Springer Proceedings in Mathematics & Statistics

Volume 188

Springer Proceedings in Mathematics & Statistics

This book series features volumes composed of selected contributions from workshops and conferences in all areas of current research in mathematics and statistics, including operation research and optimization. In addition to an overall evaluation of the interest, scientific quality, and timeliness of each proposal at the hands of the publisher, individual contributions are all refereed to the high quality standards of leading journals in the field. Thus, this series provides the research community with well-edited, authoritative reports on developments in the most exciting areas of mathematical and statistical research today.

More information about this series at http://www.springer.com/series/10533

David Loeffler · Sarah Livia Zerbes
Editors

Elliptic Curves, Modular Forms and Iwasawa Theory

In Honour of John H. Coates' 70th Birthday, Cambridge, UK, March 2015

Springer

Editors
David Loeffler
Mathematics Institute
University of Warwick
Coventry
UK

Sarah Livia Zerbes
Department of Mathematics
University College London
London
UK

ISSN 2194-1009 ISSN 2194-1017 (electronic)
Springer Proceedings in Mathematics & Statistics
ISBN 978-3-319-45031-5 ISBN 978-3-319-45032-2 (eBook)
DOI 10.1007/978-3-319-45032-2

Library of Congress Control Number: 2016948295

Mathematics Subject Classification (2010): 11F80, 11G05, 11G40, 11R23, 11S40, 14H52

Printed on acid-free paper

This Springer imprint is published by Springer Nature
The registered company is Springer International Publishing AG
The registered company address is: Gewerbestrasse 11, 6330 Cham, Switzerland

Preface

In March 2015, the Dokchitser brothers and the two of us organized two events, a workshop and a conference, in honour of John Coates' 70th birthday, in order to celebrate John's work and his mathematical heritage. Among the participants of the conference were many young mathematicians, and it is clear that John's work, in particular on Iwasawa theory, continues to be a great source of inspiration for the new generation of number theorists. It is therefore a pleasure to dedicate this volume to him, in admiration for his contributions to number theory and his influence on the subject via his many students and collaborators.

Warwick, UK
July 2016

David Loeffler
Sarah Livia Zerbes

John at the Royal Society Kavli Centre, March 2015
Published with kind permission, © John Coates

Contents

Congruences Between Modular Forms and the Birch and Swinnerton-Dyer Conjecture

Andrea Berti, Massimo Bertolini and Rodolfo Venerucci

Abstract We prove the p-part of the Birch and Swinnerton-Dyer conjecture for elliptic curves of analytic rank one for most ordinary primes.

Keywords Elliptic curves · Birch and Swinnerton-Dyer conjecture · Heegner points · Shimura curves

1 Introduction

The theory of congruences between modular forms has turned out to be a crucial player in a number of momentous results in the theory of rational points on elliptic curves. To mention only a few instances, we recall here Mazur's theory of the Eisenstein ideal [16], in which congruences between cusp forms and Eisenstein series on GL_2 are used to uniformly bound the torsion subgroups of elliptic curves over $\mathbf{Q}$. More germane to our setting, the recent work of Skinner–Urban [22] constructs classes in the p-primary Shafarevich–Tate group of an elliptic curve over $\mathbf{Q}$ (and more generally, over cyclotomic extensions) when p is ordinary and divides (the algebraic part of) the value of the associated Hasse–Weil L-series at $s = 1$. This is

A. Berti
Dipartimento di Matematica Federigo Enriques, Università di Milano,
via C. Saldini 50, Milano, Italy
e-mail: andrea85.b@gmail.com

M. Bertolini (✉) · R. Venerucci
Universität Duisburg-Essen, Fakultät Für Mathematik, Mathematikcarrée,
Thea-Leymann-Strasse 9, 45127 Essen, Germany
e-mail: massimo.bertolini@uni-due.de

R. Venerucci
e-mail: rodolfo.venerucci@uni-due.de

D. Loeffler and S.L. Zerbes (eds.), *Elliptic Curves, Modular Forms and Iwasawa Theory*, Springer Proceedings in Mathematics & Statistics 188, DOI 10.1007/978-3-319-45032-2_1

achieved by exploiting p-power congruences between cusp forms on unitary groups and Eisenstein series whose constant term encodes the special value of the L-series of the elliptic curve. On the opposite side, when this special value is non-zero, Kato's Euler system [13] arising from Steinberg symbols of modular units gives an upper bound on the p-primary Selmer group. The combination of these two results yields the validity of the p-part of the Birch and Swinnerton-Dyer conjecture for elliptic curves of analytic rank zero at almost all ordinary primes.

The goal of this paper is to present a direct proof of the p-part of the Birch and Swinnerton-Dyer conjecture for elliptic curves of analytic rank one for most ordinary primes, obtained by Zhang in [27] along a somewhat different path. More precisely, let $A/\mathbf{Q}$ be an elliptic curve of conductor N. Write $L(A/\mathbf{Q}, s)$ for the Hasse–Weil L-function of A, and $Ш(A/\mathbf{Q})$ for its Shafarevich–Tate group. When $L(A/\mathbf{Q}, s)$ has a simple zero at $s = 1$, the theorem of Gross–Zagier–Kolyvagin [11, 14] states that $A(\mathbf{Q})$ has rank one and $Ш(A/\mathbf{Q})$ is finite. Fix a modular parametrisation

$$\pi_A : X_0(N) \longrightarrow A$$

of minimal degree $\deg(\pi_A)$. For every rational point $P \in A(\mathbf{Q})$, write $h^{\mathrm{NT}}(P) \in \mathbf{R}$ for the canonical Néron–Tate height of P, and let $\Omega_A \in \mathbf{R}^*$ be the real Néron period attached to $A/\mathbf{Q}$. Set $c_A := \prod_{q|N} c_q(A)$, where $c_q(A)$ is the Tamagawa number of $A/\mathbf{Q}_q$, and denote by $a_p(A)$ the coefficient $1 + p - \bar{A}(\mathbf{F}_p)$ of A at p, and by $\mathrm{ord}_p : \mathbf{Q}^* \to \mathbf{Z}$ the p-adic valuation.

Theorem A *Assume that $A/\mathbf{Q}$ is semistable. Let $p > 7$ be a prime which does not divide* $\deg(\pi_A)$, *and is good ordinary and non-anomalous for A (i.e., $p \nmid N$ and $a_p(A) \not\equiv 0, 1 \pmod{p}$). If $L(A/\mathbf{Q}, s)$ has a simple zero at $s = 1$, then*

$$\mathrm{ord}_p\left(\frac{L'(A/\mathbf{Q}, 1)}{h^{\mathrm{NT}}(\mathbf{P}) \cdot \Omega_A}\right) = \mathrm{ord}_p\left(\#Ш(A/\mathbf{Q}) \cdot c_A\right),$$

where $\mathbf{P}$ is a generator of $A(\mathbf{Q})$ modulo torsion.

Note that the assumptions of Theorem A imply that the p-torsion of $A(\mathbf{Q})$ is trivial, and that the Tamagawa number c_A is a p-adic unit, so that it can be omitted in the statement.

By invoking the Kato–Skinner–Urban theorem mentioned above, Theorem A can be reduced (as explained in Sect. 6) to an analogous statement over an imaginary quadratic field K on which $L(A/K, s)$ has a simple zero. In light of the Gross–Zagier formula, this statement is in turn equivalent to the equality of the order of the p-primary part of the Shafarevich–Tate group of A/K and the p-part of the square of the index of a Heegner point in $A(K)$. Theorem 6.1 below proves this result by exploiting the theory of congruences between cusp forms on GL_2. In a nutshell, our strategy makes use of the explicit reciprocity laws of [3] combined with cohomological arguments and the theory of Euler systems to show that the existence of Selmer classes stated in Theorem 6.1 can be obtained from the constructive methods devised in [22] for elliptic curves of analytic rank zero.

Theorem 6.1 has been obtained independently by Zhang [27]. His method uses the reciprocity laws of loc. cit. together with [22] to prove Kolyvagin's conjecture on the non-vanishing of the cohomology classes defined in terms of Galois-theoretic derivatives of Heegner points over ring class fields. This conjecture is known to imply Theorem 6.1, thanks to prior work of Kolyvagin [15]. The method explained in this paper (a weaker version of which appears in the first author's Ph.D. thesis [1]) is more direct, insofar as it consists in an explicit comparison of Selmer groups and of special values of L-series attached to congruent modular forms.[1]

2 Modular Forms and Selmer Groups

Fix a squarefree positive integer N, a factorisation $N = N^+N^-$, and a rational prime $p > 3$ such that $p \nmid N$.

2.1 Eigenforms of Level (N^+, N^-)

Let $S_2(\Gamma_0(N))^{N^-\text{-new}}$ be the $\mathbf{C}$-vector space of weight-two cusp forms of level $\Gamma_0(N)$, which are new at every prime divisor of N^-. Write

$$\mathbb{T}_{N^+,N^-} \subset \mathrm{End}\left(S_2(\Gamma_0(N))^{N^-\text{-new}}\right)$$

for the Hecke algebra generated over $\mathbf{Z}$ by the Hecke operators T_q, for primes $q \nmid N$, and U_q for primes $q|N$.

Let R be a complete local Noetherian ring with finite residue field k_R of characteristic p. (In the following sections, R will often be chosen to be the finite ring $\mathbf{Z}/p^n\mathbf{Z}$.) An *R-valued (weight two) eigenform of level* (N^+, N^-) is a ring homomorphism

$$g : \mathbb{T}_{N^+,N^-} \longrightarrow R.$$

Denote by $S_2(N^+, N^-; R)$ the set of R-valued eigenforms of level (N^+, N^-). To every $g \in S_2(N^+, N^-; R)$ is associated—see for example [8], Sect. 2.2—a Galois representation

$$\overline{\rho}_g : G_{\mathbf{Q}} \longrightarrow \mathrm{GL}_2(k_R),$$

[1]In the recent preprint [12] the authors give a different proof of Theorem A. More precisely they combine the p-adic Gross–Zagier formula of [5] with (one divisibility in) the Iwasawa main conjecture for Rankin–Selberg convolutions [24] to deduce an analogue of Theorem 6.1 below (see also Sect. 6.6). Their approach applies only to imaginary quadratic fields in which p *splits*, but requires no assumption on the Tamagawa factors of $A/\mathbf{Q}$. In particular they remove the hypothesis $p \nmid \deg(\pi_A)$ from Theorem A.

whose semi-simplification is characterised by the following properties. Let q be a prime which does not divide Np, and let $\mathrm{Frob}_q \in G_{\mathbf{Q}}$ be an arithmetic Frobenius at q. Then $\overline{\rho}_g$ is unramified at q, and the characteristic polynomial of $\overline{\rho}_g(\mathrm{Frob}_q)$ is $X^2 - \overline{g}(T_q)X + q \in k_R[X]$, where $\overline{g} : \mathbb{T}_{N^+,N^-} \to k_R$ is the composition of g with the projection $R \twoheadrightarrow k_R$. By Théorèm 3 of *loc. cit.*, if $\overline{\rho}_g$ is *(absolutely) irreducible*, one can lift it uniquely to a Galois representation

$$\rho_g : G_{\mathbf{Q}} \longrightarrow \mathrm{GL}_2(R)$$

unramified at every prime $q \nmid Np$, and such that $\mathrm{trace}\big(\rho_g(\mathrm{Frob}_q)\big) = g(T_q)$ and $\det\big(\rho_g(\mathrm{Frob}_q)\big) = q$ for such a q. Assuming that $\overline{\rho}_g$ is irreducible, write

$$T_g \in {}_{R[G_{\mathbf{Q}}]}\mathrm{Mod}$$

for a R-module giving rise to the representation ρ_g. In other words, T_g is a free R-module of rank two, equipped with a continuous, linear action of $G_{\mathbf{Q}}$, which is unramified at every prime $q \nmid Np$, and such that Frob_q acts with characteristic polynomial $X^2 - g(T_q)X + q \in R[X]$ for every such q.

2.2 Selmer Groups

Let $g \in S_2(N^+, N^-; R)$ be an eigenform satisfying the following assumption.

Assumption 2.1 1. $\overline{\rho}_g$ is absolutely irreducible.

2. ρ_g is *ordinary* at p, i.e., there exists a short exact sequence of $G_{\mathbf{Q}_p}$-modules

$$0 \to T_g^{(p)} \to T_g \to T_g^{[p]} \to 0,$$

where $T_g^{(p)}$ (resp., $T_g^{[p]}$) is a free R-module of rank one, on which the inertia subgroup $I_{\mathbf{Q}_p} \subset G_{\mathbf{Q}_p}$ acts via the p-adic cyclotomic character $\varepsilon : G_{\mathbf{Q}_p} \twoheadrightarrow \mathrm{Gal}(\mathbf{Q}_p(\mu_{p^\infty})/\mathbf{Q}_p) \cong \mathbf{Z}_p^*$ (resp., acts via the trivial character).

3. For every prime q dividing N, there exists a unique $G_{\mathbf{Q}_q}$-submodule $T_g^{(q)} \subset T_g$, free of rank one over R, such that $G_{\mathbf{Q}_{q^2}}$ acts on $T_g^{(q)}$ via the p-adic cyclotomic character $\varepsilon : G_{\mathbf{Q}_q} \twoheadrightarrow \mathrm{Gal}(\mathbf{Q}_q(\mu_{p^\infty})/\mathbf{Q}_q) \hookrightarrow \mathbf{Z}_p^*$. (Here $\mathbf{Q}_{q^2}/\mathbf{Q}_q$ denotes the quadratic unramified extension of $\mathbf{Q}_q$.)

Let $K/\mathbf{Q}$ be an imaginary quadratic field of discriminant coprime with Np. For every (finite) prime v of K, define the *finite* and *singular* parts of the local cohomology group $H^1(K_v, T_g)$ as

$$H^1_{\mathrm{fin}}(K_v, T_g) := H^1(G_v/I_v, T_g^{I_v}); \quad H^1_{\mathrm{sing}}(K_v, T_g) := \frac{H^1(K_v, T_g)}{H^1_{\mathrm{fin}}(K_v, T_g)},$$

where I_v is the inertia subgroup of $G_v := \mathrm{Gal}(\overline{K}_v/K_v)$, and $H^1_{\mathrm{fin}}(K_v, T_g)$ is viewed as a submodule of $H^1(K_v, T_q)$ via the injective G_v/I_v-inflation map. For every prime v lying above a rational prime $q \mid Np$, define the *ordinary part* of the local cohomology $H^1(K_v, T_g)$ as

$$H^1_{\mathrm{ord}}(K_v, T_g) := \mathrm{Im}\left(H^1(K_v, T_g^{(q)}) \to H^1(K_v, T_g)\right).$$

Define the *Selmer group of* g/K as the submodule

$$\mathrm{Sel}(K, g) \subset H^1(K, T_g),$$

consisting of global cohomology classes $x \in H^1(K, T_g)$ satisfying the following conditions.

- x is *finite outside* Np: $\mathrm{res}_v(x) \in H^1_{\mathrm{fin}}(K_v, T_g)$ for every prime v of K not dividing Np.
- x is *ordinary* at every prime dividing Np: $\mathrm{res}_v(x) \in H^1_{\mathrm{ord}}(K_v, T_g)$ for every prime v of K dividing a rational prime $q|Np$.

Note that the Selmer group $\mathrm{Sel}(K, g)$ depends on g (since it depends on its level N), and not only on the representation T_g attached to it.

2.3 Admissible Primes

In this section, R will denote the finite ring $\mathbf{Z}/p^n\mathbf{Z}$, where n is a positive integer and p is a rational prime. Let $g \in S_2(N^+, N^-; \mathbf{Z}/p^n\mathbf{Z})$ be a mod-p^n eigenform of level (N^+, N^-), and let $K/\mathbf{Q}$ be an imaginary quadratic field of discriminant coprime with Np.

Following [3], we say that a rational prime ℓ is an *n-admissible* prime relative to g if the following conditions are satisfied:

A1. ℓ does not divide Np.
A2. $\ell^2 - 1$ is a unit in $\mathbf{Z}/p^n\mathbf{Z}$ (i.e. $\ell \not\equiv \pm 1 \pmod{p}$).
A3. $g(T_\ell)^2 = (\ell + 1)^2$ in $\mathbf{Z}/p^n\mathbf{Z}$.

If, in addition, ℓ is inert in K, we say that ℓ is n-admissible relative to (g, K).

For a rational prime ℓ, we say that an eigenform $g_\ell \in S_2(N^+, N^-\ell; \mathbf{Z}/p^n\mathbf{Z})$, i.e. a surjective morphism $g_\ell : \mathbb{T}_{N^+, N^-\ell} \to \mathbf{Z}/p^n\mathbf{Z}$, is an *$\ell$-level raising of* g if

$$g_\ell(T_q) = g(T_q), \qquad \text{resp. } g_\ell(U_q) = g(U_q)$$

for every prime $q \nmid N\ell$, resp. $q|N$. As recalled in loc. cit., if ℓ is n-admissible relative to g, then an ℓ-level raising g_ℓ exists.

Assume that g satisfies Assumption 2.1. Then $\overline{\rho}_g$ and $\overline{\rho}_{g_\ell}$ are isomorphic, absolutely irreducible representations of $G_{\mathbf{Q}}$ in $\mathrm{GL}_2(\mathbf{F}_p)$, and by the results recalled in Sect. 2.1, this implies that there is an isomorphism of $\mathbf{Z}/p^n\mathbf{Z}[G_{\mathbf{Q}}]$-modules

$$\mathcal{T} := T_g \cong T_{g_\ell} \in {}_{\mathbf{Z}/p^n\mathbf{Z}[G_{\mathbf{Q}}]}\mathrm{Mod}.$$

Fix such an isomorphism, that we regard as an equality from now on. The following lemma is proved by the same argument appearing in the proof of Lemma 2.6 of [3]. Write $K_\ell/\mathbf{Q}_\ell$ for the completion of K at the unique prime dividing ℓ (so $K_\ell = \mathbf{Q}_{\ell^2}$ is the quadratic unramified extension of $\mathbf{Q}_\ell$).

Lemma 2.2 *Let ℓ be an n-admissible prime relative to (g, K). Then there is a decomposition of $\mathbf{Z}/p^n\mathbf{Z}[G_{K_\ell}]$-modules*

$$\mathcal{T} = \mathbf{Z}/p^n\mathbf{Z}(\varepsilon) \oplus \mathbf{Z}/p^n\mathbf{Z},$$

where $\mathbf{Z}/p^n\mathbf{Z}(\varepsilon)$ (resp., $\mathbf{Z}/p^n\mathbf{Z}$) denotes a copy of $\mathbf{Z}/p^n\mathbf{Z}$ on which G_{K_ℓ} acts via the p-adic cyclotomic character ε (resp., acts trivially). Moreover, this decomposition induces isomorphisms

$$\begin{aligned} H^1_{\mathrm{fin}}(K_\ell, \mathcal{T}) &\cong H^1(K_\ell, \mathbf{Z}/p^n\mathbf{Z}) \cong \mathbf{Z}/p^n\mathbf{Z}; \\ H^1_{\mathrm{sing}}(K_\ell, \mathcal{T}) &\cong H^1(K_\ell, \mathbf{Z}/p^n\mathbf{Z}(\varepsilon)) \cong \mathbf{Z}/p^n\mathbf{Z}. \end{aligned} \tag{1}$$

Let $g_\ell \in S_2(N^+, N^-\ell; \mathbf{Z}/p^n\mathbf{Z})$ be an ℓ-level raising of g. One deduces that $g_\ell \in S_2(N^+, N^-\ell; \mathbf{Z}/p^n\mathbf{Z})$ satisfies Assumption 2.1 too, and (with the notations above)

$$H^1_{\mathrm{ord}}(K_\ell, T_{g_\ell}) \cong H^1_{\mathrm{sing}}(K_\ell, T_g) \cong \mathbf{Z}/p^n\mathbf{Z}. \tag{2}$$

The preceding lemma allows us to define morphisms

$$\begin{aligned} v_\ell : H^1(K, \mathcal{T}) &\longrightarrow H^1_{\mathrm{fin}}(K_\ell, \mathcal{T}) \cong \mathbf{Z}/p^n\mathbf{Z}; \\ \partial_\ell : H^1(K, \mathcal{T}) &\longrightarrow H^1_{\mathrm{ord}}(K_\ell, \mathcal{T}) \cong \mathbf{Z}/p^n\mathbf{Z}, \end{aligned}$$

defined by composing the restriction map at ℓ with the projection onto the finite and ordinary (or singular) part respectively. Given a global class $x \in H^1(K, \mathcal{T})$, we call $v_\ell(x)$ its *finite part* at ℓ, and $\partial_\ell(x)$ its *residue* at ℓ.

2.4 Raising the Level at Admissible Primes

As in the previous section, let $g \in S_2(N^+, N^-; \mathbf{Z}/p^n\mathbf{Z})$ be a mod-p^n eigenform of level (N^+, N^-).

Assumption 2.3 The data $(\overline{\rho}_g, N^+, N^-, p)$ satisfy the following conditions:

1. $N = N^+N^-$ is squarefree;
2. p does not divide N;
3. $\overline{\rho}_g : G_{\mathbf{Q}} \to \mathrm{GL}_2(\mathbf{F}_p)$ is surjective;
4. If $q \mid N^-$ and $q \equiv \pm 1 \pmod{p}$, then $\overline{\rho}_g$ is ramified at q.

The following theorem, establishing the existence of a level raising at admissible primes, comes from the work of several people, including Ribet and Diamond–Taylor.

Theorem 2.4 *Assume that Assumption 2.3 holds. Let $L = \ell_1 \cdots \ell_k$ be a product of (distinct) n-admissible primes ℓ_j relative to g. Then there exists a unique mod-p^n eigenform $g_L : \mathbb{T}_{N^+,N^-L} \longrightarrow \mathbf{Z}/p^n\mathbf{Z}$ of level (N^+, N^-L) such that*

$$g_L(T_q) = g(T_q) \ \ (\textit{for all } q \nmid NL), \qquad g_L(U_q) = g(U_q) \ \ (\textit{for all } q|N).$$

Proof We make some remarks about the references for the proof of this theorem. Assume that $N^- > 1$ and that N^- has an odd (resp., even) number of prime divisors. In this case the theorem is proved in Sect. 5 (resp., 9) of [3], working under slightly more restrictive assumptions on $(\overline{\rho}_g, N^+, N^-, p)$, subsequently removed in Sect. 4 of [18]. The method of [3] generalises previous work of Ribet (which considered the case $n = 1$), and uses Diamond–Taylor's generalisation of Ihara's Lemma (for modular curves) to Shimura curves. We refer to loc. cit. for more details and references.

Assume now that $N^- = 1$. If $n = 1$, the theorem has been proved by Ribet. If $n > 1$, the theorem can be proved by following the arguments appearing in Sect. 9 of [3] (see in particular Proposition 9.2 and Theorem 9.3), and invoking the classical Ihara Lemma (instead of Diamond–Taylor's generalisation) in the proof of Proposition 9.2. □

3 The Explicit Reciprocity Laws

In this section we recall (special cases of) the explicit reciprocity laws proved in [3], which relate Heegner points on Shimura curves to special values of Rankin L-functions (described in terms of certain *Gross points* attached to modular forms on definite quaternion algebras). Together with the proof by Kato–Skinner–Urban of the (p-part of) the Birch and Swinnerton-Dyer formula in analytic rank zero (cf. Sect. 5 below), these reciprocity laws will be at the heart of our proof of Theorem A.

Fix throughout this section a factorisation $N = N^+N^-$ of a positive integer N, a rational prime p not dividing N, and a $\mathbf{Z}_p$-valued eigenform

$$f \in S_2(N^+, N^-; \mathbf{Z}_p)$$

of level (N^+, N^-). Fix also a quadratic imaginary field $K/\mathbf{Q}$ of discriminant coprime with Np. Assume that the following hypotheses are satisfied (cf. Hypothesis CR of [18]).

Assumption 3.1 1. N^- has an *even* number of prime factors.
2. A prime divisor q of N divides N^- precisely if q is inert in $K/\mathbf{Q}$.
3. $\overline{\rho}_f : G_\mathbf{Q} \to \mathrm{GL}_2(\mathbf{F}_p)$ is surjective.
4. f is ordinary at p, i.e. $f(T_p) \in \mathbf{Z}_p^*$.
5. If $q \mid N^-$ and $q \equiv \pm 1 \pmod p$, then $\overline{\rho}_f$ is ramified at q.

3.1 Special Points on Shimura Curves

3.1.1 Shimura Curves ([3, Sect. 5])

Let $\mathscr{B} := \mathscr{B}_{N^-}$ be a quaternion algebra of discriminant N^-, let $\mathscr{R} = \mathscr{R}_{N^+}$ be an Eichler order of level N^+ in $\mathscr{B}$, and let $\mathscr{R}_{\max}$ be a maximal order of $\mathscr{B}$ containing $\mathscr{R}$. (The indefinite quaternion algebra $\mathscr{B}$ is unique up to isomorphism, while $\mathscr{R}_{\max}$ and $\mathscr{R}$ are unique up to conjugation.) Let

$$\mathcal{F}_{N^+,N^-} : \mathrm{Sch}_{/\mathbf{Z}[1/N]} \longrightarrow \mathrm{Sets}$$

be the functor attaching to a $\mathbf{Z}[1/N]$-scheme T the set of isomorphism classes of triples (A, ι, C), where

- A is an abelian scheme over T of relative dimension 2;
- ι is a morphism $\mathscr{R}_{\max} \to \mathrm{End}(A/T)$, defining an action of $\mathscr{R}_{\max}$ on A;
- C is a subgroup scheme of A, locally isomorphic to $\mathbf{Z}/N^+\mathbf{Z}$, which is stable and locally cyclic over $\mathscr{R}$.

If $N^- > 1$, the moduli problem $\mathcal{F}_{N^+,N^-}$ is coarsely represented by a smooth projective scheme

$$X_{N^+,N^-} \to \mathrm{Spec}(\mathbf{Z}[1/N]),$$

called the *Shimura curve* attached to the factorisation $N = N^+N^-$. In particular

$$X_{N^+,N^-}(F) = \mathcal{F}_{N^+,N^-}(F)$$

for every algebraically closed field F of characteristic coprime with N.

If $N^- = 1$, then the functor $\mathcal{F}_{N,1}$ can be shown to be coarsely represented by the smooth, quasi-projective modular curve $X_{N,1}^o = Y_0(N)$ over $\mathbf{Z}[1/N]$ of level $\Gamma_0(N) \subset \mathrm{SL}_2(\mathbf{Z})$. In this case we write

$$X_{N,1} = X_0(N) \to \mathrm{Spec}(\mathbf{Z}[1/N])$$

for the usual compactification obtained by adding to $X^o_{N,1}$ a finite set of cusps, which is again a smooth projective curve over $\mathbf{Z}[1/N]$.

3.1.2 Heegner Points

Under Assumption 3.1(2) $X_{N^+,N^-}(\mathbf{C})$ contains points with CM by K. More precisely, let $\mathcal{O}_K$ be the maximal order of K. Then there exists a point $\mathbb{P} = (\mathcal{A}, \iota, C) \in X_{N^+,N^-}(\mathbf{C})$ such that

$$\mathcal{O}_K \cong \mathrm{End}(\mathbb{P}), \tag{3}$$

where $\mathrm{End}(\mathbb{P}) \subset \mathrm{End}(\mathcal{A})$ denotes the ring of endomorphisms of $\mathcal{A}/\mathbf{C}$ which commute with the action of ι, and respect the level structure C. By the theory of complex multiplication,

$$\mathbb{P} \in X_{N^+,N^-}(H),$$

where $H := H_K$ is the Hilbert class field of K. Call such a $\mathbb{P} \in X_{N^+,N^-}(H)$ a *Heegner point*, and write

$$\mathrm{Heeg}_{N^+,N^-}(K) \subset X_{N^+,N^-}(H)$$

for the set of Heegner points (of conductor one).

3.1.3 Gross Points

Let $L = \ell_1 \cdots \ell_k$ be a squarefree product of an *odd* number of primes $\ell_j \nmid N$ which are *inert* in $K/\mathbf{Q}$. Let $B := B_{N^-L}$ be a definite quaternion algebra of discriminant N^-L (which is unique up to isomorphism), and let $R := R_{N^+}$ be a fixed Eichler order of level N^+ in B. The Eichler order R is not necessarily unique, even up to conjugation. Nonetheless, there are only finitely many conjugacy classes of Eichler orders of level N^+ in B, say $R_1, \ldots, R_h$. More precisely, consider the double coset space

$$\mathbb{X}_{N^+,N^-L} := \widehat{R}^* \backslash \widehat{B}^* / B^*, \tag{4}$$

where $\widehat{Z} := Z \otimes_{\mathbf{Z}} \widehat{\mathbf{Z}}$ for every ring Z, with $\widehat{\mathbf{Z}} = \prod_{q \text{ prime}} \mathbf{Z}_q$. It is a finite set, in bijection with the set of conjugacy classes of *oriented* Eichler orders of level N^+ in B, via the rule $\widehat{B}^* \ni b \mapsto R_b := b\widehat{R}b^{-1} \cap B$ (cf. [2, Sect. 1]).

Define the set of *Gross points* of level N^+ and conductor p^∞ on B as

$$\mathrm{Gr}_{N^+,N^-L}(K, \infty) := \widehat{R}^* \big\backslash \big(\mathrm{Hom}(K, B) \times \widehat{B}^*\big) \big/ B^*.$$

Here $\mathrm{Hom}(K, B)$ is the set of morphisms of algebras $f : K \to B$. (The group B^* acts on $\widehat{B}^*$ via the diagonal embedding $B^* \to \widehat{B}^*$, while it acts on $\mathrm{Hom}(K, B)$ via conjugation on B.) A Gross point $[f \times b] \in \mathrm{Gr}_{N^+,N^-L}(K, \infty)$ has *conductor one* if

$f(K) \cap b\widehat{R}b^{-1} = f(O_K)$. Denote by

$$\mathrm{Gr}_{N^+,N^-L}(K) \subset \mathrm{Gr}_{M^+,M^-L}(K,\infty)$$

the set of Gross points of conductor one. In what follows, a *Gross point (of level N^+ on B)* will always be a Gross point (of level N^+ on B) of conductor one.

3.1.4 Gross Points and Reduction of Heegner Points

With the notations of the previous section, let $L = \ell$ be a rational prime which is inert in $K/\mathbf{Q}$ and such that $\ell \nmid N$. The reduction modulo ℓ map on the Shimura curve X_{N^+,N^-} allows us to define a map

$$r_\ell : \mathrm{Heeg}_{N^+,N^-}(K) \longrightarrow \mathrm{Gr}_{N^+,N^-\ell}(K)$$

from Heegner points to Gross points. More precisely, let $\mathbb{P} = (\mathcal{A}, \iota, C) \in \mathrm{Heeg}_{N^+,N^-}(H)$. Fix a prime λ of H dividing ℓ. Since ℓ is inert in K, it is totally split in H, so that λ has associated residue field $\mathbf{F}_{\ell^2}$. The abelian variety $\mathcal{A}$ and the subgroup $C \subset \mathcal{A}$ are defined over H, and $\mathcal{A}$ has good reduction at λ. Let

$$\overline{\mathbb{P}} := \mathrm{red}_\ell(\mathbb{P}) = (\overline{\mathcal{A}}, \bar{\iota}, \overline{C}) \in X_{N^+,N^-}(\mathbf{F}_{\ell^2})$$

be the reduction of $\mathbb{P}$ modulo λ, where $\overline{\mathcal{A}}/\mathbf{F}_{\ell^2}$ and $\overline{C} \subset \overline{\mathcal{A}}$ denote the reductions of $\mathcal{A}$ and C modulo λ respectively, and $\bar{\iota}$ denotes the composition of ι with reduction of endomorphisms $\mathrm{End}(\mathcal{A}) \to \mathrm{End}(\overline{\mathcal{A}})$. Define (as above) $\mathrm{End}(\overline{\mathbb{P}}) \subset \mathrm{End}(\overline{\mathcal{A}})$ as the subring of endomorphisms of $\overline{\mathcal{A}}$ (defined over $\overline{\mathbf{F}}_\ell$) commuting with the action of ι and preserving $\overline{C}$. It turns out that $\mathrm{End}(\overline{\mathbb{P}}) \cong R_\mathbb{P}$ is isomorphic to an Eichler order $R_\mathbb{P}$ of level N^+ in $B = B_{N^-\ell}$. In light of (3), reduction of endomorphisms on $\mathcal{A}$ induces then an embedding

$$f_{\mathbb{P},\ell} : O_K \cong \mathrm{End}(\mathbb{P}) \longrightarrow \mathrm{End}(\overline{\mathbb{P}}) \cong R_\mathbb{P}.$$

Denote again by $f_{\mathbb{P},\ell} : K \to B$ the extension of scalars of $f_{\mathbb{P},\ell}$. By (4) there exists $b_\mathbb{P} \in \widehat{B}^*$ such that $R_\mathbb{P} = b_\mathbb{P}\widehat{R}b_\mathbb{P}^{-1} \cap B$. Define

$$r_\ell(\mathbb{P}) = \left[f_{\mathbb{P},\ell} \times b_\mathbb{P}\right] \in \mathrm{Gr}_{N^+,N^-\ell}(K).$$

3.1.5 Action of $\mathrm{Pic}(O_K)$

The assumptions an notations are as in the preceding sections. Write $\mathrm{Pic}(O_K)$ for the ideal class group of K, which admits the adelic description $\mathrm{Pic}(O_K) = \widehat{O}_K^* \backslash \widehat{K}^*/K^*$. Given an ideal class $\sigma \in \mathrm{Pic}(O_K)$ and a Gross point $P = [f \times b] \in \mathrm{Gr}_{N^+,N^-\ell}(K)$, define

$$P^\sigma := \left[f \times \widehat{f}(\sigma) \cdot b\right] \in \mathrm{Gr}_{N^+,N^-\ell}(K),$$

where $\widehat{f} : \widehat{K} \to \widehat{B}$ is the morphism induced on adèles by the embedding $f : K \to B$. It is easily seen that the rule $P \mapsto P^\sigma$ defines an action of $\mathrm{Pic}(O_K)$ on $\mathrm{Gr}_{N^+,N^-\ell}(K)$.

The Artin map of global class field theory gives a canonical isomorphism $\mathrm{Pic}(O_K) \cong \mathrm{Gal}(H/K)$. The set of Heegner points $\mathrm{Heeg}_{N^+,N^-}(K)$ (of conductor one) is contained in $X_{N^+,N^-}(H)$, and one obtains a natural *geometric* action of $\mathrm{Pic}(O_K)$ on $\mathrm{Heeg}_{N^+,N^-}(K)$.

With these definitions, the reduction map $r_\ell : \mathrm{Heeg}_{N^+,N^-}(K) \to \mathrm{Gr}_{N^+,N^-\ell}(K)$ defined in the preceding section is $\mathrm{Pic}(O_K)$-equivariant [2], i.e.

$$r_\ell\left(\mathbb{P}^\sigma\right) = r_\ell(\mathbb{P})^\sigma \tag{5}$$

for every ideal class $\sigma \in \mathrm{Pic}(O_K)$ and every Heegner point $\mathbb{P} \in \mathrm{Heeg}_{N^+,N^-}(K)$.

3.2 Modular Forms on Definite Quaternion Algebras

The notations and assumptions are as in Sect. 3.1.3. Let

$$\mathbb{J}_{N^+,N^-L} := \mathbf{Z}\left[\mathbb{X}_{N^+,N^-L}\right]$$

denote the group of formal divisors on the set $\mathbb{X}_{N^+,N^-L}$ defined in Eq. (4). As explained in Sect. 1.5 of [2], there is a Hecke algebra

$$\mathbf{T}_{N^+,N^-L} \subset \mathrm{End}(\mathbb{J}_{N^+,N^-L})$$

acting faithfully as a ring of endomorphisms of $\mathbb{J}_{N^+,N^-L}$, and generated over $\mathbf{Z}$ by Hecke operators t_q, for primes $q \nmid N$, and u_q, for primes $q|N$. By the Jacquet–Langlands correspondence [2, Sect. 1.6], there is an isomorphism $\mathbf{T}_{N^+,N^-L} \cong \mathbb{T}_{N^+,N^-L}$, defined by sending t_q (resp., u_q) to T_q (resp., U_q).

Let $n \in \mathbf{N} \cup \{\infty\}$, and let $g \in S_2(N^+, N^-L; \mathbf{Z}_p/p^n\mathbf{Z}_p)$ be a $\mathbf{Z}_p/p^n\mathbf{Z}_p$-valued eigenform of level (N^+, N^-L) (with the convention that $\mathbf{Z}_p/p^\infty\mathbf{Z}_p := \mathbf{Z}_p$). Then g induces a surjective morphism $g^{\mathrm{JL}} : \mathbf{T}_{N^+,N^-L} \twoheadrightarrow \mathbf{Z}_p/p^n\mathbf{Z}_p$. Let $\mathfrak{m}_g := \ker(g_{\{1\}})$ denote the maximal ideal of $\mathbf{T}_{N^+,N^-L}$ associated with (the reduction $g_{\{1\}}$ modulo p of) g, and let $\mathbb{J}_{\mathfrak{m}_g}$ and $\mathbf{T}_{\mathfrak{m}_g}$ denote the completions at $\mathfrak{m}_g$ of $\mathbb{J}_{N^+,N^-L}$ and $\mathbf{T}_{N^+,N^-L}$ respectively. According to Theorem 6.2 and Proposition 6.5 of [18], Assumption 3.1 implies that $\mathbb{J}_{\mathfrak{m}_g}$ is a free $\mathbf{T}_{\mathfrak{m}_g}$-module of rank one. As a consequence, g^{JL} induces a surjective morphism (denoted by the same symbol with a slight abuse of notation)

$$g^{\mathrm{JL}} : \mathbb{J}_{N^+,N^-L} \longrightarrow \mathbf{Z}_p/p^n\mathbf{Z}_p,$$

such that $g^{\mathrm{JL}}(h \cdot x) = g(h) \cdot g^{\mathrm{JL}}(x)$ for every $x \in \mathbb{J}_{N^+,N^-L}$ and every $h \in \mathbf{T}_{N^+,N^-L}$. Such a $\mathbf{T}_{N^+,N^-L}$-eigenform is unique up to p-adic units.

Remark 3.2 The above discussion establishes a correspondence between eigenforms in the sense of Sect. 2.1 and surjective $\mathbf{Z}_p/p^n\mathbf{Z}_p$-valued eigenforms on definite quaternion algebras. The latter is the point of view adopted in [3]; we refer the reader to Sect. 1.1 of *loc. cit.*, and in particular to Eq. (11) in the proof of Proposition 1.3, for more details.

3.2.1 Special Values Attached to Modular Forms on Definite Quaternion Algebras

There is a natural forgetful map

$$\mathrm{Gr}_{N^+,N^-L}(K) \longrightarrow \mathbb{X}_{N^+,N^-L},$$

which maps the Gross point represented by the pair $f \times b \in \mathrm{Hom}(K, B) \times \widehat{B}^*$ to the class of the idèle b in $\mathbb{X}_{N^+,N^-L}$. Any function γ defined on $\mathbb{X}_{N^+,N^-L}$ then induces a function on the set of Gross points $\mathrm{Gr}_{N^+,N^-L}(K)$, denoted again γ. Let $g : \mathbb{T}_{N^+,N^-L} \to \mathbf{Z}_p/p^n\mathbf{Z}_p$ be as above. Thanks to the Jacquet–Langlands correspondence recalled in the preceding section, one can define the *special value attached to* (g, K) by

$$\mathscr{L}_p(g/K) := \sum_{\sigma \in \mathrm{Pic}(O_K)} g^{\mathrm{JL}}\left(x^\sigma\right) \in \mathbf{Z}_p/p^n\mathbf{Z}_p, \tag{6}$$

where $x \in \mathrm{Gr}_{N^+,N^-L}(K)$ is any fixed Gross point of level N^+ on B. The special value $\mathscr{L}_p(g/K)$ is well defined up to multiplication by a p-adic unit. (Once g^{JL} is fixed, $\mathscr{L}_p(g/K)$ can be shown to be independent, up to sign, of the choice of the Gross point x fixed to define it. We refer to Sect. 3 of [2] for more details.)

When $n = \infty$, so that g arises from a classical modular form, $\mathscr{L}_p(g/K)$ is essentially equal to the square-root of the special value $L(g/K, 1)$, as explained in Sect. 4 below.

3.3 *The Reciprocity Laws*

Fix throughout this section a positive integer n, and denote by

$$f_{\{n\}} \in S_2(N^+, N^-; \mathbf{Z}/p^n\mathbf{Z})$$

the reduction of f modulo p^n (i.e. the composition of $f : \mathbb{T}_{N^+,N^-} \to \mathbf{Z}_p$ with the natural projection $\mathbf{Z}_p \twoheadrightarrow \mathbf{Z}/p^n\mathbf{Z}$). An n-admissible prime relative to $(f_{\{n\}}, K)$ is also said to be *n-admissible relative to* (f, K).

3.3.1 The Graph of Modular Forms

Let $\mathcal{L} = \mathcal{L}_n$ denote the set of squarefree products $L = \ell_1 \cdots \ell_r$ of n-admissible primes ℓ_j relative to (f, K). One can decompose $\mathcal{L} = \mathcal{L}^{\mathrm{def}} \coprod \mathcal{L}^{\mathrm{indef}}$, where $L \in \mathcal{L}^{\mathrm{def}}$ is a *definite vertex* (resp., $L \in \mathcal{L}^{\mathrm{indef}}$ is an *indefinite vertex*) if the number r of primes dividing L is *odd* (resp., *even*).

According to Theorem 2.4 (and recalling Assumption 3.1), to every $L \in \mathcal{L}$ is associated a unique mod-p^n eigenform

$$f_L \in S_2(N^+, N^- L; \mathbf{Z}/p^n\mathbf{Z})$$

of level $(N^+, N^- L)$, such that $f_L(T_q) = f_{\{n\}}(T_q)$ for every prime $q \nmid NL$ and $f_L(U_q) = f_{\{n\}}(U_q)$ for every prime $q|N$.

3.3.2 Construction of Cohomology Classes

Let $L \in \mathcal{L}^{\mathrm{indef}}$ be an indefinite vertex. Let $X_L := X_{N^+, N^- L}/\mathbf{Q}$ be the Shimura curve of level $(N^+, N^- L)$, let $J_L/\mathbf{Q}$ be the Jacobian variety of X_L, and let $\mathrm{Ta}_p(J_L)$ be its p-adic Tate module. As explained e.g. in [2], the Hecke algebra $\mathbb{T}_{N^+, N^- L}$ acts faithfully as a ring of $\mathbf{Q}$-rational endomorphisms of J_L. Theorem 5.17 of [3], as generalised in Proposition 4.4 of [18], states that there is an isomorphism of $\mathbf{Z}/p^n\mathbf{Z}[G_\mathbf{Q}]$-modules

$$\pi_L : \mathrm{Ta}_p(J_L)/I_L \cong T_{f_L} \cong T_{f,n}, \tag{7}$$

where $I_L \subset \mathbb{T}_{N^+, N^- L}$ denotes the kernel of $f_L \in S_2(N^+, N^- L; \mathbf{Z}/p^n\mathbf{Z})$, $T_{f_L} \in {}_{\mathbf{Z}/p^n\mathbf{Z}[G_\mathbf{Q}]}\mathrm{Mod}$ is the Galois representation attached in Sect. 2.1 to the eigenform f_L, and $T_{f,n} := T_f \otimes_\mathbf{Z} \mathbf{Z}/p^n\mathbf{Z}$ (so that $T_{f,n} \cong T_{f_{\{n\}}}$). Let Pic_L denote the Picard variety of X_L. Since I_L is not an Eisenstein ideal, the natural map $J_L(K) \hookrightarrow \mathrm{Pic}_L(K)$ induces an isomorphism $J_L(K)/I_L \cong \mathrm{Pic}_L(K)/I_L$. One can then define the morphism

$$\mathrm{k}_L : \mathrm{Pic}_L(K)/I_L \cong J_L(K)/I_L \xrightarrow{\delta} H^1(K, \mathrm{Ta}_p(J_L)/I_L) \overset{\pi_L}{\cong} H^1(K, T_{f,n}),$$

where δ denotes the map induced by the global Kummer map $J_L(K)\widehat{\otimes}\mathbf{Z}_p \hookrightarrow H^1(K, \mathrm{Ta}_p(J_L))$ after taking the quotients by I_L. Fix now a Heegner point $\mathbb{P}(L) \in \mathrm{Heeg}_{N^+, N^- L}(K) \subset X_L(H)$, let

$$\mathbf{P}(L) := \sum_{\sigma \in \mathrm{Gal}(H/K)} \mathbb{P}(L)^\sigma \in \mathrm{Pic}_L(K),$$

and define the global cohomology class

$$\kappa(L) := \mathrm{k}_L\big(\mathbf{P}(L)\big) \in H^1(K, T_{f,n}).$$

The class $\kappa(L)$ is uniquely determined, up to sign, by the choice of the isomorphism π_L in (7) [3]. It is then naturally associated with the pair (f, L) up to multiplication by a p-adic unit.

3.3.3 The Special Values

The constructions of Sects. 3.2.1 and 3.3.1 attach to a definite vertex $L \in \mathcal{L}^{\mathrm{def}}$ the quaternionic special value

$$\mathscr{L}_p(L) := \mathscr{L}_p(f_L/K) \in \mathbf{Z}/p^n\mathbf{Z}.$$

This is canonically attached to the pair (f, L) up to multiplication by a p-adic unit.

3.3.4 The First Reciprocity Law

Let $L \in \mathcal{L}^{\mathrm{def}}$, and let $\ell \in \mathcal{L}^{\mathrm{def}}$ be a n-admissible prime relative to (f, K) such that $\ell \nmid L$. Recall the residue map $\partial_\ell : H^1(K, T_{f,n}) \to H^1_{\mathrm{ord}}(K_\ell, T_{f,n}) \cong \mathbf{Z}/p^n\mathbf{Z}$ introduced in Sect. 2.3. The following theorem is a special case of [3, Theorem 4.1].

Theorem 3.3 *The equality*

$$\partial_\ell\big(\kappa(L\ell)\big) = \mathscr{L}_p(L)$$

holds in $\mathbf{Z}/p^n\mathbf{Z}$, up to multiplication by a p-adic unit.

3.3.5 The Second Reciprocity Law

Let $L \in \mathcal{L}^{\mathrm{indef}}$ be an indefinite vertex, and let $\ell \in \mathcal{L}^{\mathrm{def}}$ be a n-admissible prime which does not divide L. Recall the morphism $v_\ell : H^1(K, T_{f,n}) \to H^1_{\mathrm{fin}}(K_\ell, T_{f,n}) \cong \mathbf{Z}/p^n\mathbf{Z}$. The following theorem is a special case of [3, Theorem 4.2].

Theorem 3.4 *The equality*

$$v_\ell\big(\kappa(L)\big) = \mathscr{L}_p(L\ell)$$

holds in $\mathbf{Z}/p^n\mathbf{Z}$, up to multiplication by a p-adic unit.

Proof This is proved in Sect. 9 of [3] when $N^- \neq 1$ (i.e. X_{N^+,N^-} is not the classical modular curve of level $\Gamma_0(N)$), using Diamond–Taylor's generalisation of Ihara's Lemma to Shimura curves. On the other hand, making use of the classical Ihara's Lemma, the argument of loc. cit. also applies to the case $N^- = 1$. To handle the case $N^- = 1$, one may alternately go through the argument of Vatsal in [23, Sect. 6], where the case $N^- = 1$ and $n = 1$ of Theorem 3.4 is proved, and note that the proof applies also to the case $n > 1$. $\square$

4 Gross' Special Value Formula

In this section only, let $N = N^+N^-$ be a squarefree integer coprime with p, such that N^- is a product of an *odd* number of primes. Let $K/\mathbf{Q}$ be a quadratic imaginary field of discriminant coprime with Np. Let $g \in S_2(N^+, N^-; \mathbf{Z}_p)$ be a $\mathbf{Z}_p$-valued eigenform of level (N^+, N^-). We impose in this section the following hypotheses (cf. Assumption 3.1):

Assumption 4.1 The data $(\overline{\rho}_g, K, N^+, N^-)$ satisfy the following conditions:
1. N^- has an *odd* number of prime factors.
2. A prime divisor q of N divides N^- precisely if q is inert in $K/\mathbf{Q}$.
3. $\overline{\rho}_g : G_{\mathbf{Q}} \to \mathrm{GL}_2(\mathbf{F}_p)$ is surjective.
4. If $q \mid N^-$ and $q \equiv \pm 1 \pmod{p}$, then $\overline{\rho}_g$ is ramified at q.

Section 3.2.1 (see Eq. (6) and the discussion following it) attached to g and K a special value

$$\mathscr{L}_p(g/K) \in \mathbf{Z}_p,$$

well defined up to multiplication by a p-adic unit. Gross' formula compares this *quaternionic special value* to the *algebraic part* of the *complex special value* of g/K, defined as

$$L^{\mathrm{alg}}(g/K, 1) := \frac{L(g/K, 1)}{\Omega_g} \in \mathbf{Z}_p.$$

Here $L(g/K, s) := L(g, s) \cdot L(g, \epsilon_K, s)$ is the product of the Hecke complex L-series of g with that of the twist $g \otimes \epsilon_K$ of g by the quadratic character $\epsilon_K : (\mathbf{Z}/D\mathbf{Z})^* \to \{\pm 1\}$ of K. Moreover $\Omega_g \in \mathbf{C}^*$ is the *canonical Shimura period* of g. In order to define it, we briefly recall the definition of congruence numbers, referring to [18] for more details. Given a positive integer M and a factorisation $M = M^+ \cdot M^-$, write $\widehat{\mathbb{T}}_{M^+,M^-}$ for the p-adic completion of $\mathbb{T}_{M^+,M^-}$. For every eigenform $\phi \in S_2(M^+, M^-; \mathbf{Z}_p)$, define the *congruence ideal*

$$\eta_\phi(M^+, M^-) := \hat{\phi}\Big(\mathrm{Ann}_{\widehat{\mathbb{T}}_{M^+,M^-}}\big(\ker(\hat{\phi})\big)\Big) \subset \mathbf{Z}_p,$$

where $\hat{\phi} : \widehat{\mathbb{T}}_{M^+,M^-} \to \mathbf{Z}_p$ is the morphism induced by ϕ. One identifies $\eta_\phi(M^+, M^-)$ with the non-negative power of p that generates it, in other words we regard it as a positive integer. Then $\eta_\phi(M^+, M^-) = 1$ precisely if there is no non-trivial congruence modulo p between ϕ and eigenforms of level (L, M^-), for some divisor $L|M^+$. The canonical Shimura period mentioned above is defined as

$$\Omega_g := \frac{(g, g)}{\eta_g(N, 1)},$$

where (g, g) is the Petersson norm of $g \in S_2(\Gamma_0(N))$, and where we write again g to denote the composition of $g : \mathbb{T}_{N^+,N^-} \to \mathbf{Z}_p$ with the natural projection $\mathbb{T}_{N,1} \to \mathbb{T}_{N^+,N^-}$ in order to define $\eta_g(N, 1)$.

Before stating Gross' formula, we also need to introduce the Tamagawa exponents attached to g at primes dividing N. Let ϕ denote either g or its quadratic twist $g \otimes \epsilon_K$. Write as usual $T_\phi \in {}_{\mathbf{Z}_p[G_\mathbf{Q}]}\mathrm{Mod}$ for the p-adic representation attached to ϕ, and $A_\phi := T_\phi \otimes_{\mathbf{Z}_p} \mathbf{Q}_p/\mathbf{Z}_p$. Given a prime $q \mid N$, the *Tamagawa factor* $c_q(\phi)$ is defined to be the cardinality of (the finite group) $H^1(\mathrm{Frob}_q, A_\phi^{I_q})$, where I_q is the inertia subgroup of $G_{\mathbf{Q}_q}$. The *Tamagawa exponent* $t_q(g) = t_q(g/K)$ of g at q is the p-adic valuation of $c_q(g) \cdot c_q(g \otimes \epsilon_K)$. (If $q|N^-$ then $t_q(g)$ is the largest integer $n \geqslant 0$ such that the $G_\mathbf{Q}$-module $A_g[p^n]$ is unramified at q, cf. [18, Definition 3.3].)

The following result is due the the work many people, including Gross, Daghigh, Hatcher, Hui Xue, Ribet–Takahashi, and Pollack–Weston. We refer to [18] and Sect. 3.1 of [4] for more details and precise references.

Theorem 4.2 *The equality*

$$L^{\mathrm{alg}}(g/K, 1) = \mathscr{L}_p(g/K)^2 \cdot \prod_{q|N^-} p^{t_q(g)}$$

holds in $\mathbf{Z}_p$, up to multiplication by a p-adic unit.

Proof Combine Lemma 2.2 and Theorem 6.8 of [18]. □

5 A Theorem of Kato and Skinner–Urban

This section states the result of Kato–Skinner–Urban mentioned in the Introduction, proving the validity of the p-part of the Birch and Swinnerton-Dyer conjecture for weight-two newforms of analytic rank zero (under some technical conditions). Let $g \in S_2(1, N; \mathbf{Z}_p)$ be a weight-two newform with Fourier coefficients in $\mathbf{Z}_p$. Let $K/\mathbf{Q}$ be a quadratic imaginary field of discriminant coprime with Np. Consider as in the preceding section the algebraic part $L^{\mathrm{alg}}(g/K, 1) \in \mathbf{Z}_p$ of the complex special value of g/K. On the algebraic side, write as usual $A_g := T_g \otimes_{\mathbf{Z}_p} \mathbf{Q}_p/\mathbf{Z}_p \in_{\mathbf{Z}_p[G_\mathbf{Q}]} \mathrm{Mod}$ for the discrete representation attached to g. Assume that $p \nmid N$ is a prime of *good ordinary reduction* for g, i.e. that $g(T_p) \in \mathbf{Z}_p^*$. This implies that A_g fits into a short exact sequence of $\mathbf{Z}_p[G_{\mathbf{Q}_p}]$-modules

$$0 \to A_g^+ \to A_g \to A_g^- \to 0,$$

where $A_g^\pm \cong \mathbf{Q}_p/\mathbf{Z}_p$ as $\mathbf{Z}_p$-modules, and $G_{\mathbf{Q}_p}$ acts on A_g^+ via $\varepsilon \cdot \gamma_{g,p}^{-1}$, where $\varepsilon : G_{\mathbf{Q}_p} \to \mathbf{Z}_p^*$ denotes the p-adic cyclotomic character, and $\gamma_{g,p} : G_\mathbf{Q} \twoheadrightarrow G_{\mathbf{Q}_p}/I_{\mathbf{Q}_p} \to$

$\mathbf{Z}_p^*$ is the unramified character of $G_{\mathbf{Q}_p}$ sending an arithmetic Frobenius in $G_{\mathbf{Q}_p}/I_{\mathbf{Q}_p}$ to $g(U_p)$. Hence $A_g^- \cong \mathbf{Q}_p/\mathbf{Z}_p(\gamma_{g,p})$ is unramified, with $G_{\mathbf{Q}_p}$ acting via $\gamma_{g,p}$. Define the p-primary *Greenberg (strict) Selmer group* of g/K by

$$\mathrm{Sel}_{p^\infty}(K, g) := \ker\left(H^1(K_{Np}/K, A_g) \xrightarrow{\mathrm{res}_{Np}} \prod_{v|p} \frac{H^1(K_v, A_g)}{H^1_{\mathrm{ord}}(K_v, A_g)_{\mathrm{div}}} \times \prod_{v|N} H^1(K_v, A_g) \right),$$

where K_{Np}/K denotes the maximal algebraic extension of K which is unramified outside Np, and $H^1(K_{Np}/K, A_g)$ stands for $H^1(\mathrm{Gal}(K_{Np}/K), A_g)$. Moreover, the map res_{Np} denotes the direct sum of the restriction maps at v, running over the primes v of K which divide Np. Finally, for every prime $v|p$ of K, $H^1_{\mathrm{ord}}(K_v, A_g) \subset H^1(K_v, A_g)$ is the image of $H^1(K_v, A_g^+)$, and $H^1_{\mathrm{ord}}(K_v, A_g)_{\mathrm{div}}$ is its maximal p-divisible subgroup.

The following theorem combines the work of Kato [13] and Skinner–Urban [22] on the Iwasawa main conjecture for GL_2. More precisely, it follows from Theorem 3.29 of [22], applied to g and its quadratic twist $g \otimes \epsilon_K$, taking into account the algebraic Birch and Swinnerton-Dyer formulae proved by Mazur. For the precise statement in the level of generality required here, we refer to Theorem B in Skinner's preprint [21].

Recall the Tamagawa exponent $t_q(g) = t_q(g/K)$ attached to every prime $q|N$ in the preceding section.

Theorem 5.1 *Assume that*

1. *$p \nmid N$ and g is p-ordinary,*
2. *the residual representation $\overline{\rho}_g : G_{\mathbf{Q}} \to \mathrm{GL}_2(\mathbf{F}_p)$ is irreducible,*
3. *there exists a prime $q \mid N$ such that $\overline{\rho}_g$ is ramified at q.*

Then $L^{\mathrm{alg}}(g/K, 1) \neq 0$ if and only if $\mathrm{Sel}_{p^\infty}(K, g)$ is finite. In this case, the equality

$$L^{\mathrm{alg}}(g/K, 1) = \#\mathrm{Sel}_{p^\infty}(K, g) \prod_{q|N} p^{t_q(g)}$$

holds in $\mathbf{Z}_p$, up to multiplication by p-adic units.

6 Heegner Points and Shafarevich–Tate Groups

Let $A/\mathbf{Q}$ be an elliptic curve of conductor N. Fix a modular parametrisation

$$\pi_A : X_0(N) \longrightarrow A$$

of minimal degree $\deg(\pi_A)$. Let $K/\mathbf{Q}$ be a quadratic imaginary field of discriminant coprime with Np, satisfying the Heegner hypothesis that every prime divisor of N *splits* in $K/\mathbf{Q}$. Fix a Heegner point $\mathbb{P} \in \mathrm{Heeg}_{N,1}(H) \subset X_0(N)(H)$ (see Sect. 3.1.2, recalling that $X_0(N) = X_{N,1}$ and H/K is the Hilbert class field of K). Define the Heegner point over K

$$P_K := \mathrm{Trace}_{H/K}\big(\pi_A(\mathbb{P})\big) \in A(K).$$

The theorem of Gross–Zagier [11] states that P_K is a non-torsion point in $A(K)$ if and only the Hasse–Weil L-function $L(A/K, s)$ of A/K has a simple zero at $s = 1$. Moreover, according to the work of Kolyvagin [14], if P_K is a non-torsion point, the Mordell–Weil group $A(K)$ has rank one and the Shafarevich–Tate group $Ш(A/K)$ is finite. In this case, denote by

$$I_p(P_K) := p^{\mathrm{ord}_p[A(K):\, \mathbf{Z}P_K]}$$

the p-part of the index of $\mathbf{Z}P_K$ in $A(K)$. Write, as customary, $Ш(A/K)_{p^\infty}$ for the p-primary part of the Shafarevich–Tate group of A/K. The following theorem is the main result of this note and will imply Theorem A of the Introduction.

Theorem 6.1 *Assume that $A/\mathbf{Q}$ is semistable, and that $p > 7$ is a prime which does not divide* $\deg(\pi_A)$. *Assume furthermore that* $a_p(A) \not\equiv 0, 1 \pmod p$, *resp.* $a_p(A) \not\equiv 0, \pm 1 \pmod p$ *when p is split, resp. inert in K, and that all primes dividing N are split in K. If $L(A/K, s)$ has a simple zero at $s = 1$, then*

$$I_p(P_K)^2 = \#Ш(A/K)_{p^\infty}.$$

The proof of Theorem 6.1 is given in Sect. 6.5.

6.1 Setting and Notations

Assume from now on that the assumptions of Theorem 6.1 are satisfied, and fix a positive integer n such that

$$n > 2 \cdot \max\Big\{\mathrm{ord}_p\big(I_p(P_K)\big),\ \mathrm{ord}_p\big(\#Ш(A/K)_{p^\infty}\big)\Big\}. \quad (8)$$

Let $f = f_A \in S_2(\Gamma_0(N), \mathbf{Z})$ be the weight-two newform of level N attached to $A/\mathbf{Q}$ by the modularity theorem. With the notations of Sect. 2.1, one considers

$$f \in S_2(N, 1; \mathbf{Z}_p);\ \ N^+ := N;\ \ N^- := 1.$$

Note that, since f is q-new at every prime $q \mid N$, one can consider $f \in S_2(N/m, m; \mathbf{Z}_p)$, for every positive divisor m of N. In other words, $f : \mathbb{T}_N := \mathbb{T}_{N,1} \to \mathbf{Z}_p$ factorises through the m-new quotient $\mathbb{T}_{N/m,m}$ of $\mathbb{T}_N$, for every positive divisor m of N. As in Sect. 3.3, for every $m \in \mathbf{N} \cup \{\infty\}$ let $f_{\{m\}} \in S_2(N, 1; \mathbf{Z}_p/p^m\mathbf{Z}_p)$ denote the reduction of f modulo p^m.

Lemma 6.2 1. *The data* (f, N^+, N^-, K, p) *satisfy Assumption 3.1.*
2. $f_{\{m\}}$ *satisfies Assumption 2.1 for every* $m \in \mathbf{N} \cup \{\infty\}$.

Proof Parts 1, 2, 4 and 5 of Assumption 3.1 are satisfied since A is ordinary at p and $N^- = 1$. As $A/\mathbf{Q}$ is semistable and $p > 7$, Assumption 3.1(3) holds by a result of Mazur [16]. Moreover, the representation $T_{f,m} = T_f \otimes \mathbf{Z}_p/p^m\mathbf{Z}_p$ associated with $f_{\{m\}}$ is ordinary at p, hence Assumption 2.1(2) holds. Finally, since $p \nmid \deg(\pi_A)$, Assumption 2.1(3) holds by a result of Ribet [19], as explained in Lemma 2.2 of [3]. □

With the notations of Sect. 3.3.1, write $\mathcal{L}_m$ for the graph associated to $f_{\{m\}}$, for $m \in \mathbf{N}$. Let $L \in \mathcal{L}_m$ and let $f_L \in S_2(N, L; \mathbf{Z}/p^m\mathbf{Z})$ be the L-level raising of $f_{\{m\}}$ (cf. Sect. 3.3.1). We say that f_L *can be lifted to a true modular form* if there exists a $\mathbf{Z}_p$-valued eigenform $g = g_L \in S_2(N, L; \mathbf{Z}_p)$ of level (N, L) whose reduction modulo p^m equals f_L (i.e. such that $f_L = g_{\{m\}}$).

6.2 Level Raising at One Prime

Let $\ell \in \mathcal{L}_n^{\mathrm{def}}$ be an n-admissible prime relative to (f, K). The next result shows that the conclusion of Theorem 6.1 holds under certain assumptions.

Proposition 6.3 *Assume that* f_ℓ *can be lifted to a true modular form. Moreover, assume that the map* $A(K) \otimes \mathbf{Z}/p^n\mathbf{Z} \to A(K_\ell) \otimes \mathbf{Z}/p^n\mathbf{Z}$ *(induced by the natural inclusion* $A(K) \hookrightarrow A(K_\ell)$*) is injective. Then*

$$I_p(P_K)^2 = \#\text{Ш}(A/K)_{p^\infty}.$$

The rest of this section will be devoted to the proof Proposition 6.3. Section 3.3.2 attaches to $f_{\{n\}}$ and $1 \in \mathcal{L}^{\mathrm{indef}}$ a global cohomology class $\kappa(1) \in H^1(K, T_{f,n})$. The representation $T_{f,n}$ attached to $f_{\{n\}}$ is nothing but the p^n-torsion submodule A_{p^n} of $A = A(\overline{\mathbf{Q}})$. Since $\overline{\rho}_f$ is irreducible, π_A induces isomorphisms of $\mathbf{Z}_p[G_\mathbf{Q}]$-modules $\pi_A : \mathrm{Ta}_p(J)/I_f \cong T_f$ and $\pi_{A,n} : \mathrm{Ta}_p(J)/I_{f,n} \cong T_{f,n}$, where $J/\mathbf{Q} = \mathrm{Jac}(X_0(N))$ is the Jacobian variety of $X_0(N)$, $I_f := \ker(f)$ and $I_{f,n} := \ker\left(f_{\{n\}}\right)$. One can then take $\pi_{A,n} = \pi_1$ in (7), and retracing definitions it follows that, up to multiplication by p-adic units,

$$\kappa(1) = \delta(P_K) \in \mathrm{Sel}(K, f_{\{n\}}), \tag{9}$$

where δ denotes the global Kummer map $A(K)/p^n \hookrightarrow H^1(K, A_{p^n})$. We observe that the class $\kappa(1)$ belongs to the Selmer group $\mathrm{Sel}(K, f_{\{n\}})$, defined in Sect. 2.2 by ordinary conditions (which can be imposed in the current context in light of Lemma 6.2), since this Selmer group coincides with the usual p^n-Selmer group of A in which the local conditions are described in terms of the local Kummer maps. Indeed, our assumption on $a_p(A)$ being $\not\equiv 1$ or $\not\equiv \pm 1 \pmod p$ implies that the local Selmer conditions at p agree (see for example [9]). As for the primes dividing N, this is a direct consequence of the theory of non-archimedean uniformisation for A. This yields the equality up to units

$$I_p(P_K) = v_\ell\big(\kappa(1)\big) \in H^1_{\text{fin}}(K_\ell, T_{f,n}) \cong \mathbf{Z}/p^n\mathbf{Z} \tag{10}$$

(see Sect. 2.3 for the last isomorphism). To see this, consider the composition

$$A(K) \otimes \mathbf{Z}/p^n\mathbf{Z} \to A(K_\ell) \otimes \mathbf{Z}/p^n\mathbf{Z} \overset{\delta}{\hookrightarrow} H^1_{\text{fin}}(K_\ell, T_{f,n}) \cong \mathbf{Z}/p^n\mathbf{Z}. \tag{11}$$

Since $p > 7$, one has $A(K)_p = 0$ by Mazur's theorem. Moreover, as we are assuming $\text{ord}_{s=1}L(A/K, s) = 1$, the Gross–Zagier–Kolyvagin theorem gives that $A(K)$ has rank one. It follows that $A(K) \otimes \mathbf{Z}/p^n\mathbf{Z} \cong \mathbf{Z}/p^n\mathbf{Z}$. Since by assumption the first map in (11) is injective, the composition (11) is an isomorphism, and the claim (10) follows. Theorem 3.4 then yields the equality

$$I_p(P_K) = \mathscr{L}_p(\ell) \in \mathbf{Z}/p^n\mathbf{Z}, \tag{12}$$

up to multiplication by p-adic units. Let now $g \in S_2(N, \ell; \mathbf{Z}_p)$ be a $\mathbf{Z}_p$-valued eigenform of level (N, ℓ) lifting f_ℓ. Combining Theorem 4.2 with Theorem 5.1 yields (up to p-adic units)

$$\mathscr{L}_p(g/K)^2 \cdot p^{t_\ell(g)} \overset{\text{Theorem 4.2}}{=} L^{\text{alg}}(g/K, 1) \overset{\text{Theorem 5.1}}{=} \#\text{Sel}_{p^\infty}(K, g) \cdot p^{t_\ell(g)}. \tag{13}$$

More precisely, note that g satisfies the assumptions of Theorems 4.2 and 5.1 by Lemma 6.2. Moreover, as explained in the proof of Lemma 2.2 of [3], the assumption $p \nmid \deg(\pi_A)$ and Ribet's lowering the level theorem [19] imply that $A_p \cong A_{g,1}$ is ramified at every prime $q \mid N$. By the definition of $t_q(g)$, this gives $\prod_{q|N\ell} p^{t_q(g)} = p^{t_\ell(g)}$, and the first equality in (13). Since by construction $\mathscr{L}_p(g/K) \equiv \mathscr{L}_p(\ell) \pmod{p^n}$, and $I_p(P_K)$ is non-zero in $\mathbf{Z}/p^n\mathbf{Z}$ by (8), $\mathscr{L}_p(g/K) \neq 0$ by (12), hence $L^{\text{alg}}(g/K, 1) \neq 0$, and the second equality in (13) follows by Theorem 5.1. Combining Eqs. (12) and (13) give the identity

$$I_p(P_K)^2 = \#\text{Sel}_{p^\infty}(K, g). \tag{14}$$

It then remains to compare the cardinality of the p-primary Selmer group $\text{Sel}_{p^\infty}(K, g)$ with that of the p-primary part of the Shafarevich–Tate group $\text{Ш}(A/K)$. In order to do that, one first notes that

$$\text{Sel}(K, f_\ell) \cong \text{Sel}_{p^\infty}(K, g), \tag{15}$$

where $\text{Sel}(K, f_\ell)$ is the p^n-Selmer group attached in Sect. 2.1 to $f_\ell = g_{\{n\}}$. (Note that f_ℓ satisfies Assumption 2.1, thanks to Lemma 6.2.) By the irreducibility of A_p and our assumptions on $a_p(A)$, it is easily seen that the natural map $\text{Sel}(K, f_\ell) \to \text{Sel}_{p^\infty}(K, g)[p^n]$ is an isomorphism (cf. [9]). On the other hand, Eqs. (12), (13) and

(8) imply that $p^n > \#\mathrm{Sel}_{p^\infty}(K, g)$, hence (15) follows. One is thus reduced to compare the cardinality of $\mathrm{Sel}(K, f_\ell)$ to that of $Ш(A/K)_{p^\infty}$. Kummer theory inserts $Ш(A/K)_{p^\infty}$ in a short exact sequence

$$0 \to A(K) \otimes \mathbf{Z}/p^n\mathbf{Z} \to \mathrm{Sel}(K, f_{\{n\}}) \to Ш(A/K)_{p^\infty} \to 0$$

(one uses again $p^n > \#Ш(A/K)_{p^\infty}$, which follows by (8)). By the discussion above this gives

$$\#Ш(A/K)_{p^\infty} = p^{-n} \cdot \#\mathrm{Sel}(K, f_{\{n\}}). \tag{16}$$

We claim that

$$\mathrm{Sel}^{(\ell)}(K, f_{\{n\}}) = \mathrm{Sel}(K, f_{\{n\}}). \tag{17}$$

where the suffix (ℓ) indicates condition at ℓ relaxed. To prove this, let $x \in \mathrm{Sel}^{(\ell)}(K, f_n)$ be a Selmer class relaxed at ℓ; we have to show that $x \in \mathrm{Sel}(K, f_{\{n\}})$, i.e. that its residue $\partial_\ell(x)$ at ℓ vanishes. Since (11) is an isomorphism, there exists a class $y \in A(K)/p^n \hookrightarrow \mathrm{Sel}(K, f_{\{n\}})$ such that $\mathrm{res}_\ell(y) = v_\ell(y) \in H^1_{\mathrm{fin}}(K_\ell, T_{f,n}) \cong \mathbf{Z}/p^n\mathbf{Z}$ is a unit modulo p^n. For every prime v of K, let $\langle -, - \rangle_v : H^1(K_v, T_{f,n}) \times H^1(K_v, T_{f,n}) \to H^2(K_v, \mu_{p^n}) \cong \mathbf{Z}/p^n\mathbf{Z}$ be the perfect local Tate pairing attached to the Weil pairing $T_{f,n} \times T_{f,n} \to \mu_{p^n}$. The subspace $H^1_{\mathrm{fin}}(K_v, T_{f,n})$ (resp., $H^1_{\mathrm{ord}}(K_v, T_{f,n})$ for $v | \ell N p$) is maximal isotropic for $\langle -, - \rangle_v$, i.e. it is equal to its own orthogonal complement under $\langle -, - \rangle_v$. By the reciprocity law of global class field theory and the definition of $\mathrm{Sel}^{(\ell)}(K, f_{\{n\}})$:

$$0 = \sum_v \langle \mathrm{res}_v(x), \mathrm{res}_v(y) \rangle_v = \langle \mathrm{res}_\ell(x), \mathrm{res}_\ell(y) \rangle_\ell = \langle \partial_\ell(x), v_\ell(y) \rangle_\ell ,$$

where the first sum runs over all primes of K. Since $v_\ell(y)$ generates $H^1_{\mathrm{fin}}(K_\ell, T_{f,n})$ by assumption and the Tate local duality induces a perfect pairing between this finite part and the ordinary (or singular) part $H^1_{\mathrm{ord}}(K_\ell, T_{f,n})$, this implies $\partial_\ell(x) = 0$, as was to be shown.

Using again that (11) is an isomorphism, together with Eq. (2), one deduces the exact sequence

$$0 \to \mathrm{Sel}(K, f_\ell) \to \mathrm{Sel}^{(\ell)}(K, f_{\{n\}}) \xrightarrow{v_\ell} H^1_{\mathrm{fin}}(K_\ell, T_{f,n}) \cong \mathbf{Z}/p^n\mathbf{Z} \to 0. \tag{18}$$

This allows us to conclude the proof of the proposition, as it gives

$$\begin{aligned} I_p(P_K)^2 &\overset{(14)}{=} \#\mathrm{Sel}_{p^\infty}(K, g) \overset{(15)}{=} \#\mathrm{Sel}(K, f_\ell) \overset{(18)}{=} p^{-n} \cdot \#\mathrm{Sel}^{(\ell)}(K, f_{\{n\}}) \\ &\overset{(17)}{=} p^{-n} \cdot \#\mathrm{Sel}(K, f_{\{n\}}) \overset{(16)}{=} \#Ш(A/K)_{p^\infty}. \end{aligned}$$

6.3 Level Raising at Three n-admissible Primes

Write in this section $\mathcal{L} = \mathcal{L}_{2n}$. Fix three primes ℓ_1, ℓ_2 and ℓ_3 in $\mathcal{L}$ (so that ℓ_1, ℓ_2 and ℓ_3 are $2n$-admissible primes relative to (f, K)).

Since Assumption 3.1 is satisfied by Lemma 6.2, Sect. 3.3.2 attaches to $(f, 1)$ and $(f, \ell_1\ell_2)$ Selmer classes

$$\kappa(1) = \delta(P_K) \in \mathrm{Sel}(K, f_{\{2n\}}); \quad \kappa(\ell_1\ell_2) \in \mathrm{Sel}(K, f_{\ell_1\ell_2}).$$

The fact that the first class belongs to $\mathrm{Sel}(K, f_{\{2n\}})$ was explained after Eq. (9). A similar argument applies to the second class, recalling that it arises as the Kummer image of a Heegner point on the Shimura curve $X_{N,\ell_1\ell_2}$ and invoking the Cerednik–Drinfeld theory of non-archimedean uniformisation for this curve at the primes ℓ_1 and ℓ_2 (see [3] for more details). If $\kappa(\ell_1\ell_1) \neq 0$, set

$$\widetilde{\kappa}(\ell_1\ell_2) := p^{t-1} \cdot \kappa(\ell_1\ell_2) \in H^1(K, T_{f,1}),$$

where $t \leqslant 2n$ is the smallest positive integer such that $p^t \cdot \kappa(\ell_1\ell_2) = 0$ (and we identify $H^1(K, T_{f,1})$ with $H^1(K, T_{f,2n})[p]$, which is possible since $T_{f,1}^{G_K} = 0$). If $\kappa(\ell_1\ell_2) = 0$, set $\widetilde{\kappa}(\ell_1\ell_2) := 0$ (in $H^1(K, T_{f,1})$). Recall the morphisms $v_{\ell_j} : H^1(K, T_{f,k}) \to H^1_{\mathrm{fin}}(K_{\ell_j}, T_{f,k}) \cong \mathbf{Z}/p^k\mathbf{Z}$ ($k \geqslant 1$). The aim of this section is to prove the following proposition.

Proposition 6.4 *Assume that $f_{\ell_1\ell_2\ell_3}$ can be lifted to a true modular form of level $(N, \ell_1\ell_2\ell_3)$. Assume moreover that the restriction map $A(K)/p^n \to A(K_{\ell_1})/p^n$ at ℓ_1 is injective, and that $v_{\ell_3}\big(\widetilde{\kappa}(\ell_1\ell_2)\big) \neq 0$. Then*

$$I_p(P_K)^2 = \#\text{Ш}(A/K)_{p^\infty}.$$

The rest of this section will be devoted to the proof of Proposition 6.4. In particular, assume from now on that the assumptions of the proposition are satisfied.

Let $r \leqslant 2n$ be a positive integer. Since ℓ_1, ℓ_2 and ℓ_3 are $2n$-admissible primes, they are also r-admissible primes relative to (f, K, p). For every divisor m of $\ell_1\ell_2\ell_3$ write

$$\mathrm{Sel}_{p^r}(K, f_m) \subset H^1(K, T_{f,r}); \quad \mathrm{Sel}_{p^r}(K, f) := \mathrm{Sel}(K, f_{\{r\}})$$

to denote the Selmer group attached to the reduction modulo p^r of the mod-p^{2n} form f_m. For every $L \in \mathcal{L}$, let

$$\mathrm{Sel}_{p^r}^{(L)}(K, f_m) \subset H^1(K, T_{f,r})$$

be the relaxed Selmer group at L, i.e. the Selmer group defined by the same local conditions used to define $\mathrm{Sel}_{p^r}(K, f_m)$ at every prime of K which does not divide L, and by imposing *no* local condition at every prime of K dividing L. As explained in Sect. 3 of [3] (see in particular Proposition 3.3 and the references therein), we can enlarge $\ell_1\ell_2\ell_3$ to an integer $L \in \mathcal{L}$ which *controls* the Selmer group. More precisely, there exists $L \in \mathcal{L}$, divisible by $\ell_1\ell_2\ell_3$, such that the restriction map $\mathrm{Sel}_{p^{2n}}(K, f_{\{2n\}}) \to \bigoplus_{\ell|L} H^1(K_\ell, T_{f,2n})$ is injective and

$$\mathrm{Sel}_{p^{2n}}^{(L)}(K, f) \cong \left(\mathbf{Z}/p^{2n}\mathbf{Z}\right)^{\#L}$$

is free of rank $\#L$ over $\mathbf{Z}/p^{2n}\mathbf{Z}$, where $\#L := \#\{\ell : \ell \text{ prime and } \ell|L\}$. Fix from now on such an L. For every element $0 \neq x \in \mathrm{Sel}_{p^{2n}}^{(L)}(K, f)$, denote by $\mathrm{ord}_p(x)$ the largest integer such that $x \in p^{\mathrm{ord}_p(x)} \cdot \mathrm{Sel}_{p^{2n}}^{(L)}(K, f)$.

Theorems 3.3 and 3.4 yield the equality (up to multiplication by p-adic units)

$$I_p(P_K) = v_{\ell_1}\left(\kappa(1)\right) \overset{\text{Theorem 3.4}}{=} \mathscr{L}_p(\ell_1) \overset{\text{Theorem 3.3}}{=} \partial_{\ell_2}\left(\kappa(\ell_1\ell_2)\right) \in \mathbf{Z}/p^{2n}\mathbf{Z}, \qquad (19)$$

the first equality being a consequence of the injectivity of the localisation map $A(K)/p^n \hookrightarrow A(K_{\ell_1})/p^n$, as explained in the proof Proposition 6.3 (see (10)). By (8) one deduces

$$\xi(\ell_1\ell_2) := \mathrm{ord}_p\left(\kappa(\ell_1\ell_2)\right) \leqslant \mathrm{ord}_p\left(\partial_{\ell_2}(\kappa(\ell_1\ell_2))\right) = \mathrm{ord}_p\left(I_p(P_K)\right) < n. \qquad (20)$$

Let $\widehat{\kappa}(\ell_1\ell_2) \in \mathrm{Sel}_{p^{2n}}^{(L)}(K, f)$ be such that $p^{\xi(\ell_1\ell_2)} \cdot \widehat{\kappa}(\ell_1\ell_2) = \kappa(\ell_1\ell_2) \in \mathrm{Sel}_{p^{2n}}^{(L)}(K, f)$. Consider the natural map

$$\mathrm{Sel}_{p^{2n}}^{(L)}(K, f) \longrightarrow \mathrm{Sel}_{p^n}^{(L)}(K, f) \qquad (21)$$

induced by the projection $T_{f,2n} \twoheadrightarrow T_{f,n}$, and write $\kappa'(\ell_1\ell_2) \in \mathrm{Sel}_{p^n}^{(L)}(K, f)$ for the image of $\widehat{\kappa}(\ell_1\ell_2)$. Note that, while $\widehat{\kappa}(\ell_1\ell_2)$ is well-defined only up to elements in $\mathrm{Sel}_{p^{2n}}^{(L)}(K, f)[p^{\xi(\ell_1\ell_2)}]$, $\kappa'(\ell_1\ell_2)$ depends only on $\kappa(\ell_1\ell_2)$.

Lemma 6.5 *The class $\kappa'(\ell_1\ell_2)$ enjoys the following properties:*

1. *$\kappa'(\ell_1\ell_2) \in \mathrm{Sel}_{p^n}(K, f_{\ell_1\ell_2})$;*
2. *$\kappa'(\ell_1\ell_2)$ has exact order p^n;*
3. *$\partial_{\ell_2}\left(\kappa(\ell_1\ell_2)\right) \pmod{p^n} = p^{\xi(\ell_1\ell_2)} \cdot \partial_{\ell_2}\left(\kappa'(\ell_1\ell_2)\right) \in \mathbf{Z}/p^n\mathbf{Z}$, up to multiplication by units in $(\mathbf{Z}/p^n\mathbf{Z})^*$;*
4. *$v_{\ell_3}\left(\kappa'(\ell_1\ell_2)\right) \in (\mathbf{Z}/p^n\mathbf{Z})^*$ and $v_{\ell_3}\left(\kappa(\ell_1\ell_2)\right) \pmod{p^n} = p^{\xi(\ell_1\ell_2)}$, up to units in $(\mathbf{Z}/p^n\mathbf{Z})^*$.*

Proof Since $\mathrm{Sel}_{p^{2n}}^{(L)}(K, f)$ is free over $\mathbf{Z}/p^{2n}\mathbf{Z}$, $\widehat{\kappa}(\ell_1\ell_2)$ has order p^{2n}. If $x \in \mathrm{Sel}_{p^{2n}}^{(L)}(K, f) \subset H^1(K, T_{f,2n})$ belongs to the kernel of the map (21), then x comes from a class in $H^1(K, p^n \cdot T_{f,2n})$, hence is killed by p^n. It follows that $\kappa'(\ell_1\ell_2)$ has order p^n, thus proving part 2. To show part 1, i.e. that $\kappa'(\ell_1\ell_2)$ belongs to $\mathrm{Sel}_{p^n}(K, f_{\ell_1\ell_2})$, one has to prove that $v_q\left(\kappa'(\ell_1\ell_2)\right) = 0$ for $q|\ell_1\ell_2$, and that $\partial_\ell\left(\kappa'(\ell_1\ell_2)\right) = 0$ for every ℓ dividing $L/\ell_1\ell_2$. This follows by the fact that $p^{\xi(\ell_1\ell_2)} \cdot \widehat{\kappa}(\ell_1\ell_2)$ already satisfies these properties, and by the fact that $\xi(\ell_1\ell_2) < n$ (see 20). Indeed, for $? \in \{\mathrm{fin}, \mathrm{sing}\}$ and

$k \in \{n, 2n\}$, there is an isomorphism $H^1_?(K, T_{f,k}) \cong \mathbf{Z}/p^k\mathbf{Z}$ (cf. Sect. 2.3), and the morphism $H^1_?(K_\ell, T_{f,2n}) \to H^1_?(K_\ell, T_{f,n})$ induced by $T_{f,2n} \twoheadrightarrow T_{f,n}$ corresponds to the canonical projection $\mathbf{Z}/p^{2n}\mathbf{Z} \to \mathbf{Z}/p^n\mathbf{Z}$. Part 3 also follows by the last argument. Finally, let t be the order of $\kappa(\ell_1\ell_2) \in \mathrm{Sel}^{(L)}_{p^{2n}}(K, f_{\ell_1\ell_2})$, so that $p^{\xi(\ell_1\ell_2)+t-1} \cdot \widehat{\kappa}(\ell_1\ell_2) = \widetilde{\kappa}(\ell_1\ell_2)$, and $\xi(\ell_1\ell_2) + t = 2n$. By assumption, $v_{\ell_3}\big(\widetilde{\kappa}(\ell_1\ell_2)\big) \neq 0$, which implies that $v_{\ell_3}\big(\widehat{\kappa}(\ell_1\ell_2)\big)$ has order p^{2n} in $\mathbf{Z}/p^{2n}\mathbf{Z}$, i.e. it is a unit modulo p^{2n}. Since, as remarked above, $v_{\ell_3}\big(\kappa'(\ell_1\ell_2)\big)$ is the image of $v_{\ell_3}\big(\widehat{\kappa}(\ell_1\ell_2)\big)$ under the projection $\mathbf{Z}/p^{2n}\mathbf{Z} \twoheadrightarrow \mathbf{Z}/p^n\mathbf{Z}$, Part 4 follows. □

Thanks to part 3 of the preceding lemma, (19) can be rewritten in term of the class $\kappa'(\ell_1\ell_2)$, i.e.

$$I_p(P_K) = p^{\xi(\ell_1\ell_2)} \cdot \partial_{\ell_2}\big(\kappa'(\ell_1\ell_2)\big) \in \mathbf{Z}/p^n\mathbf{Z}.$$

(The latter equality, valid up to multiplication by p-adic units, takes place now in $\mathbf{Z}/p^n\mathbf{Z}$, while (19) was an equality in $\mathbf{Z}/p^{2n}\mathbf{Z}$.) Moreover, Theorem 3.4 and Part 4 of the preceding lemma give

$$\mathscr{L}_p(\ell_1\ell_2\ell_3) \pmod{p^n} \overset{\text{Theorem 3.4}}{=} v_{\ell_3}\big(\kappa(\ell_1\ell_2)\big) \pmod{p^n} = p^{\xi(\ell_1\ell_2)} \in \mathbf{Z}/p^n\mathbf{Z}$$

(as usual up to p-adic units). We now make use of the assumption that $f_{\ell_1\ell_2\ell_3}$ can be lifted to a true modular form $g \in S_2(N, \ell_1\ell_2\ell_3; \mathbf{Z}_p)$. Using Theorem 4.2 and Theorem 5.1 one proves, by the same argument used in the proof of Proposition 6.3, that up to p-adic units

$$\mathscr{L}_p(g/K)^2 = \#\mathrm{Sel}_{p^\infty}(K, g) = \#\mathrm{Sel}_{p^n}(K, f_{\ell_1\ell_2\ell_3}).$$

(To justify the second equality, note that $\mathscr{L}_p(\ell_1\ell_2\ell_3) \pmod{p^n} = p^{\xi(\ell_1\ell_2)}$ is non-zero in $\mathbf{Z}/p^n\mathbf{Z}$, as follows by (20), and proceed as in the proof of Eq. (15) in the proof of Proposition 6.3.) The preceding three equations combine to give

$$I_p(P_K)^2 = p^{2\cdot\mathrm{ord}_p\left(\partial_{\ell_2}\left(\kappa'(\ell_1\ell_2)\right)\right)} \cdot \#\mathrm{Sel}_{p^n}(K, f_{\ell_1\ell_2\ell_3}). \tag{22}$$

(Here, given $0 \neq x \in \mathbf{Z}/p^n\mathbf{Z}$, $\mathrm{ord}_p(x)$ denotes the positive integer s.t. $\big(p^{\mathrm{ord}_p(x)}\big) = (x)$ as ideals of $\mathbf{Z}/p^n\mathbf{Z}$.) The proof of Proposition 6.4 will then result combining Eq. (22) with the following lemma.

Lemma 6.6

$$\#\text{Ш}(A/K)_{p^\infty} = p^{2\cdot\mathrm{ord}_p\left(\partial_{\ell_2}\left(\kappa'(\ell_1\ell_2)\right)\right)} \cdot \#\mathrm{Sel}_{p^n}(K, f_{\ell_1\ell_2\ell_3}).$$

Proof Recall that by assumption the localisation map $A(K)/p^n \hookrightarrow A(K_{\ell_1})/p^n$ is injective. As in the proof of Proposition 6.3, this implies

$$\#\mathrm{Sel}_{p^n}(K, f_{\ell_1}) = \#\text{Ш}(A/K)_{p^\infty}. \tag{23}$$

Given this, the proof naturally breaks into two parts. One first compares the Selmer groups $\mathrm{Sel}_{p^n}(K, f_{\ell_1})$ and $\mathrm{Sel}_{p^n}(K, f_{\ell_1\ell_2})$, and proves the equality

$$\#\mathrm{Sel}_{p^n}(K, f_{\ell_1\ell_2}) = p^{n-2\cdot\mathrm{ord}_p\left(\partial_{\ell_2}\left(\kappa'(\ell_1\ell_2)\right)\right)} \cdot \#\mathrm{Sel}_{p^n}(K, f_{\ell_1}). \tag{24}$$

One then compares the Selmer groups $\mathrm{Sel}_{p^n}(K, f_{\ell_1\ell_2})$ and $\mathrm{Sel}_{p^n}(K, f_{\ell_1\ell_2\ell_3})$, and shows that

$$\#\mathrm{Sel}_{p^n}(K, f_{\ell_1\ell_2}) = p^n \cdot \#\mathrm{Sel}_{p^n}(K, f_{\ell_1\ell_2\ell_3}). \tag{25}$$

The lemma will then follow by combining the preceding three equations.

By Poitou–Tate duality, as formulated e.g. in [20, Theorem 1.7.3] (see also [17, Chap. I]), and the very definitions of the Selmer groups (see Sect. 2.2), there is an exact sequence

$$\begin{aligned} 0 \to \mathrm{Sel}_{p^n}(K, f_{\ell_1}) &\to \mathrm{Sel}_{p^n}^{(\ell_2)}(K, f_{\ell_1\ell_2}) \\ &\xrightarrow{\partial_{\ell_2}} H^1_{\mathrm{sing}}(K_{\ell_2}, T_{f,n}) \cong H^1_{\mathrm{fin}}(K_{\ell_2}, T_{f,n})^\vee \xrightarrow{v_{\ell_2}^\vee} \mathrm{Sel}_{p^n}(K, f_{\ell_1})^\vee, \end{aligned}$$

where $(\cdot)^\vee := \mathrm{Hom}(\cdot, \mathbf{Z}/p^n\mathbf{Z})$, the isomorphism is induced by the local Tate pairing (cf. the proof of Proposition 6.3), and $v_{\ell_2}^\vee$ refers to the dual of the morphism $v_{\ell_2} = \mathrm{res}_{\ell_2} : \mathrm{Sel}_{p^n}(K, f_{\ell_1}) \to H^1_{\mathrm{fin}}(K, T_{f,n})$. Similarly, one has the exact sequence

$$\begin{aligned} 0 \to \mathrm{Sel}_{p^n}(K, f_{\ell_1\ell_2}) &\to \mathrm{Sel}_{p^n}^{(\ell_2)}(K, f_{\ell_1\ell_2}) \\ &\xrightarrow{v_{\ell_2}} H^1_{\mathrm{fin}}(K_{\ell_2}, T_{f,n}) \cong H^1_{\mathrm{sing}}(K_{\ell_2}, T_{f,n})^\vee \xrightarrow{\partial_{\ell_2}^\vee} \mathrm{Sel}_{p^n}(K, f_{\ell_1\ell_2})^\vee. \end{aligned}$$

The existence of these exact sequences yields

$$\begin{aligned} \#\partial_{\ell_2}\left(\mathrm{Sel}_{p^n}(K, f_{\ell_1\ell_2})\right) \cdot \#v_{\ell_2}\left(\mathrm{Sel}_{p^n}^{(\ell_2)}(K, f_{\ell_1\ell_2})\right) &= p^n \\ = \#\partial_{\ell_2}\left(\mathrm{Sel}_{p^n}^{(\ell_2)}(K, f_{\ell_1\ell_2})\right) \cdot \#v_{\ell_2}\left(\mathrm{Sel}_{p^n}(K, f_{\ell_1})\right). \end{aligned} \tag{26}$$

We claim that

$$\partial_{\ell_2}\left(\mathrm{Sel}_{p^n}(K, f_{\ell_1\ell_2})\right) = \partial_{\ell_2}\left(\kappa'(\ell_1\ell_2)\right) \cdot \mathbf{Z}/p^n\mathbf{Z} = \partial_{\ell_2}\left(\mathrm{Sel}_{p^n}^{(\ell_2)}(K, f_{\ell_1\ell_2})\right). \tag{27}$$

This would easily imply Eq. (24). Indeed, Eqs. (26) and (27) would then give

$$\frac{\#\partial_{\ell_2}\left(\mathrm{Sel}_{p^n}^{(\ell_2)}(K,f_{\ell_1\ell_2})\right)}{\#v_{\ell_2}\left(\mathrm{Sel}_{p^n}^{(\ell_2)}(K,f_{\ell_1\ell_2})\right)} \overset{(27)}{=} \frac{\left(\#\partial_{\ell_2}\left(\mathrm{Sel}_{p^n}(K,f_{\ell_1\ell_2})\right)\right)^2}{\#\partial_{\ell_2}\left(\mathrm{Sel}_{p^n}(K,f_{\ell_1\ell_2})\right)\cdot\#v_{\ell_2}\left(\mathrm{Sel}_{p^n}^{(\ell_2)}(K,f_{\ell_1\ell_2})\right)}$$
$$\overset{(26)\text{ and }(27)}{=} p^{n-2\mathrm{ord}_p\left(\partial_{\ell_2}\left(\kappa'(\ell_1\ell_2)\right)\right)}.$$

On the other hand, the (trivial part of the) exact sequences above show that the first term in the previous equation is equal to the ratio $\#\mathrm{Sel}_{p^n}(K,f_{\ell_1\ell_2})/\#\mathrm{Sel}_{p^n}(K,f_{\ell_1})$, and Eq. (24) would follow. In order to prove Eq. (27), note that

$$\mathrm{Sel}_{p^n}^{(\ell_2)}(K,f_{\ell_1\ell_2}) = \kappa'(\ell_1\ell_2)\cdot\mathbf{Z}/p^n\mathbf{Z}\oplus Ш_{\ell_1\ell_2}$$

for a certain direct summand $Ш_{\ell_1\ell_2}$. This follows by Parts 1 and 2 of Lemma 6.5. Assume *ad absurdum* that there is a class (β,α), with $\beta\in\kappa'(\ell_1\ell_2)\cdot\mathbf{Z}/p^n\mathbf{Z}$ and $0\neq\alpha\in Ш_{\ell_1\ell_2}$, such that

$$\partial_{\ell_2}\left(\kappa'(\ell_1\ell_2)\right)\cdot\mathbf{Z}/p^n\mathbf{Z}\subsetneq\partial_{\ell_2}(\beta,\alpha)\cdot\mathbf{Z}/p^n\mathbf{Z}=\partial_{\ell_2}\left(\mathrm{Sel}_{p^n}^{(\ell_2)}(K,f_{\ell_1\ell_2})\right).$$

Without loss of generality, one can assume $\beta=0$. Say $\partial_{\ell_2}(\alpha)=u_1\cdot p^t$ and $\partial_{\ell_2}\left(\kappa'(\ell_1\ell_2)\right)=u_2\cdot p^{t'}$, for units $u_j\in(\mathbf{Z}/p^n\mathbf{Z})^*$, and integers $t<t'<n$. Since the images of $p^{t'-t}\cdot\alpha$ and $\kappa'(\ell_1\ell_2)$ under ∂_{ℓ_2} generate the same ideal of $\mathbf{Z}/p^n\mathbf{Z}$, there exists a unit $u\in(\mathbf{Z}/p^n\mathbf{Z})^*$ such that $u\cdot p^{t'-t}\cdot\alpha-\kappa'(\ell_1\ell_2)$ belongs to the kernel of ∂_{ℓ_2}. In other words $u\cdot p^{t'-t}\cdot\alpha-\kappa'(\ell_1\ell_2)\in\mathrm{Sel}_{p^n}(K,f_{\ell_1})$. Let C be the smallest non-negative integer such that p^C kills $Ш(A/K)_{p^\infty}$. Equation (23) implies that p^C kills $u\cdot p^{t'-t}\cdot\alpha-\kappa'(\ell_1\ell_2)$, so that $p^C\cdot\kappa'(\ell_1\ell_2)=u\cdot p^{C+t'-t}\cdot\alpha=0$. Since $\kappa'(\ell_1\ell_2)$ has order p^n by Lemma 6.5(2), this implies $C\geqslant n$, which is impossible by the choice (8) of n. This contradiction proves the second equality in (27), and since $\kappa'(\ell_1\ell_2)\in\mathrm{Sel}_{p^n}(K,f_{\ell_1\ell_2})$ by Lemma 6.5(1), the first equality follows too. As explained above, this also proves Eq. (24).

To conclude the proof of the proposition, one is left with the proof of Eq. (25). By Parts 1 and 4 of Lemma 6.5, $\kappa'(\ell_1\ell_2)\in\mathrm{Sel}_{p^n}(K,f_{\ell_1\ell_2})$, and $v_{\ell_3}\left(\kappa'(\ell_1\ell_2)\right)$ generates $\mathbf{Z}/p^n\mathbf{Z}$. In particular

$$v_{\ell_3}\left(\mathrm{Sel}_{p^n}^{(\ell_3)}(K,f_{\ell_1\ell_2\ell_3})\right)=\mathbf{Z}/p^n\mathbf{Z}=v_{\ell_3}\left(\mathrm{Sel}_{p^n}(K,f_{\ell_1\ell_2})\right).$$

As for Eq. (17) in the proof of Proposition 6.3, this implies (via Poitou–Tate duality)

$$\mathrm{Sel}_{p^n}^{(\ell_3)}(K,f_{\ell_1\ell_2\ell_3})=\mathrm{Sel}_{p^n}(K,f_{\ell_1\ell_2}),$$

and then Eq. (25) follows from the short exact sequence

$$0 \to \mathrm{Sel}_{p^n}(K, f_{\ell_1\ell_2\ell_3}) \to \mathrm{Sel}_{p^n}^{(\ell_3)}(K, f_{\ell_1\ell_2\ell_3}) \xrightarrow{v_{\ell_3}} H^1_{\mathrm{fin}}(K, T_{f,n}) \cong \mathbf{Z}/p^n\mathbf{Z} \to 0. \qquad \square$$

6.4 A Lifting Theorem

In order to apply Propositions 6.3 and 6.4 to the proof of Theorem 6.1, we need the following *lifting theorem*, proved in Part 3 of [6]. The notations and assumptions are as in the previous sections; in particular $p \nmid \deg(\pi_A)$.

Theorem 6.7 *Let ℓ_1 be a $2n$-admissible prime relative to (f, K, p). Assume that f_{ℓ_1}* cannot *be lifted to a true modular form. Then there exists infinitely many pairs (ℓ_2, ℓ_3) of $2n$-admissible primes such that:*

1. *$f_{\ell_1\ell_2\ell_3} \in S_2(N, \ell_1\ell_2\ell_3; \mathbf{Z}/p^{2n}\mathbf{Z})$ can be lifted to a true modular form $g := g_{\ell_1\ell_2\ell_3} \in S_2(N, \ell_1\ell_2\ell_3; \mathbf{Z}_p)$,*
2. *$v_{\ell_3}\left(\tilde{\kappa}(\ell_1\ell_2)\right) \neq 0$ if and only if $\tilde{\kappa}(\ell_1\ell_2) \neq 0$.*

6.5 Proof of Theorem 6.1

The following proposition is a consequence of Theorem 3.2 of [3].

Proposition 6.8 *For every positive integer t, there exist infinitely many t admissible primes ℓ relative to (f, K) such that the natural map $\iota_{\ell,t} : A(K) \otimes \mathbf{Z}/p^t\mathbf{Z} \to A(K_\ell) \otimes \mathbf{Z}/p^t\mathbf{Z}$ is an isomorphism.*

Proof As noted in the proof of Proposition 6.3, under our assumptions $A(K) \otimes \mathbf{Z}/p^t\mathbf{Z} \cong \mathbf{Z}/p^t\mathbf{Z} \cdot \mathbb{P}$, for every generator $\mathbb{P}$ of $A(K)$ modulo torsion. Similarly, for every t-admissible prime ℓ, the local Kummer map gives an isomorphism $A(K_\ell) \otimes \mathbf{Z}/p^t\mathbf{Z} \cong H^1_{\mathrm{fin}}(K_\ell, A_{p^t}) \cong \mathbf{Z}/p^t\mathbf{Z}$ (cf. Sect. 2.3). Let $\kappa_p \in H^1(K, A_p)$ be the image of $\mathbb{P} \pmod p \in A(K) \otimes \mathbf{F}_p$ under the global Kummer map $A(K) \otimes \mathbf{F}_p \hookrightarrow H^1(K, A_p)$. Theorem 3.2 of [3] shows that there exist infinitely many t-admissible primes ℓ relative to (f, K) such that $v_\ell(\kappa_p) \neq 0$ in $H^1_{\mathrm{fin}}(K_\ell, A_p)$. Since $v_\ell(\kappa_p)$ is the image of $\iota_{\ell,t}(\mathbb{P}) \pmod p \in A(K_\ell) \otimes \mathbf{F}_p$ under the local Kummer map $A(K_\ell) \otimes \mathbf{F}_p \cong H^1_{\mathrm{fin}}(K_\ell, A_p)$, this implies that $\iota_{\ell,t}(\mathbb{P})$ is not divisible by p in $A(K_\ell) \otimes \mathbf{Z}/p^t\mathbf{Z}$, i.e. that $\iota_{\ell,t}$ is an isomorphism, hence proving the proposition. $\square$

We are now ready to prove Theorem 6.1. Thanks to the preceding proposition, one can fix a $2n$-admissible prime ℓ_1 relative to (f, K) such that $A(K) \otimes \mathbf{Z}/p^{2n}\mathbf{Z} \cong A(K_{\ell_1}) \otimes \mathbf{Z}/p^{2n}\mathbf{Z}$. Let $f_{\ell_1} \in S_2(N, \ell_1; \mathbf{Z}/p^{2n}\mathbf{Z})$ be a level raising at ℓ_1 of the reduction of f modulo p^{2n}. If f_{ℓ_1} can be lifted to a true modular form of level (N, ℓ_1), apply Proposition 6.3 to conclude the proof of Theorem 6.1. Assume, on the contrary, that f_{ℓ_1} cannot be lifted to a true modular form. Then Theorem 6.7 guarantees the existence of infinitely many pairs (ℓ_2, ℓ_3) of $2n$-admissible primes such that: (i) the level raising $f_{\ell_1\ell_2\ell_3} \in S_2(N, \ell_1\ell_2\ell_3; \mathbf{Z}/p^{2n}\mathbf{Z})$ at $\ell_1\ell_2\ell_3$ of $f_{\{2n\}}$ can be lifted to a true modular form, and (ii) the image of $\widetilde{\kappa}(\ell_1\ell_2) \in H^1(K, A_p)$ under the map $v_{\ell_3} : H^1(K, A_p) \to H^1_{\mathrm{fin}}(K_{\ell_3}, A_p)$ is non-zero. Indeed Eq. (20) implies that $\widetilde{\kappa}(\ell_1\ell_2) \neq 0$ for every $2n$-admissible prime ℓ_2, thanks to the injectivity of the localisation map $A(K)/p^n \hookrightarrow A(K_{\ell_1})/p^n$ at ℓ_1, so that (ii) holds true. In this case, Theorem 6.1 is a consequence of Proposition 6.4.

6.6 Generalisations

The statements of the results of the previous sections do not strive for a maximal degree of generality, but rather aim at keeping technicalities and notations as simple as possible. In this section, we briefly point at possible ways of generalising our results.

Semistability. Theorem 6.1 (and Theorem A of the Introduction) is stated under the assumption that the elliptic curve A is semistable. This makes it possible to consider in our arguments Selmer groups defined in terms of ordinary local conditions at the bad primes, and therefore to compare Selmer groups attached to different modular forms in a direct and elementary way. When N is not squarefree, the lack of natural ordinary conditions at the non-semistable primes may be obviated by imposing non-self dual local conditions. For example, one may view the cohomology classes $\kappa(1)$ and $\kappa(\ell_1\ell_2)$ as belonging to Selmer groups with relaxed local conditions at these primes, and keep track of the appearance of the restricted counterparts of these Selmer groups in the Poitou–Tate sequences of the proofs of Sect. 6.

The Heegner hypothesis. Section 7 below deduces Theorem A from Theorem 6.1 by choosing an auxiliary imaginary quadratic field K in which all prime divisors of the conductor of A are split, and hence the Heegner hypothesis of Sect. 6 is satisfied. A more general version of Theorem 6.1 can be proved along the same lines when K satisfies the generalised Heegner hypothesis Assumption 3.1(1,2). In this case, the Heegner point P_K arises on the Shimura curve X_{N^+,N^-} and the class $\kappa(\ell_1\ell_2)$ comes from a Heegner point on $X_{N^+,N^-\ell_1\ell_2}$. Since the results of the previous sections hold at this level of generality, and the Gross–Zagier formula has been generalised to Shimura curves by Zhang [26], the proof of Theorem 6.1 goes through unchanged.

The non-anomalous condition. Our main results depend on the assumption that p is a non-anomalous ordinary prime for A/K. This implies that the local Selmer condition at p arising from the local Kummer map coincides with the ordinary condition. The latter condition is defined solely in terms of the Galois representation A_{p^n}. As a consequence, both classes $\kappa(1)$ and $\kappa(\ell_1\ell_2)$, which are defined as the Kummer images of Heegner points on *different* Shimura curves, satisfy the *same* local condition at p. When p is anomalous, one faces the need of directly comparing the images of the two different local Kummer maps. This requires a more sophisticated analysis of the models for A_{p^n} over the ring of integers of $K \otimes \mathbf{Z}_p$, as is carried out for example in Sect. 4 of [10].

The ordinary condition. The technical heart of our proof of Theorem 6.1 is represented by the explicit reciprocity laws of Sect. 3, which hold without the ordinary assumption. (Note that this hypothesis is imposed in [3] in order to obtain results over the anticyclotomic $\mathbf{Z}_p$-extension of K, and not just over the base.) In order to extend Theorem 6.1 (and its consequence Theorem A) to an elliptic curve having supersingular reduction at p, one considers Selmer groups where the local condition at p is defined to be the Kummer condition. As above, the comparison of Selmer conditions at p can be done following [10]. In order to complete the proofs, one needs an extension of Theorem 5.1 to the supersingular setting, similar to that announced in [25].

7 Proof of Theorem A

In this section we prove Theorem A stated in the Introduction.

Thus, as in the Introduction and in Sect. 6, let $A/\mathbf{Q}$ be a semistable elliptic curve of conductor N, let $p > 7$ be a non-anomalous prime of good ordinary reduction, and fix a modular parametrisation $\pi_A : X_0(N) \to A$ of minimal degree $\deg(\pi_A)$. Assume moreover that p does not divide $\deg(\pi_A)$ and that $L(A/\mathbf{Q}, s)$ has a simple zero at $s = 1$.

`Step I.` Thanks to the results of [7], there exists a quadratic imaginary field $K/\mathbf{Q}$ such that

(α) the discriminant of $K/\mathbf{Q}$ is coprime with $6Np$, and every prime divisor of Np splits in $K/\mathbf{Q}$;
(β) the Hasse–Weil L-function $L(A/K, s)$ of A/K has a simple zero at $s = 1$.

Writing $A^K/\mathbf{Q}$ for the K-quadratic twist of A, one has $L(A/K, s) = L(A/\mathbf{Q}, s) \cdot L(A^K/\mathbf{Q}, s)$, so that ($\beta$) is equivalent to $L(A^K/\mathbf{Q}, 1) \neq 0$. In particular

$$L'(A/K, 1) = L'(A/\mathbf{Q}, 1) \cdot L(A^K/\mathbf{Q}, 1). \tag{28}$$

`Step II.` The Gross–Zagier formula [11, Sect. 5, Theorem 2.1] states that

$$\frac{D_K^{\frac{1}{2}} \cdot L'(A/K, 1)}{c^2 \cdot \Omega_{A/K} \cdot h^{\mathrm{NT}}(\mathbf{P}_K)} = [A(K) : \mathbf{Z}P_K]^2.$$

Here c is the *Manin constant* associated with the strong Weil curve in the isogeny class of $A/\mathbf{Q}$, D_K is the absolute value of the discriminant of $K/\mathbf{Q}$, and $\Omega_{A/K} \in \mathbf{C}^*$ is the Néron period of A/K. Moreover, $\mathbf{P}_K$ denotes a generator of $A(K)/\text{torsion}$, $h^{\mathrm{NT}}(\mathbf{P}_K) \in \mathbf{R}$ its Néron–Tate canonical height, and $P_K \in A(K)$ the Heegner point attached to π_A (cf. Sect. 6). A result of Mazur [16] states that $p^2|4N$ if $p|c$, hence c^2 is a p-adic unit in our setting. Moreover, $\Omega_{A/K} = D_K^{1/2} \cdot \Omega_A \cdot \Omega_{A^K}$, where Ω_* is the real Néron period of the elliptic curve $*/\mathbf{Q}$ [11, P. 312]. Using (28), the preceding equation gives

$$\frac{L'(A/\mathbf{Q}, 1)}{\Omega_A \cdot h^{\mathrm{NT}}(\mathbf{P})} \cdot \frac{L(A^K/\mathbf{Q}, 1)}{\Omega_{A^K}} \doteq [A(K) : \mathbf{Z}P_K]^2, \tag{29}$$

where $\doteq$ denotes equality up to multiplication by a p-adic unit, and $\mathbf{P}$ is a generator of $A(\mathbf{Q})/\text{torsion}$. (Note that in our setting $P_K \in A(\mathbf{Q})$, as the sign in the functional equation satisfied by $L(A/\mathbf{Q}, s)$ is -1.)

`Step III.` As explained in the proof of Proposition 6.3 (see in particular the discussion following Eq. (13)), the residual representation $\overline{\rho}_{A,p} = \overline{\rho}_f$ is ramified at every prime $q|N$, hence $\overline{\rho}_{A^K,p}$ is also ramified at every prime $q|N$. Then the local Tamagawa number $c_q(A) = c_q(A^K)$ is a p-adic unit for every $q|N$, so that every local Tamagawa number of $A^K/\mathbf{Q}$ is a p-adic unit. (Indeed, if a prime $q \nmid N$ divides the conductor of $A^K/\mathbf{Q}$, then q divides the absolute discriminant of K, and $A^K/\mathbf{Q}$ has additive reduction at q and $c_q(A^K) \leqslant 4$). According to Theorem 2 of [22] (cf. Theorem 5.1) the p-part of the Birch and Swinnerton-Dyer formula holds for $A^K/\mathbf{Q}$:

$$\frac{L(A^K/\mathbf{Q}, 1)}{\Omega_{A^K}} \doteq \#\text{Ш}(A^K/\mathbf{Q})_{p^\infty}. \tag{30}$$

(Note that loc. cit. requires $\overline{\rho}_{A^K,p}$ to be surjective. On the other hand, as proved in [21, Theorem B], the irreducibility of $\overline{\rho}_{A^K,p}$ is sufficient for the arguments of [22].)

`Step IV.` According to Theorem 6.1

$$[A(K) : \mathbf{Z}P_K]^2 \doteq \#\text{Ш}(A/K)_{p^\infty} \doteq \#\text{Ш}(A/\mathbf{Q})_{p^\infty} \cdot \#\text{Ш}(A^K/\mathbf{Q})_{p^\infty}.$$

Since $c_A := \prod_{q|N} c_q(A)$ is a p-adic unit, combining the preceding equation with (29) and (30) yields

$$\frac{L'(A/\mathbf{Q}, 1)}{\Omega_A \cdot h^{\mathrm{NT}}(\mathbf{P})} \doteq \#\text{Ш}(A/\mathbf{Q})_{p^\infty} \cdot c_A,$$

concluding the proof of Theorem A.

References

1. Berti, A.: On the Birch and Swinnerton-Dyer conjecture for elliptic curves of analytic rank one. Ph.D. Thesis, University of Milan (2014)
2. Bertolini, M., Darmon, H.: Heegner points on Mumford–Tate curves. Invent. Math. **126**(3) (1996)
3. Bertolini, M., Darmon, H.: Iwasawa's main conjecture for elliptic curves over anticyclotomic $\mathbb{Z}_p$-extensions. Ann. Math. **162** (2005)

4. Bertolini, M., Darmon, H.: Hida families and rational points on elliptic curves. Invent. Math. **168**(2) (2007)
5. Bertolini, M., Darmon, H., Prasanna, K.: Generalized Heegner cycles and p-adic Rankin L-series. With an appendix by Brian Conrad. Duke Math. J. **162**(6) (2013)
6. Bertolini, M., Venerucci, R.: The anticyclotomic Iwasawa main conjectures. Preprint (2015)
7. Bump, D., Friedberg, S., Hoffstein, J.: Nonvanishing theorems for L-functions of modular forms and their derivatives. Invent. Math. **102**(3) (1990)
8. Carayol, H.: Formes modulaires et représentations galoisiennes à valeurs dans un anneau local complet. In: Mazur, B., Stevens, G. (eds.) P-adic monodromy and the Birch and Swinnerton-Dyer conjecture. American Mathematical Society (1994)
9. Greenberg, R.: Iwasawa Theory for Elliptic Curves. Springer, New York, Inc. (1997)
10. Gross, B., Parson, J.: On the local divisibility of Heegner points. In: Number Theory, Analysis and Geometry. Springer, New York (2012)
11. Gross, B., Zagier, D.: Heegner points and derivatives of L-series. Invent. Math. **86**(2) (1986)
12. Jetchev, D., Skinner, C., Wan, X.: The Birch and Swinnerton-Dyer formula for elliptic curves of analytic rank one. Preprint (2015)
13. Kato, K.: p-adic Hodge theory and values of zeta functions of modular forms. Astérisque **295** (2004)
14. Kolyvagin, V.A.: Euler systems. In: The Grothendieck Festschrift, vol. II. Progr. Math., 87, Birkhäuser Boston, Boston, MA (1990)
15. Kolyvagin, V.A.: On the structure of Selmer groups. Math. Ann. **291**(2) (1991)
16. Mazur, B.: Rational isogenies of prime degree. Invent. Math. **44**(2) (1978) (With an appendix by D. Goldfeld)
17. Milne, J.: Arithmetic Duality Theorems. Kea Books (2004)
18. Pollack, R., Weston, T.: On anticyclotomic μ-invariants of modular forms. Compositio Math. **147**(5) (2011)
19. Ribet, K.: On modular representations of $\mathrm{Gal}(\overline{\mathbf{Q}}/\mathbf{Q})$ arising from modular forms. Invent. Math. **100**(2) (1990)
20. Rubin, K.: Euler Systems (Hermann Weyl Lectures). In: Annals of Mathematics Studies, vol. 147. Princeton University Press (2000)
21. Skinner, C.: Multiplicative reduction and the cyclotomic main conjecture for GL_2. Preprint (2014)
22. Skinner, C., Urban, E.: The Iwasawa main conjecture for GL_2. Invent. Math. **195**(1) (2014)
23. Vatsal, V.: Special values of anticyclotomic L-functions. Duke Math. J. **116**(2) (2003)
24. Wan, X.: Iwasawa main conjecture for Rankin-Selberg p-adic L-functions. Preprint (2013)
25. Wan, X.: Iwasawa main conjecture for supersingular elliptic curves. Preprint, arXiv:1411.6352 (2014)
26. Zhang, S.: Heights of Heegner points on Shimura curves. Ann. Math. **153**(1) (2001)
27. Zhang, W.: Selmer groups and the indivisibility of Heegner points. Cambridge J. Math. **2**(2) (2014)

p-adic Measures for Hermitian Modular Forms and the Rankin–Selberg Method

Thanasis Bouganis

Abstract In this work we construct p-adic measures associated to an ordinary Hermitian modular form using the Rankin–Selberg method.

Keywords Hermitian modular forms · Special L-values · p-adic measures

1 Introduction

p-adic measures are known to play an important role in Iwasawa theory, since they constitute the analytic part of the various Main Conjectures. In this paper we are interested in p-adic measures attached to an ordinary Hermitian modular form $\mathbf{f}$. There has been work on the subject by Harris et al. [20, 21], where the first steps towards the construction of p-adic measures associated to ordinary Hermitian modular forms were made. Actually in their work they construct a p-adic Eisenstein measure (see also the works of Eischen [15, 16] on this), and provide a sketch of the construction of a p-adic measure associated to an ordinary Hermitian modular form. We also mention here our work [4], where we constructed p-adic measures associated to Hermitian modular forms of definite unitary groups of one and two variables. All these works impose the following assumption on the prime number p: if we denote by K the CM field associated to the Hermitian modular form $\mathbf{f}$ and let F be the maximal totally real subfield of K, then all the primes in F above p must be split in K. One of the main motivation of this work is to consider the case where p does not satisfy this condition.

Actually this work differs from the once mentioned above on the method used to obtain the p-adic measures. Indeed the previous works utilize the doubling method in order to construct the p-adic measures, where in this work we will use the

This work is dedicated with gratefulness and admiration to Professor John Coates in occasion of his 70th birthday.

T. Bouganis (✉)
Department of Mathematical Sciences, Durham University, Science Laboratories, South Rd., Durham DH1 3LE, UK
e-mail: athanasios.bouganis@durham.ac.uk

D. Loeffler and S.L. Zerbes (eds.), *Elliptic Curves, Modular Forms and Iwasawa Theory*, Springer Proceedings in Mathematics & Statistics 188, DOI 10.1007/978-3-319-45032-2_2

Rankin–Selberg method. In the Rankin–Selberg method one obtains an integral representation of the L-values as a Petersson inner product of $\mathbf{f}$ with a product of a theta series and a Siegel-type Eisenstein series, where in the doubling method the L-values can be represented as a Petersson inner product of $\mathbf{f}$ with another Hermitian form, which is obtained by pulling back a Siegel-type Eisenstein series of a larger unitary group. Of course one should remark right away that the use of the Rankin–Selberg method puts some serious restrictions on the unitary groups which may be considered. In particular, the archimedean components of the unitary group must be of the form $U(n, n)$, where the doubling method allows situations of the form $U(n, m)$ with $n \neq m$. However, we believe that it reasonable to expect, with the current stage of knowledge at least, to relax the splitting assumption only in the cases of $U(n, n)$. The reason being that in the cases of $U(n, m)$ with $n \neq m$, in order to obtain the special L-values, one needs to evaluate Siegel-type Eisenstein series on CM points, and in the p-adic setting, one needs that this CM points correspond to abelian varieties with complex multiplication, which are ordinary at p, and hence the need for the splitting assumption. For example, even in the "simplest" case of the definite $U(1) = U(1, 0)$, which is nothing else than the case of p-adic measures for Hecke characters of a CM field K considered by Katz in [24], even today, in this full generality, it is not known how to remove the assumption on the primes above p in F being split in K. We need to remark here that in some special cases (for example elliptic curves over $\mathbb{Q}$ with CM by imaginary quadratic fields), there are results which provide some p-adic distributions associated to Hecke characters of CM fields.

In this work we make some assumptions, which will simplify various technicalities, and we postpone to a later work [7] for a full account. In particular, we fix an odd prime p, and write $\mathfrak{P}_i$ for the prime ideals in F above p, which are inert in K. We write $\mathfrak{p}_i$ for the prime ideal of K above $\mathfrak{P}_i$, and denote by S the set of these primes. We will assume that $S \neq \emptyset$. Then our aim is to construct p-measures for the Galois group $\mathrm{Gal}(K(\prod_i \mathfrak{p}_i^{\infty})/K)$, where $K(\prod_i \mathfrak{p}_i^{\infty})$ denotes the maximal abelian extension of K unramified outside the prime ideals $\mathfrak{p}_i$. As we said already our techniques can also handle the situation of primes split in K, and this will be done in [7]. The other simplifying assumptions which we impose in this work, which will be lifted in [7], are

1. we assume that the class number of the CM field K is equal to the class number of the underlying unitary group with repsect to the standard congruence subgroup. This for example happens when the class number of F is taken equal to one,
2. we will investigate the interpolation properties of the p-adic measures only for the special values for which the corresponding Eisenstein series in the Rankin–Selberg method are holomorphic, and not just nearly-holomorphic.

We should also remark that this present work should be seen as the unitary analogue of the work of Panchishkin [27], and Courtieu and Panchishkin [12] in the Siegel modular form case. We should say here that the second assumption above can be lifted by developing the techniques of Courtieu and Panchishkin on the holomorphic projection in the unitary case. Actually the techniques of this present work grew out of the efforts of the author to extend the work of Courtieu and Panchishkin in the following directions, which is also one of the aims of [7],

1. to consider the situation of totally real fields (they consider the case of $\mathbb{Q}$),
2. to obtain the interpolation properties also for Hecke characters which are not totally ramified.
3. to construct the measures also for symplectic groups of odd genus. In their work they consider the case of even genus, and hence no half-integral theta, and Eisenstein series appear in the construction. We remark here that, over $\mathbb{Q}$, the work of Böcherer and Schmidt [2], provides the existence of these p-adic measures, in both odd and even genus. However their techniques seem to be hard to extend to the totally real field situation.

Indeed in this paper we work completely adelically, which allow us to work over any field. Moreover, we use a more precise form of the so-called Adrianov–Kalinin identity, shown by Shimura, which allows us to obtain a better understanding of the bad Euler factors above p. And finally, we work here the interpolation properties for characters that may be unramified at some of the primes of the set S. Note that only at these primes one sees the needed modification of the Euler factors above p at the interpolation properties.

Notation: Since our main references for this work are the two books of Shimura [29, 30] our notation is the one used by Shimura in his books.

2 Hermitian Modular Forms

In this section, which is similar to the corresponding section in [6], we introduce the notion of a Hermitian modular form, both classically and adelically. We follow closely the books of Shimura [29, 30], and we remark that we adopt the convention done in the second book with respect to the weight of Hermitian modular forms (see the discussion on p. 32, Sect. 5.4 in [30]).

Let K be an algebra equipped with an involution ρ. For a positive integer $n \in \mathbb{N}$ we define the matrix $\eta := \eta_n := \begin{pmatrix} 0 & -1_n \\ 1_n & 0 \end{pmatrix} \in GL_{2n}(K)$, and the group $G := U(n,n) := \{\alpha \in GL_{2n}(K) | \alpha^* \eta \alpha = \eta\}$, where $\alpha^* := {}^t\alpha^\rho$. Moreover we define $\hat{\alpha} := (\alpha^*)^{-1}$ and $S := S^n := \{s \in M_n(K) | s^* = s\}$ for the set of Hermitian matrices with entries in K. If we take $K = \mathbb{C}$ and let ρ to denote the complex conjugation then the group $G(\mathbb{R}) = \{\alpha \in GL_{2n}(\mathbb{C}) | \alpha^* \eta \alpha = \eta\}$ acts on the symmetric space (Hermitian upper half space)

$$\mathbb{H}_n := \{z \in M_n(\mathbb{C}) | i(z^* - z) > 0\},$$

by linear fractional transformations,

$$\alpha \cdot z := (a_\alpha z + b_\alpha)(c_\alpha z + d_\alpha)^{-1} \in \mathbb{H}_n, \quad \alpha = \begin{pmatrix} a_\alpha & b_\alpha \\ c_\alpha & d_\alpha \end{pmatrix} \in G(\mathbb{R}), \quad z \in \mathbb{H}_n,$$

where the $a_\alpha, b_\alpha, c_\alpha, d_\alpha$ are taken in $M_n(\mathbb{C})$.

Let now K be a CM field of degree $2d := [K : \mathbb{Q}]$ and we write F for its maximal totally real subfield. Moreover we write $\mathfrak{r}$ for the ring of integers of K, $\mathfrak{g}$ for that of F, D_F and D_K for their discriminants and $\mathfrak{d}$ for the different ideal of F. We write $\mathbf{a}$ for the set of archimedean places of F. We now pick a CM type $(K, \{\tau_v\}_{v\in\mathbf{a}})$ of K, where $\tau_v \in Hom(K, \mathbb{C})$. For an element $a \in K$ we set $a_v := \tau_v(a) \in \mathbb{C}$. We will also regard $\mathbf{a}$ as the archimedean places of K corresponding to the embeddings τ_v of the selected CM type. Finally we let $\mathbf{b}$ be the set of all complex embeddings of K, and we note that $\mathbf{b} = \{\tau_v, \tau_v\rho | v \in \mathbf{a}\}$, where ρ denotes complex conjugation acting on the CM field K. By abusing the notation we may also write $\mathbf{b} = \mathbf{a} \coprod \mathbf{a}\rho$.

We write $G_{\mathbb{A}}$ for the adelic group of G, and $G_{\mathbf{h}} = \prod'_v G_v$ (restricted product) for its finite part, and $G_{\mathbf{a}} = \prod_{v\in\mathbf{a}} G_v$ for its archimedean part. Note that we understand G as an algebraic group over F, and hence the finite places v above are finite places of F, which will be denoted by $\mathbf{h}$. For a description of G_v at a finite place we refer to [29, Chap. 2]. Given two fractional ideals $\mathfrak{a}$ and $\mathfrak{b}$ of F such that $\mathfrak{a}\mathfrak{b} \subseteq \mathfrak{g}$, we define following Shimura the subgroup of $G_{\mathbb{A}}$,

$$D[\mathfrak{a}, \mathfrak{b}] := \left\{ \begin{pmatrix} a_x & b_x \\ c_x & d_x \end{pmatrix} \in G_{\mathbb{A}} | a_x \prec \mathfrak{g}_v, b_x \prec \mathfrak{a}_v, c_x \prec \mathfrak{b}_v, d_x \prec \mathfrak{g}_v, \quad \forall v \in \mathbf{h} \right\},$$

where we use the notation $\prec$ in [30, p. 11], where $x \prec \mathfrak{b}_v$ means that the v-component of the matrix x has are all its entries in $\mathfrak{b}_v$. Again we take a_x, b_x, c_x, d_x to be n by n matrices. For a finite adele $q \in G_{\mathbf{h}}$ we define $\Gamma^q = \Gamma^q(\mathfrak{b}, \mathfrak{c}) := G \cap qD[\mathfrak{b}^{-1}, \mathfrak{b}\mathfrak{c}]q^{-1}$, a congruence subgroup of G. Given a finite order Hecke character ψ of K of conductor dividing $\mathfrak{c}$ we define a character on $D[\mathfrak{b}^{-1}, \mathfrak{b}\mathfrak{c}]$ by $\psi(x) = \prod_{v|\mathfrak{c}} \psi_v(det(a_x)_v)^{-1}$, where ψ_v denotes the local component of ψ at the finite place v, and a character ψ_q on Γ^q by $\psi_q(\gamma) = \psi(q^{-1}\gamma q)$.

We write $\mathbb{Z}^{\mathbf{a}} := \prod_{v\in\mathbf{a}} \mathbb{Z}$, $\mathbb{Z}^{\mathbf{b}} := \prod_{v\in\mathbf{b}} \mathbb{Z}$ and $\mathcal{H} := \prod_{v\in\mathbf{a}} \mathbb{H}_n$. We embed $\mathbb{Z} \hookrightarrow \mathbb{Z}^{\mathbf{a}}$ diagonally and for an $m \in \mathbb{Z}$ we write $m\mathbf{a} \in \mathbb{Z}^{\mathbf{a}}$ for its image. We will simply write $\mathbf{a}$ for $1\mathbf{a}$. We define an action of $G_{\mathbb{A}}$ on $\mathcal{H}$ by $g \cdot z := g_{\mathbf{a}} \cdot z := (g_v \cdot z_v)_{v\in\mathbf{a}}$, with $g \in G_{\mathbb{A}}$ and $z = (z_v)_{v\in\mathbf{a}} \in \mathcal{H}$. For a function $f : \mathcal{H} \to \mathbb{C}$ and an element $k \in \mathbb{Z}^{\mathbf{b}}$ we define

$$(f|_k\alpha)(z) := j_\alpha(z)^{-k} f(\alpha \cdot z), \ \alpha \in G_{\mathbb{A}}, \ z \in \mathcal{H},$$

where,

$$j_\alpha(z)^{-k} := \prod_{v\in\mathbf{a}} det(c_{\alpha_v} z_v + d_{\alpha_v})^{-k_v} det(c^{\rho}_{\alpha_v}\, {}^t z_v + d^{\rho}_{\alpha_v})^{-k_{v\rho}}, \quad z = (z_v)_{v\in\mathbf{a}} \in \mathcal{H}.$$

For fixed $\mathfrak{b}$ and $\mathfrak{c}$ as above, and $q \in G_{\mathbf{h}}$ and a Hecke character ψ of K, we define,

Definition 2.1 [30, p. 31] A function $f : \mathcal{H} \to \mathbb{C}$ is called a Hermitian modular form for the congruence subgroup Γ^q of weight $k \in \mathbb{Z}^{\mathbf{b}}$ and nebentype ψ_q if:

1. f is holomorphic,
2. $f|_k\gamma = \psi_q(\gamma) f$ for all $\gamma \in \Gamma^q$,
3. f is holomorphic at cusps (see [30, p. 31] for this notion).

The space of Hermitian modular forms of weight k for the congruences group Γ^q and nebentype ψ_q will be denoted by $\mathcal{M}_k(\Gamma^q, \psi_q)$. For any $\gamma \in G$ we have a Fourier expansion of the form (see [30, p. 33])

$$(f|_k\gamma)(z) = \sum_{s\in\mathfrak{S}} c(s,\gamma;f)e_{\mathbf{a}}(sz),\ \ c(s,\gamma;f)\in\mathbb{C}, \tag{1}$$

where $\mathfrak{S}$ a lattice in $S_+ := \{s \in S|\ s_v \geq 0,\ \ \forall v \in \mathbf{a}\}$, and

$$e_{\mathbf{a}}(x) := exp(2\pi i \sum_v tr(x_v)).$$

An f is called a cusp form if $c(s,\gamma;f)=0$ for any $\gamma\in G$ and s with $det(s)=0$. The space of cusp forms we will be denoted by $\mathcal{S}_k(\Gamma^q,\psi_q)$. When we do not wish to determine the nebentype we will be writing $f\in\mathcal{M}_k(\Gamma^q)$, and this should be understood that there exists some ψ_q as above such that $f\in\mathcal{M}_k(\Gamma^q,\psi_q)$.

We now turn to the adelic Hermitian modular forms. If we write D for a group of the form $D[\mathfrak{b}^{-1},\mathfrak{bc}]$, and ψ a Hecke character of finite order then we define,

Definition 2.2 [30, p. 166] A function $\mathbf{f}: G_{\mathbb{A}} \to \mathbb{C}$ is called an adelic Hermitian modular form if

1. $\mathbf{f}(\alpha x w) = \psi(w) j_w^k(\mathbf{i})\mathbf{f}(x)$ for $\alpha\in G$, $w\in D$ with $w_{\mathbf{a}}(\mathbf{i})=\mathbf{i}$,
2. For every $p\in G_{\mathbf{h}}$ there exists $f_p\in\mathcal{M}_k(\Gamma^p,\psi_p)$, where $\Gamma^p := G\cap pCp^{-1}$ such that $\mathbf{f}(py) = (f_p|_k y)(\mathbf{i})$ for every $y\in G_{\mathbf{a}}$.

Here we write $\mathbf{i} := (i1_n,\ldots,i1_n)\in\mathcal{H}$. We denote this space by $\mathcal{M}_k(D,\psi)$, and the space of cusp forms by $\mathcal{S}_k(D,\psi)$. As in the classical case above, we will write just $\mathcal{M}_k(D)$ if we do not wish to determine the nebentype. A simple computation shows, if $\mathbf{f}\in\mathcal{M}_k(D,\psi)$ then the form $\mathbf{f}^*(x) := \mathbf{f}(x\eta_{\mathbf{h}}^{-1})$ belongs to $\mathcal{M}_k(D',\psi^{-c})$ where $D' := D[\mathfrak{bc},\mathfrak{b}^{-1}]$ and $\psi^{-c}(x) := \psi(x^\rho)^{-1}$.

By [29, Chap. 2] there exists a finite set $\mathcal{B}\subset G_{\mathbf{h}}$ such that $G_{\mathbb{A}} = \coprod_{b\in\mathcal{B}} GbD$ and an isomorphism $\mathcal{M}_k(D,\psi)\cong\oplus_{b\in\mathcal{B}}\mathcal{M}_k(\Gamma^b,\psi_b)$ (see [29, Chap. 2]). We note here that for the congruence subgroups $D[\mathfrak{b}^{-1},\mathfrak{bc}]$ the cardinality of the set $\mathcal{B}$ does not depend on the ideal $\mathfrak{c}$ and its elements can be selected to be of the form $\begin{pmatrix}\hat{q} & 0\\ 0 & q\end{pmatrix}$ with $q\in GL_n(K)_{\mathbf{h}}$, and $q_v=1$ for $v|\mathfrak{c}$, (see for example [6, Lemma 2.6]). For a $q\in GL_n(K)_{\mathbb{A}}$ and an $s\in S_{\mathbb{A}}$ we have

$$\mathbf{f}\left(\begin{pmatrix} q & s\hat{q}\\ 0 & \hat{q}\end{pmatrix}\right) = \sum_{\tau\in S_+} c_{\mathbf{f}}(\tau,q)e_{\mathbb{A}}(\tau s).$$

For the properties of $c_{\mathbf{f}}(\tau,q)$ we refer to the [30, Proposition 20.2] and for the definition of $e_{\mathbb{A}}$ to [30, p. 127]. We also note that sometimes we may write $c(\tau,q;\mathbf{f})$ for $c_{\mathbf{f}}(\tau,q)$.

For a subfield L of $\mathbb{C}$ we will be writing $\mathcal{M}_k(\Gamma^q, \psi, L)$ for the Hermitian modular forms in $\mathcal{M}_k(\Gamma^q, \psi)$ whose Fourier expansion at infinity, that is γ is the identity in Eq. 1, has coefficients in L. For a fixed set $\mathcal{B}$ as above we will be writing $\mathcal{M}_k(D, \psi, L)$ for the subspace of $\mathcal{M}_k(D, \psi)$ consisting of elements whose image under the above isomorphism lies in $\oplus_{b\in\mathcal{B}}\mathcal{M}_k(\Gamma^b, \psi_b, L)$. Finally we define the adelic cusp forms $\mathcal{S}_k(D, \psi)$ to be the subspace of $\mathcal{M}_k(D, \psi)$, which maps to $\oplus_{b\in\mathcal{B}}\mathcal{S}_k(\Gamma^b, \psi_b)$. As above, when we do not wish to determine the nebentype we simply write $\mathcal{M}_k(\Gamma^q, L)$ and $\mathcal{M}_k(D, L)$.

We fix an embedding $\overline{\mathbb{Q}} \hookrightarrow \mathbb{C}$ and write F^{cl} for the Galois closure of F over $\mathbb{Q}$. Then by [30, Chap. II, Sect. 10] we have a well-defined action of the absolute Galois group $\mathrm{Gal}(\overline{\mathbb{Q}}/F^{cl})$ on $\mathcal{M}_k(\Gamma^q, \overline{\mathbb{Q}})$ given by an action on the Fourier-coefficients of the expansion at infinity. This action will be denoted by f^σ for an $f \in \mathcal{M}_k(\Gamma^q, \overline{\mathbb{Q}})$ and $\sigma \in \mathrm{Gal}(\overline{\mathbb{Q}}/F^{cl})$. A similar action can be defined on the space $\mathcal{M}_k(D, \overline{\mathbb{Q}})$ (see [30, p. 193, Lemma 23.14]), and will be also denoted by $\mathbf{f}^\sigma$ for an $\mathbf{f} \in \mathcal{M}_k(D, \overline{\mathbb{Q}})$. In both cases (classical and adelic) the action of the absolute Galois group preserves the space of cusp forms.

We close this section with a final remark concerning Hecke characters. Given an (adelic) Hecke character χ of K (or F), we will be abusing the notation and write χ also for the corresponding ideal character.

3 Eisenstein and Theta Series

3.1 Eisenstein Series

In this section we collect some facts concerning Siegel-type Eisenstein series. We closely follow [30, Chap. IV].

We consider a $k \in \mathbb{Z}^{\mathbf{b}}$, an integral ideal $\mathfrak{c}$ in F and a unitary Hecke character χ of K with infinity component of the form $\chi_{\mathbf{a}}(x) = x_{\mathbf{a}}^{\ell}|x_{\mathbf{a}}|^{-\ell}$, where $\ell = (k_v - k_{v\rho})_{v\in\mathbf{a}}$ and of conductor dividing $\mathfrak{c}$. For a fractional ideal $\mathfrak{b}$ we write C for $D[\mathfrak{b}^{-1}, \mathfrak{b}\mathfrak{c}]$. Then for a pair $(x, s) \in G_{\mathbb{A}} \times \mathbb{C}$, we denote by $E_{\mathbb{A}}(x, s)$ or $E_{\mathbb{A}}(x, s; \chi, \mathfrak{c})$ the Siegel type Eisenstein series associated to the character χ and the weight k. We recall here its definition, taken from [30, p. 131],

$$E_{\mathbb{A}}(x, s) = \sum_{\gamma\in P\backslash G} \mu(\gamma x)\epsilon(\gamma x)^{-s}, \quad \Re(s) >> 0,$$

where P is the standard Siegel parabolic subgroup and the function $\mu : G_{\mathbb{A}} \to \mathbb{C}$ is supported on $P_{\mathbb{A}}C \subset G_{\mathbb{A}}$, defined by,

$$\mu(x) = \chi_{\mathbf{h}}(det(d_p))^{-1}\chi_{\mathfrak{c}}(det(d_w))^{-1} j_x(\mathbf{i})^{-k}|j_x(\mathbf{i})|^m,$$

where $x = pw$ with $p \in P_{\mathbb{A}}$ and $w \in C$, and $m = (k_v + k_{v\rho})_v$. Here we define $|j_x(\mathbf{i})|^m := \prod_{v \in \mathbf{a}} |j_{x_v}(i 1_n)|^{m_v}$. The function $\epsilon : G_{\mathbb{A}} \to \mathbb{C}$ is defined as $\epsilon(x) = |det(d_p d_p^*)|_{\mathbb{A}}$ where $x = pw$ with $p \in P_{\mathbb{A}}$ and $w \in D[\mathfrak{b}^{-1}, \mathfrak{b}]$. Here for an adele $x \in F_{\mathbb{A}}^{\times}$ we write $|x|_{\mathbb{A}}$ for the adele norm normalized as in [29, 30]. Moreover we define the normalized Eisenstein series

$$D_{\mathbb{A}}(x, s) = E_{\mathbb{A}}(x, s) \prod_{i=0}^{n-1} L_{\mathfrak{c}}(2s - i, \chi_1 \theta^i),$$

where θ is the non-trivial character associated to K/F and χ_1 is the restriction of the Hecke character χ to $F_{\mathbb{A}}^{\times}$. We note that since we consider unitary characters the infinity part of such a character is of the form $(\chi_1)_{\mathbf{a}}(x) = \prod_{v \in \mathbf{a}} \left(\frac{x_v}{|x_v|}\right)^{\ell_v}$, and it will be often denoted by $sgn(x_{\mathbf{a}})^{\ell}$. Moreover for a Hecke character ϕ of F, we write $L_{\mathfrak{c}}(s, \phi)$ for the Dirichlet series associated to ϕ with the Euler factors at the primes dividing $\mathfrak{c}$ removed.

For a $q \in GL_n(K)_{\mathbf{h}}$ we define $D_q(z, s; k, \chi, \mathfrak{c})$, a function on $(z, s) \in \mathcal{H} \times \mathbb{C}$, associated to $D_{\mathbb{A}}(x, s)$ by the rule (see [30, p. 146]),

$$D_q(x \cdot \mathbf{i}, s; k, \chi, \mathfrak{c}) = j_x^k(\mathbf{i}) D_{\mathbb{A}}(diag[q, \hat{q}]x, s).$$

We now introduce yet another Eisenstein series for which we have explicit information about their Fourier expansion. In particular we define the $E_{\mathbb{A}}^*(x, s) := E_{\mathbb{A}}(x\eta_{\mathbf{h}}^{-1}, s)$ and $D_{\mathbb{A}}^*(x, s) := D_{\mathbb{A}}(x\eta_{\mathbf{h}}^{-1}, s)$, and as before we write $D_q^*(z, s; k, \chi, \mathfrak{c})$ for the series associated to $D_{\mathbb{A}}^*(x, s)$. We now write the Fourier expansion of $E_{\mathbb{A}}^*(x, s)$ as,

$$E_{\mathbb{A}}^* \left(\begin{pmatrix} q & \sigma \hat{q} \\ 0 & \hat{q} \end{pmatrix}, s \right) = \sum_{h \in S} c(h, q, s) \mathbf{e}_{\mathbb{A}}(h\sigma), \tag{2}$$

where $q \in GL_n(K)_{\mathbb{A}}$ and $\sigma \in S_{\mathbb{A}}$. We now state a result of Shimura on the coefficients $c(h, q, s)$. We first define an $\mathfrak{r}$-lattice in $S := S^n$, by

$$T := T^n := \{x \in S | tr(xy) \subset \mathfrak{g}, \quad \forall y \in S(\mathfrak{r})\},$$

where $S(\mathfrak{r}) := S \cap M_n(\mathfrak{r})$. T is usually called the dual lattice to $S(\mathfrak{r})$. For a finite place v of F we write T_v for $T \otimes_{\mathfrak{r}} \mathfrak{r}_v$.

Proposition 3.1 (Shimura, Proposition 18.14 and Proposition 19.2 in [29]). *Suppose that $\mathfrak{c} \neq \mathfrak{g}$. Then $c(h, q, s) \neq 0$ only if $({}^t q h q)_v \subset (\mathfrak{d}\mathfrak{b}^{-1}\mathfrak{c}^{-1})_v T_v^n$ for every $v \subset \mathbf{h}$. In this case*

$$c(h, q, s) = C(S)\chi(det(-q))^{-1} |det(qq^*)|_{\mathbf{h}}^{n-s} |det(qq^*)|_{\mathbf{a}}^{s} N(\mathfrak{b}\mathfrak{c})^{-n^2} \times$$

$$\alpha_{\mathfrak{c}}(\omega \cdot {}^t q h q, 2s, \chi_1) \prod_{v \in \mathbf{a}} \xi(q_v q_v^*, h_v; s + (k_v + k_{v\rho})/2, s - (k_v + k_{v\rho})/2)),$$

where $N(\cdot)$ denotes the norm from F to $\mathbb{Q}$, $|x|_{\mathbf{h}} := \prod_{v\in\mathbf{h}} |x_v|_v$ with $|\cdot|_v$ the normalized absolute value at the finite place v, ω is a finite idele such that $\omega\mathfrak{r} = \mathfrak{b}\mathfrak{d}$, and

$$C(S) := 2^{n(n-1)d}|D_F|^{-n/2}|D_K|^{-n(n-1)/4}.$$

For the function $\xi(g_v, h_v, s, s')$ with $0 < g_v \in S_v$, $h_v \in S_v$, $s, s' \in \mathbb{C}$, $v \in \mathbf{a}$ we refer to [30, p. 134].

Moreover if we write r for the rank of h and let $g \in GL_n(F)$ such that $g^{-1}hg = diag[h', 0]$ with $h' \in S^r$. Then

$$\alpha_{\mathfrak{c}}(\omega \cdot {}^t qhq, 2s, \chi_1) = \Lambda_{\mathfrak{c}}(s)^{-1}\Lambda_h(s) \prod_{v\in\mathbf{c}} f_{h,q,v}\left(\chi(\pi_v)|\pi_v|^{2s}\right),$$

where

$$\Lambda_{\mathfrak{c}}(s) = \prod_{i=0}^{n-1} L_{\mathfrak{c}}(2s - i, \chi_1\theta^i), \quad \Lambda_h(s) = \prod_{i=0}^{n-r+1} L_{\mathfrak{c}}(2s - n - i, \chi_1\theta^{n+i-1}).$$

Here $f_{h,q,v}$ are polynomials with constant term 1 and coefficients in $\mathbb{Z}$; they are independent of χ. The set $\mathbf{c}$ is determined as follows: $\mathbf{c} = \emptyset$ if $r = 0$. If $r > 0$, then take $g_v \in GL_n(\mathfrak{r}_v)$ for each $v \nmid \mathfrak{c}$ so that $(\omega q^* hq)_v = g_v^* diag[\xi_v, 0]g_v$ with $\xi_v \in T_v^r$. Then $\mathbf{c}$ consists of all the v prime to $\mathfrak{c}$ of the following two types: (i) v is ramified in K and (ii) v is unramified in K and $det(\xi_v) \notin \mathfrak{g}_v^\times$.

For a number field W, a $k \in \mathbb{Z}^{\mathbf{b}}$ and $r \in \mathbb{Z}^{\mathbf{a}}$ we follow [30] and write $\mathcal{N}_k^r(W)$ for the space of W-rational nearly holomorphic modular forms of weight k (see [30, p. 103 and p. 110] for the definition). Regarding the near holomorphicity of the Eisenstein series $D_q(z, s; \chi, \mathfrak{c})$ we have the following theorem of Shimura,

Theorem 3.2 (Shimura, Theorem 17.12 in [30]) *We set $m := (k_v + k_{v\rho})_{v\in\mathbf{a}} \in \mathbb{Z}^{\mathbf{a}}$. Let K' be the reflex field of K with respect to the selected CM type and K_χ the field generated over K' by the values of χ. Let Φ be the Galois closure of K over $\mathbb{Q}$ and $\mu \in \mathbb{Z}$ with $2n - m_v \leqslant \mu \leqslant m_v$ and $m_v - \mu \in 2\mathbb{Z}$ for every $v \in \mathbf{a}$. Then $D_q(z, \mu/2; k, \chi, \mathfrak{c})$ belongs to $\pi^\beta \mathcal{N}_k^r(\Phi K_\chi \mathbb{Q}_{ab})$, except when $0 \leqslant \mu < n$, $\mathfrak{c} = \mathfrak{g}$, and $\chi_1 = \theta^\mu$, where $\beta = (n/2)\sum_{v\in\mathbf{a}}(m_v + \mu) - dn(n-1)/2$. Moreover $r = n(m - \mu + 2)/2$ if $\mu = n + 1$, $F = \mathbb{Q}$ and $\chi_1 = \theta^{n+1}$. In all other cases we have $r = (n/2)(m - |\mu - n|\mathbf{a} - n\mathbf{a})$.*

We now work out the positivity of the Fourier expansion of some holomoprhic Eisenstein series. In particular we assume that $m = \mu\mathbf{a}$ and we consider the series $D_{\mathbb{A}}^*(x, s)$ for $s = \frac{\mu}{2}$ and for $s = n - \frac{\mu}{2}$. For an $h \in S$, and $c(h, q, s)$ as in Eq. 2, we define $c(h, s) := \prod_{i=0}^{n-1} L_{\mathfrak{c}}(2s - i, \chi_1\theta^i)c(h, q, s)$, that is the hth Fourier coefficient of $D_{\mathbb{A}}^*(x, s)$. Then we have the following,

Proposition 3.3 (Shimura, Proposition 17.6 in [30]) *Exclude the case where $\mu = n + 1$, $F = \mathbb{Q}$ and $\chi = \theta^{n+1}$. Then we have that $c(h, \frac{\mu}{2}) \neq 0$ only in the following situations*

1. $h = 0$, *and* $\mu = n$,
2. $h \neq 0$, $\mu > n$ *and* $h_v > 0$ *for all* $v \in \mathbf{a}$,
3. $h \neq 0$, $\mu = n$ *and* $h_v \geqslant 0$ *for all* $v \in \mathbf{a}$.

Proof This follows directly from [30, Proposition 17.6], where the positivity of $c(h, q, \frac{\mu}{2})$ is considered, after observing that $\Lambda_c(\mu/2) = \prod_{i=0}^{n-1} L_{\mathfrak{c}}(\mu - i, \chi_1\theta^i) \neq 0$ for $\mu > n$. For $\mu = n$ we need to observe that $L(s, \chi_1\theta^{n-1})$ does not have a pole at $s = 1$, since $\chi_1\theta^{n-1}$ is not the trivial character, since $(\chi_1)_{\mathbf{a}}(x) = sgn(x_{\mathbf{a}})^{n\mathbf{a}}$, and hence $(\chi_1\theta^{n-1})_{\mathbf{a}}(x) = sgn(x_{\mathbf{a}})$. hence not trivial. □

The other holomorphic Eisenstein series, i.e. $s = n - \frac{\mu}{2}$, has a completely different behaviour. Namely, independently of μ, it may have non-trivial Fourier coefficients even for $h \geqslant 0$ not of full rank, that is with $det(h) = 0$. Let us explain this. By Proposition 3.1 we observe that $c(h, s)$ is equal to a finite non-vanishing factor times

$$f(s)\Lambda_h(s)\prod_{v\in\mathbf{a}} \xi\left(y_v, h_v; s+\mu/2, s-\mu/2\right), \quad y_v := q_v q_v^*,$$

where $f(s) := \prod_{v\in\mathbf{c}} f_{h,q,v}\left(\chi(\pi_v)|\pi_v|^{2s}\right)$, and for the function ξ we have (see [30, p. 140]) that

$$\xi(y_v, h_v; a, b) = i^{nb-na}2^{\tau}\pi^{\epsilon}\frac{\Gamma_t(a+b-n)}{\Gamma_{n-q}(a)\Gamma_{n-p}(b)}det(y_v)^{n-a-b}\times$$

$$\delta_+(h_v y_v)^{a-n+q/2}\delta_-(h_v y_v)^{b-n+p/2}\omega(2\pi y_v, h_v; a, b),$$

where p (resp. q) is the number of positive (resp. negative) eigenvalues of h_v and $t = n - p - q$; $\delta_+(x)$ is the product of all positive eigenvalues of x and $\delta_-(x) = \delta_+(-x)$, and

$$\Gamma_n(s) := \pi^{n(n-1)/2}\prod_{\nu=0}^{n-1}\Gamma(s-\nu).$$

For the quantities τ, ϵ and the function $\omega(\cdot)$ we refer to [30, p. 140], since they do not play any role in the argument below. We are interested in the values

$$f(n-\mu/2)\Lambda_h(n-\mu/2)\prod_{v\in\mathbf{a}}\xi\left(y_v, h_v; n, n-\mu\right),$$

with $\mu \geqslant n$.

Let us write r for the rank of h, then $\Lambda_h(s) = \prod_{i=0}^{n-1-r} L_{\mathfrak{c}}(2s - n - i, \chi_1\theta^{n+i-1})$ and hence $\Lambda_h(n - \mu/2) = \prod_{i=0}^{n-1-r} L_{\mathfrak{c}}(n - \mu - i, \chi_1\theta^{n+i-1})$. We now note that $(\chi_1)_{\mathbf{a}}(x) = sgn(x_{\mathbf{a}})^{\mu\mathbf{a}}$ and hence after setting $\psi_i := \chi_1\theta^{n+i-1}$ we obtain $(\psi_i)_{\mathbf{a}}(x) = sgn(x_{\mathbf{a}})^{(\mu+n+i-1)\mathbf{a}}$. We now conclude that the quantity $\Lambda_h(n - \mu/2)$ may not be zero since by [30, Lemma 17.5] we have that $L(n - \mu - i, \psi_i) = 0$ if $n - \mu - i \equiv \mu + n + i - 1 \mod 2$ (the so-called trivial zeros), which never holds. For the gamma factors we have for $h = 0$,

$$\prod_{v\in\mathbf{a}} \frac{\Gamma_n(n-\mu)}{\Gamma_n(n)\Gamma_n(n-\mu)} = \prod_{v\in\mathbf{a}} \frac{1}{\Gamma_n(n)} \neq 0.$$

Suppose that $h \neq 0$ and let $r = rank(h)$. Then

$$\prod_{v\in\mathbf{a}} \frac{\Gamma_{n-r}(n-\mu)}{\Gamma_n(n)\Gamma_{n-r}(n-\mu)} = \prod_{v\in\mathbf{a}} \frac{1}{\Gamma_n(n)} \neq 0.$$

In particular we conclude that in the case of $s = n - \frac{\mu}{2}$ we may have non-trivial Fourier coefficients even if the matrix h is not positive definite.

3.2 Theta Series

We start by recalling some results of Shimura in (the appendices of) [29, 30] regarding Hermitian theta series. We set $V := M_n(K)$ and we let $\mathcal{S}(V_{\mathbf{h}})$ to denote the space of Schwartz–Bruhat functions on $V_{\mathbf{h}} := \prod'_{v\in\mathbf{h}} V_v$. We consider an element $\lambda \in \mathcal{S}(V_{\mathbf{h}})$ and an $\mu \in \mathbb{Z}^{\mathbf{b}}$ such that $\mu_v\mu_{v\rho} = 0$ for all $v \in \mathbf{a}$ and $\mu_v \geqslant 0$ for all $v \in \mathbf{b}$. For a $\tau \in S_+ \cap GL_n(K)$ we then consider the theta series defined in [30, p. 277]),

$$\theta(z,\lambda) := \sum_{\xi\in V} \lambda(\xi) det(\xi)^{\mu\rho} \mathbf{e}^n_{\mathbf{a}}(\xi^*\tau\xi), \quad z \in \mathcal{H},$$

where $det(\xi)^{\mu\rho} := \left(\prod_{v\in\mathbf{b}} det(\xi_v)^{\mu_v}\right)^{\rho}$. We fix a Hecke character ϕ of K with infinity type $\phi_{\mathbf{a}}(y) = y^{-\mathbf{a}}|y|^{\mathbf{a}}$ and such that $\phi_1 = \theta$, where we recall that we write θ for the non-trivial character of K/F. Such a character ϕ always exists, [30, Lemma A5.1], but may not be unique. We now let ω be a Hecke character of K and we write $\mathfrak{f}$ for its conductor and define $\mathfrak{h} = \mathfrak{f} \cap \mathfrak{g}$. Following Shimura we introduce the notation,

$$R^* = \{w \in M_n(K)_{\mathbb{A}} | w_v \prec \mathfrak{r}_v, \forall v \in \mathbf{h}\},$$

and we fix an element $r \in GL_n(K)_{\mathbf{h}}$. Then we define the function $\lambda \in \mathcal{S}(V_{\mathbf{h}})$ by

$$\lambda(x) := \omega(det(r)^{-1}) \prod_{v|\mathfrak{h}} \omega_v(det(r_v x_v^{-1})),$$

if $r^{-1}x \in R^*$ and $r_v^{-1}x_v \in GL_n(\mathfrak{r}_v)$ for all $v|\mathfrak{h}$, and we set $\lambda(x) = 0$ otherwise. As it is explained in Shimura [30, Theorem A5.4] there is an action of $G_{\mathbb{A}}$ on $\mathcal{S}(V_{\mathbf{h}})$, which will be denoted by ${}^x\ell$ for $x \in G_{\mathbb{A}}$ and $\ell \in \mathcal{S}(V_{\mathbf{h}})$. Then we define the adelic theta function $\theta_{\mathbb{A}}$ on $G_{\mathbb{A}}$ by

$$\theta_{\mathbb{A}}(x,\omega) := \theta_{\mathbb{A}}(x,\lambda) := j_x^l(\mathbf{i})\theta(x\cdot\mathbf{i}, {}^x\lambda), \quad x \in G_{\mathbb{A}},$$

where $l = \mu + n\mathbf{a} \in \mathbb{Z}^{\mathbf{b}}$. Then Shimura shows that

$$\theta_{\mathbb{A}}(\alpha x w, \lambda) = j_w^l(\mathbf{i})^{-1}\theta_{\mathbb{A}}(x, {}^w\lambda), \quad \alpha \in G, w \in G_{\mathbb{A}}, \text{ and } w \cdot \mathbf{i} = \mathbf{i}. \tag{3}$$

and,

Theorem 3.4 (Shimura, Sect. A5.5 in [30] and Proposition A7.16 in [29]) $\theta_{\mathbb{A}}(x, \omega)$ *is an element in* $M_l(C, \omega')$ *with* $C = D[\mathfrak{b}^{-1}, \mathfrak{b}\mathfrak{c}]$ *and* $\omega' = \omega\phi^{-n}$, *and* $l = \mu + n\mathbf{a}$. *Moreover* $\theta_{\mathbb{A}}(x, \omega)$ *is a cusp form if* $\mu \neq 0$. *The ideals* $\mathfrak{b}$ *and* $\mathfrak{c}$ *are given as follows. We define a fractional ideals* $\mathfrak{y}$ *and* $\mathfrak{t}$ *in* F *such that* $g^*\tau g \in \mathfrak{y}$ *and* $h^*\tau^{-1}h \in \mathfrak{t}^{-1}$ *for all* $g \in r\mathfrak{g}^n$ *and* $h \in \mathfrak{r}^n$. *Then we can take*

$$(\mathfrak{b}, \mathfrak{b}\mathfrak{c}) = (\mathfrak{d}\mathfrak{y}, \mathfrak{d}(\mathfrak{t}\mathfrak{e}\mathfrak{f}^{\rho}\mathfrak{f} \cap \mathfrak{y}\mathfrak{e} \cap \mathfrak{y}\mathfrak{f})),$$

where $\mathfrak{e}$ *is the relative discriminant of* K *over* F. *For an element* $q \in GL_n(K)_{\mathbf{h}}$ *we have that the* q*th component of the theta series is given by*

$$\theta_{q,\omega}(z) = \omega'(det(q)^{-1})|det(q)|_K^{n/2} \times$$

$$\sum_{\xi \in V \cap r R^* q^{-1}} \omega_{\mathbf{a}}(det(\xi))\omega(det(r^{-1}\xi q)\mathfrak{r})det(\xi)^{\mu\rho}\mathbf{e}_{\mathbf{a}}(\xi^*\tau\xi z).$$

where $\xi \in V \cap rR^*q^{-1}$ *such that* $\xi^*\tau\xi = \sigma$.

For our later applications we now work out the functional equation with respect to the action of the element $\eta = \eta_n = \begin{pmatrix} 0 & -1_n \\ 1_n & 0 \end{pmatrix}$. In particular we are interested in the theta series $\theta^*_{\mathbb{A}}(x, \omega) := \theta_{\mathbb{A}}(x\eta_{\mathbf{h}}^{-1}, \omega)$. We note that by Eq. 3 we have that

$$\theta^*_{\mathbb{A}}(x, \omega) = \theta_{\mathbb{A}}(x\eta_{\mathbf{h}}^{-1}, \lambda) =$$

$$\theta_{\mathbb{A}}((-1)_{\mathbf{h}}x\eta_{\mathbf{h}}, \lambda) = \theta_{\mathbb{A}}((-1)_{\mathbf{h}}x, {}^{\eta}\lambda) = \omega'_{\mathfrak{c}}(-1)\theta_{\mathbb{A}}(x, {}^{\eta}\lambda),$$

and by [30, Theorem A5.4 (6)] we have that

$${}^{\eta}\lambda(x) = i^p|N_{F/\mathbb{Q}}(det(2\tau^{-1}))|^n \int_{V_{\mathbf{h}}} \lambda(y)\mathbf{e}_{\mathbf{h}}(-2^{-1}Tr_{K/F}(tr(y^*\tau x)))dy,$$

where $p = n^2[F : \mathbb{Q}]$ and dy is the Haar measure on $V_{\mathbf{h}}$ such that the volume of $M_n(\mathfrak{r})_{\mathbf{h}}$ is $|D_K|^{-n^2/2}$. We now compute the integral

$$I(x) := \int_{V_{\mathbf{h}}} \lambda(y)\mathbf{e}_{\mathbf{h}}(-2^{-1}Tr_{K/F}(tr(y^*\tau x)))dy.$$

We have

$$I(x) = \omega(det(r))^{-1}\left(\prod_{v\nmid\mathfrak{h}}\int_{r_vM_n(\mathfrak{r}_v)}\mathbf{e}_v(-2^{-1}Tr_{K_v/F_v}(tr(y_v^*\tau_vx_v)))d_vy\right)\times$$

$$\left(\prod_{v|\mathfrak{h}}\int_{r_vGL_n(\mathfrak{r}_v)}\omega(det(r_v^{-1}y_v))^{-1}\mathbf{e}_v(-2^{-1}Tr_{K_v/F_v}(tr(y_v^*\tau_vx_v)))d_vy\right).$$

We compute the local integrals separately. For a prime $v \nmid \mathfrak{h}$ we have

$$\int_{r_vM_n(\mathfrak{r}_v)}\mathbf{e}_v(-2^{-1}Tr_{K_v/F_v}(tr(y_v^*\tau_vx_v)))d_vy =$$

$$|det(r)|_v\int_{M_n(\mathfrak{r}_v)}\mathbf{e}_v(-2^{-1}Tr_{K_v/F_v}(tr(y_v^*r_v^*\tau_vx_v)))d_vy =$$

$$|det(r)|_v\int_{M_n(\mathfrak{r}_v)}\mathbf{e}_v(-2^{-1}Tr_{K_v/F_v}(tr(x_v^*\tau_v^*r_vy_v)))d_vy =$$

$$\begin{cases} 0, & \text{if } x_v^*\tau_v^*r_v \notin T; \\ |det(r)|_v|D_{K_v}|_v^{n^2/2}, & \text{otherwise.}\end{cases},$$

where $T := \{x \in M_n(K_v)|tr(xy) \in \mathfrak{d}_v^{-1},\ \forall y \in M_n(\mathfrak{r}_v))\}$ and D_{K_v} is the discriminant of K_v. For the other finite places, we obtain generalized Gauss sums. We have

$$\int_{r_vGL_n(\mathfrak{r}_v)}\omega(det(r_v^{-1}y_v))^{-1}\mathbf{e}_v(-2^{-1}Tr_{K_v/F_v}(tr(y_v^*\tau_vx_v)))d_vy =$$

$$|det(r)|_v\int_{GL_n(\mathfrak{r}_v)}\omega(det(y_v))^{-1}\mathbf{e}_v(-2^{-1}Tr_{K_v/F_v}(tr(y_v^*r_v^*\tau_vx_v)))d_vy =$$

$$|det(r)|_v\int_{GL_n(\mathfrak{r}_v)}\omega(det(y_v))^{-1}\mathbf{e}_v(-2^{-1}Tr_{K_v/F_v}(tr(x_v^*\tau_v^*r_vy_v)))d_vy.$$

By a standard argument (see for example [22, pp. 259–260]), this integral is zero, if $x_v^*\tau_v^*r_v\mathfrak{r}_v \neq (\mathfrak{f}\mathfrak{d}_K)^{-1}T_v^\times$, where $T_v^\times := T_v \cap GL_n(\mathfrak{r}_v)$. If $x_v^*\tau_v^*r_v\mathfrak{r}_v = (\mathfrak{f}\mathfrak{d}_K)^{-1}T_v^\times$, then after the change of variable $y_v \mapsto (x_v^*\tau_v^*r_v)^{-1}y_v$we have that the integral is equal to

$$|det(r)|_v|det(x_v^*\tau_v^*r_v)|^{-1}\times$$

$$\int_{\mathfrak{f}^{-1}\mathfrak{d}_K^{-1}GL_n(\mathfrak{r}_v)}\omega(det((x_v^*\tau_v^*r_v)^{-1}y_v))^{-1}\mathbf{e}_v(-2^{-1}Tr_{K_v/F_v}(tr(y_v)))d_vy =$$

$$|det(x_v^*\tau_v^*)|^{-1}\omega(det(\tau_v^* r_v x_v^*))\times$$

$$\int_{\mathfrak{f}^{-1}\mathfrak{d}_K^{-1}GL_n(\mathfrak{r}_v)} \omega(y_v)^{-1}\mathbf{e}_v(-2^{-1}Tr_{K_v/F_v}(tr(y_v)))d_v y$$

We then have that $|det(x_v^*\tau_v^*)|^{-1} = |det(r_v)|_v N(\mathfrak{f}\mathfrak{d})^{-n}$ and hence we can rewrite the above expression as

$$|det(r_v)|_v N(\mathfrak{f}\mathfrak{d})^{-n}\omega(det((fd)^n\tau_v^* r_v x_v^*))\omega(fd)^{-n}\times$$

$$\int_{\mathfrak{f}^{-1}\mathfrak{d}_K^{-1}GL_n(\mathfrak{r}_v)} \omega(y_v)^{-1}\mathbf{e}_v(-2^{-1}Tr_{K_v/F_v}(tr(y_v)))d_v y$$

for some elements f, d such that $(f) = \mathfrak{f}_v$ and $(d) = \mathfrak{d}_v$. By a standard argument (see for example [22, p. 259]), we obtain

$$N(\mathfrak{d})^{-n}\omega(fd)^{-n}\int_{\mathfrak{f}^{-1}\mathfrak{d}_K^{-1}GL_n(\mathfrak{r}_v)} \omega(y_v)^{-1}\mathbf{e}_v(-2^{-1}Tr_{K_v/F_v}(tr(y_v)))d_v y =$$

$$\sum_{y\in(M_n(\mathfrak{r}_v)/M_n(\mathfrak{d}\mathfrak{f}_v)} \omega_v(det(y))^{-1}\mathbf{e}_v(-tr(y)).$$

We set $\tau_n(\omega^{-1}) := \sum_{y\in(M_n(\mathfrak{r}_v)/M_n(\mathfrak{d}\mathfrak{f}_v)}\omega_v(det(y))^{-1}\mathbf{e}_v(-tr(y))$, and we note that in the case that ω is primitive we have that the last integral can be related to one-dimensional standard Gauss sums (see for example [2, p. 1410]). In particular in such a case we have $\tau_n(\omega^{-1}) = N(\mathfrak{d})^{\frac{n(n-1)}{2}}\tau(\omega^{-1})^n$ where $\tau(\omega^{-1})$ the standard one dimensional Gauss sum, associated to the character ω^{-1}. We summarize the above calculations in the following Proposition.

Proposition 3.5 *Let ω be a primitive character of conductor $\mathfrak{f}$. For the theta series $\theta_{\mathbb{A}}^*(x,\omega) \in \mathcal{M}_l(C', \omega'^{-c})$ with $C' := D[\mathfrak{b}\mathfrak{c}, \mathfrak{b}^{-1}]$ we have*

$$\theta_{\mathbb{A}}^*(x,\omega) = i^{n^2[F:\mathbb{Q}]}|N(2det(\tau)^{-1})|^n\omega'_{\mathfrak{c}}(-1)|det(r)|_{\mathbf{h}}N(\mathfrak{f})^{-n}N(\mathfrak{d})^{(\frac{-n}{2})}\times$$

$$\prod_{v|\mathfrak{f}} N(\mathfrak{d}_v)^{\frac{n^2}{2}}\tau(\omega^{-1})^n\theta_{\mathbb{A}}(x,\lambda^*),$$

where $\lambda^*(x) = \omega_{\mathfrak{f}}((fd)^n det(\tau_v r_v x_v^*))$ for $x \in T$ and $x_v^*\tau_v^* r_v \in \mathfrak{f}\mathfrak{d}^{-1}T_v^\times$ for all $v|\mathfrak{f}$, and zero otherwise.

We close this section by making a remark on the support of the q-expansion of θ^*. We first set,

$$C(\omega) := i^{n^2[F:\mathbb{Q}]}|N(2det(\tau)^{-1})|^n\omega'_{\mathfrak{c}}(-1)|det(r)|_{\mathbf{h}}N(\mathfrak{f})^{-n}N(\mathfrak{d})^{\frac{-n}{2}}\prod_{v|\mathfrak{f}}N(\mathfrak{d}_v)^{\frac{n^2}{2}}. \quad (4)$$

and take some $q \in GL_n(K)_{\mathbf{h}}$. Then, the qth component of θ^* is given by

$$\theta_q^*(z) = i^{n^2d}|det(q)|_{\mathbf{h}}^{n/2}\phi(det(q))^n\sum_{\xi\in V}I(\xi q)det(\xi)^{\mu\rho}e_{\mathbf{a}}(\xi^*\tau\xi z)$$

If $det(\xi)\neq 0$, then $I(\xi q)\neq 0$ only when $(\tau^* r q^*\xi^*)_v \in (\mathfrak{f}\mathfrak{d})^{-1}T_v^{\times}$ for all $v|\mathfrak{f}\mathfrak{d}$. That is,

$$\theta_q^*(z) = C(\omega)\tau(\omega^{-1})^n \sum_{\xi\in\widehat{\mathfrak{f}\mathfrak{d}}R_{\mathfrak{f}\mathfrak{d}}^{\times}\tau^{-1}\hat{r}q^{-1}\cap V}\omega_{\mathfrak{f}}((fd)^n\tau^* r q^*\xi^*)det(\xi)^{\mu\rho}e_{\mathbf{a}}(\xi^*\tau\xi z).$$

In particular we have that $(\xi^*\tau\xi)_v \in (\mathfrak{f}\mathfrak{d})^{-1}\hat{q}r^{-1}\hat{\tau}T_v^{\times}\tau T_v^{\times}\tau^{-1}\hat{r}q^{-1}\widehat{\mathfrak{f}\mathfrak{d}}$ for all $v|\mathfrak{f}$.

4 The L-function Attached to a Hermitian Modular Form

4.1 The Standard L-function

We fix a fractional ideal $\mathfrak{b}$ and an integral ideal $\mathfrak{c}$ of F. We set $C = D[\mathfrak{b}^{-1}, \mathfrak{b}\mathfrak{c}]$. For the fixed group C and for an integral ideal $\mathfrak{a}$ of K we write $T(\mathfrak{a})$ for the Hecke operator associated to it as it is defined for example in [30, p. 162].

We consider a non-zero adelic Hermitian modular form $\mathbf{f} \in \mathcal{M}_k(C,\psi)$ and assume that we have $\mathbf{f}|T(\mathfrak{a}) = \lambda(\mathfrak{a})\mathbf{f}$ with $\lambda(\mathfrak{a}) \in \mathbb{C}$ for all integral ideals $\mathfrak{a}$. If χ denotes a Hecke character of K of conductor $\mathfrak{f}$, for $s \in \mathbb{C}$ with $\Re(s) >> 0$ we consider the Dirichlet series

$$Z(s,\mathbf{f},\chi) := \left(\prod_{i=1}^{2n}L_{\mathfrak{c}}(2s-i+1,\chi_1\theta^{i-1})\right)\times\sum_{\mathfrak{a}}\lambda(\mathfrak{a})\chi(\mathfrak{a})N(\mathfrak{a})^{-s}, \quad (5)$$

where the sum runs over all integral ideals of K. It is shown in [30, p. 171] that this series has an Euler product representation, which we write as $Z(s,\mathbf{f},\chi) = \prod_{\mathfrak{q}}Z_{\mathfrak{q}}\left(\chi(\mathfrak{q})N(\mathfrak{q})^{-s}\right)$, where the product is over all prime ideals of K. Here we remind the reader (see introduction) that we abuse the notation and write χ also for the ideal character associated to the Hecke character χ. For the description of the Euler factors $Z_{\mathfrak{q}}$ at the prime ideal $\mathfrak{q}$ of K we have (see [30, p. 171]),

1. $Z_{\mathfrak{q}}(X) = \prod_{i=1}^n\left((1-N(\mathfrak{q})^{n-1}t_{\mathfrak{q},i}X)(1-N(\mathfrak{q})^n t_{\mathfrak{q},i}^{-1}X)\right)^{-1}$, if $\mathfrak{q}^{\rho}=\mathfrak{q}$ and $\mathfrak{q}\nmid\mathfrak{c}$,
2. $$Z_{\mathfrak{q}_1}(X_1)Z_{\mathfrak{q}_2}(X_2) =$$

$$\prod_{i=1}^{2n}\left((1-N(\mathfrak{q}_1)^{2n}t^{-1}_{\mathfrak{q}_1\mathfrak{q}_2,i}X_1)(1-N(\mathfrak{q}_2)^{-1}t_{\mathfrak{q}_1\mathfrak{q}_2,i}X_2)\right)^{-1},$$

if $\mathfrak{q}_1 \neq \mathfrak{q}_2$, $\mathfrak{q}_1^\rho = \mathfrak{q}_2$ and $\mathfrak{q}_i \nmid \mathfrak{c}$ for $i = 1, 2$,

3. $Z_{\mathfrak{q}}(X) = \prod_{i=1}^{n}\left((1-N(\mathfrak{q})^{n-1}t_{\mathfrak{q},i}X)\right)^{-1}$, if $\mathfrak{q}^\rho = \mathfrak{q}$ and $\mathfrak{q}|\mathfrak{c}$,
4.
$$Z_{\mathfrak{q}_1}(X_1)Z_{\mathfrak{q}_2}(X_2) =$$

$$\prod_{i=1}^{2n}\left((1-N(\mathfrak{q}_1)^{n-1}t_{\mathfrak{q}_1\mathfrak{q}_2,i}X_1)(1-N(\mathfrak{q}_2)^{n-1}t_{\mathfrak{q}_1\mathfrak{q}_2,n+i}X_2)\right)^{-1},$$

if $\mathfrak{q}_1 \neq \mathfrak{q}_2$, $\mathfrak{q}_1^\rho = \mathfrak{q}_2$ and $\mathfrak{q}_i|\mathfrak{c}$ for $i = 1, 2$,

where the $t_{?,i}$ above for $? = \mathfrak{q}, \mathfrak{q}_1\mathfrak{q}_2$ are the Satake parameters associated to the eigenform $\mathbf{f}$. We also introduce the L-function,

$$L(s, \mathbf{f}, \chi) := \prod_{\mathfrak{q}} Z_{\mathfrak{q}}\left(\chi(\mathfrak{q})(\psi/\psi_{\mathfrak{c}})(\pi_{\mathfrak{q}})N(\mathfrak{q})^{-s}\right), \quad \Re(s) >> 0 \tag{6}$$

where $\pi_{\mathfrak{q}}$ a uniformizer of $K_{\mathfrak{q}}$. We note here that we may obtain the Dirichelt series in Eq. 5 from the one in Eq. 6, up to a finite number of Euler factors, by setting $\chi\psi^{-1}$ for χ. Moreover if ψ is trivial then the two series coincide.

4.2 The Rankin–Selberg Integral Representation

We recall that in Sect. 3.2 we have fixed a Hecke character ϕ of K of infinity part $\phi_{\mathbf{a}}(y) = y_{\mathbf{a}}^{-\mathbf{a}}|y_{\mathbf{a}}|^{\mathbf{a}}$ and the restriction of ϕ to $F_{\mathbb{A}}^\times$ is the non-trivial Hecke character θ corresponding to the extension K/F. Keeping the notations from above we define $t \in \mathbb{Z}^{\mathbf{a}}$ to be the infinity type of χ, that is $\chi_{\mathbf{a}}(x) = x_{\mathbf{a}}^{-t}|x_{\mathbf{a}}|^t$. We then define $\mu \in \mathbb{Z}^{\mathbf{b}}$ by

$$\mu_v := t_v - k_{v\rho} + k_v, \text{ and } \mu_{v\rho} := 0 \text{ if } t_v \geqslant k_{v\rho} - k_v,$$

and

$$\mu_v := 0, \text{ and } \mu_{v\rho} := k_{v\rho} - k_v - t_v \text{ if } t_v < k_{v\rho} - k_v.$$

We moreover set $l := \mu + n\mathbf{a}$, $\psi' := \chi^{-1}\phi^{-n}$ and $h := 1/2(k_v + k_{v\rho} + l_v + l_{v\rho})_{v\in\mathbf{a}}$. Given μ, ϕ, τ and χ as above we write $\theta_\chi(x) := \theta_{\mathbb{A}}(x, \lambda) \in \mathcal{M}_l(C', \psi')$ for the theta series that we can associate to $(\mu, \phi, \tau, \chi^{-1})$ by taking $\omega := \chi^{-1}$ in Theorem 3.4. We write $\mathfrak{c}'$ for the integral ideal defined by $C' = D[\mathfrak{b}'^{-1}, \mathfrak{b}'\mathfrak{c}']$.

We now fix a decomposition $GL_n(K)_{\mathbb{A}} = \coprod_{q\in Q} GL_n(K)qEGL_n(K)_{\mathbf{a}}$, where $E = \prod_{v\in\mathbf{h}} GL_n(\mathfrak{r}_v)$. In particular the size of the set Q is nothing else than the class number of K. Given an element $f \in \mathcal{S}_k(\Gamma^q, \psi_q)$, and a function g on $\mathcal{H}$ such that $g|_k\gamma = \psi_q(\gamma)f$ for all $\gamma \in \Gamma^q$ we define the Petersson inner product

$$< f, g > := < f, g >_{\Gamma^q} := \int_{\Gamma^q \backslash \mathcal{H}} f(z)\overline{g(z)}\delta(z)^m dz,$$

where $\delta(z) := det(\frac{i}{2}(z^* - z))$ and dz a measure on $\Gamma^q \setminus \mathcal{H}$ defined as in [30, Lemma 3.4] and $m = (m_v)_{v\in\mathbf{a}}$ with $m_v = k_v + k_{v\rho}$.

The following theorem (see also [25, Theorem 7.8]) is obtained by combining results of Shimura [30] and Klosin [25]. For details we refer to [6, Sect. 4].

Theorem 4.1 (Shimura, Klosin) *Let $0 \neq \mathbf{f} \in \mathcal{M}_k(C, \psi))$ such that $\mathbf{f}|T(\mathfrak{a}) = \lambda(\mathfrak{a})\mathbf{f}$ for every $\mathfrak{a}$, and assume that $k_v + k_{v\rho} \geqslant n$ for some $v \in \mathbf{a}$, then there exists $\tau \in S_+ \cap GL_n(K)$ and $r \in GL_n(K)_{\mathbf{h}}$ such that*

$$\Gamma((s))\psi_{\mathfrak{c}}(det(r))c_{\mathbf{f}}(\tau, r)L(s + 3n/2, \mathbf{f}, \chi) =$$

$$\Lambda_{\mathfrak{c}}(s + 3n/2, \theta(\psi\chi)_1) \cdot \left(\prod_{v\in\mathbf{b}} g_v(\chi(\pi_{\mathfrak{p}})N(\mathfrak{p})^{-2s-3n})\right) det(\tau)^{s\mathbf{a}+h}|det(r)|_K^{-s-n/2} \times$$

$$C_0 \sum_{q\in Q} |det(qq^*)|_F^{-n} < f_q(z), \theta_{q,\chi}(z)E_q(z, \bar{s} + n; k - l, (\psi'/\psi)^c, \mathfrak{c}'') >_{\Gamma^q(\mathfrak{c}'')},$$

where

$$\Gamma((s)) := \prod_{v\in\mathbf{a}} (4\pi)^{-n(s+h_v)}\Gamma_n(s + h_v), \quad and \quad C_0 := \frac{[\Gamma_0(\mathfrak{c}'') : \Gamma]A}{\sharp X}.$$

where $\mathfrak{c}''$ any non-trivial integral ideal of F such that $\mathfrak{c}\mathfrak{c}'|\mathfrak{c}''$, $\Gamma^q(\mathfrak{c}'') := G \cap qD[\mathfrak{e}, \mathfrak{e}\mathfrak{h}]q^{-1}$, with $\mathfrak{e} = \mathfrak{b} + \mathfrak{b}'$ and $\mathfrak{h} = \mathfrak{e}^{-1}(\mathfrak{b}\mathfrak{c}'' \cap \mathfrak{b}'\mathfrak{c}'')$. Moreover $g_v(\cdot)$ are Siegel-series related to the polynomials $f_{\tau,r,v}(x)$ mentioned in Proposition 3.1 above, and we refer to [30, Theorem 20.4] for the precise definition. Finally X denotes the set of Hecke characters of infinity type t and conductor dividing $\mathfrak{f}_\chi$, Γ is a congruence subgroup of $SU(n, n)$ which appears in the [30, p. 179], and A some fixed rational number times some powers of π, and is independent of χ.

We will make the following assumption (see also the introduction):

Assumption. We assume that the class number of K is equal to the class number of $U(n, n)/F$ with respect to the full congruence subgroup $D[\mathfrak{b}^{-1}, \mathfrak{b}]$. For example this holds when the class number of F is taken equal to one [29, p. 66].
From the above assumption it follows that

$$\sum_{q\in Q} |det(qq^*)|_F^{-n} < f_q(z), \theta_{q,\chi}(z)E_q(z, \bar{s} + n; k - l, (\psi'/\psi)^c, \mathfrak{c}'') >_{\Gamma^q(\mathfrak{c}'')} =$$

$$< \mathbf{f}(x), \theta_{\mathbb{A},\chi}(x)\widetilde{E}_{\mathbb{A}}(x, \bar{s}+n; k-l, (\psi'/\psi)^c, \mathfrak{c}'' >_{\mathfrak{c}''},$$

where $\widetilde{E}_{\mathbb{A}}(x, \bar{s}+n; k-l, (\psi'/\psi)^c, \mathfrak{c}'')$ is the adelic Eisenstein series with q-component $|det(qq^*)|_F^{-n} E_q(z, \bar{s}+n; k-l, (\psi'/\psi)^c, \mathfrak{c}'')$, and $< \cdot, \cdot >_{\mathfrak{c}''}$ is the adelic Petersson inner product associated to the group $D[\mathfrak{e}, \mathfrak{e}\mathfrak{h}]$ as defined for example in [29, Eq. 10.9.6], but not normalized, and hence depends on the level. Moreover we define,

$$\widetilde{D}_{\mathbb{A}}(x, \bar{s}+n; k-l, \Psi, \mathfrak{c}'') := \overline{\Lambda_{\mathfrak{c}}(s+3n/2, \theta(\psi\chi)_1)}\widetilde{E}_{\mathbb{A}}(x, \bar{s}+n; k-l, \Psi, \mathfrak{c}''), \tag{7}$$

where $\Psi := (\psi'/\psi)^c$.

5 Algebraicity of Special L-Values

In this section we present some algebraicity results on the special values of the L-function introduced above, which were obtained in [6]. Results of this kind have been obtained by Shimura [30], but over the algebraic closure of $\mathbb{Q}$, and in [6] we worked out the precise field of definition, as well as, the reciprocity properties. There is also work by Harris [18, 19] and we refer to [6] for a discussion of how the results there compare with the ones presented here.

We consider a cuspidal Hecke eigenform $0 \neq \mathbf{f} \in \mathcal{S}_k(C, \psi; \overline{\mathbb{Q}})$ with $C := D[\mathfrak{b}^{-1}, \mathfrak{b}\mathfrak{c}]$ for some fractional ideal $\mathfrak{b}$ and integral ideal $\mathfrak{c}$ of F. We start by introducing some periods associated to $\mathbf{f}$. These periods are the analogue in the unitary case of periods introduced by Sturm in [31], and generalized in [3, 5], in the symplectic case (i.e. Siegel modular forms). In the following theorem we write $< \cdot, \cdot >$ for the adelic inner product associated to the group C.

Theorem 5.1 *Let $\mathbf{f} \in \mathcal{S}_k(D, \psi, \overline{\mathbb{Q}})$ be an eigenform, and define $m_v := k_v + k_{\rho v}$ for all $v \in \mathbf{a}$. Let Φ be the Galois closure of K over $\mathbb{Q}$ and write W for the extension of Φ generated by the Fourier coefficients of $\mathbf{f}$ and their complex conjugation. Assume $m_0 := min_v(m_v) > 3n+2$. Then there exists a period $\Omega_{\mathbf{f}} \in \mathbb{C}^\times$ and a finite extension Ψ of Φ such that for any $\mathbf{g} \in \mathcal{S}_k(\overline{\mathbb{Q}})$ we have*

$$\left(\frac{< \mathbf{f}, \mathbf{g} >}{\Omega_{\mathbf{f}}}\right)^\sigma = \frac{< \mathbf{f}^\sigma, \mathbf{g}^{\sigma'} >}{\Omega_{\mathbf{f}^\sigma}},$$

for all $\sigma \in \mathrm{Gal}(\overline{\mathbb{Q}}/\Psi)$, with $\sigma' := \rho\sigma\rho$. Here $\Omega_{\mathbf{f}^\sigma}$ is the period attached to the eigenform $\mathbf{f}^\sigma$. Moreover $\Omega_{\mathbf{f}}$ depends only on the eigenvalues of $\mathbf{f}$ and we have $\frac{<\mathbf{f},\mathbf{f}>}{\Omega_{\mathbf{f}}} \in (W\Psi)^\times$. In particular we have $\frac{<\mathbf{f},\mathbf{g}>}{<\mathbf{f},\mathbf{f}>} \in (W\Psi)(\mathbf{g}, \mathbf{g}^\rho)$, where $(W\Psi)(\mathbf{g}, \mathbf{g}^\rho)$ denotes the extension of $W\Psi$ obtained by adjoining the values of the Fourier coefficients of $\mathbf{g}$ and $\mathbf{g}^\rho$.

We note that the extension Ψ does not depend on $\mathbf{f}$, but only on K and n. We refer to [6] for more details on this. The following two theorems were obtained in [6].

Theorem 5.2 *Let $\mathbf{f} \in \mathcal{S}_k(C, \psi; \overline{\mathbb{Q}})$ be an eigenform for all Hecke operators, and assume that $m_0 \geqslant 3n+2$. Let χ be a character of K such that $\chi_{\mathbf{a}}(x) = x_{\mathbf{a}}^t |x_{\mathbf{a}}|^{-t}$ with $t \in \mathbb{Z}^{\mathbf{a}}$, and define $\mu \in \mathbb{Z}^{\mathbf{b}}$ by $\mu_v := -t_v - k_{v\rho} + k_v$ and $\mu_{v\rho} = 0$ if $k_{v\rho} - k_v + t_v \leqslant 0$, and $\mu_v = 0$ and $\mu_{v\rho} = k_{v\rho} - k_v + t_v$, if $k_{v\rho} - k_v + t_v > 0$. Assume moreover that either*

1. *there exists $v, v' \in \mathbf{a}$ such that $m_v \neq m_{v'}$, or*
2. *$m_v = m_0$ for all v and $m_0 > 4n-2$, or*
3. *$\mu \neq 0$.*

Then let $\sigma_0 \in \frac{1}{2}\mathbb{Z}$ such that

$$4n - m_v + |k_v - k_{v\rho} - t_v| \leqslant 2\sigma_0 \leqslant m_v - |k_v - k_{v\rho} - t_v|,$$

and,

$$2\sigma_0 - t_v \in 2\mathbb{Z}, \quad \forall v \in \mathbf{a}.$$

We exclude the following cases: For $n \leqslant 2\sigma_0 < 2n$, if we write $\mathfrak{f}'$ for the conductor of the character χ_1, then there is no choice of the integral ideal $\mathfrak{c}''$ as in Theorem 4.1 such that for any prime ideal $\mathfrak{q}$ of F, $\mathfrak{q}|\mathfrak{c}''\mathfrak{c}^{-1}$ implies either $\mathfrak{q}|\mathfrak{f}'$ or $\mathfrak{q}$ ramifies in K.

We let W be a number field such that $\mathbf{f}, \mathbf{f}^{\rho} \in \mathcal{S}_k(W)$ and $\Psi\Phi \subset W$, where Φ is the Galois closure of K in $\overline{\mathbb{Q}}$, and Ψ as in the Theorem 5.1 then

$$\frac{L(\sigma_0, \mathbf{f}, \chi)}{\pi^{\beta}\tau(\chi_1^n\psi_1^n\theta^{n^2})^{\rho} i^{n\sum_{v\in\mathbf{a}} p_v} < \mathbf{f}, \mathbf{f} >} \in \mathcal{W} := W(\chi),$$

where $\beta = n(\sum_v m_v) + d(2n\sigma_0 - 2n^2 + n)$, $W(\chi)$ obtained from W by adjoining the values of χ on finite adeles, and $p \in \mathbb{Z}^{\mathbf{a}}$ is defined for $v \in \mathbf{a}$ as $p_v = \frac{m_v - |k_v - k_{v\rho} - t_v| - 2\sigma_0}{2}$ if $\sigma_0 \geqslant n$, and $p_v = \frac{m_v - |k_v - k_{v\rho} - t_v| - 4n + 2\sigma_0}{2}$ if $\sigma_0 < n$.

Theorem 5.3 *Let $\mathbf{f} \in \mathcal{S}_k(C, \psi; \overline{\mathbb{Q}})$ be an eigenform for all Hecke operators. With notation as before we take $m_0 > 3n+2$. Let χ be a Hecke character of K such that $\chi_{\mathbf{a}}(x) = x_{\mathbf{a}}^t |x_{\mathbf{a}}|^{-t}$ with $t \in \mathbb{Z}^{\mathbf{a}}$. Define $\mu \in \mathbb{Z}^{\mathbf{b}}$ as in the previous theorem. With the same assumptions as in the previous theorem and with $\Omega_{\mathbf{f}} \in \mathbb{C}^{\times}$ as defined in Theorem 5.1 we have for all $\sigma \in \mathrm{Gal}(\overline{\mathbb{Q}}/\Psi_Q)$ that*

$$\left(\frac{L(\sigma_0, \mathbf{f}, \chi)}{\pi^{\beta}\tau(\chi_1^n\psi_1^n\theta^{n^2})^{\rho} i^{n\sum_{v\in\mathbf{a}} p_v}\Omega_{\mathbf{f}}}\right)^{\sigma} = \frac{L(\sigma_0, \mathbf{f}^{\sigma}, \chi^{\sigma})}{\pi^{\beta}\tau((\chi_1^n\psi_1^n\theta^{n^2})^{\sigma})^{\rho} i^{n\sum_{v\in\mathbf{a}} p_v}\Omega_{\mathbf{f}^{\sigma}}},$$

where $\Psi_Q = \Psi$ if $\sigma_0 \in \mathbb{Z}$ and it is the algebraic extension of Ψ obtained by adjoining $|det(qq^*)|_{\mathbf{h}}^{1/2}$ for all $q \in Q$, if $\sigma_0 \in \frac{1}{2}\mathbb{Z}$, where the set Q is defined in Sect. 4.

6 The Euler Factors Above p and the Trace Operator

We now fix an odd prime p and write S for the set of prime ideals in K above p such that they are inert with respect to the totally real subfield F. We assume of course that $S \neq \emptyset$. A typical element in this set will be denoted by $\mathfrak{p}$.

For a fractional ideal $\mathfrak{b}$ and an integral ideal $\mathfrak{c}$ of F, which are taken prime to the ideals in the set S, we define $C := D[\mathfrak{b}^{-1}, \mathfrak{b}\mathfrak{c}]$. We consider a non-zero $\mathbf{f} \in S_k(C, \psi)$, which we take to be an eigenform for all Hecke operators with respect to C. Furthermore we let χ be a Hecke character of K of conductor $\mathfrak{f}_\chi$ (or simply $\mathfrak{f}$ if there is no danger of confusion), supported in the set S. As we mentioned in the introduction our aim is to obtain measures that interpolate special values of $L(s, \mathbf{f}, \chi)$ such that the Eisenstein series involved in the Theorem 4.1 are holomorphic. In particular if we write $t \in \mathbb{Z}^{\mathbf{a}}$ for the infinite type of the character χ and define $\mu \in \mathbb{Z}^{\mathbf{b}}$ as in Sect. 4, then we will assume that

$$(k_v - \mu_v - n) + (k_{v\rho} - \mu_{v\rho}) = r, \quad \forall v \in \mathbf{a},$$

for some $r \geqslant n$, where we exclude the case of $r = n+1$, $F = \mathbb{Q}$ and $\chi_1 = \theta$. For a fixed character χ we define

1. $\Theta_\chi := \Theta := \theta_{\mathbb{A}}(x, \chi^{-1})$, where we put some special condition on the element $r \in GL_n(K)_{\mathbf{h}}$ in the definition of the theta series. Namely we pick the element $r \in GL_n(K)_{\mathbf{h}}$ such that $r_v = \pi_v r_v'$ with $r_v' \in GL_n(\mathfrak{r}_v)$ for v not dividing the conductor and $v \in S$, and $r_v \in GL_n(\mathfrak{r}_v)$ for $v \in S$ and dividing the conductor. For τ we assume that $\tau_v \in GL_n(\mathfrak{r}_v)$.
2. $\Theta^*_\chi := \Theta^* := \theta^*_{\mathbb{A}}(x, \chi^{-1})$, with similar conditions on r and τ as above.
3. $\mathbf{E}_{\chi,+} := \mathbf{E}_+ := \widetilde{D}_{\mathbb{A}}(x, \frac{r}{2}; k - l, \Psi, \mathfrak{c}'')$,
4. $\mathbf{E}^*_{\chi,+} := \mathbf{E}^*_+ := \widetilde{D}^*_{\mathbb{A}}(x, \frac{r}{2}; k - l, \Psi, \mathfrak{c}'')$,
5. $\mathbf{E}_{\chi,-} := \mathbf{E}_- := \widetilde{D}_{\mathbb{A}}(x, n - \frac{r}{2}; k - l, \Psi, \mathfrak{c}'')$,
6. $\mathbf{E}^*_{\chi,-} := \mathbf{E}^*_- := \widetilde{D}^*_{\mathbb{A}}(x, n - \frac{r}{2}; k - l, \Psi, \mathfrak{c}'')$,

where $\Psi := (\chi^{-1}\phi^{-n}\psi^{-1})^c$, $\mathfrak{c}''$ is as in Theorem 4.1 and the Eisenstein series $\widetilde{D}_{\mathbb{A}}$ was introduced in Eq. 7.

We now recall some facts about Hecke operators taken from [29, 30]. The action of the Hecke operator $T_C(\xi) := T(\xi) := C\xi C$ for some $\xi \in G_{\mathbf{h}}$, such that $C\xi C = \bigsqcup_{y \in Y} Cy$ for a finite set Y, is defined by,

$$(\mathbf{f}|C\xi C)(x) := \sum_{y \subset Y} \psi_{\mathfrak{c}}(det(a_y))^{-1}\mathbf{f}(xy^{-1}).$$

Following Shimura, we introduce the notation $E := \prod_{v \in \mathbf{h}} GL_n(\mathfrak{r}_v)$ and $B := \{x \in GL_n(K)_{\mathbf{h}} | x \prec \mathfrak{r}\}$. We have,

Lemma 6.1 (Shimura, Lemma 19.2 in [30]) *Let* $\sigma = diag[\hat{q}, q] \in G_v$ *with* $q \in B_v$ *and* $v|\mathfrak{c}$. *Then*

$$C_v \sigma C_v = \bigsqcup_{d,b} C_v \begin{pmatrix} \hat{d} & \widehat{db} \\ 0 & d \end{pmatrix},$$

with $d \in E_v \setminus E_v q E_v$ *and* $b \in S(\mathfrak{b}^{-1})_v / d^* S(\mathfrak{b}^{-1})_v d$, *where* $S(\mathfrak{b}^{-1}) := S \cap M_n(\mathfrak{b}^{-1})$.

We now introduce the following notation. Let $v \in \mathbf{h}$ be a finite place of F which correspond to a prime ideal of F, that is inert in K. We write $\mathfrak{p}$ for the ideal in K corresponding to the place in K above v, and π_v (or π when there is no fear of confusion) for a uniformizer of $\mathfrak{p}$. Since the choice of v determines uniquely a place of K (since we deal with the inert situation) we will often abuse the notation and write v also for this place of K.

For an integral ideal $\mathfrak{c}$ such that $v|\mathfrak{c}$ we write $U(\pi_i)$, for an $i = 1, \ldots, n$, for the operator $C\xi C$ defined by taking $\xi_{v'} = 1_{2n}$ for v' not equal to v and $\xi_v = diag[\hat{q}, q]$ with $q = diag[\pi, \ldots, \pi, 1, \ldots, 1]$ where there are i-many π's. Sometimes, we will also write $U(\pi)$ or $U(\mathfrak{p})$ for $U(\pi_n)$.

6.1 The Unramified Part of the Character

We now describe how we can choose the elements d in Lemma 6.1 for the operators $U(\pi_i)$. We have,

Lemma 6.2 *Let* $q = diag[\pi, \pi, \ldots \pi, 1, \ldots, 1]$ *with* m *many* π*'s. Then we have that in the decomposition*

$$E_v q E_v = \bigsqcup_d E_v d,$$

the representatives $d = (d_{ij})_{i,j}$*'s are all the lower triangular matrices such that,*

1. *there exist* $n - m$ *many* 1 *on the diagonal and the rest elements of the diagonal are equal to* π. *Write* S *for the subset of* $\{1, \ldots, n\}$ *such that* $i \in S$ *if and only if* $d_{ii} = \pi$.
2. *For any* $i > j$, *we have*

$$d_{ij} = \begin{cases} 0 & \text{if } j \notin S \text{ and } i \in S \\ 0 & \text{if } j \in S \text{ and } i \in S \\ \alpha & \text{if } j \in S \text{ and } i \notin S \end{cases},$$

where $\alpha \in \mathfrak{r}_v$ *runs over some fixed representatives of* $\mathfrak{r}_v/\mathfrak{p}_v$, *where* $\mathfrak{p}_v$ *the maximal ideal of* $\mathfrak{r}_v$.

Proof See [8, pp. 55–56] □

We now let λ_i be the eigenvalues of $\mathbf{f}$ with respect to the operators $U(\pi_i)$. For the fixed prime ideal $\mathfrak{p}$ as above we write t_i for the Satake parameters $t_{\mathfrak{p},i}$ associated to $\mathbf{f}$ as introduced in Sect. 4.

Lemma 6.3 *We have the identity*

$$\lambda_n^{-1}\left(\sum_{i=0}^{n}(-1)^i N(\mathfrak{p})^{\frac{i(i-1)}{2}+ni+\frac{n(n-1)}{2}}\lambda_i X^i\right) =$$

$$(-1)^n N(\mathfrak{p})^{n(2n-1)} X^n \prod_{i=1}^{n}(1-t_i^{-1}N(\mathfrak{p})^{1-n}X^{-1}).$$

Proof We first note that,

$$\sum_{i=0}^{n}(-1)^i N(\mathfrak{p})^{\frac{i(i-1)}{2}+ni}\lambda_i X^i = \prod_{i=1}^{n}(1-N(\mathfrak{p})^{n-1}t_i X). \tag{8}$$

This follows from [30, Lemma 19.13] and the fact that (see [30, p. 163])

$$\sum_{d\in E_v\backslash B_v}\omega_0(E_v d)|det(d)|_v^{-n}X^{v_{\mathfrak{p}}(det(d))} = \prod_{i=1}^{n}(1-N(\mathfrak{p})^{n-1}t_i X)^{-1},$$

where $v_{\mathfrak{p}}(\cdot)$ is the discrete valuation corresponding to the prime $\mathfrak{p}$, $|\cdot|_v$ the absolute value at v normalized as $|\pi|_v = N(\mathfrak{p})^{-1}$. For the definition of $\omega_0(E_v d)$, we first find an upper triangular matrix g so that $E_v d = E_v g$ and then we define $\omega_0(E_v d) := \prod_{i=1}^{n}\left(N(\mathfrak{p})^{-2i}t_i\right)^{e_i}$, where the $e_i \in \mathbb{Z}$ are so that $g_{ii} = \pi^{e_i}$ for $g = (g_{ij})$.

We can rewrite the right hand side of Eq. 8 as

$$\prod_{i=1}^{n}(1-N(\mathfrak{p})^{n-1}t_i X) =$$

$$N(\mathfrak{p})^{n(n-1)}(-1)^n(t_1t_2\dots t_n)X^n\prod_{i=1}^{n}(1-t_i^{-1}N(\mathfrak{p})^{1-n}X^{-1}).$$

Moreover we have by Eq. 8 that $\lambda_n = N(\mathfrak{p})^{-\frac{n(n+1)}{2}}t_1t_2\dots t_n$. So we conclude that

$$\lambda_n^{-1}\left(\sum_{i=0}^{n}(-1)^i N(\mathfrak{p})^{\frac{i(i-1)}{2}+ni}\lambda_i X^i\right) =$$

$$(-1)^n N(\mathfrak{p})^{n(n-1)+\frac{n(n+1)}{2}} X^n \prod_{i=1}^{n}(1-t_i^{-1}N(\mathfrak{p})^{1-n}X^{-1}),$$

or,

$$\lambda_n^{-1}\left(\sum_{i=0}^{n}(-1)^i N(\mathfrak{p})^{\frac{i(i-1)}{2}+ni+\frac{n(n-1)}{2}}\lambda_i X^i\right)=$$

$$(-1)^n N(\mathfrak{p})^{n(2n-1)}X^n\prod_{i=1}^{n}(1-t_i^{-1}N(\mathfrak{p})^{1-n}X^{-1}).$$

□

In particular if χ is a Hecke character of K which is taken unramified at $\mathfrak{p}$ and we set $X := \chi(\mathfrak{p})N(\mathfrak{p})^{s_+}$ with $s_+ := -\frac{n+r}{2}$ for some $r \in \mathbb{Z}$ we obtain,

$$\lambda_n^{-1}\left(\sum_{i=0}^{n}(-1)^i N(\mathfrak{p})^{\frac{(n-i)(n-i-1)}{2}-\frac{r-3n+2}{2}}\lambda_i \chi(\mathfrak{p})^i\right)=$$

$$(-1)^n N(\mathfrak{p})^{n(2n-1)-n(\frac{r+n}{2})}\chi(\mathfrak{p})^n\prod_{i=1}^{n}(1-\chi(\mathfrak{p})^{-1}t_i^{-1}N(\mathfrak{p})^{\frac{r-n+2}{2}}),$$

or

$$\lambda_n^{-1}\left(\sum_{i=0}^{n}(-1)^i N(\mathfrak{p})^{\frac{(n-i)(n-i-1)}{2}-\frac{r-3n+2}{2}}\lambda_i \chi(\mathfrak{p})^{i-n}\right)= \tag{9}$$

$$(-1)^n N(\mathfrak{p})^{n(2n-1)-n(\frac{r+n}{2})}\prod_{i=1}^{n}(1-\chi(\mathfrak{p})^{-1}t_i^{-1}N(\mathfrak{p})^{\frac{r-n+2}{2}}),$$

and if we set $X := \chi(\mathfrak{p})N(\mathfrak{p})^{s_-}$ with $s_- := -\frac{3n-r}{2}$, we obtain,

$$\lambda_n^{-1}\left(\sum_{i=0}^{n}(-1)^i N(\mathfrak{p})^{\frac{(n-i)(n-i-1)}{2}-\frac{-n-r+2}{2}}\lambda_i \chi(\mathfrak{p})^{i-n}\right)= \tag{10}$$

$$(-1)^n N(\mathfrak{p})^{n(2n-1)-n(\frac{3n-r}{2})}\prod_{i=1}^{n}(1-\chi(\mathfrak{p})^{-1}t_i^{-1}N(\mathfrak{p})^{\frac{n-r+2}{2}}).$$

We also make a general remark about the adjoint operator of the Hecke operators introduced in Lemma 6.1. First we note that,

$$\begin{pmatrix}0&1\\-1&0\end{pmatrix}\begin{pmatrix}a&b\\c&d\end{pmatrix}\begin{pmatrix}0&-1\\1&0\end{pmatrix}=\begin{pmatrix}d&-c\\-b&a\end{pmatrix}$$

In particular we have

$$\eta_{\mathbf{h}}^{-1} D[\mathfrak{b}^{-1}, \mathfrak{b}\mathfrak{c}]\eta_{\mathbf{h}} = D[\mathfrak{b}\mathfrak{c}, \mathfrak{b}^{-1}].$$

Now if we write W for the operator $(\mathbf{f}|W)(x) := \mathbf{f}(x\eta_{\mathbf{h}}^{-1})$ we have,

Lemma 6.4 *For* $\mathbf{f}, \mathbf{g} \in \mathcal{M}_k(C, \psi)$ *we have*

$$< \mathbf{f}|C\sigma C, \mathbf{g} >_{\mathfrak{c}} = < \mathbf{f}, \mathbf{g}|W\widetilde{C}\tilde{\sigma}\widetilde{C}W^{-1} >_{\mathfrak{c}}$$

where $\widetilde{C} := D[\mathfrak{b}\mathfrak{c}, \mathfrak{b}^{-1}]$, *and* $\tilde{\sigma} := diag[\widehat{q^*}, q^*]$ *if* $\sigma = diag[\hat{q}, q]$.

Proof By Proposition 11.7 in [29] we have that $< \mathbf{f}|C\sigma C, \mathbf{g} > = < \mathbf{f}, \mathbf{g}|C\sigma^{-1}C >$. Of course we have $\sigma^{-1} = diag[q^*, q^{-1}]$. Moreover we have that

$$C\sigma^{-1}C = WW^{-1}CWW^{-1}\sigma^{-1}WW^{-1}CWW^{-1}$$

and we have that $W\sigma^{-1}W^{-1} = diag[q^{-1}, q^*] = diag[\widehat{q^*}, q^*]$. Moreover the group $W^{-1}CW = D[\mathfrak{b}\mathfrak{c}, \mathfrak{b}^{-1}]$ if $C = D[\mathfrak{b}^{-1}, \mathfrak{b}\mathfrak{c}]$. Moreover we note that we may write $D[\mathfrak{b}\mathfrak{c}, \mathfrak{b}^{-1}] = D[\tilde{\mathfrak{b}}^{-1}, \tilde{\mathfrak{b}}\mathfrak{c}]$ by taking $\tilde{\mathfrak{b}} = \mathfrak{b}^{-1}\mathfrak{c}^{-1}$. □

For the fixed ideal $\mathfrak{p}$, and an $s \in \mathbb{C}$, we define the operator $J(\mathfrak{p}, s)$ on $\mathcal{M}_k(C, \psi)$ as

$$J(\mathfrak{p}, s) := \sum_{i=0}^{n} (-1)^i N(\mathfrak{p})^{\frac{i(i-1)}{2} + i(n+s) + \frac{n(n-1)}{2}} (\chi)(\mathfrak{p})^{i-n} U(\pi_i).$$

We now note by Lemma 6.3 we have that for the eigenform $\mathbf{f}$

$$\mathbf{f}|J(\mathfrak{p}, s) = \lambda_n (-1)^n N(\mathfrak{p})^{n(2n-1)} N(\mathfrak{p})^{ns} \prod_{i=1}^{n} (1 - N(\mathfrak{p})^{1-n}\chi(\mathfrak{p})^{-1} t_i^{-1} N(\mathfrak{p})^{-s})\mathbf{f}$$

We will need to consider the adjoint operator of $J(\mathfrak{p}, s)$ with respect to the Petersson inner product. In particular if we write

$$< \mathbf{f}|J(\mathfrak{p}, s), \mathbf{g} > = < \mathbf{f}, \mathbf{g}|W\widetilde{J(\mathfrak{p}, s)}W^{-1} >,$$

then by Lemma 6.4 we have that

$$\widetilde{J(\mathfrak{p}, s)} = \sum_{i=0}^{n} (-1)^i N(\mathfrak{p})^{\frac{i(i-1)}{2} + i(n+s) + \frac{n(n-1)}{2}} \chi(\mathfrak{p})^{n-i} U(\pi_i),$$

where we keep writing $U(\pi_i)$ for the Hecke operator

$$D[\mathfrak{b}\mathfrak{c}, \mathfrak{b}^{-1}] diag[\pi, \pi, \dots, \pi, 1 \dots, 1] D[\mathfrak{b}\mathfrak{c}, \mathfrak{b}^{-1}].$$

We note here that

$$D[\mathfrak{bc}, \mathfrak{b}^{-1}]diag[\pi, \pi, \dots, \pi, 1 \dots, 1]D[\mathfrak{bc}, \mathfrak{b}^{-1}] =$$

$$D[\mathfrak{bc}, \mathfrak{b}^{-1}]diag[\pi^{\rho}, \pi^{\rho}, \dots, \pi^{\rho}, 1 \dots, 1]D[\mathfrak{bc}, \mathfrak{b}^{-1}]$$

Of particular interest for us are the operators $\widetilde{J(\mathfrak{p}, s_{\pm})}$ where we recall we have defined $s_{+} := -\frac{r+n}{2}$ and $s_{-} := \frac{3n-r}{2}$. We set $\mathfrak{m}_0 := \mathfrak{c}\mathfrak{p}\mathfrak{p}^{\rho}$. We note that $\Theta^{*}\mathbf{E}^{*}_{\pm} \in D[\mathfrak{b}\mathfrak{a}\mathfrak{m}_0, \mathfrak{b}^{-1}]$ for some ideals $\mathfrak{a}, \mathfrak{b}$ prime to $\mathfrak{q}$. This is clear for the Eisenstein series by its definition, and for the theta series we need to observe that since we are taking an $r \in GL_n(K)_{\mathbf{h}}$ of the form $\pi\pi^{\rho}r'$ for some $r' \in GL_n(K)_{\mathbf{h}}$ with $r_v \in GL_n(\mathfrak{r}_v)$ we have that the ideals $\mathfrak{t}$ and $\mathfrak{y}$ are equal to $\mathfrak{q}\mathfrak{q}^{\rho}$. Hence we have that $\theta \in \mathcal{M}_l(D[(\mathfrak{d}\mathfrak{q}\mathfrak{q}^{\rho})^{-1}, \mathfrak{d}\mathfrak{q}\mathfrak{q}^{\rho}\mathfrak{e}\mathfrak{f}\mathfrak{f}^{\rho}])$. Hecne $\theta^{*} \in \mathcal{M}_l(D[\mathfrak{d}\mathfrak{q}\mathfrak{q}^{\rho}\mathfrak{e}\mathfrak{f}\mathfrak{f}^{\rho}, (\mathfrak{d}\mathfrak{q}\mathfrak{q}^{\rho})^{-1}]) \subset \mathcal{M}_l(D[\mathfrak{d}\mathfrak{q}\mathfrak{q}^{\rho}\mathfrak{e}\mathfrak{f}\mathfrak{f}^{\rho}, \mathfrak{d}^{-1}]) \subset \mathcal{M}_l(D[\mathfrak{d}\mathfrak{c}\mathfrak{q}\mathfrak{q}^{\rho}\mathfrak{e}\mathfrak{f}\mathfrak{f}^{\rho}, \mathfrak{d}^{-1}])$. We then take $\mathfrak{b} = \mathfrak{d}^{-1}$ and $\mathfrak{a} = \mathfrak{e}\mathfrak{f}\mathfrak{f}^{\rho}$.
Before we go further, we collect some facts which will be needed in the proof of the following Theorem. We start by recalling the so-called generalized Möbius function as for example defined by Shimura in [30, pp. 163–164]. We restrict ourselves to the local version of it, since this will be enough for our purposes. We have fixed a finite place v of the filed K (recall here our abusing of notation explained above), and write K_v for the completion at v and $\mathfrak{r}_v$ for its ring of integers. We continue writing $\mathfrak{p}$ for the prime ideal of $\mathfrak{r}$ corresponding to the finite place v, and $\mathfrak{p}_v$ for the maximal ideal of $\mathfrak{r}_v$. Finally we write π for a fixed uniformizer of $\mathfrak{r}_v$.
The generalized Möbius function will be denoted by μ, and it is defined on the set of $\mathfrak{r}_v$-submodules of a torsion $\mathfrak{r}_v$-module. In particular we cite the following lemma [30, Lemma 19.10].

Lemma 6.5 *To every finitely generated torsion $\mathfrak{r}_v$-module A we can uniquely assign an integer $\mu(A)$ so that*

$$\sum_{B \subset A} \mu(B) = \begin{cases} 1 & \text{if } A = \{0\} \\ 0 & \text{if } A \neq \{0\} \end{cases}.$$

We also recall two properties (see [30] for a proof) of this generalized Möbius function, which will play an important role later. We have

1. $\mu((\mathfrak{r}_v/\mathfrak{p}_v)^r) = (-1)^r N(\mathfrak{p})^{r(r-1)/2}$ if $0 \leqslant r \in \mathbb{Z}$.
2. $\mu(A) \neq 0$ if and only if A is annihilated by a square free integral ideal of K_v.

Let us now denote by $\mathcal{L} := \mathcal{L}_{\ell}$ the set of $\mathfrak{r}_v$-lattices in K_v^{ℓ}. Given an $y \in GL_{\ell}(K_v)$ and an $L \in \mathcal{L}$ we define a new lattice by $yL := \{yx | x \in L\} \in \mathcal{L}$. Conversely it is clear that given two lattices $M, L \in \mathcal{L}$ there exists a $y \in GL_{\ell}(K_v)$ such that $M = yL$. We also note that if $L, M \in \mathcal{L}$ and $M \subset L$ then we can write $\mu(L/M)$. Let us now take $L := \mathfrak{r}_v^{\ell} \subset K_v^{\ell}$. Then by [30, Lemma 19.13] we have

$$\sum_{L \supset M \in \mathcal{L}} \mu(L/M) X^{v_{\mathfrak{p}}(det(y))} = \prod_{i=1}^{\ell} (1 - N(\mathfrak{p})^{i-1} X), \tag{11}$$

where the sum runs over all lattices $M \in \mathcal{L}$ contained in L, and y is defined so that $M = yL$. Here we write $v_{\mathfrak{p}}(\cdot)$ for the normalized discrete valuation of K_v.
We will now use the above equality to obtain a relation between the number of left cosets in the decomposition of Lemma 6.2. We set $E_\ell := GL_\ell(\mathfrak{r}_v)$ and for an $m \leqslant \ell$ we set $\pi_m^{(\ell)} := diag[\pi, \pi, \ldots, \pi, 1, \ldots, 1] \in GL_\ell(K_v)$ with m-many π's. As we have seen in Lemma 6.2 we have a decomposition

$$E_\ell \pi_m^{(\ell)} E_\ell = \bigsqcup_{d_m^{(\ell)}} E_\ell d_m^{(\ell)},$$

for some $d_m^{(\ell)} \in GL_\ell(K_v) \cap M_\ell(\mathfrak{r}_v)$. We write $\mu_m^{(\ell)}$ for the number of the cosets in the above decomposition. Then we have,

Lemma 6.6 *With notation as above,*

$$\sum_{i=0}^{\ell} (-1)^i N(\mathfrak{p})^{\frac{i(i-1)}{2}} \mu_i^{(\ell)} = 0.$$

Proof We first note that by taking the transpose of the decomposition above we may also work with right cosets, that is $E_\ell \pi_m^{(\ell)} E_\ell = \bigsqcup_{d_m^{(\ell)}} {}^t d_m^{(\ell)} E_\ell$. We now let $L := \mathfrak{r}_v^\ell$, and we see that to every coset ${}^t d_m^{(\ell)} E_\ell$ for $0 \leqslant m \leqslant \ell$ we can associate a lattice $M \in \mathcal{L}$ by $M := {}^t d_m^{(\ell)} L$. Since ${}^t d_m^{(\ell)}$ are integral we have $M \subset L$. Moreover in the sum $\sum_{L \supset M \in \mathcal{L}}' \mu(L/M) X^{v_{\mathfrak{p}}(det(y))}$, because of property (ii) of the Möbius function, we have that the $y's$ have square free elementary divisors. Indeed it is enough to notice (see for example [9, Theorem 1.4.1]) that for the lattice $M = yL$ we have that L/M is isomorphic to $\oplus_{0 \leqslant i \leqslant r} (\mathfrak{r}_v/\mathfrak{p}_v)^{e_i}$ where e_i are the (powers) of the elementary divisors of y, and r its rank. In particular we can conclude that each y in the sum $\sum_{L \supset M \in \mathcal{L}} \mu(L/M) X^{v_{\mathfrak{p}}(det(y))}$ belongs to some ${}^t d_m^{(\ell)} E_\ell$ for m equal to $v_{\mathfrak{p}}(det(y))$. That is, we may write

$$\sum_{L \supset M \in \mathcal{L}} \mu(L/M) X^{v_{\mathfrak{p}}(det(y))} = \sum_{i=0}^{\ell} (-1)^i N(\mathfrak{p})^{\frac{i(i-1)}{2}} \mu_i^{(\ell)} X^i,$$

where we have used property (i) of the Möbius function. We now set $X = 1$ and use Eq. 11 to conclude the lemma. □

We are now ready to prove the following theorem.

Theorem 6.7 *Let* $\mathfrak{p} \in S$ *and write* v *for the finite place of* F *corresponding to* $\mathfrak{p}$ *as above. Consider a Hecke character* χ *of* K *unramified at the prime* $\mathfrak{p}$. *Let* $\mathbf{F}_\pm := \Theta^* \mathbf{E}_\pm^*$ *and write*

$$\mathbf{g}_\pm := \mathbf{F}_\pm | \widetilde{J(\mathfrak{p}, s_\pm)}.$$

Then, for $q \in GL(K)_{\mathbf{h}}$, with $q_v \in GL_n(\mathfrak{r}_v)$, we have

$$\mathbf{g}_\pm\left(\begin{pmatrix} q & s\hat{q} \\ 0 & \hat{q} \end{pmatrix}\right) = C(\mathfrak{p}, s_\pm) \sum_{\tau \in S_+} c(\tau, q; \mathbf{g}_\pm)\mathbf{e}^n_{\mathbb{A}}(\tau s)$$

with $C(\mathfrak{p}, s_\pm) := (-1)^n N(\mathfrak{p})^{n(n-1)+n(n+s_\pm)}\psi(\mathfrak{p})^{-n}$ and,

$$c(\tau, q; \mathbf{g}_\pm) = \sum_{\tau_1+\tau_2=\tau} c(\tau_1, q\pi, \Theta^*)c(\tau_2, q\pi, \mathbf{E}^*_\pm),$$

where $(\tau_1)_v \in (\pi_v\pi_v^\rho)^{-1}T_v^\times$, where $T_v^\times = T_v \cap GL_n(\mathfrak{r}_v)$ and we recall that

$$T = \{x \in S | tr(S(\mathfrak{r})x) \subset \mathfrak{g}\},$$

where $S(\mathfrak{r}) = S \cap M_n(\mathfrak{r})$, and $T_v := T \otimes_{\mathfrak{r}} \mathfrak{r}_v$.

Proof We will show the Theorem when $\mathbf{F} := \mathbf{F}_+ = \Theta^*\mathbf{E}^*_+$, and a similar proof shows also the case of $\mathbf{F}_- = \Theta^*\mathbf{E}^*_-$. We set $\mathbf{g} := \mathbf{g}_+$, and we note that the Nebentype of $\Theta^*\mathbf{E}^*_+$ is ψ^{-c}. We then have,

$$\mathbf{g}\left(\begin{pmatrix} q & s\hat{q} \\ 0 & \hat{q} \end{pmatrix}\right) =$$

$$N(\mathfrak{p})^{\frac{n(n-1)}{2}} \sum_{i=0}^{n} B_i \sum_{d_i} \psi_v(det(d_i))^{-1} \sum_{b_i} \mathbf{F}\left(\begin{pmatrix} q & s\hat{q} \\ 0 & \hat{q} \end{pmatrix}\begin{pmatrix} \hat{d}_i^{-1} & -b_i d_i^{-1} \\ 0 & d_i^{-1} \end{pmatrix}\right),$$

where here we write d_i and b_i for the d's and b's corresponding to the Hecke operator $U(\pi_i)$ as described in Lemma 6.1, and in order to make the formulas a bit shorter we have introduced the notation $B_i := (-1)^i N(\mathfrak{p})^{\frac{i(i-1)}{2}+i(n+s_+)}\chi(\mathfrak{p})^{i-n}$. In particular we have that

$$N(\mathfrak{p})^{-\frac{n(n-1)}{2}}\mathbf{g}\left(\begin{pmatrix} q & s\hat{q} \\ 0 & \hat{q} \end{pmatrix}\right) =$$

$$\sum_{i=0}^{n} B_i \sum_{d_i} \psi_v(det(d_i))^{-1} \sum_{b_i} \mathbf{F}\left(\begin{pmatrix} q\hat{d}_i^{-1} & -qb_i d_i^{-1} + s\hat{q}d_i^{-1} \\ 0 & \hat{q}d_i^{-1} \end{pmatrix}\right) =$$

$$\sum_{i=0}^{n} B_i \sum_{d_i} \psi_v(det(d_i))^{-1} \sum_{b_i} \mathbf{F}\left(\begin{pmatrix} qd_i^* & (-qb_i q^* + s)\widehat{qd_i^*} \\ 0 & \widehat{qd_i^*} \end{pmatrix}\right) =$$

$$\sum_{i=0}^{n} B_i \sum_{d_i} \psi_v(det(d_i))^{-1} \sum_{b_i} \sum_{\tau\in S_+} c(\tau, qd_i^*; \mathbf{F})\mathbf{e}_{\mathbb{A}}^n(\tau(-qb_iq^* + s)) =$$

$$\sum_{i=0}^{n} B_i \sum_{d_i} \psi_v(det(d_i))^{-1} \sum_{\tau\in S_+} c(\tau, qd_i^*; \mathbf{F}) \left(\sum_{b_i} \mathbf{e}_{\mathbf{h}}^n(-\tau qb_iq^*)\right) \mathbf{e}_{\mathbb{A}}^n(\tau s)$$

Since $b_i \in S(\mathfrak{bc})_v$, we have by [30, Lemma 19.6] that

$$\sum_{b_i} \mathbf{e}_{\mathbf{h}}^n(-\tau qb_iq^*) = |det(d_i)|_{\mathbb{A}}^{-n},$$

if $(q^*\tau q)_v \in \mathfrak{d}^{-1}\mathfrak{b}^{-1}\mathfrak{c}^{-1}T_v$ for all $v \in \mathbf{h}$ and zero otherwise. We now write

$$c(\tau, qd_i^*; \mathbf{F}) = \sum_{\tau_1+\tau_2=\tau} c(\tau_1, qd_i^*; \Theta^*)c(\tau_2, qd_i^*; \mathbf{E}_+^*),$$

and from above we have that $(q^*\tau q)_v \in \mathfrak{d}^{-1}\mathfrak{b}^{-1}\mathfrak{c}^{-1}T_v$ for all $v \in \mathbf{h}$. Moreover we have that $c(\tau_1, qd_i^*; \Theta^*) \neq 0$ only if $(q^*\tau_1 q)_v \in \mathfrak{d}^{-1}\mathfrak{b}^{-1}\mathfrak{c}^{-1}d_i^{-1}T_v\widehat{d_i}$ for all $v \in \mathbf{h}$ and $c(\tau_2, qd_i^*; \mathbf{E}_+^*) \neq 0$, only if $(q^*\tau_2 q)_v \in \mathfrak{d}^{-1}\mathfrak{b}^{-1}\mathfrak{c}^{-1}d_i^{-1}T_v\widehat{d_i}$ for all $v \in \mathbf{h}$. In the above sum we run over all possible pairs of positive semi-definite hermitian matrices τ_1, τ_2 with $\tau_1 + \tau_2 = \tau$, and set $c(\tau_1, qd_i^*; \Theta^*) = c(\tau_2, qd_i^*; \mathbf{E}_+^*) = 0$ if τ_1, τ_2 are not in the set described above.
From now on we will be writing v for the finite place of F corresponding to the prime ideal $\mathfrak{p}$. We introduce the notation

$$S_i := \{s \in S : q^*sq \in \mathfrak{d}^{-1}\mathfrak{b}^{-1}\mathfrak{c}^{-1}d_i^{-1}T\widehat{d_i}, ord_{\mathfrak{p}}(\mathfrak{dbc}\nu(s)) = 2i\},$$

where $\nu(s)$ is the so-called denominator ideal associated to a matrix s, as for example defined in [29, Chap. I, Sect. 3]. That is, the valuation at $\mathfrak{p}$ of the denominator-ideal of the symmetric matrix q^*sq is exactly i, after clearing powers of $\mathfrak{p}$ coming from $\mathfrak{dcb}$. We note that since $\tau \in S_0$ we have that $\tau_1 \in S_i$ if and only if $\tau_2 \in S_i$ if $\tau_1 + \tau_2 = \tau$. We now rewrite the Fourier expansion of $\mathbf{g}$ as

$$\sum_{\tau\in S_+} N(\mathfrak{p})^{\frac{n(n-1)}{2}} \chi(\mathfrak{p})^{-n} \sum_{i=0}^{n} (-1)^i N(\mathfrak{p})^{\frac{i(i-1)}{2}+i(n+s_+)} \chi(\mathfrak{p}^i) \sum_{d_i} \psi_v(det(d_i))^{-1} \times$$

$$\sum_{\tau_1+\tau_2=\tau} c(\tau_1, qd_i^*; \Theta^*)c(\tau_2, qd_i^*; \mathbf{E}_+^*)|det(d_i)|_v^{-n}\mathbf{e}_{\mathbf{a}}^n(\mathbf{1}\hat{A}'{}^tq\tau q)\mathbf{e}_{\mathbb{A}}^n(\tau s),$$

where we have used the fact that $|det(d_i)|_{\mathbb{A}} = |det(d_i)|_v$ since $(d_i)_{v'} = 1_n$ for any finite place v' not equal to v. We now work the inner sum for any fixed τ. That is,

$$\sum_{i=0}^{n}(-1)^i N(\mathfrak{p})^{\frac{i(i-1)}{2}+i(n+s_+)}\chi(\mathfrak{p}^i)\sum_{d_i}\psi_v(det(d_i))^{-1}\times$$

$$\sum_{\tau_1+\tau_2=\tau} c(\tau_1, qd_i^*; \Theta^*)c(\tau_2, qd_i^*; \mathbf{E}_+^*)|det(d_i)|_v^{-n}, \tag{12}$$

or

$$\sum_{\tau_1+\tau_2=\tau}\sum_{i=0}^{n}(-1)^i N(\mathfrak{p})^{\frac{i(i-1)}{2}} N(\mathfrak{p})^{i(n+s_+)}\chi(\mathfrak{p}^i)\times$$

$$\sum_{d_i}\psi_v(det(d_i))^{-1}c(\tau_1, qd_i^*; \Theta^*)c(\tau_2, qd_i^*; \mathbf{E}_+^*)|det(d_i)|_v^{-n} \tag{13}$$

We claim that this sum is equal to

$$N(\mathfrak{p})^{\frac{n(n-1)}{2}+n(n+s_+)}(-1)^n\chi(\mathfrak{p})^n\psi(\mathfrak{p})^{-n}\sum_{\tau_1+\tau_2=\tau} c(\tau_1, q\pi, \Theta^*)c(\tau_2, q\pi, \mathbf{E}_+^*), \tag{14}$$

where $(q^*\tau_1 q)_v, (q^*\tau_2 q)_v \in \pi^{-2}\mathfrak{d}^{-1}\mathfrak{b}^{-1}\mathfrak{c}^{-1}T_v^{\times} = S_n$. Note that this is enough in order to establish the claim of the Theorem.
To show this, we consider the nth term of the Eq. 12, that is the summand with $i = n$ and we recall that the d_n's run over the single element πI_n . That is, the nth term is equal to

$$N(\mathfrak{p})^{\frac{n(n-1)}{2}+n(n+s_+)}(-1)^n\chi(\mathfrak{p})^n\psi(\mathfrak{p})^{-n}\sum_{\tau_1+\tau_2=\tau} c(\tau_1, q\pi, \Theta^*)c(\tau_2, q\pi, \mathbf{E}_+^*), \tag{15}$$

where $(q^*\tau_1 q)_v, (q^*\tau_2 q)_v \in \mathfrak{d}^{-1}\mathfrak{b}^{-1}\mathfrak{c}^{-1}d_n^{-1}T_v\widehat{d_n} = \mathfrak{d}^{-1}\mathfrak{b}^{-1}\mathfrak{c}^{-1}\pi^{-2}T_v$.

Note that the difference of the expression in Eq. 15, and the claimed sum in Eq. 14 is the difference of the support of the Fourier coefficients. Indeed note that in Eq. 15 (or better say in the line right after) we write T_v where in Eq. 14 we write $T_v^{\times}$, and of course $T_v^{\times} \subset T_v$. Hence our aim is to prove that for every pair (τ_1, τ_2) with $\tau_1 + \tau_2 = \tau$ and $\tau_i \in S_j$ with $j < n$ that contributes a non-trivial term in Eq. 15, its contribution will be cancelled out by the lower terms (i.e. $i < n$) that appear in Eq. 13. So the only terms that "survive" the cancellation will be the ones with $\tau_1, \tau_2 \in S_n$. Moreover all lower terms will be cancelled out.

We note that if we consider a $\tau_1 \in S_j$ (hence $\tau_2 \in S_j$) with $j < n$, then we observe that given such a τ_1 and τ_2, we have that $c(\tau_1, qd_m^*; \Theta^*)c(\tau_2, qd_m^*; \mathbf{E}_+^*) \neq 0$ implies that $m \geqslant j$. Indeed since $\tau_1, \tau_2 \in S_j$ we have for any $m < j$ that $(q^*\tau_1 q)_v, (q^*\tau_2 q)_v \notin \mathfrak{d}^{-1}\mathfrak{b}^{-1}\mathfrak{c}^{-1}d_i^{-1}T_v\widehat{d_i}$.

So in what follows we fix a pair τ_1 and τ_2 in S_j for some $j \geqslant 0$ with $j < n$. By [29, Lemma 13.3] since we are interested in the question whether τ_1, τ_2 belong to a particular lattice, we may assume without loss of generality that our τ_1, τ_2, locally at v, are of the form $diag[s_1, \dots, s_n]$ for $s_i \in K_v$. After reordering the s_i's we may assume that $s_{j+1}, \dots, s_n$ are integral, while the rest have non-trivial denominators. That means, that the d_m's for $j \leqslant m \leqslant n$ with $c(\tau_1, qd_m^*; \Theta^*)c(\tau_2, qd_m^*; \mathbf{E}_+^*) \neq 0$ can be taken of a very particular form, namely we can take them to be lower triangular matrices (by Lemma 6.2) with the diagonal of the form $diag[\pi, \dots, \pi, \pi^{e_{j+1}}, \dots, \pi^{e_n}]$, where $e_{j+1}, \dots, e_n \in \{0, 1\}$ and $e_{j+1} + \ldots + e_n = m - j$. Indeed the first j many π's on the diagonal are imposed to us in order $d_j \tau 1 d_j^*, d_j \tau_2 d_j^*$ to have integral coefficients along the diagonal. Given such a pair of indices m and j, with $m \geqslant j$ we will write $\lambda_m^{(j)}$ for the number of left cosets $E_v d_m$ with diagonal of d_m as just described. From now on when we write a d_m or d_j it will be always one of this particular form (i.e. lower diagonal and with the above mentioned description of the diagonal).
We now claim that we may write

$$c(\tau_1, qd_n^*, \Theta^*) = \alpha_{n,j} c(\tau_1, qd_j^*, \Theta^*),$$

and

$$c(\tau_2, qd_n^*, \mathbf{E}_+^*) = \beta_{n,j} c(\tau_2, qd_j^*, \mathbf{E}_+^*),$$

for some $\alpha_{n,j}$ and $\beta_{n,j}$, and any d_j. The terms $c(\tau_1, qd_j^*, \Theta^*)$, $c(\tau_2, qd_j^*, \mathbf{E}_+^*)$ are not trivially zero since $(d_j q^* \tau_i d_j^*)_v \in \mathfrak{b}\mathfrak{d}\mathfrak{c}^{-1} T_v$. Actually for any m with $n \geqslant m \geqslant j$, and for any d_m and d_j of the form mentioned in the previous paragraph regarding their diagonal we may write

$$c(\tau_1, qd_m^*, \Theta^*) = \alpha_{m,j} c(\tau_1, qd_j^*, \Theta^*),$$

and

$$c(\tau_2, qd_m^*, \mathbf{E}_+^*) = \beta_{m,j} c(\tau_2, qd_j^*, \mathbf{E}_+^*),$$

for $\tau_1, \tau_2 \in S_j$. We now compute the $\alpha_{m,j}$, $\beta_{m,j}$. We have by the explicit description of the Fourier coefficients in Proposition 3.1 that,

$$c(\tau_2, qd_m^*, \mathbf{E}_+^*) = (\psi\chi)(det(d_m d_j^{-1}))\phi(det(d_m d_j^{-1}))^n \times$$

$$|det(d_m d_m^\rho) d_j^{-1} d_j^{-\rho}|_v^{n-r/2} c(\tau_2, qd_j^*, \mathbf{E}_+^*).$$

Now we consider the theta series. We first notice that in order to compute the coefficients $c(\tau_1, qd_i^*; \Theta_\chi^*)$ for any i with $0 \leqslant i \leqslant n$ it is enough to compute the Fourier coefficients of $\theta_\chi(xw)$ with $w = diag[d_i^*, d_i^{-1}]_{\mathbf{h}}$. We now note that by [30, Eq. (A5.7)] we have that

$$\theta_\chi(xw) = |det(d_i^\rho)|_v^{n/2} \phi_{\mathbf{h}}(det(d_i^\rho)^n \chi_{\mathfrak{f}_\chi}(det(d_i))\theta_\chi(x),$$

where we have used [30, Theorem A5.4] and the definition of the theta series. In particular we conclude that

$$c(\tau_1, qd_m^*, \Theta^*) = |det(d_m^\rho)det(d_j^{-\rho})|_v^{n/2}\phi_{\mathbf{h}}(det(d_m^\rho)det(d_j^{-\rho}))^n c(\tau_1, qd_j^*, \Theta^*).$$

where we have used the fact that the character χ is unramified at $\mathfrak{p}$, and hence $\chi_{\mathfrak{f}_\chi}$ can bc ignored.

We now note that $det(d_m) = \pi^m$ and $det(d_j) = \pi^j$. In particular we have

$$\beta_{m,j} = (\chi\psi)(\pi^{m-j})\phi(\pi^{m-j})^n|\pi^{m-j}|_v^{n-r/2},$$

and

$$\alpha_{m,j} = |\pi^{m-j}|_v^{n/2}\phi(\pi^{(m-j)\rho})^n$$

In particular we observe that the $\alpha_{m,j}$ and $\beta_{m,j}$ do not depend on the specific class of Ed_m and Ed_j.

Now we remark that the coefficients $c(\tau_1, qd_j^*, \Theta^*)$ and $c(\tau_2, qd_j^*, \mathbf{E}_+^*)$ depend only on the determinant of d_j, and not on the particular choice of the d_j, as it follows from the explicit description of the Fourier coefficients of the Eisenstein series in Propositions 3.1 and of the theta series in 3.5. Especially for the theta series we remark that it is important here that the character χ is unramified at $\mathfrak{p}$. Hence going back to the Eq. 13, we observe that we can factor the term $c(\tau_1, qd_j^*, \Theta^*)c(\tau_2, qd_j^*, \mathbf{E}_+^*)$ since it does not depend on a particular choice of d_j. Here we remind the reader the convention done above, that the d_j's are taken of a particular form, i.e. lower diagonal and a condition on the diagonal are described above. So for the fixed choice of the pair τ_1 and τ_2, we see that in order to establish the cancellation of the contribution of the fixed pair (τ_1, τ_2) in the sum, we need to show, that

$$\sum_{m=j}^{n}(-1)^m N(\mathfrak{p})^{\frac{m(m-1)}{2}+m(n+s_+)+\frac{n(n-1)}{2}}\chi(\mathfrak{p})^{n-m}\alpha_{m,j}\beta_{m,j}|det(d_m)|_v^{-n}\lambda_m^{(j)} = 0.$$

(We remark one more time here that the outer summation runs from j to n, since for the fixed choice of τ_1 and τ_2 we have that $c(\tau_1, qd_i^*; \Theta^*)c(\tau_2, qd_i^*; \mathbf{E}_+^*) = 0$ for $i < j$.)

Using the fact that $\phi((\pi\pi^\rho)^{m-j})^n$ is equal to $\phi_1(\pi\pi^\rho)^{(m-j)n}$ and the restriction $\phi_1 = \theta$, a quadratic character, we obtain $\phi((\pi\pi^\rho)^{m-j})^n = 1$. Hence we may rewrite the above sum as

$$(\chi^{n-j}\psi^{-j})(\mathfrak{p})\sum_{m=j}^{n}(-1)^m N(\mathfrak{p})^{\frac{m(m-1)}{2}+m(n+s_+)+\frac{n(n-1)}{2}}|\pi^{m-j}|_v^{n-r/2}|\pi^{m-j}|_v^{n/2}\times$$

$$|det(d_m)|_v^{-n}\lambda_m^{(j)} = 0.$$

Of course $|\pi|_v = N(\mathfrak{p})^{-1}$ and hence we have

$$\sum_{m=j}^{n}(-1)^m N(\mathfrak{p})^{\frac{m(m-1)}{2}+m(n+s_+)+\frac{n(n-1)}{2}+(j-m)(n-r/2)+(j-m)n/2+mn}\lambda_m^{(j)} =$$

$$\sum_{m=j}^{n}(-1)^m N(\mathfrak{p})^{\frac{m(m-1)}{2}+m(n+s_+)+\frac{n(n-1)}{2}+(j-m)(n-r/2)+(j-m)n/2+mn}\lambda_m^{(j)} =$$

$$N(\mathfrak{p})^{j(n-r/2+n/2)}\sum_{m=j}^{n}(-1)^m N(\mathfrak{p})^{\frac{2(\frac{m(m-1)}{2}+m(n+s_+)+\frac{n(n-1)}{2})+mr-mn}{2}}\lambda_m^{(j)}.$$

That is, we need to establish that

$$\sum_{m=j}^{n}(-1)^m N(\mathfrak{p})^{\frac{2(\frac{m(m-1)}{2}+m(n+s_+)+\frac{n(n-1)}{2})+mr-mn}{2}}\lambda_m^{(j)} = 0,$$

which is equivalent to

$$\sum_{m=j}^{n}(-1)^m N(\mathfrak{p})^{\frac{m(m-1+n+2s_++r)}{2}}\lambda_m^{(j)} = 0,$$

and since $s_+ = -\frac{r+n}{2}$ we get that we need to show that,

$$\sum_{m=j}^{n}(-1)^m N(\mathfrak{p})^{\frac{m(m-1)}{2}}\lambda_m^{(j)} = 0. \tag{16}$$

We now recall that we are considering d_m's of very particular form, namely lower diagonal matrices where the diagonal is of the form $diag[\pi,\ldots,\pi,\pi^{e_{j+1}},\ldots,\pi^{e_n}]$, where $e_{j+1},\ldots,e_n \in \{0,1\}$ and $e_{j+1}+\ldots+e_n = m-j$. We wrote $\lambda_m^{(j)}$ for the number of them. Recalling now the notation introduced in Lemma 6.6, we claim that

$$\lambda_m^{(j)} = \mu_{m-j}^{(n-j)} \times N(\mathfrak{p})^{(n-m)j}. \tag{17}$$

We first recall that by Lemma 6.2 we may pick the d_m's in the decomposition $E_v\pi_m E_v = \bigsqcup_{dm} E_v d_m$ such that, if we write $d_m = (a_{ik})$ we have that $a_{ik} = 0$ for $i < k$ (i.e. lower triangular), and for $i > k$ we have that a_{ik} could be any representative in $\mathfrak{r}_v$ of $\mathfrak{r}_v/\mathfrak{p}_v$ for $k \in S$ and $i \notin S$ and zero otherwise, where S is the subset of $\{1,\ldots,n\}$ of cardinality m indicating the indices of the π's in the diagonal of d_m. Since we consider d_m's with π in the first j entries of the diagonal we have that $a_{ik} = 0$ for $1 \leqslant k < i \leqslant j$. Moreover the number of choices for the lower right $n-j \times n-j$ part of d_m is equal to $\mu_{m-j}^{(n-j)}$ since we are putting $m-j$ many π on a diagonal of length $n-j$. We can conclude the claimed equality after observing that we are free to pick for the entry a_{ik} with $i>k$ and $j+1\leqslant i\leqslant n$, and $1\leqslant k\leqslant j$ (i.e. the lower left $(n-j)\times j$ part) any representative of $\mathfrak{r}_v/\mathfrak{p}_v$ as long as $a_{ii}=1$. That is we have $N(\mathfrak{p})^{(n-m)j}$ many choices, since we place $n-m$ many ones in the $n-j$ many lower entries of the diagonal of d_m.

By Lemma 6.6 we have,

$$\sum_{i=0}^{n-j}(-1)^i N(\mathfrak{p})^{\frac{i(i-1)}{2}}\mu_i^{(n-j)}=0,$$

and using Eq. 17 we obtain

$$\sum_{i=0}^{n-j}(-1)^i N(\mathfrak{p})^{\frac{i(i-1)}{2}}N(\mathfrak{p})^{-(n-(i+j))j}\lambda_{i+j}^{(j)}=0,$$

or,

$$\sum_{m=j}^{n}(-1)^{m-j}N(\mathfrak{p})^{\frac{(m-j)(m-j-1)}{2}-(n-m)j}\lambda_m^{(j)}=0,$$

or,

$$\sum_{m=j}^{n}(-1)^{m}N(\mathfrak{p})^{\frac{m(m-1)}{2}}\lambda_m^{(j)}=0,$$

which establishes Equality (16), and hence concludes the proof. □

6.2 *The Ramified Part of the Character*

We now fix two integral ideals $\mathfrak{c}_1$ and $\mathfrak{c}_2$ of F with $\mathfrak{c}_1|\mathfrak{c}_2$. We write $C_i:=D[\mathfrak{b}^{-1},\mathfrak{b}\mathfrak{c}_i]$, for $i=1,2$ and define the trace operator $Tr_{\mathfrak{c}_1}^{\mathfrak{c}_2}:\mathcal{M}_k(C_2,\psi)\to\mathcal{M}_k(C_1,\psi)$ by

$$\mathbf{f}\mapsto Tr_{\mathfrak{c}_1}^{\mathfrak{c}_2}(\mathbf{f})(x):=\sum_{r\in R}\psi_{\mathfrak{c}_2}(det(a_r))^{-1}\mathbf{f}(xr),$$

where R is a set of left coset representatives of $D[\mathfrak{b}^{-1},\mathfrak{b}\mathfrak{c}_2]\setminus D[\mathfrak{b}^{-1},\mathfrak{b}\mathfrak{c}_1]$. We note that for a Hermitian cusp form $\mathbf{g}\in\mathcal{S}_k(C_1,\psi)$ we have the well known identity

$$<\mathbf{g},\mathbf{f}>_{\mathfrak{c}_2}=<\mathbf{g},Tr_{\mathfrak{c}_1}^{\mathfrak{c}_2}(\mathbf{f})>_{\mathfrak{c}_1},\qquad(18)$$

where $< \cdot, \cdot >_{\mathfrak{c}_i}$ denotes the adelic inner product with respect to the group $D[\mathfrak{b}^{-1}, \mathfrak{b}\mathfrak{c}_i]$. We now give an explicit description of the trace operator $Tr_{\mathfrak{c}_1}^{\mathfrak{c}_2}$ in the case of $supp(\mathfrak{c}_1) = supp(\mathfrak{c}_2)$, where by $supp(\mathfrak{m})$ of an ideal $\mathfrak{m}$ is defined to be the set of prime ideals $\mathfrak{q}$ of F with $\mathfrak{q}|\mathfrak{m}$. We note that this is similar to the description given in [27, p. 91, p. 136]. We write $\mathfrak{c}_2\mathfrak{c}_1^{-1} = \mathfrak{c}$ for some integral ideal $\mathfrak{c}$ and we fix elements $c, c_1, c_2 \in F_{\mathbb{A}}^{\times}$ such that $c_? \mathfrak{g} = \mathfrak{c}_?$ as well as $b \in F_{\mathbb{A}}^{\times}$ such that $b\mathfrak{g} = \mathfrak{b}$. We first show the following lemma.

Lemma 6.8 *Let $\mathfrak{a}$ be an integral ideal prime to $\mathfrak{c}_2$. Then we have the decomposition*

$$D[\mathfrak{b}^{-1}, \mathfrak{b}\mathfrak{a}\mathfrak{c}_1] = \bigsqcup_{r \in R} D[\mathfrak{b}^{-1}, \mathfrak{b}\mathfrak{a}\mathfrak{c}_2]r,$$

where

$$R = \{\begin{pmatrix} 1 & 0 \\ bac_1u & 1 \end{pmatrix} |u \in S(\mathfrak{g})_{\mathbf{h}} \mod \mathfrak{c}\},$$

with $a \in F_{\mathbb{A}}^{\times}$ such that $a\mathfrak{g} = \mathfrak{a}$.

Proof Clearly without loss of generality we can set $\mathfrak{a} = \mathfrak{g}$. Moreover it is clear that the right hand side of the claimed decomposition is included into the left. To prove the other inclusion we consider an element $\begin{pmatrix} A & B \\ C & D \end{pmatrix} \in D[\mathfrak{b}^{-1}, \mathfrak{b}\mathfrak{c}_1]$ and show that there exists an $r \in R$ such that $\begin{pmatrix} A & B \\ C & D \end{pmatrix} r^{-1} \in D[\mathfrak{b}^{-1}, \mathfrak{b}\mathfrak{c}_2]$ or otherwise there exists $u \in S(\mathfrak{g})_{\mathbf{h}} \mod \mathfrak{c}$ such that

$$\begin{pmatrix} A & B \\ C & D \end{pmatrix} \begin{pmatrix} 1 & 0 \\ bc_1u & 1 \end{pmatrix} \in D[\mathfrak{b}^{-1}, \mathfrak{b}\mathfrak{c}_2].$$

That is, we need to prove that there exists such a u as above so that $C + bc_1Du \prec \mathfrak{b}\mathfrak{c}_2$. Since $C \prec \mathfrak{b}\mathfrak{c}_1\mathfrak{r}$ we can write it as $C = bc_1C_0$ with $C_0 \prec \mathfrak{r}$, and hence we need to show that $bc_1(C_0 + Du) \prec \mathfrak{b}\mathfrak{c}_2\mathfrak{r}$. By our assumption that $supp(\mathfrak{c}_1) = supp(\mathfrak{c}_2)$ we have that $DA^* \equiv 1_n \mod \left(\prod_{\mathfrak{q}|\mathfrak{c}} \mathfrak{q}\right)\mathfrak{r}$. For a prime ideal $\mathfrak{q}$ that divides $\mathfrak{c}$ we write $e_{\mathfrak{q}}$ for the largest power of it that divides $\mathfrak{c}$ and we define $e := max(e_{\mathfrak{q}})$. Then we have that $(DA^* - 1_n)^e \prec \mathfrak{c}\mathfrak{r}$. That means that there exists an element $\tilde{D} \prec \mathfrak{r}$ such that $D\tilde{D} \equiv 1 \mod \mathfrak{c}\mathfrak{r}$ and $\tilde{D}C_0 \in S(\mathfrak{g})_{\mathbf{h}}$. Indeed we have that

$$(DA^* - 1_n)^e = DA^*DA^* \cdots DA^* + ...(-1)^e I_n \prec \mathfrak{c}\mathfrak{r},$$

or equivalently

$$D\left(A^*DA^* \cdots DA^* + \cdots + A^*\right) \equiv (-1)^{e-1} I_n \mod \mathfrak{c}\mathfrak{r}.$$

So we need only to check that the matrix

$$\left(A^*DA^*\cdots DA^* + \cdots + A^*\right)C_0 = A^*DA^*\cdots DA^*C_0 + \ldots + A^*C_0$$

is hermitian. But we know that A^*C is hermitian and since $bc_1 \in F_{\mathbb{A}}^\times$ we have that also A^*C_0 is hermitian. The same reasoning holds for the product DC_0^*. In particular we have

$$(A^*DA^*\cdots DA^*C_0)^* = C_0^*AD^*A\cdots D^*A = A^*C_0D^*A\cdots DA^* =$$

$$A^*DC_0^*A\cdots DA^* =$$

$$\cdots = A^*DA^*\cdots C_0A = A^*DA^*\cdots DA^*C_0.$$

This establishes the claim. Then we can take $u = (-1)^e\tilde{D}C_0$ to conclude the proof. □

Let us now assume that the deal $\mathfrak{c} = \mathfrak{c}_2\mathfrak{c}_1^{-1}$ above is the norm of an integral ideal $\mathfrak{c}_0$ of K, that is $\mathfrak{c} = N_{K/F}(\mathfrak{c}_0)$. We also pick an element $c_0 \in K_{\mathbb{A}}^\times$ such that $c_0\mathfrak{r} = \mathfrak{c}_0$. We consider now the Hecke operator $T_C(\mathfrak{c}) := T(\mathfrak{c}) := \prod_{v|\mathfrak{c}} T(\sigma_v)$ for $\sigma_v = diag[\widehat{c_{0v}}1_n, c_{0v}1_n]$, where we take $C = D[\mathfrak{b}\mathfrak{c}_1, \mathfrak{b}^{-1}\mathfrak{c}]$. Note that this group is of the form $D[\tilde{\mathfrak{b}}^{-1}, \tilde{\mathfrak{b}}\tilde{\mathfrak{c}}]$ with $\tilde{\mathfrak{b}} = (\mathfrak{b}\mathfrak{c}_1)^{-1}$ and $\tilde{\mathfrak{c}} = \mathfrak{c}\mathfrak{c}_1 = \mathfrak{c}_2$. By Lemma 6.1 we have that

$$C_v\sigma_vC_v = \coprod_b C_v\begin{pmatrix}\widehat{c_{0v}}1_n & \widehat{c_{0v}}b\\ 0 & c_{0v}1_n\end{pmatrix},$$

where $b \in S(\mathfrak{b}\mathfrak{c}_1)_v/\mathfrak{c}S(\mathfrak{b}\mathfrak{c}_1)_v$. We now observe the identity

$$\begin{pmatrix}0 & -1_n\\ 1_n & 0\end{pmatrix}\begin{pmatrix}c_0^*1_n & -c_0^{-1}b\\ 0 & c_0{}^{-1}1_n\end{pmatrix}\begin{pmatrix}\widehat{c_0}1_n & 0\\ 0 & c_01_n\end{pmatrix}\begin{pmatrix}0 & 1_n\\ -1_n & 0\end{pmatrix} = \begin{pmatrix}1_n & 0\\ b & 1_n\end{pmatrix}.$$

We now write $V(\mathfrak{c}_0) : \mathcal{M}_k(D[\mathfrak{b}\mathfrak{c}_2, \mathfrak{b}^{-1}]) \to \mathcal{M}_k(D[\mathfrak{b}\mathfrak{c}_1, \mathfrak{c}\mathfrak{b}^{-1}])$ for the operator defined by $\mathbf{f}(x) \mapsto \mathbf{f}\left(x\begin{pmatrix}\widehat{c_0}1_n & 0\\ 0 & c_01_n\end{pmatrix}\right)$. We can conclude from the above calculation that the trace operator can be decomposed as

$$Tr_{\mathfrak{c}_1}^{\mathfrak{c}_2} = W \circ V(\mathfrak{c}_0) \circ T(\mathfrak{c}) \circ W^{-1},$$

where the operators are operating from the right. We note that in general the image of the right hand side is in $\mathcal{M}_k(D[\mathfrak{b}^{-1}\mathfrak{c}, \mathfrak{b}\mathfrak{c}_1])$ which contains of course $\mathcal{M}_k(D[\mathfrak{b}^{-1}, \mathfrak{b}\mathfrak{c}_1])$, where the image of the trace operator lies. We summarize the above calculations to the following lemma.

Lemma 6.9 *With notation as above, and assuming that there exists a* $\mathfrak{c}_0$ *such that* $\mathfrak{c} = N_{K/F}(\mathfrak{c}_0)$ *we have*

$$Tr_{\mathfrak{c}_1}^{\mathfrak{c}_2} = W \circ V(\mathfrak{c}_0) \circ T(\mathfrak{c}) \circ W^{-1}.$$

The effect of $T(\mathfrak{c})$ **on the** q**-expansion.** We now study the effect of the operator $T(\mathfrak{c})$ and of $V(\mathfrak{c}_0)$ on the q-expansion of an automorphic form $\mathbf{F}$ which we take in $\mathcal{M}_k([D[\mathfrak{b}\mathfrak{c}_1, \mathfrak{b}^{-1}\mathfrak{c}]], \psi^{-c})$. We write

$$\mathbf{F}\left(\begin{pmatrix} q & s\hat{q} \\ 0 & \hat{q} \end{pmatrix}\right) = \sum_{\tau \in S_+} c(\tau, q; \mathbf{F}) \mathbf{e}^n_{\mathbb{A}}(\tau s).$$

Setting $\mathbf{G} := \mathbf{F}|T(\mathbf{c})$ we have,

$$\mathbf{G}\left(\begin{pmatrix} q & s\hat{q} \\ 0 & \hat{q} \end{pmatrix}\right) =$$

$$\psi(det(c_0))^{-1} \sum_b \mathbf{F}\left(\begin{pmatrix} q & s\hat{q} \\ 0 & \hat{q} \end{pmatrix}\begin{pmatrix} {c_0}^* 1_n & -{c_0}^{-1} b \\ 0 & {c_0}^{-1} 1_n \end{pmatrix}\right) =$$

where b runs over the set $S(\mathfrak{b}\mathfrak{c}_1)_v / \mathfrak{c}S(\mathfrak{b}\mathfrak{c}_1)$. In particular

$$\mathbf{G}\left(\begin{pmatrix} q & s\hat{q} \\ 0 & \hat{q} \end{pmatrix}\right) =$$

$$\psi(det(c_0))^{-1} \sum_b \mathbf{F}\left(\begin{pmatrix} qc_0^* & -qc_0^{-1} b + s\hat{q}c_0^{-1} \\ 0 & \hat{q}c_0^{-1} \end{pmatrix}\right) =$$

$$\sum_b \mathbf{F}\left(\begin{pmatrix} qc_0^* & (-qbq^* + s)\widehat{qc_0^*} \\ 0 & \widehat{qc_0^*} \end{pmatrix}\right) =$$

$$\psi(det(c_0))^{-1} \sum_b \sum_{\tau \in S_+} c(\tau, qc_0^*; \mathbf{F}) \mathbf{e}^n_{\mathbb{A}}(\tau(-qbq^* + s)) =$$

$$\psi(det(c_0))^{-1} \sum_{\tau \in S_+} \left(\sum_b \mathbf{e}^n_{\mathbb{A}}(\tau q b q^*) \right) c(\tau, qc_0^*; \mathbf{F}) \mathbf{e}^n_{\mathbb{A}}(\tau s) =$$

$$\psi(det(c_0))^{-1} \sum_{\tau \in S_+} \left(\sum_b \mathbf{e}^n_{\mathbf{h}}(\tau q b q^*) \right) c(\tau, qc_0^*; \mathbf{F}) \mathbf{e}^n_{\mathbb{A}}(\tau s).$$

Note that the inner sum is well-defined since by [30, Proposition 20.2] we have that $c(\tau, qc_0^*; \mathbf{f}) = 0$ unless $e_{\mathbf{h}}^n(q^* c_0 \tau q c_0^* s) = 1$ for every $s \in S(\mathfrak{b}\mathfrak{c}_1)_{\mathbf{h}}$. Moreover (see for example [30, Lemma 19.6]) we have that

$$\sum_b \mathbf{e}_{\mathbf{h}}^n(\tau q b q^*) = |c_0|_K^{-n^2},$$

if $\tau \in \Lambda := qT(\mathfrak{b}\mathfrak{c}_1)q^*$, and zero otherwise. Here $T(\mathfrak{b}\mathfrak{c}_1)$ denotes the dual lattice of $S(\mathfrak{b}\mathfrak{c}_1) := S \cap M(\mathfrak{b}\mathfrak{c}_1))$. That is,

$$\mathbf{G}\left(\begin{pmatrix} q & s\hat{q} \\ 0 & \hat{q} \end{pmatrix}\right) = |c_0|_K^{-n^2} \psi(det(c_0))^{-1} \sum_{\tau \in \Lambda} c(\tau, qc_0^*; \mathbf{F}) \mathbf{e}_{\mathbb{A}}^n(\tau s). \tag{19}$$

The effect of $V(\mathfrak{c}_0)$ on the q-expansion. Now we turn to the operator $V(\mathfrak{c}_0)$. With $\mathbf{F}$ we now consider $\mathbf{G} = \mathbf{F}|V(\mathfrak{c}_0)$. Then for the q-expansion of $\mathbf{G}$ we have,

$$\mathbf{G}\left(\begin{pmatrix} q & s\hat{q} \\ 0 & \hat{q} \end{pmatrix}\right) = \mathbf{F}\left(\begin{pmatrix} q & s\hat{q} \\ 0 & \hat{q} \end{pmatrix}\begin{pmatrix} \widehat{c_0} 1_n & 0 \\ 0 & c_0 1_n \end{pmatrix}\right) = \mathbf{F}\left(\begin{pmatrix} \widehat{c_0} q & s\widehat{q} c_0 \\ 0 & \widehat{q} c_0 \end{pmatrix}\right) =$$

$$\sum_{\tau \in S_+} c(\tau, \widehat{c_0} q; \mathbf{F}) \mathbf{e}_{\mathbb{A}}^n(\tau s).$$

We now take $\mathbf{F}$ of a particular dorm, namely we take $\mathbf{F} = \Theta^* \mathbf{E} *_+$, and assume that the conductor $\mathfrak{f}_\chi$ of the character χ has the property that $\mathfrak{f}_\chi | \mathfrak{c}_0$. We have that

$$c(\tau, q\widehat{c_0}, \Theta^* \mathbf{E}_+^*) = \sum_{\tau_1 + \tau_2 = \tau} c(\tau_1, q\widehat{c_0}, \Theta^*) c(\tau_2, q\widehat{c_0}, \mathbf{E}_+^*)$$

We now note that

$$c(\tau_1, q\widehat{c_0}, \Theta^*) = |c_0^\rho|_K^{-\frac{n}{2}} \phi_{\mathbf{h}}(c_0^\rho)^{n^2} \chi_{\mathfrak{f}_\chi}(c_0)^n c(\tau, q, \Theta^*),$$

and

$$c(\tau_2, q\widehat{c_0}, \mathbf{E}_+^*) = (\psi\chi)(c_0)^{-n} \phi(c_0)^{-n^2} |c_0|^{-n(n-r/2)} c(\tau_2, q, \mathbf{E}_+^*).$$

We then conclude that,

$$c(\tau, q\widehat{c_0}, \Theta^* \mathbf{E}_+^*) = \psi(c_0)^{-n} |c_0|_K^{-\frac{n}{2} - n(n-r/2)} c(\tau, q, \Theta^* \mathbf{E}_+^*). \tag{20}$$

In particular we have that $c(\tau, q\widehat{c_0}, \Theta^* \mathbf{E}_+^*) \neq 0$ only if

$$(c_0^{-\rho} q^* \tau c_0^{-1} q)_v \in (\mathfrak{f}_\chi^{-1} \mathfrak{f}_\chi^{-\rho})_v T_v,$$

for all $v | p$.

Similarly we have for $\Theta^*\mathbf{E}^*_-$ but we need to replace $\frac{r}{2}$ with $n-\frac{r}{2}$ in the above equations. That is

$$c(\tau, q\widehat{c_0}, \Theta^*\mathbf{E}^*_-) = \psi(c_0)^{-n}|c_0|_K^{-\frac{n}{2}-n(r/2)} c(\tau, q, \Theta^*\mathbf{E}^*_-). \tag{21}$$

6.3 Rewriting the Rankin–Selberg Integral

We now use the above identities to rewrite the Rankin–Selberg integrals. We consider a $\mathfrak{p} \in S$ and we let $\mathbf{f}_0 \in S_k(C, \psi)$ be an eigenform for the Hecke operator $U(\mathfrak{p})$, of eigenvalue $\alpha(\mathfrak{p})$. We take $C := D[\mathfrak{b}^{-1}, \mathfrak{b}\mathfrak{m}_0]$ where $\mathfrak{m}_0 := \mathfrak{c}'\mathfrak{p}\mathfrak{p}^\rho$ for some $\mathfrak{c}'$ prime to $\mathfrak{p}$. We now consider a Hecke character χ of K, of some conductor $\mathfrak{f}_\chi$, and write $\mathfrak{m}_\chi$ for the ideal $\mathfrak{c}'\mathfrak{p}^{n_\mathfrak{p}}\mathfrak{p}^{n_\mathfrak{p}\rho}$ where $\mathfrak{p}^{n_\mathfrak{p}}$ is the smallest power-$\mathfrak{p}$ ideal contained in the conductor $\mathfrak{f}_\chi$. Moreover we take $\mathfrak{c}'$ small enough so that it includes the prime to $\mathfrak{p}$ level of Θ. Note that by Theorem 3.4 the level of Θ supported at $\mathfrak{p}$ is exactly $\mathfrak{p}^{n_\mathfrak{p}}\mathfrak{p}^{n_\mathfrak{p}\rho}$. We then show,

Proposition 6.10 *Consider any $c_\mathfrak{p} \in \mathbb{N}$ with $c_\mathfrak{p} \geqslant n_\mathfrak{p} \geqslant 1$. Then we have*

$$\alpha(\mathfrak{p})^{-n_\mathfrak{p}-1} < \mathbf{f}_0, \Theta\mathbf{E}_\pm >_{\mathfrak{m}_\chi} =$$

$$\alpha(\mathfrak{p})^{-c_\mathfrak{p}-1} < \mathbf{f}_0|W, \Theta^*\mathbf{E}^*_\pm|V(\mathfrak{p})^{n_\mathfrak{p}-1} \circ U(\mathfrak{p})^{c_\mathfrak{p}-1} >_{\mathfrak{m}_0}.$$

Proof

$$< \mathbf{f}_0, \Theta\mathbf{E}_\pm >_{\mathfrak{m}_\chi} =$$

$$< \mathbf{f}_0, \Theta\mathbf{E}_\pm|Tr^{\mathfrak{m}_\chi}_{\mathfrak{m}_0} >_{\mathfrak{m}_0} = < \mathbf{f}_0, \Theta\mathbf{E}_\pm|W \circ V(\mathfrak{p})^{n_\mathfrak{p}-1} \circ U(\mathfrak{p})^{n_\mathfrak{p}-1} \circ W^{-1} >_{\mathfrak{m}_0} =$$

$$\alpha(\mathfrak{p})^{n_\mathfrak{p}-c_\mathfrak{p}} < \mathbf{f}_0|U(\mathfrak{p})^{c_\mathfrak{p}-n_\mathfrak{p}}, \Theta\mathbf{E}_\pm|W \circ V(\mathfrak{p})^{n_\mathfrak{p}-1} \circ U(\mathfrak{p})^{n_\mathfrak{p}-1} \circ W^{-1} >_{\mathfrak{m}_0} =$$

$$\frac{< \mathbf{f}_0, \Theta\mathbf{E}_\pm|W \circ V(\mathfrak{p})^{n_\mathfrak{p}-1} \circ U(\mathfrak{p})^{n_\mathfrak{p}-1} \circ W^{-1} \circ W \circ U(\mathfrak{p})^{c_\mathfrak{p}-n_\mathfrak{p}} \circ W^{-1} >_{\mathfrak{m}_0}}{\alpha(\mathfrak{p})^{-n_\mathfrak{p}+c_\mathfrak{p}}} =$$

$$\alpha(\mathfrak{p})^{n_\mathfrak{p}-c_\mathfrak{p}} < \mathbf{f}_0, \Theta\mathbf{E}_\pm|W \circ V(\mathfrak{p})^{n_\mathfrak{p}-1} \circ U(\mathfrak{p})^{c_\mathfrak{p}-1} \circ W^{-1} >_{\mathfrak{m}_0} =$$

$$\alpha(\mathfrak{p})^{n_\mathfrak{p}-c_\mathfrak{p}} < \mathbf{f}_0|W, \Theta^*\mathbf{E}^*_\pm|V(\mathfrak{p})^{n_\mathfrak{p}-1} \circ U(\mathfrak{p})^{c_\mathfrak{p}-1} >_{\mathfrak{m}_0}.$$

Hence

$$< \mathbf{f}_0, \Theta\mathbf{E}_\pm >_{\mathfrak{m}_\chi} = \alpha(\mathfrak{p})^{n_\mathfrak{p}-c_\mathfrak{p}} < \mathbf{f}_0|W, \Theta^*\mathbf{E}^*_\pm|V(\mathfrak{p})^{n_\mathfrak{p}-1} \circ U(\mathfrak{p})^{c_\mathfrak{p}-1} >_{\mathfrak{m}_0},$$

or

$$\alpha(\mathfrak{p})^{-n_\mathfrak{p}} < \mathbf{f}_0, \Theta\mathbf{E}_\pm >_{\mathfrak{m}_\chi} = \alpha(\mathfrak{p})^{-c_\mathfrak{p}} < \mathbf{f}_0|W, \Theta^*\mathbf{E}^*_\pm|V(\mathfrak{p})^{n_\mathfrak{p}-1} \circ U(\mathfrak{p})^{c_\mathfrak{p}-1} >_{\mathfrak{m}_0}.$$

□

7 The p-stabilization

Let us consider $C := D[\mathfrak{b}^{-1}, \mathfrak{b}\mathfrak{c}]$, where we take the integral ideal $\mathfrak{c}$ prime to the ideals in the fixed set S. We consider a Hermitian cusp form $\mathbf{f}$ in $\mathcal{S}_k(C, \psi)$ which we take to be an eigenform for all the "good" Hecke operators in $\prod_{v \nmid \mathfrak{c}} \mathfrak{R}(C_v, \mathfrak{X}_v)$, where $\mathfrak{R}(C_v, \mathfrak{X}_v)$ is the local Hecke algebra at v defined in [30, Chap. IV]. Our aim in this section is to construct a Hermitian cusp form $\mathbf{f}_0$, of level $\mathfrak{c} \prod_{\mathfrak{p} \in S} \mathfrak{p}\mathfrak{p}^{\rho} =: \mathfrak{c}\mathfrak{m}$ which is an eigenform for all the "good" Hecke operators away from $\mathfrak{c}\mathfrak{m}$ and for the operators $U(\pi_{v,i})$ for all finite places v corresponding to prime ideals in the set S. Our construction is the unitary analogue of the symplectic situation considered in [2, Sect. 9]. It is important to mention here that our construction is adelic, so it can be used to generalize the one in [2] to the totally real field situation. Here, as we mentioned in the introduction, we restrict ourselves to the case where all prime ideals in S are inert, but our arguments generalize also to the split case. We will consider this in [7].

We write $\mathbf{M}_S$ for the submodule of $\mathcal{S}_k(D[\mathfrak{b}^{-1}, \mathfrak{b}\mathfrak{c}\mathfrak{m}], \psi)$ generated by $\mathbf{f}$ under the action of the Hecke algebra $\prod_{v \in S} \mathfrak{R}(C'_v, \mathfrak{X}_v)$, where $C' = D(\mathfrak{b}^{-1}, \mathfrak{b}\mathfrak{c}\mathfrak{m})$. We let $\mathbf{f}_0 \in \mathbf{M}_S$ to be a non-trivial eigenform of all the Hecke operators in $\prod_{v \in S} \mathfrak{R}(C'_v, \mathfrak{X}_v)$. In particular $\mathbf{f}_0 \neq 0$. We write the adelic q-expansion of $\mathbf{f}$ as

$$\mathbf{f}\left(\begin{pmatrix} q & s\hat{q} \\ 0 & \hat{q} \end{pmatrix}\right) = \sum_{\tau \in S_+} c(\tau, q; \mathbf{f}) \mathbf{e}^n_{\mathbb{A}}(\tau s).$$

and of $\mathbf{f}_0$ as,

$$\mathbf{f}_0\left(\begin{pmatrix} q & s\hat{q} \\ 0 & \hat{q} \end{pmatrix}\right) = \sum_{\tau \in S_+} c(\tau, q; \mathbf{f}_0) \mathbf{e}^n_{\mathbb{A}}(\tau s).$$

We pick a $\tau \in S_+ \cap GL_n(K)$ and $q \in GL_n(K)_{\mathbf{h}}$ such that $c(\tau, q; \mathbf{f}_0) \neq 0$. In particular that means that we have $q^* \tau q \in T$, where as always T denotes the dual lattice to $S(\mathfrak{b}^{-1}) := S \cap M_n(\mathfrak{b}^{-1})$. Then for any finite place v corresponding to a prime ideal $\mathfrak{p} \in S$ we have [30, Eq. (20.15)]

$$Z_v(\mathbf{f}_0, X) c(\tau, q; \mathbf{f}_0) = \sum_d \psi_{\mathfrak{c}}(det(d^*)) |det(d)^*|_v^{-n} X^{v_{\mathfrak{p}}(det(d^*))} c(\tau, q d^*; \mathbf{f}_0), \quad (22)$$

where $d \in E_v \setminus E_v q E_v$, and $Z_v(\mathfrak{f}_0, X)$ denotes the Euler factor $Z_{\mathfrak{p}}(X)$ of Sect. 4.1. Moreover $v_{\mathfrak{p}}(\cdot)$ is the valuation associated to the ideal $\mathfrak{p}$, and $|\cdot|_v$ the normalized norm.

Following Böcherer and Schmidt [2] we now try to describe the right hand side of (22) using the Satake parameters of the form $\mathbf{f}$. As in [loc. cit.] we start with the Andrianov type identity generalized by Shimura [30, Theorem 20.4]. For the selected $\tau \in S_+ \cap GL_n(K)$ and $q \in GL_n(K)_{\mathbf{h}}$ we define (this is the local version at v of the series $D(\tau, q; \mathbf{f})$ considered in [30, p. 169])

$$D_v(\tau, q : \mathbf{f}, X) := \sum_{x \in B_v/E_v} \psi_{\mathfrak{c}}(det(qx))|det(x)|_v^{-n} c(\tau, qx; \mathbf{f}) X^{v_{\mathfrak{p}}(det(x))},$$

where $B_v = GL_n(K_v) \cap M_n(\mathfrak{r}_v)$. We will employ now what may be considered as a local version of the Andrianov–Kalinin equality in the unitary case. Namely we will relate the above series $D_v(\tau, q : \mathbf{f}, X)$ to the Euler factor $Z_v(\mathbf{f}_0, X)$.

We first introduce some notation. We let $\mathcal{L}_\tau$ be the set of $\mathfrak{r}$-lattices L in K^n such that $\ell^* \tau \ell \in \mathfrak{b}\mathfrak{d}^{-1}$ for all $\ell \in L$. Moreover for the chosen ideal $\mathfrak{c}$ above, and for two $\mathfrak{r}$ lattices M, N we write $M < N$ if $M \subset N$ and $M \otimes_{\mathfrak{r}} \mathfrak{r}_v = N \otimes_{\mathfrak{r}} \mathfrak{r}_v$ for every $v \mid \mathfrak{c}$. We now set $L := q\mathfrak{r}^n$. Then we have the following local version of [30, Theorem 20.7],

$$D_v(\tau, q; \mathbf{f}, X) \cdot \mathfrak{L}_{0,v}(X) \cdot g_v(X) =$$

$$Z_v(\mathbf{f}, X) \cdot \sum_{L_v < M_v \in \mathcal{L}_\tau} \mu(M_v/L_v) \psi_{\mathfrak{c}}(det(y)) X^{v_{\mathfrak{p}}(det(q^* \hat{y})} c(\tau, y; \mathbf{f}),$$

where $\mathfrak{L}_{0,v}(X) := \prod_{i=0}^{n-1}(1 - (-1)^{i-1} N(\mathfrak{p})^{n+i} X)^{-1}$, and $g_v(X)$ is a polynomial in X with integers coefficients and constant term equal to 1. In the sum over the M's, we take $y \in GL_n(K_v)$ such that $M_v = y\mathfrak{r}^n$ and $y^{-1}q \in B_v$. Furthermore $\mu(\cdot)$ is the generalized Möbius function introduced in the previous section, and as in the last section we write $v_{\mathfrak{p}}(\cdot)$ for the discrete valuation associated to the prime ideal $\mathfrak{p}$. We now cite the following lemma regarding $g_v(X)$ (see [23, Lemma 5.2.4]).

Lemma 7.1 *Write* $(q^* \tau q)_v = diag[1_{n-r}, \pi_v s_1]$ *with* $s_1 \in S^r(\mathfrak{r}_v)$. *Then we have*

$$g_v(X) = \prod_{i=0}^{r-1} (1 - (-1)^{i-1} N(\mathfrak{p})^{n+i} X).$$

In particular we conclude that if $(q^* \tau q)_v$ *is divisible by* π_v *(i.e.* $r = n$*) then we have that* $g_v(X)$ *is equal to* $\mathfrak{L}_{0,v}^{-1}(X)$.

Our next step is to rewrite the expression

$$\sum_{L_v < M_v \in \mathcal{L}_{\tau,v}} \mu(M_v/L_v) \psi_{\mathfrak{c}}(det(y)) X^{v_{\mathfrak{p}}(det(q^* \hat{y})} c(\tau, y; \mathbf{f}),$$

in terms of the action of the Hecke algebra. By the above lemma if we take $\pi_v q$ instead of q we obtain,

$$D_v(\tau, \pi q; \mathbf{f}, X) = Z_v(\mathbf{f}, X) \times$$

$$\sum_{L_v < M_v \in \mathcal{L}_\tau} \mu(M_v/L_v) \psi_{\mathfrak{c}}(det(y)) X^{v_{\mathfrak{p}}(det(q^* \pi^* \hat{y})} c(\tau, y; \mathbf{f}),$$

where now $L_v = \pi_v \mathfrak{r}_v^n$ and $M = y\mathfrak{r}^n$. Since the y's are supported only at v and we are taking $\mathfrak{p} \not| \mathfrak{c}$ we have $\psi_{\mathfrak{c}}(det(y)) = 1$. That is,

$$D_v(\tau, \pi q; \mathbf{f}, X) = Z_v(\mathbf{f}, X) \cdot \sum_{L_v < M_v \in \mathcal{L}_\tau} \mu(M_v/L_v) X^{v_{\mathfrak{p}}(det(q^* \pi^* \hat{y}))} c(\tau, y; \mathbf{f}).$$

Now we rewrite the above expression in terms of the Hecke operators $U(\pi_j)$. In particular we have (see [30, proof of Theorem 19.8]),

$$D_v(\tau, \pi q; \mathbf{f}, X) = Z_v(\mathbf{f}, X) \times$$

$$c\left(\tau, q; \mathbf{f} \middle| \left(\sum_{i=0}^{n} (-1)^n N(\mathfrak{p})^{i(i-1)/2} \psi_v(\pi^{i-n}) N(\mathfrak{p})^{-n(n-i)} U(\pi_{n-i}) X^i \right)\right),$$

where recall that we write the action of the Hecke operators from the right. Using the fact that $\mathbf{f}_0$ is obtained from $\mathbf{f}$ by using the Hecke operators at the prime $\mathfrak{p}$, and the fact that the Hecke algebra is commutative we obtain that the above relation holds also for $\mathbf{f}_0$. That is, we have

$$D_v(\tau, \pi q; \mathbf{f}_0, X) = Z_v(\mathbf{f}, X) \times \tag{23}$$

$$c\left(\tau, q; \mathbf{f}_0 \middle| \left(\sum_{i=0}^{n} (-1)^n N(\mathfrak{p})^{i(i-1)/2} \psi_v(\pi^{i-n}) N(\mathfrak{p})^{-n(n-i)} U(\pi_{n-i}) X^i \right)\right).$$

We first rewrite the left hand side of the above equation. We recall that

$$D_v(\tau, \pi q; \mathbf{f}_0, X) = \sum_{x \in B_v/E_v} \psi_{\mathfrak{c}}(det(qx)) |det(x)|_v^{-n} c(\tau, \pi q x; \mathbf{f}_0) X^{v_{\mathfrak{p}}(det(x))}.$$

Now we use the fact that $\mathbf{f}_0$ is an eigenform for the operators $U(\pi_i)$. We write λ_i for the eigenvalues. Then we have that

$$c(\tau, \pi q x, \mathbf{f}_0) = N(\mathfrak{p})^{-n^2} \psi_v(\pi)^{-n} \lambda_n c(\tau, q x, \mathbf{f}_0).$$

That is we obtain,

$$D_v(\tau, \pi q; \mathbf{f}_0, X) = Z_v(\mathbf{f}_0, X) \lambda_n N(\mathfrak{p})^{-n^2} c(\tau, q, \mathbf{f}_0),$$

and so we can rewrite Eq. 23 as,

$$Z_v(\mathbf{f}_0, X) \lambda_n N(\mathfrak{p})^{-n^2} c(\tau, q, \mathbf{f}_0) =$$

$$Z_v(\mathbf{f}, X)c(\tau, q; \mathbf{f}_0)\left(\sum_{i=0}^{n}(-1)^i N(\mathfrak{p})^{i(i-1)/2}\psi_v(\pi^i)N(\mathfrak{p})^{-n(n-i)}\lambda_{n-i}X^i\right).$$

We note that we have

$$N(\mathfrak{p})^{\frac{i(i-1)}{2}+ni}\lambda_i = N(\mathfrak{p})^{i(n-1)}E_i(t_1, \ldots, t_n) \tag{24}$$

and $(t_1 \ldots t_n)^{-1}E_{n-i}(t_1, \ldots, t_n) = E_i(t_1^{-1}, \ldots, t_n^{-1})$ where E_i is the ith symmetric polynomial. Indeed Eq. 24 is the unitary analogue of the formula employed in [2, pp. 1429–1430] of how to obtain the eigenvalues of the Hecke operators $U(\pi_i)$ from the Satake parameters at $\mathfrak{p}$, and it can be shown in the same way. Hence we conclude that after picking τ and q such that $c(\tau, q, \mathbf{f}_0) \neq 0$ we have

$$\lambda_n N(\mathfrak{p})^{-n^2}Z_v(\mathbf{f}_0, X) = Z_v(\mathbf{f}, X)\times$$

$$(\sum_{i=0}^{n}(-1)^i N(\mathfrak{p})^{i(i-1)/2}\psi_v(\pi^i)X^i N(\mathfrak{p})^{-n^2}N(\mathfrak{p})^{-\frac{i(i-1)}{2}+2ni-\frac{n(n+1)}{2}}\times$$

$$(t_1 \ldots t_n)E_i(t_1^{-1}, \ldots, t_n^{-1})),$$

and using the fact that $N(\mathfrak{p})^{\frac{n(n+1)}{2}}\lambda_n = t_1 \ldots t_n$ we get

$$Z_v(\mathbf{f}_0, X) = Z_v(\mathbf{f}, X)\left(\sum_{i=0}^{n}(-1)^i\psi_v(\pi^{n-i})X^i N(\mathfrak{p})^{2ni}E_i(t_1^{-1}, \ldots, t_n^{-1})\right) =$$

$$Z_v(\mathbf{f}, X)\left(\sum_{i=0}^{n}(-1)^i\psi_v(\pi^i)N(\mathfrak{p})^{2ni}E_i(t_1^{-1}, \ldots, t_n^{-1})X^i\right),$$

and so

$$Z_v(\mathbf{f}_0, X) = Z_v(\mathbf{f}, X)\left(\sum_{i=0}^{n}(-1)^i\psi_v(\pi^i)X^i N(\mathfrak{p})^{2ni}E_i(t_1^{-1}, \ldots, t_n^{-1})\right).$$

Equivalently

$$Z_v(\mathbf{f}_0, X) = Z_v(\mathbf{f}, X)\prod_{i=0}^{n}\left(1 - N(\mathfrak{p})^{2n}\psi_v(\pi)^i t_i^{-1}X^i\right),$$

and so we conclude that

$$Z_v(\mathbf{f}_0, X)\prod_{i=0}^{n}\left(1 - N(\mathfrak{p})^{2n}\psi(\pi)^i t_i^{-1}X^i\right)^{-1} = Z_v(\mathbf{f}, X).$$

We now make the following definition

Definition 7.2 Let $\mathbf{f} \in \mathcal{S}_k(C, \psi)$ be a Hecke eigenform for $C = D[\mathfrak{b}^{-1}, \mathfrak{b}\mathfrak{c}]$. Let $\mathfrak{p}$ be a prime of K prime to $\mathfrak{c}$, which is inert over F. Then we say that $\mathbf{f}$ is ordinary at $\mathfrak{p}$ if there exists an eigenform $0 \neq \mathbf{f}_0 \in \mathbf{M}_{\{\mathfrak{p}\}} \subset \mathcal{S}_k(D[\mathfrak{b}^{-1}, \mathfrak{b}\mathfrak{c}\mathfrak{p}\mathfrak{p}^{\rho}], \psi)$ with Satake parameters $t_{\mathfrak{p},i}$ such that

$$\left\| \left(\prod_{i=1}^{n} t_{\mathfrak{p},i}\right) N(\mathfrak{p})^{-\frac{n(n+1)}{2}} \right\|_p = 1,$$

where $\| \cdot \|_p$ the normalized absolute value at p.

Summarizing the computations of this section we have,

Theorem 7.3 *Let* $\mathbf{f}$ *be an cuspidal Hecke eigenform. Assume that* $\mathbf{f}$ *is ordinary for all primes in* K *above* p *that are inert from* F. *Then we can associate to it a cuspidal Hecke eigenform* $\mathbf{f}_0$ *such that its Euler factors above* p *are related by the equation*

$$Z_{\mathfrak{p}}(\mathbf{f}_0, X)\prod_{i=0}^{n}\left(1 - N(\mathfrak{p})^{2n}\psi_v(\pi)^i t_i^{-1}X^i\right)^{-1} = Z_{\mathfrak{p}}(\mathbf{f}, X),$$

where $Z_{\mathfrak{p}}(\mathbf{f}, X)$ *and* $Z_{\mathfrak{p}}(\mathbf{f}_0, X)$ *are given by* (i) *and* (iii) *respectively of the Euler factors described at the beginning of Sect. 4. Moreover the eigenvalues of* $\mathbf{f}_0$ *with respect to the Hecke operators* $U(\mathfrak{p})$ *are* p*-adic units. For all other primes* $\mathfrak{q}$ *we have* $Z_{\mathfrak{q}}(\mathbf{f}, X) = Z_{\mathfrak{q}}(\mathbf{f}_0, X)$.

8 p-adic Measures for Ordinary Hermitian Modular Forms

We recall that for a fixed odd prime p we write S for the set of all prime ideals above p in K, that are inert from F, and we assume that $S \neq \emptyset$. Moreover we denote by $\mathfrak{v}$ the ideal $\prod_{\mathfrak{p}\in S}\mathfrak{p}$. We denote by $K(S)$ the maximal abelian extension of K unramified outside the set S, and we write G for the Galois group of the extension $K(S)/K$. We consider a Hecke eigenform $\mathbf{f} \in \mathcal{S}_k(C, \psi)$ with $C = D[\mathfrak{b}^{-1}, \mathfrak{b}\mathfrak{c}]$ for some ideals $\mathfrak{b}$ and $\mathfrak{c}$ of F which are prime to p. We assume that $m_0 \geqslant 3n + 2$, where we recall that $m_0 := min_{v\in\mathbf{a}}(m_v)$ with $m_v := k_v + k_{v\rho}$. Moreover we take $\mathbf{f}$ to be ordinary at every prime $\mathfrak{p}$ in the set S in the sense defined in the previous section. By Theorem 7.3 we can associate to it a Hermitian modular form $\mathbf{f}_0$. In particular the eigenvalues of $\mathbf{f}_0$ with respect to the Hecke operators $U(\mathfrak{p})$ for all $\mathfrak{p} \in S$ are p-adic units, where we recall that we write $U(\mathfrak{p})$ for the Hecke operator $U(\pi_n)$ where π is a uniformizer corresponding to the prime ideal $\mathfrak{p}$. In this section we write $\alpha(\mathfrak{p})$ for $U(\mathfrak{p})\mathbf{f}_0 = \alpha(\mathfrak{p})\mathbf{f}_0$. We also write $\{t_{i,\mathfrak{p}}\}$ for the Satake parameters of $\mathbf{f}_0$ at the prime $\mathfrak{p}$.

Given a $k \in \mathbb{Z}^{\mathbf{b}}$ and a $t \in \mathbb{Z}^{\mathbf{a}}$ we define a $\mu \in \mathbb{Z}^{\mathbf{b}}$ as in Sect. 4. Since in this paper we have been working with unitary Hecke characters so far we need to establish a correspondence between Galois characters and unitary Hecke characters. We start by recalling the definition of a Grössencharacter of type A_0 for the CM field K. In the following for an integral ideal $\mathfrak{m}$ of K we write $I(\mathfrak{m})$ for the free abelian group generated by all prime ideals of K prime to $\mathfrak{m}$.

Definition 8.1 A Grössencharacter of type A_0, in the sense of Weil, of conductor dividing a given integral ideal $\mathfrak{m}$ of K, is a homomorphism $\chi : I(\mathfrak{m}) \to \overline{\mathbb{Q}}$ such that there exist integers $\lambda(\tau)$ for each $\tau : K \hookrightarrow \mathbb{C}$, such that for each $\alpha \in K^\times$ we have

$$\chi((\alpha)) = \prod_{\tau} \tau(\alpha)^{\lambda(\tau)}, \quad \text{if } \alpha \equiv 1 \mod {}^\times\mathfrak{m}.$$

Here the condition $\alpha \equiv 1 \mod {}^\times\mathfrak{m}$ means that if we write $\mathfrak{m} = \prod_{\mathfrak{q}} \mathfrak{q}^{n_{\mathfrak{q}}}$ with $\mathfrak{q}$ distinct prime ideals and $n_{\mathfrak{q}} \in \mathbb{N}$ then $v_{\mathfrak{q}}(\alpha - 1) \geqslant n_{\mathfrak{q}}$, where $v_{\mathfrak{q}}$ the standard discrete valuation associated to the prime ideal $\mathfrak{q}$.

It is well known (see for example [24]) if since we are taking K to be a CM field then the above $\lambda(\tau)$ must satisfy some conditions. In particular if we select a CM type of K, which we identify with the places $\mathbf{a}$ of F, then there exists integers d_v for each $v \in \mathbf{a}$ and an integer k such that

$$\chi((\alpha)) = \prod_{v \in \mathbf{a}} \left(\frac{1}{\alpha_v^k} \left(\frac{\alpha_v^\rho}{\alpha_v} \right)^{d_v} \right), \quad \text{if } \alpha \equiv 1 \mod {}^\times\mathfrak{m}.$$

We now keep writing χ for the associated, by class field theory, adelic character to χ. As it is explained in [24, p. 286] the infinity type is of the form,

$$\chi_{\mathbf{a}}(x) = \prod_{v \in \mathbf{a}} \left(\frac{x_v^{k+d_v}}{x_v^{\rho d_v}} \right). \tag{25}$$

We now consider the unitary character $\chi^1 := \chi|\cdot|_{\mathbb{A}_K}^{-k/2}$, where $|\cdot|_{\mathbb{A}_K}$ the adelic absolute value with archimedean part $|x|_{\mathbf{a}} = \prod_{v \in \mathbf{a}} |x_v|_v$, where $|\cdot|_v$ is the standard absolute value of $\mathbb{C}$. We then have that

$$\chi_{\mathbf{a}}^1(x) = \prod_{v \in \mathbf{a}} \left(\frac{x_v^{k/2+d_v}}{\bar{x}_v^{\,k/2+d_v}} \right) = \prod_{v \in \mathbf{a}} \left(\frac{x_v^{k+2d_v}}{(x_v \bar{x}_v)^{k/2+d_v}} \right) = \prod_{v \in \mathbf{a}} \left(\frac{x_v^{k+2d_v}}{|x_v|^{k+2d_v}} \right).$$

In particular to a Grössencharacter χ of type A_0 of infinity type as in Eq. 25 we can associate a unitary character χ^1 of infinity type $\{m_v\}_{v\in\mathbf{a}}$ with $m_v := k + 2d_v$. The relation between the associated L functions is given by

$$L(s,\chi) = L(s + k/2, \chi^1).$$

In particular, in what follows, when we say that we consider a character χ of G of infinite type $t \in \mathbb{Z}^{\mathbf{a}}$ we shall mean that the corresponding unitary character, in the way we explained above, is of infinity type t. And we will keep writing χ, instead of χ^1 for this corresponding unitary character.

Now we return to the general setting introduced at the beginning of this section. Given a character χ of G we write $\mathfrak{f}_\chi = \prod_{\mathfrak{p}\in S}\mathfrak{p}^{n_\mathfrak{p}}$ for its conductor and define the ideal $\mathfrak{m}_\chi := \mathfrak{a}\prod_j(\mathfrak{p}\mathfrak{p}^\rho)^{m_\mathfrak{p}}$ where $m_\mathfrak{p} = n_\mathfrak{p}$ for $n_\mathfrak{p} \neq 0$ and $m_\mathfrak{p} = 1$ for $n_\mathfrak{p} = 0$, and $\mathfrak{a}$ is a small enough ideal so that it is included in $\mathfrak{c}$ and the prime to S level of the theta series Θ_χ, where Θ_χ is defined at the beginning of Sect. 6. Moreover we define $\mathfrak{m}_0 := \mathfrak{a}\prod_{\mathfrak{p}\in S}\mathfrak{p}\mathfrak{p}^\rho$ and

$$A^+(\chi) := C(\chi^{-1})^{-1}C(S)^{-1}N(\mathfrak{f}_\chi)^{n^2-\frac{n}{2}-n(n-\frac{r}{2})}N(\mathfrak{v}),$$

where $C(\chi^{-1})$ was defined in Eq. 4, $C(S)$ in Proposition 3.1, and we recall that $\mathfrak{v} = \prod_{\mathfrak{p}\in S}\mathfrak{p}$. We also define

$$A^-(\chi) := C(\chi^{-1})^{-1}C(S)^{-1}N(\mathfrak{f}_\chi)^{n^2-\frac{n}{2}-\frac{nr}{2}}N(\mathfrak{v}),$$

$$B^+(\chi) := \prod_{\mathfrak{p}\nmid\mathfrak{f}_\chi} N(\mathfrak{p})^{n(2n-1)-n(\frac{r}{2}+\frac{3n}{2}-1)-n^2}\left(\prod_{\mathfrak{p}\nmid\mathfrak{f}_\chi} C(\mathfrak{p}, -\frac{n+r}{2})\right)^{-\rho},$$

$$B^-(\chi) := \prod_{\mathfrak{p}\nmid\mathfrak{f}_\chi} N(\mathfrak{p})^{n(2n-1)-n(-\frac{r}{2}+\frac{5n}{2}-1)-n^2}\left(\prod_{\mathfrak{p}\nmid\mathfrak{f}_\chi} C(\mathfrak{p}, -\frac{3n-r}{2})\right)^{-\rho},$$

where $C(\mathfrak{p}, s)$ was defined in Theorem 6.7. We also write $C_0(\mathfrak{m}_\chi)$ for the quantity appearing in Theorem 4.1 by taking $\mathfrak{c}''$ equal to $\mathfrak{m}_\chi$ there. We then have the following theorems,

Theorem 8.2 *Assume we are given a $t \in \mathbb{Z}^{\mathbf{a}}$ such that*

$$(k_v - \mu_v - n) + (k_{v\rho} - \mu_{v\rho}) = r, \quad \forall v \in \mathbf{a}$$

for some $r \geqslant n$. Moreover assume that $r > n$ if $\psi_1 = 1$ or $\mathfrak{c} = \mathfrak{g}$. Then there exists a measure $\mu^+_{\mathbf{f},t}$ of G such that for any primitive Hecke character χ of conductor $\mathfrak{f}_\chi = \prod_\mathfrak{p}\mathfrak{p}^{n_\mathfrak{p}}$ of infinite type $\chi_\mathbf{a}(x) = x_\mathbf{a}^{-t}|x_\mathbf{a}|^t$ we have

$$\int_G \chi d\mu^+_{\mathbf{f},t} = \frac{A^+(\chi)B^+(\chi)}{C_0(\mathfrak{m}_\chi)} \left(\prod_{\mathfrak{p}|\mathfrak{f}_\chi} \alpha(\mathfrak{p})^{-n_\mathfrak{p}}\right) \tau(\chi)^{-n\rho} \times$$

$$\prod_{\mathfrak{p}\nmid\mathfrak{f}_\chi} \prod_{i=1}^{n} \left(\frac{1-\chi(\mathfrak{p})^{-1}t_{i,\mathfrak{p}}^{-1}N(\mathfrak{p})^{\frac{r-n+2}{2}}}{1-\chi(\mathfrak{p})t_{i,\mathfrak{p}}N(\mathfrak{p})^{\frac{n-r}{2}-1}} \right) \times \frac{L_\mathfrak{v}(\frac{r+n}{2},\mathbf{f},\chi)}{\pi^\beta \Omega_{\mathbf{f}_0}},$$

where β is as in Theorem 5.2, and $\Omega_{\mathbf{f}_0} \in \mathbb{C}^\times$ is the period defined in Theorem 5.1 corresponding to the eigenform $\mathbf{f}_0$. In the case of $r = n+1$ and $F = \mathbb{Q}$ we exclude the characters χ such that $(\chi\psi)_1 = \theta$.

We remark here that on the left hand side, χ denotes a Galois character to which by class field theory we can associate a Hecke character of A_0 type, and by the process described above we can further associate to it a unitary character χ^1. Then as it was indicated above it is our convention that in the right hand side of the above theorem we write χ for this χ^1. Moreover we recall that we declared the infinite type of χ to be the infinite type of χ^1.

Furthermore we remark that the archimedean periods we use for our interpolation properties are the ones related to $\mathbf{f}_0$. However it is not hard to see by the definition of these periods in [6] that they are related to $\Omega_\mathbf{f}$ by some algebraic factor, which can be made very precise. For the cases excluded in the above theorem we have the following theorem.

Theorem 8.3 *We let $\mathfrak{q}$ be a prime ideal of F, prime to p. Assume that $r = n$ and further that $\psi_1 = 1$ or $\mathfrak{c} = \mathfrak{g}$ there exists a measure $\mu^+_{\mathbf{f},\mathfrak{q},t}$ such that for all characters χ of G of infinite type t we have*

$$\int_G \chi d\mu^+_{\mathbf{f},\mathfrak{q},t} = \frac{A^+(\chi)B^+(\chi)}{C_0(\mathfrak{m}_\chi)} \left(\prod_{\mathfrak{p}|\mathfrak{f}_\chi} \alpha(\mathfrak{p})^{-n_\mathfrak{p}}\right) \tau(\chi)^{-n\rho} \times$$

$$\prod_{\substack{i=0\\ i+n\equiv 1\, mod\, 2}}^{n-1} (1-(\chi\psi)_1(\mathfrak{q})N(\mathfrak{q})^{i+1}) \prod_{\mathfrak{p}\nmid\mathfrak{f}_\chi} \prod_{i=1}^{n} \left(\frac{1-\chi(\mathfrak{p})^{-1}t_{i,\mathfrak{p}}^{-1}N(\mathfrak{p})}{1-\chi(\mathfrak{p})t_{i,\mathfrak{p}}N(\mathfrak{p})^{-1}} \right) \frac{L_\mathfrak{v}(n,\mathbf{f},\chi)}{\pi^\beta \Omega_{\mathbf{f}_0}},$$

where $A^+(\chi)$ and $B^+(\chi)$ are defined by taking $r = n$ there.

For the other critical value, which does not involve nearly-holomorphic Eisenstein series we have the following theorem.

Theorem 8.4 *Assume that $\psi_1 \neq 1$, $\mathfrak{c} \neq \mathfrak{g}$ and $r \geqslant n$. Then there exists a measure $\mu^-_{\mathbf{f},t}$ on G such that for all characters χ of G of infinite type t we have,*

$$\int_G \chi d\mu^-_{\mathbf{f},t} = \frac{A^-(\chi)B^-(\chi)}{C_0(\mathfrak{m}_\chi)} \left(\prod_{\mathfrak{p}|\mathfrak{f}_\chi} \alpha(\mathfrak{p})^{-n_\mathfrak{p}}\right) \tau(\chi)^{-n\rho}$$

$$\prod_{\mathfrak{p}\nmid\mathfrak{f}_\chi}\prod_{i=1}^{n}\left(\frac{1-\chi(\mathfrak{p})t_{i,\mathfrak{p}}^{-1}N(\mathfrak{p})^{\frac{n-r+2}{2}}}{1-\chi(\mathfrak{p})t_{i,\mathfrak{p}}N(\mathfrak{p})^{\frac{r-n}{2}-1}}\right)\frac{L_{\mathfrak{v}}(\frac{3n-r}{2},\mathbf{f},\chi)}{\pi^{\beta}\Omega_{\mathbf{f}_0}},$$

And finally,

Theorem 8.5 *Assume that $\psi_1 = 1$ or $\mathfrak{c} = \mathfrak{g}$, and moreover $r \geqslant n$. Let $\mathfrak{q}$ be an ideal prime to p. Then there exist a measure $\mu^{-}_{\mathbf{f},\mathfrak{q},t}$ such that*

$$\int_G \chi d\mu^{-}_{\mathbf{f},\mathfrak{q}} = \frac{A^{-}(\chi)B^{-}(\chi)}{C_0(\mathfrak{m}_\chi)}\left(\prod_{\mathfrak{p}|\mathfrak{f}_\chi}\alpha(\mathfrak{p})^{-n_\mathfrak{p}}\right)\tau(\chi)^{-n\rho}\times$$

$$\prod_{\substack{i=0\\ n+i\equiv 1 \bmod 2}}^{n-1}(1-(\chi\psi)_1(\mathfrak{q})N(\mathfrak{q})^{r+i+1-n})\prod_{\mathfrak{p}\nmid\mathfrak{f}_\chi}\prod_{i=1}^{n}\left(\frac{1-\chi(\mathfrak{p})^{-1}t_{i,\mathfrak{p}}^{-1}N(\mathfrak{p})^{\frac{n-r+2}{2}}}{1-\chi(\mathfrak{p})t_{i,\mathfrak{p}}N(\mathfrak{p})^{\frac{r-n}{2}-1}}\right)\times$$

$$\frac{L_{\mathfrak{v}}(\frac{3n-r}{2},\mathbf{f},\chi)}{\pi^{\beta}\Omega_{\mathbf{f}_0}}.$$

Remark 8.6 We remark that in the interpolation properties above, at the modified Euler factors above p, we use the Satake parameters of the Hermitian form $\mathbf{f}_0$, and not of $\mathbf{f}$. However Theorem 7.3 provides a relation between them.

The rest of this section is devoted to proving the above theorems. We will establish in details the proof of Theorem 8.2 and then comment on the needed modifications to establish the rest.

We define,

$$\mathcal{F}^{+}_{\chi} := \Theta^{*}\mathbf{E}^{*}_{+}|\left(\prod_{\mathfrak{p}|\mathfrak{f}_\chi}V(\pi_\mathfrak{p})^{n_\mathfrak{p}-1}\right)\left(\prod_{\mathfrak{p}\nmid\mathfrak{f}_\chi}C(\mathfrak{p},s_{+})^{-1}\widetilde{J(\mathfrak{p},s_{+})}\right),$$

and

$$\mathcal{F}^{-}_{\chi} := \Theta^{*}\mathbf{E}^{*}_{-}|\left(\prod_{\mathfrak{p}|\mathfrak{f}_\chi}V(\pi_\mathfrak{p})^{n_\mathfrak{p}-1}\right)\left(\prod_{\mathfrak{p}\nmid\mathfrak{f}_\chi}C(\mathfrak{p},s_{-})^{-1}\widetilde{J(\mathfrak{p},s_{-})}\right),$$

where Θ^* and $\mathbf{E}^*_{\pm}$ are the series defined at the beginning of Sect. 6, associated to the character χ, and $C(\mathfrak{p}, s_{\pm})$ is defined in Theorem 6.7. We now define the following distribution on G, which later we will show it is actually a measure. For the definition of the distribution it is enough to give the values at each character χ of infinite type t.

$$\int_G \chi d\mu'_{\mathbf{f},+,t} := \frac{1}{\pi^\beta \Omega_{\mathbf{f}_0}} A^+(\chi) \left(\prod_{\mathfrak{p}|\mathfrak{f}_\chi} \alpha(\mathfrak{p})^{-n_\mathfrak{p}}\right) \left(\prod_{\mathfrak{p}\nmid\mathfrak{f}_\chi} \alpha(\mathfrak{p})^{-2}\right) \tau(\chi)^{-n\rho} \times$$

$$< \mathbf{f}_0 | W, \mathcal{F}^+_\chi | \prod_{\mathfrak{p}|\mathfrak{f}_\chi} U(\mathfrak{p})^{n_\mathfrak{p}-1} >_{\mathfrak{m}_0},$$

We now show that $\mu'_{\mathbf{f},+,t}$ is actually a measure. We start by recalling the classical Kummer congruences (see [24]). Let Y be a profinite topological space, and R a p-adic ring.

Proposition 8.7 (abstract Kummer congruences) *Suppose R is flat over $\mathbb{Z}_p$, and let $\{f_i\}_{i\in I}$ be a collection of elements of $Cont(Y, R)$, whose $R[1/p]$-span is uniformly dens in $Cont(Y, R[1/p])$. Let $\{a_i\}_{i\in I}$ be a family elements of R with the same indexing set I. Then there exists an R-valued p-adic measure μ on Y such that*

$$\int_Y f_i d\mu = a_i, \quad \forall i \in I$$

if and only if the a_i's satisfy the following "Kummer congruences":

for every collection $\{b_i\}_{i\in I}$ of elements in $R[1/p]$ which are zero for all but finitely many i, and every integer n such that

$$\sum_i b_i f_i(y) \in p^n R, \quad \forall y \in Y,$$

we have

$$\sum_i b_i a_i \in p^n R.$$

Proof [24] □

Proposition 8.8 *The distribution $\mu'_{\mathbf{f},+,t}$ is a measure.*

Proof We establish the Kummer congruences. We first start with a remark. For a character χ of conductor $\mathfrak{f}_\chi = \prod_{\mathfrak{p}\in S} \mathfrak{p}^{n_\mathfrak{p}}$ we consider any vector $c = (c_\mathfrak{p})_{\mathfrak{p}\in S}$ with $c_\mathfrak{p} \in \mathbb{Z}$, and $c_\mathfrak{p} \geqslant max(n_\mathfrak{p}, 1)$ for all $\mathfrak{p} \in S$. Then, by the same considerations as in the proof of Proposition 6.10, we have that

$$\left(\prod_{\mathfrak{p}|\mathfrak{f}_\chi} \alpha(\mathfrak{p})^{-n_\mathfrak{p}}\right) \left(\prod_{\mathfrak{p}\nmid\mathfrak{f}_\chi} \alpha(\mathfrak{p})^{-2}\right) < \mathbf{f}_0 | W, \mathcal{F}^+_\chi | \prod_{\mathfrak{p}|\mathfrak{f}_\chi} U(\mathfrak{p})^{n_\mathfrak{p}-1} >_{\mathfrak{m}_0}$$

$$\left(\prod_{\mathfrak{p}|\mathfrak{f}_\chi} \alpha(\mathfrak{p})^{-n_\mathfrak{p}-1}\right) \left(\prod_{\mathfrak{p}\nmid\mathfrak{f}_\chi} \alpha(\mathfrak{p})^{-2}\right) < \mathbf{f}_0 | W, \mathcal{F}^+_\chi | \prod_{\mathfrak{p}|\mathfrak{f}_\chi} U(\mathfrak{p})^{n_\mathfrak{p}} >_{\mathfrak{m}_0} =$$

$$\left(\prod_{\mathfrak{p}|\mathfrak{f}_\chi} \alpha(\mathfrak{p})^{-c_\mathfrak{p}-1}\right)\left(\prod_{\mathfrak{p}\nmid\mathfrak{f}_\chi} \alpha(\mathfrak{p})^{-c_\mathfrak{p}-1}\right) \times$$

$$< \mathbf{f}_0|W, \mathcal{F}_\chi^+| \left(\prod_{\mathfrak{p}|\mathfrak{f}_\chi} U(\mathfrak{p})^{c_\mathfrak{p}}\right)\left(\prod_{\mathfrak{p}\nmid\mathfrak{f}_\chi} U(\mathfrak{p})^{c_\mathfrak{p}-1}\right) >_{\mathfrak{m}_0} .$$

We now consider a finite set of characters χ_i with $i = 1, \ldots \ell$ of conductors $\mathfrak{f}_{\chi_i} = \prod_{\mathfrak{p}\in S} \mathfrak{p}^{n_{\mathfrak{p},i}}$. We define $c = (c_\mathfrak{p})_{\mathfrak{p}\in S}$ with $c_\mathfrak{p} := max(max_i(n_{\mathfrak{p},i}), 1)$. We now let $\mathcal{O}$ be a large enough p-adic ring and take elements $a_i \in \mathcal{O}[1/p]$ such that

$$\sum_{i=0}^{\ell} a_i \chi_i \in p^m \mathcal{O}$$

for some $m \in \mathbb{N}$. We then establish the congruences

$$\sum_{i=0}^{\ell} a_i A^+(\chi_i)\tau(\chi)^{-n\rho}\mathcal{F}_{\chi_i}^+| \left(\prod_{\mathfrak{p}|\mathfrak{f}_{\chi_i}} U(\mathfrak{p})^{c_\mathfrak{p}}\right)\left(\prod_{\mathfrak{p}\nmid\mathfrak{f}_\chi} U(\mathfrak{p})^{c_\mathfrak{p}-1}\right) \in p^m \mathcal{O}[[q]].$$

The above statement should be understood that the q-expansion of the Hermitian modular form on the left has coefficients in $p^m\mathcal{O}$.

The first observation here is that by Theorem 6.7 and by the discussion right after Proposition 3.5, the Fourier expansion for all

$$\mathcal{G}_i := \mathcal{F}_{\chi_i}^+| \left(\prod_{\mathfrak{p}|\mathfrak{f}_{\chi_i}} U(\mathfrak{p})^{c_\mathfrak{p}}\right)\left(\prod_{\mathfrak{p}\nmid\mathfrak{f}_\chi} U(\mathfrak{p})^{c_\mathfrak{p}-1}\right),$$

is supported at the same Hermitian matrices. That is, the sets $Supp_i := \{(\tau, q) : c(\tau, q; \mathcal{G}_i) \neq 0\}$ for $i = 1, \ldots, \ell$, are the same.

We note here that we need to apply one power less of the Hecke operators $U(\mathfrak{p})$ at the primes $\mathfrak{p}$ which divide $\mathfrak{f}_\chi$, since for the rest we have already applied $U(\mathfrak{p})$ as the n'th term of the operator $J(\mathfrak{p}, s_+)$.

It now follows from the explicit description of the Fourier coefficients given in Propositions 3.1 and 3.5 and by Eq. 20 that the coefficients of $A^+(\chi_i)\tau(\chi_i)^{-n\rho}\mathcal{F}_{\chi_i}^+$ are all p-integral and that we have the congruences

$$\sum_{i=0}^{\ell} a_i A^+(\chi_i)\tau(\chi)^{-\rho}\mathcal{F}_{\chi_i}| \left(\prod_{\mathfrak{p}|\mathfrak{f}_{\chi_i}} U(\mathfrak{p})^{c_\mathfrak{p}}\right)\left(\prod_{\mathfrak{p}\in S} U(\mathfrak{p})^{c_\mathfrak{p}-1}\right) \in p^m \mathcal{O}[[q]].$$

Indeed, let us write R for the "polynomial" ring $\mathcal{O}[\mathfrak{q}|\mathfrak{q} \in \mathcal{P}_S]$, in the variables $\mathfrak{q} \in \mathcal{P}_S$. where $\mathcal{P}_S$ is the set of prime ideals of K not in the set S. A character χ of G, induces then a ring homomorphism $\chi_R : R \to \overline{\mathbb{Q}}_p$, where we have extended $\mathcal{O}$- linear the multiplicative map $\chi : \mathcal{P}_S \to \overline{\mathbb{Q}}_p^\times$. Given an element $P \in R$ we write $P(\chi)$ for $\chi_R(P) \in \overline{\mathbb{Q}}_p$. Then by Propositions 3.5 and 3.1 we have that the Fourier coefficients of $A^+(\chi_i)\tau(\chi_i)^{-n\rho}\mathcal{F}^+_{\chi_i}$ at any given Hermitian matrix τ are of the form $P_{\tau_1}(\chi_i)P_{\tau_2}(\chi_i) = P_\tau(\chi_i)$ for some P_{τ_i}, $P_\tau \in R$, with $P_\tau = P_{\tau_1}P_{\tau_2}$. In particular if we have $\sum_i a_i\chi_i \in p^m\mathcal{O}$ then $\sum_i a_i P_\tau(\chi_i) \in p^m\mathcal{O}$. We also remark here that we need to use also Proposition 3.3, which guarantees that the coefficients of the Eisenstein series are supported only at full rank Hermitian matrices, and hence no L-values of Dirichlet series appear in the Fourier coefficients (and so the polynomial description above is enough). Moreover we also use the fact that the operator $U(\mathfrak{p})$ is p-integral as it was shown using the q-expansion in Eq. 19, where in the notation there $U(\mathfrak{p})^m = T(\mathfrak{p}^m)$ for any $m \in \mathbb{N}$ and $\mathfrak{p} \in S$.

It is now a standard argument using the finite dimension of the space of cusp forms of a particular level (see for example [2, Lemma 9.7] or [11, p. 134]) to show that by taking projection to $\mathbf{f}_0|W$ we obtain a measure. For this of course we use also by Theorem 5.1, $\Omega(\mathbf{f}_0)$ is up to algebraic factor equal to $< \mathbf{f}_0, \mathbf{f}_0 >$. Hence we conclude that $\mu'_{\mathbf{f},t,+}$ is indeed a measure. □

We now define the measure $\mu_{\mathbf{g}}$ on G by

$$\int_G \chi d\mu_{\mathbf{g}} := \prod_{v \in \mathbf{b}} g_v(\chi(\pi_v)|\pi_v|^{r+n}),$$

where $g_v(X)$ are the polynomials appearing in Theorem 4.1. Note that $g_v \in \mathbb{Z}[X]$ with $g_v(0) = 1$, and hence since we evaluate then at places prime to p, we have that $\mu_{\mathbf{g}}$ is indeed a measure. We now define are measure $\mu^+_{\mathbf{f},2}$ as the convolution of $\mu'_{\mathbf{f},+}$ with $\mu_{\mathbf{g}}$. In particular we now obtain after evaluating at a character χ that,

$$\int_G \chi d\mu^+_{\mathbf{f},2} = \left(\int_G \chi d\mu'_{\mathbf{f},+}\right)\left(\int_G \chi d\mu_{\mathbf{g}}\right) =$$

$$\frac{1}{\pi^\beta \Omega_{\mathbf{f}_0}} A^+(\chi) \left(\prod_{\mathfrak{p}|\mathfrak{f}_\chi} \alpha(\mathfrak{p})^{-n_\mathfrak{p}}\right)\left(\prod_{\mathfrak{p}\nmid\mathfrak{f}_\chi} \alpha(\mathfrak{p})^{-2}\right) \tau(\chi)^{-n\rho} \times$$

$$< \mathbf{f}_0|W, \mathcal{F}^+_\chi| \prod_{\mathfrak{p}|\mathfrak{f}_\chi} U(\mathfrak{p})^{n_\mathfrak{p}-1} >_{\mathfrak{m}_0} \prod_{v \in \mathbf{b}} g_v(\chi(\pi_v)|\pi_v|^{r+n}).$$

However we have by Proposition 6.10, by taking there $n_{\mathfrak{p}} = c_{\mathfrak{p}}$ for all $\mathfrak{p}|\mathfrak{f}_{\chi}$ that

$$\left(\prod_{\mathfrak{p}|\mathfrak{f}_{\chi}} \alpha(\mathfrak{p})^{-n_{\mathfrak{p}}}\right)\left(\prod_{\mathfrak{p}\nmid\mathfrak{f}_{\chi}} \alpha(\mathfrak{p})^{-2}\right) < \mathbf{f}_0|W, \mathcal{F}_{\chi}^{+} >_{\mathfrak{m}_0} =$$

$$\left(\prod_{\mathfrak{p}|\mathfrak{f}_{\chi}} \alpha(\mathfrak{p})^{-n_{\mathfrak{p}}}\right)\left(\prod_{\mathfrak{p}\nmid\mathfrak{f}_{\chi}} \alpha(\mathfrak{p})^{-2}\right) < \mathbf{f}_0|\prod_{\mathfrak{p}\nmid\mathfrak{f}_{\chi}} J(\mathfrak{p}, s_{+}), \Theta_{\chi}\mathbf{E}_{+,\chi} >_{\mathfrak{m}_{\chi}},$$

and using Lemma 6.3, and in particular Eq. 9, we get that the above is equal to

$$\left(\prod_{\mathfrak{p}|\mathfrak{f}_{\chi}} \alpha(\mathfrak{p})^{-n_{\mathfrak{p}}}\right)\left(\prod_{\mathfrak{p}\nmid\mathfrak{f}_{\chi}} \alpha(\mathfrak{p})^{-1}\right)\prod_{\mathfrak{p}\nmid\mathfrak{f}_{\chi}}(-1)^{n} N(\mathfrak{p})^{n(2n-1)-n(\frac{n+r}{2})} \times$$

$$\prod_{i=1}^{n}(1-\chi(\mathfrak{p})^{-1}t_i^{-1}N(\mathfrak{p})^{\frac{r-n+2}{2}}) < \mathbf{f}_0, \Theta_{\chi}\mathbf{E}_{\chi,+} >_{\mathfrak{m}_{\chi}} .$$

We now use Theorem 4.1, where we pick an invertible τ such that $c(\tau, r, \mathbf{f}_0) \neq 0$, which is of course always possible since $\mathbf{f}_0$ is a cusp form. Moreover after using the fact that $c(\tau, \pi r, \mathbf{f}_0) = N(\mathfrak{p})^{-n^2}\alpha(\mathfrak{p})c(\tau, r, \mathbf{f}_0)$ we have that

$$\int_G \chi d\mu_{\mathbf{f},2}^{+} = B \times C_0^{-1}\left(\prod_{\mathfrak{p}|\mathfrak{f}_{\chi}} \alpha(\mathfrak{p})^{-n_{\mathfrak{p}}}\right) A^{+}(\chi)\tau(\chi)^{-n\rho} \times$$

$$\left(\prod_{\mathfrak{p}\nmid\mathfrak{f}_{\chi}} N(\mathfrak{p})^{n(2n-1)-n(\frac{n+r}{2})-n^2}\prod_{i=1}^{n}(1-\chi(\mathfrak{p})^{-1}t_i^{-1}N(\mathfrak{p})^{\frac{r-n+2}{2}})\right)\frac{L(\frac{r+n}{2}, \mathbf{f}_0, \chi)}{\pi^{\beta}\Omega_{\mathbf{f}_0}},$$

where B is some non-zero algebraic constant independent of χ. We then define the measure $\mu_{\mathbf{f},t,+} := B^{-1}\mu_{\mathbf{f},2}$. Using the fact that $\mathbf{f}$ and $\mathbf{f}_0$ have the same Satake parameters away from p, we obtain the claimed interpolation properties of Theorem 8.2.

The proofs of Theorems 8.3, 8.4 and 8.5 are similar, we just need to take some extra care for the fact that in the Fourier coefficients of the Eisenstein series involve values of various Dirichlet series. In order to establish the congruences we use the Barsky, Cassou-Noguès, Deligne–Ribet p-adic L-function [1, 10, 14]. Let us write $F(p^{\infty})$ for the maximal abelian extension of F unramified outside p and infinity. Then it is known that if we pick an ideal $\mathfrak{q}$ of F prime to p, then there exists a measure $\mu_{F,\mathfrak{q}}$ of the Galois group $G' := \mathrm{Gal}(F(p^{\infty})/F)$, such that for any $k \geqslant 1$ we have,

$$\int_{G'} \chi N^k d\mu_{F,\mathfrak{q}} = \left(1-\chi(\mathfrak{q})N(\mathfrak{q})^k\right) L_{(p)}(1-k, \chi),$$

where $\mathcal{N}$ denotes the cyclotomic character. Moreover if we select some primitive character ψ, of some non-trivial conductor prime to p, then we can define a twisted measure $\mu_{F,\psi}$ on G' such that for any $k \geqslant 1$ we have,

$$\int_{G'} \chi \mathcal{N}^k d\mu_{F,\psi} = L_{(p)}(1-k, \chi\psi),$$

where in both equations $L_{(p)}(1-k, ?)$ means that we remove the Euler factors above p. Now we are ready to deal with the proof of the theorems. We explain it for Theorem 8.3, and similarly we argue for the rest. The main difference is the fact that the τ'th Fourier expansion of $\Theta^*\mathbf{E}^*$ is of the form $P_{\tau_1}(\chi)P_{\tau_2}(\chi)$ (with notation as before) multiplied by the L-values $\prod_{i=0}^{n-1-r_2} L_c(-i, \chi_1\theta^{n+i-1})$, where r_2 is the rank of the matrix τ_2. That is we need to establish congruences of the form

$$\sum_i a_i P_{\tau_1}(\chi_i)P_{\tau_2}(\chi_i) \prod_{\substack{i=0 \\ i+n\equiv 1 \bmod 2}}^{n-1} (1-\chi_{i,1}^{-1}(\mathfrak{q})\mathcal{N}(\mathfrak{q})^{i+1})\times$$

$$\prod_{i=0}^{n-1-r_2} L_c(-i, \chi_{i,1}^{-1}\theta^{n+i-1}) \in p^n\mathcal{O}.$$

But now the congruences follow from the existence of the Cassou-Nogues, Deligne–Ribet p-adic measure since the above congruences can be understood as convolution (which we denote as product below) of the measures

$$\left(\prod_{\substack{i=0 \\ i+n\equiv 1 \mod 2}}^{n-1} \mathcal{N}^{i+1}\mu_{F,\mathfrak{q}}\right) \star \left(\prod_{i=0}^{n-1-r_2} \mathcal{N}^{i+1}\mu_{F,\theta^{n+i-1}}\right) \star P,$$

where P is the measure in the Iwasawa algebra represented by the polynomial $P_{\tau_1} \times P_{\tau_2} \in R$, where the Iwasawa algebra. The rest of the proof is entirely identical where of course we need to replace the quantities $A^+(\chi)$ and $B^+(\chi)$ with $A^-(\chi)$ and $B^-(\chi)$ respectively.

9 The Values of the p-adic Measures

We now obtain a result regarding the values of the p-adic measures constructed above. We show the following theorem.

Theorem 9.1 *Write μ for any of the measures constructed in Theorems 8.2, 8.3, 8.4 and 8.5. Define the normalized measure*

$$\mu' := \left(\tau(\psi_1\theta^{n^2})\right)^{-\rho} i^{-n\sum_{v\in\mathfrak{a}} p_v}\mu,$$

where the p_v's are defined as in Theorem 5.2. Assume that one of the cases of Theorem 5.2 occurs. Then μ' is W-valued, where W is the field appearing in the Theorem 5.2.

Proof By comparing the interpolation properties of the measure μ' and the reciprocity law shown in Theorem 5.3, we need only to establish that the Gauss sums $\tau(\chi_1)$ and $\tau(\chi)$ have the same reciprocity properties, namely $\left(\frac{\tau(\chi)}{\tau(\chi_1)}\right)^\sigma = \frac{\tau(\chi^\sigma)}{\tau(\chi_1^\sigma)}$ for any $\sigma \in \mathrm{Gal}(\overline{\mathbb{Q}}/W)$, and any character χ of G. For the proof we follow the strategy sketched in [17, p. 33] and [28, p. 105].
We first recall a property (see [26, p. 36]) of the transfer map ,

$$det(\rho) = \theta \cdot \chi \circ Ver = \theta \cdot \chi_1,$$

where $\rho := Ind_F^K(\chi)$ is the two-dimensional representation induced from K to F, and for the second equakity we used the fact that the restriction $F_{\mathbb{A}}^\times \hookrightarrow K_{\mathbb{A}}^\times$ on the automorphic side is the transfer map (Ver) on the Galois side. We note here that the result in [26] is more general but we have applied it to our special case (i.e. χ is a one-dimensional representation and the extension K/F is quadratic). Recalling that the gauss sum attached to a character is closely related to the Deligne–Langlands epsilon factor attached to the same character, we have that

$$\tau(det(\rho)) = \tau(\theta\chi_1) = \pm\tau(\chi_1)\tau(\theta),$$

where we have used the fact that K/F is unramified above p, χ_1 can be ramified only above p, θ is a quadratic character, and the property [32, p. 15, Eq. (3.4.6)]. Now we note that by [13, p. 330, Eq. 5.5.1 and 5.5.2] we have that

$$\left(\frac{\tau(\rho)}{\tau(det(\rho))}\right)^\sigma = \frac{\tau(\rho^\sigma)}{\tau(det(\rho^\sigma))}$$

for all $\sigma \in \mathrm{Gal}(\overline{\mathbb{Q}}/\mathbb{Q})$. We note here that we write $\tau(\rho)$ for the Deligne–Langlands epsilon factor associated to the representation ρ. In particular since $\tau(\theta) \in W$ we have that

$$\left(\frac{\tau(\rho)}{\tau(\chi_1)}\right)^\sigma = \frac{\tau(\rho^\sigma)}{\tau(\chi_1^\sigma)},$$

and now using the fact that also $\tau(\rho) = \tau(\chi)$ up to elements in $W^\times$ we conclude that

$$\left(\frac{\tau(\chi)}{\tau(\chi_1)}\right)^\sigma = \frac{\tau(\chi^\sigma)}{\tau(\chi_1^\sigma)}, \quad \sigma \in \mathrm{Gal}(\overline{\mathbb{Q}}/W),$$

which concludes the proof of the theorem. □

References

1. Barsky, D.: Fonctions zêta p-adiques d'une classe de rayon des corps de nombres totalement réels. In: Amice, Y., Barskey, D., Robba, P. (eds.) Groupe d'Etude d'Analyse Ultramétrique (5e année) (1977/78)
2. Böcherer, S., Schmidt, C.-G.: p-adic measures attached to Siegel modular forms. Annales de l' institut Fourier, tome **50**(5), 1375–1443 (2000)
3. Bouganis, Th: On special L-values attached to Siegel modular forms. In: Bouganis, Th., Venjakob, O. (eds.) Iwasawa Theory 2012-state of the art and recent developments. Springer (2014)
4. Bouganis, Th: Non-abelian p-adic L-functions and Eisenstein series of unitary groups; the CM method. Ann. Inst. Fourier (Grenoble) **64**(2), 793–891 (2014)
5. Bouganis, Th.: On special L-values attached to metaplectic modular forms (submitted)
6. Bouganis, Th: On the algebraicity of special L-values of Hermitian modular forms. Documenta Mathematica **20**, 1293–1329 (2015)
7. Bouganis, Th.: On p-adic measures for Hermitian and Siegel modular forms (in preparation)
8. Bump, D.: Hecke Algebras (Notes available on-line)
9. Bump, D.: Automorphic Forms and Representations. Cambridge Stud. Adv. Math. **53**, CUP (1998)
10. Cassou-Noguès, P.: Valeurs aux entiers négatifs des fonctions zêta et fonctions zêta p-adiques. Inventiones Mathematicae **51**(1), 29–59 (1979)
11. Coates, J., Schmidt, C.-G.: Iwasawa Theory for the symmetric square of an elliptic curve. J. Reine Angew. Math. 104–156
12. Courtieu, M., Panchishkin, A.: Non-Archimedean L-functions and Arithmetical Siegel Aodular Forms. 2nd edn. Lecture Notes in Mathematics, 1471. Springer, Berlin (2004)
13. Deligne, P.: Valeurs de Fonctions L et Périodes d'Intégrales. In: Proceedings of Symposia in Pure Mathematics, vol. 33, part 2, pp. 313–346 (1979)
14. Deligne, P., Ribet, K.: Values of abelian L-functions at negative integers over totally real fields. Inventiones Mathematicae **59**(3) (1980)
15. Eischen, E.: A p-adic Eisenstein measure for unitary groups. J. Reine Angew. Math. **699**, 111–142 (2015)
16. Eischen, E.: p-adic differential operators on automorphic forms on unitary groups. Ann. Inst. Fourier (Grenoble) **62**(1), 177–243 (2012)
17. Harder, G., Schappacher, N.: Special values of Hecke L-functions and abelian integrals. Springer Lecture Notes in Mathematics, vol. 1111 (1985)
18. Harris, M.: L-functions and periods of polarized regular motives. J. reine angew. Math. **483**, 75–161 (1997)
19. Harris, M.: A simple proof of rationality of Siegel–Weil Eisenstein series, Eisenstein series and applications, pp. 149–185, Progr. Math., 258, Birkhäuser Boston, Boston, MA (2008)
20. Harris, M., Li, J.-S., Skinner, C.: The Rallis inner product formula and p-adic L-functions. In: Rallis. S. (eds.) Automorphic Representations, L-Functions and Applications: Progress and Prospects. de Gruyter, Berlin, pp. 225–255 (2005)
21. Harris, M., Li, J.-S., Skinner, C.: p-adic L-functions for unitary Shimura varieties. In: Construction of the Eisenstein measure, Documenta Math., Extra Volume: John H.Coates' Sixtieth Birthday pp. 393–464 (2006)
22. Hida, H.: Elementary theory of L-functions and Eisenstein series. London Mathematical Society, Student Texts 26, CUP (1993)
23. Katsurada, H.: On the period of the Ikeda lift for $U(m, n)$. http://arxiv.org/abs/1102.4393
24. Katz, N.: p-adic L-functions for CM fields. Inventiones Math. **49**, 199–297 (1978)
25. Klosin, K.: Maass spaces on U(2,2) and the Bloch-Kato conjecture for the symmetric square motive of a modular form. J. Math. Soc. Jpn. **67**(2), 797–860 (2015)
26. Martinet, J.: Character theory and Artin L-functions. In: Fröhlich, A. (ed.) Algebraic Number Fields, Proceedings of LMS Symposium Durham, Academic Press (1977)
27. Panchishkin, A.: Non-Archimedean L-functions of Siegel and Hilbert Modular Forms. 1st edn. Lecture Notes in Mathematics, 1471. Springer, Berlin (1991)

28. Schappacher, N.: Periods of Hecke Characters. Lecture Notes in Mathematics, Springer, 1301 (1988)
29. Shimura, G.: Euler Products and Eisenstein Series. In: CBMS Regional Conference Series in Mathematics, No. 93. American Mathematical Society (1997)
30. Shimura, G.: Arithmeticity in the Theory of Automorphic Forms, Mathematical Surveys and Monographs, vol. 82. American Mathematical Society (2000)
31. Sturm, J.: The critical values of zeta functions associated to the symplectic group. Duke Math. J. **48**(2) (1981)
32. Tate, J.: Number Theoretic Background. In: Proceedings of Symposia in Pure Mathematics, vol. 33, part 2, pp. 3–26 (1979)

Big Image of Galois Representations Associated with Finite Slope p-adic Families of Modular Forms

Andrea Conti, Adrian Iovita and Jacques Tilouine

Abstract We prove that the Lie algebra of the image of the Galois representation associated with a finite slope family of modular forms contains a congruence subalgebra of a certain level. We interpret this level in terms of congruences with CM forms.

Keywords Modular forms · CM forms · Galois representations · Lie algebras · Relative Sen theory · Eigenvarieties

1 Introduction

Let f be a non-CM cuspidal eigenform and let ℓ be a prime integer. By the work of Ribet [15, 17] and Momose [13], it is known that the ℓ-adic Galois representation $\rho_{f,\ell}$ associated with f has large image for every ℓ and that for almost every ℓ it satisfies

(cong$_\ell$) Im $\rho_{f,\ell}$ contains the conjugate of a principal congruence subgroup $\Gamma(\ell^m)$ of $\mathrm{SL}_2(\mathbb{Z}_\ell)$.

For instance if Im $\rho_{f,\ell}$ contains an element with eigenvalues in $\mathbb{Z}_\ell^\times$ distinct modulo ℓ then (cong$_\ell$) holds.

The work of Conti and Tilouine is supported by the Programs ArShiFo ANR-10-BLAN-0114 and PerColaTor ANR-14-CE25-0002-01.

A. Conti (✉) · J. Tilouine
Université Paris 13, Sorbonne Paris Cité, LAGA, CNRS (UMR 7539),
99, Avenue J.-B. Clément, 93430 Villetaneuse, France
e-mail: conti@math.univ-paris13.fr

J. Tilouine
e-mail: tilouine@math.univ-paris13.fr

A. Iovita
Department of Mathematics and Statistics, Concordia University, Montreal, Canada
e-mail: adrian.iovita@concordia.ca

A. Iovita
Dipartimento di Matematica, Universita Degli Studi di Padova, Padova, Italy

D. Loeffler and S.L. Zerbes (eds.), *Elliptic Curves, Modular Forms and Iwasawa Theory*, Springer Proceedings in Mathematics & Statistics 188, DOI 10.1007/978-3-319-45032-2_3

In [9], Hida proved an analogous statement for p-adic families of non-CM ordinary cuspidal eigenforms, where p is any odd prime integer. We fix once and for all an embedding $\overline{\mathbb{Q}} \hookrightarrow \overline{\mathbb{Q}}_p$, identifying $\mathrm{Gal}(\overline{\mathbb{Q}}_p/\mathbb{Q}_p)$ with a decomposition subgroup G_p of $\mathrm{Gal}(\overline{\mathbb{Q}}/\mathbb{Q})$. We also choose a topological generator u of $\mathbb{Z}_p^\times$. Let $\Lambda = \mathbb{Z}_p[[T]]$ be the Iwasawa algebra and let $\mathfrak{m} = (p, T)$ be its maximal ideal. A special case of Hida's first main theorem ([9, Theorem I]) is the following.

Theorem 1.1 *Let* $\mathbf{f}$ *be a non-CM Hida family of ordinary cuspidal eigenforms defined over a finite extension* $\mathbb{I}$ *of* Λ *and let* $\rho_{\mathbf{f}}\colon \mathrm{Gal}(\overline{\mathbb{Q}}/\mathbb{Q}) \to \mathrm{GL}_2(\mathbb{I})$ *be the associated Galois representation. Assume that* $\rho_{\mathbf{f}}$ *is residually irreducible and that there exists an element* d *in its image with eigenvalues* $\alpha, \beta \in \mathbb{Z}_p^\times$ *such that* $\alpha^2 \not\equiv \beta^2 \pmod{p}$. *Then there exists a nonzero ideal* $\mathfrak{l} \subset \Lambda$ *and an element* $g \in \mathrm{GL}_2(\mathbb{I})$ *such that*

$$g\Gamma(\mathfrak{l})g^{-1} \subset \mathrm{Im}\,\rho_{\mathbf{f}},$$

where $\Gamma(\mathfrak{l})$ *denotes the principal congruence subgroup of* $\mathrm{SL}_2(\Lambda)$ *of level* $\mathfrak{l}$.

Under mild technical assumptions it is also shown in [9, Theorem II] that if the image of the residual representation of $\rho_{\mathbf{f}}$ contains a conjugate of $\mathrm{SL}_2(\mathbb{F}_p)$ then $\mathfrak{l}$ is trivial or $\mathfrak{m}$-primary, and if the residual representation is dihedral "of CM type" the height one prime factors P of $\mathfrak{l}$ are exactly those of the g.c.d. of the adjoint p-adic L function of $\mathbf{f}$ and the anticyclotomic specializations of Katz's p-adic L functions associated with certain Hecke characters of an imaginary quadratic field. This set of primes is precisely the set of congruence primes between the given non-CM family and the CM families.

In her Ph.D. dissertation (see [12]), J. Lang improved on Hida's Theorem I. Let $\mathbb{T}$ be Hida's big ordinary cuspidal Hecke algebra; it is finite and flat over Λ. Let $\mathrm{Spec}\,\mathbb{I}$ be an irreducible component of $\mathbb{T}$. It corresponds to a surjective Λ-algebra homomorphism $\theta\colon \mathbb{T} \to \mathbb{I}$ (a Λ-adic Hecke eigensystem). We also call θ a Hida family. Assume that it is not residually Eisenstein. It gives rise to a residually irreducible continuous Galois representation $\rho_\theta\colon G_{\mathbb{Q}} \to \mathrm{GL}_2(\mathbb{I})$ that is p-ordinary. We suppose for simplicity that $\mathbb{I}$ is normal. Consider the Λ-algebra automorphisms σ of $\mathbb{I}$ for which there exists a finite order character $\eta_\sigma\colon G_{\mathbb{Q}} \to \mathbb{I}^\times$ such that for every prime ℓ not dividing the level, $\sigma \circ \theta(T_\ell) = \eta_\sigma(\ell)\theta(T_\ell)$ (see [12, 17]). These automorphisms form a finite abelian 2-group Γ. Let $\mathbb{I}_0$ be the subring of $\mathbb{I}$ fixed by Γ. Let $H_0 = \bigcap_{\sigma\in\Gamma} \ker\, \eta_\sigma$; it is a normal open subgroup of $G_{\mathbb{Q}}$. One may assume, up to conjugation by an element of $\mathrm{GL}_2(\mathbb{I})$, that $\rho_\theta|_{H_0}$ takes values in $\mathrm{GL}_2(\mathbb{I}_0)$.

Theorem 1.2 [12, Theorem 2.4] *Let* $\theta\colon \mathbb{T} \to \mathbb{I}$ *be a non-CM Hida family such that* $\overline{\rho}_\theta$ *is absolutely irreducible. Assume that* $\overline{\rho}_\theta|_{H_0}$ *is an extension of two distinct characters. Then there exists a nonzero ideal* $\mathfrak{l} \subset \mathbb{I}_0$ *and an element* $g \in \mathrm{GL}_2(\mathbb{I})$ *such that*

$$g\Gamma(\mathfrak{l})g^{-1} \subset \mathrm{Im}\,\rho_\theta,$$

where $\Gamma(\mathfrak{l})$ *denotes the principal congruence subgroup of* $\mathrm{SL}_2(\mathbb{I}_0)$ *of level* $\mathfrak{l}$.

For all of these results it is important to assume the ordinarity of the family, as it implies the ordinarity of the Galois representation and in particular that some element of the image of inertia at p is conjugate to the matrix

$$C_T = \begin{pmatrix} u^{-1}(1+T) & * \\ 0 & 1 \end{pmatrix}.$$

Conjugation by the element above defines a Λ-module structure on the Lie algebra of a pro-p subgroup of $\mathrm{Im}\,\rho_\theta$ and this is used to produce the desired ideal $\mathfrak{l}$. Hida and Lang use Pink's theory of Lie algebras of pro-p subgroups of $\mathrm{SL}_2(\mathbb{I})$.

In this paper we propose a generalization of Hida's work to the finite slope case. We establish analogues of Hida's Theorems I and II. These are Theorems 6.2, 7.1 and 7.4 in the text. Moreover, we put ourselves in the more general setting considered in Lang's work. In the positive slope case the existence of a normalizing matrix analogous to C_T above is obtained by applying relative Sen theory ([19, 21]) to the expense of extending scalars to the completion $\mathbb{C}_p$ of an algebraic closure of $\mathbb{Q}_p$.

More precisely, for every $h \in (0, \infty)$, we define an Iwasawa algebra $\Lambda_h = \mathcal{O}_h[[t]]$ (where $t = p^{-s_h}T$ for some $s_h \in \mathbb{Q} \cap]\frac{1}{p-1}, \infty[$ and $\mathcal{O}_h$ is a finite extension of $\mathbb{Z}_p$ containing p^{s_h} such that its fraction field is Galois over $\mathbb{Q}_p$) and a finite torsion free Λ_h-algebra $\mathbb{T}_h$ (see Sect. 3.1), called an adapted slope $\leqslant h$ Hecke algebra. Let $\theta\colon \mathbb{T}_h \to \mathbb{I}^\circ$ be an irreducible component; it is finite and torsion-free over Λ_h. The notation $\mathbb{I}^\circ$ is borrowed from the theory of Tate algebras, but $\mathbb{I}^\circ$ is not a Tate or an affinoid algebra. We write $\mathbb{I} = \mathbb{I}^\circ[p^{-1}]$. We assume for simplicity that $\mathbb{I}^\circ$ is normal. The finite slope family θ gives rise to a continuous Galois representation $\rho_\theta\colon G_\mathbb{Q} \to \mathrm{GL}_2(\mathbb{I}^\circ)$. We assume that the residual representation $\overline{\rho_\theta}$ is absolutely irreducible. We introduce the finite abelian 2-group Γ as above, together with its fixed ring $\mathbb{I}_0$ and the open normal subgroup $H_0 \subset G_\mathbb{Q}$. In Sect. 5.1 we define a ring $\mathbb{B}_r$ (with an inclusion $\mathbb{I}_0 \hookrightarrow \mathbb{B}_r$) and a Lie algebra $\mathfrak{H}_r \subset \mathfrak{sl}_2(\mathbb{B}_r)$ attached to the image of ρ_θ. In the positive slope case CM families do not exist (see Sect. 3.3) hence no "non-CM" assumption is needed in the following. As before we can assume, after conjugation by an element of $\mathrm{GL}_2(\mathbb{I}^\circ)$, that $\rho_\theta(H_0) \subset \mathrm{GL}_2(\mathbb{I}_0^\circ)$. Let $P_1 \subset \Lambda_h$ be the prime $(u^{-1}(1+T)-1)$.

Theorem 1.3 (Theorem 6.2) *Let $\theta\colon \mathbb{T}_h \to \mathbb{I}^\circ$ be a positive slope family such that $\overline{\rho}_\theta|_{H_0}$ is absolutely irreducible. Assume that there exists $d \in \rho_\theta(H_0)$ with eigenvalues $\alpha, \beta \in \mathbb{Z}_p^\times$ such that $\alpha^2 \not\equiv \beta^2 \pmod{p}$. Then there exists a nonzero ideal $\mathfrak{l} \subset \mathbb{I}_0[P_1^{-1}]$ such that*

$$\mathfrak{l} \cdot \mathfrak{sl}_2(\mathbb{B}_r) \subset \mathfrak{H}_r.$$

The largest such ideal $\mathfrak{l}$ is called the Galois level of θ.

We also introduce the notion of fortuitous CM congruence ideal for θ (see Sect. 3.4). It is the ideal $\mathfrak{c} \subset \mathbb{I}$ given by the product of the primary ideals modulo which a congruence between θ and a slope $\leqslant h$ CM form occurs. Following the proof of Hida's Theorem II we are able to show (Theorem 7.1) that the set of primes of $\mathbb{I}_0 = \mathbb{I}_0^\circ[p^{-1}]$ containing $\mathfrak{l}$ coincides with the set of primes containing $\mathfrak{c} \cap \mathbb{I}_0$, except possibly for the primes of $\mathbb{I}_0$ above P_1 (the weight 1 primes).

Several generalizations of the present work are currently being studied by one of the authors.[1] They include a generalization of [10], where the authors treated the ordinary case for GSp_4 with a residual representation induced from the one associated with a Hilbert modular form, to the finite slope case and to bigger groups and more types of residual representations.

Acknowledgements. This paper owes much to Hida's recent paper [9]. We also thank Jaclyn Lang for making her dissertation [12] available to us and for some very useful remarks pertaining to Sect. 4. We thank the referee of this article for the careful reading of the manuscript and for useful suggestions which hopefully led to improvements.

2 The Eigencurve

2.1 The Weight Space

Fix a prime integer $p > 2$. We call *weight space* the rigid analytic space over $\mathbb{Q}_p$, $\mathcal{W}$, canonically associated with the formal scheme over $\mathbb{Z}_p$, $\mathrm{Spf}(\mathbb{Z}_p[[\mathbb{Z}_p^\times]])$. The $\mathbb{C}_p$-points of $\mathcal{W}$ parametrize continuous homomorphisms $\mathbb{Z}_p^\times \to \mathbb{C}_p^\times$.

Let X be a rigid analytic space defined over some finite extension $L/\mathbb{Q}_p$. We say that a subset S of $X(\mathbb{C}_p)$ is Zariski-dense if the only closed analytic subvariety Y of X satisfying $S \subset Y(\mathbb{C}_p)$ is X itself.

For every $r > 0$, we denote by $\mathcal{B}(0, r)$, respectively $\mathcal{B}(0, r^-)$, the closed, respectively open, disc in $\mathbb{C}_p$ of centre 0 and radius r. The space $\mathcal{W}$ is isomorphic to a disjoint union of $p-1$ copies of the open unit disc $\mathcal{B}(0, 1^-)$ centre in 0 and indexed by the group $\mathbb{Z}/(p-1)\mathbb{Z} = \widehat{\mu}_{p-1}$. If u denotes a topological generator of $1 + p\mathbb{Z}_p$, then an isomorphism is given by

$$\mathbb{Z}/(p-1)\mathbb{Z} \times \mathcal{B}(0, 1^-) \to \mathcal{W}, \quad (i, v) \mapsto \chi_{i,v},$$

where $\chi_{i,v}((\zeta, u^x)) = \zeta^i (1+v)^x$. Here we wrote an element of $\mathbb{Z}_p^\times$ uniquely as a pair (ζ, u^x) with $\zeta \in \mu_{p-1}$ and $x \in \mathbb{Z}_p$. We make once and for all the choice $u = 1 + p$.

We say that a point $\chi \in \mathcal{W}(\mathbb{C}_p)$ is classical if there exists $k \in \mathbb{N}$ and a finite order character $\psi\colon \mathbb{Z}_p^\times \to \mathbb{C}_p^\times$ such that χ is the character $z \mapsto z^k \psi(z)$. The set of classical points is Zariski-dense in $\mathcal{W}(\mathbb{C}_p)$.

If $\mathrm{Spm}\, R \subset \mathcal{W}$ is an affinoid open subset, we denote by $\kappa = \kappa_R\colon \mathbb{Z}_p^\times \to R^\times$ its tautological character given by $\kappa(t)(\chi) = \chi(t)$ for every $\chi \in \mathrm{Spm}\, R$. Recall ([3, Proposition 8.3]) that κ_R is r-analytic for every sufficiently small radius $r > 0$ (by which we mean that it extends to a rigid analytic function on $\mathbb{Z}_p^\times \mathcal{B}(1, r)$).

[1] A. Conti.

2.2 Adapted Pairs and the Eigencurve

Let N be a positive integer prime to p. We recall the definition of the spectral curve Z^N and of the cuspidal eigencurve C^N of tame level $\Gamma_1(N)$. These objects were constructed in [6] for $p > 2$ and $N = 1$ and in [3] in general. We follow the presentation of [3, Part II]. Let $\mathrm{Spm}\, R \subset \mathcal{W}$ be an affinoid domain and let $r = p^{-s}$ for $s \in \mathbb{Q}$ be a radius smaller than the radius of analyticity of κ_R. We denote by $M_{R,r}$ the R-module of r-overconvergent modular forms of weight κ_R. It is endowed it with a continuous action of the Hecke operators T_ℓ, $\ell \nmid Np$, and U_p. The action of U_p on $M_{R,r}$ is completely continuous, so we can consider its associated Fredholm series $F_{R,r}(T) = \det(1 - U_p T | M_{R,r}) \in R\{\{T\}\}$. These series are compatible when R and r vary, in the sense that there exists $F \in \Lambda\{\{T\}\}$ that restricts to $F_{R,r}(T)$ for every R and r.

The series $F_{R,r}(T)$ converges everywhere on the R-affine line $\mathrm{Spm}\, R \times \mathbb{A}^{1,an}$, so it defines a rigid curve $Z^N_{R,r} = \{F_{R,r}(T) = 0\}$ in $\mathrm{Spm}\, R \times \mathbb{A}^{1,an}$. When R and r vary, these curves glue into a rigid space Z^N endowed with a quasi-finite and flat morphism $w_Z \colon Z^N \to \mathcal{W}$. The curve Z^N is called the spectral curve associated with the U_p-operator. For every $h \geqslant 0$, let us consider

$$Z_R^{N,\leqslant h} = Z_R^N \cap \left(\mathrm{Spm}\, R \times B(0, p^h)\right).$$

By [3, Lemma 4.1] $Z_R^{N,\leqslant h}$ is quasi-finite and flat over $\mathrm{Spm}\, R$.

We now recall how to construct an admissible covering of Z^N.

Definition 2.1 We denote by C the set of affinoid domains $Y \subset Z$ such that:

- there exists an affinoid domain $\mathrm{Spm}\, R \subset \mathcal{W}$ such that Y is a union of connected components of $w_Z^{-1}(\mathrm{Spm}\, R)$;
- the map $w_Z|_Y \colon Y \to \mathrm{Spm}\, R$ is finite.

Proposition 2.2 [3, Theorem 4.6] *The covering C is admissible.*

Note in particular that an element $Y \in C$ must be contained in $Z_R^{N,\leqslant h}$ for some h.

For every R and r as above and every $Y \in C$ such that $w_Z(Y) = \mathrm{Spm}\, R$, we can associate with Y a direct factor M_Y of $M_{R,r}$ by the construction in [3, Sect. I.5]. The abstract Hecke algebra $\mathcal{H} = \mathbb{Z}[T_\ell]_{\ell \nmid Np}$ acts on $M_{R,r}$ and M_Y is stable with respect to this action. Let $\mathbb{T}_Y$ be the R-algebra generated by the image of $\mathcal{H}$ in $\mathrm{End}_R(M_Y)$ and let $C_Y^N = \mathrm{Spm}\, \mathbb{T}_Y$. Note that it is reduced as all Hecke operators are self-adjoint for a certain pairing and mutually commute.

For every Y the finite covering $C_Y^N \to \mathrm{Spm}\, R$ factors through $Y \to \mathrm{Spm}\, R$. The eigencurve C^N is defined by gluing the affinoids C_Y^N into a rigid curve, endowed with a finite morphism $C^N \to Z^N$. The curve C^N is reduced and flat over $\mathcal{W}$ since it is so locally.

We borrow the following terminology from Bellaïche.

Definition 2.3 [1, Definition II.1.8] Let Spm $R \subset \mathcal{W}$ be an affinoid open subset and $h > 0$ be a rational number. The couple (R, h) is called adapted if $Z_R^{N,\leqslant h}$ is an element of C.

By [1, Corollary II.1.13] the sets of the form $Z_R^{N,\leqslant h}$ are sufficient to admissibly cover the spectral curve.

Now we fix a finite slope h. We want to work with families of slope $\leqslant h$ which are finite over a wide open subset of the weight space. In order to do this it will be useful to know which pairs (R, h) in a connected component of $\mathcal{W}$ are adapted. If Spm $R' \subset$ Spm R are affinoid subdomains of $\mathcal{W}$ and (R, h) is adapted then (R', h) is also adapted by [1, Proposition II.1.10]. By [3, Lemma 4.3], the affinoid Spm R is adapted to h if and only if the weight map $Z_R^{N,\leqslant h} \to$ Spm R has fibres of constant degree.

Remark 2.4 Given a slope h and a classical weight k, it would be interesting to have a lower bound for the radius of a disc of centre k adapted to h. A result of Wan ([24, Theorem 2.5]) asserts that for a certain radius r_h depending only on h, N and p, the degree of the fibres of $Z_{\mathcal{B}(k,r_h)}^{N,\leqslant h} \to$ Spm $\mathcal{B}(k, r_h)$ at classical weights is constant. Unfortunately we do not know whether the degree is constant at all weights of $\mathcal{B}(k, r_h)$, so this is not sufficient to answer our question. Estimates for the radii of adapted discs exist in the case of eigenvarieties for groups different than GL_2; see for example the results of Chenevier on definite unitary groups ([4, Sect. 5]).

2.3 Pseudo-characters and Galois Representations

Let K be a finite extension of $\mathbb{Q}_p$ with valuation ring O_K. Let X be a rigid analytic variety defined over K. We denote by $O(X)$ the ring of global analytic functions on X equipped with the coarsest locally convex topology making the restriction map $O(X) \to O(U)$ continuous for every affinoid $U \subset X$. It is a Fréchet space isomorphic to the inverse limit over all affinoid domains U of the K-Banach spaces $O(U)$. We denote by $O(X)^\circ$ the O_K-algebra of functions bounded by 1 on X, equipped with the topology induced by that on $O(X)$. The question of the compactness of this ring is related to the following property of X.

Definition 2.5 [2, Definition 7.2.10] We say that a rigid analytic variety X defined over K is nested if there is an admissible covering $X = \bigcup X_i$ by open affinoids X_i defined over K such that the maps $O(X_{i+1}) \to O(X_i)$ induced by the inclusions are compact.

We equip the ring $O(X)^\circ$ with the topology induced by that on $O(X) = \varprojlim_i O(X_i)$.

Lemma 2.6 *[2, Lemma 7.2.11(ii)] If X is reduced and nested, then $O(X)^\circ$ is a compact (hence profinite) O_K-algebra.*

We will be able to apply Lemma 2.6 to the eigenvariety thanks to the following.

Proposition 2.7 [2, Corollary 7.2.12] *The eigenvariety C^N is nested for $K = \mathbb{Q}_p$.*

Given a reduced nested subvariety X of C^N defined over a finite extension K of $\mathbb{Q}_p$ there is a pseudo-character on X obtained by interpolating the classical ones. Let $\mathbb{Q}^{N_p}$ be the maixmal extension of $\mathbb{Q}$ uniamified outside N_p and let $G_{\mathbb{Q}}$, $N_p = Gal(\mathbb{Q}^{N_p}/\mathbb{Q})$.

Proposition 2.8 [1, Theorem IV.4.1] *There exists a unique pseudo-character*

$$\tau\colon G_{\mathbb{Q},Np} \to \mathcal{O}(X)^\circ$$

of dimension 2 such that for every ℓ prime to Np, $\tau(\mathrm{Frob}_\ell) = \psi_X(T_\ell)$, where ψ_X is the composition of $\psi\colon \mathcal{H} \to \mathcal{O}(C^N)^\circ$ with the restriction map $\mathcal{O}(C^N)^\circ \to \mathcal{O}(X)^\circ$.

Remark 2.9 One can take as an example of X a union of irreducible components of C^N in which case $K = \mathbb{Q}_p$. Later we will consider other examples where $K \neq \mathbb{Q}_p$.

3 The Fortuitous Congruence Ideal

In this section we will define families with slope bounded by a finite constant and coefficients in a suitable profinite ring. We will show that any such family admits at most a finite number of classical specializations which are CM modular forms. Later we will define what it means for a point (not necessarily classical) to be CM and we will associate with a family a congruence ideal describing its CM points. Contrary to the ordinary case, the non-ordinary CM points do not come in families so the points detected by the congruence ideal do not correspond to a crossing between a CM and a non-CM family. For this reason we call our ideal the "fortuitous congruence ideal".

3.1 The Adapted Slope $\leqslant h$ Hecke Algebra

Throughout this section we fix a slope $h > 0$. Let $C^{N,\leqslant h}$ be the subvariety of C^N whose points have slope $\leqslant h$. Unlike the ordinary case treated in [9] the weight map $w^{\leqslant h}\colon C^{N,\leqslant h} \to \mathcal{W}$ is not finite which means that a family of slope $\leqslant h$ is not in general defined by a finite map over the entire weight space. The best we can do in the finite slope situation is to place ourselves over the largest possible wide open subdomain U of $\mathcal{W}$ such that the restricted weight map $w^{\leqslant h}|_U\colon C^{N,\leqslant h} \times_{\mathcal{W}} U \to U$ is finite. This is a domain "adapted to h" in the sense of Definition 2.3 where only affinoid domains were considered. The finiteness property will be necessary in order to apply going-up and going-down theorems.

Let us fix a rational number s_h such that for $r_h = p^{-s_h}$ the closed disc $B(0, r_h)$ is adapted for h. We assume that $s_h > \frac{1}{p-1}$ (this will be needed later to assure the convergence of the exponential map). Let $\eta_h \in \overline{\mathbb{Q}}_p$ be an element of p-adic valuation s_h. Let K_h be the Galois closure (in $\mathbb{C}_p$) of $\mathbb{Q}_p(\eta_h)$ and let $\mathcal{O}_h$ be its valuation ring. Recall that T is the variable on the open disc of radius 1. Let $t = \eta_h^{-1}T$ and $\Lambda_h = \mathcal{O}_h[[t]]$. This is the ring of analytic functions, with $\mathcal{O}_h$-coefficients and bounded by one, on the wide open disc $\mathcal{B}_h$ of radius p^{-s_h}. There is a natural map $\Lambda \to \Lambda_h$ corresponding to the restriction of analytic functions on the open disc of radius 1, with $\mathbb{Z}_p$ coefficients and bounded by 1, to the open disc of radius r_h. The image of this map is the ring $\mathbb{Z}_p[[\eta t]] \subset \mathcal{O}_h[[t]]$.

For $i \geqslant 1$, let $s_i = s_h + 1/i$ and $\mathcal{B}_i = \mathcal{B}(0, p^{-s_i})$. The open disc $\mathcal{B}_h$ is the increasing union of the affinoid discs $\mathcal{B}_i$. For each i a model for $\mathcal{B}_i$ over K_h is given by Berthelot's construction of $\mathcal{B}_h$ as the rigid space associated with the $\mathcal{O}_h$-formal scheme Spf Λ_h. We recall it briefly following [7, Sect. 7]. Let

$$A_{r_i}^\circ = \mathcal{O}_h\langle t, X_i\rangle/(pX_i - t^i).$$

We have $\mathcal{B}_i = \mathrm{Spm}\, A_{r_i}^\circ[p^{-1}]$ as rigid space over K_h. For every i we have a morphism $A_{r_{i+1}}^\circ \to A_{r_i}^\circ$ given by

$$X_{i+1} \mapsto X_i t$$

$$t \mapsto t$$

We have induced compact morphisms $A_{r_{i+1}}^\circ[p^{-1}] \to A_{r_i}^\circ[p^{-1}]$, hence open immersions $\mathcal{B}_i \to \mathcal{B}_{i+1}$ defined over K_h. The wide open disc $\mathcal{B}_h$ is defined as the inductive limit of the affinoids $\mathcal{B}_i$ with these transition maps. We have $\Lambda_h = \varprojlim_i A_{r_i}^\circ$.

Since the s_i are strictly bigger than s_h for each i, $\mathcal{B}(0, p^{-s_i}) = \mathrm{Spm}\, A_{r_i}^\circ[p^{-1}]$ is adapted to h. Therefore for every $r > 0$ sufficiently small and for every $i \geqslant 1$ the image of the abstract Hecke algebra acting on $M_{A_{r_i},r}$ provides a finite affinoid $A_{r_i}^\circ$-algebra $\mathbb{T}_{A_{r_i}^\circ,r}^{\leqslant h}$. The morphism $w_{A_{r_i}^\circ,r}\colon \mathrm{Spm}\,\mathbb{T}_{A_{r_i}^\circ,r}^{\leqslant h} \to \mathrm{Spm}\, A_{r_i}^\circ$ is finite. For $i < j$ we have natural open immersions $\mathrm{Spm}\,\mathbb{T}_{A_{r_j}^\circ,r}^{\leqslant h} \to \mathrm{Spm}\,\mathbb{T}_{A_{r_i}^\circ,r}^{\leqslant h}$ and corresponding restriction maps $\mathbb{T}_{A_{r_i}^\circ,r}^{\leqslant h} \to \mathbb{T}_{A_{r_j}^\circ,r}^{\leqslant h}$. We call C_h the increasing union $\bigcup_{i\in\mathbb{N},r>0} \mathrm{Spm}\,\mathbb{T}_{A_{r_i}^\circ,r}^{\leqslant h}$; it is a wide open subvariety of C^N. We denote by $\mathbb{T}_h$ the ring of rigid analytic functions bounded by 1 on C_h. We have $\mathbb{T}_h = \mathcal{O}(C_h)^\circ = \varprojlim_{i,r} \mathbb{T}_{A_{r_i}^\circ,r}^{\leqslant h}$. There is a natural weight map $w_h\colon C_h \to \mathcal{B}_h$ that restricts to the maps $w_{A_{r_i}^\circ,r}$. It is finite because the closed ball of radius r_h is adapted to h.

3.2 The Galois Representation Associated with a Family of Finite Slope

Since $\mathcal{O}(B_h)^\circ = \Lambda_h$, the map w_h gives $\mathbb{T}_h$ the structure of a finite Λ_h-algebra; in particular $\mathbb{T}_h$ is profinite.

Let $\mathfrak{m}$ be a maximal ideal of $\mathbb{T}_h$. The residue field $k = \mathbb{T}_h/\mathfrak{m}$ is finite. Let $\mathbb{T}_\mathfrak{m}$ denote the localization of $\mathbb{T}_h$ at $\mathfrak{m}$. Since Λ_h is henselian, $\mathbb{T}_\mathfrak{m}$ is a direct factor of

$\mathbb{T}_h$, hence it is finite over Λ_h; it is also local noetherian and profinite. It is the ring of functions bounded by 1 on a connected component of C_h. Let $W = W(k)$ be the ring of Witt vectors of k. By the universal property of W, $\mathbb{T}_\mathfrak{m}$ is a W-algebra. The affinoid domain Spm $\mathbb{T}_\mathfrak{m}$ contains a zarisiki-dense set of points x corresponding to cuspidal eigenforms f_x of weight $w(x) = k_x \geqslant 2$ and level Np. The Galois representations ρ_{f_x} associated with the f_x give rise to a residual representation $\overline{\rho}\colon G_{\mathbb{Q},Np} \to \mathrm{GL}_2(k)$ that is independent of f_x. By Proposition 2.8, we have a pseudo-character

$$\tau_{\mathbb{T}_\mathfrak{m}}: G_{\mathbb{Q},Np} \to \mathbb{T}_\mathfrak{m}$$

such that for every classical point $x\colon \mathbb{T}_\mathfrak{m} \to L$, defined over some finite extension $L/\mathbb{Q}_p$, the specialization of $\tau_{\mathbb{T}_\mathfrak{m}}$ at x is the trace of Lf_x.

Proposition 3.1 *If $\overline{\rho}$ is absolutely irreducible there exists a unique continuous irreducible Galois representation*

$$\rho_{\mathbb{T}_\mathfrak{m}}: G_{\mathbb{Q},Np} \to \mathrm{GL}_2(\mathbb{T}_\mathfrak{m}),$$

lifting $\overline{\rho}$ and whose trace is $\tau_{\mathbb{T}_\mathfrak{m}}$.

This follows from a result of Nyssen and Rouquier ([14], [18, Corollary 5.2]), since $\mathbb{T}_\mathfrak{m}$ is local henselian.

Let $\mathbb{I}^\circ$ be a finite torsion-free Λ_h-algebra. We call *family* an irreducible component of Spec $\mathbb{T}_h$ defined by a surjective morphism $\theta\colon \mathbb{T}_h \to \mathbb{I}^\circ$ of Λ_h-algebras. Since such a map factors via $\mathbb{T}_\mathfrak{m} \to \mathbb{I}^\circ$ for some maximal ideal $\mathfrak{m}$ of $\mathbb{T}_h$, we can define a residual representation $\overline{\rho}$ associated with θ. Suppose that $\overline{\rho}$ is irreducible. By Proposition 3.1 we obtain a Galois representation $\rho\colon G_\mathbb{Q} \to \mathrm{GL}_2(\mathbb{I}^\circ)$ associated with θ.

Remark 3.2 If $\eta_h \notin \mathbb{Q}_p$, Λ_h is not a power series ring over $\mathbb{Z}_p$.

3.3 *Finite Slope CM Modular Forms*

In this section we study non-ordinary finite slope CM modular forms. We say that a family is CM if all its classical points are CM. We prove that for every $h > 0$ there are no CM families with positive slope $\leqslant h$. However, contrary to the ordinary case, every family of finite positive slope may contain classical CM points of weight $k \geqslant 2$. Let F be an imaginary quadratic field, $\mathfrak{f}$ an integral ideal in F, $I_\mathfrak{f}$ the group of fractional ideals prime to $\mathfrak{f}$. Let σ_1, σ_2 be the embeddings of F into $\mathbb{C}$ (say that $\sigma_1 = \mathrm{Id}_F$) and let $(k_1, k_2) \in \mathbb{Z}^2$. A Grössencharacter ψ of infinity type (k_1, k_2) defined modulo $\mathfrak{f}$

is a homomorphism $\psi : I_{\mathfrak{f}} \to \mathbb{C}^*$ such that $\psi((\alpha)) = \sigma_1(\alpha)^{k_1}\sigma_2(\alpha)^{k_2}$ for all $\alpha \equiv 1 \pmod{^\times \mathfrak{f}}$. Consider the q-expansion

$$\sum_{\mathfrak{a} \subset O_F, (\mathfrak{a}, \mathfrak{f})=1} \psi(\mathfrak{a}) q^{N(\mathfrak{a})},$$

where the sum is over ideals $\mathfrak{a} \subset O_F$ and $N(\mathfrak{a})$ denotes the norm of $\mathfrak{a}$. Let $F/\mathbb{Q}$ be an imaginary quadratic field of discriminant D and let ψ be a Grössencharacter of exact conductor $\mathfrak{f}$ and infinity type $(k-1, 0)$. By [22, Lemma 3] the expansion displayed above defines a cuspidal newform $f(F, \psi)$ of level $N(\mathfrak{f})D$.

Ribet proved in [16, Theorem 4.5] that if a newform g of weight $k \geqslant 2$ and level N has CM by an imaginary quadratic field F, one has $g = f(F, \psi)$ for some Grössencharacter ψ of F of infinity type $(k-1, 0)$.

Definition 3.3 We say that a classical modular eigenform g of weight k and level Np has CM by an imaginary quadratic field F if its Hecke eigenvalues for the operators T_ℓ, $\ell \nmid Np$, coincide with those of $f(F, \psi)$ for some Grössencharacter ψ of F of infinity type $(k-1, 0)$. We also say that g is CM without specifying the field.

Remark 3.4 For g as in the definition the Galois representations $\rho_g, \rho_{f(F,\psi)} : G_{\mathbb{Q}} \to \mathrm{GL}_2(\overline{\mathbb{Q}}_p)$ associated with g and $f(F, \psi)$ are isomorphic, hence the image of the representation ρ_g is contained in the normalizer of a torus in GL_2.

Proposition 3.5 *Let g be a CM modular eigenform of weight k and level Np^m with N prime to p and $m \geqslant 0$. Then its p-slope is either* 0, $\frac{k-1}{2}$, $k-1$ *or infinite.*

Proof Let F be the quadratic imaginary field and ψ the Grössencharacter of F associated with the CM form g by Definition 3.3. Let $\mathfrak{f}$ be the conductor of ψ.

We assume first that g is p-new, so that $g = f(F, \psi)$. Let a_p be the U_p-eigenvalue of g. If p is inert in F we have $a_p = 0$, so the p-slope of g is infinite. If p splits in F as $\mathfrak{p}\bar{\mathfrak{p}}$, then $a_p = \psi(\mathfrak{p}) + \psi(\bar{\mathfrak{p}})$. We can find an integer n such that $\mathfrak{p}^n$ is a principal ideal (α) with $\alpha \equiv 1 \pmod{^\times \mathfrak{f}}$. Hence $\psi((\alpha)) = \alpha^{k-1}$. Since α is a generator of $\mathfrak{p}^n$ we have $\alpha \in \mathfrak{p}$ and $\alpha \notin \bar{\mathfrak{p}}$; moreover $\alpha^{k-1} = \psi((\alpha)) = \psi(\mathfrak{p})^n$, so we also have $\psi(\mathfrak{p}) \in \mathfrak{p} - \bar{\mathfrak{p}}$. In the same way we find $\psi(\bar{\mathfrak{p}}) \in \bar{\mathfrak{p}} - \mathfrak{p}$. We conclude that $\psi(\mathfrak{p}) + \psi(\bar{\mathfrak{p}})$ does not belong to $\mathfrak{p}$, so its p-adic valuation is 0.

If p ramifies as $\mathfrak{p}^2$ in F, then $a_p = \psi(\mathfrak{p})$. As before we find n such that $\mathfrak{p}^n = (\alpha)$ with $\alpha \equiv 1 \pmod{^\times \mathfrak{f}}$. Then $(\psi(\mathfrak{p}))^n \psi(\mathfrak{p}^n) = \psi((\alpha)) = \alpha^{k-1} = \mathfrak{p}^{n(k-1)}$. By looking at p-adic valuations we find that the slope is $\frac{k-1}{2}$.

If g is not p-new, it is the p-stabilization of a CM form $f(F, \psi)$ of level prime to p. If a_p is the T_p-eigenvalue of $f(F, \psi)$, the U_p-eigenvalue of g is a root of the Hecke polynomial $X^2 - a_p X + \zeta p^{k-1}$ for some root of unity ζ. By our discussion of the p-new case, the valuation of a_p belongs to the set $\{0, \frac{k-1}{2}, k-1\}$. Then it is easy to see that the valuations of the roots of the Hecke polynomial belong to the same set. □

We state a useful corollary.

Corollary 3.6 *There are no CM families of strictly positive slope.*

Proof We show that the eigencurve C_h contains only a finite number of points corresponding to classical CM forms. It will follow that almost all classical points of a family in C_h are non-CM. Let f be a classical CM form of weight k and positive slope. By Proposition 3.5 its slope is at least $\frac{k-1}{2}$. If f corresponds to a point of C_h its slope must be $\leqslant h$, so we obtain an inequality $\frac{k-1}{2} \leqslant h$. The set of weights $\mathcal{K}$ satisfying this condition is finite. Since the weight map $C_h \to B_h$ is finite, the set of points of C_h whose weight lies in $\mathcal{K}$ is finite. Hence the number of CM forms in C_h is also finite. □

We conclude that, in the finite positive slope case, classical CM forms can appear only as isolated points in an irreducible component of the eigencurve C_h. In the ordinary case, the congruence ideal of a non-CM irreducible component is defined as the intersection ideal of the CM irreducible components with the given non-CM component. In the case of a positive slope family $\theta\colon \mathbb{T}_h \to \mathbb{I}^\circ$, we need to define the congruence ideal in a different way.

3.4 Construction of the Congruence Ideal

Let $\theta\colon \mathbb{T}_h \to \mathbb{I}^\circ$ be a family. We write $\mathbb{I} = \mathbb{I}^\circ[p^{-1}]$.

Fix an imaginary quadratic field F where p is inert or ramified; let $-D$ be its discriminant. Let $\mathfrak{Q}$ be a primary ideal of $\mathbb{I}$; then $\mathfrak{q} = \mathfrak{Q} \cap \Lambda_h$ is a primary ideal of Λ_h. The projection $\Lambda_h \to \Lambda_h/\mathfrak{q}$ defines a point of $\mathcal{B}_h$ (possibly non-reduced) corresponding to a weight $\kappa_{\mathfrak{Q}}\colon \mathbb{Z}_p^* \to (\Lambda_h/\mathfrak{q})^*$. For $r > 0$ we denote by $\mathcal{B}_r$ the ball of centre 1 and radius r in $\mathbb{C}_p$. By [3, Proposition 8.3] there exists $r > 0$ and a character $\kappa_{\mathfrak{Q},r}\colon \mathbb{Z}_p^\times \cdot \mathcal{B}_r \to (\Lambda_h/\mathfrak{q})^\times$ extending $\kappa_{\mathfrak{Q}}$.

Let σ be an embedding $F \hookrightarrow \mathbb{C}_p$. Let r and $\kappa_{\mathfrak{Q},r}$ be as above. For m sufficiently large $\sigma(1 + p^m O_F)$ is contained in $\mathbb{Z}_p^\times \cdot \mathcal{B}_r$, the domain of definition of $\kappa_{\mathfrak{Q},r}$.

For an ideal $\mathfrak{f} \subset O_F$ let $I_\mathfrak{f}$ be the group of fractional ideals prime to $\mathfrak{f}$. For every prime ℓ not dividing Np we denote by $a_{\ell,\mathfrak{Q}}$ the image of the Hecke operator T_ℓ in $\mathbb{I}^\circ/\mathfrak{Q}$. We define here a notion of non-classical CM point of θ (hence of the eigencurve C_h) as follows.

Definition 3.7 Let $F, \sigma, \mathfrak{Q}, r, \kappa_{\mathfrak{Q},r}$ be as above. We say that $\mathfrak{Q}$ defines a CM point of weight $\kappa_{\mathfrak{Q},r}$ if there exist an integer $m > 0$, an ideal $\mathfrak{f} \subset O_F$ with norm $N(\mathfrak{f})$ such that $DN(\mathfrak{f})$ divides N, a quadratic extension $(\mathbb{I}/\mathfrak{Q})'$ of $\mathbb{I}/\mathfrak{Q}$ and a homomorphism $\psi\colon I_{\mathfrak{f}p^m} \to (\mathbb{I}/\mathfrak{Q})'^\times$ such that:

1. $\sigma(1 + p^m O_F) \subset \mathbb{Z}_p^\times \cdot \mathcal{B}_r$;
2. for every $\alpha \in O_F$ with $\alpha \equiv 1 \,(\mathrm{mod}^\times \mathfrak{f}p^m)$, $\psi((\alpha)) = \kappa_{\mathfrak{Q},r}(\alpha)\alpha^{-1}$;
3. $a_{\ell,\mathfrak{Q}} = 0$ if L is a prime inert in F and not dividing Np;
4. $a_{\ell,\mathfrak{Q}} = \psi(\mathfrak{l}) + \psi(\bar{\mathfrak{l}})$ if ℓ is a prime splitting as $\mathfrak{l}\bar{\mathfrak{l}}$ in F and not dividing Np.

Note that $\kappa_{\mathfrak{Q},r}(\alpha)$ is well defined thanks to condition 1.

Remark 3.8 If $\mathfrak{P}$ is a prime of $\mathbb{I}$ corresponding to a classical form f then $\mathfrak{P}$ is a CM point if and only if f is a CM form in the sense of Sect. 3.3.

Proposition 3.9 *The set of CM points in* $\operatorname{Spec}\mathbb{I}$ *is finite.*

Proof By contradiction assume it is infinite. Then we have an injection $\mathbb{I} \hookrightarrow \prod_{\mathfrak{P}} \mathbb{I}/\mathfrak{P}$ where $\mathfrak{P}$ runs over the set of CM prime ideals of $\mathbb{I}$. One can assume that the imaginary quadratic field of complex multiplication is constant along $\mathbb{I}$. We can also assume that the ramification of the associated Galois characters $\lambda_{\mathfrak{P}}\colon G_F \to (\mathbb{I}/\mathfrak{P})^\times$ is bounded (in support and in exponents). On the density one set of primes of F prime to $\mathfrak{f}p$ and of degree one, they take values in the image of $\mathbb{I}^\times$ hence they define a continuous Galois character $\lambda\colon G_F \to \mathbb{I}^\times$ such that $\rho_\theta = \operatorname{Ind}_{G_F}^{G_\mathbb{Q}}\lambda$, which is absurd (by Corallary 3.6 and specialization at non-CM classical points which do exist). □

Definition 3.10 The (fortuitous) congruence ideal $\mathfrak{c}_\theta$ associated with the family θ is defined as the intersection of all the primary ideals of $\mathbb{I}$ corresponding to CM points.

Remark 3.11 (Characterizations of the CM locus)

1. Assume that $\overline{\rho}_\theta = \operatorname{Ind}_{G_K}^{G_\mathbb{Q}}\overline{\lambda}$ for a unique imaginary quadratic field K. Then the closed subscheme $V(\mathfrak{c}_\theta) = \operatorname{Spec}\mathbb{I}/\mathfrak{c}_\theta \subset \operatorname{Spec}\mathbb{I}$ is the largest subscheme on which there is an isomorphism of Galois representations $\rho_\theta \cong \rho_\theta \otimes \left(\frac{K/\mathbb{Q}}{\bullet}\right)$. Indeed, for every artinian $\mathbb{Q}_p$-algebra A, a CM point $x\colon \mathbb{I} \to A$ is characterized by the conditions $x(T_\ell) = x(T_\ell)\left(\frac{K/\mathbb{Q}}{\ell}\right)$ for all primes ℓ not dividing Np.
2. Note that N is divisible by the discriminant D of K. Assume that $\mathbb{I}$ is N-new and that D is prime to N/D. Let W_D be the Atkin-Lehner involution associated with D. Conjugation by W_D defines an automorphism ι_D of $\mathbb{T}_h$ and of $\mathbb{I}$. Then $V(\mathfrak{c}_\theta)$ coincides with the (schematic) invariant locus $(\operatorname{Spec}\mathbb{I})^{\iota_D=1}$.

4 The Image of the Representation Associated with a Finite Slope Family

It is shown by Lang in [12, Theorem 2.4] that, under some technical hypotheses, the image of the Galois representation $\rho\colon G_\mathbb{Q} \to \mathrm{GL}_2(\mathbb{I}^\circ)$ associated with a non-CM ordinary family $\theta\colon \mathbb{T} \to \mathbb{I}^\circ$ contains a congruence subgroup of $\mathrm{SL}_2(\mathbb{I}_0^\circ)$, where $\mathbb{I}_0^\circ$ is the subring of $\mathbb{I}^\circ$ fixed by certain "symmetries" of the representation ρ. In order to study the Galois representation associated with a non-ordinary family we will adapt some of the results in [12] to this situation. Since the crucial step ([12, Theorem 4.3]) requires the Galois ordinarity of the representation (as in [9, Lemma 2.9]), the results of this section will not imply the existence of a congruence subgroup of $\mathrm{SL}_2(\mathbb{I}_0^\circ)$ contained in the image of ρ. However, we will prove in later sections the existence of a "congruence Lie subalgebra" of $\mathfrak{sl}_2(\mathbb{I}_0^\circ)$ contained in a suitably defined Lie algebra of the image of ρ by means of relative Sen theory.

For every ring R we denote by $Q(R)$ its total ring of fractions.

4.1 The Group of Self-twists of a Family

We follow [12, Sect. 2] in this subsection. Let $h \geqslant 0$ and $\theta\colon \mathbb{T}_h \to \mathbb{I}^\circ$ be a family of slope $\leqslant h$ defined over a finite torsion free Λ_h-algebra $\mathbb{I}^\circ$. Recall that there is a natural map $\Lambda \to \Lambda_h$ with image $\mathbb{Z}_p[[\eta t]]$.

Definition 4.1 We say that $\sigma \in \mathrm{Aut}_{Q(\mathbb{Z}_p[[\eta t]])}(Q(\mathbb{I}^\circ))$ is a conjugate self-twist for θ if there exists a Dirichlet character $\eta_\sigma\colon G_\mathbb{Q} \to \mathbb{I}^{\circ,\times}$ such that

$$\sigma(\theta(T_\ell)) = \eta_\sigma(\ell)\theta(T_\ell)$$

for all but finitely many primes ℓ.

Any such σ acts on $\Lambda_h = O_h[[t]]$ by restriction, trivially on t and by a Galois automorphism on O_h. The conjugates self-twists for θ form a subgroup of $\mathrm{Aut}_{Q(\mathbb{Z}_p[[\eta t]])}(Q(\mathbb{I}^\circ))$. We recall the following result which holds without assuming the ordinarity of θ.

Lemma 4.2 [12, Lemma 7.1] *Γ is a finite abelian 2-group.*

We suppose from now on that $\mathbb{I}^\circ$ is normal. The only reason for this hypothesis is that in this case $\mathbb{I}^\circ$ is stable under the action of Γ on $Q(\mathbb{I}^\circ)$, which is not true in general. This makes it possible to define the subring $\mathbb{I}_0^\circ$ of elements of $\mathbb{I}^\circ$ fixed by Γ.

Remark 4.3 The hypothesis of normality of $\mathbb{I}^\circ$ is just a simplifying one. We could work without it by introducing the Λ_h-order $\mathbb{I}^{\circ,\prime} = \Lambda_h[\theta(T_\ell), \ell \nmid Np] \subset \mathbb{I}^\circ$: this is an analogue of the Λ-order $\mathbb{I}'$ defined in [12, Sect. 2] and it is stable under the action of Γ. We would define $\mathbb{I}_0^\circ$ as the fixed subring of $\mathbb{I}^{\circ,\prime}$ and the arguments in the rest of the article could be adapted to this setting.

The subring of Λ_h fixed by Γ is an $O_{h,0}$ form of Λ_h for some subring $O_{h,0}$ of O_h. We denote it by $\Lambda_{h,0}$ the field of fractions of $O_{h,0}$.

Remark 4.4 By definition Γ fixes $\mathbb{Z}_p[[\eta t]]$, so we have $\mathbb{Z}_p[[\eta t]] \subset \Lambda_{h,0}$. In particular it makes sense to speak about the ideal $P_k\Lambda_{h,0}$ for every arithmetic prime $P_k = (1+\eta t - u^k) \subset \mathbb{Z}_p[[\eta t]]$. Note that $P_k\Lambda_h$ defines a prime ideal of Λ_h if and only if the weight k belongs to the open disc B_h, otherwise $P_k\Lambda_h = \Lambda_h$. We see immediately that the same statement is true if we replace Λ_h by $\Lambda_{h,0}$.

Note that $\mathbb{I}_0^\circ$ is a finite extension of $\Lambda_{h,0}$ because $\mathbb{I}^\circ$ is a finite Λ_h-algebra. Moreover, we have $K_h^\Gamma = K_{h,0}$ (although the inclusion $\Lambda_h \cdot \mathbb{I}_0^\circ \subset \mathbb{I}^\circ$ may not be an equality).

We define two open normal subgroups of $G_\mathbb{Q}$ by:

- $H_0 = \bigcap_{\sigma \in \Gamma} \ker \eta_\sigma$;
- $H = H_0 \cap \ker(\det \overline{\rho})$.

Note that H_0 is an open normal subgroup of $G_\mathbb{Q}$ and that H is a n open normal subgroup of H_0 and $G_\mathbb{Q}$.

4.2 The Level of a General Ordinary Family

We recall the main result of [12]. Denote by $\mathbb{T}$ the big ordinary Hecke algebra, which is finite over $\Lambda = \mathbb{Z}_p[[T]]$. Let $\theta\colon \mathbb{T} \to \mathbb{I}^\circ$ be an ordinary family with associated Galois representation $\rho\colon G_\mathbb{Q} \to \mathrm{GL}_2(\mathbb{I}^\circ)$. The representation ρ is p-ordinary, which means that its restriction $\rho|_{D_p}$ to a decomposition subgroup $D_p \subset G_\mathbb{Q}$ is reducible. There exist two characters $\varepsilon, \delta\colon D_p \to \mathbb{I}^{\circ,\times}$, with δ unramified, such that $\rho|_{D_p}$ is an extension of ε by δ.

Denote by $\mathbb{F}$ the residue field of $\mathbb{I}^\circ$ and by $\overline{\rho}$ the representation $G_\mathbb{Q} \to \mathrm{GL}_2(\mathbb{F})$ obtained by reducing ρ modulo the maximal ideal of $\mathbb{I}^\circ$. Lang introduces the following technical condition.

Definition 4.5 The p-ordinary representation $\overline{\rho}$ is called H_0-regular if $\overline{\varepsilon}|_{D_p \cap H_0} \neq \overline{\delta}|_{D_p \cap H_0}$.

The following result states the existence of a Galois level for ρ.

Theorem 4.6 [12, Theorem 2.4] *Let $\rho\colon G_\mathbb{Q} \to \mathrm{GL}_2(\mathbb{I}^\circ)$ be the representation associated with an ordinary, non-CM family $\theta\colon \mathbb{T} \to \mathbb{I}^\circ$. Assume that $p > 2$, the cardinality of $\mathbb{F}$ is not 3 and the residual representation $\overline{\rho}$ is absolutely irreducible and H_0-regular. Then there exists $\gamma \in \mathrm{GL}_2(\mathbb{I}^\circ)$ such that $\gamma \cdot \mathrm{Im}\,\rho \cdot \gamma^{-1}$ contains a congruence subgroup of* $\mathrm{SL}_2(\mathbb{I}_0^\circ)$.

The proof relies on the analogous result proved by Ribet [15] and Momose [13] for the p-adic representation associated with a classical modular form.

4.3 An Approximation Lemma

In this subsection we prove an analogue of [10, Lemma 4.5]. It replaces in our approach the use of Pink's Lie algebra theory, which is relied upon in the case of ordinary representations in [9, 12]. Let $\mathbb{I}_0^\circ$ be a local domain that is finite torsion free over Λ_h. It does not need to be related to a Hecke algebra for the moment.

Let N be an open normal subgroup of $G_\mathbb{Q}$ and let $\rho\colon N \to \mathrm{GL}_2(\mathbb{I}_0^\circ)$ be an arbitrary continuous representation. We denote by $\mathfrak{m}_{\mathbb{I}_0^\circ}$ the maximal ideal of $\mathbb{I}_0^\circ$ and by $\mathbb{F} = \mathbb{I}_0^\circ/\mathfrak{m}_{\mathbb{I}_0^\circ}$ its residue field of cardinality q. In the lemma we do not suppose that ρ comes from a family of modular forms. We will only assume that it satisfies the following technical condition:

Definition 4.7 Keep notations as above. We say that the representation $\rho\colon N \to \mathrm{GL}_2(\mathbb{I}_0^\circ)$ is $\mathbb{Z}_p$-regular if there exists $d \in \mathrm{Im}\,\rho$ with eigenvalues $d_1, d_2 \in \mathbb{Z}_p$ such that $d_1^2 \not\equiv d_2^2 \pmod{p}$. We call d a $\mathbb{Z}_p$-regular element. If N' is an open normal subgroup of N then we say that ρ is $(N', \mathbb{Z}_p)$-regular if $\rho|_{N'}$ is $\mathbb{Z}_p$-regular.

Let $B^\pm$ denote the Borel subgroups consisting of upper, respectively lower, triangular matrices in GL_2. Let $U^\pm$ be the unipotent radical of $B^\pm$.

Proposition 4.8 *Let $\mathbb{I}_0^\circ$ be a finite torsion free $\Lambda_{h,0}$-algebra, N an open normal subgroup of $G_\mathbb{Q}$ and $\rho\colon N \to \mathrm{GL}_2(\mathbb{I}_0^\circ)$ a continuous representation that is $\mathbb{Z}_p$-regular. Suppose (upon replacing ρ by a conjugate) that a $\mathbb{Z}_p$-regular element is diagonal. Let $\mathbf{P}$ be an ideal of $\mathbb{I}_0^\circ$ and $\rho_\mathbf{P}\colon N \to \mathrm{GL}_2(\mathbb{I}_0^\circ/\mathbf{P})$ be the representation given by the reduction of ρ modulo $\mathbf{P}$. Let $U^\pm(\rho)$, and $U^\pm(\rho_\mathbf{P})$ be the upper and lower unipotent subgroups of $\mathrm{Im}\,\rho$, and $\mathrm{Im}\,\rho_\mathbf{P}$, respectively. Then the natural maps $U^+(\rho) \to U^+(\rho_\mathbf{P})$ and $U^-(\rho) \to U^-(\rho_\mathbf{P})$ are surjective.*

Remark 4.9 The ideal $\mathbf{P}$ in the proposition is not necessarily prime. At a certain point we will need to take $\mathbf{P} = P\mathbb{I}_0^\circ$ for a prime ideal P of $\Lambda_{h,0}$.

As in [10, Lemma 4.5] we need two lemmas. Since the argument is the same for U^+ and U^-, we will only treat here the upper triangular case $U = U^+$ and $B = B^+$.

For $* = U, B$ and every $j \geqslant 1$ we define the groups

$$\Gamma_*(\mathbf{P}^j) = \{x \in \mathrm{SL}_2(\mathbb{I}_0^\circ) \mid x \pmod{\mathbf{P}^j} \in *(\mathbb{I}_0^\circ/\mathbf{P}^j)\}.$$

Let $\Gamma_{\mathbb{I}_0^\circ}(\mathbf{P}^j)$ be the kernel of the reduction morphism $\pi_j\colon \mathrm{SL}_2(\mathbb{I}_0^\circ) \to \mathrm{SL}_2(\mathbb{I}_0^\circ/\mathbf{P}^j)$. Note that $\Gamma_U(\mathbf{P}^j) = \Gamma_{\mathbb{I}_0^\circ}(\mathbf{P}^j)U(\mathbb{I}_0^\circ)$ consists of matrices $\begin{pmatrix} a & b \\ c & d \end{pmatrix}$ such that $a, d \equiv 1 \pmod{\mathbf{P}^j}$, $c \equiv 0 \pmod{\mathbf{P}^j}$. Let $K = \mathrm{Im}\,\rho$ and

$$K_U(\mathbf{P}^j) = K \cap \Gamma_U(\mathbf{P}^j), \quad K_B(\mathbf{P}^j) = K \cap \Gamma_B(\mathbf{P}^j).$$

Since $U(\mathbb{I}_0^\circ)$ and $\Gamma_{\mathbb{I}_0^\circ}(\mathbf{P})$ are p-profinite, the groups $\Gamma_U(\mathbf{P}^j)$ and $K_U(\mathbf{P}^j)$ for all $j \geqslant 1$ are also p-profinite. Note that

$$\left[\begin{pmatrix} a & b \\ c & -a \end{pmatrix}, \begin{pmatrix} e & f \\ g & -e \end{pmatrix}\right] = \begin{pmatrix} bg-cf & 2(af-be) \\ 2(ce-ag) & cf-bg \end{pmatrix}.$$

From this we obtain the following.

Lemma 4.10 *If $X, Y \in \mathfrak{sl}_2(\mathbb{I}_0^\circ) \cap \begin{pmatrix} \mathbf{P}^j & \mathbf{P}^k \\ \mathbf{P}^i & \mathbf{P}^j \end{pmatrix}$ with $i \geqslant j \geqslant k$, then $[X, Y] \in \begin{pmatrix} \mathbf{P}^{i+k} & \mathbf{P}^{j+k} \\ \mathbf{P}^{i+j} & \mathbf{P}^{i+k} \end{pmatrix}$.*

We denote by $\mathrm{D}\Gamma_U(\mathbf{P}^j)$ the topological commutator subgroup $(\Gamma_U(\mathbf{P}^j), \Gamma_U(\mathbf{P}^j))$. Lemma 4.10 tells us that

$$\mathrm{D}\Gamma_U(\mathbf{P}^j) \subset \Gamma_B(\mathbf{P}^{2j}) \cap \Gamma_U(\mathbf{P}^j). \tag{1}$$

By the $\mathbb{Z}_p$-regularity assumption, there exists a diagonal element $d \in K$ with eigenvalues in $\mathbb{Z}_p$ and distinct modulo p. Consider the element $\delta = \lim_{n\to\infty} d^{p^n}$, which belongs to K since this is p-adically complete. In particular δ normalizes K. It is also diagonal with coefficients in $\mathbb{Z}_p$, so it normalizes $K_U(\mathbf{P}^j)$ and $\Gamma_B(\mathbf{P}^j)$. Since $\delta^p = \delta$, the eigenvalues δ_1 and δ_2 of δ are roots of unity of order dividing $p-1$. They still satisfy $\delta_1^2 \neq \delta_2^2$ as $p \neq 2$.

Set $\alpha = \delta_1/\delta_2 \in \mathbb{F}_p^\times$ and let a be the order of α as a root of unity. We see α as an element of $\mathbb{Z}_p^\times$ via the Teichmüller lift. Let H be a p-profinite group normalized by δ. Since H is p-profinite, every $x \in H$ has a unique a-th root. We define a map $\Delta\colon H \to H$ given by

$$\Delta(x) = [x \cdot \mathrm{ad}(\delta)(x)^{\alpha^{-1}} \cdot \mathrm{ad}(\delta^2)(x)^{\alpha^{-2}} \cdots \mathrm{ad}(\delta^{a-1})(x)^{\alpha^{1-a}}]^{1/a}$$

Lemma 4.11 *If $u \in \Gamma_U(\mathbf{P}^j)$ for some $j \geqslant 1$, then $\Delta^2(u) \in \Gamma_U(\mathbf{P}^{2j})$ and $\pi_j(\Delta(u)) = \pi_j(u)$.*

Proof If $u \in \Gamma_U(\mathbf{P}^j)$, we have $\pi_j(\Delta(u)) = \pi_j(u)$ as Δ is the identity map on $U(\mathbb{I}_0^\circ/\mathbf{P}^j)$. Let $\mathrm{D}\Gamma_U(\mathbf{P}^j)$ be the topological commutator subgroup of $\Gamma_U(\mathbf{P}^j)$. Since Δ induces the projection of the $\mathbb{Z}_p$-module $\Gamma_U(\mathbf{P}^j)/\mathrm{D}\Gamma_U(\mathbf{P}^j)$ onto its α-eigenspace for $\mathrm{ad}(d)$, it is a projection onto $U(\mathbb{I}_0^\circ)\mathrm{D}\Gamma_U(\mathbf{P}^j)/\mathrm{D}\Gamma_U(\mathbf{P}^j)$. The fact that this is exactly the α-eigenspace comes from the Iwahori decomposition of $\Gamma_U(\mathbf{P}^j)$, hence a similar direct sum decomposition holds in the abelianization $\Gamma_U(\mathbf{P}^j)/\mathrm{D}\Gamma_U(\mathbf{P}^j)$.

By (1), we have $\mathrm{D}\Gamma_U(\mathbf{P}^j) \subset \Gamma_B(\mathbf{P}^{2j}) \cap \Gamma_U(\mathbf{P}^j)$. Since the α-eigenspace of $\Gamma_U(\mathbf{P}^j)/\mathrm{D}\Gamma_U(\mathbf{P}^j)$ is inside $\Gamma_B(\mathbf{P}^{2j})$, Δ projects $u\Gamma_U(\mathbf{P}^j)$ to

$$\overline{\Delta}(u) \in (\Gamma_B(\mathbf{P}^{2j}) \cap \Gamma_U(\mathbf{P}^j))/\mathrm{D}\Gamma_U(\mathbf{P}^j).$$

In particular, $\Delta(u) \in \Gamma_B(\mathbf{P}^{2j}) \cap \Gamma_U(\mathbf{P}^j)$. Again apply Δ. Since $\Gamma_B(\mathbf{P}^{2j})/\Gamma_{\mathbb{I}_0^\circ}(\mathbf{P}^{2j})$ is sent to $\Gamma_U(\mathbf{P}^{2j})/\Gamma_{\mathbb{I}_0^\circ}(\mathbf{P}^{2j})$ by Δ, we get $\Delta^2(u) \in \Gamma_U(\mathbf{P}^{2j})$ as desired. □

Proof We can now prove Proposition 4.8. Let $\overline{u} \in U(\mathbb{I}_0^\circ/\mathbf{P}) \cap \mathrm{Im}(\rho_\mathbf{P})$. Since the reduction map $\mathrm{Im}(\rho) \to \mathrm{Im}(\rho_\mathbf{P})$ induced by π_1 is surjective, there exists $v \in \mathrm{Im}(\rho)$ such that $\pi_1(v) = \overline{u}$. Take $u_1 \in U(\mathbb{I}_0^\circ)$ such that $\pi_1(u_1) = \overline{u}$ (this is possible since $\pi_1\colon U(\Lambda_h) \to U(\Lambda_h/P)$ is surjective). Then $vu_1^{-1} \in \Gamma_{\mathbb{I}_0^\circ}(\mathbf{P})$, so $v \in K_U(\mathbf{P})$.

By compactness of $K_U(\mathbf{P})$ and by Lemma 4.11, starting with v as above, we see that $\lim_{m\to\infty} \Delta^m(v)$ converges $\mathbf{P}$-adically to $\Delta^\infty(v) \in U(\mathbb{I}_0^\circ) \cap K$ with $\pi_1(\Delta^\infty(v)) = \overline{u}$. □

Remark 4.12 Proposition 4.8 is true with the same proof if we replace $\Lambda_{h,0}$ by Λ_h and $\mathbb{I}_0^\circ$ by a finite torsion free Λ_h-algebra.

As a first application of Proposition 4.8 we give a result that we will need in the next subsection. Given a representation $\rho\colon G_\mathbb{Q} \to \mathrm{GL}_2(\mathbb{I}^\circ)$ and every ideal $\mathbf{P}$ of $\mathbb{I}^\circ$ we define $\rho_\mathbf{P}$, $U^\pm(\rho)$ and $U^\pm(\rho_\mathbf{P})$ as above, by replacing $\mathbb{I}_0^\circ$ by $\mathbb{I}^\circ$.

Proposition 4.13 *Let $\theta\colon \mathbb{T}_h \to \mathbb{I}^\circ$ be a family of slope $\leqslant h$ and $\rho_\theta\colon G_\mathbb{Q} \to \mathrm{GL}_2(\mathbb{I}^\circ)$ be the representation associated with θ. Suppose that ρ_θ is $(H_0, \mathbb{Z}_p)$-regular and let*

ρ be a conjugate of ρ_θ such that $\operatorname{Im}\rho|_{H_0}$ *contains a diagonal $\mathbb{Z}_p$-regular element. Then $U^+(\rho)$ and $U^-(\rho)$ are both nontrivial.*

Proof By density of classical points in $\mathbb{T}_h$ we can choose a prime ideal $\mathbf{P}\subset\mathbb{I}^\circ$ corresponding to a classical modular form f. The modulo $\mathbf{P}$ representation $\rho_{\mathbf{P}}$ is the p-adic representation classically associated with f. By the results of [13, 15] and the hypothesis of $(H_0,\mathbb{Z}_p)$-regularity of L, there exists an ideal $\mathfrak{l}_{\mathbf{P}}$ of $\mathbb{Z}_p$ such that $\operatorname{Im}\rho_{\mathbf{P}}$ contains the congruence subgroup $\Gamma_{\mathbb{Z}_p}(\mathfrak{l}_{\mathbf{P}})$. In particular $U^+(\rho_{\mathbf{P}})$ and $U^-(\rho_{\mathbf{P}})$ are both nontrivial. Since the maps $U^+(\rho)\to U^+(\rho_{\mathbf{P}})$ and $U^-(\rho)\to U^-(\rho_{\mathbf{P}})$ are surjective we find nontrivial elements in $U^+(\rho)$ and $U^-(\rho)$. □

We adapt the work in [12, Sect. 7] to show the following.

Proposition 4.14 *Suppose that the representation $\rho\colon G_{\mathbb{Q}}\to\mathrm{GL}_2(\mathbb{I}^\circ)$ is $(H_0,\mathbb{Z}_p)$-regular. Then there exists $g\in\mathrm{GL}_2(\mathbb{I}^\circ)$ such that the conjugate representation $g\rho g^{-1}$ satisfies the following two properties:*

1. *the image of $g\rho g^{-1}|_{H_0}$ is contained in $\mathrm{GL}_2(\mathbb{I}_0^\circ)$;*
2. *the image of $g\rho g^{-1}|_{H_0}$ contains a diagonal $\mathbb{Z}_p$-regular element.*

Proof As usual we choose a $\mathrm{GL}_2(\mathbb{I}^\circ)$-conjugate of ρ such that a $\mathbb{Z}_p$-regular element d is diagonal. We still write ρ for this conjugate representation and we show that it also has property (1).

Recall that for every $\sigma\in\Gamma$ there is a character $\eta_\sigma\colon G_{\mathbb{Q}}\to(\mathbb{I}^\circ)^\times$ and an equivalence $\rho^\sigma\cong\rho\otimes\eta_\sigma$. Then for every $\sigma\in\Gamma$ there exists $\mathbf{t}_\sigma\in\mathrm{GL}_2(\mathbb{I}^\circ)$ such that, for all $g\in G_{\mathbb{Q}}$,

$$\rho^\sigma(g)=\mathbf{t}_\sigma\eta_\sigma(g)\rho(g)\mathbf{t}_\sigma^{-1}. \tag{2}$$

We prove that the matrices $\mathbf{t}_\sigma$ are diagonal. Let $\rho(t)$ be a non-scalar diagonal element in $\operatorname{Im}\rho$ (for example d). Evaluating (2) at $g=t$ we find that $\mathbf{t}_\sigma$ must be either a diagonal or an antidiagonal matrix. Now by Proposition 4.13 there exists a nontrivial element $\rho(u^+)\in\operatorname{Im}\rho\cap U^+(\mathbb{I}^\circ)$. Evaluating (2) at $g=u^+$ we find that $\mathbf{t}_\sigma$ cannot be antidiagonal.

It is shown in [12, Lemma 7.3] that there exists an extension A of $\mathbb{I}^\circ$, at most quadratic, and a function $\zeta\colon\Gamma\to A^\times$ such that $\sigma\to\mathbf{t}_\sigma\zeta(\sigma)^{-1}$ defines a cocycle with values in $\mathrm{GL}_2(A)$. The proof of this result does not require the ordinarity of ρ. Equation (2) remains true if we replace $\mathbf{t}_\sigma$ with $\mathbf{t}_\sigma\zeta(\sigma)^{-1}$, so we can and do suppose from now on that $\mathbf{t}_\sigma$ is a cocycle with values in $\mathrm{GL}_2(A)$. In the rest of the proof we assume for simplicity that $A=\mathbb{I}^\circ$, but everything works in the same way if A is a quadratic extension of $\mathbb{I}^\circ$ and $\mathbb{F}$ is the residue field of A.

Let $V=(\mathbb{I}^\circ)^2$ be the space on which $G_{\mathbb{Q}}$ acts via ρ. As in [12, Sect. 7] we use the cocycle $\mathbf{t}_\sigma$ to define a twisted action of Γ on $(\mathbb{I}^\circ)^2$. For $v=(v_1,v_2)\in V$ we denote by v^σ the vector (v_1^σ,v_2^σ). We write $v^{[\sigma]}$ for the vector $\mathbf{t}_\sigma^{-1}v^\sigma$. Then $v\to v^{[\sigma]}$ gives an action of Γ since $\sigma\mapsto\mathbf{t}_\sigma$ is a cocycle. Note that this action is $\mathbb{I}_0^\circ$-linear.

Since $\mathbf{t}_\sigma$ is diagonal for every $\sigma\in\Gamma$, the submodules $V_1=\mathbb{I}^\circ(1,0)$ and $V_2=\mathbb{I}^\circ(0,1)$ are stable under the action of Γ. We show that each V_i contains an element fixed by Γ. We denote by $\mathbb{F}$ the residue field $\mathbb{I}^\circ/\mathfrak{m}_{\mathbb{I}}^\circ$. Note that the action of Γ on V_i induces an action of Γ on the one-dimensional $\mathbb{F}$-vector space $V_i \otimes \mathbb{I}^\circ/\mathfrak{m}_{\mathbb{I}^\circ}$. We show that for each i the space $V_i \otimes \mathbb{I}^\circ/\mathfrak{m}_{\mathbb{I}^\circ}$ contains a nonzero element $\overline{v}_i$ fixed by Γ. This is a consequence of the following argument, a form of which appeared in an early preprint of [12]. Let w be any nonzero element of $V_i \otimes \mathbb{I}^\circ/\mathfrak{m}_{\mathbb{I}^\circ}$ and let a be a variable in $\mathbb{F}$. The sum

$$S_{aw} = \sum_{\sigma\in\Gamma}(aw)^{[\sigma]}$$

is clearly Γ-invariant. We show that we can choose a such that $S_{aw} \neq 0$. Since $V_i \otimes \mathbb{I}^\circ/\mathfrak{m}_{\mathbb{I}^\circ}$ is one-dimensional, for every $\sigma \in \Gamma$ there exists $\alpha_\sigma \in \mathbb{F}$ such that $w^{[\sigma]} = \alpha_\sigma w$. Then

$$S_{aw} = \sum_{\sigma\in\Gamma}(aw)^{[\sigma]} = \sum_{\sigma\in\Gamma} a^\sigma w^{[\sigma]} = \sum_{\sigma\in\Gamma} a^\sigma \alpha_\sigma w = \left(\sum_{\sigma\in\Gamma} a^\sigma \alpha_\sigma a^{-1}\right) aw.$$

By Artin's lemma on the independence of characters, the function $f(a) = \sum_{\sigma\in\Gamma} a^\sigma \alpha_\sigma a^{-1}$ cannot be identically zero on $\mathbb{F}$. By choosing a value of a such that $f(a) \neq 0$ we obtain a nonzero element $\overline{v}_i = S_{aw}$ fixed by Γ.

We show that $\overline{v}_i$ lifts to an element $v_i \in V_i$ fixed by Γ. Let $\sigma_0 \in \Gamma$. By Lemma 4.2 Γ is a finite abelian 2-group, so the minimal polynomial $P_m(X)$ of $[\sigma_0]$ acting on V_i divides $X^{2^k} - 1$ for some integer k. In particular the factor $X - 1$ appears with multiplicity at most 1. We show that its multiplicity is exactly 1. If $\overline{P_m}$ is the reduction of P_m modulo $\mathfrak{m}_{\mathbb{I}^\circ}$ then $\overline{P_m}([\sigma_0]) = 0$ on $V_i \otimes \mathbb{I}^\circ/\mathfrak{m}_{\mathbb{I}^\circ}$. By our previous argument there is an element of $V_i \otimes \mathbb{I}^\circ/\mathfrak{m}_{\mathbb{I}^\circ}$ fixed by Γ (hence by $[\sigma_0]$) so we have $(X - 1) \mid \overline{P_m(X)}$. Since $p > 2$ the polynomial $X^{2^k} - 1$ has no double roots modulo $\mathfrak{m}_{\mathbb{I}^\circ}$, so neither does $\overline{P_m}$. By Hensel's lemma the factor $X - 1$ lifts to a factor $X - 1$ in P_m and $\overline{v}_i$ lifts to an element $v_i \in V_i$ fixed by $[\sigma_0]$. Note that $\mathbb{I}^\circ \cdot v_i = V_i$ by Nakayama's lemma since $\overline{v}_i \neq 0$.

We show that v_i is fixed by all of Γ. Let $W_{[\sigma_0]} = \mathbb{I}^\circ v_i$ be the one-dimensional eigenspace for $[\sigma_0]$ in V_i. Since Γ is abelian $W_{[\sigma_0]}$ is stable under Γ. Let $\sigma \in \Gamma$. Since σ has order 2^k in Γ for some $k \geqslant 0$ and $v_i^{[\sigma]} \in W_{[\sigma_0]}$, there exists a root of unity ζ_σ of order 2^k satisfying $v_i^{[\sigma]} = \zeta_\sigma v_i$. Since $\overline{v}_i^{[\sigma]} = \overline{v}_i$, the reduction of ζ_σ modulo $\mathfrak{m}_{\mathbb{I}^\circ}$ must be 1. As before we conclude that $\zeta_\sigma = 1$ since $p \neq 2$.

We found two elements $v_1 \in V_1$, $v_2 \in V_2$ fixed by Γ. We show that every element of $v \in V$ fixed by Γ must belong to the $\mathbb{I}_0^\circ$-submodule generated by v_1 and v_2. We proceed as in the end of the proof of [12, Theorem 7.5]. Since V_1 and V_2 are Γ-stable we must have $v \in V_1$ or $v \in V_2$. Suppose without loss of generality that $v \in V_1$. Then $v = \alpha v_1$ for some $\alpha \in \mathbb{I}^\circ$. If $\alpha \in \mathbb{I}_0^\circ$ then $v \in \mathbb{I}_0^\circ v_1$, as desired. If $\alpha \notin \mathbb{I}_0^\circ$ then there exists $\sigma \in \Gamma$ such that $\alpha^\sigma \neq \alpha$. Since v_1 is $[\sigma]$-invariant we obtain $(\alpha v_1)^{[\sigma]} = \alpha^\sigma v_1^{[\sigma]} = \alpha^\sigma v_1 \neq \alpha v_1$, so αv_1 is not fixed by $[\sigma]$, a contradiction.

Now (v_1, v_2) is a basis for V over $\mathbb{I}^\circ$, so the $\mathbb{I}_0^\circ$ submodule $V_0 = \mathbb{I}_0^\circ v_1 + \mathbb{I}_0^\circ v_2$ is an $\mathbb{I}_0^\circ$-lattice in V. Recall that $H_0 = \bigcap_{\sigma\in\Gamma} \ker \eta_\sigma$. We show that V_0 is stable under the action of H_0 via $\rho|_{H_0}$, i.e. that if $v \in V$ is fixed by Γ, so is $\rho(h)v$ for every $h \in H_0$. This is a consequence of the following computation, where v and h are as before and $\sigma \in \Gamma$:

$$(\rho(h)v)^{[\sigma]} = \mathbf{t}_\sigma^{-1}\rho(h)^\sigma v^\sigma = \mathbf{t}_\sigma^{-1}\eta_\sigma(h)\rho(h)^\sigma v^\sigma = \mathbf{t}_\sigma^{-1}\mathbf{t}_\sigma\rho(h)\mathbf{t}_\sigma^{-1}v^\sigma = \rho(h)v^{[\sigma]}.$$

Since V_0 is an $\mathbb{I}_0^\circ$-lattice in V stable under $\rho|_{H_0}$, we conclude that $\mathrm{Im}\,\rho|_{H_0} \subset \mathrm{GL}_2(\mathbb{I}_0^\circ)$. □

4.4 Fullness of the Unipotent Subgroups

From now on we write ρ for the element in its $\mathrm{GL}_2(\mathbb{I}^\circ)$ conjugacy class such that $\rho|_{H_0} \in \mathrm{GL}_2(\mathbb{I}_0^\circ)$. Recall that H is the open subgroup of H_0 defined by the condition $\det \overline{\rho}(h) = 1$ for every $h \in H$. As in [12, Sect. 4] we define a representation $H \to \mathrm{SL}_2(\mathbb{I}_0^\circ)$ by

$$\rho_0 = \rho|_H \otimes (\det \rho|_H)^{-\frac{1}{2}}.$$

We can take the square root of the determinant thanks to the definition of H. We will use the results of [12] to deduce that the $\Lambda_{h,0}$-module generated by the unipotent subgroups of the image of ρ_0 is big. We will later deduce the same for ρ.

We fix from now on a height one prime $P \subset \Lambda_{h,0}$ with the following properties:

1. there is an arithmetic prime $P_k \subset \mathbb{Z}_p[[\eta t]]$ satisfying $k > h+1$ and $P = P_k\Lambda_{h,0}$;
2. every prime $\mathfrak{P} \subset \mathbb{I}^\circ$ lying above P corresponds to a non-CM point.

Such a prime always exists. Indeed, by Remark 4.4 every classical weight $k > h+1$ contained in the disc B_h defines a prime $P = P_k\Lambda_{h,0}$ satisfying (1), so such primes are Zariski-dense in $\Lambda_{h,0}$, while the set of CM primes in $\mathbb{I}^\circ$ is finite by Proposition 3.9.

Remark 4.15 Since $k > h+1$, every point of $\mathrm{Spec}\,\mathbb{T}_h$ above P_k is classical by [5, Theorem 6.1]. Moreover the weight map is étale at every such point by [11, Theorem 11.10]. In particular the prime $P\mathbb{I}_0^\circ = P_k\mathbb{I}_0^\circ$ splits as a product of distinct primes of $\mathbb{I}_0^\circ$.

Make the technical assumption that the order of the residue field $\mathbb{F}$ of $\mathbb{I}^\circ$ is not 3. For every ideal $\mathbf{P}$ of $\mathbb{I}_0^\circ$ over P we let $\pi_\mathbf{P}$ be the projection $\mathrm{SL}_2(\mathbb{I}_0^\circ) \to \mathrm{SL}_2(\mathbb{I}_0^\circ/\mathbf{P})$. We still denote by $\pi_\mathbf{P}$ the restricted maps $U^\pm(\mathbb{I}_0^\circ) \to U^\pm(\mathbb{I}_0^\circ/\mathbf{P})$.

Let $G = \mathrm{Im}\,\rho_0$. For every ideal $\mathbf{P}$ of $\mathbb{I}_0^\circ$ we denote by $\rho_{0,\mathbf{P}}$ the representation $\pi_\mathbf{P}(\rho_0)$ and by $G_\mathbf{P}$ the image of $\rho_\mathbf{P}$, so that $G_\mathbf{P} = \pi_\mathbf{P}(G)$. We state two results from Lang's work that come over unchanged to the non-ordinary setting.

Proposition 4.16 [12, Corollary 6.3] *Let $\mathfrak{P}$ be a prime of $\mathbb{I}_0^\circ$ over P. Then $G_\mathfrak{P}$ contains a congruence subgroup $\Gamma_{\mathbb{I}_0^\circ/\mathfrak{P}}(\mathfrak{a}) \subset \mathrm{SL}_2(\mathbb{I}_0^\circ/\mathfrak{P})$. In particular $G_\mathfrak{P}$ is open in $\mathrm{SL}_2(\mathbb{I}_0^\circ/\mathfrak{P})$.*

Proposition 4.17 [12, Proposition 5.1] *Assume that for every prime $\mathfrak{P} \subset \mathbb{I}_0^\circ$ over P the subgroup $G_\mathfrak{P}$ is open in $\mathrm{SL}_2(\mathbb{I}_0^\circ/\mathfrak{P})$. Then the image of G in $\prod_{\mathfrak{P}|P} \mathrm{SL}_2(\mathbb{I}_0^\circ/\mathfrak{P})$ through the map $\prod_{\mathfrak{P}|P} \pi_\mathfrak{P}$ contains a product of congruence subgroups $\prod_{\mathfrak{P}|P} \Gamma_{\mathbb{I}_0^\circ/\mathfrak{P}}(\mathfrak{a}_\mathfrak{P})$.*

Remark 4.18 The proofs of Propositions 4.16 and 4.17 rely on the fact that the big ordinary Hecke algebra is étale over Λ at every arithmetic point. In order for these proofs to adapt to the non-ordinary setting it is essential that the prime P satisfies the properties above Remark 4.15.

We let $U^\pm(\rho_0) = G \cap U^\pm(\mathbb{I}_0^\circ)$ and $U^\pm(\rho_\mathbf{P}) = G_\mathbf{P} \cap U^\pm(\mathbb{I}_0^\circ/\mathbf{P})$. We denote by $U(\rho_\mathbf{P})$ either the upper or lower unipotent subgroups of $G_\mathbf{P}$ (the choice will be fixed throughout the proof). By projecting to the upper right element we identify $U^+(\rho_0)$ with a $\mathbb{Z}_p$-submodule of $\mathbb{I}_0^\circ$ and $U^+(\rho_{0,\mathbf{P}})$ with a $\mathbb{Z}_p$-submodule of $\mathbb{I}_0^\circ/\mathbf{P}$. We make analogous identifications for the lower unipotent subgroups. We will use Propositions 4.17 and 4.8 to show that, for both signs, $U^\pm(\rho)$ spans $\mathbb{I}_0^\circ$ over $\Lambda_{h,0}$.

First we state a version of [12, Lemma 4.10], with the same proof. Let A and B be Noetherian rings with B integral over A. We call *A-lattice* an A-submodule of B generated by the elements of a basis of $Q(B)$ over $Q(A)$.

Lemma 4.19 *Any A-lattice in B contains a nonzero ideal of B. Conversely, every nonzero ideal of B contains an A-lattice.*

We prove the following proposition by means of Proposition 4.8. We could also use Pink theory as in [12, Sect. 4].

Proposition 4.20 *Consider $U^\pm(\rho_0)$ as subsets of $Q(\mathbb{I}_0^\circ)$. For each choice of sign the $Q(\Lambda_{h,0})$-span of $U^\pm(\rho_0)$ is $Q(\mathbb{I}_0^\circ)$. Equivalently the $\Lambda_{h,0}$-span of $U^\pm(\rho_0)$ contains a $\Lambda_{h,0}$-lattice in $\mathbb{I}_0^\circ$.*

Proof Keep notations as above. We omit the sign when writing unipotent subgroups and we refer to either the upper or lower ones (the choice is fixed throughout the proof). Let P be the prime of $\Lambda_{h,0}$ chosen above. By Remark 4.15 the ideal $P\mathbb{I}_0^\circ$ splits as a product of distinct primes in $\mathbb{I}_0^\circ$. When $\mathfrak{P}$ varies among these primes, the map $\bigoplus_{\mathfrak{P}|P} \pi_\mathfrak{P}$ gives embeddings of $\Lambda_{h,0}/P$-modules $\mathbb{I}_0^\circ/P\mathbb{I}_0^\circ \hookrightarrow \bigoplus_{\mathfrak{P}|P} \mathbb{I}_0^\circ/\mathfrak{P}$ and $U(\rho_{P\mathbb{I}_0^\circ}) \hookrightarrow \bigoplus_{\mathfrak{P}|P} U(\rho_\mathfrak{P})$. The following diagram commutes:

$$\begin{array}{ccc} U(\rho_{P\mathbb{I}_0^\circ}) & \xhookrightarrow{\bigoplus_{\mathfrak{P}|P} \pi_\mathfrak{P}} & \bigoplus_{\mathfrak{P}|P} U(\rho_\mathfrak{P}) \\ \big\downarrow & & \big\downarrow \\ \mathbb{I}_0^\circ/P\mathbb{I}_0^\circ & \xhookrightarrow{\bigoplus_{\mathfrak{P}|P} \pi_\mathfrak{P}} & \bigoplus_{\mathfrak{P}|P} \mathbb{I}_0^\circ/\mathfrak{P} \end{array} \tag{3}$$

By Proposition 4.17 there exist ideals $\mathfrak{a}_\mathfrak{P} \subset \mathbb{I}_0^\circ/\mathfrak{P}$ such that $(\bigoplus_{\mathfrak{P}|P} \pi_\mathfrak{P})(G_{P\mathbb{I}_0^\circ}) \supset \bigoplus_{\mathfrak{P}|P} \Gamma_{\mathbb{I}_0^\circ/\mathfrak{P}}(\mathfrak{a}_\mathfrak{P})$. In particular $(\bigoplus_{\mathfrak{P}|P} \pi_\mathfrak{P})(U(\rho_{P\mathbb{I}_0^\circ})) \supset \bigoplus_{\mathfrak{P}|P}(\mathfrak{a}_\mathfrak{P})$. By Lemma 4.19 each ideal $\mathfrak{a}_\mathfrak{P}$ contains a basis of $Q(\mathbb{I}_0^\circ/\mathfrak{P})$ over $Q(\Lambda_{h,0}/P)$, so that the

$Q(\Lambda_{h,0}/P)$-span of $\bigoplus_{\mathfrak{P}|P} \mathfrak{a}_{\mathfrak{P}}$ is the whole $\bigoplus_{\mathfrak{P}|P} Q(\mathbb{I}_0^\circ/\mathfrak{P})$. Then the $Q(\Lambda_{h,0}/P)$-span of $(\bigoplus_{\mathfrak{P}|P} \pi_{\mathfrak{P}})(G_{\mathfrak{P}} \cap U(\rho_{\mathfrak{P}}))$ is also $\bigoplus_{\mathfrak{P}|P} Q(\mathbb{I}_0^\circ/\mathfrak{P})$. By commutativity of diagram (3) we deduce that the $Q(\Lambda_{h,0}/P)$-span of $G_P \cap U(\rho_{P\mathbb{I}_0^\circ})$ is $Q(\mathbb{I}_0^\circ/P\mathbb{I}_0^\circ)$. In particular $G_{P\mathbb{I}_0^\circ} \cap U(\rho_{P\mathbb{I}_0^\circ})$ contains a $\Lambda_{h,0}/P$-lattice, hence by Lemma 4.19 a nonzero ideal $\mathfrak{a}_P$ of $\mathbb{I}_0^\circ/P\mathbb{I}_0^\circ$.

Note that the representation $\rho_0\colon H \to \mathrm{SL}_2(\mathbb{I}_0^\circ)$ satisfies the hypotheses of Proposition 4.8. Indeed we assumed that $\rho\colon G_{\mathbb{Q}} \to \mathrm{GL}_2(\mathbb{I})$ is $(H_0, \mathbb{Z}_p)$-regular, so the image of $\rho|_{H_0}$ contains a diagonal $\mathbb{Z}_p$-regular element d. Since H is a normal subgroup of H_0, $\rho(H)$ is a normal subgroup of $\rho(H_0)$ and it is normalized by d. By a trivial computation we see that the image of $\rho_0 = \rho|_H \otimes (\det \rho|_H)^{-1/2}$ is also normalized by d.

Let $\mathfrak{a}$ be an ideal of $\mathbb{I}_0^\circ$ projecting to $\mathfrak{a}_P \subset U(\rho_{0,P\mathbb{I}_0^\circ})$. By Proposition 4.8 applied to ρ_0 we obtain that the map $U(\rho_0) \to U(\rho_{0,P\mathbb{I}_0^\circ})$ is surjective, so the $\mathbb{Z}_p$-module $\mathfrak{a} \cap U(\rho_0)$ also surjects to $\mathfrak{a}_P$. Since $\Lambda_{h,0}$ is local we can apply Nakayama's lemma to the $\Lambda_{h,0}$-module $\Lambda_{h,0}(\mathfrak{a} \cap U(\rho_0)$ to conclude that it coincides with $\mathfrak{a}$. Hence $\mathfrak{a} \subset \Lambda_{h,0} \cdot U(\rho_0)$, so the $\Lambda_{h,0}$-span of $U(\rho_0)$ contains a $\Lambda_{h,0}$-lattice in $\mathbb{I}_0^\circ$ by lemma 4.19. □

We show that Proposition 4.20 is true if we replace ρ_0 by $\rho|_H$. This will be a consequence of the description of the subnormal sugroups of $\mathrm{GL}_2(\mathbb{I}^\circ)$ presented in [23], but we need a preliminary step because we cannot induce a $\Lambda_{h,0}$-module structure on the unipotent subgroups of G. For a subgroup $\mathcal{G} \subset \mathrm{GL}_2(\mathbb{I}_0^\circ)$ define $\mathcal{G}^p = \{g^p,\ g \in G\}$ and $\widetilde{\mathcal{G}} = \mathcal{G}^p \cap (1 + p\mathrm{M}_2(\mathbb{I}_0^\circ))$. Let $\widetilde{\mathcal{G}}^{\Lambda_{h,0}}$ be the subgroup of $\mathrm{GL}_2(\mathbb{I}^\circ)$ generated by the set $\{g^\lambda\colon g \in \widetilde{\mathcal{G}}, \lambda \in \Lambda_{h,0}\}$ where $g^\lambda = \exp(\lambda \log g)$. We have the following.

Lemma 4.21 *The group $\widetilde{\mathcal{G}}^{\Lambda_{h,0}}$ contains a congruence subgroup of* $\mathrm{SL}_2(\mathbb{I}_0^\circ)$ *if and only if both of the unipotent subgroups $\mathcal{G} \cap U^+(\mathbb{I}_0^\circ)$ and $\mathcal{G} \cap U^-(\mathbb{I}_0^\circ)$ contain a basis of a $\Lambda_{h,0}$-lattice in $\mathbb{I}_0^\circ$.*

Proof It is easy to see that $\mathcal{G} \cap U^+(\mathbb{I}_0^\circ)$ contains the basis of a $\Lambda_{h,0}$-lattice in $\mathbb{I}_0^\circ$ if and only if the same is true for $\widetilde{\mathcal{G}} \cap U^+(\mathbb{I}_0^\circ)$. The same is true for U^-. By a standard argument, used in the proofs of [9, Lemma 2.9] and [12, Proposition 4.2], $\mathcal{G}^{\Lambda_{h,0}} \subset \mathrm{GL}_2(\mathbb{I}_0^\circ)$ contains a congruence subgroup of $\mathrm{SL}_2(\mathbb{I}_0^\circ)$ if and only if both its upper and lower unipotent subgroup contain an ideal of $\mathbb{I}_0^\circ$. We have $U^+(\mathbb{I}_0^\circ) \cap \mathcal{G}^{\Lambda_{h,0}} = \Lambda_{h,0}(\mathcal{G} \cap U^+(\mathbb{I}_0^\circ))$, so by Lemma 4.19 $U^+(\mathbb{I}_0^\circ) \cap \mathcal{G}^{\Lambda_{h,0}}$ contains an ideal of $\mathbb{I}_0^\circ$ if and only if $\mathcal{G} \cap U^+(\mathbb{I}_0^\circ)$ contains a basis of a $\Lambda_{h,0}$-lattice in $\mathbb{I}_0^\circ$. We proceed in the same way for U^-. □

Now let $G_0 = \mathrm{Im}\, \rho|_H$, $G = \mathrm{Im}\, \rho_0$. Note that $G_0 \cap \mathrm{SL}_2(\mathbb{I}_0^\circ)$ is a normal subgroup of G. Let $f\colon \mathrm{GL}_2(\mathbb{I}_0^\circ) \to \mathrm{SL}_2(\mathbb{I}_0^\circ)$ be the homomorphism sending g to $\det(g)^{-1/2}g$. We have $G = f(G_0)$ by definition of ρ_0. We show the following.

Proposition 4.22 *The subgroups $G_0 \cap U^\pm(\mathbb{I}_0^\circ)$ both contain the basis of a $\Lambda_{h,0}$-lattice in $\mathbb{I}_0^\circ$ if and only if $G \cap U^\pm(\mathbb{I}_0^\circ)$ both contain the basis of a $\Lambda_{h,0}$-lattice in $\mathbb{I}_0^\circ$.*

Proof Since $G = f(G_0)$ we have $\widetilde{G} = f(\widetilde{G_0})$. This implies that $\widetilde{G}^{\Lambda_{h,0}} = f(\widetilde{G_0}^{\Lambda_{h,0}})$. We remark that $\widetilde{G_0}^{\Lambda_{h,0}} \cap \mathrm{SL}_2(\mathbb{I}_0^\circ)$ is a normal subgroup of $\widetilde{G}^{\Lambda_{h,0}}$. Indeed

$\widetilde{G_0}^{\Lambda_{h,0}} \cap \mathrm{SL}_2(\mathbb{I}_0^\circ)$ is normal in $\widetilde{G_0}^{\Lambda_{h,0}}$, so its image $f(G_0^{\Lambda_{h,0}} \cap \mathrm{SL}_2(\mathbb{I}_0^\circ)) = G_0^{\Lambda_{h,0}} \cap \mathrm{SL}_2(\mathbb{I}_0^\circ)$ is normal in $f(G_0^{\Lambda_{h,0}}) = \widetilde{G}^{\Lambda_{h,0}}$.

By [23, Corollary 1] a subgroup of $\mathrm{GL}_2(\mathbb{I}_0^\circ)$ contains a congruence subgroup of $\mathrm{SL}_2(\mathbb{I}_0^\circ)$ if and only if it is subnormal in $\mathrm{GL}_2(\mathbb{I}_0^\circ)$ and it is not contained in the centre. We note that $\widetilde{G_0}^{\Lambda_{h,0}} \cap \mathrm{SL}_2(\mathbb{I}_0^\circ) = (\widetilde{G_0} \cap \mathrm{SL}_2(\mathbb{I}_0^\circ))^{\Lambda_{h,0}}$ is not contained in the subgroup $\{\pm 1\}$. Otherwise also $\widetilde{G_0} \cap \mathrm{SL}_2(\mathbb{I}_0^\circ)$ would be contained in $\{\pm 1\}$ and $\mathrm{Im}\,\rho \cap \mathrm{SL}_2(\mathbb{I}_0^\circ)$ would be finite, since $\widetilde{G_0}$ is of finite index in G_0^p. This would give a contradiction: indeed if $\mathfrak{P}$ is an arithmetic prime of $\mathbb{I}^\circ$ of weight greater than 1 and $\mathfrak{P}' = \mathfrak{P} \cap \mathbb{I}_0^\circ$, the image of ρ modulo $\mathfrak{P}'$ contains a congruence subgroup of $\mathrm{SL}_2(\mathbb{I}_0^\circ/\mathfrak{P}')$ by the result of [15].

Since $\widetilde{G_0}^{\Lambda_{h,0}} \cap \mathrm{SL}_2(\mathbb{I}_0^\circ)$ is a normal subgroup of $\widetilde{G}^{\Lambda_{h,0}}$, we deduce by [23, Corollary 1] that $\widetilde{G_0}^{\Lambda_{h,0}} \cap \mathrm{SL}_2(\mathbb{I}_0^\circ)$ (hence $\widetilde{G_0}^{\Lambda_{h,0}}$) contains a congruence subgroup of $\mathrm{SL}_2(\mathbb{I}_0^\circ)$ if and only if $\widetilde{G}^{\Lambda_{h,0}}$ does. By applying Lemma 4.21 to $\mathcal{G} = G_0$ and $\mathcal{G} = G$ we obtain the desired equivalence. □

By combining Propositions 4.20 and 4.22 we obtain the following.

Corollary 4.23 *The $\Lambda_{h,0}$-span of each of the unipotent subgroups* $\mathrm{Im}\,\rho \cap U^\pm$ *contains a $\Lambda_{h,0}$-lattice in $\mathbb{I}_0^\circ$.*

Unlike in the ordinary case we cannot deduce from the corollary that $\mathrm{Im}\,\rho$ contains a congruence subgroup of $\mathrm{SL}_2(\mathbb{I}_0^\circ)$, since we are working over $\Lambda_h \neq \Lambda$ and we cannot induce a Λ_h-module structure (not even a Λ-module structure) on $\mathrm{Im}\,\rho \cap U^\pm$. The proofs of [9, Lemma 2.9] and [12, Proposition 4.3] rely on the existence, in the image of the Galois group, of an element inducing by conjugation a Λ-module structure on $\mathrm{Im}\,\rho \cap U^\pm$. In their situation this is predicted by the condition of Galois ordinarity of ρ. In the non-ordinary case we will find an element with a similar property via relative Sen theory. In order to do this we will need to work with a suitably defined Lie algebra rather than with the group itself.

5 Relative Sen Theory

We recall the notations of Sect. 3.1. In particular $r_h = p^{-s_h}$, with $s_h \in \mathbb{Q}$, is the h-adapted radius (which we also take smaller than $p^{-\frac{1}{p-1}}$), η_h is an element in $\mathbb{C}_p$ of norm r_h, K_h is the Galois closure in $\mathbb{C}_p$ of $\mathbb{Q}_p(\eta_h)$ and O_h is the ring of integers in K_h. The ring Λ_h of analytic functions bounded by 1 on the open disc $\mathcal{B}_h = \mathcal{B}(0, r_h^-)$ is identified to $O_h[[t]]$. We take a sequence of radii $r_i = p^{-s_h - 1/i}$ converging to r_h and denote by $A_{r_i} = K_h\langle t, X_i\rangle/(pX_i - t^i)$ the K_h-algebra defined in Sect. 3.1 which is a form over K_h of the $\mathbb{C}_p$-algebra of analytic functions on the closed ball $\mathcal{B}(0, r_i)$ (its Berthelot model). We denote by $A_{r_i}^\circ$ the O_h-subalgebra of functions bounded by 1. Then $\Lambda_h = \varprojlim_i A_{r_i}^\circ$ where $A_{r_j}^\circ \to A_{r_i}^\circ$ for $i < j$ is the restriction of analytic functions.

We defined in Sect. 4.1 a subring $\mathbb{I}_0^\circ \subset \mathbb{I}^\circ$, finite over $\Lambda_{h,0} \subset \Lambda_h$. For r_i as above, we write $A_{0,r_i}^\circ = \mathcal{O}_{h,0}\langle t, X_i\rangle/(pX_i - t^i)$ with maps $A_{0,r_j}^\circ \to A_{0,r_i}^\circ$ for $i < j$, so that $\Lambda_{h,0} = \varprojlim_i A_{0,r_i}^\circ$. Let $\mathbb{I}_{r_i}^\circ = \mathbb{I}^\circ \widehat{\otimes}_{\Lambda_h} A_{r_i}^\circ$ and $\mathbb{I}_{0,r_i}^\circ = \mathbb{I}_0^\circ \widehat{\otimes}_{\Lambda_{h,0}} A_{0,r_i}^\circ$, both endowed with their p-adic topology. Note that $(\mathbb{I}_{r_i}^\circ)^\Gamma = \mathbb{I}_{r_i,0}^\circ$.

Consider the representation $\rho\colon G_\mathbb{Q} \to \mathrm{GL}_2(\mathbb{I}^\circ)$ associated with a family $\theta\colon \mathbb{T}_h \to \mathbb{I}^\circ$. We observe that ρ is continuous with respect to the profinite topology of $\mathbb{I}^\circ$ but not with respect to the p-adic topology. For this reason we fix an arbitrary radius r among the r_i defined above and consider the representation $\rho_r\colon G_\mathbb{Q} \to \mathrm{GL}_2(\mathbb{I}_r^\circ)$ obtained by composing ρ with the inclusion $\mathrm{GL}_2(\mathbb{I}^\circ) \hookrightarrow \mathrm{GL}_2(\mathbb{I}_r^\circ)$. This inclusion is continuous, hence the representation ρ_r is continuous with respect to the p-adic topology on $\mathrm{GL}_2(\mathbb{I}_{0,r}^\circ)$.

Recall from Proposition 4.14 that, after replacing ρ by a conjugate, there is an open normal subgroup $H_0 \subset G_\mathbb{Q}$ such that the restriction $\rho|_{H_0}$ takes values in $\mathrm{GL}_2(\mathbb{I}_0^\circ)$ and is $(H_0, \mathbb{Z}_p)$-regular. Then the restriction $\rho_r|_{H_0}$ gives a representation $H_0 \to \mathrm{GL}_2(\mathbb{I}_{0,r}^\circ)$ which is continuous with respect to the p-adic topology on $\mathrm{GL}_2(\mathbb{I}_{0,r}^\circ)$.

5.1 *Big Lie Algebras*

Recall that $G_p \subset G_\mathbb{Q}$ denotes our chosen decomposition group at p. Let G_r and G_r^{loc} be the images respectively of H_0 and $G_p \cap H_0$ under the representation $\rho_r|_{H_0}\colon H_0 \to \mathrm{GL}_2(\mathbb{I}_{0,r}^\circ)$. Note that they are actually independent of r since they coincide with the images of H_0 and $G_p \cap H_0$ under ρ.

For every ring R and ideal $I \subset R$ we denote by $\Gamma_{\mathrm{GL}_2(R)}(I)$ the GL_2-congruence subgroup consisting of elements $g \in \mathrm{GL}_2(R)$ such that $g \equiv \mathrm{Id}_2 \pmod{I}$. Let $G_r' = G_r \cap \Gamma_{\mathrm{GL}_2(\mathbb{I}_{0,r}^\circ)}(p)$ and $G_r^{\prime,\mathrm{loc}} = G_r^{\mathrm{loc}} \cap \Gamma_{\mathrm{GL}_2(\mathbb{I}_{0,r}^\circ)}(p)$, so that G_r' and $G_r^{\prime,\mathrm{loc}}$ are pro-p groups. Note that the congruence subgroups $\Gamma_{\mathrm{GL}_2(\mathbb{I}_{0,r})}(p^m)$ are open in $\mathrm{GL}_2(\mathbb{I}_{0,r})$ for the p-adic topology. In particular G_r' and $G_r^{\prime,\mathrm{loc}}$ can be identified with the images under ρ of the absolute Galois groups of finite extensions of $\mathbb{Q}$ and respectively $\mathbb{Q}_p$.

Remark 5.1 We remark that we can choose an arbitrary r_0 and set, for every r, $G_r' = G_r \cap \Gamma_{\mathrm{GL}_2(\mathbb{I}_{0,r_0}^\circ)}(p)$. Then G_r' is a pro-p subgroup of G_r for every r and it is independent of r since G_r is. This will be important in Theorem 7.1 where we will take projective limits over r of various objects.

We set $A_{0,r} = A_{0,r}^\circ[p^{-1}]$ and $\mathbb{I}_{0,r} = \mathbb{I}_{0,r}^\circ[p^{-1}]$. We consider from now on G_r' and $G_r^{\prime,\mathrm{loc}}$ as subgroups of $\mathrm{GL}_2(\mathbb{I}_{0,r})$ through the inclusion $\mathrm{GL}_2(\mathbb{I}_{0,r}^\circ) \hookrightarrow \mathrm{GL}_2(\mathbb{I}_{0,r})$.

We want to define big Lie algebras associated with the groups G_r' and $G_r^{\prime,\mathrm{loc}}$. For every nonzero ideal $\mathfrak{a}$ of the principal ideal domain $A_{0,r}$, we denote by $G_{r,\mathfrak{a}}'$ and $G_{r,\mathfrak{a}}^{\prime,\mathrm{loc}}$ the images respectively of G_r' and $G_r^{\prime,\mathrm{loc}}$ under the natural projection $\mathrm{GL}_2(\mathbb{I}_{0,r}) \to \mathrm{GL}_2(\mathbb{I}_{0,r}/\mathfrak{a}\mathbb{I}_{0,r})$. The pro-$p$ groups $G_{r,\mathfrak{a}}'$ and $G_{r,\mathfrak{a}}^{\prime,\mathrm{loc}}$ are topologically of finite type so we can define the corresponding $\mathbb{Q}_p$-Lie algebras $\mathfrak{H}_{r,\mathfrak{a}}$ and $\mathfrak{H}_{r,\mathfrak{a}}^{\mathrm{loc}}$ using the p-adic logarithm map: $\mathfrak{H}_{r,\mathfrak{a}} = \mathbb{Q}_p \cdot \mathrm{Log}\, G_{r,\mathfrak{a}}'$ and $\mathfrak{H}_{r,\mathfrak{a}}^{\mathrm{loc}} = \mathbb{Q}_p \cdot \mathrm{Log}\, G_{r,\mathfrak{a}}^{\prime,\mathrm{loc}}$. They are closed Lie subalgebras of the finite dimensional $\mathbb{Q}_p$-Lie algebra $\mathrm{M}_2(\mathbb{I}_{0,r}/\mathfrak{a}\mathbb{I}_{0,r})$.

Let $B_r = \varprojlim_{(\mathfrak{a},P_1)=1} A_{0,r}/\mathfrak{a}A_{0,r}$ where the inverse limit is taken over nonzero ideals $\mathfrak{a} \subset A_{0,r}$ prime to $P_1 = (u^{-1}(1+T)-1)$ (the reason for excluding P_1 will become clear later). We endow B_r with the projective limit topology coming from the p-adic topology on each quotient. We have a topological isomorphism of $K_{h,0}$-algebras

$$B_r \cong \prod_{P \neq P_1} \widehat{(A_{0,r})_P},$$

where the product is over primes P and $\widehat{(A_{0,r})_P} = \varprojlim_{m\geqslant 1} A_{0,r}/P^m A_{0,r}$ denotes the $K_{h,0}$-Fréchet space inverse limit of the finite dimensional $K_{h,0}$-vector spaces $A_{0,r}/P^m A_{0,r}$. Similarly, let $\mathbb{B}_r = \varprojlim_{(\mathfrak{a},P_1)=1} \mathbb{I}_{0,r}/\mathfrak{a}\mathbb{I}_{0,r}$, where as before $\mathfrak{a}$ varies over all nonzero ideals of $A_{0,r}$ prime to P_1. We have

$$\mathbb{B}_r \cong \prod_{P \neq P_1} \widehat{(\mathbb{I}_{0,r})_{P\mathbb{I}_{0,r}}} \cong \prod_{\mathfrak{P} \nmid P_1} \widehat{(\mathbb{I}_{0,r})_{\mathfrak{P}}} \cong \varprojlim_{(\mathfrak{Q},P_1)=1} \mathbb{I}_{0,r}/\mathfrak{Q},$$

where the second product is over primes $\mathfrak{P}$ of $\mathbb{I}_{0,r}$ and the projective limit is over primary ideals $\mathfrak{Q}$ of $\mathbb{I}_{0,r}$. Here $\widehat{(\mathbb{I}_{0,r})_{\mathfrak{P}}}$ denotes the projective limit of finite dimensional $K_{h,0}$-algebras (endowed with the p-adic topology). The last isomorphism follows from the fact that $\mathbb{I}_{0,r}$ is finite over $A_{0,r}$, so that there is an isomorphism $\mathbb{I}_{0,r} \otimes \widehat{(A_{0,r})_P} = \prod_{\mathfrak{P}} \widehat{(\mathbb{I}_{0,r})_{\mathfrak{P}}}$ where P is a prime of $A_{0,r}$ and $\mathfrak{P}$ varies among the primes of $\mathbb{I}_{0,r}$ above P. We have natural continuous inclusions $A_{0,r} \hookrightarrow B_r$ and $\mathbb{I}_{0,r} \hookrightarrow \mathbb{B}_r$, both with dense image. The map $A_{0,r} \hookrightarrow \mathbb{I}_{0,r}$ induces an inclusion $B_r \hookrightarrow \mathbb{B}_r$ with closed image. Note however that $\mathbb{B}_r$ is not finite over B_r. We will work with $\mathbb{B}_r$ for the rest of this section, but we will need B_r later.

For every $\mathfrak{a}$ we have defined Lie algebras $\mathfrak{H}_{r,\mathfrak{a}}$ and $\mathfrak{H}_{r,\mathfrak{a}}^{\mathrm{loc}}$ associated with the finite type Lie groups $G'_{r,\mathfrak{a}}$ and $G'^{,\mathrm{loc}}_{r,\mathfrak{a}}$. We take the projective limit of these algebras to obtain Lie subalgebras of $\mathrm{M}_2(\mathbb{B}_r)$.

Definition 5.2 The Lie algebras associated with G'_r and $G'^{,\mathrm{loc}}_r$ are the closed $\mathbb{Q}_p$-Lie subalgebras of $\mathrm{M}_2(\mathbb{B}_r)$ given respectively by

$$\mathfrak{H}_r = \varprojlim_{(\mathfrak{a},P_1)=1} \mathfrak{H}_{r,\mathfrak{a}}$$

and

$$\mathfrak{H}_r^{\mathrm{loc}} = \varprojlim_{(\mathfrak{a},P_1)=1} \mathfrak{H}_{r,\mathfrak{a}}^{\mathrm{loc}},$$

where as usual the products are taken over nonzero ideals $\mathfrak{a} \subset A_{0,r}$ prime to P_1.

For every ideal $\mathfrak{a}$ prime to P_1, we have continuous homomorphisms $\mathfrak{H}_r \to \mathfrak{H}_{r,\mathfrak{a}}$ and $\mathfrak{H}_r^{\mathrm{loc}} \to \mathfrak{H}_{r,\mathfrak{a}}^{\mathrm{loc}}$. Since the transition maps are surjective these homomorphisms are surjective.

Remark 5.3 The limits in Definition 5.2 can be replaced by limits over primary ideals of $\mathbb{I}_{0,r}$. Explicitly, let $\mathfrak{Q}$ be a primary ideal of $\mathbb{I}_{0,r}$. Let $G'_{r,\mathfrak{Q}}$ be the image of G'_r via the natural projection $\mathrm{GL}_2(\mathbb{I}_{0,r}) \to \mathrm{GL}_2(\mathbb{I}_{0,r}/\mathfrak{Q})$ and let $\mathfrak{H}_{r,\mathfrak{Q}}$ be the Lie algebra associated with $G'_{r,\mathfrak{Q}}$ (which is a finite type Lie group). We have an isomorphism of topological Lie algebras

$$\mathfrak{H}_r = \varprojlim_{(\mathfrak{Q},P_1)=1} \mathfrak{H}_{r,\mathfrak{Q}},$$

where the limit is taken over primary ideals $\mathfrak{Q}$ of $\mathbb{I}_{0,r}$. This is naturally a subalgebra of $\mathrm{M}_2(\mathbb{B}_r)$ since $\mathbb{B}_r \cong \varprojlim_{(\mathfrak{Q},P_1)=1} \mathbb{I}_{0,r}/\mathfrak{Q}$. The same goes for the local algebras.

5.2 The Sen Operator Associated with a Galois Representation

Recall that there is a finite extension $K/\mathbb{Q}_p$ such that $G_r^{\prime,\mathrm{loc}}$ is the image of $\rho|_{\mathrm{Gal}(\overline{K}/K)}$ and, for an ideal $P \subset A_{0,r}$ and $m \geqslant 1$, $G_{r,P^m}^{\prime,\mathrm{loc}}$ is the image of $\rho_{r,P^m}|_{\mathrm{Gal}(\overline{K}/K)}$. Following [19, 21] we can define a Sen operator associated with $\rho_r|_{\mathrm{Gal}(\overline{K}/K)}$ and $\rho_{r,P^m}|_{\mathrm{Gal}(\overline{K}/K)}$ for every ideal $P \subset A_{0,r}$ and every $m \geqslant 1$. We will see that these operators satisfy a compatibility property. We write for the rest of the section ρ_r and ρ_{r,P^m} while implicitly taking the domain to be $\mathrm{Gal}(\overline{K}/K)$.

We begin by recalling the definition of the Sen operator associated with a representation $\tau\colon \mathrm{Gal}(\overline{K}/K) \to \mathrm{GL}_m(\mathcal{R})$ where $\mathcal{R}$ is a Banach algebra over a p-adic field L. We follow [21]. We can suppose $L \subset K$; if not we just restrict τ to the open subgroup $\mathrm{Gal}(\overline{K}/KL) \subset \mathrm{Gal}(\overline{K}/K)$.

Let L_∞ be a totally ramified $\mathbb{Z}_p$-extension of L. Let γ be a topological generator of $\Gamma = \mathrm{Gal}(L_\infty/L)$, $\Gamma_n \subset \Gamma$ the subgroup generated by γ^{p^n} and $L_n = L_\infty^{\gamma^{p^n}}$, so that $L_\infty = \cup_n L_n$. Let $L'_n = L_n K$ and $G'_n = \mathrm{Gal}(\overline{L}/L'_n)$. If $\mathcal{R}^m$ is the $\mathcal{R}$-module over which $\mathrm{Gal}(\overline{K}/K)$ acts via τ, define an action of $\mathrm{Gal}(\overline{K}/K)$ on $\mathcal{R}\widehat{\otimes}_L\mathbb{C}_p$ by letting $\sigma \in \mathrm{Gal}(\overline{K}/K)$ map $x \otimes y$ to $\tau(\sigma)(x) \otimes \sigma(y)$. Then by the results of [19, 21] there is a matrix $M \in \mathrm{GL}_m\left(\mathcal{R}\widehat{\otimes}_L\mathbb{C}_p\right)$, an integer $n \geqslant 0$ and a representation $\delta\colon \Gamma_n \to \mathrm{GL}_m(\mathcal{R} \otimes_L L'_n)$ such that for all $\sigma \in G'_n$

$$M^{-1}\tau(\sigma)\sigma(M) = \delta(\sigma).$$

Definition 5.4 The Sen operator associated with τ is

$$\phi = \lim_{\sigma \to 1} \frac{\log\bigl(\delta(\sigma)\bigr)}{\log(\chi(\sigma))} \in \mathrm{M}_m(\mathcal{R}\widehat{\otimes}_L\mathbb{C}_p).$$

The limit exists as for σ close to 1 the map $\sigma \mapsto \dfrac{\log\bigl(\delta(\sigma)\bigr)}{\log(\chi(\sigma))}$ is constant. It is proved in [21, Sect. 2.4] that ϕ does not depend on the choice of δ and M.

If $L=\mathcal{R}=\mathbb{Q}_p$, we define the Lie algebra $\mathfrak{g}$ associated with $\tau(\mathrm{Gal}(\overline{K}/K))$ as the $\mathbb{Q}_p$-vector space generated by the image of the logarithm map in $\mathrm{M}_m(\mathbb{Q}_p)$. In this situation the Sen operator ϕ associated with τ has the following property.

Theorem 5.5 [19, Theorem 1] *For a continuous representation $\tau\colon G_K\to\mathrm{GL}_m(\mathbb{Q}_p)$, the Lie algebra $\mathfrak{g}$ of the group $\tau(\mathrm{Gal}(\overline{K}/K))$ is the smallest $\mathbb{Q}_p$-subspace of $\mathrm{M}_m(\mathbb{Q}_p)$ such that $\mathfrak{g}\otimes_{\mathbb{Q}_p}\mathbb{C}_p$ contains ϕ.*

This theorem is valid in the absolute case above, but relies heavily on the fact that the image of the Galois group is a finite dimensional Lie group. In the relative case it is doubtful that its proof can be generalized.

5.3 The Sen Operator Associated with ρ_r

Set $\mathbb{I}_{0,r,\mathbb{C}_p}=\mathbb{I}_{0,r}\widehat{\otimes}_{K_{h,0}}\mathbb{C}_p$. It is a Banach space for the natural norm. Let $\mathbb{B}_{r,\mathbb{C}_p}=\mathbb{B}_r\widehat{\otimes}_{K_{h,0}}\mathbb{C}_p$; it is the topological $\mathbb{C}_p$-algebra completion of $\mathbb{B}_r\otimes_{K_{h,0}}\mathbb{C}_p$ for the (uncountable) set of nuclear seminorms $p_{\mathfrak{a}}$ given by the norms on $\mathbb{I}_{0,r,\mathbb{C}_p}/\mathfrak{a}\mathbb{I}_{0,r,\mathbb{C}_p}$ via the specialization morphisms $\pi_{\mathfrak{a}}\colon\mathbb{B}_r\otimes_{K_{h,0}}\mathbb{C}_p\to\mathbb{I}_{0,r,\mathbb{C}_p}/\mathfrak{a}\mathbb{I}_{0,r,\mathbb{C}_p}$. Let $\mathfrak{H}_{r,\mathfrak{a},\mathbb{C}_p}=\mathfrak{H}_{r,\mathfrak{a}}\otimes_{K_{h,0}}\mathbb{C}_p$ and $\mathfrak{H}^{\mathrm{loc}}_{r,\mathfrak{a},\mathbb{C}_p}=\mathfrak{H}^{\mathrm{loc}}_{r,\mathfrak{a},}\otimes_{K_{h,0}}\mathbb{C}_p$. Then we define $\mathfrak{H}_{r,\mathbb{C}_p}=\mathfrak{H}_r\widehat{\otimes}_{K_{h,0}}\mathbb{C}_p$ as the topological $\mathbb{C}_p$-Lie algebra completion of $\mathfrak{H}_r\otimes_{K_{0,h}}\mathbb{C}_p$ for the (uncountable) set of seminorms $p_{\mathfrak{a}}$ given by the norms on $\mathfrak{H}_{r,\mathfrak{a},\mathbb{C}_p}$ and similar specialization morphisms $\pi_{\mathfrak{a}}\colon\mathfrak{H}_{r,}\otimes_{K_{h,0}}\mathbb{C}_p\to\mathfrak{H}_{r,\mathfrak{a},\mathbb{C}_p}$. We define in the same way $\mathfrak{H}^{\mathrm{loc}}_{r,\mathbb{C}_p}$ in terms of the norms on $\mathfrak{H}^{\mathrm{loc}}_{r,\mathfrak{a},\mathbb{C}_p}$. Note that by definition we have

$$\mathfrak{H}_{r,\mathbb{C}_p}=\varprojlim_{(\mathfrak{a},P_1)=1}\mathfrak{H}_{r,\mathfrak{a},\mathbb{C}_p},\quad\text{and }\mathfrak{H}^{\mathrm{loc}}_{r,\mathbb{C}_p}=\varprojlim_{(\mathfrak{a},P_1)=1}\mathfrak{H}^{\mathrm{loc}}_{r,\mathfrak{a},\mathbb{C}_p}.$$

We apply the construction of the previous subsection to $L=K_{h,0},\mathcal{R}=\mathbb{I}_{0,r}$ which is a Banach L-algebra with the p-adic topology, and $\tau=\rho_r$. We obtain an operator $\phi_r\in\mathrm{M}_2(\mathbb{I}_{0,r,\mathbb{C}_p})$. Recall that we have a natural continuous inclusion $\mathbb{I}_{0,r}\hookrightarrow\mathbb{B}_r$, inducing inclusions $\mathbb{I}_{0,r,\mathbb{C}_p}\hookrightarrow\mathbb{B}_{r,\mathbb{C}_p}$ and $\mathrm{M}_2(\mathbb{I}_{0,r,\mathbb{C}_p})\hookrightarrow\mathrm{M}_2(\mathbb{B}_{r,\mathbb{C}_p})$. We denote all these inclusions by $\iota_{\mathbb{B}_r}$ since it will be clear each time to which we are referring to. We will prove in this section that $\iota_{\mathbb{B}_r}(\phi_r)$ is an element of $\mathfrak{H}^{\mathrm{loc}}_{r,\mathbb{C}_p}$.

Let $\mathfrak{a}$ be a nonzero ideal of $A_{0,r}$. Let us apply Sen's construction to $L=K_{h,0}$, $\mathcal{R}=\mathbb{I}_{0,r}/\mathfrak{a}\mathbb{I}_{0,r}$ and $\tau=\rho_{r,\mathfrak{a}}\colon\mathrm{Gal}(\overline{K}/K)\to\mathrm{GL}_2(\mathbb{I}_{0,r}/\mathfrak{a}\mathbb{I}_{0,r})$; we obtain an operator $\phi_{r,\mathfrak{a}}\in\mathrm{M}_2(\mathbb{I}_{0,r}/\mathfrak{a}\mathbb{I}_{0,r}\widehat{\otimes}_{K_{h,0}}\mathbb{C}_p)$.

Let

$$\pi_{\mathfrak{a}}\colon\mathrm{M}_2(\mathbb{I}_{0,r}\widehat{\otimes}_{K_{h,0}}\mathbb{C}_p)\to\mathrm{M}_2(\mathbb{I}_{0,r}/\mathfrak{a}\mathbb{I}_{0,r}\widehat{\otimes}_{K_{h,0}}\mathbb{C}_p)$$

and

$$\pi_{\mathfrak{a}}^{\times}\colon\mathrm{GL}_2(\mathbb{I}_{0,r}\widehat{\otimes}_{K_{h,0}}\mathbb{C}_p)\to\mathrm{GL}_2(\mathbb{I}_{0,r}/\mathfrak{a}\mathbb{I}_{0,r}\widehat{\otimes}_{K_{h,0}}\mathbb{C}_p)$$

be the natural projections.

Proposition 5.6 *We have* $\phi_{r,\mathfrak{a}} = \pi_{\mathfrak{a}}(\phi_r)$ *for all* $\mathfrak{a}$.

Proof Recall from the construction of ϕ_r that there exist $M \in \mathrm{GL}_2\left(\mathbb{I}_{0,r,\mathbb{C}_p}\right)$, $n \geqslant 0$ and $\delta\colon \Gamma_n \to \mathrm{GL}_2(\mathbb{I}_{0,r}\widehat{\otimes}_{K_{h,0}} K'_{h,0,n})$ such that for all $\sigma \in G'_n$ we have

$$M^{-1}\rho_r(\sigma)\sigma(M) = \delta(\sigma) \tag{4}$$

and

$$\phi_r = \lim_{\sigma\to 1} \frac{\log(\delta(\sigma))}{\log(\chi(\sigma))}. \tag{5}$$

Let $M_{\mathfrak{a}} = \pi_{\mathfrak{a}}^{\times}(M) \in \mathrm{GL}_2(\mathbb{I}_{0,r,\mathbb{C}_p}/\mathfrak{a}\mathbb{I}_{0,r,\mathbb{C}_p})$ and

$$\delta_{\mathfrak{a}} = \pi_{\mathfrak{a}}^{\times} \circ \delta\colon \Gamma_n \to \mathrm{GL}_2((\mathbb{I}_{0,r}/\mathfrak{a}\mathbb{I}_{0,r})\widehat{\otimes}_{K_{h,0}} K'_{h,0,n}).$$

Denote by $\phi_{r,\mathfrak{a}} \in \mathrm{M}_2((\mathbb{I}_{0,r}/\mathfrak{a}\mathbb{I}_{0,r})\widehat{\otimes}_{K_{h,0}} K'_{h,0,n})$ the Sen operator associated with $\rho_{r,\mathfrak{a}}$. Now (4) gives

$$M_{\mathfrak{a}}^{-1}\rho_{r,\mathfrak{a}}(\sigma)\sigma(M_{\mathfrak{a}}) = \delta_{\mathfrak{a}}(\sigma) \tag{6}$$

so we can calculate $\phi_{r,\mathfrak{a}}$ as

$$\phi_{r,\mathfrak{a}} = \lim_{\sigma\to 1} \frac{\log(\delta_{\mathfrak{a}}(\sigma))}{\log(\chi(\sigma))} \in \mathrm{M}_2(\mathcal{R}\widehat{\otimes}_L \mathbb{C}_p). \tag{7}$$

By comparing this with (5) we see that $\phi_{r,\mathfrak{a}} = \pi_{\mathfrak{a}}(\phi_r)$. □

Let $\phi_{r,\mathbb{B}_r} = \iota_{\mathbb{B}_r}(\phi_r)$. For a nonzero ideal $\mathfrak{a}$ of $A_{0,r}$ let $\pi_{\mathbb{B}_r,\mathfrak{a}}$ be the natural projection $\mathbb{B}_r \to \mathbb{I}_{0,r}/\mathfrak{a}\mathbb{I}_{0,r}$. Clearly $\pi_{\mathbb{B}_r,\mathfrak{a}}(\phi_{r,\mathbb{B}_r}) = \pi_{\mathfrak{a}}(\phi_r)$ and $\phi_{r,\mathfrak{a}} = \pi_{\mathfrak{a}}(\phi_r)$ by Proposition 5.6, so we have $\phi_{r,\mathbb{B}_r} = \varprojlim_{(\mathfrak{a},P_1)=1} \phi_{r,\mathfrak{a}}$.

We apply Theorem 5.5 to show the following.

Proposition 5.7 *Let* $\mathfrak{a}$ *be a nonzero ideal of* $A_{0,r}$ *prime to* P_1. *The operator* $\phi_{r,\mathfrak{a}}$ *belongs to the Lie algebra* $\mathfrak{H}^{\mathrm{loc}}_{r,\mathfrak{a},\mathbb{C}_p}$.

Proof Let n be the dimension over $\mathbb{Q}_p$ of $\mathbb{I}_{0,r}/\mathfrak{a}\mathbb{I}_{0,r}$; by choosing a basis $(\omega_1,\dots,\omega_n)$ of this algebra as a $\mathbb{Q}_p$-vector space, we can define an injective ring homomorphism $\alpha\colon \mathrm{M}_2(\mathbb{I}_{0,r}/\mathfrak{a}\mathbb{I}_{0,r}) \hookrightarrow \mathrm{M}_{2n}(\mathbb{Q}_p)$ and an injective group homomorphism $\alpha^{\times}\colon \mathrm{GL}_2(\mathbb{I}_{0,r}/\alpha\mathbb{I}_{0,r}) \hookrightarrow \mathrm{GL}_{2n}(\mathbb{Q}_p)$. In fact, an endomorphism f of the $(\mathbb{I}_{0,r}/\mathfrak{a}\mathbb{I}_{0,r})$-module $(\mathbb{I}_{0,r}/\mathfrak{a}\mathbb{I}_{0,r})^2 = (\mathbb{I}_{0,r}/\mathfrak{a}\mathbb{I}_{0,r})\cdot e_1 \oplus (\mathbb{I}_{0,r}/\mathfrak{a}\mathbb{I}_{0,r})\cdot e_2$ is $\mathbb{Q}_p$-linear, so it induces an endomorphism $\alpha(f)$ of the $\mathbb{Q}_p$-vector space $(\mathbb{I}_{0,r}/\mathfrak{a}\mathbb{I}_{0,r})^2 = \bigoplus_{i,j}\mathbb{Q}_p\cdot\omega_i e_j$; furthermore if α is an automorphism then $\alpha(f)$ is one too. In particular $\rho_{r,\mathfrak{a}}$ induces a representation $\rho_{r,\mathfrak{a}}^{\alpha} = \alpha^{\times}\circ\rho_{r,\mathfrak{a}}\colon \mathrm{Gal}(\overline{K}/K) \to \mathrm{GL}_{2n}(\mathbb{Q}_p)$. The image of $\rho_{r,\mathfrak{a}}^{\alpha}$ is the group $G_{r,\mathfrak{a}}^{\mathrm{loc},\alpha} = \alpha^{\times}(G_{r,\mathfrak{a}}^{\mathrm{loc}})$. We consider its Lie algebra $\mathfrak{H}_{r,\mathfrak{a}}^{\mathrm{loc},\alpha} = \mathbb{Q}_p\cdot\mathrm{Log}\,(G_{r,\mathfrak{a}}^{\mathrm{loc},\alpha}) \subset \mathrm{M}_{2n}(\mathbb{Q}_p)$. The p-adic logarithm commutes with α in the sense that $\alpha(\mathrm{Log}\,x) = \mathrm{Log}\,(\alpha^{\times}(x))$ for every $x \in \Gamma_{\mathbb{I}_{0,r}/\mathfrak{a}\mathbb{I}_{0,r}}(p)$, so we have $\mathfrak{H}_{r,\mathfrak{a}}^{\mathrm{loc},\alpha} = \alpha(\mathfrak{H}_{r,\mathfrak{a}}^{\mathrm{loc}})$ (recall that $\mathfrak{H}_{r,\mathfrak{a}}^{\mathrm{loc}} = \mathbb{Q}_p\cdot\mathrm{Log}\,G_{r,\mathfrak{a}}^{\mathrm{loc}}$).

Let $\phi_{r,\mathfrak{a}}^{\alpha}$ be the Sen operator associated with $\rho_{r,\mathfrak{a}}^{\alpha}\colon \mathrm{Gal}(\overline{K}/K) \to \mathrm{GL}_{2n}(\mathbb{Q}_p)$. By Theorem 5.5 we have $\phi_{r,\mathfrak{a}}^{\alpha} \in \mathfrak{H}_{r,\mathfrak{a},\mathbb{C}_p}^{\mathrm{loc},\alpha} = \mathfrak{H}_{r,\mathfrak{a}}^{\mathrm{loc},\alpha}\widehat{\otimes}\mathbb{C}_p$. Denote by $\alpha_{\mathbb{C}_p}$ the map $\alpha\widehat{\otimes}1\colon \mathrm{M}_2(\mathbb{I}_{0,r,\mathbb{C}_p}/\mathfrak{a}\mathbb{I}_{0,r,\mathbb{C}_p}) \hookrightarrow \mathrm{M}_{2n}(\mathbb{C}_p)$. We show that $\phi_{r,\mathfrak{a}}^{\alpha_{\mathbb{C}_p}} = \alpha_{\mathbb{C}_p}(\phi_{r,\mathfrak{a}})$, from which it follows that $\phi_{r,\mathfrak{a}} \in \mathfrak{H}_{r,\mathfrak{a},\mathbb{C}_p}^{\mathrm{loc}}$ since $\mathfrak{H}_{r,\mathfrak{a},\mathbb{C}_p}^{\mathrm{loc},\alpha_{\mathbb{C}_p}} = \alpha_{\mathbb{C}_p}(\mathfrak{H}_{r,\mathfrak{a},\mathbb{C}_p}^{\mathrm{loc}})$ and $\alpha_{\mathbb{C}_p}$ is injective. Now let $M_{\mathfrak{a}}$, $\delta_{\mathfrak{a}}$ be as in (6) and $M_{\mathfrak{a}}^{\alpha_{\mathbb{C}_p}} = \alpha_{\mathbb{C}_p}(M_{\mathfrak{a}})$, $\delta_{\mathfrak{a}}^{\alpha_{\mathbb{C}_p}} = \alpha_{\mathbb{C}_p} \circ \delta_{\mathfrak{a}}$. By applying α_C to (4) we obtain $(M_{\mathfrak{a}}^{\alpha_{\mathbb{C}_p}})^{-1}\rho_{r,\mathfrak{a}}^{\alpha_{\mathbb{C}_p}}(\sigma)\sigma(M_{\mathfrak{a}}^{\alpha_{\mathbb{C}_p}}) = \delta_{\mathfrak{a}}^{\alpha_{\mathbb{C}_p}}(\sigma)$ for every $\sigma \in G_n'$, so we can calculate

$$\phi_{r,\mathfrak{a}}^{\alpha_{\mathbb{C}_p}} = \lim_{\sigma\to 1} \frac{\log(\delta_{\mathfrak{a}}^{\alpha_{\mathbb{C}_p}}(\sigma))}{\log(\chi(\sigma))},$$

which coincides with $\alpha_{\mathbb{C}_p}(\phi_{r,\mathfrak{a}})$. □

Proposition 5.8 *The element $\phi_{r,\mathbb{B}_r}$ belongs to $\mathfrak{H}_{r,\mathbb{C}_p}^{\mathrm{loc}}$, hence to $\mathfrak{H}_{r,\mathbb{C}_p}$.*

Proof By definition of the space $\mathfrak{H}_{r,\mathbb{C}_p}^{\mathrm{loc}}$ as completion of the space $\mathfrak{H}_r^{\mathrm{loc}} \otimes_{K_{h,0}} \mathbb{C}_p$ for the seminorms $p_{\mathfrak{a}}$ given by the norms on $\mathfrak{H}_{r,\mathfrak{a},\mathbb{C}_p}^{\mathrm{loc}}$, we have $\mathfrak{H}_{r,\mathbb{C}_p}^{\mathrm{loc}} = \varprojlim_{(\mathfrak{a},P_1)=1} \mathfrak{H}_{r,\mathfrak{a},\mathbb{C}_p}^{\mathrm{loc}}$. By Proposition 5.6, we have $\phi_{r,\mathbb{B}_r} = \varprojlim_{\mathfrak{a}} \phi_{r,\mathfrak{a}}$ and by Proposition 5.7 we have, for every $\mathfrak{a}$, $\phi_{r,\mathfrak{a}} \in \mathfrak{H}_{r,\mathfrak{a},\mathbb{C}_p}^{\mathrm{loc}}$. We conclude that $\phi_{r,\mathbb{B}_r} \in \mathfrak{H}_{r,\mathbb{C}_p}^{\mathrm{loc}}$. □

Remark 5.9 In order to prove that our Lie algebras are "big" it will be useful to work with primary ideals of A_r, as we did in this subsection. However, in light of Remark 5.3, all of the results can be rewritten in terms of primary ideals $\mathfrak{Q}$ of $\mathbb{I}_{0,r}$. This will be useful in the next subsection, when we will interpolate the Sen operators corresponding to the attached to the classical modular forms representations.

From now on we identify $\mathbb{I}_{0,r,\mathbb{C}_p}$ with a subring of $\mathbb{B}_{r,\mathbb{C}_p}$ via $\iota_{\mathbb{B}_r}$, so we also identify $\mathrm{M}_2(\mathbb{I}_{0,r})$ with a subring of $\mathrm{M}_2(\mathbb{B}_r)$ and $\mathrm{GL}_2(\mathbb{I}_{0,r,\mathbb{C}_p})$ with a subgroup of $\mathrm{GL}_2(\mathbb{B}_{r,\mathbb{C}_p})$. In particular we identify ϕ_r with $\phi_{r,\mathbb{B}_r}$ and we consider ϕ_r as an element of $\mathfrak{H}_{r,\mathbb{C}_p} \cap \mathrm{M}_2(\mathbb{I}_{0,r,\mathbb{C}_p})$.

5.4 The Characteristic Polynomial of the Sen Operator

Sen proved the following result.

Theorem 5.10 *Let L_1 and L_2 be two p-adic fields. Assume for simplicity that L_2 contains the normal closure of L_1. Let $\tau\colon \mathrm{Gal}(\overline{L}_1/L_1) \to \mathrm{GL}_m(L_2)$ be a continuous representation. For each embedding $\sigma\colon L_1 \to L_2$, there is a Sen operator $\phi_{\tau,\sigma} \in \mathrm{M}_m(\mathbb{C}_p \otimes_{L_1,\sigma} L_2)$ associated with τ and σ. If τ is Hodge-Tate and its Hodge-Tate weights with respect to σ are $h_{1,\sigma},\ldots,h_{m,\sigma}$ (with multiplicities, if any), then the characteristic polynomial of $\phi_{\tau,\sigma}$ is $\prod_{i=1}^m(X - h_{i,\sigma})$.*

Now let $k \in \mathbb{N}$ and $P_k = (u^{-k}(1+T) - 1)$ be the corresponding arithmetic prime of $A_{0,r}$. Let $\mathfrak{P}_f$ be a prime of $\mathbb{I}_r$ above P, associated with the system of Hecke eigenvalues of a classical modular form f. Let $\rho_r \colon \mathbb{G}_Q \to \mathrm{GL}_2(\mathbb{I}_r)$ be as usual. The specialization of ρ_r modulo $\mathfrak{P}$ is the representation $\rho_f \colon G_{\mathbb{Q}} \to \mathrm{GL}_2(\mathbb{I}_r/\mathfrak{P})$ classically associated with f, defined over the field $K_f = \mathbb{I}_r/\mathfrak{P}_f\mathbb{I}_r$. By a theorem of Faltings [8], when the weight of the form f is k, the representation ρ_f is Hodge-Tate of Hodge-Tate weights 0 and $k-1$. Hence by Theorem 5.10 the Sen operator ϕ_f associated with ρ_f has characteristic polynomial $X(X-(k-1))$. Let $\mathfrak{P}_{f,0} = \mathfrak{P}_f \cap \mathbb{I}_{0,r}$. With the notations of the previous subsection, the specialization of ρ_r modulo $\mathfrak{P}_{f,0}$ gives a representation $\rho_{r,\mathfrak{P}_{f,0}} \colon \mathrm{Gal}(\overline{K}/K) \to \mathrm{GL}_2(\mathbb{I}_{0,r}/\mathfrak{P}_{f,0})$, which coincides with $\rho_f|_{\mathrm{Gal}(\overline{K}/K)}$. In particular the Sen operator $\phi_{r,\mathfrak{P}_{f,0}}$ associated with $\rho_{r,\mathfrak{P}_{f,0}}$ is ϕ_f.

By Proposition 5.6 and Remark 5.9, the Sen operator $\phi_r \in \mathrm{M}_2(\mathbb{I}_{0,r,\mathbb{C}_p})$ specializes modulo $\mathfrak{P}_{f,0}$ to the Sen operator $\phi_{r,\mathfrak{P}_{f,0}}$ associated with $\rho_{r,\mathfrak{P}_{f,0}}$, for every f as above. Since the primes of the form $\mathfrak{P}_{f,0}$ are dense in $\mathbb{I}_{0,r,\mathbb{C}_p}$, the eigenvalues of $\phi_{r,Q}$ are given by the unique interpolation of those of $\rho_{r,\mathfrak{P}_{f,0}}$. This way we will recover an element of $\mathrm{GL}_2(\mathbb{B}_{r,\mathbb{C}_p})$ with the properties we need.

Given $f \in A_{0,r}$ we define its p-adic valuation by $v_p'(f) = \inf_{x \in \mathcal{B}(0,r)} v_p(f(x))$, where v_p is our chosen valuation on $\mathbb{C}_p$. Then if $v'(f-1) \leqslant p^{-\frac{1}{p-1}}$ there are well-defined elements $\log(f)$ and $\exp(\log(f))$ in $A_{0,r}$, and $\exp(\log(f)) = f$.

Let $\phi_r' = \log(u)\phi_r$. Note that ϕ_r' is a well-defined element of $\mathrm{M}_2(\mathbb{B}_{r,\mathbb{C}_p})$ since $\log(u) \in \mathbb{Q}_p$. Recall that we denote by C_T the matrix $\mathrm{diag}(u^{-1}(1+T), 1)$. We have the following.

Proposition 5.11 *1. The eigenvalues of ϕ_r' are* $\log(u^{-1}(1+T))$ *and* 0. *In particular the exponential $\Phi_r = \exp(\phi_r')$ is defined in* $\mathrm{GL}_2(\mathbb{B}_{r,\mathbb{C}_p})$. *Moreover Φ_r' is conjugate to C_T in* $\mathrm{GL}_2(\mathbb{B}_{r,\mathbb{C}_p})$.
2. The element Φ_r' of part (1) normalizes $\mathfrak{H}_{r,\mathbb{C}_p}$.

Proof For every $\mathfrak{P}_{f,0}$ as in the discussion above, the element $\log(u)\phi_r$ specializes to $\log(u)\phi_{r,\mathfrak{P}_{f,0}}$ modulo $\mathfrak{P}_{f,0}$. If $\mathfrak{P}_{f,0}$ is a divisor of P_k, the eigenvalues of $\log(u)\phi_{r,\mathfrak{P}_{f,0}}$ are $\log(u)(k-1)$ and 0. Since $1+T = u^k$ modulo $\mathfrak{P}_{f,0}$ for every prime $\mathfrak{P}_{f,0}$ dividing P_k, we have $\log(u^{-1}(1+T)) = \log(u^{k-1}) = (k-1)\log(u)$ modulo $\mathfrak{P}_{f,0}$. Hence the eigenvalues of $\log(u)\phi_{r,\mathfrak{P}_{f,0}}$ are interpolated by $\log(u^{-1}(1+T))$ and 0.

Recall that in Sect. 3.1 we chose r_h smaller than $p^{-\frac{1}{p-1}}$. Since $r < r_h$, $v_p'(T) < p^{-\frac{1}{p-1}}$. In particular $\log(u^{-1}(1+T))$ is defined and $\exp(\log(u^{-1}(1+T))) = u^{-1}(1+T)$, so $\Phi_r = \exp(\phi_r')$ is also defined and its eigenvalues are $u^{-1}(1+T)$ and 1. The difference between the two is $u^{-1}(1+T)-1$; this elements belongs to P_1, hence it is invertible in $\mathbb{B}_r$. This proves (1).

By Proposition 5.8, $\phi_r \in \mathfrak{H}_{r,\mathbb{C}_p}$. Since $\mathfrak{H}_{r,\mathbb{C}_p}$ is a $\mathbb{Q}_p$-Lie algebra, $\log(u)\phi_r$ is also an element of $\mathfrak{H}_{r,\mathbb{C}_p}$. Hence its exponential Φ_r' normalizes $\mathfrak{H}_{r,\mathbb{C}_p}$. □

6 Existence of the Galois Level for a Family with Finite Positive Slope

Let $r_h \in p^{\mathbb{Q}} \cap]0, p^{-\frac{1}{p-1}}[$ be the radius chosen in Sect. 3. As usual we write r for any one of the radii r_i of Sect. 3.1. Recall that $\mathfrak{H}_r \subset \mathrm{M}_2(\mathbb{B}_r)$ is the Lie algebra attached to the image of ρ_r (see Definition 5.2) and $\mathfrak{H}_{r,\mathbb{C}_p} = \mathfrak{H}_r \widehat{\otimes} \mathbb{Q}_p \mathbb{C}_p$. Let $\mathfrak{u}^{\pm}$ and $\mathfrak{u}^{\pm}_{\mathbb{C}_p}$ be the upper and lower nilpotent subalgebras of $\mathfrak{H}_r$, and $\mathfrak{H}_{r,\mathbb{C}_p}$ respectively.

Remark 6.1 The commutative Lie algebra $\mathfrak{u}^{\pm}$ is independent of r because it is equal to $\mathbb{Q}_p \cdot \mathrm{Log}(U(\mathbb{I}_0^\circ) \cap G'_r)$ which is independent of r, provided $r_1 \leqslant r < r_h$.

We fix $r_0 \in p^{\mathbb{Q}} \cap]0, r_h[$ arbitrarily and we work from now on with radii r satisfying $r_0 \leqslant r < r_h$. As in Remark 5.1 this fixes a finite extension of $\mathbb{Q}$ corresponding to the inclusion $G'_r \subset G_r$. For $r < r'$ we have a natural inclusion $\mathbb{I}_{0,r'} \hookrightarrow \mathbb{I}_{0,r}$. Since $\mathbb{B}_r = \varprojlim_{(\mathfrak{a}P_1)=1} \mathbb{I}_{0,r}/\mathfrak{a}\mathbb{I}_{0,r}$ this induces an inclusion $\mathbb{B}_{r'} \hookrightarrow \mathbb{B}_r$. We will consider from now on $\mathbb{B}_{r'}$ as a subring of $\mathbb{B}_r$ for every $r < r'$. We will also consider $\mathrm{M}_2(\mathbb{I}_{0,r',\mathbb{C}_p})$ and $\mathrm{M}_2(\mathbb{B}_{r'})$ as subsets of $\mathrm{M}_2(\mathbb{I}_{0,r,\mathbb{C}_p})$ and $\mathrm{M}_2(\mathbb{B}_r)$ respectively. These inclusions still hold after taking completed tensor products with $\mathbb{C}_p$.

Recall the elements $\phi'_r = \log(u)\phi_r \in \mathrm{M}_2(\mathbb{B}_{r,\mathbb{C}_p})$ and $\Phi'_r = \exp(\phi'_r) \in \mathrm{GL}_2(\mathbb{B}_{r,\mathbb{C}_p})$ defined at the end of the previous section. The Sen operator ϕ_r is independent of r in the following sense: if $r < r' < r_h$ and $\mathbb{B}_{r',\mathbb{C}_p} \to \mathbb{B}_{r,\mathbb{C}_p}$ is the natural inclusion then the image of $\phi_{r'}$ under the induced map $\mathrm{M}_2(\mathbb{B}_{r',\mathbb{C}_p}) \to \mathrm{M}_2(\mathbb{B}_{r,\mathbb{C}_p})$ is ϕ_r. We deduce that ϕ'_r and Φ'_r are also independent of r (in the same sense).

By Proposition 5.11, for every $r < r_h$ there exists an element $\beta_r \in \mathrm{GL}_2(\mathbb{B}_{r,\mathbb{C}_p})$ such that $\beta_r \Phi'_r \beta_r^{-1} = C_T$. Since Φ'_r normalizes $\mathfrak{H}_{r,\mathbb{C}_p}$, $C_T = \beta_r \Phi'_r \beta_r^{-1}$ normalizes $\beta_r \mathfrak{H}_{r,\mathbb{C}_p} \beta_r^{-1}$.

We denote by $\mathfrak{U}^{\pm}$ the upper and lower nilpotent subalgebras of $\mathfrak{sl}_2$. The action of C_T on $\mathfrak{H}_{r,\mathbb{C}_p}$ by conjugation is semisimple, so we can decompose $\beta_r \mathfrak{H}_{r,\mathbb{C}_p} \beta_r^{-1}$ as a sum of eigenspaces for C_T:

$$\beta_r \mathfrak{H}_{r,\mathbb{C}_p} \beta_r^{-1} = \left(\beta_r \mathfrak{H}_{r,\mathbb{C}_p} \beta_r^{-1}\right)[1] \oplus \left(\beta_r \mathfrak{H}_{r,\mathbb{C}_p} \beta_r^{-1}\right)[u^{-1}(1+T)] \oplus \left(\beta_r \mathfrak{H}_{r,\mathbb{C}_p} \beta_r^{-1}\right)[u(1+T)^{-1}]$$

with $\left(\beta_r \mathfrak{H}_{r,\mathbb{C}_p} \beta_r^{-1}\right)[u^{-1}(1+T)] \subset \mathfrak{U}^+(\mathbb{B}_{r,\mathbb{C}_p})$ and $\left(\beta_r \mathfrak{H}_{r,\mathbb{C}_p} \beta_r^{-1}\right)[u(1+T)^{-1}] \subset \mathfrak{U}^-(\mathbb{B}_{r,\mathbb{C}_p})$.

Moreover, the formula

$$\begin{pmatrix} u^{-1}(1+T) & 0 \\ 0 & 1 \end{pmatrix} \begin{pmatrix} 1 & \lambda \\ 0 & 1 \end{pmatrix} \begin{pmatrix} u^{-1}(1+T) & 0 \\ 0 & 1 \end{pmatrix}^{-1} = \begin{pmatrix} 1 & u^{-1}(1+T)\lambda \\ 0 & 1 \end{pmatrix}$$

shows that the action of C_T by conjugation coincides with multiplication by $u^{-1}(1+T)$. By linearity this gives an action of the polynomial ring $\mathbb{C}_p[T]$ on $\beta_r \mathfrak{H}_{r,\mathbb{C}_p} \beta_r^{-1} \cap \mathfrak{U}^+(\mathbb{B}_{r,\mathbb{C}_p})$, compatible with the action of $\mathbb{C}_p[T]$ on $\mathfrak{U}^+(\mathbb{B}_{r,\mathbb{C}_p})$ given by the inclusions

$\mathbb{C}_p[T] \subset \Lambda_{h,0,\mathbb{C}_p} \subset B_{r,\mathbb{C}_p} \subset \mathbb{B}_{r,\mathbb{C}_p}$. Since $\mathbb{C}_p[T]$ is dense in $A_{h,0,\mathbb{C}_p}$ for the p-adic topology, it is also dense in $B_{r,\mathbb{C}_p}$. Since $\mathfrak{H}_{r,\mathbb{C}_p}$ is a closed Lie subalgebra of $\mathrm{M}_2(\mathbb{B}_{r,\mathbb{C}_p})$, we can define by continuity a $B_{r,\mathbb{C}_p}$-module structure on $\beta_r\mathfrak{H}_{r,\mathbb{C}_p}\beta_r^{-1} \cap \mathfrak{U}^+(\mathbb{B}_{r,\mathbb{C}_p})$, compatible with that on $\mathfrak{U}^+(\mathbb{B}_{r,\mathbb{C}_p})$. Similarly we have

$$\begin{pmatrix} u^{-1}(1+T) & 0 \\ 0 & 1 \end{pmatrix} \begin{pmatrix} 1 & 0 \\ \mu & 1 \end{pmatrix} \begin{pmatrix} u^{-1}(1+T) & 0 \\ 0 & 1 \end{pmatrix}^{-1} = \begin{pmatrix} 1 & 0 \\ u(1+T)^{-1}\mu & 1 \end{pmatrix}.$$

We note that $1+T$ is invertible in $A_{0,r}$ since $T = p^{s_h}t$ where $r_h = p^{-s_h}$. Therefore C_T is invertible and by twisting by $(1+T) \mapsto (1+T)^{-1}$ we can also give $\beta_r\mathfrak{H}_{r,\mathbb{C}_p}\beta_r^{-1} \cap \mathfrak{U}^-(\mathbb{B}_{r,\mathbb{C}_p})$ a structure of $B_{r,\mathbb{C}_p}$-module compatible with that on $\mathfrak{U}^-(\mathbb{B}_{r,\mathbb{C}_p})$.

By combining the previous remarks with Corollary 4.23, we prove the following "fullness" result for the big Lie algebra $\mathfrak{H}_r$.

Theorem 6.2 *Suppose that the representation ρ is $(H_0, \mathbb{Z}_p)$-regular. Then there exists a nonzero ideal $\mathfrak{l}$ of $\mathbb{I}_0$, independent of $r < r_h$, such that for every such r the Lie algebra $\mathfrak{H}_r$ contains $\mathfrak{l} \cdot \mathfrak{sl}_2(\mathbb{B}_r)$.*

Proof Since $U^{\pm}(\mathbb{B}_r) \cong \mathbb{B}_r$, we can and shall identify $\mathfrak{u}^+ = \mathbb{Q}_p \cdot \mathrm{Log}\, G'_r \cap \mathfrak{U}^+(\mathbb{B}_r)$ with a $\mathbb{Q}_p$-vector subspace of $\mathbb{B}_r$ (actually of $\mathbb{I}_0$), and $\mathfrak{u}^+_{\mathbb{C}_p}$ with a $\mathbb{C}_p$-vector subspace of $\mathbb{B}_{r,\mathbb{C}_p}$. We repeat that these spaces are independent of r since G'_r is, provided that $r_0 \leqslant r < r_h$ (see Remark 5.1). By Corollary 4.23, $\mathfrak{u}^{\pm} \cap \mathbb{I}_0$ contains a basis $\{e_{i,\pm}\}_{i\in I}$ for $Q(\mathbb{I}_0)$ over $Q(\Lambda_{h,0})$. The set $\{e_{i,+}\}_{i\in I} \subset \mathfrak{u}^+$ is a basis for $Q(\mathbb{I}_0)$ over $Q(\Lambda_{h,0})$, so $\mathfrak{u}^+$ contains the basis of a $\Lambda_{h,0}$-lattice in $\mathbb{I}_0$. By Lemma 4.19 we deduce that $\Lambda_{h,0}\mathfrak{u}^+$ contains a nonzero ideal $\mathfrak{a}^+$ of $\mathbb{I}_0$. Hence we also have $B_{r,\mathbb{C}_p}\mathfrak{u}^+_{\mathbb{C}_p} \supset B_{r,\mathbb{C}_p}\mathfrak{a}^+$. Now $\mathfrak{a}^+$ is an ideal of $\mathbb{I}_0$ and $B_{r,\mathbb{C}_p}\mathbb{I}_{0,\mathbb{C}_p} = \mathbb{B}_{r,\mathbb{C}_p}$, so $B_{r,\mathbb{C}_p}\mathfrak{a}^+ = \mathfrak{a}^+\mathbb{B}_{r,\mathbb{C}_p}$ is an ideal in $\mathbb{B}_{r,\mathbb{C}_p}$. We conclude that $B_{r,\mathbb{C}_p} \cdot \mathfrak{u}^+ \supset \mathfrak{a}^+\mathbb{B}_{r,\mathbb{C}_p}$ for a nonzero ideal $\mathfrak{a}^+$ of $\mathbb{I}_0$. We proceed in the same way for the lower unipotent subalgebra, obtaining $B_{r,\mathbb{C}_p} \cdot \mathfrak{u}^- \supset \mathfrak{a}^-\mathbb{B}_{r,\mathbb{C}_p}$ for some nonzero ideal $\mathfrak{a}^-$ of $\mathbb{I}_0$.

Consider now the Lie algebra $B_{r,\mathbb{C}_p}\mathfrak{H}_{\mathbb{C}_p} \subset \mathrm{M}_2(\mathbb{B}_{r,\mathbb{C}_p})$. Its nilpotent subalgebras are $B_{r,\mathbb{C}_p}\mathfrak{u}^+$ and $B_{r,\mathbb{C}_p}\mathfrak{u}^-$, and we showed $B_{r,\mathbb{C}_p}\mathfrak{u}^+ \supset \mathfrak{a}^+\mathbb{B}_{r,\mathbb{C}_p}$ and $B_{r,\mathbb{C}_p}\mathfrak{u}^- \supset \mathfrak{a}^-\mathbb{B}_{r,\mathbb{C}_p}$. Denote by $\mathfrak{t} \subset \mathfrak{sl}_2$ the subalgebra of diagonal matrices over $\mathbb{Z}$. By taking the Lie bracket, we see that $[\mathfrak{U}^+(\mathfrak{a}^+\mathbb{B}_{r,\mathbb{C}_p}), \mathfrak{U}^-(\mathfrak{a}^-\mathbb{B}_{r,\mathbb{C}_p})]$ spans $\mathfrak{a}^+ \cdot \mathfrak{a}^- \cdot \mathfrak{t}(\mathbb{B}_{r,\mathbb{C}_p})$ over $B_{r,\mathbb{C}_p}$. We deduce that $B_{r,\mathbb{C}_p}\mathfrak{H}_{\mathbb{C}_p} \supset \mathfrak{a}^+ \cdot \mathfrak{a}^- \cdot \mathfrak{sl}_2(\mathbb{B}_{r,\mathbb{C}_p})$. Let $\mathfrak{a} = \mathfrak{a}^+ \cdot \mathfrak{a}^-$. Now $\mathfrak{a} \cdot \mathfrak{sl}_2(\mathbb{B}_{r,\mathbb{C}_p})$ is a $\mathbb{B}_{r,\mathbb{C}_p}$-Lie subalgebra of $\mathfrak{sl}_2(\mathbb{B}_{r,\mathbb{C}_p})$. Recall that $\beta_r \in \mathrm{GL}_2(\mathbb{B}_{r,\mathbb{C}_p})$; hence by stability by conjugation we have $\beta_r\left(\mathfrak{a} \cdot \mathfrak{sl}_2(\mathbb{B}_{r,\mathbb{C}_p})\right)\beta_r^{-1} = \mathfrak{a} \cdot \mathfrak{sl}_2(\mathbb{B}_{r,\mathbb{C}_p})$. Thus, we constructed $\mathfrak{a}$ such that $B_{r,\mathbb{C}_p}\left(\beta_r\mathfrak{H}_{r,\mathbb{C}_p}\beta_r^{-1}\right) \supset \mathfrak{a} \cdot \mathfrak{sl}_2(\mathbb{B}_{r,\mathbb{C}_p})$. In particular, if $\mathfrak{u}^{\pm,\beta_r}_{\mathbb{C}_p}$ denote the unipotent subalgebras of $\beta_r\mathfrak{H}_{r,\mathbb{C}_p}\beta_r^{-1}$, we have $B_{r,\mathbb{C}_p}\mathfrak{u}^{\pm,\beta_r}_{\mathbb{C}_p} \supset \mathfrak{a}\mathbb{B}_{r,\mathbb{C}_p}$ for both signs. By the discussion preceding the proposition the subalgebras $\mathfrak{u}^{\pm,\beta_r}_{\mathbb{C}_p}$ have a structure of $B_{r,\mathbb{C}_p}$-modules, which means that $\mathfrak{u}^{\pm,\beta_r}_{\mathbb{C}_p} = B_{r,\mathbb{C}_p}\mathfrak{u}^{\pm,\beta_r}_{\mathbb{C}_p}$. We conclude that $\mathfrak{u}^{\pm,\beta_r}_{\mathbb{C}_p} \supset \beta_r\left(\mathfrak{a} \cdot \mathfrak{U}^{\pm}(\mathbb{B}_{r,\mathbb{C}_p})\right)\beta_r^{-1}$ for both signs. By the usual argument of taking the bracket, we obtain $\beta_r\mathfrak{H}_{r,\mathbb{C}_p}\beta_r^{-1} \supset \mathfrak{a}^2 \cdot \mathfrak{sl}_2(\mathbb{B}_{r,\mathbb{C}_p})$. We can untwist by the invertible matrix β_r to conclude that, for $\mathfrak{l} = \mathfrak{a}^2$, we have $\mathfrak{H}_{r,\mathbb{C}_p} \supset \mathfrak{l} \cdot \mathfrak{sl}_2(\mathbb{B}_{r,\mathbb{C}_p})$.

Let us get rid of the completed extension of scalars to $\mathbb{C}_p$. For every ideal $\mathfrak{a} \subset \mathbb{I}_{0,r}$ not dividing P_1, let $\mathfrak{H}_{r,\mathfrak{a}}$ be the image of $\mathfrak{H}_r$ in $\mathrm{M}_2(\mathbb{I}_{0,r}/\mathfrak{a}\mathbb{I}_{0,r})$. Consider the two finite dimensional $\mathbb{Q}_p$-vector spaces $\mathfrak{H}_{r,\mathfrak{a}}$ and $\mathfrak{l} \cdot \mathfrak{sl}_2(\mathbb{I}_{0,r}/\mathfrak{a}\mathbb{I}_{0,r})$. Note that they are both subspaces of the finite dimensional $\mathbb{Q}_p$-vector space $\mathrm{M}_2(\mathbb{I}_{0,r}/\mathfrak{a}\mathbb{I}_{0,r})$. After extending scalars to $\mathbb{C}_p$, we have

$$\mathfrak{l} \cdot \mathfrak{sl}_2(\mathbb{I}_{0,r}/\mathfrak{a}\mathbb{I}_{0,r}) \otimes \mathbb{C}_p \subset \mathfrak{H}_{r,\mathfrak{a}} \otimes \mathbb{C}_p. \tag{8}$$

Let $\{e_i\}_{i\in I}$ be an orthonormal basis of the Banach space $\mathbb{C}_p$ over $\mathbb{Q}_p$, with I some index set, such that $1 \in \{e_i\}_{i\in I}$. Let $\{v_j\}_{j=1,\ldots,n}$ be a $\mathbb{Q}_p$-basis of $\mathrm{M}_2(\mathbb{I}_{0,r}/\mathfrak{a}\mathbb{I}_{0,r})$ such that, for some $d \leqslant n$, $\{v_j\}_{j=1,\ldots,d}$ is a $\mathbb{Q}_p$-basis of $\mathfrak{H}_{r,\mathfrak{a}}$.

Let v be an element of $\mathfrak{l} \cdot \mathfrak{sl}_2(\mathbb{I}_{0,r}/\mathfrak{a}\mathbb{I}_{0,r})$. Then $v \otimes 1 \in \mathfrak{l} \cdot \mathfrak{sl}_2(\mathbb{I}_{0,r}/\mathfrak{a}\mathbb{I}_{0,r}) \otimes \mathbb{C}_p$ and by (8) we have $v \otimes 1 \in \mathfrak{H}_{r,\mathfrak{a}} \otimes \mathbb{C}_p$. As $\{v_j \otimes e_i\}_{1\leqslant j\leqslant d, i\in I}$, and $\{v_j \otimes e_i\}_{1\leqslant j\leqslant n, i\in I}$ are orthonormal bases of $\mathfrak{H}_{r,\mathfrak{a}} \otimes \mathbb{C}_p$, and $\mathrm{M}_2(\mathbb{I}_{0,r}/\mathfrak{a}\mathbb{I}_{0,r}) \otimes \mathbb{C}_p$ over $\mathbb{Q}_p$, respectively there exist $\lambda_{j,i} \in \mathbb{Q}_p$, $(j, i) \in \{1, 2, \ldots, d\} \times I$ converging to 0 in the filter of complements of finite subsets of $\{1, 2, \ldots, d\} \times I$ such that $v \otimes 1 = \sum_{j=1,\ldots,d;\, i\in I} \lambda_{j,i}(v_j \otimes e_i)$.

But $v \otimes 1 \in \mathrm{M}_2(\mathbb{I}_{0,r}/\mathfrak{a}\mathbb{I}_{0,r}) \otimes 1 \subset \mathrm{M}_2(\mathbb{I}_{0,r}/\mathfrak{a}\mathbb{I}_{0,r}) \otimes \mathbb{C}_p$ and therefore $v \otimes 1 = \sum_{1\leqslant j\leqslant n} a_j(v_j \otimes 1)$, for some $a_j \in \mathbb{Q}_p$, $j = 1, \ldots, n$. By the uniqueness of a representation of an element in a $\mathbb{Q}_p$-Banach space in terms of a given orthonormal basis we have

$$v \otimes 1 = \sum_{j=1}^{d} a_j(v_j \otimes 1), \quad \text{i.e.} \quad v = \sum_{j=1}^{d} a_j v_j \in \mathfrak{H}_{r,\mathfrak{a}}.$$

By taking the projective limit over $\mathfrak{a}$, we conclude that

$$\mathfrak{l} \cdot \mathfrak{sl}_2(\mathbb{B}_r) \subset \mathfrak{H}_r.$$

□

Definition 6.3 The Galois level of the family $\theta\colon \mathbb{T}_h \to \mathbb{I}^\circ$ is the largest ideal $\mathfrak{l}_\theta$ of $\mathbb{I}_0[P_1^{-1}]$ such that $\mathfrak{H}_r \supset \mathfrak{l}_\theta \cdot \mathfrak{sl}_2(\mathbb{B}_r)$ for all $r < r_h$.

It follows by the previous remarks that $\mathfrak{l}_\theta$ is nonzero.

7 Comparison Between the Galois Level and the Fortuitous Congruence Ideal

Let $\theta\colon \mathbb{T}_h \to \mathbb{I}^\circ$ be a slope $\leqslant h$ family. We keep all the notations from the previous sections. In particular $\rho\colon G_{\mathbb{Q}} \to \mathrm{GL}_2(\mathbb{I}^\circ)$ is the Galois representation associated with θ. We suppose that the restriction of ρ to H_0 takes values in $\mathrm{GL}_2(\mathbb{I}_0^\circ)$. Recall that

$\mathbb{I} = \mathbb{I}^\circ[p^{-1}]$ and $\mathbb{I}_0 = \mathbb{I}_0^\circ[p^{-1}]$. Also recall that P_1 is the prime of $\Lambda_{h,0}$ generated by $u^{-1}(1+T)-1$. Let $\mathfrak{c} \subset \mathbb{I}$ be the congruence ideal associated with θ. Set $\mathfrak{c}_0 = \mathfrak{c} \cap \mathbb{I}_0$ and $\mathfrak{c}_1 = \mathfrak{c}_0\mathbb{I}_0[P_1^{-1}]$. Let $\mathfrak{l} = \mathfrak{l}_\theta \subset \mathbb{I}_0[P_1^{-1}]$ be the Galois level of θ. For an ideal $\mathfrak{a}$ of $\mathbb{I}_0[P_1^{-1}]$ we denote by $V(\mathfrak{a})$ the set of prime ideals of $\mathbb{I}_0[P_1^{-1}]$ containing $\mathfrak{a}$. We prove the following.

Theorem 7.1 *Suppose that*

1. *ρ is $(H_0, \mathbb{Z}_p)$-regular;*
2. *there exists no pair (F, ψ), where F is a real quadratic field and $\psi\colon \mathrm{Gal}(\overline{F}/F) \to \mathbb{F}^\times$ is a character, such that $\overline{\rho}\colon G_\mathbb{Q} \to \mathrm{GL}_2(\mathbb{F}) \cong \mathrm{Ind}_F^\mathbb{Q}\psi$.*

Then we have $V(\mathfrak{l}) = V(\mathfrak{c}_1)$.

Before giving the proof we make some remarks. Let P be a prime of $\mathbb{I}_0[P_1^{-1}]$ and Q be a prime factor of $P\mathbb{I}[P_1^{-1}]$. We consider ρ as a representation $G_\mathbb{Q} \to \mathrm{GL}_2(\mathbb{I}[P_1^{-1}])$ by composing it with the inclusion $\mathrm{GL}_2(\mathbb{I}) \hookrightarrow \mathrm{GL}_2(\mathbb{I}[P_1^{-1}])$. We have a representation $\rho_Q\colon G_\mathbb{Q} \to \mathrm{GL}_2(\mathbb{I}[P_1^{-1}]/Q)$ obtained by reducing ρ modulo Q. Its restriction $\rho_Q|_{H_0}$ takes values in $\mathrm{GL}_2(\mathbb{I}_0[P_1^{-1}]/(Q \cap \mathbb{I}_0[P_1^{-1}])) = \mathrm{GL}_2(\mathbb{I}_0[P_1^{-1}]/P)$ and coincides with the reduction ρ_P of $\rho|_{H_0}\colon H_0 \to \mathrm{GL}_2(\mathbb{I}_0[P_1^{-1}])$ modulo P. In particular $\rho_Q|_{H_0}$ is independent of the chosen prime factor Q of $P\mathbb{I}[P_1^{-1}]$.

We say that a subgroup of $\mathrm{GL}_2(A)$ for some algebra A finite over a p-adic field K is *small* if it admits a finite index abelian subgroup. Let P, Q be as above, G_P be the image of $\rho_P\colon H_0 \to \mathrm{GL}_2(\mathbb{I}_0[P_1^{-1}]/P)$ and G_Q be the image of $\rho_Q\colon G_\mathbb{Q} \to \mathrm{GL}_2(\mathbb{I}[P_1^{-1}]/Q)$. By our previous remark ρ_P coincides with the restriction $\rho_Q|_{H_0}$, so G_P is a finite index subgroup of G_Q for every Q. In particular G_P is small if and only if G_Q is small for all prime factors Q of $P\mathbb{I}[P_1^{-1}]$.

Now if Q is a CM point the representation ρ_Q is induced by a character of $\mathrm{Gal}(F/\mathbb{Q})$ for an imaginary quadratic field F. Hence G_Q admits an abelian subgroup of index 2 and G_P is also small.

Conversely, if G_P is small, $G_{Q'}$ is small for every prime Q' above P. Choose any such prime Q'; by the argument in [16, Proposition 4.4] $G_{Q'}$ has an abelian subgroup of index 2. It follows that $\rho_{Q'}$ is induced by a character of $\mathrm{Gal}(\overline{F}_{Q'}/F_{Q'})$ for a quadratic field $F_{Q'}$. If $F_{Q'}$ is imaginary then Q' is a CM point. In particular, if we suppose that the residual representation $\bar{\rho}\colon G_\mathbb{Q} \to \mathrm{GL}_2(\mathbb{F})$ is not induced by a character of $\mathrm{Gal}(\overline{F}/F)$ for a real quadratic field $F/\mathbb{Q}$, then $F_{Q'}$ is imaginary and Q' is CM. The above argument proves that G_P is small if and only if all points $Q' \subset \mathbb{I}[P_1^{-1}]$ above P are CM.

Proof We prove first that $V(\mathfrak{c}_1) \subset V(\mathfrak{l})$. Fix a radius $r < r_h$. By contradiction, suppose that a prime P of $\mathbb{I}_0[P_1^{-1}]$ contains $\mathfrak{c}_0$ but P does not contain $\mathfrak{l}$. Then there exists a prime factor Q of $P\mathbb{I}[P_1^{-1}]$ such that $\mathfrak{c} \subset Q$. By definition of $\mathfrak{c}$ we have that Q is a CM point in the sense of Sect. 3.4, hence the representation $\rho_{\mathbb{I}[P_1^{-1}],Q}$ has small image in $\mathrm{GL}_2(\mathbb{I}[P_1^{-1}]/Q)$. Then its restriction $\rho_{\mathbb{I}[P_1^{-1}],Q}|_{H_0} = \rho_P$ also has small image in $\mathrm{GL}_2(\mathbb{I}_0[P_1^{-1}]/P)$. We deduce that there is no nonzero ideal $\mathfrak{J}_P$ of $\mathbb{I}_0[P_1^{-1}]/P$ such that the Lie algebra $\mathfrak{H}_{r,P}$ contains $\mathfrak{J}_P \cdot \mathfrak{sl}_2(\mathbb{I}_0[P_1^{-1}]/P)$.

Now by definition of $\mathfrak{l}$ we have $\mathfrak{l}\cdot\mathfrak{sl}_2(\mathbb{B}_r)\subset\mathfrak{H}_r$. Since reduction modulo P gives a surjection $\mathfrak{H}_r\to\mathfrak{H}_{r,P}$, by looking at the previous inclusion modulo P we find $\mathfrak{l}\cdot\mathfrak{sl}_2(\mathbb{I}_{0,r}[P_1^{-1}]/P\mathbb{I}_{0,r}[P_1^{-1}])\subset\mathfrak{H}_{r,P}$. If $\mathfrak{l}\not\subset P$ we have $\mathfrak{l}/P\neq 0$, which contradicts our earlier statement. We deduce that $\mathfrak{l}\subset P$.

We prove now that $V(\mathfrak{l})\subset V(\mathfrak{c}_1)$. Let $P\subset\mathbb{I}_0[P_1^{-1}]$ be a prime containing $\mathfrak{l}$. Recall that $\mathbb{I}_0[P_1^{-1}]$ has Krull dimension one, so $\kappa_P=\mathbb{I}_0[P_1^{-1}]/P$ is a field. Let Q be a prime of $\mathbb{I}[P_1^{-1}]$ above P. As before ρ reduces to representations $\rho_Q\colon G_\mathbb{Q}\to\mathrm{GL}_2(\mathbb{I}[P_1^{-1}]/Q)$ and $\rho_P\colon H_0\to\mathrm{GL}_2(\mathbb{I}_0[P_1^{-1}]/P)$. Let $\mathfrak{P}\subset\mathbb{I}_0[P_1^{-1}]$ be the P-primary component of $\mathfrak{l}$ and let $\mathfrak{A}$ be an ideal of $\mathbb{I}_0[P_1^{-1}]$ containing $\mathfrak{P}$ such that the localization at P of $\mathfrak{A}/\mathfrak{P}$ is one-dimensional over κ_P. Choose any $r<r_h$. Let $\mathfrak{s}=\mathfrak{A}/\mathfrak{P}\cdot\mathfrak{sl}_2(\mathbb{I}_{0,r}[P_1^{-1}]/\mathfrak{P})\cap\mathfrak{H}_{r,\mathfrak{P}}$, that is a Lie subalgebra of $\mathfrak{A}/\mathfrak{P}\cdot\mathfrak{sl}_2(\mathbb{I}_{0,r}[P_1^{-1}]/\mathfrak{P})$.

We show that $\mathfrak{s}$ is stable under the adjoint action $\mathrm{Ad}(\rho_Q)$ of $G_\mathbb{Q}$. Let $\mathfrak{Q}$ be the Q-primary component of $\mathfrak{l}\cdot\mathbb{I}[P_1^{-1}]$. Recall that $\mathfrak{H}_{r,\mathfrak{P}}$ is the Lie algebra associated with the pro-p group $\mathrm{Im}\,\rho_{r,\mathfrak{Q}}|_{H_0}\cap\Gamma_{\mathrm{GL}_2(\mathbb{I}_{0,r_0}[P_1^{-1}]/\mathfrak{P})}(p)\subset\mathrm{GL}_2(\mathbb{I}_{0,r}[P_1^{-1}]/\mathfrak{P})$. Since this group is open in $\mathrm{Im}\,\rho_{r,\mathfrak{Q}}\subset\mathrm{GL}_2(\mathbb{I}_r[P_1^{-1}]/\mathfrak{Q})$, the Lie algebra associated with $\mathrm{Im}\,\rho_{r,\mathfrak{Q}}$ is again $\mathfrak{H}_{r,\mathfrak{P}}$. In particular $\mathfrak{H}_{r,\mathfrak{P}}$ is stable under $\mathrm{Ad}(\rho_Q)$. Since $\mathfrak{H}_{r,\mathfrak{P}}\subset\mathfrak{sl}_2(\mathbb{I}_{0,r}[P_1^{-1}]/\mathfrak{P})$ we have $\mathfrak{A}/\mathfrak{P}\cdot\mathfrak{sl}_2(\mathbb{I}_{0,r}[P_1^{-1}]/\mathfrak{P})\cap\mathfrak{H}_{r,\mathfrak{P}}=\mathfrak{A}/\mathfrak{P}\cdot\mathfrak{sl}_2(\mathbb{I}_r[P_1^{-1}]/\mathfrak{Q})\cap\mathfrak{H}_{r,\mathfrak{P}}$. Now $\mathfrak{A}/\mathfrak{P}\cdot\mathfrak{sl}_2(\mathbb{I}_r[P_1^{-1}]/\mathfrak{Q})$ is clearly stable under $\mathrm{Ad}(\rho_Q)$, so the same is true for $\mathfrak{A}/\mathfrak{P}\cdot\mathfrak{sl}_2(\mathbb{I}_r[P_1^{-1}]/\mathfrak{Q})\cap\mathfrak{H}_{r,\mathfrak{P}}$, as desired.

We consider from now on $\mathfrak{s}$ as a Galois representation via $\mathrm{Ad}(\rho_Q)$. By the proof of Theorem 6.2 we can assume, possibly considering a sub-Galois representation, that $\mathfrak{H}_r$ is a $\mathbb{B}_r$-submodule of $\mathfrak{sl}_2(\mathbb{B}_r)$ containing $\mathfrak{l}\cdot\mathfrak{sl}_2(\mathbb{B}_r)$ but not $\mathfrak{a}\cdot\mathfrak{sl}_2(\mathbb{B}_r)$ for any $\mathfrak{a}$ strictly bigger than $\mathfrak{l}$. This allows us to speak of the localization $\mathfrak{s}_P$ of $\mathfrak{s}$ at P. Note that, since $\mathfrak{P}$ is the P-primary component of $\mathfrak{l}$ and $\mathfrak{A}_P/\mathfrak{P}_P\cong\kappa_P$, when P-localizing we find $\mathfrak{H}_{r,P}\supset\mathfrak{P}_P\cdot\mathfrak{sl}_2(\mathbb{B}_{r,P})$ and $\mathfrak{H}_{r,P}\not\supset\mathfrak{A}_P\cdot\mathfrak{sl}_2(\mathbb{B}_{r,P})$.

The localization at P of $\mathfrak{a}/\mathfrak{P}\cdot\mathfrak{sl}_2(\mathbb{I}_{0,r}[P_1^{-1}]/\mathfrak{P})$ is $\mathfrak{sl}_2(\kappa_P)$, so $\mathfrak{s}_P$ is contained in $\mathfrak{sl}_2(\kappa_P)$. It is a κ_P-representation of $G_\mathbb{Q}$ (via $\mathrm{Ad}(\rho_Q)$) of dimension at most 3. We distinguish various cases following its dimension.

We cannot have $\mathfrak{s}_P=0$. By exchanging the quotient with the localization we would obtain $(\mathfrak{A}_P\cdot\mathfrak{sl}_2(\mathbb{B}_{r,P})\cap\mathfrak{H}_{r,P})/\mathfrak{P}_P=0$. By Nakayama's lemma $\mathfrak{A}_P\cdot\mathfrak{sl}_2(\mathbb{B}_{r,P})\cap\mathfrak{H}_{r,P}=0$, which is absurd since $\mathfrak{A}_P\cdot\mathfrak{sl}_2(\mathbb{B}_{r,P})\cap\mathfrak{H}_{r,P}\supset\mathfrak{P}_P\cdot\mathfrak{sl}_2(\mathbb{B}_{r,P})\neq 0$.

We also exclude the three-dimensional case. If $\mathfrak{s}_P=\mathfrak{sl}_2(\kappa_P)$, by exchanging the quotient with the localization we obtain $(\mathfrak{A}_P\cdot\mathfrak{sl}_2(\mathbb{B}_{r,P})\cap\mathfrak{H}_{r,P})/\mathfrak{P}_P=(\mathfrak{A}_P\cdot\mathfrak{sl}_2(\mathbb{I}_{0,r,P}[P_1^{-1}]))/\mathfrak{P}_P\mathbb{I}_{0,r,P}[P_1^{-1}]$, because we have $\mathfrak{A}_P\mathbb{I}_{0,r,P}[P_1^{-1}]/\mathfrak{P}_P\mathbb{I}_{0,r,P}[P_1^{-1}]=\left(\mathbb{I}_{0,r,P}[P_1^{-1}]/\mathfrak{P}_P\mathbb{I}_{0,r,P}[P_1^{-1}]\right)$ and this is isomorphic to κ_P. By Nakayama's lemma we would conclude that $\mathfrak{H}_{r,P}\supset\mathfrak{A}\cdot\mathfrak{sl}_2(\mathbb{B}_{r,P})$, which is absurd.

We are left with the one and two-dimensional cases. If $\mathfrak{s}_P$ is two-dimensional we can always replace it by its orthogonal in $\mathfrak{sl}_2(\kappa_P)$ which is one-dimensional; indeed the action of $G_\mathbb{Q}$ via $\mathrm{Ad}(\rho_Q)$ is isometric with respect to the scalar product $\mathrm{Tr}(XY)$ on $\mathfrak{sl}_2(\kappa_P)$.

Suppose that $\mathfrak{sl}_2(\kappa_P)$ contains a one-dimensional stable subspace. Let ϕ be a generator of this subspace over κ_P. Let $\chi\colon G_\mathbb{Q}\to\kappa_P$ denote the character satisfying $\rho_Q(g)\phi\rho_Q(g)^{-1}=\chi(g)\phi$ for all $g\in G_\mathbb{Q}$. Now ϕ induces a nontrivial morphism of representations $\rho_Q\to\rho_Q\otimes\chi$. Since ρ_Q and $\rho_Q\otimes\chi$ are irreducible, by Schur's lemma ϕ must be invertible. Hence we obtain an isomorphism $\rho_Q\cong\rho_Q\otimes\chi$. By taking determinants we see that χ must be quadratic. If $F_0/\mathbb{Q}$ is the quadratic extension fixed by $\ker\chi$, then ρ_Q is induced by a character ψ of $\mathrm{Gal}(\overline{F_0}/F_0)$. By assumption the residual representation $\rho_{\mathfrak{m}_{\mathbb{I}}}\colon G_{\mathbb{Q}} \to \mathrm{GL}_2(\mathbb{F})$ is not of the form $\mathrm{Ind}_F^{\mathbb{Q}}\psi$ for a real quadratic field F and a character $\mathrm{Gal}(\overline{F}/F) \to \mathbb{F}^\times$. We deduce that F_0 must be imaginary, so Q is a CM point by Remark 3.11(1). By construction of the congruence ideal $\mathfrak{c} \subset Q$ and $\mathfrak{c}_0 \subset Q \cap \mathbb{I}_0[P_1^{-1}] = P$. □

We prove a corollary.

Corollary 7.2 *If the residual representation $\overline{\rho}\colon G_{\mathbb{Q}} \to \mathrm{GL}_2(\mathbb{F})$ is not dihedral then $\mathfrak{l} = 1$.*

Proof Since $\overline{\rho}$ is not dihedral there cannot be any CM point on the family $\theta\colon \mathbb{T}_h \to \mathbb{I}^\circ$. By Theorem 7.1 we deduce that $\mathfrak{l}$ has no nontrivial prime factor, hence it is trivial. □

Remark 7.3 Theorem 7.1 gives another proof of Proposition 3.9. Indeed the CM points of a family $\theta\colon \mathbb{T}_h \to \mathbb{I}^\circ$ correspond to the prime factors of its Galois level, which are finite in number.

We also give a partial result about the comparison of the exponents of the prime factors in $\mathfrak{c}_1$ and $\mathfrak{l}$. This is an analogous of what is proved in [9, Theorem 8.6] for the ordinary case; our proof also relies on the strategy there. For every prime P of $\mathbb{I}_0[P_1^{-1}]$ we denote by $\mathfrak{c}_1^P$ and $\mathfrak{l}^P$ the P-primary components of $\mathfrak{c}_1$ and $\mathfrak{l}$ respectively.

Theorem 7.4 *Suppose that $\overline{\rho}$ is not induced by a character of G_F for a real quadratic field $F/\mathbb{Q}$. We have $(\mathfrak{c}_1^P)^2 \subset \mathfrak{l}^P \subset \mathfrak{c}_1^P$.*

Proof The inclusion $\mathfrak{l}^P \subset \mathfrak{c}_1^P$ is proved in the same way as the first inclusion of Theorem 7.1.

We show that the inclusion $(\mathfrak{c}_1^P)^2 \subset \mathfrak{l}^P$ holds. If $\mathfrak{c}_1^P$ is trivial this reduces to Theorem 7.1, so we can suppose that P is a factor of $\mathfrak{c}_1$. Let Q denote any prime of $\mathbb{I}[P_1^{-1}]$ above P. Let $\mathfrak{c}_1^Q$ be a Q-primary ideal of $\mathbb{I}[P_1^{-1}]$ satisfying $\mathfrak{c}_1^Q \cap \mathbb{I}_0[P_1^{-1}] = \mathfrak{c}_1^P$. Since P divides $\mathfrak{c}_1$, Q is a CM point, so we have an isomorphism $\rho_P \cong \mathrm{Ind}_F^{\mathbb{Q}}\psi$ for an imaginary quadratic field $F/\mathbb{Q}$ and a character $\psi\colon G_F \to \mathbb{C}_p^\times$. Choose any $r < r_h$. Consider the κ_P-vector space $\mathfrak{s}_{\mathfrak{c}_1^P} = \mathfrak{H}_r \cap \mathfrak{c}_1^P \cdot \mathfrak{sl}_2(\mathbb{I}_{0,r})/\mathfrak{H}_r \cap \mathfrak{c}_1^P P \cdot \mathfrak{sl}_2(\mathbb{I}_{0,r})$. We see it as a subspace of $\mathfrak{sl}_2(\mathfrak{c}_1^P/\mathfrak{c}_1^P P) \cong \mathfrak{sl}_2(\kappa_P)$. By the same argument as in the proof of Theorem 7.1, $\mathfrak{s}_{\mathfrak{c}_1^P}$ is stable under the adjoint action $\mathrm{Ad}(\rho_{\mathfrak{c}_1^Q Q})\colon G_{\mathbb{Q}} \to \mathrm{Aut}(\mathfrak{sl}_2(\kappa_P))$.

Let $\chi_{F/\mathbb{Q}}\colon G_{\mathbb{Q}} \to \mathbb{C}_p^\times$ be the quadratic character defined by the extension $F/\mathbb{Q}$. Let $\varepsilon \in G_{\mathbb{Q}}$ be an element projecting to the generator of $\mathrm{Gal}(F/\mathbb{Q})$. Let $\psi^\varepsilon\colon G_F \to \mathbb{C}_p^\times$ be given by $\psi^\varepsilon(\tau) = \psi(\varepsilon\tau\varepsilon^{-1})$. Set $\psi^- = \psi/\psi^\varepsilon$. Since $\rho_Q \cong \mathrm{Ind}_F^{\mathbb{Q}}\psi$, we have a decomposition $\mathrm{Ad}(\rho_Q) \cong \chi_{F/\mathbb{Q}} \oplus \mathrm{Ind}_F^{\mathbb{Q}}\psi^-$, where the two factors are irreducible. Now we have three possibilities for the Galois isomorphism class of $\mathfrak{s}_{\mathfrak{c}_1^P}$: it is either that of $\mathrm{Ad}(\rho_Q)$ or that of one of the two irreducible factors.

If $\mathfrak{s}_{\mathfrak{c}_1^P} \cong \mathrm{Ad}(\rho_Q)$, then as κ_P-vector spaces $\mathfrak{s}_{\mathfrak{c}_1^P} = \mathfrak{sl}_2(\kappa_P)$. By Nakayama's lemma $\mathfrak{H}_r \supset \mathfrak{c}_1^P \cdot \mathfrak{sl}_2(\mathbb{B}_r)$. This implies $\mathfrak{c}_1^P \subset \mathfrak{l}^P$, hence $\mathfrak{c}_1^P = \mathfrak{l}^P$ in this case.

If $\mathfrak{s}_{\mathfrak{c}_1^P}$ is one-dimensional then we proceed as in the proof of Theorem 7.1 to show that $\rho_{\mathfrak{c}_1^Q Q}\colon G_{\mathbb{Q}} \to \mathrm{GL}_2(\mathbb{I}_r[P_1^{-1}]/\mathfrak{c}_1^Q Q\mathbb{I}_r[P_1^{-1}])$ is induced by a character $\psi_{\mathfrak{c}_1^Q Q}\colon G_F \to \mathbb{C}_p^\times$. In particular the image of $\rho_{\mathfrak{c}_1^P P}\colon H \to \mathrm{GL}_2(\mathbb{I}_{0,r}[P_1^{-1}]/\mathfrak{c}_1^P P\mathbb{I}_{0,r})$ is small. This is a contradiction, since $\mathfrak{c}_1^P$ is the P-primary component of $\mathfrak{c}_1$, hence it is the smallest P-primary ideal $\mathfrak{A}$ of $\mathbb{I}_{0,r}[P_1^{-1}]$ such that the image of $\rho_{\mathfrak{A}}\colon G_{\mathbb{Q}} \to \mathrm{GL}_2(\mathbb{I}_r[P_1^{-1}]/\mathfrak{A}\mathbb{I}_r[P_1^{-1}])$ is small.

Finally, suppose that $\mathfrak{s}_{\mathfrak{c}_1^P} \cong \mathrm{Ind}_F^{\mathbb{Q}}\psi^-$. Let $d = \mathrm{diag}(d_1, d_2) \in \rho(G_{\mathbb{Q}})$ be the image of a $\mathbb{Z}_p$-regular element. Since d_1 and d_2 are nontrivial modulo the maximal ideal of $\mathbb{I}_0^\circ$, the image of d modulo $\mathfrak{c}_1^Q Q$ is a nontrivial diagonal element $d_{\mathfrak{c}_1^Q Q} = \mathrm{diag}(d_{1,\mathfrak{c}_1^Q Q}, d_{2,\mathfrak{c}_1^Q Q}) \in \rho_{\mathfrak{c}_1^Q Q}(G_{\mathbb{Q}})$. We decompose $\mathfrak{s}_{\mathfrak{c}_1^P}$ in eigenspaces for the adjoint action of $d_{\mathfrak{c}_1^Q Q}$: we write $\mathfrak{s}_{\mathfrak{c}_1^P} = \mathfrak{s}_{\mathfrak{c}_1^P}[a] \oplus \mathfrak{s}_{\mathfrak{c}_1^P}[1] \oplus \mathfrak{s}_{\mathfrak{c}_1^P}[a^{-1}]$, where $a = d_{1,\mathfrak{c}_1^Q Q}/d_{2,\mathfrak{c}_1^Q Q}$. Now $\mathfrak{s}_{\mathfrak{c}_1^P}[1]$ is contained in the diagonal torus, on which the adjoint action of $G_{\mathbb{Q}}$ is given by the character $\chi_{F/\mathbb{Q}}$. Since $\chi_{F/\mathbb{Q}}$ does not appear as a factor of $\mathfrak{s}_{\mathfrak{c}_1^P}$, we must have $\mathfrak{s}_{\mathfrak{c}_1^P}[1] = 0$. This implies that $\mathfrak{s}_{\mathfrak{c}_1^P}[a] \neq 0$ and $\mathfrak{s}_{\mathfrak{c}_1^P}[a^{-1}] \neq 0$. Since $\mathfrak{s}_{\mathfrak{c}_1^P}[a] = \mathfrak{s}_{\mathfrak{c}_1^P} \cap \mathfrak{u}^+(\kappa_P)$ and $\mathfrak{s}_{\mathfrak{c}_1^P}[a^{-1}] = \mathfrak{s}_{\mathfrak{c}_1^P} \cap \mathfrak{u}^-(\kappa_P)$, we deduce that $\mathfrak{s}_{\mathfrak{c}_1^P}$ contains nontrivial upper and lower nilpotent elements $\overline{u^+}$ and $\overline{u^-}$. Then $\overline{u^+}$ and $\overline{u^-}$ are the images of some elements u^+ and u^- of $\mathfrak{H}_r \cap \mathfrak{c}_1^P \cdot \mathfrak{sl}_2(\mathbb{I}_{0,r}[P_1^{-1}])$ nontrivial modulo $\mathfrak{c}_1^P P$. The Lie bracket $t = [u^+, u^-]$ is an element of $\mathfrak{H}_r \cap \mathfrak{t}(\mathbb{I}_{0,r}[P_1^{-1}])$ (where $\mathfrak{t}$ denotes the diagonal torus) and it is nontrivial modulo $(\mathfrak{c}_1^P)^2 P$. Hence the κ_P-vector space $\mathfrak{s}_{(\mathfrak{c}_1^P)^2} = \mathfrak{H}_r \cap (\mathfrak{c}_1^P)^2 \cdot \mathfrak{sl}_2(\mathbb{I}_{0,r,\mathbb{C}_p}[P_1^{-1}])/\mathfrak{H}_r \cap (\mathfrak{c}_1^P)^2 P \cdot \mathfrak{sl}_2(\mathbb{I}_{0,r,\mathbb{C}_p}[P_1^{-1}])$ contains nontrivial diagonal, upper nilpotent and lower nilpotent elements, so it is three-dimensional. By Nakayama's lemma we conclude that $\mathfrak{H}_r \supset (\mathfrak{c}_1^P)^2 \cdot \mathfrak{sl}_2(\mathbb{I}_{0,r}[P_1^{-1}])$, so $(\mathfrak{c}_1^P)^2 \subset \mathfrak{l}^P$. □

References

1. Bellaïche, J.: Eigenvarieties and adjoint p-adic L-functions (preprint)
2. Bellaïche, J., Chenevier, G.: Families of Galois Representations and Selmer groups. Astérisque 324, Soc. Math, France (2009)
3. Buzzard, K.: Eigenvarieties, in L-functions and Galois representations. In: Proceedings of Conference Durham: LMS Lect. Notes Series 320. Cambridge University Press 2007, pp. 59–120 (2004)
4. Chenevier, G.: Familles p-adiques de formes automorphes pour GL(n). J. Reine Angew. Math. **570**, 143–217 (2004)
5. Coleman, R.: Classical and overconvergent modular forms. Invent. Math. **124**, 214–241 (1996)
6. Coleman, R., Mazur, B.: The eigencurve. In: Galois Representations in Arithmetic Algebraic Geometry, London Math. Soc. Lecture Note Ser. 254, Cambridge University Press, Cambridge, pp. 1–113 (1998)
7. de Jong, A.J.: Crystalline Dieudonné theory via formal and rigid geometry. Publ. Math. Inst. Hautes Études Sci. **82**(1), 5–96 (1995)
8. Faltings, G.: Hodge-Tate structures and modular forms. Math. Ann. **278**, 133–149 (1987)
9. Hida, H.: Big Galois representations and p-adic L-functions. Compos. Math. **151**, 603–654 (2015)
10. Hida, H., Tilouine, J.: Big image of Galois representations and congruence ideals. In: Dieulefait, L., Heath-Brown, D.R., Faltings, G., Manin, Y.I., Moroz, B.Z., Wintenberger, J.-P. (eds.)

Arithmetic Geometry, Proceedings of Workshop on Serre's Conjecture, Hausdorff Inst. Math., Bonn. Cambridge University Press, pp. 217–254 (2015)
11. Kisin, M.: Overconvergent modular forms and the Fontaine-Mazur conjecture. Invent. Math. **153**, 363–454 (2003)
12. Lang, J.: On the image of the Galois representation associated to a non-CM Hida family. Algebra Number Theory **10**(1), 155–194 (2016)
13. Momose, F.: On the ℓ-adic representations attached to modular forms. J. Fac. Sci. Univ. Tokyo Sect. IA Math. **28**(1), 89–109 (1981)
14. Nyssen, L.: Pseudo-representations. Math. Ann. **306**(2), 257–284 (1996)
15. Ribet, K.: On ℓ-adic representations attached to modular forms. Invent. Math. **28**, 245–275 (1975)
16. Ribet, K.: Galois representations attached to modular forms with Nebentypus. In: Modular functions of one variable V, Lecture Notes in Math., vol. 601, pp. 17–51. Springer (1977)
17. Ribet, K.: On ℓ-adic representations attached to modular forms. II. Glasgow Math. J. **27**, 185–194 (1985)
18. Rouquier, R.: Caractérisation des caractères et pseudo-caractères. J. Algebra **180**, 571–586 (1996)
19. Sen, S.: Lie algebras of Galois groups arising from Hodge-Tate modules. Ann. Math. **97**(1), 160–170 (1973)
20. Sen, S.: Continuous cohomology and p-adic Hodge theory. Invent. Math. **62**(1), 89–116 (1980)
21. Sen, S.: An infinite dimensional Hodge-Tate theory. Bull. Soc. Math. France **121**, 13–34 (1993)
22. Shimura, G.: On elliptic curves with complex multiplication as factors of the Jacobians of modular function fields. Nagoya Math. J. **43**, 199–208 (1971)
23. Tazhetdinov, S.: Subnormal structure of two-dimensional linear groups over local rings. Algebra i Logika **22**(6), 707–713 (1983)
24. Wan, D.: Dimension variation of classical and p-adic modular forms. Invent. Math. **133**, 449–463 (1998)

Behaviour of the Order of Tate–Shafarevich Groups for the Quadratic Twists of $X_0(49)$

Andrzej Dąbrowski, Tomasz Jędrzejak and Lucjan Szymaszkiewicz

Abstract We present the results of our search for the orders of Tate–Shafarevich groups for the quadratic twists of $E = X_0(49)$.

Keywords Elliptic curves · Quadratic twists · Tate–Shafarevich group · Distribution of central L-values · Cohen–Lenstra heuristics

MSCs 11G05 · 11G40 · 11Y50

1 Introduction

Let E be an elliptic curve defined over $\mathbb{Q}$ of conductor N_E, and let $L(E, s)$ denote its L-series. Let $Ш(E)$ be the Tate–Shafarevich group of E, $E(\mathbb{Q})$ the group of rational points, and $R(E)$ the regulator, with respect to the Néron–Tate height pairing. Finally, let Ω_E be the least positive real period of the Néron differential on E, and define $C_\infty(E) = \Omega_E$ or $2\Omega_E$ according as $E(\mathbb{R})$ is connected or not, and let $C_{\text{fin}}(E)$ denote the product of the Tamagawa factors of E at the bad primes. The Euler product defining $L(E, s)$ converges for $\text{Re}\, s > 3/2$. The modularity conjecture, proven by Wiles–Taylor–Diamond–Breuil–Conrad, implies that $L(E, s)$ has an analytic continuation to an entire function. The Birch and Swinnerton-Dyer conjecture relates the arithmetic data of E to the behaviour of $L(E, s)$ at $s = 1$.

Dedicated to John Coates on his seventieth birthday.

A. Dąbrowski (✉) · T. Jędrzejak · L. Szymaszkiewicz
Institute of Mathematics, University of Szczecin, Wielkopolska 15, 70-451 Szczecin, Poland
e-mail: dabrowskiandrzej7@gmail.com; dabrowsk@wmf.univ.szczecin.pl

T. Jędrzejak
e-mail: tjedrzejak@gmail.com

L. Szymaszkiewicz
e-mail: lucjansz@gmail.com

D. Loeffler and S.L. Zerbes (eds.), *Elliptic Curves, Modular Forms and Iwasawa Theory*, Springer Proceedings in Mathematics & Statistics 188, DOI 10.1007/978-3-319-45032-2_4

Conjecture 1 (Birch and Swinnerton-Dyer) *(i) L-function $L(E,s)$ has a zero of order $r = \operatorname{rank} E(\mathbb{Q})$ at $s = 1$,*
(ii) Ш(E) is finite, and

$$\lim_{s\to 1} \frac{L(E,s)}{(s-1)^r} = \frac{C_\infty(E) C_{fin}(E) R(E) |Ш(E)|}{|E(\mathbb{Q})_{tors}|^2}.$$

If Ш(E) is finite, the work of Cassels and Tate shows that its order must be a square.

The first general result in the direction of this conjecture was proven for elliptic curves E with complex multiplication by Coates and Wiles in 1976 [3], who showed that if $L(E,1) \neq 0$, then the group $E(\mathbb{Q})$ is finite. Gross and Zagier [12] showed that if $L(E,s)$ has a first-order zero at $s = 1$, then E has a rational point of infinite order. Rubin [23] proves that if E has complex multiplication and $L(E,1) \neq 0$, then Ш(E) is finite. Let g_E be the rank of $E(\mathbb{Q})$ and let r_E the order of the zero of $L(E,s)$ at $s = 1$. Then Kolyvagin [15] proved that, if $r_E \leqslant 1$, then $r_E = g_E$ and Ш(E) is finite. The work [11] completed Rubin's verification of the Birch and Swinnerton-Dyer conjecture for the quadratic twists of $X_0(49)$ when the complex L-series of the twist does not vanish at $s = 1$. Coates et al. [1, 2] showed that there is a large class of explicit quadratic twists of $X_0(49)$ whose complex L-series does not vanish at $s = 1$, and for which the full Birch and Swinnerton-Dyer conjecture is valid. We recall that $E = X_0(49)$ has a minimal Weierstrass equation $y^2 + xy = x^3 - x^2 - 2x - 1$. Its Néron differential $\omega = \frac{dx}{2y+x}$ has fundamental real period $\Omega_E = \frac{\Gamma(1/7)\Gamma(2/7)\Gamma(4/7)}{2\pi\sqrt{7}} = 1.9333117\ldots$ In what follows we shall study numerical data arising from the conjecture of Birch and Swinnerton-Dyer for the quadratic twists of E. Our reason for considering the quadratic twists of this particular curve is that, in our present state of knowledge, one can prove more cases of the full Birch–Swinnerton-Dyer conjecture for these quadratic twists than for the quadratic twists of any other elliptic curve over $\mathbb{Q}$ with small conductor.

The numerical studies and conjectures by Conrey–Keating–Rubinstein–Snaith [5], Delaunay [8, 9], Quattrini [20, 21], Watkins [25], Radziwiłł–Soundararajan [22] (see also the paper [7] and references therein) substantially extend the systematic tables given by Cremona. Our present computations are over a considerably larger range of quadratic twists, and support all previous conjectures, as well as giving rise to some new ones (see our Conjectures 7 and 8 below).

In this paper we present the results of our search for the orders of Tate–Shafarevich groups for the quadratic twists of E for rather large ranges of the index. Our calculations may be served as an appendix to the following beautiful results obtained by Gonzalez-Avilés ([11], Theorem B), and Coates et al. ([2], Theorems 1.2 and 1.4). If d is the discriminant of a quadratic field, E_d will denote the twist of E by $\mathbb{Q}(\sqrt{d})/\mathbb{Q}$.

Theorem 2 ([11], Theorem B) *If $L(E_d, 1) \neq 0$, then the full Birch and Swinnerton-Dyer conjecture is true for E_d.*

Theorem 3 ([2], Theorem 1.2) *Let $d = p_1 \dots p_l$ be a product of $\geqslant 0$ distinct primes, which are $\equiv 1$ mod 4 and inert in $\mathbb{Q}(\sqrt{-7})$. Then $L(E_d, 1) \neq 0$, $E_d(\mathbb{Q})$ is finite, the Tate–Shafarevich group of E_d is finite of odd cardinality, and the full Birch–Swinnerton-Dyer conjecture is valid for E_d.*

Theorem 4 ([2], a special case of Theorem 1.4) *Let l_0 be a prime number > 3, which is $\equiv 3$ mod 4 and inert in $\mathbb{Q}(\sqrt{-7})$. Assume that $q_1, \dots q_r$ are distinct rational primes, which are $\equiv 1$ mod 4 and inert in both the fields $\mathbb{Q}(\sqrt{-7})$ and $\mathbb{Q}(\sqrt{-l_0})$. Put $d = -l_0 q_1 \dots q_r$. Then $L(E_d, s)$ has a simple zero at $s = 1$, $E_d(\mathbb{Q})$ has rank 1, and the Tate–Shafarevich group of E_d is finite of odd cardinality.*

Our numerical data gives the order of $\text{Ш}(E_d)$ for all 5598893691 odd positive discriminants d, prime to 7, for which $d < 32 \cdot 10^9$ and $L(E_d, 1) \neq 0$. The calculations suggest that for any positive integer k there is square-free positive integer d, $(d, 7) = 1$ (or even infinitely many such d's), such that E_d has rank zero and $|\text{Ш}(E_d)| = k^2$ (Sects. 3 and 4); in Sect. 11 we propose asymptotical formulae for the number of such d's. The numerical data gives strong information, discussed in Sect. 7, for the asymptotic behaviour of the sum of the orders of the $\text{Ш}(E_d)$ for odd positive d prime to 7, with $L(E_d, 1) \neq 0$ over all such d with $d \leqslant X$ as $X \to \infty$. It turns out that both distributions of $L(E_d, 1)$ and $\log(|\text{Ш}(E_d)|/\sqrt{d})$ follow an approximate normal distribution (Sect. 9). In the last section we numerically confirm that $|\text{Ш}(E_d)| = 1$ is about as common as $L(E_d, 1) = 0$ when $\epsilon(E_d) = 1$. The attached table contains, for each positive integer $k \leqslant 1793$ (and for selected integers up to 2941), an elliptic curve E_{d_k} with $|\text{Ш}(E_{d_k})| = k^2$.

Of course, all the experiments concerning statistics of the L-values of quadratic twists of $X_0(49)$, and related orders of Tate–Shafarevich groups, can be repeated for quadratic twists of other elliptic curves (see [6]).

At the end of December 2013, John Coates asked one of us (A. D.) to establish some results about large orders of Ш for the quadratic twists of $X_0(49)$ (using Theorem 1.2 in [2]). It was the starting point for us to make extensive calculations reported in this article. We heartily thank John for his constant support, suggestions (i.e., he proposed to find and/or test asymptotic formulae in Sect. 7) and many corrections. Finally, we thank him for sending us a manuscript by Radziwiłł and Soundararajan [22] and some comments by Heath-Brown [13]. We thank Bjorn Poonen for sending us his comments and, especially, for his question (see Sect. 11). We thank the anonymous referees for their comments which improved the final version of this paper.

The main part of the computations was carried out in 2015 on desktop computers Core(TM) 2 Quad Q8300 4GB/8GB. For the calculations of examples in Sect. 5 we also used the HPC cluster HAL9000. All machines are located at the Department of Mathematics and Physics of Szczecin University.

2 Formula for the Order of $Ш(E_d)$ When $L(E_d, 1) \neq 0$

We can compute $L(E_d, 1)$ when it is non-zero for a huge range of positive discriminants $d \equiv 1 (\text{mod } 4)$ thanks to the remarkable ideas discovered by Waldspurger, and worked out explicitly in this particular case by Lehman [17]. These ideas show that $L(E_d, 1)$, when it is non-zero, is essentially equal to the d-th Fourier coefficient of an explicit modular form of weight 3/2, and we now recall the precise result which Lehman proves.

Notation. Let $q := e^{2\pi i z}$, $\Theta(z) = \sum_{n\in\mathbb{Z}} q^{n^2}$, $\Theta_t(z) = \Theta(tz) = \sum_{n\in\mathbb{Z}} q^{tn^2}$. Let d be a positive square free integer, prime to 7, and $\equiv 1 (\text{mod } 4)$. Let $l_1(d)$ (resp. $l_2(d)$) denote the number of odd prime divisors p of d such that $(p/7) = 1$ (resp. $(p/7) = -1$). Define $l(d) = l_1(d) + \frac{1}{2} l_2(d)$ if $l_2(d)$ is even, and $l(d) = l_1(d) + \frac{1}{2}(l_2(d) - 1)$ if $l_2(d)$ is odd.

Let $g = g_1 + \cdots + g_6$, where

$$g_1 = \sum [q^{(14m+1)^2+(14n)^2} - q^{(14m+7)^2+(14n+6)^2}]$$

$$g_2 = \sum [q^{(14m+3)^2+(14n)^2} - q^{(14m+7)^2+(14n+4)^2}]$$

$$g_3 = \sum [q^{(14m+5)^2+(14n)^2} - q^{(14m+7)^2+(14n+2)^2}]$$

$$g_4 = \sum [q^{(14m+1)^2+(14n+2)^2} - q^{(14m+5)^2+(14n+6)^2}]$$

$$g_5 = \sum [q^{(14m+3)^2+(14n+6)^2} - q^{(14m+1)^2+(14n+4)^2}]$$

$$g_6 = \sum [q^{(14m+5)^2+(14n+4)^2} - q^{(14m+3)^2+(14n+2)^2}]$$

and all sums are taken over all $m, n \in \mathbb{Z}$.

Let $g\Theta_{28} = \sum a_n q^n$. Let Ω_d denote the least positive real period of the Néron differential on E_d. Then for d as above, we have $\Omega_d = \frac{\Omega_E}{\sqrt{d}}$, and Lehman ([17], Theorem 2) proves, in particular, the following result: $L(E_d, 1) = \Omega_d a_d^2$ if $(d/7) = -1$, and $L(E_d, 1) = \frac{1}{2}\Omega_d a_d^2$ if $(d/7) = 1$.

Assume that $a_d \neq 0$. Then $L(E_d, 1) \neq 0$. In this case the full Birch and Swinnerton-Dyer conjecture is valid ([11], Theorem B), hence using ([17], p. 268) we obtain the following result.

Corollary 1 *Assume d is positive, square free integer, prime to 7, and $\equiv 1 (\text{mod } 4)$. If $a_d \neq 0$, then* $|Ш(E_d)| = \frac{a_d^2}{4^{l(d)}}$.

Note that the weight 3/2 modular form $g\Theta_{28}$ may be constructed using ternary quadratic forms. This construction will be used in our algorithm (see the Appendix), hence we give some details. We introduce the following notations: for a positive definite integral quadratic form $f(x_1, \ldots, x_m)$ define $\theta(f)$ to be the power series

$$\theta(f) := \sum_{(k_1,\ldots,k_m)\in\mathbb{Z}^m} q^{f(k_1,\ldots,k_m)}.$$

We denote the form $f(x, y, z) = ax^2 + by^2 + cz^2 + ryz + szx + txy$ by the array $\begin{bmatrix} a\ b\ c \\ r\ s\ t \end{bmatrix}$, and put $\theta(f_1, f_2) := \frac{1}{2}(\theta(f_1) - \theta(f_2))$. Then (see [17], p. 259 for details) we get

$$g\Theta_{28} = g'\Theta_{28} + g''\Theta_{28},$$

where

$$g'\Theta_{28} = \theta\left(\begin{bmatrix} 1 & 28 & 196 \\ 0 & 0 & 0 \end{bmatrix}, \begin{bmatrix} 4 & 28 & 49 \\ 0 & 0 & 0 \end{bmatrix}\right),$$

$$g''\Theta_{28} = \theta\left(\begin{bmatrix} 5 & 40 & 28 \\ 0 & 0 & 4 \end{bmatrix}, \begin{bmatrix} 13 & 17 & 28 \\ 0 & 0 & 10 \end{bmatrix}\right).$$

Definition 5 (i) We say that a positive integer d satisfies condition (*), if $d = p_1 \cdots p_l$ is a product of distinct primes which are $\equiv 1 \bmod 4$ and $(p_i/7) = -1$ for all $i = 1, \ldots, l$. (ii) We say that a positive integer d satisfies condition (**), if d is square-free, $d \equiv 1 \bmod 4$, $(d, 7) = 1$, and $a_d \neq 0$.

Note that any d satisfying the condition (*) satisfies the condition (**) as well (use Theorems 2 and 3).

3 Examples of Rank Zero Elliptic Curves E_d with $|Ш(E_d)| = k^2$ for all $k \leqslant 1793$

Our data contains values of $|Ш(E_d)|$ for 5598893691 values of $d \leqslant 32 \cdot 10^9$ satisfying (**) (and for 715987381 values of d satisfying (*)).

In the attached table we exhibit, for each positive integer $k \leqslant 1793$ (and for selected integers up to 2941), an elliptic curve E_{d_k} with $|Ш(E_{d_k})| = k^2$. Note that for each odd positive integer $l \leqslant 2357$ there is an elliptic curve E_{d_l} with $|Ш(E_{d_l})| = l^2$. Our calculations strongly support the following

Conjecture 6 *Let $E = X_0(49)$. For any positive integer k there is square-free positive integer d, $(d, 7) = 1$, such that E_d has rank zero and $|Ш(E_d)| = k^2$.*

4 Frequency of Orders of Ш

Let $N_k(x)$ (resp. $N_k^*(x)$) denote the number of integers $d \leqslant x$ satisfying (*) (resp. (**)), and such that $|Ш(E_d)| = k^2$. Let $M_k(x) := \frac{N_k(2x)}{N_k(x)}$ (resp. $M_k^*(x) := \frac{N_k^*(2x)}{N_k^*(x)}$). Using our data, we obtain the following tables.

x	$M_1(x)$	$M_3(x)$	$M_5(x)$	$M_7(x)$	$M_9(x)$	$M_{11}(x)$
$1 \cdot 10^9$	1.670002	1.673796	1.678710	1.682080	1.695299	1.702050
$2 \cdot 10^9$	1.669373	1.673490	1.677806	1.682654	1.692469	1.696480
$3 \cdot 10^9$	1.668695	1.672879	1.677747	1.681843	1.688599	1.692675
$4 \cdot 10^9$	1.669351	1.672975	1.676227	1.680354	1.686862	1.691462
$5 \cdot 10^9$	1.670374	1.673585	1.676169	1.679956	1.685623	1.690646
$6 \cdot 10^9$	1.670751	1.673366	1.675929	1.679536	1.684822	1.690233
$7 \cdot 10^9$	1.671123	1.673133	1.675905	1.679332	1.683803	1.689585
$8 \cdot 10^9$	1.670935	1.673007	1.676074	1.679353	1.683255	1.689208
$9 \cdot 10^9$	1.670890	1.672742	1.675902	1.679188	1.683266	1.688139
$10 \cdot 10^9$	1.670861	1.672626	1.675780	1.679025	1.683358	1.687904
$11 \cdot 10^9$	1.670619	1.672538	1.675438	1.678478	1.683040	1.687505
$12 \cdot 10^9$	1.670764	1.673039	1.675417	1.678316	1.682781	1.687242
$13 \cdot 10^9$	1.670597	1.673224	1.675475	1.678079	1.682533	1.687109
$14 \cdot 10^9$	1.670479	1.673145	1.675411	1.677997	1.682744	1.686674
$15 \cdot 10^9$	1.670658	1.673080	1.675425	1.677969	1.682986	1.685881
$16 \cdot 10^9$	1.670893	1.673113	1.675090	1.677817	1.682823	1.685623

x	$M_1^*(x)$	$M_2^*(x)$	$M_3^*(x)$	$M_4^*(x)$	$M_5^*(x)$	$M_6^*(x)$
$1 \cdot 10^9$	1.728915	1.756191	1.742642	1.778071	1.758349	1.794058
$2 \cdot 10^9$	1.727257	1.752530	1.739237	1.772804	1.753243	1.785324
$3 \cdot 10^9$	1.726643	1.751384	1.737529	1.769437	1.750071	1.781009
$4 \cdot 10^9$	1.726260	1.750318	1.736594	1.767249	1.748203	1.777811
$5 \cdot 10^9$	1.725806	1.749001	1.735493	1.765948	1.746848	1.775631
$6 \cdot 10^9$	1.725426	1.748400	1.735025	1.764595	1.745359	1.773905
$7 \cdot 10^9$	1.724843	1.747711	1.734246	1.763498	1.744371	1.772759
$8 \cdot 10^9$	1.724452	1.747431	1.733720	1.762646	1.743659	1.771485
$9 \cdot 10^9$	1.724231	1.746896	1.733419	1.761800	1.743276	1.770443
$10 \cdot 10^9$	1.724024	1.746636	1.733082	1.761021	1.742697	1.769579
$11 \cdot 10^9$	1.723739	1.746219	1.732533	1.760323	1.742181	1.768670
$12 \cdot 10^9$	1.723712	1.745862	1.732306	1.759965	1.741811	1.767875
$13 \cdot 10^9$	1.723749	1.745593	1.732126	1.759491	1.741372	1.767234
$14 \cdot 10^9$	1.723679	1.745369	1.731859	1.759140	1.740965	1.766635
$15 \cdot 10^9$	1.723582	1.744934	1.731700	1.758824	1.740612	1.766010
$16 \cdot 10^9$	1.723609	1.744755	1.731335	1.758363	1.740290	1.765564

The values $M_k(x)$ and $M_l^*(x)$ ($x \to \infty$) oscillate very closely near (or converge to) some constants c_k and c_l^*.

Our calculations therefore suggest the following

Conjecture 7 *Let $E = X_0(49)$. For any positive integer k there are infinitely many positive square-free integers d, $(d, 7) = 1$, such that E_d has rank zero and $|Ш(E_d)| = k^2$.*

In the last section we state a more precise conjecture, which suggests, in particular, that all the constants c_k and c_l^* are equal to $2^{3/4} \approx 1.68179283$.

5 Large Orders of Ш

The article [7] presents results of search for elliptic curves with exceptionally large (analytic) orders of the Tate–Shafarevich group. It contains, in particular, 134 examples of rank zero elliptic curves E with $|Ш(E)| > 1832^2$, with the record $|Ш(E)| = 63408^2$.

Our data gives 5102 examples of rank zero elliptic curves E_d with $|Ш(E_d)| > 1832^2$. Note that we obtain 30 elliptic curves E_d with $|Ш(E_d)| > 2500^2$, with the record $|Ш(E_{28715939033})| = 2941^2$.

Using the approximations to $|Ш(E_d)|$ (by evaluating $L(E_d, 1)$ with sufficiently accuracy as in [7], p. 411) we were able to find two examples of E_d with much larger orders of Tate–Shafarevich groups: $7440^2 \leqslant |Ш(E_{10^{14}+7521})| \leqslant 7560^2$, and $7000^2 \leqslant |Ш(E_{10^{14}+7857})| \leqslant 7160^2$ (using 1.75×10^{13} terms of the L-series). The values $d_1 = 10^{14} + 7521$ and $d_2 = 10^{14} + 7857$ are primes satisfying the condition (*), hence the groups $Ш(E_{d_1})$ and $Ш(E_{d_2})$ both have odd order.

Finally, let us propose two candidates E_d with $|Ш(E_d)| > 15000^2$ (Fig. 1) and one candidate E_d with $|Ш(E_d)| > 100000^2$ (Fig. 2).

6 Large Primes Dividing the Orders of Ш

Another open problem about Ш of elliptic curves defined over $\mathbb{Q}$ is the following one: do exist arbitrarily large primes p such that there exists some elliptic curve E over $\mathbb{Q}$ with $Ш(E)(p) \neq 0$ ([1], p. 2)?

From our tables it follows that for $E = X_0(49)$, we have $|Ш(E_{25306669001})| = 2851^2$, with 2851 a prime (the largest prime dividing $|Ш(E_d)|$ at the moment). Also, for any prime $p \leqslant 2357$ there is an elliptic curve E_{d_p} such that $|Ш(E_{d_p})| = p^2$.

Note that from the Table 1 on page 415 of the article [7], we obtain the following (analytic) order of Ш: $|Ш(E(16, 472))| = 3119^2$, where $E(n, m) : y^2 = x(x + m)(x + m - 4 \cdot 3^{2n+1})$. The prime 3119 is good ordinary for $E(16, 472)$, and one may use [24] to prove that the analytic order of Ш coincides with the actual order of Ш in this case.

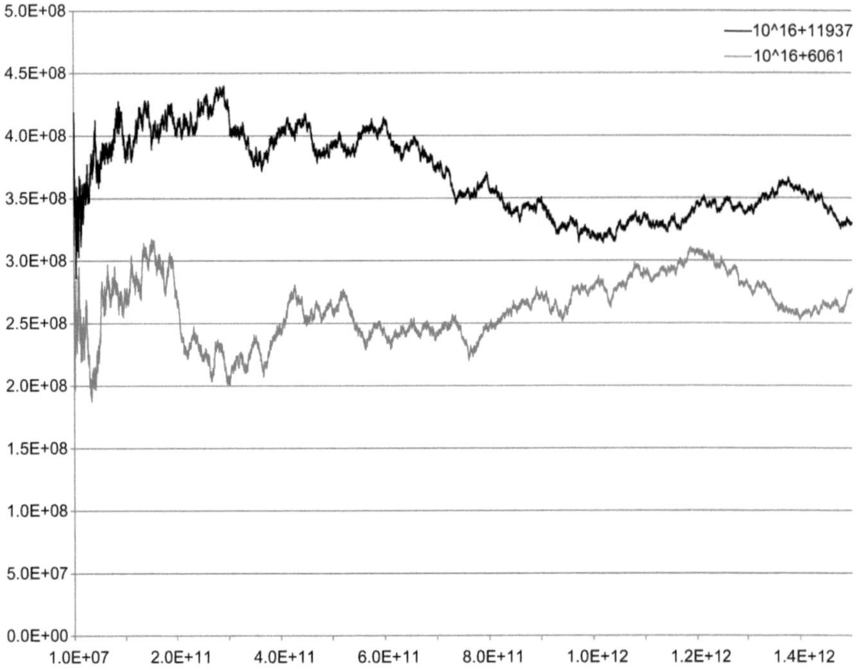

Fig. 1 Beginning of the approximation to $|\text{Ш}(E_d)|$, for $d = 10^{16} + 11937$ and $d = 10^{16} + 6061$ (sum of 1.5×10^{12} terms of the L-series)

Fig. 2 Beginning of the approximation to $|\text{Ш}(E_d)|$, for $d = 10^{20} + 1537$ (sum of 10^{12} terms of the L-series)

7 Asymptotic Formulae

7.1 Quadratic Twists of Rank Zero

Let $V(X) := \{d \leqslant X : d \text{ satisfies the condition } (*)\}$, and $W(X) := \{d \leqslant X : d \text{ satisfies the condition } (**)\}$. We put

$$S(X) := \sum_{d \in V(X)} |Ш(E_d)|, \qquad Z(X) := \sum_{d \in W(X)} |Ш(E_d)|,$$

$$s(X) := \frac{S(X)}{X^{3/2}}, \qquad z(X) := \frac{Z(X)}{X^{3/2}},$$

$$s^*(X) := \frac{(\log X)^{1/8} S(X)}{X^{3/2}}, \qquad z^*(X) := \frac{(\log X)^{1/8} Z(X)}{X^{3/2}}.$$

We have numerically checked that the sequences $s(X)$ and $z(X)$ oscillate very closely (or converge to) some positive constants. Therefore, we proposed the following asymptotic formulae:

$$\sum_{d \in T(X)} |Ш(E_d)| \sim A_E(T) X^{3/2}, \quad X \to \infty, \tag{1}$$

where $A_E(T)$ ($T = V$ or W) are constants depending on E.

R. Heath-Brown ([13]) has proposed a variant of (1):

$$\sum_{d \in V(X)} |Ш(E_d)| \sim B_E X^{3/2} (\log X)^{-1/8}, \quad X \to \infty. \tag{2}$$

It is indeed intriguing which asymptotic formula should be correct. As the following pictures (Figs. 3 and 4) show, it may be difficult numerically to decide ...

Delaunay [8] has used predictions on L-functions coming from random matrix theory (see [4]) to give conjectures for the first leading order asymptotic for

$$M_E(k, T) := \frac{1}{T^*} \sum |Ш(E_d)|^k,$$

for any fixed elliptic curve E over $\mathbb{Q}$ and real positive number k, where the sum is over all fundamental discriminants $d < 0$ coprime with N_E (and satisfying some restrictions) such that $|d| \leqslant T$ and $L(E_d, 1) \neq 0$, and T^* denotes the number of terms in the sum (see Conjecture 6.1 in [8]).

Consider the case $k = 1$, and take $E = X_0(49)$. In this case $\epsilon(E_d) = 1$ if and only if $d > 0$ and $(d, 7) = 1$ or $d < 0$ and $7|d$, hence we cannot apply Conjecture 6.1 directly to E. Instead, we can take $F = E_{-1} : y^2 = x^3 - 35x + 98$. Then $F_{-d} = E_d$, and in our situation Conjecture 6.1 reads as follows:

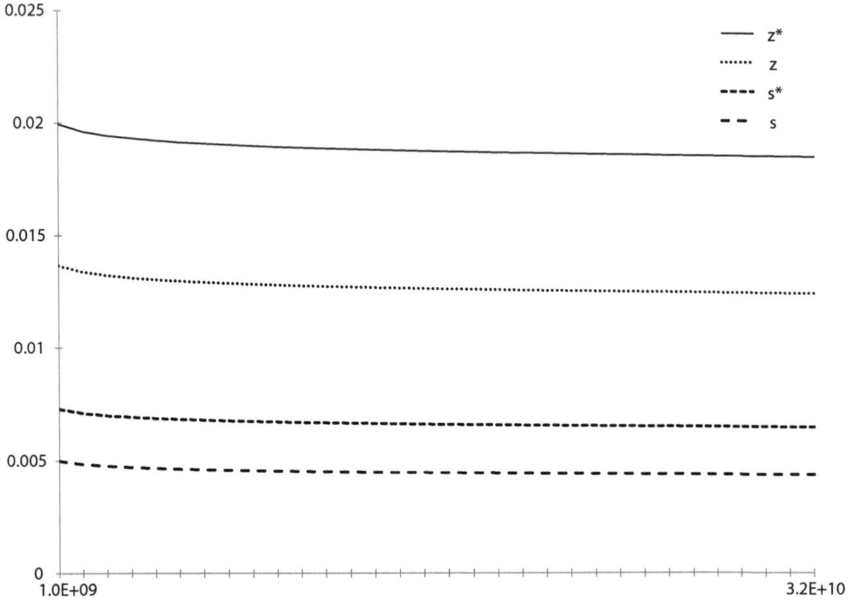

Fig. 3 Numerical evidence for the asymptotic formulas (1) and (2), using the arithmetic sequence of arguments

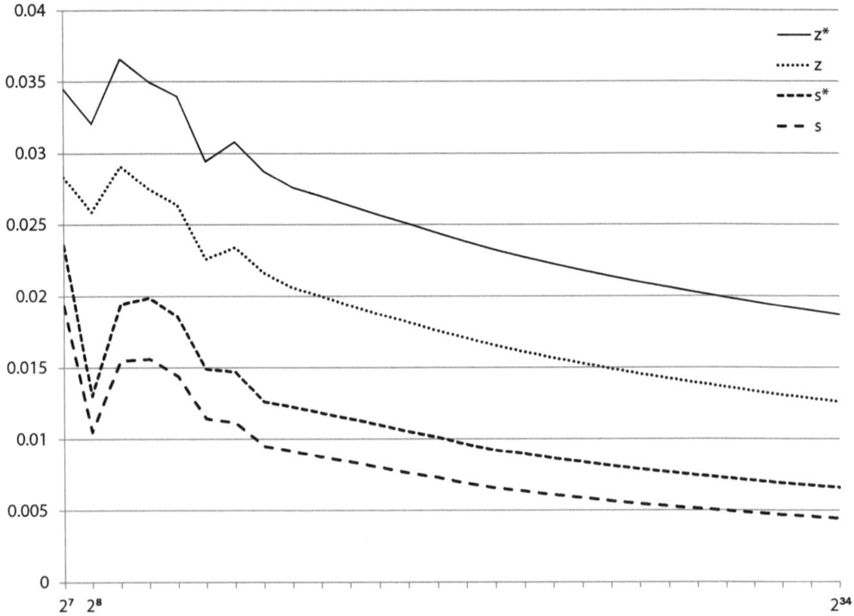

Fig. 4 Numerical evidence for the asymptotic formulas (1) and (2), using the geometric sequence of arguments

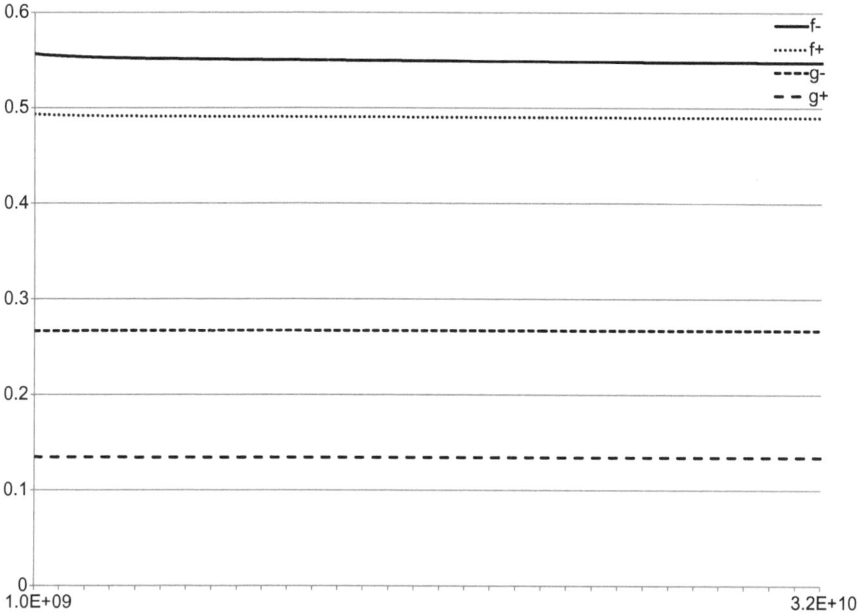

Fig. 5 Numerical evidence for the conjectures 6.1 and 4.2 in [8] in the case $E = X_0(49)$, using the arithmetic sequence of arguments

$$M_E(1,T)^{\pm} \sim C_E^{\pm} T^{1/2} (\log T)^{-5/8}, \quad T \to \infty,$$

for some $C_E^{\pm} > 0$, where $M_E(1,T)^{\pm}$ denotes the subsum of $M_E(1,T)$ restricted to $d \in W(T)$ satisfying $(d/7) = +1$ (or $(d/7) = -1$, respectively). If we restrict to prime discriminants, then we obtain a similar conjecture, but without the log term (Conjecture 4.2 in [8]).

Let $N_E(1,T)^{\pm}$ be a subsum of $M_E(1,T)^{\pm}$, restricted to prime discriminants. Let $f^{\pm}(T) := \frac{(\log T)^{5/8} M_E(1,T)^{\pm}}{T^{1/2}}$, and $g^{\pm}(T) := \frac{N_E(1,T)^{\pm}}{T^{1/2}}$. We obtain the following pictures confirming the Conjectures 6.1 and 4.2 in [8] in the case $E = X_0(49)$ (Figs. 5 and 6).

7.2 Quadratic Twists of Rank One

A general conjecture of Le Boudec ([16], (1.6)), when applied to $E = X_0(49)$, asserts that, if $\Sigma(X)$ denotes the set of odd negative square-free integers d prime to 7, with absolute value at most X, such that $L(E_d, s)$ has a zero of order 1 at $s = 1$, we should have the asymptotic formula

$$\sum_{d \in \Sigma(X)} |\text{Ш}(E_d)|\, R(E_d) \sim C_E X^{3/2} \log X \quad \text{as} \quad X \to \infty. \tag{3}$$

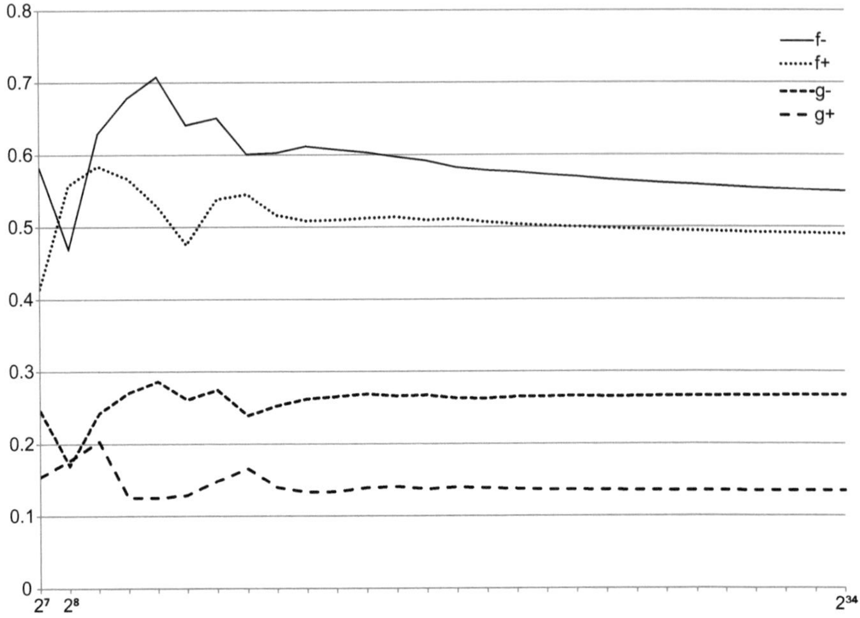

Fig. 6 Numerical evidence for the conjectures 6.1 and 4.2 in [8] in the case $E = X_0(49)$, using the geometric sequence of arguments

At present, we do not unfortunately know if the exact Birch–Swinnerton-Dyer conjecture for the order of $\text{Ш}(E_d)$ is valid for $d \in \Sigma(X)$. However, in what follows, we simply have carried out calculations which use the conjectural analytic order of $\text{Ш}(E_d)$. We now give some numerical data in support of Le Boudec's conjecture in the special case of Theorem 4 (Theorem 1.4 in [2]). More precisely, we fix a prime $l > 3$ which is congruent to 3 mod 4 and inert in the field $\mathbb{Q}(\sqrt{-7})$, and let R be a product of distinct primes, which are congruent to 1 mod 4 and inert in both of the fields $\mathbb{Q}(\sqrt{-7})$ and $\mathbb{Q}(\sqrt{-l})$. Take $d = -lR$, and let $V_l(X)$ denote the set of all such d with absolute value at most X. Then, inserting the precise values for the Tamagawa factors in this case, the above asymptotic formula leads naturally to the conjecture that, for each fixed choice of l, we should also have an asymptotic formula

$$\sum_{d \in V_l(X)} \frac{L'(E_d, 1)\sqrt{-d}}{\Omega_E 2^{r(R)}} \sim C_l X^{3/2} \log X \quad \text{as} \quad X \to \infty, \tag{4}$$

where $r(R)$ denotes the number of prime factors of R, and C_l is a positive constant. Writing $T_l(X)$ for the left hand side of this proposed asymptotic formula, we define

$$t_l(X) := \frac{T_l(X)}{X^{3/2} \log X}.$$

Then, using PARI/GP ([18]) for computations of $L'(E_d, 1)$, we obtain the following data:

X	$t_{19}(X)$	$t_{31}(X)$	$t_{47}(X)$	$t_{59}(X)$	$t_{83}(X)$
250000	0.00013902	0.00008012	0.00006212	0.00003896	0.00003245
500000	0.00011921	0.00007196	0.00006106	0.00003640	0.00003059
750000	0.00011147	0.00006782	0.00005860	0.00003660	0.00003231
1000000	0.00010830	0.00006689	0.00005706	0.00003821	0.00002984
1250000	0.00010783	0.00006785	0.00005560	0.00003814	0.00002996
1500000	0.00010778	0.00006985	0.00005396	0.00003708	0.00002964
1750000	0.00010860	0.00006767	0.00005027	0.00003650	0.00003010
2000000	0.00010621	0.00006648	0.00005119	0.00003567	0.00003079
2250000	0.00010566	0.00006736	0.00005087	0.00003545	0.00003046
2500000	0.00010501	0.00006739	0.00005191	0.00003518	0.00002851
2750000	0.00010359	0.00006582	0.00005205	0.00003353	0.00002812
3000000	0.00010342	0.00006580	0.00005118	0.00003353	0.00002695
3250000	0.00010292	0.00006436	0.00005069	0.00003323	0.00002778
3500000	0.00010190	0.00006371	0.00005027	0.00003337	0.00002710
3750000	0.00010135	0.00006277	0.00004977	0.00003286	0.00002702
4000000	0.00010013	0.00006263	0.00004938	0.00003205	0.00002710
4250000	0.00009997	0.00006175	0.00004914	0.00003268	0.00002653
4500000	0.00009872	0.00006253	0.00004927	0.00003252	0.00002655
4750000	0.00009777	0.00006236	0.00004850	0.00003274	0.00002676
5000000	0.00009764	0.00006165	0.00004870	0.00003248	0.00002759
5250000	0.00009707	0.00006179	0.00004834	0.00003305	0.00002755
5500000	0.00009696	0.00006205	0.00004814	0.00003327	0.00002741
5750000	0.00009708	0.00006225	0.00004798	0.00003326	0.00002717
6000000	0.00009654	0.00006233	0.00004762	0.00003298	0.00002733

and the following picture (Fig. 7).

8 Cohen–Lenstra Heuristics for the Order of Ш

Delaunay [9] has considered Cohen–Lenstra heuristics for the order of Tate–Shafarevich group. He predicts, among others, that in the rank zero case, the probability that $|Ш(E)|$ of a given elliptic curve E over $\mathbb{Q}$ is divisible by a prime p should be $f_0(p) := 1 - \prod_{j=1}^{\infty}(1 - p^{1-2j}) = \frac{1}{p} + \frac{1}{p^3} + \cdots$. Hence, $f_0(2) \approx 0.580577$, $f_0(3) \approx 0.360995$, $f_0(5) \approx 0.206660$, $f_0(7) \approx 0.145408$, and so on. The papers of Quattrini ([20, 21]) make a correction to Delaunay's heuristics for p-divisibility of $|Ш(E_d)|$ in the family of quadratic twists of a given elliptic curve E of square-free conductor for odd primes dividing the order of $E(\mathbb{Q})_{tors}$. The author gives an explanation of why and when the original Cohen–Lenstra heuristics should be used for the prediction of the p-divisibility of $|Ш(E_d)|$.

Let $F(X)$ (resp. $G(X)$) denote the number of $d \leqslant X$ satisfying (*) (resp. (**)). Let $F_p(X)$ (resp. $G_p(X)$) denote the number of $d \leqslant X$ satisfying (*) (resp. satisfying

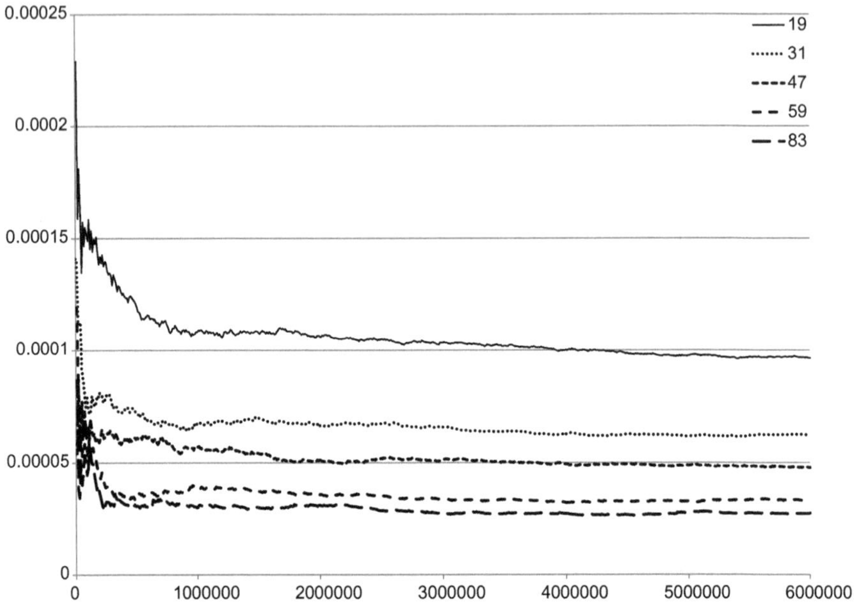

Fig. 7 Numerical evidence for the asymptotic formula (4)

(**)), and such that $|Ш(E_d)|$ is divisible by p. Let $f_p(X) := \frac{F_p(X)}{F(X)}$, and $g_p(X) := \frac{G_p(X)}{G(X)}$. We obtain the following table (see the next page).

The functions $g_3(X)$ and $g_5(X)$ both tend to the Delaunay numbers $f_0(3)$ and $f_0(5)$, respectively. Additionally restricting to the twists satisfying (*) (i.e., considering the functions $f_3(X)$ and $f_5(X)$) tends to speed the convergence. The function $g_2(X)$ tends (slowly) to $f_0(2)$. Finally, the table shows that the probability that $|Ш(E_d)|$ is divisible by 7 deviates from Delaunay's prediction. Note that $N_{X_0(49)} = 49$ is not square-free, hence the papers of Quattrini do not explain this situation.

9 Distributions of $L(E_d, 1)$ and $|Ш(E_d)|$

9.1 Distribution of $L(E_d, 1)$

It is a classical result (due to Selberg) that the values of $\log|\zeta(\frac{1}{2}+it)|$ follow a normal distribution.

Let E be any elliptic curve defined over $\mathbb{Q}$. Let $\mathcal{E}$ denote the set of all fundamental discriminants d with $(d, 2N_E) = 1$ and $\epsilon_E(d) = \epsilon_E \chi_d(-N_E) = 1$, where ϵ_E is the root number of E and $\chi_d = (d/\cdot)$. Keating and Snaith [14] have conjectured that, for

X	$g_2(X)$	$f_3(X)$	$g_3(X)$	$f_5(X)$	$g_5(X)$	$f_7(X)$	$g_7(X)$
$1 \cdot 10^9$	0.524765	0.359655	0.343200	0.206042	0.186796	0.162955	0.142212
$2 \cdot 10^9$	0.529699	0.359866	0.345360	0.206251	0.189156	0.163044	0.144630
$3 \cdot 10^9$	0.532425	0.359882	0.346472	0.206308	0.190392	0.163055	0.145932
$4 \cdot 10^9$	0.534302	0.359993	0.347244	0.206389	0.191231	0.163065	0.146795
$5 \cdot 10^9$	0.535716	0.360069	0.347810	0.206375	0.191835	0.163086	0.147442
$6 \cdot 10^9$	0.536861	0.360112	0.348264	0.206414	0.192318	0.163115	0.147960
$7 \cdot 10^9$	0.537804	0.360147	0.348629	0.206418	0.192714	0.163116	0.148387
$8 \cdot 10^9$	0.538615	0.360193	0.348945	0.206425	0.193046	0.163110	0.148740
$9 \cdot 10^9$	0.539317	0.360219	0.349216	0.206442	0.193343	0.163121	0.149050
$10 \cdot 10^9$	0.539944	0.360237	0.349461	0.206444	0.193599	0.163134	0.149321
$11 \cdot 10^9$	0.540497	0.360248	0.349663	0.206451	0.193820	0.163140	0.149564
$12 \cdot 10^9$	0.541004	0.360266	0.349853	0.206454	0.194025	0.163141	0.149782
$13 \cdot 10^9$	0.541465	0.360269	0.350021	0.206464	0.194209	0.163143	0.149977
$14 \cdot 10^9$	0.541890	0.360272	0.350182	0.206472	0.194382	0.163150	0.150158
$15 \cdot 10^9$	0.542281	0.360285	0.350322	0.206479	0.194538	0.163153	0.150322
$16 \cdot 10^9$	0.542646	0.360290	0.350456	0.206487	0.194681	0.163161	0.150478
$17 \cdot 10^9$	0.542984	0.360302	0.350580	0.206493	0.194817	0.163169	0.150618
$18 \cdot 10^9$	0.543301	0.360320	0.350695	0.206497	0.194940	0.163168	0.150753
$19 \cdot 10^9$	0.543601	0.360322	0.350803	0.206498	0.195057	0.163173	0.150879
$20 \cdot 10^9$	0.543883	0.360330	0.350903	0.206496	0.195165	0.163175	0.150997
$21 \cdot 10^9$	0.544151	0.360331	0.350995	0.206494	0.195268	0.163171	0.151108
$22 \cdot 10^9$	0.544404	0.360342	0.351086	0.206500	0.195368	0.163167	0.151211
$23 \cdot 10^9$	0.544647	0.360358	0.351174	0.206503	0.195464	0.163170	0.151309
$24 \cdot 10^9$	0.544877	0.360366	0.351258	0.206510	0.195552	0.163171	0.151404
$25 \cdot 10^9$	0.545100	0.360371	0.351334	0.206513	0.195635	0.163174	0.151494
$26 \cdot 10^9$	0.545312	0.360374	0.351408	0.206512	0.195715	0.163166	0.151578
$27 \cdot 10^9$	0.545513	0.360387	0.351478	0.206510	0.195791	0.163167	0.151659
$28 \cdot 10^9$	0.545707	0.360386	0.351541	0.206501	0.195864	0.163168	0.151738
$29 \cdot 10^9$	0.545894	0.360395	0.351606	0.206499	0.195934	0.163171	0.151813
$30 \cdot 10^9$	0.546074	0.360408	0.351669	0.206499	0.196001	0.163174	0.151885
$31 \cdot 10^9$	0.546248	0.360413	0.351726	0.206501	0.196067	0.163173	0.151955
$32 \cdot 10^9$	0.546416	0.360411	0.351784	0.206508	0.196131	0.163172	0.152022

$d \in \mathcal{E}$, the quantity $\log L(E_d, 1)$ has a normal distribution with mean $-\frac{1}{2} \log\log |d|$ and variance $\log\log |d|$; see [5] for numerical data towards this conjecture.

Below we consider the case $E = X_0(49)$. Our data allow to confirm that the values $\log L(E_d, 1)$ indeed follow an approximate normal distribution.

Here is some explanation for the next figures. Let $B = 32 \cdot 10^9$, $V = \{d \leqslant B : d \text{ satisfies } (*)\}$, $W = \{d \leqslant B : d \text{ satisfies } (**)\}$ and $I_x = [x, x + 0.1)$ for $x \in \{-10, -9.9, -9.8, \ldots, 10\}$. We create a histogram with bins I_x from the data $\left\{\left(\log L(E_d, 1) + \frac{1}{2} \log\log d\right) / \sqrt{\log\log d} : d \in V\right\}$ and normalize it in such a way that the total area of bars is equal to 1. Below we picture this histogram together with a graph of the standard normal density function (Fig. 8).

Next, we do the same, but with W in place of V (Fig. 9).

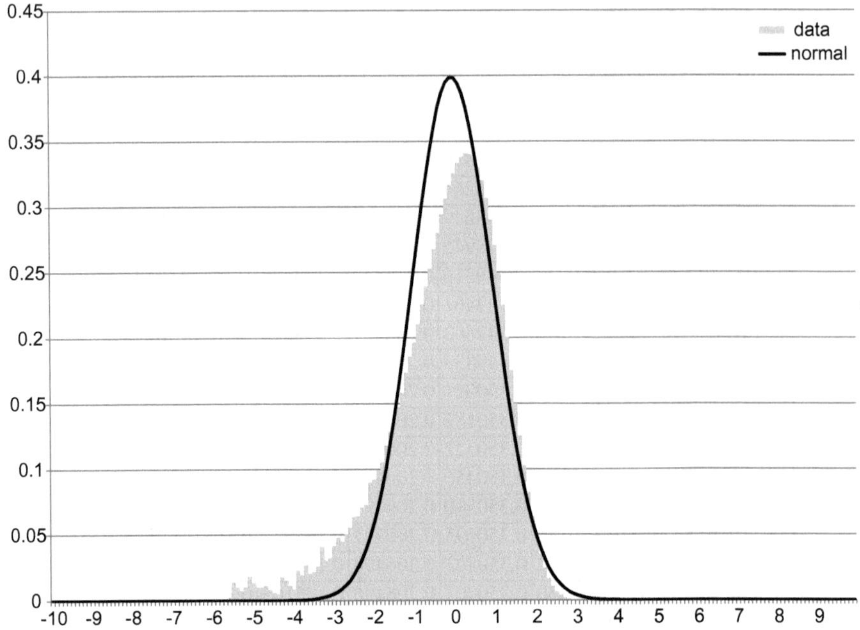

Fig. 8 Histogram of values $\left(\log L(E_d, 1) + \frac{1}{2}\log\log d\right)/\sqrt{\log\log d}$ for $d \leqslant B$ satisfying (*). The *black line* depicts a graph of the standard normal density function

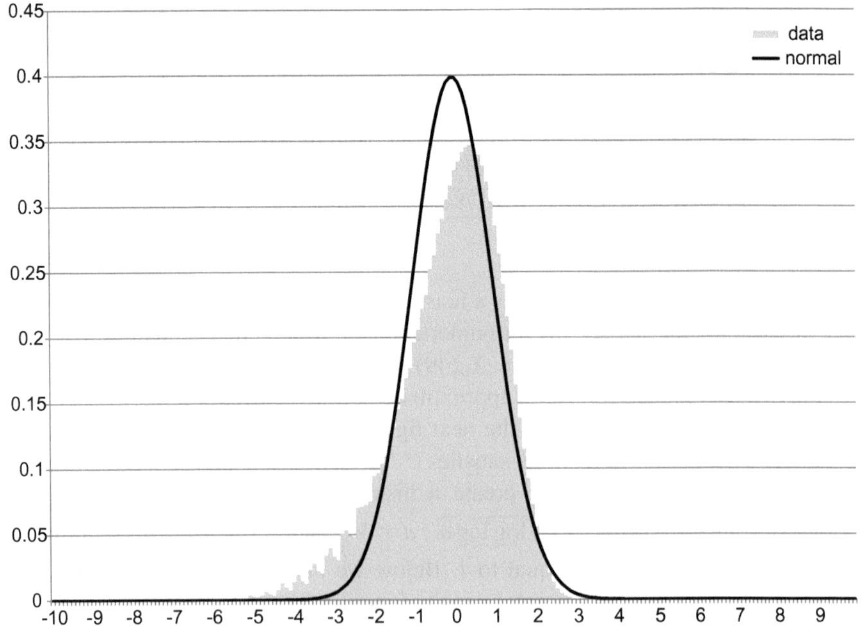

Fig. 9 Histogram of values $\left(\log L(E_d, 1) + \frac{1}{2}\log\log d\right)/\sqrt{\log\log d}$ for $d \leqslant B$ satisfying (**). The *black line* depicts a graph of the standard normal density function

The paper of Conrey et al. [5] contains similar data (millions of quadratic twists for thousands of elliptic curves). Their data compares not just against the limiting Gaussian (as in our paper), but against the distribution suggested from random matrix theory (which tends to the standard Gaussian); for more details see Sect. 6 in [5].

9.2 Distribution of $|Ш(E_d)|$

It is an interesting question to find results (or at least a conjecture) on distribution of the order of the Tate–Shafarevich group for rank zero quadratic twists of an elliptic curve over $\mathbb{Q}$.

It turns out that the values of $\log(|Ш(E_d)|/\sqrt{d})$ are the more natural ones (compare Conjecture 1 in [22]). Let $\mu = -\frac{1}{2} - \frac{3}{2}\log 2$, $\sigma^2 = 1 + \frac{5}{2}(\log 2)^2$ (it is the case for $[K : \mathbb{Q}] = 2$ in Conjecture 1 in [22]). We create a histogram from the data $\left\{\left(\log(|Ш(E_d)|/\sqrt{d}) - \mu \log\log d\right)/\sqrt{\sigma^2 \log\log d} : d \in V\right\}$ and normalize it in such a way that the total area of bars is equal to 1. Below we picture this histogram together with a graph of the standard normal density function (Fig. 10).

Next, we do the same, but with W in place of V (Fig. 11).

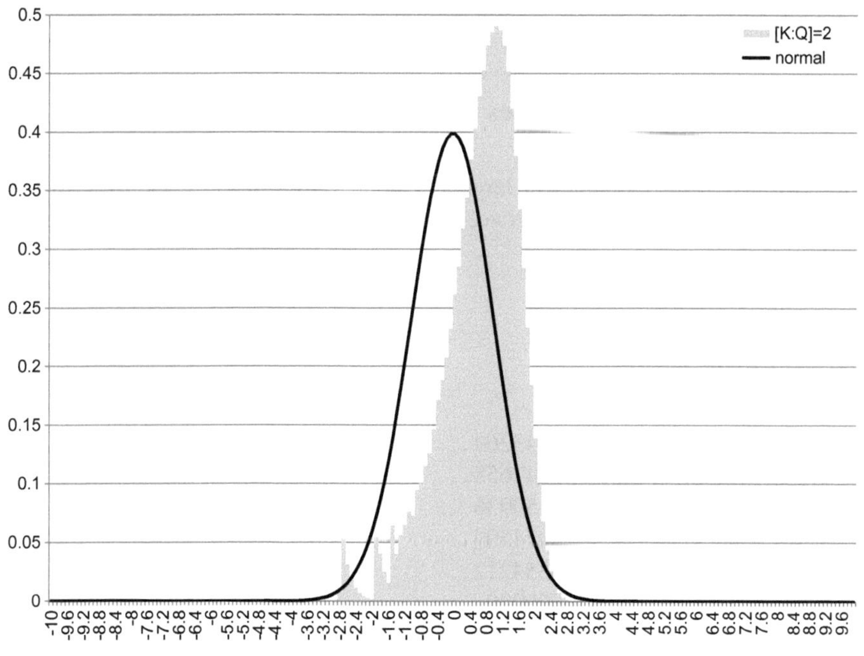

Fig. 10 Histogram of values $\left(\log(|Ш(E_d)|/\sqrt{d}) - \mu \log\log d\right)/\sqrt{\sigma^2 \log\log d}$ for $d \leqslant B$ satisfying (*). The *black line* depicts a graph of the standard normal density function

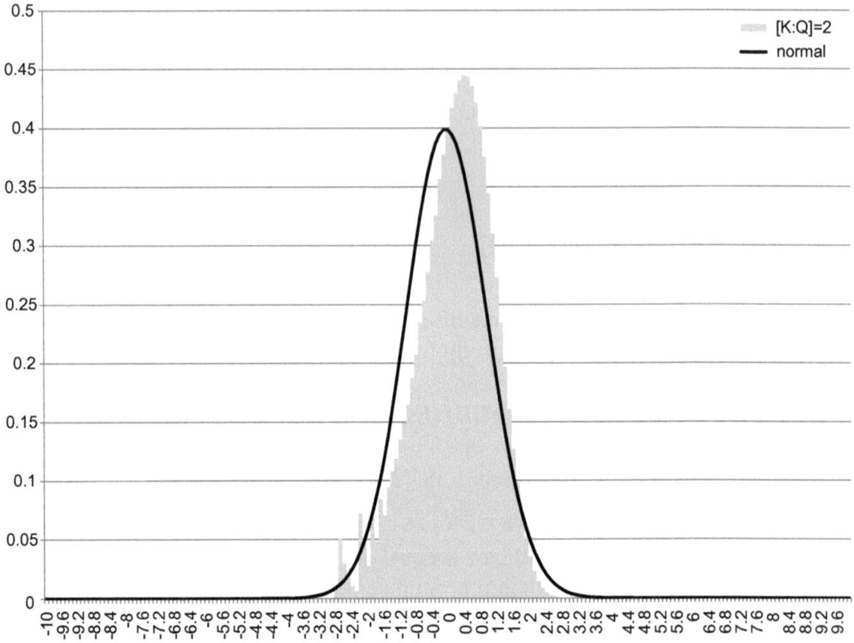

Fig. 11 Histogram of values $\left(\log(|\text{Ш}(E_d)|/\sqrt{d}) - \mu \log\log d\right) / \sqrt{\sigma^2 \log\log d}$ for $d \leqslant B$ satisfying (**). The *black line* depicts a graph of the standard normal density function

10 Large and Small Values of $L(E_d, 1)$

Here we give some examples of large and small values of $L(E_d, 1)$. We also give some examples of small gaps between the values of $L(E_d, 1)$ (compare Sect. 3 in [7]).

10.1 Large Values

$$L(E_{12010333305}, 1) = 139.0972543269\ldots$$
$$L(E_{24320258169}, 1) = 130.2497841658\ldots$$
$$L(E_{30942205545}, 1) = 130.0598150936\ldots$$
$$L(E_{21502242105}, 1) = 129.4879974509\ldots$$
$$L(E_{26284959705}, 1) = 128.3672354212\ldots$$
$$L(E_{17391204345}, 1) = 127.8286009701\ldots$$
$$L(E_{24406185945}, 1) = 127.3116124586\ldots$$
$$L(E_{18840415665}, 1) = 127.0854001988\ldots$$

10.2 Small Values

$L(E_{31999908701}, 1) = 0.0000108075564\ldots$
$L(E_{31999917269}, 1) = 0.0000108075549\ldots$
$L(E_{31999918117}, 1) = 0.0000108075548\ldots$
$L(E_{31999937569}, 1) = 0.0000108075515\ldots$
$L(E_{31999943197}, 1) = 0.0000108075505\ldots$
$L(E_{31999952249}, 1) = 0.0000108075490\ldots$
$L(E_{31999975069}, 1) = 0.0000108075451\ldots$
$L(E_{31999994129}, 1) = 0.0000108075419\ldots$

10.3 Small Gaps Between L-values

We expect that there are infinitely many d's with trivial $Ш(E_d)$, and hence we expect $L(E_d, 1)$ may take arbitrarily small values. Anyway, it may be of some interest to have examples of small gaps between L-values in case of non-trivial Sha's.

d_1	d_2	$\lvert L(E_{d_1}, 1) - L(E_{d_2}, 1)\rvert$	$\lvert Ш(E_{d_1})\rvert$	$\lvert Ш(E_{d_2})\rvert$
31999874185	31999874189	$1.08 \cdot 10^{-14}$	1^2	2^2
31999576809	31999576813	$2.16 \cdot 10^{-14}$	2^2	2^2
31999771129	31999771133	$4.86 \cdot 10^{-14}$	3^2	3^2
31999662013	31999662017	$1.35 \cdot 10^{-13}$	5^2	5^2
31999835293	31999835297	$2.97 \cdot 10^{-13}$	21^2	21^2
31999908217	31999908221	$7.78 \cdot 10^{-13}$	6^2	12^2
31999945877	31999945881	$9.13 \cdot 10^{-13}$	13^2	26^2
31999535093	31999535101	$1.55 \cdot 10^{-12}$	24^2	6^2

11 $|Ш(E_d)| = 1$ is About as Common as $L(E_d, 1) = 0$

Poonen [19] has recently asked one of us, whether our data show that $|Ш(E_d)|=1$ is about as common as $L(E_d, 1) = 0$? It turns out that the literature contains conflicting predictions about this (see Sect. 4.5 of [25] for a discussion).

Our computational evidence resolves this problem. Let $f(x)$ denote the number of positive square-free integers $d \leqslant x$, congruent to 1 modulo 4, such that $(d, 7) = 1$, $L(E_d, 1) \neq 0$, and $|Ш(E_d)| = 1$. Let $g(x)$ denote the number of positive square-free integers $d \leqslant x$, congruent to 1 modulo 4, such that $(d, 7) = 1$, and $L(E_d, 1) = 0$. We obtain the following graph (Fig. 12).

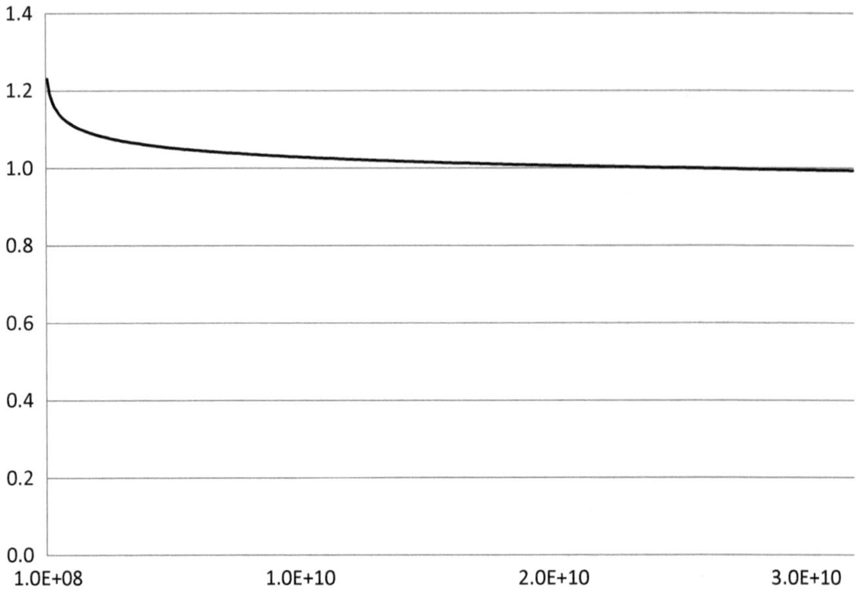

Fig. 12 Graph of the function $f(x)/g(x)$

We expect (Delaunay–Watkins [10], Heuristics 1.1):

$$g(x) \sim cx^{3/4}(\log x)^{3/8+\sqrt{2}/2}, \quad x \to \infty,$$

hence we may expect a similar asymptotic formula for $f(x)$ as well.

Now let $f_k(x)$ denote the number of positive square-free integers $d \leqslant x$, congruent to 1 modulo 4, such that $(d, 7) = 1$, $L(E_d, 1) \neq 0$, and $|Ш(E_d)| = k^2$. Let

$$F_k(x) := \frac{x^{3/4}(\log x)^{3/8+\sqrt{2}/2}}{f_k(x)},$$

$$G(x) := \frac{x^{3/4}(\log x)^{3/8+\sqrt{2}/2}}{g(x)}.$$

The above calculations and the next graph (Fig. 13) suggest the following

Conjecture 8 *For any positive integer k there are constants $c_k > 0$ and d_k, such that*

$$f_k(x) \sim c_k x^{3/4}(\log x)^{d_k}, \quad x \to \infty.$$

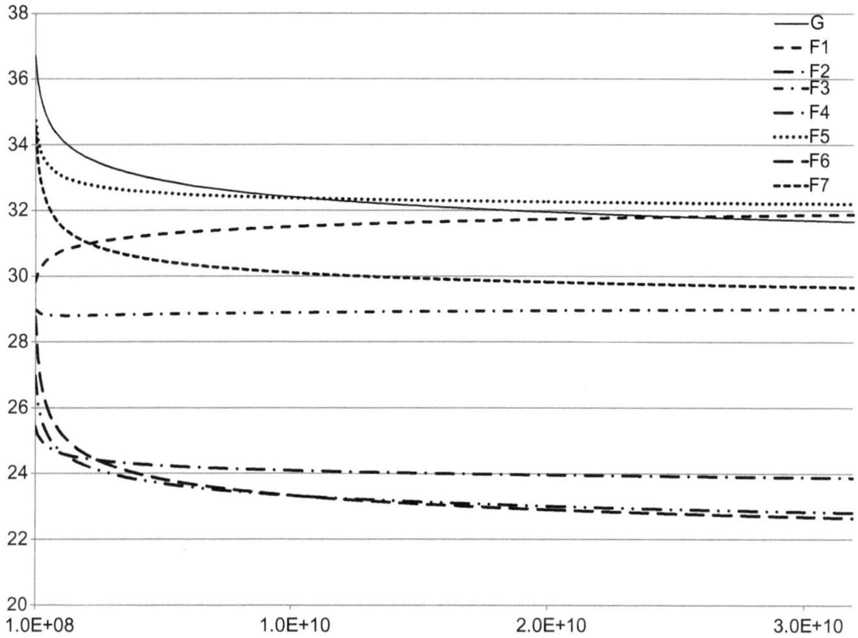

Fig. 13 Graphs of the functions $G(x)$ and $F_k(x)$, $k = 1, 2, 3, 4, 5, 6, 7$

Appendix: The Algorithm and the Implementation

The strategy is to use the construction described in the end of Sect. 2 to compute the coefficients a_d for d satisfying condition (**) up to $32 \cdot 10^9$, and use Corollary 1.

We present our algorithm using computer algebra system PARI/GP [18].

```
sha(B) =
{
   /* define quadratic forms */
   f1 = [1,0,0;  0,28,0; 0,0,196];
   f2 = [4,0,0;  0,28,0;  0,0,49];
   f3 = [5,2,0;  2,40,0;  0,0,28];
   f4 = [13,5,0; 5,17,0;  0,0,28];

   /* compute a_d for all d ∈ {1,...,B} */
   a = Vec(qfrep(f1,B)) - Vec(qfrep(f2,B)) +
       Vec(qfrep(f3,B)) - Vec(qfrep(f4,B));

   /* enumerate all d satisfying (**) */
   forstep(d = 1, B, 4,
      if(Mod(d,7)!=0 && issquarefree(d),
         f = factor(d);
         l1 = 0; l2 = 0;
         for(i = 1, omega(d),
```

```
            if(kronecker(f[i,1],7)==1, l1=l1+1, l2=l2+1)
          );
          if(Mod(l2,2)==1, l=l1+(l2-1)/2, l=l1+l2/2);
          print(d, "", abs(a[d]/2^l));
        )
    )
}
```

The key point of the above implementation is using the `qfrep` function. Recall that `qfrep` (q, B) returns the vector whose i-th entry ($1 \leqslant i \leqslant B$) is half the number of vectors v such that $q(v) = i$. Routine `qfrep` relies on `qfminim` function which enumerates, using the Fincke–Pohst algorithm, the vectors v for which $q(v) \leqslant B$.

We used the above PARI/GP script (with small modifications) to compute `sha`(B) for $B = 5 \cdot 10^7$ (on standard desktop PC). We made further progress implementing `qfrep` function in `C++` language. Our routine directly enumerates all vectors v for which $f_i(v) \leqslant B$ for $i = 1, 2, 3, 4$. Such a straightforward approach is more effective for forms f_1, f_2, f_3, f_4 than the sophisticated Fincke–Pohst algorithm used by PARI/GP. Moreover, we made some optimizations. Most important among them are:

- time optimization—exploiting symmetries of f_1, f_2, f_3, f_4,
- memory optimization—storing in memory only values of a_d for $d \equiv 1 \pmod 4$.

Enumerating numbers satisfying condition (**) (and factoring them) also takes some time. One can be speed it up by filtering out numbers which are not square-free. We did it by using a modified sieve of Eratosthenes. However, the real bottleneck is computing `qfrep`.

Let us add that our algorithm is quite easily parallelizable. It appears that performing computation in parallel will make possible to increase B substantially compared to our achievement.

Tables

For each positive integer $k \leqslant 1793$, the column headed d_1 gives the smallest integer d for which $|\text{Ш}(E_d)| = k^2$. One interesting observation is that all odd orders $|\text{Ш}(E_{d_1})|$ are realized by the integers d_1 satisfying the condition (*).

Selected values $1795 \leqslant k \leqslant 2941$ are also considered.

k	d_1	k	d_1	k	d_1	k	d_1
1	1	2	93	3	73	4	177
5	257	6	933	7	929	8	4337
9	2281	10	6073	11	3169	12	6609
13	5897	14	14177	15	12241	16	20497
17	10937	18	19713	19	5641	20	52257
21	18793	22	40769	23	31513	24	63473
25	26249	26	55617	27	23369	28	63849
29	62929	30	121881	31	49993	32	152769
33	65609	34	100857	35	62401	36	167073
37	98257	38	322921	39	96353	40	226913
41	133769	42	206273	43	151273	44	734001
45	110977	46	337681	47	129457	48	498129
49	253553	50	549817	51	152953	52	518137
53	152249	54	702353	55	291457	56	612529
57	247369	58	673817	59	368857	60	953313
61	365249	62	964793	63	626377	64	847793
65	290657	66	1319649	67	527729	68	1217049
69	536017	70	1091841	71	957361	72	2060353
73	637297	74	1501329	75	423097	76	1135649
77	1465469	78	1707729	79	955769	80	1827193
81	570113	82	2874369	83	682009	84	1234137
85	1101593	86	2827553	87	1899481	88	2229529
89	1885673	90	2341817	91	1323689	92	2799217
93	1381337	94	3018513	95	1242169	96	2904801
97	1917697	98	4294313	99	1790897	100	3567881
101	1625321	102	4518273	103	1866857	104	5884041
105	1781569	106	4184049	107	2915713	108	6165329
109	2182249	110	4479897	111	3647689	112	4909017
113	1465313	114	5427489	115	2761841	116	6469849
117	2687257	118	6350073	119	3393449	120	4884177
121	3524041	122	7486329	123	3485513	124	7240809
125	3613193	126	4935001	127	4229657	128	7353921
129	3486257	130	7753601	131	4459601	132	4110177
133	4693177	134	8832657	135	3247313	136	8314777
137	4296977	138	10538889	139	5507297	140	11180073
141	3688081	142	10114889	143	6025801	144	6302409
145	3653369	146	14245449	147	5294833	148	11250761
149	3106921	150	10362081	151	5946337	152	13688313
153	6790073	154	10240521	155	7491361	156	15464089
157	4103641	158	10692817	159	7016777	160	11928953
161	6718193	162	15799897	163	6645721	164	13927593
165	10108297	166	12971121	167	6120929	168	19275657
169	10688753	170	21326937	171	6078337	172	9310449
173	8860361	174	16962969	175	9539281	176	23387809
177	9032609	178	18940849	179	9425113	180	24630321
181	9843529	182	21259921	183	5634809	184	20181561
185	15130393	186	18589729	187	10534921	188	26802777

k	d_1	k	d_1	k	d_1	k	d_1
189	9128969	190	18017673	191	15028201	192	27317361
193	17238961	194	27932529	195	13267873	196	18475521
197	19233889	198	32768481	199	18412817	200	17722329
201	14262433	202	35158337	203	17371513	204	31450409
205	15861473	206	55430673	207	17282137	208	25378833
209	16847953	210	20905977	211	13179577	212	30805273
213	17764913	214	47121369	215	14565193	216	24423177
217	17119073	218	31973313	219	20660377	220	30225721
221	20663593	222	32356041	223	24736529	224	31522353
225	17100961	226	44057561	227	23429761	228	61238433
229	20035217	230	30310809	231	17521937	232	45713721
233	26153209	234	53720529	235	19521001	236	30965713
237	17479313	238	45580921	239	19624729	240	59076249
241	24796313	242	46666337	243	15196457	244	47964921
245	24126161	246	67880649	247	15737417	248	59498961
249	27527393	250	67310681	251	31900529	252	71401089
253	21488809	254	42480201	255	21141041	256	62559121
257	21436001	258	44968137	259	23661529	260	68143553
261	28188257	262	83482809	263	45616297	264	74407953
265	33502577	266	76802441	267	45721681	268	56817777
269	30511001	270	80564961	271	42257857	272	65262849
273	30407369	274	51619593	275	34562401	276	86165913
277	28530241	278	100543353	279	29771201	280	55275609
281	33775801	282	102490809	283	38382041	284	89933937
285	27594521	286	94586473	287	41793233	288	84939537
289	47313209	290	94270929	291	25854097	292	74695377
293	49080337	294	129084873	295	48796537	296	97992497
297	41571113	298	108653521	299	63138337	300	114844137
301	28987073	302	117318657	303	70938377	304	120142353
305	54726241	306	106517777	307	62983121	308	108212241
309	54211177	310	132195057	311	57847201	312	107424529
313	64804081	314	136571681	315	67153729	316	72567049
317	39044641	318	103714801	319	68778097	320	124306361
321	67515881	322	99630969	323	42683441	324	126667249
325	81725521	326	128132673	327	22476089	328	69421713
329	82804153	330	131429937	331	35634569	332	103211529
333	53915137	334	127330809	335	58071737	336	145127361
337	82549193	338	127179537	339	81256817	340	212734713
341	94701017	342	120707049	343	74528177	344	180657537
345	82682417	346	163723673	347	83092201	348	127062681
349	64266313	350	158078897	351	45363041	352	165215121
353	86228729	354	150696393	355	74605177	356	158006489
357	96923641	358	168255201	359	101236001	360	199147433
361	61967953	362	208401153	363	75284753	364	179223529
365	81361481	366	190068321	367	128132273	368	233660249
369	104815049	370	166966409	371	78813817	372	166934289
373	85236353	374	259338137	375	100124561	376	150853497
377	86855849	378	195302193	379	117023129	380	170908593

k	d_1	k	d_1	k	d_1	k	d_1
381	119282593	382	229290681	383	124208657	384	250172553
385	84921329	386	242076657	387	81585241	388	198355873
389	111613529	390	317098497	391	68312473	392	164772969
393	137121073	394	317478617	395	184317257	396	267280841
397	157495033	398	198455217	399	95602057	400	201599049
401	141095993	402	315033657	403	162887129	404	252130649
405	129060721	406	239496657	407	75453481	408	309730961
409	145253513	410	263668161	411	114535441	412	258161649
413	70924577	414	201012681	415	156669217	416	348700969
417	133510961	418	255577281	419	133337329	420	243488697
421	163103713	422	349629081	423	129344561	424	248961057
425	113424457	426	158340513	427	161716729	428	457284881
429	149221609	430	306979737	431	188287097	432	263466921
433	113174249	434	379096881	435	116677553	436	341102721
437	138979921	438	344625297	439	142217729	440	402660633
441	173153249	442	337717857	443	124106569	444	303693153
445	187920529	446	348974817	447	141120313	448	374142729
449	169920161	450	415195257	451	119896241	452	456636161
453	172998929	454	370842873	455	186067649	456	284126217
457	211471489	458	322781001	459	175500121	460	424009721
461	175321193	462	383282377	463	227914553	464	411552969
465	151632193	466	405886881	467	160286201	468	493254001
469	133377289	470	381535529	471	150114793	472	467868809
473	199481561	474	367048641	475	231417217	476	428448673
477	216300353	478	464970993	479	182712193	480	377366529
481	203950673	482	481370849	483	208906417	484	466546089
485	221357009	486	611992833	487	213593561	488	445591689
489	152807297	490	365071281	491	227448577	492	324398649
493	152292209	494	327557561	495	158411977	496	455620881
497	235272161	498	298850577	499	159619337	500	487175313
501	87873329	502	433020569	503	209599633	504	468120129
505	233102873	506	680179593	507	248435521	508	409342217
509	167807489	510	499217601	511	201167833	512	502073881
513	256670657	514	601398561	515	265990297	516	591746313
517	225790673	518	474463697	519	244563961	520	560288513
521	191110121	522	724247857	523	259344209	524	455548713
525	263635321	526	426422649	527	313004473	528	484397841
529	231144521	530	556803769	531	257557873	532	652226129
533	328620697	534	579162081	535	335578081	536	618274897
537	312412721	538	888968217	539	291056657	540	577412049
541	243334657	542	871280337	543	389624233	544	484212369
545	320529089	546	874368489	547	260262113	548	784207257
549	298280401	550	598627929	551	359759921	552	374213393
553	217628393	554	840530793	555	258609433	556	789059793
557	376545137	558	899077953	559	310206713	560	849373977
561	276498809	562	720256769	563	359961713	564	605796249
565	505198489	566	660711921	567	374576513	568	693060153
569	285955057	570	582042129	571	194086553	572	971076633

k	d_1	k	d_1	k	d_1	k	d_1
573	269512417	574	812238113	575	394565449	576	531793497
577	376637449	578	558769737	579	177118241	580	800651713
581	314235617	582	971046561	583	278549137	584	713190633
585	343632217	586	910562577	587	299414377	588	806580569
589	339441217	590	969882153	591	404910817	592	1180596561
593	445593769	594	746080449	595	366680009	596	588145881
597	499709489	598	1000317921	599	453223081	600	857027697
601	482163793	602	709173529	603	464499577	604	1085969329
605	583993777	606	611140737	607	477113129	608	767941113
609	487518937	610	1205784057	611	345190481	612	1042100481
613	433569953	614	759143361	615	295290449	616	1060039817
617	271819777	618	916876041	619	399513761	620	833937729
621	351775609	622	1014335921	623	461803457	624	1063477353
625	604768433	626	1195237257	627	355313929	628	1130731233
629	356217217	630	1088179233	631	523841737	632	1296431537
633	373251833	634	1046573089	635	533158321	636	1129661633
637	437185369	638	1001125833	639	690863353	640	1049542113
641	393691721	642	1219662481	643	442113409	644	866810121
645	473179657	646	918013177	647	422803841	648	925311633
649	628298753	650	908022641	651	416028409	652	668155161
653	639545801	654	788481633	655	456778633	656	1253196393
657	553788233	658	1011782193	659	559500833	660	1095831129
661	502398097	662	1556083761	663	540229913	664	1511104233
665	487247689	666	1530974993	667	419408113	668	1006678033
669	557895281	670	1502333529	671	748804241	672	954448401
673	431472649	674	1503531033	675	567004513	676	1390863849
677	486471529	678	845750049	679	563530553	680	1401556137
681	472466377	682	1131586201	683	774729497	684	1153585497
685	424046153	686	1292386929	687	384060241	688	969826881
689	677852257	690	1146387201	691	472867673	692	1284311337
693	590461073	694	1448799433	695	552263017	696	1430856177
697	487192513	698	1065709793	699	789946601	700	1272089361
701	617717609	702	1305583953	703	622046417	704	1815425929
705	510979577	706	886583769	707	571233809	708	1259122593
709	634656713	710	1451816193	711	769553489	712	1902558001
713	919520761	714	1671253809	715	555591601	716	1737287913
717	690268577	718	1691545857	719	758546561	720	1417870169
721	626618809	722	2027470713	723	645226937	724	1660673337
725	357461201	726	1159545633	727	803008321	728	1172959881
729	645947833	730	1532369217	731	786522889	732	1220017089
733	822906313	734	1812159633	735	723749137	736	1677838089
737	797421113	738	1334772609	739	533625049	740	2156546177
741	898149521	742	1908161193	743	777702721	744	1475258537
745	764284097	746	1932838161	747	774980849	748	1597045321
749	805117801	750	1649813817	751	912356273	752	1922828889
753	535552681	754	2049963393	755	618088673	756	1439459169
757	1012636441	758	1734170937	759	704129537	760	2013014697
761	673372577	762	2328013713	763	832572401	764	1695155961

k	d_1	k	d_1	k	d_1	k	d_1
765	920887169	766	1176556089	767	831201097	768	1843728657
769	786715537	770	2185812897	771	637134697	772	1860573801
773	623324777	774	2428781793	775	854043041	776	1309574841
777	413798201	778	2198450769	779	790908361	780	1722152721
781	1058446409	782	1562389249	783	673866737	784	1600255897
785	859210817	786	2201361121	787	845971769	788	2347753809
789	1009504417	790	2403252273	791	780662161	792	2485110489
793	991564633	794	1670427609	795	988737601	796	1304526953
797	851169289	798	2039945329	799	909129017	800	1739103809
801	936222409	802	1575219297	803	1063972649	804	2389858689
805	810098873	806	2374269969	807	838474873	808	1497563313
809	980245633	810	1902373617	811	1228354273	812	1950931833
813	636479009	814	1263787449	815	1314825769	816	2641549089
817	691199521	818	2555306121	819	839732129	820	2539277841
821	880782593	822	2059914081	823	1530012697	824	1619827449
825	1079910089	826	2651407089	827	877717273	828	2528252409
829	1123787593	830	2115598521	831	879991633	832	2757188417
833	910187017	834	2576804673	835	1162667929	836	2967468729
837	1194444817	838	1496186553	839	1092092489	840	2828807529
841	1168598161	842	1741340841	843	1217448433	844	2961741017
845	1055512121	846	2086860777	847	1176846961	848	2857144161
849	1145287009	850	3129250569	851	1359540241	852	2734116361
853	1270901369	854	2849667881	855	1316031401	856	4109398617
857	1192513873	858	2589301401	859	1747673497	860	2374378953
861	1665814057	862	2457866009	863	864297857	864	2597941641
865	1105681657	866	2126163873	867	1474091393	868	2422942097
869	813330241	870	2587316257	871	1215280433	872	2864765121
873	1055546617	874	2955604521	875	1588324849	876	2540162769
877	1166140897	878	1740197649	879	1071471217	880	3251124937
881	1121666929	882	2896790241	883	1071890161	884	2631081489
885	1370348857	886	2229607857	887	1314894769	888	2641504209
889	1094391521	890	3179094297	891	1421740409	892	2424853489
893	2003784313	894	2601422553	895	1324413313	896	3435753297
897	1457825417	898	3103542033	899	1602399217	900	4071247833
901	798447257	902	3530623273	903	1367841121	904	2494992417
905	1568185337	906	1507151409	907	1403055569	908	3180171369
909	720843289	910	3779810841	911	1697289089	912	2537937033
913	1444968089	914	3481002897	915	1492811897	916	3819843393
917	1201733153	918	2923319177	919	1520806369	920	3225574697
921	1219732793	922	3645265017	923	1480790041	924	2561150937
925	1303714201	926	2471502009	927	1672411249	928	3086409481
929	1134207353	930	3259442041	931	1342180313	932	3616278537
933	1823027209	934	3784692993	935	1818849121	936	3746518561
937	1919893561	938	4533867321	939	1944481873	940	2560471321
941	1902900761	942	2760382353	943	1998745937	944	4549819233
945	1727638009	946	2346199689	947	1791347633	948	3866224849
949	1776360337	950	2284988073	951	1859912113	952	3095542209
953	1968641401	954	4091400681	955	1242925193	956	4225706769

k	d_1	k	d_1	k	d_1	k	d_1
957	1892915281	958	4045337521	959	1343608489	960	3763601697
961	1814149921	962	4314724129	963	1530725993	964	3609639993
965	2095192361	966	2436844953	967	1351773289	968	3652908681
969	1829988737	970	3910470657	971	1267624417	972	3916603681
973	1608509857	974	3902482497	975	2267415617	976	4570911097
977	1857529337	978	4374788097	979	2397032977	980	5057369153
981	1670595169	982	4467133353	983	1846662313	984	4018295033
985	1762663633	986	3335288433	987	1284164017	988	4156050761
989	1285009081	990	4916862897	991	2476064449	992	4493035633
993	1271996969	994	4194691593	995	1940749649	996	4604658081
997	1711463473	998	4304953977	999	1545011873	1000	5834979417
1001	1847473897	1002	4000561161	1003	2587251313	1004	4444371537
1005	1970441177	1006	3405718369	1007	1382096873	1008	5178973193
1009	1676958233	1010	3411899313	1011	1764844217	1012	3933861297
1013	2665391921	1014	4085187033	1015	2155796977	1016	4976692977
1017	1727206529	1018	4202383529	1019	2238334649	1020	4709707473
1021	1906163689	1022	4994064329	1023	2099589449	1024	3672154313
1025	2398817921	1026	4792036593	1027	2531008121	1028	6935325609
1029	2054505073	1030	3141709761	1031	2776699313	1032	5715222753
1033	2134171513	1034	4244558761	1035	1927111289	1036	5127372177
1037	1838393969	1038	4227048849	1039	2140121833	1040	2998130601
1041	2288746241	1042	4985786569	1043	2278021241	1044	3622165017
1045	1723885337	1046	4033170177	1047	1948369009	1048	4866766521
1049	2181026153	1050	4926721809	1051	1899948697	1052	4513174257
1053	1780107257	1054	6008035881	1055	1981477217	1056	4505164217
1057	2128430041	1058	3772506593	1059	2822718281	1060	4274440017
1061	1704499681	1062	4783736721	1063	1541303833	1064	4850259177
1065	2098010809	1066	5239266033	1067	2637658841	1068	4392626921
1069	2149662329	1070	5038266297	1071	2158972121	1072	4546632633
1073	2086733353	1074	3829269273	1075	2998234721	1076	6655000377
1077	2648174233	1078	5672683977	1079	2252184889	1080	5032809849
1081	1771959913	1082	5685794529	1083	2053479553	1084	5816631921
1085	1511663233	1086	4288434441	1087	2787474721	1088	4952525601
1089	2244439553	1090	6708573953	1091	3044027353	1092	4837226241
1093	2746126073	1094	7094498513	1095	2877152249	1096	6873218697
1097	1772069249	1098	4869092089	1099	2817189961	1100	4743326977
1101	3022586257	1102	4954226601	1103	2498748409	1104	6842596993
1105	2487064193	1106	7970709993	1107	2353272241	1108	5216995617
1109	1926922913	1110	4832534281	1111	2792473897	1112	5454046977
1113	3116936497	1114	6769807377	1115	2517392177	1116	5703735801
1117	2671437473	1118	7489651409	1119	2477257033	1120	5843140473
1121	2207569633	1122	6967177233	1123	3071942201	1124	5210538337
1125	2860318481	1126	4725738361	1127	2158369337	1128	4380257841
1129	2955926249	1130	6026522217	1131	2854320649	1132	8687265337
1133	3132253633	1134	3941363721	1135	3007750849	1136	8538508113
1137	2660576873	1138	4614141057	1139	2714667961	1140	5295320529
1141	2520918889	1142	5026383321	1143	3100841593	1144	5145053313
1145	3892302041	1146	6648477441	1147	3307165321	1148	7713100497

k	d_1	k	d_1	k	d_1	k	d_1
1149	3459028681	1150	8423248593	1151	3855433433	1152	6303556081
1153	4081831337	1154	8181165073	1155	2370744913	1156	5505018033
1157	3265573513	1158	6115577073	1159	3420560993	1160	8620240729
1161	2074438369	1162	6685243161	1163	3091638289	1164	6461731329
1165	2557770689	1166	8386327569	1167	3909415849	1168	7717795649
1169	3346828609	1170	7157766009	1171	2546895473	1172	6856502937
1173	2793810361	1174	6047040057	1175	2157996097	1176	7838284001
1177	3455089697	1178	6647141153	1179	2842924577	1180	3590709753
1181	2476686001	1182	6121270401	1183	3243393473	1184	8151936081
1185	2750240993	1186	8253927273	1187	4210292057	1188	7049052497
1189	3727000777	1190	6917832177	1191	1880429633	1192	8525295417
1193	2945848097	1194	7813345929	1195	4223004449	1196	7625410321
1197	2935348321	1198	8133389889	1199	3453727913	1200	9033894193
1201	3900859073	1202	7929221217	1203	4421113777	1204	5713891737
1205	3245362609	1206	8728800369	1207	2985830417	1208	9863502889
1209	3775474961	1210	8046600153	1211	3447639473	1212	4400227329
1213	2407830617	1214	8811977777	1215	2882908073	1216	7791324753
1217	2283976033	1218	4647909473	1219	3339966881	1220	9564956097
1221	2923551601	1222	6811015929	1223	3830663849	1224	7028709697
1225	4063280401	1226	6696778449	1227	3844261441	1228	8576958993
1229	3535227961	1230	7620325329	1231	3092311897	1232	8908428153
1233	3776364689	1234	9188637753	1235	4696847321	1236	7724221737
1237	3856164217	1238	8480502209	1239	3861511657	1240	9144421801
1241	3395460017	1242	9294254817	1243	4180275881	1244	7201797177
1245	4078091713	1246	8112886233	1247	4499842297	1248	7881839529
1249	3775972633	1250	7019585673	1251	4105962433	1252	5848850001
1253	3991021849	1254	6065628801	1255	3564439913	1256	10740230169
1257	3463552753	1258	9561871929	1259	3185608033	1260	6169512273
1261	4293851609	1262	9078298377	1263	3156206177	1264	8582985849
1265	3897792881	1266	10075921769	1267	4139956313	1268	10380813601
1269	4570493801	1270	9356528801	1271	3890225513	1272	9442108113
1273	5438791777	1274	9367475217	1275	4870511089	1276	11355293009
1277	4570534433	1278	5457766953	1279	4582329937	1280	7966743801
1281	3303470393	1282	10603149393	1283	2858681489	1284	8329309041
1285	4838698793	1286	7925188569	1287	4214651369	1288	6365154713
1289	4383924977	1290	7700745369	1291	4783441193	1292	10899763737
1293	3183701393	1294	7947641353	1295	4216864601	1296	10375292481
1297	4884197377	1298	8881826633	1299	3896076089	1300	12504176953
1301	4842107689	1302	5756693721	1303	2723480233	1304	10029103001
1305	5075648857	1306	10231827369	1307	4881446321	1308	10362529857
1309	4663594841	1310	8630540241	1311	4361721577	1312	6785253561
1313	4473287153	1314	11339103273	1315	5213216977	1316	9030929009
1317	3765257761	1318	8397845481	1319	5921712193	1320	10518869337
1321	3553813609	1322	10411769897	1323	4880881361	1324	6586436201
1325	4041084001	1326	9637389353	1327	4263481897	1328	10794914769
1329	4020541969	1330	8861462817	1331	4655094673	1332	10320158377
1333	4225091729	1334	12687213841	1335	4849883897	1336	8635792713
1337	5056156097	1338	11247364209	1339	3533357441	1340	12338265921

k	d_1	k	d_1	k	d_1	k	d_1
1341	6188174921	1342	10179842689	1343	4931635369	1344	10717061457
1345	4992350177	1346	9914338641	1347	3640667057	1348	10149929057
1349	5178899449	1350	11908911953	1351	3790717769	1352	9688139169
1353	2687142617	1354	11319565489	1355	5446670857	1356	10184514657
1357	4413137921	1358	11811948801	1359	5491510897	1360	14981176697
1361	6845363297	1362	11520637489	1363	5726707801	1364	9844492793
1365	4074339769	1366	13189267281	1367	4590877273	1368	10263989121
1369	6090656801	1370	11577597441	1371	5360341369	1372	8969163897
1373	3729738193	1374	10030041849	1375	5790874417	1376	13672932177
1377	3354967201	1378	9721180921	1379	3593612993	1380	11572057281
1381	2975326049	1382	11990306353	1383	5409192193	1384	12071006817
1385	6181709449	1386	14069494849	1387	6109066921	1388	11024766993
1389	5274319777	1390	9642295441	1391	6249679649	1392	11173712089
1393	3920782057	1394	14970384881	1395	4291103633	1396	14873830449
1397	4970225489	1398	14282140161	1399	3113018473	1400	9907899873
1401	5602055801	1402	13224194153	1403	4388529953	1404	9080218569
1405	5874339553	1406	14392574233	1407	5313687361	1408	8070174249
1409	5340878497	1410	8445957249	1411	5799147161	1412	11307387921
1413	6256545641	1414	14603912313	1415	5182399633	1416	11291079633
1417	5761864793	1418	8611063521	1419	4283780297	1420	11965703817
1421	4919896369	1422	13148498001	1423	5418737449	1424	11819709609
1425	6452481929	1426	8849047089	1427	6464797601	1428	10724557281
1429	5850358097	1430	11668121193	1431	4450743881	1432	14117182377
1433	7530295477	1434	11005822041	1435	4781099249	1436	16464319257
1437	5625246553	1438	15200800833	1439	6232068121	1440	11543895273
1441	5582017241	1442	11467212681	1443	6162816449	1444	12587849569
1445	4505553209	1446	9179582801	1447	4759466801	1448	12803192537
1449	5190934153	1450	13376175873	1451	5026362641	1452	17479263937
1453	6196053953	1454	12803941329	1455	6038713553	1456	14943995697
1457	5826184769	1458	12952966161	1459	8336832073	1460	12575735841
1461	6697005209	1462	12862360281	1463	6953992769	1464	14337419209
1465	5551779049	1466	13519359177	1467	5740147537	1468	11113911169
1469	6633294233	1470	10581863009	1471	6089924993	1472	14453700969
1473	6641047097	1474	13050281721	1475	6676376873	1476	13635058089
1477	4879320689	1478	11540902641	1479	5218748209	1480	15361738681
1481	5142288889	1482	8955885801	1483	8016755057	1484	12068317977
1485	5782637041	1486	14976660297	1487	7880353297	1488	16452839881
1489	6484799473	1490	9905393521	1491	5707371641	1492	17674712841
1493	6689379689	1494	12127943433	1495	6879464849	1496	16182669297
1497	4306434697	1498	14691262881	1499	8152320377	1500	15122451873
1501	6323595697	1502	14457595441	1503	6009027289	1504	13185180849
1505	5685159793	1506	14022943169	1507	8097404209	1508	16948680681
1509	6547557281	1510	14908391969	1511	7678220273	1512	19449660561
1513	6075080353	1514	17372940729	1515	6823802473	1516	10667702049
1517	7286535961	1518	16155325929	1519	6453592841	1520	14756469729
1521	7044532081	1522	11995984617	1523	6981504113	1524	17180091273
1525	7773449641	1526	19556487329	1527	6239985569	1528	18678344721
1529	7319054153	1530	9633634761	1531	8359447409	1532	19243693401

k	d_1	k	d_1	k	d_1	k	d_1
1533	5383103641	1534	15336632257	1535	5435424889	1536	16518110361
1537	6493576057	1538	11799669513	1539	6261258617	1540	11453876897
1541	7007402449	1542	16811621457	1543	7119113321	1544	15084189201
1545	5032367849	1546	19853096433	1547	6206696401	1548	18095827017
1549	5585570233	1550	12228799353	1551	8651999977	1552	13545865481
1553	7061734417	1554	15351766761	1555	7881434881	1556	19088671361
1557	6358784129	1558	17521254633	1559	8866491401	1560	14777854617
1561	5406250657	1562	16055379201	1563	8111629553	1564	18455413729
1565	7557590969	1566	19354854017	1567	6376897561	1568	11162316201
1569	10325243801	1570	14174249937	1571	8346350809	1572	14805992793
1573	8639085097	1574	19876647729	1575	7310427353	1576	13383109041
1577	7692909481	1578	17923856217	1579	7980327521	1580	13791208929
1581	4445597953	1582	14074036089	1583	5656117361	1584	17916753801
1585	9527118401	1586	14217349353	1587	7082431481	1588	16185164297
1589	6205767769	1590	16566006801	1591	9414671033	1592	19394644593
1593	9484650257	1594	10596671121	1595	8488474417	1596	13021079817
1597	7823658209	1598	24921801969	1599	9490053809	1600	19530979913
1601	9941854033	1602	11917095513	1603	8980051961	1604	17863582801
1605	8408412017	1606	19317873369	1607	9030034649	1608	21150606513
1609	10040135537	1610	22641182337	1611	7404361369	1612	18619595009
1613	11724971569	1614	14844290169	1615	8054991409	1616	18427052337
1617	7654546177	1618	20158634841	1619	8074096649	1620	10101987681
1621	6833900393	1622	21247408473	1623	9205065929	1624	19442890041
1625	9320044529	1626	20777329497	1627	7474670633	1628	16407640929
1629	8062549801	1630	19745737113	1631	5708233177	1632	19947287721
1633	7857319273	1634	24913676649	1635	10897495537	1636	18635575281
1637	9765563569	1638	22715248593	1639	6518086921	1640	11710935681
1641	10877325889	1642	24782078073	1643	7997553217	1644	17307309497
1645	9688815713	1646	20375827873	1647	8266590337	1648	21286015017
1649	6771151313	1650	21418425473	1651	7759261313	1652	17058280017
1653	7055863681	1654	23824112489	1655	7797935281	1656	22941611697
1657	6898160657	1658	19080932649	1659	8573955281	1660	21767826777
1661	11803489417	1662	20306822649	1663	10557367441	1664	19351782129
1665	9643518041	1666	23188938609	1667	8053982201	1668	17951834217
1669	9542343233	1670	26324496353	1671	8320139033	1672	14965707817
1673	9223372409	1674	16030498793	1675	12397060721	1676	21302932753
1677	7388864993	1678	20675197713	1679	8424181121	1680	15850391313
1681	9979688393	1682	23169114809	1683	9821011049	1684	21532687521
1685	9032961017	1686	17768279433	1687	8223478961	1688	19766229081
1689	12157301161	1690	18187374489	1691	10753068737	1692	18319821657
1693	9909960329	1694	26639467017	1695	10258117313	1696	21471116241
1697	13692281329	1698	15424707057	1699	6648674609	1700	22105438041
1701	10328316337	1702	23340311481	1703	7672509353	1704	19865714313
1705	7345957961	1706	20083328049	1707	12110334193	1708	12603492609
1709	14665471217	1710	21153612153	1711	12214804297	1712	16901015217
1713	7883012129	1714	20265748617	1715	11526296753	1716	17900661153
1717	9022814057	1718	23173716129	1719	8558039537	1720	18317261649
1721	9755539897	1722	26642566113	1723	11295454553	1724	21852465793

k	d_1	k	d_1	k	d_1	k	d_1
1725	9820146593	1726	25820071817	1727	13386871657	1728	19234668121
1729	10200354113	1730	22356100321	1731	10496099857	1732	15563628777
1733	10280518337	1734	19458436089	1735	10344401849	1736	15334875609
1737	12478800689	1738	22729799577	1739	13185137881	1740	31556733609
1741	11977301593	1742	13991205561	1743	9726851417	1744	26735056217
1745	7959615929	1746	20732047473	1747	13544803529	1748	23934761337
1749	10109738497	1750	15138620841	1751	10531936889	1752	26307135033
1753	11742293489	1754	15293415993	1755	11740768681	1756	21444330777
1757	11408438473	1758	26855399481	1759	9561059017	1760	19279833721
1761	9953890057	1762	25903870689	1763	13747239089	1764	23244127729
1765	10512541913	1766	29607980417	1767	14566400561	1768	31222130153
1769	13852180417	1770	29675608953	1771	9052924193	1772	21842571921
1773	12653890049	1774	21060861537	1775	10345905937	1776	22913920929
1777	11299193849	1778	23773736001	1779	12959578369	1780	27545395593
1781	11502866521	1782	27768982497	1783	12903652081	1784	28861680633
1785	12271976993	1786	24459731697	1787	10862200793	1788	15961497577
1789	10612904713	1790	24324206537	1791	10638082801	1792	31848889713
1793	10500588257						

k	d_1	k	d_1	k	d_1	k	d_1
1795	14302640609	1796	27358359081	1797	11675814881	1798	27273950169
1799	12288253057	1800	29409340097	1801	10747223609	1802	17116275873
1803	10761662033	1805	11795108209	1806	24360921033	1807	14006501017
1809	11315214497	1810	30452439257	1811	10525057129	1812	28348764753
1813	8922565193	1814	31795855249	1815	10367096249	1816	28742795673
1817	8638659049	1818	24484322193	1819	13819473449	1820	15894922737
1821	11682471617	1822	28719137049	1823	10534815857	1824	17666542041
1825	10081327513	1827	12077934553	1828	22366484121	1829	14616252913
1830	20798720121	1831	11063929649	1832	28289473809	1833	8002632881
1834	25611350433	1835	9212193857	1836	27665418993	1837	16898429161
1838	26484256857	1839	9016016953	1840	24896737497	1841	15427422041
1842	26136546297	1843	10786881689	1844	28223186289	1845	16636782473
1847	12676735273	1849	7747069097	1850	28252003081	1851	10879109849
1852	26361529041	1853	13897278121	1854	21989559201	1855	10695553697
1856	27284626329	1857	14003640817	1858	27657269313	1859	12177360529
1860	21356618529	1861	11888816113	1862	30328356601	1863	9906836593
1864	21976435497	1865	14472946801	1866	20748624513	1867	11552645537
1868	27402172977	1869	11303255617	1870	24913897737	1871	8443601753
1872	31756304409	1873	16026449393	1874	29842152657	1875	12307826081
1876	27713889569	1877	9654472721	1879	15210808849	1880	28870006929
1881	15669411673	1882	28934387553	1883	16555237537	1885	11660985689
1886	27292507377	1887	14953069561	1889	10128518657	1891	17239580153
1892	30427087497	1893	11212691801	1894	30501258393	1895	15299539457
1896	27590751057	1897	13077468809	1898	31827861641	1899	11483109641
1901	12606717017	1903	15584020681	1905	11813267161	1906	16367835009
1907	18412667969	1909	11836986241	1910	21635667673	1911	13272833377
1912	30472364553	1913	12531171217	1914	23353861521	1915	8202143393

k	d_1	k	d_1	k	d_1	k	d_1
1917	14755875137	1918	25708307601	1919	15066358481	1920	22043395081
1921	9023034793	1923	11690362561	1925	14786098913	1927	13989520417
1929	10757106673	1930	26996413497	1931	11823613913	1933	16867973321
1935	15576283369	1937	16795975817	1939	17334712729	1941	15871885753
1942	31072472337	1943	12023446697	1945	10890996689	1946	22747002657
1947	16774867913	1948	27535083801	1949	14984974273	1950	23125134057
1951	15365245153	1952	28463088977	1953	14557374529	1954	29968558449
1955	16991135209	1956	29772869289	1957	18647799593	1958	30016450209
1959	17536828609	1961	16387251809	1963	16939173761	1964	24017100321
1965	17519090473	1966	25475162953	1967	18063913321	1969	19189797049
1971	16094600017	1973	19185850561	1975	22692362201	1977	13792212017
1979	16926239609	1981	18654590113	1982	31863559753	1983	11258242889
1984	26625714369	1985	17771644441	1987	19770734129	1989	16941181249
1991	16964769281	1993	18166946593	1995	19893527657	1997	10175019889
1999	18426841921	2001	16120899073	2003	10683271289	2005	19279072217
2007	20008876177	2009	15135767321	2011	12772102033	2013	15387373321
2015	17258610449	2016	26923237881	2017	15488878849	2019	20066054209
2021	14898157433	2022	25753502769	2023	19001943929	2025	17314824481
2027	11829608209	2029	16940925433	2031	19334270129	2033	23685116761
2035	18311081017	2037	14836645081	2039	16373229137	2041	17298071521
2043	14748148673	2045	10982533169	2047	21082442393	2049	23866962377
2051	18769206193	2053	19125553481	2054	29534671497	2055	14107004761
2057	21482977393	2059	14756819873	2061	12912128689	2063	10837182401
2065	18331106561	2067	10481469337	2069	17415046001	2071	20282274193
2073	18819832793	2075	17545036193	2077	15578650289	2078	27729367257
2079	24057117089	2081	16148176609	2082	25882365777	2083	12490953649
2085	10832312753	2087	19455168137	2089	18270610081	2091	21856626049
2093	23238238169	2095	24068136049	2097	16522756057	2099	15962858257
2101	25476168961	2103	24847607321	2105	18015016057	2106	31923588929
2107	20803781729	2109	19673142289	2111	18392434289	2113	23220719273
2115	12636915289	2117	19315124009	2119	28994112073	2121	20050974761
2123	29088481481	2125	18655061473	2127	19574918641	2129	18895809337
2131	23112080113	2133	14523603209	2135	22249389529	2137	21773555593
2139	18025003393	2141	22630967561	2142	29046167529	2143	18693988129
2145	18877428953	2147	18136409801	2149	19695859753	2151	19861422377
2153	24196460777	2155	23555454961	2156	29713956801	2157	20859254233
2159	20632640297	2161	27332078161	2162	28975285761	2163	18060492217
2165	21744616193	2167	20857806913	2169	14622455329	2171	20886550177
2173	14434855249	2175	16846410769	2177	15977018641	2179	23282095961
2181	25607323801	2183	26851308977	2185	19995487417	2187	26886280081
2189	27014177657	2191	14464864153	2193	20535271249	2195	26841314281
2197	22297034849	2199	23926888417	2201	23201739361	2203	16987173097
2205	18142454081	2207	16398494993	2209	27148857433	2211	22374953593
2213	23897368169	2215	20780753537	2217	24137876041	2219	21973542001
2221	21305351633	2222	31449231969	2223	25203670673	2224	31563684681
2225	19548976033	2227	19151122577	2229	17388825793	2231	20015710609
2233	23774802841	2235	17229928721	2237	25054037273	2239	28215927697

k	d_1	k	d_1	k	d_1	k	d_1
2241	25364314097	2243	24693406769	2245	19663680161	2247	23166334049
2249	22198449089	2251	21782856049	2253	23058697577	2255	23490722401
2257	23182069777	2259	25375332953	2261	23321397889	2263	23134466353
2265	21099281657	2267	20500043689	2269	24136990729	2271	26367890993
2273	23874674297	2275	18425670833	2277	26640229873	2279	23580652217
2281	18777503057	2282	28804400601	2283	25006505753	2285	28135395473
2287	25885481009	2289	24003072217	2291	25217386561	2293	22590552449
2295	24478979353	2297	29279333513	2299	21622666817	2301	28206525689
2303	29479981529	2305	30915727681	2307	23347446401	2309	30173417497
2311	17259667001	2313	20830862281	2315	28875641633	2317	22175499113
2319	20768662297	2321	22111529257	2323	22985300057	2325	21970530497
2327	21542492929	2329	25253712697	2331	24055982809	2333	27344154281
2335	31659517921	2337	26380992137	2339	24173328793	2341	23702333329
2343	22874829473	2345	27503106937	2347	23997306689	2349	25080505033
2351	27005350529	2353	19594199089	2355	26290364593	2357	31187585617
2361	21535123417	2363	30890690297	2365	17997494969	2367	25941664313
2368	30521729001	2369	21152347649	2373	28640771921	2375	25445748593
2377	24642382537	2379	27949170833	2383	29346656233	2385	27353459369
2387	22092862993	2389	26541069889	2391	27922751849	2395	25679645297
2397	25567788161	2401	26806417097	2403	25471907233	2411	23201469721
2413	27471244057	2415	28729140457	2417	31918397593	2423	19416040537
2427	27250644433	2429	28702862873	2431	18452796697	2433	31076018153
2437	19044233393	2443	28724687897	2447	24087157561	2449	24340659377
2451	24310203641	2453	29613412849	2455	25877124769	2457	20105114921
2459	31572674153	2465	31474193953	2469	31751925329	2473	25877913169
2479	25781498417	2483	30687948241	2485	29489657473	2487	23214266969
2489	24403608241	2493	28496723993	2495	26242884937	2499	29141913769
2511	27983986649	2523	31630888169	2531	30568914073	2533	29836994353
2545	30815861849	2551	29163166121	2553	18839920273	2555	25973619241
2561	27573697457	2565	20167085041	2567	30338840489	2575	24854975473
2579	29969335969	2623	29674805977	2627	26057264561	2645	22700098081
2667	26463497129	2683	28569879721	2705	30513902753	2713	29668713889
2735	28004847841	2757	20013907409	2783	31014739937	2801	31532536313
2851	25306669001	2869	30730146737	2941	28715939033		

References

1. Coates, J.: Lectures on the Birch-Swinnerton-Dyer Conjecture. Not. ICCM **1**, 29–46 (2013)
2. Coates, J., Li, Y., Tian, Y., Zhai, S.: Quadratic twists of elliptic curves. Proc. Lond. Math. Soc. **110**, 357–394 (2015)
3. Coates, J., Wiles, A.: On the conjecture of Birch and Swinnerton-Dyer. Invent. Math. **39**, 223–251 (1977)
4. Conrey, J.B., Farmer, D.W., Keating, J.P., Rubinstein, M.O., Snaith, N.C.: Integral moments of L-functions. Proc. Lond. Math. Soc. **91**, 33–104 (2005)
5. Conrey, J.B., Keating, J.P., Rubinstein, M.O., Snaith, N.C.: Random matrix theory and the Fourier coefficients of half-integral weight forms. Exp. Math. **15**, 67–82 (2006)
6. Dąbrowski, A., Szymaszkiewicz, L.: Behaviour of the order of Tate–Shafarevich groups for the quadratic twists of elliptic curves (in preparation)
7. Dąbrowski, A., Wodzicki, M.: Elliptic curves with large analytic order of Ш, In: Algebra, Arithmetic and Geometry (in honour of Yu.I. Manin, vol. I). Progress Math. **269**, 407–421 (2009)
8. Delaunay, C.: Moments of the orders of Tate-Shafarevich groups. Int. J. Number Theory **1**, 243–264 (2005)
9. Delaunay, C., Heuristics on class groups and on Tate–Shafarevich groups: the magic of the Cohen–Lenstra heuristics. In: Ranks of elliptic curves and random matrix theory. Lond. Math. Soc. Lect. Ser. **341**, 323–340 (2007)
10. Delaunay, C., Watkins, M.: The powers of logarithm for quadratic twists. In: Ranks of elliptic curves and random matrix theory. Lond. Math. Soc. Lect. Ser. **341**, 189–193 (2007)
11. Gonzalez-Avilés, C.D.: On the conjecture of Birch and Swinnerton-Dyer. Trans. Am. Math. Soc. **349**, 4181–4200 (1997)
12. Gross, B., Zagier, D.: Heegner points and derivatives of L-series. Invent. Math. **84**, 225–320 (1986)
13. Heath-Brown, R.: Letter to John Coates (2015)
14. Keating, J.P., Snaith, N.C.: Random matrix theory and $\zeta(1/2+it)$. Commun. Math. Phys. **214**(1), 57–89 (2000)
15. Kolyvagin, V.: Finiteness of $E(\mathbb{Q})$ and Ш for a class of Weil curves. Math. USSR Izv. **32**, 523–541 (1989)
16. Le Boudec, P.: Height of rational points on quadratic twists of a given elliptic curve. arXiv:1404.7738v1 [math.NT] 30 April 2014
17. Lehman, J.L.: Rational points on elliptic curves with complex multiplication by the ring of integers in $\mathbb{Q}(\sqrt{-7})$. J. Number Theory **27**, 253–272 (1987)
18. The PARI Group, PARI/GP version 2.7.2, Bordeaux (2014). http://pari.math.u-bordeaux.fr/
19. Poonen, B.: Letter to A. Dąbrowski (2015)
20. Quattrini, P.: On the distribution of analytic $\sqrt{Ш}$ values on quadratic twists of elliptic curves. Exp. Math. **15**, 355–365 (2006)
21. Quattrini, P.: The effect of torsion on the distribution of Ш among quadratic twists of an elliptic curve. J. Number Theory **131**, 195–211 (2011)
22. Radziwiłł, M., Soundararajan, K.: Moments and distribution of central L-values of quadratic twists of elliptic curves. Invent. Math. **202**, 1029–1068 (2015)
23. Rubin, K.: Tate-Shafarevich groups and L-functions of elliptic curves with complex multiplication. Invent. Math. **89**, 527–560 (1987)
24. Skinner, Ch., Urban, E.: The Iwasawa main conjectures for GL_2. Invent. Math. **195**, 1–277 (2014)
25. Watkins, M.: Some heuristics about elliptic curves. Exp. Math. **17**, 105–125 (2008)

Compactifications of S-arithmetic Quotients for the Projective General Linear Group

Takako Fukaya, Kazuya Kato and Romyar Sharifi

Abstract Let F be a global field, let S be a nonempty finite set of places of F which contains the archimedean places of F, let $d \geqslant 1$, and let $X = \prod_{v \in S} X_v$ where X_v is the symmetric space (resp., Bruhat-Tits building) associated to $\mathrm{PGL}_d(F_v)$ if v is archimedean (resp., non-archimedean). In this paper, we construct compactifications $\Gamma\backslash\bar{X}$ of the quotient spaces $\Gamma\backslash X$ for S-arithmetic subgroups Γ of $\mathrm{PGL}_d(F)$. The constructions make delicate use of the maximal Satake compactification of X_v (resp., the polyhedral compactification of X_v of Gérardin and Landvogt) for v archimedean (resp., non-archimedean). We also consider a variant of $\bar{X}$ in which we use the standard Satake compactification of X_v (resp., the compactification of X_v due to Werner).

MSCs Primary 14M25 · Secondary 14F20

1 Introduction

1.1 Let $d \geqslant 1$, and let $X = \mathrm{PGL}_d(\mathbb{R})/\mathrm{PO}_d(\mathbb{R}) \cong \mathrm{SL}_d(\mathbb{R})/\mathrm{SO}_d(\mathbb{R})$. The Borel–Serre space (resp., reductive Borel–Serre space) $\bar{X}$ contains X as a dense open subspace [3] (resp., [26]). If Γ is a subgroup of $\mathrm{PGL}_d(\mathbb{Z})$ of finite index, this gives rise to a compactification $\Gamma\backslash\bar{X}$ of $\Gamma\backslash X$.

Dedicated to Professor John Coates on the occasion of his 70th birthday.

T. Fukaya · K. Kato
Department of Mathematics, University of Chicago, 5734 S. University Ave., Chicago, IL 60637, USA
e-mail: takako@math.uchicago.edu

K. Kato
e-mail: kkato@math.uchicago.edu

R. Sharifi (✉)
Department of Mathematics, UCLA, Los Angeles, CA 90095-1555, USA
e-mail: sharifi@math.ucla.edu

D. Loeffler and S.L. Zerbes (eds.), *Elliptic Curves, Modular Forms and Iwasawa Theory*, Springer Proceedings in Mathematics & Statistics 188, DOI 10.1007/978-3-319-45032-2_5

1.2 Let F be a global field, which is to say either a number field or a function field in one variable over a finite field. For a place v of F, let F_v be the local field of F at v. Fix $d \geqslant 1$.

In this paper, we will consider the space X_v of all homothety classes of norms on F_v^d and a certain space $\bar{X}_{F,v}$ which contains X_v as a dense open subset. For $F = \mathbb{Q}$ and v the real place, X_v is identified with $\mathrm{PGL}_d(\mathbb{R})/\mathrm{PO}_d(\mathbb{R})$, and $\bar{X}_{F,v}$ is identified with the reductive Borel–Serre space associated to $\mathrm{PGL}_d(F_v)$. We have the following analogue of 1.1.

Theorem 1.3 *Let F be a function field in one variable over a finite field, let v be a place of F, and let O be the subring of F consisting of all elements which are integral outside v. Then for any subgroup Γ of $\mathrm{PGL}_d(O)$ of finite index, the quotient $\Gamma\backslash\bar{X}_{F,v}$ is a compact Hausdorff space which contains $\Gamma\backslash X_v$ as a dense open subset.*

1.4 Our space $\bar{X}_{F,v}$ is not a very new object. In the case that v is non-archimedean, X_v is identified as a topological space with the Bruhat-Tits building of $\mathrm{PGL}_d(F_v)$. In this case, $\bar{X}_{F,v}$ is similar to the polyhedral compactification of X_v of Gérardin [7] and Landvogt [19], which we denote by $\bar{X}_v$. To each element of $\bar{X}_v$ is associated a parabolic subgroup of PGL_{d,F_v}. We define $\bar{X}_{F,v}$ as the subset of $\bar{X}_v$ consisting of all elements for which the associated parabolic subgroup is F-rational. We endow $\bar{X}_{F,v}$ with a topology which is different from its topology as a subspace of $\bar{X}_v$.

In the case $d = 2$, the boundary $\bar{X}_v \setminus X_v$ of $\bar{X}_v$ is $\mathbb{P}^1(F_v)$, whereas the boundary $\bar{X}_{F,v} \setminus X_v$ of $\bar{X}_{F,v}$ is $\mathbb{P}^1(F)$. Unlike $\bar{X}_v$, the space $\bar{X}_{F,v}$ is not compact, but the arithmetic quotient as in 1.1 and 1.3 is compact (see 1.6).

1.5 In §4, we derive the following generalization of 1.1 and 1.3.

Let F be a global field. For a nonempty finite set S of places of F, let $\bar{X}_{F,S}$ be the subspace of $\prod_{v\in S} \bar{X}_{F,v}$ consisting of all elements $(x_v)_{v\in S}$ such that the F-parabolic subgroup associated to x_v is independent of v. Let X_S denote the subspace $\prod_{v\in S} X_v$ of $\bar{X}_{F,S}$.

Let S_1 be a nonempty finite set of places of F containing all archimedean places of F, let S_2 be a finite set of places of F which is disjoint from S_1, and let $S = S_1 \cup S_2$. Let O_S be the subring of F consisting of all elements which are integral outside S.

Our main result is the following theorem (see Theorem 4.1.4).

Theorem 1.6 *Let Γ be a subgroup of $\mathrm{PGL}_d(O_S)$ of finite index. Then the quotient $\Gamma\backslash(\bar{X}_{F,S_1} \times X_{S_2})$ is a compact Hausdorff space which contains $\Gamma\backslash X_S$ as a dense open subset.*

1.7 If F is a number field and S_1 coincides with the set of archimedean places of F, then the space $\bar{X}_{F,S_1}$ is the maximal Satake space of the Weil restriction of $\mathrm{PGL}_{d,F}$ from F to $\mathbb{Q}$. In this case, the theorem is known for $S = S_1$ through the work of Satake [23] and in general through the work of Ji et al. [14, 4.4].

1.8 We also consider a variant $\bar{X}^\flat_{F,v}$ of $\bar{X}_{F,v}$ and a variant $\bar{X}^\flat_{F,S}$ of $\bar{X}_{F,S}$ with continuous surjections

$$\bar{X}_{F,v} \to \bar{X}^\flat_{F,v}, \quad \bar{X}_{F,S} \to \bar{X}^\flat_{F,S}.$$

In the case v is non-archimedean (resp., archimedean), $\bar{X}^{\flat}_{F,v}$ is the part with "F-rational boundary" in Werner's compactification (resp., the standard Satake compactification) $\bar{X}^{\flat}_{v}$ of X_v [24, 25] (resp., [22]), endowed with a new topology. We will obtain an analogue of 1.6 for this variant.

To grasp the relationship with the Borel–Serre compactification [3], we also consider a variant $\bar{X}^{\sharp}_{F,v}$ of $\bar{X}_{F,v}$ which has a continuous surjection $\bar{X}^{\sharp}_{F,v} \to \bar{X}_{F,v}$, and we show that in the case that $F = \mathbb{Q}$ and v is the real place, $\bar{X}^{\sharp}_{\mathbb{Q},v}$ coincides with the Borel–Serre space associated to $\mathrm{PGL}_{d,\mathbb{Q}}$ (3.7.4). If v is non-archimedean, the space $\bar{X}^{\sharp}_{F,v}$ is not Hausdorff (3.7.6) and does not seem useful.

1.9 What we do in this paper is closely related to what Satake did in [22, 23]. In [22], he defined a compactification of a symmetric Riemannian space. In [23], he took the part of this compactification with "rational boundary" and endowed it with the Satake topology. Then he showed that the quotient of this part by an arithmetic group is compact. We take the part $\bar{X}_{F,v}$ of $\bar{X}_v$ with "F-rational boundary" to have a compact quotient by an arithmetic group. So, the main results and their proofs in this paper might be evident to the experts in the theory of Bruhat-Tits buildings, but we have not found them in the literature.

1.10 We intend to apply the compactification 1.3 to the construction of toroidal compactifications of the moduli space of Drinfeld modules of rank d in a forthcoming paper. In Sect. 4.7, we give a short explanation of this plan, along with two other potential applications, to asymptotic behavior of heights of motives and to modular symbols over function fields.

1.11 We plan to generalize the results of this paper from PGL_d to general reductive groups in another forthcoming paper. The reason why we separate the PGL_d-case from the general case is as follows. For PGL_d, we can describe the space $\bar{X}_{F,v}$ via norms on finite-dimensional vector spaces over F_v (this method is not used for general reductive groups), and these norms play an important role in the analytic theory of toroidal compactifications.

1.12 In §2, we review the compactifications of Bruhat-Tits buildings in the non-archimedean setting and symmetric spaces in the archimedean setting. In §3 and §4, we discuss our compactifications.

1.13 We plan to apply the results of this paper to the study of Iwasawa theory over a function field F. We dedicate this paper to John Coates, who has played a leading role in the development of Iwasawa theory.

2 Spaces Associated to Local Fields

In this section, we briefly review the compactification of the symmetric space (resp., of the Bruhat-Tits building) associated to PGL_d of an archimedean (resp., non-archimedean) local field. See the papers of Satake [22] and Borel–Serre [3] (resp., Gérardin [7], Landvogt [19], and Werner [24, 25]) for details.

Let E be a local field. This means that E is a locally compact topological field with a non-discrete topology. That is, E is isomorphic to $\mathbb{R}$, $\mathbb{C}$, a finite extension of $\mathbb{Q}_p$ for some prime number p, or $\mathbb{F}_q((T))$ for a finite field $\mathbb{F}_q$.

Let $|\ |\colon E \to \mathbb{R}_{\geqslant 0}$ be the normalized absolute value. If $E \cong \mathbb{R}$, this is the usual absolute value. If $E \cong \mathbb{C}$, this is the square of the usual absolute value. If E is non-archimedean, this is the unique multiplicative map $E \to \mathbb{R}_{\geqslant 0}$ such that $|a| = \sharp(O_E/aO_E)^{-1}$ if a is a nonzero element of the valuation ring O_E of E.

Fix a positive integer d and a d-dimensional E-vector space V.

2.1 Norms

2.1.1 We recall the definitions of norms and semi-norms on V.

A norm (resp., semi-norm) on V is a map $\mu\colon V \to \mathbb{R}_{\geqslant 0}$ for which there exist an E-basis $(e_i)_{1\leqslant i\leqslant d}$ of V and an element $(r_i)_{1\leqslant i\leqslant d}$ of $\mathbb{R}^d_{>0}$ (resp., $\mathbb{R}^d_{\geqslant 0}$) such that

$$\mu(a_1e_1 + \cdots + a_de_d) = \begin{cases} (r_1^2|a_1|^2 \cdots + r_d^2|a_d|^2)^{1/2} & \text{if } E \cong \mathbb{R}, \\ r_1|a_1| + \cdots + r_d|a_d| & \text{if } E \cong \mathbb{C}, \\ \max(r_1|a_1|, \ldots, r_d|a_d|) & \text{otherwise.} \end{cases}$$

for all $a_1, \ldots, a_d \in E$.

2.1.2 We will call the norm (resp., semi-norm) μ in the above, the norm (resp., semi-norm) given by the basis $(e_i)_i$ and by $(r_i)_i$.

2.1.3 We have the following characterizations of norms and semi-norms.

(1) If $E \cong \mathbb{R}$ (resp., $E \cong \mathbb{C}$), then there is a one-to-one correspondence between semi-norms on V and symmetric bilinear (resp., Hermitian) forms $(\ ,\ )$ on V such that $(x, x) \geqslant 0$ for all $x \in V$. The semi-norm μ corresponding to $(\ ,\ )$ is given by $\mu(x) = (x, x)^{1/2}$ (resp., $\mu(x) = (x, x)$). This restricts to a correspondence between norms and forms that are positive definite.

(2) If E is non-archimedean, then (as in [9]) a map $\mu\colon V \to \mathbb{R}_{\geqslant 0}$ is a norm (resp., semi-norm) if and only if μ satisfies the following (i)–(iii) (resp., (i) and (ii)):

   (i) $\mu(ax) = |a|\mu(x)$ for all $a \in E$ and $x \in V$,
   (ii) $\mu(x + y) \leqslant \max(\mu(x), \mu(y))$ for all $x, y \in V$, and
   (iii) $\mu(x) > 0$ if $x \in V \setminus \{0\}$.

These well-known facts imply that if μ is a norm (resp., semi-norm) on V and V' is an E-subspace of V, then the restriction of μ to V' is a norm (resp., semi-norm) on V'.

2.1.4 We say that two norms (resp., semi-norms) μ and μ' on V are equivalent if $\mu' = c\mu$ for some $c \in \mathbb{R}_{>0}$.

2.1.5 The group $\mathrm{GL}_V(E)$ acts on the set of all norms (resp., semi-norms) on V: for $g \in \mathrm{GL}_V(E)$ and a norm (resp., semi-norm) μ on V, $g\mu$ is defined as $\mu \circ g^{-1}$. This action preserves the equivalence in 2.1.4.

2.1.6 Let V^* be the dual space of V. Then there is a bijection between the set of norms on V and the set of norms on V^*. That is, for a norm μ on V, the corresponding norm μ^* on V^* is given by

$$\mu^*(\varphi) = \sup\left(\frac{|\varphi(x)|}{\mu(x)} \mid x \in V \setminus \{0\}\right) \quad \text{for } \varphi \in V^*.$$

For a norm μ on V associated to a basis $(e_i)_i$ of V and $(r_i)_i \in \mathbb{R}^d_{>0}$, the norm μ^* on V^* is associated to the dual basis $(e_i^*)_i$ of V^* and $(r_i^{-1})_i$. This proves the bijectivity.

2.1.7 For a norm μ on V and for $g \in \mathrm{GL}_V(E)$, we have

$$(\mu \circ g)^* = \mu^* \circ (g^*)^{-1},$$

which is to say $(g\mu)^* = (g^*)^{-1}\mu^*$, where $g^* \in \mathrm{GL}_{V^*}(E)$ is the transpose of g.

2.2 Definitions of the Spaces

2.2.1 Let X_V denote the set of all equivalence classes of norms on V (as in 2.1.4). We endow X_V with the quotient topology of the subspace topology on the set of all norms on V inside $\mathbb{R}^V$.

2.2.2 In the case that E is archimedean, we have

$$X_V \cong \begin{cases} \mathrm{PGL}_d(\mathbb{R})/\,\mathrm{PO}_d(\mathbb{R}) \cong \mathrm{SL}_d(\mathbb{R})/\,\mathrm{SO}_d(\mathbb{R}) & \text{if } E \cong \mathbb{R} \\ \mathrm{PGL}_d(\mathbb{C})/\,\mathrm{PU}(d) \cong \mathrm{SL}_d(\mathbb{C})/\,\mathrm{SU}(d) & \text{if } E \cong \mathbb{C}. \end{cases}$$

In the case E is non-archimedean, X_V is identified with (a geometric realization of) the Bruhat-Tits building associated to PGL_V [4] (see also [5, Sect. 2]).

2.2.3 Recall that for a finite-dimensional vector space $H \neq 0$ over a field I, the following four objects are in one-to-one correspondence:

(i) a parabolic subgroup of the algebraic group GL_H over I,
(ii) a parabolic subgroup of the algebraic group PGL_H over I,
(iii) a parabolic subgroup of the algebraic group SL_H over I, and
(iv) a flag of I-subspaces of H (i.e., a set of subspaces containing $\{0\}$ and H and totally ordered under inclusion).

The bijections (ii) $\mapsto$ (i) and (i) $\mapsto$ (iii) are the taking of inverse images. The bijection (i) $\mapsto$ (iv) sends a parabolic subgroup P to the set of all P-stable I-subspaces of H, and the converse map takes a flag to its isotropy subgroup in GL_H.

2.2.4 Let $\bar{X}_V$ be the set of all pairs (P, μ), where P is a parabolic subgroup of the algebraic group PGL_V over E and, if

$$0 = V_{-1} \subsetneq V_0 \subsetneq \cdots \subsetneq V_m = V$$

denotes the flag corresponding to P (2.2.3), then μ is a family $(\mu_i)_{0 \leqslant i \leqslant m}$, where μ_i is an equivalence class of norms on V_i / V_{i-1}.

We have an embedding $X_V \hookrightarrow \bar{X}_V$ which sends μ to (PGL_V, μ).

2.2.5 Let $\bar{X}_V^\flat$ be the set of all equivalence classes of nonzero semi-norms on the dual space V^* of V (2.1.4). We have an embedding $X_V \hookrightarrow \bar{X}_V^\flat$ which sends μ to μ^* (2.1.6).

This set $\bar{X}_V^\flat$ is also identified with the set of pairs (W, μ) with W a nonzero E-subspace of V and μ an equivalence class of a norm on W. In fact, μ corresponds to an equivalence class μ^* of a norm on the dual space W^* of W (2.1.6), and μ^* is identified via the projection $V^* \to W^*$ with an equivalence class of semi-norms on V^*.

We call the understanding of $\bar{X}_V^\flat$ as the set of such pairs (W, μ) the definition of $\bar{X}_V^\flat$ in the second style. In this interpretation of $\bar{X}_V^\flat$, the above embedding $X_V \to \bar{X}_V^\flat$ is written as $\mu \mapsto (V, \mu)$.

2.2.6 In the case that E is non-archimedean, $\bar{X}_V$ is the polyhedral compactification of the Bruhat-Tits building X_V by Gérardin [7] and Landvogt [19] (see also [11, Proposition 19]), and $\bar{X}_V^\flat$ is the compactification of X_V by Werner [24, 25]. In the case that E is archimedean, $\bar{X}_V$ is the maximal Satake compactification, and $\bar{X}_V^\flat$ is the minimal Satake compactification for the standard projective representation of $\mathrm{PGL}_V(E)$, as constructed by Satake in [22] (see also [2, 1.4]). The topologies of $\bar{X}_V$ and $\bar{X}_V^\flat$ are reviewed in Sect. 2.3 below.

2.2.7 We have a canonical surjection $\bar{X}_V \to \bar{X}_V^\flat$ which sends (P, μ) to (V_0, μ_0), where V_0 is as in 2.2.4, and where we use the definition of $\bar{X}_V^\flat$ of the second style in 2.2.5. This surjection is compatible with the inclusion maps from X_V to these spaces.

2.2.8 We have the natural actions of $\mathrm{PGL}_V(E)$ on X_V, $\bar{X}_V$ and $\bar{X}_V^\flat$ by 2.1.5. These actions are compatible with the canonical maps between these spaces.

2.3 *Topologies*

2.3.1 We define a topology on $\bar{X}_V$.

Take a basis $(e_i)_i$ of V. We have a commutative diagram

$$\begin{array}{ccc} \mathrm{PGL}_V(E) \times \mathbb{R}_{>0}^{d-1} & \longrightarrow & X_V \\ \downarrow & & \downarrow \\ \mathrm{PGL}_V(E) \times \mathbb{R}_{\geqslant 0}^{d-1} & \longrightarrow & \bar{X}_V. \end{array}$$

Here the upper arrow is $(g,t) \mapsto g\mu$, where μ is the class of the norm on V associated to $((e_i)_i, (r_i)_i)$ with $r_i = \prod_{1 \leqslant j < i} t_j^{-1}$, and where $g\mu$ is defined by the action of $\mathrm{PGL}_V(E)$ on X_V (2.2.8). The lower arrow is $(g,t) \mapsto g(P,\mu)$, where $(P,\mu) \in \bar{X}_V$ is defined as follows, and $g(P,\mu)$ is then defined by the action of $\mathrm{PGL}_V(E)$ on $\bar{X}_V$ (2.2.8). Let

$$I = \{j \mid t_j = 0\} \subset \{1, \ldots, d-1\},$$

and write

$$I = \{c(i) \mid 0 \leqslant i \leqslant m-1\},$$

where $m = \sharp I$ and $1 \leqslant c(0) < \cdots < c(m-1) \leqslant d-1$. If we also let $c(-1) = 0$ and $c(m) = d$, then the set of

$$V_i = \sum_{j=1}^{c(i)} F e_j$$

with $-1 \leqslant i \leqslant m$ forms a flag in V, and P is defined to be the corresponding parabolic subgroup of PGL_V (2.2.3). For $0 \leqslant i \leqslant m$, we take μ_i to be the equivalence class of the norm on V_i/V_{i-1} given by the basis $(e_j)_{c(i-1)<j\leqslant c(i)}$ and the sequence $(r_j)_{c(i-1)<j\leqslant c(i)}$ with $r_j = \prod_{c(i-1)<k<j} t_k^{-1}$.

Both the upper and the lower horizontal arrows in the diagram are surjective, and the topology on X_V coincides with the topology as a quotient space of $\mathrm{PGL}_V(E) \times \mathbb{R}_{>0}^{d-1}$ via the upper horizontal arrow. The topology on $\bar{X}_V$ is defined as the quotient topology of the topology on $\mathrm{PGL}_V(E) \times \mathbb{R}_{\geqslant 0}^{d-1}$ via the lower horizontal arrow. It is easily seen that this topology is independent of the choice of the basis $(e_i)_i$.

2.3.2 The space $\bar{X}_V^\flat$ has the following topology: the space of all nonzero semi-norms on V^* has a topology as a subspace of the product $\mathbb{R}^{V^*}$, and $\bar{X}_V^\flat$ has a topology as a quotient of it.

2.3.3 Both $\bar{X}_V$ and $\bar{X}_V^\flat$ are compact Hausdorff spaces containing X_V as a dense open subset. This is proved in [7, 19, 24, 25] in the case that E is non-archimedean and in [2, 22] in the archimedean case.

2.3.4 The topology on $\bar{X}_V^\flat$ coincides with the image of the topology on $\bar{X}_V$. In fact, it is easily seen that the canonical map $\bar{X}_V \to \bar{X}_V^\flat$ is continuous (using, for instance, [25, Theorem 5.1]). Since both spaces are compact Hausdorff and this continuous map is surjective, the topology on $\bar{X}_V^\flat$ is the image of that of $\bar{X}_V$.

3 Spaces Associated to Global Fields

Let F be a global field, which is to say, either a number field or a function field in one variable over a finite field. We fix a finite-dimensional F-vector space V of dimension $d \geqslant 1$. For a place v of F, let $V_v = F_v \otimes_F V$. We set $X_v = X_{V_v}$ and $X_v^\flat = X_{V_v}^\flat$ for brevity. If v is non-archimedean, we let O_v, k_v, q_v, ϖ_v denote the valuation ring of F_v, the residue field of O_v, the order of k_v, and a fixed uniformizer in O_v, respectively.

In this section, we define sets $\bar{X}^{\star}_{F,v}$ containing X_v for $\star \in \{\sharp, \ , \flat\}$, which serve as our rational partial compactifications. Here, $\bar{X}_{F,v}$ (resp., $\bar{X}^{\flat}_{F,v}$) is defined as a subset of X_v (resp., $\bar{X}^{\flat}_v$), and $\bar{X}^{\sharp}_{F,v}$ has $\bar{X}_{F,v}$ as a quotient. In §3.2, by way of example, we describe these sets and various topologies on them in the case that $d = 2$, $F = \mathbb{Q}$, and v is the real place. For $\star \neq \sharp$, we construct more generally sets $\bar{X}^{\star}_{F,S}$ for a nonempty finite set S of places of F. In §3.1, we describe $\bar{X}^{\star}_{F,S}$ as a subset of $\prod_{v \in S} \bar{X}^{\star}_{F,v}$.

In §3.3 and §3.4, we define topologies on these sets. That is, in §3.3, we define the "Borel–Serre topology", while in §3.4, we define the "Satake topology" on $\bar{X}_{F,v}$ and, assuming S contains all archimedean places of F, on $\bar{X}^{\flat}_{F,S}$. In §3.5, we prove results on $\bar{X}_{F,v}$. In §3.6, we compare the following topologies on $\bar{X}_{F,v}$ (resp., $\bar{X}^{\flat}_{F,v}$): the Borel–Serre topology, the Satake topology, and the topology as a subspace of $\bar{X}_v$ (resp., $\bar{X}^{\flat}_v$). In §3.7, we describe the relationship between these spaces and Borel–Serre and reductive Borel–Serre spaces.

3.1 Definitions of the Spaces

3.1.1 Let $\bar{X}_{F,v} = \bar{X}_{V,F,v}$ be the subset of $\bar{X}_v$ consisting of all elements (P, μ) such that P is F-rational. If P comes from a parabolic subgroup P' of PGL_V over F, we also denote (P, μ) by (P', μ).

3.1.2 Let $\bar{X}^{\flat}_{F,v}$ be the subset of $\bar{X}^{\flat}_v$ consisting of all elements (W, μ) such that W is F-rational (using the definition of $\bar{X}^{\flat}_v$ in the second style in 2.2.5). If W comes from an F-subspace W' of V, we also denote (W, μ) by (W', μ).

3.1.3 Let $\bar{X}^{\sharp}_{F,v}$ be the set of all triples (P, μ, s) such that $(P, \mu) \in \bar{X}_{F,v}$ and s is a splitting

$$s\colon \bigoplus_{i=0}^{m} (V_i/V_{i-1})_v \xrightarrow{\sim} V_v$$

over F_v of the filtration $(V_i)_{-1 \leqslant i \leqslant m}$ of V corresponding to P.

We have an embedding $X_v \hookrightarrow \bar{X}^{\sharp}_{F,v}$ that sends μ to (PGL_V, μ, s), where s is the identity map of V_v.

3.1.4 We have a diagram with a commutative square

$$\begin{array}{ccccc} \bar{X}^{\sharp}_{F,v} & \twoheadrightarrow & \bar{X}_{F,v} & \twoheadrightarrow & \bar{X}^{\flat}_{F,v} \\ & & \downarrow & & \downarrow \\ & & \bar{X}_v & \twoheadrightarrow & \bar{X}^{\flat}_v. \end{array}$$

Here, the first arrow in the upper row forgets the splitting s, and the second arrow in the upper row is $(P, \mu) \mapsto (V_0, \mu_0)$, as is the lower arrow (2.2.7).

3.1.5 The group $\mathrm{PGL}_V(F)$ acts on the sets $\bar{X}_{F,v}$, $\bar{X}^{\flat}_{F,v}$ and $\bar{X}^{\sharp}_{F,v}$ in the canonical manner.

3.1.6 Now let S be a nonempty finite set of places of F.

- Let $\bar{X}_{F,S}$ be the subset of $\prod_{v\in S} \bar{X}_{F,v}$ consisting of all elements $(x_v)_{v\in S}$ such that the parabolic subgroup of $G = \mathrm{PGL}_V$ associated to x_v is independent of v.
- Let $\bar{X}^{\flat}_{F,S}$ be the subset of $\prod_{v\in S} \bar{X}^{\flat}_{F,v}$ consisting of all elements $(x_v)_{v\in S}$ such that the F-subspace of V associated to x_v is independent of v.

We will denote an element of $\bar{X}_{F,S}$ as (P, μ), where P is a parabolic subgroup of G and $\mu \in \prod_{v\in S, 0\leqslant i\leqslant m} X_{(V_i/V_{i-1})_v}$ with $(V_i)_i$ the flag corresponding to P. We will denote an element of $\bar{X}^{\flat}_{F,S}$ as (W, μ), where W is a nonzero F-subspace of V and $\mu \in \prod_{v\in S} X_{W_v}$. We have a canonical surjective map

$$\bar{X}_{F,S} \to \bar{X}^{\flat}_{F,S}$$

which commutes with the inclusion maps from X_S to these spaces.

3.2 *Example: Upper Half-Plane*

3.2.1 Suppose that $F = \mathbb{Q}$, v is the real place, and $d = 2$.

In this case, the sets X_v, $\bar{X}_v = \bar{X}^{\flat}_v$, $\bar{X}_{\mathbb{Q},v} = \bar{X}^{\flat}_{\mathbb{Q},v}$, and $\bar{X}^{\sharp}_{\mathbb{Q},v}$ are described by using the upper half-plane. In §2, we discussed topologies on the first two spaces. The remaining spaces also have natural topologies, as will be discussed in §3.3 and §3.4: the space $\bar{X}^{\sharp}_{\mathbb{Q},v}$ is endowed with the Borel–Serre topology, and $\bar{X}_{\mathbb{Q},v}$ has two topologies, the Borel–Serre topology and Satake topology, which are both different from its topology as a subspace of $\bar{X}_v$. In this section, as a prelude to §3.3 and §3.4, we describe what the Borel–Serre and Satake topologies look like in this special case.

3.2.2 Let $\mathfrak{H} = \{x + yi \mid x, y \in \mathbb{R}, y > 0\}$ be the upper half-plane. Fix a basis $(e_i)_{i=1,2}$ of V. For $z \in \mathfrak{H}$, let μ_z denote the class of the norm on V corresponding to the class of the norm on V^* given by $ae_1^* + be_2^* \mapsto |az + b|$ for $a, b \in \mathbb{R}$. Here $(e_i^*)_{1\leqslant i\leqslant d}$ is the dual basis of $(e_i)_i$, and $|\ \ |$ denotes the usual absolute value (not the normalized absolute value) on $\mathbb{C}$. We have a homeomorphism

$$\mathfrak{H} \xrightarrow{\sim} X_v, \quad z \mapsto \mu_z$$

which is compatible with the actions of $\mathrm{SL}_2(\mathbb{R})$.

For the square root $i \in \mathfrak{H}$ of -1, the norm $ae_1 + be_2 \mapsto (a^2 + b^2)^{1/2}$ has class μ_i. For $z = x + yi$, we have

$$\mu_z = \begin{pmatrix} y & x \\ 0 & 1 \end{pmatrix} \mu_i.$$

The action of $\begin{pmatrix} -1 & 0 \\ 0 & 1 \end{pmatrix} \in \mathrm{GL}_2(\mathbb{R})$ on X_v corresponds to $x + yi \mapsto -x + yi$ on $\mathfrak{H}$.

3.2.3 The inclusions

$$X_v \subset \bar{X}_{\mathbb{Q},v} \subset \bar{X}_v$$

can be identified with

$$\mathfrak{H} \subset \mathfrak{H} \cup \mathbb{P}^1(\mathbb{Q}) \subset \mathfrak{H} \cup \mathbb{P}^1(\mathbb{R}).$$

Here $z \in \mathbb{P}^1(\mathbb{R}) = \mathbb{R} \cup \{\infty\}$ corresponds to the class in $\bar{X}_v^\flat = \bar{X}_v$ of the semi-norm $ae_1^* + be_2^* \mapsto |az + b|$ (resp., $ae_1^* + be_2^* \mapsto |a|$) on V^* if $z \in \mathbb{R}$ (resp., $z = \infty$). These identifications are compatible with the actions of $\mathrm{PGL}_V(\mathbb{Q})$.

The topology on $\bar{X}_v$ of 2.3.1 is the topology as a subspace of $\mathbb{P}^1(\mathbb{C})$.

3.2.4 Let B be the Borel subgroup of PGL_V consisting of all upper triangular matrices for the basis $(e_i)_i$, and let $0 = V_{-1} \subsetneq V_0 = \mathbb{Q}e_1 \subsetneq V_1 = V$ be the corresponding flag. Then $\infty \in \mathbb{P}^1(\mathbb{Q})$ is understood as the point (B, μ) of $\bar{X}_{\mathbb{Q},v}$, where μ is the unique element of $X_{(V_0)_v} \times X_{(V/V_0)_v}$.

Let $\bar{X}_{\mathbb{Q},v}(B) = \mathfrak{H} \cup \{\infty\} \subset \bar{X}_{\mathbb{Q},v}$ and let $\bar{X}^\sharp_{\mathbb{Q},v}(B)$ be the inverse image of $\bar{X}_{\mathbb{Q},v}(B)$ in $\bar{X}^\sharp_{\mathbb{Q},v}$. Then for the Borel–Serre topology defined in §3.3, we have a homeomorphism

$$\bar{X}^\sharp_{\mathbb{Q},v}(B) \cong \{x + yi \mid x \in \mathbb{R}, 0 < y \leqslant \infty\} \supset \mathfrak{H}.$$

Here $x + \infty i$ corresponds to (B, μ, s) where s is the splitting of the filtration $(V_{i,v})_i$ given by the embedding $(V/V_0)_v \to V_v$ that sends the class of e_2 to $xe_1 + e_2$.

The Borel–Serre topology on $\bar{X}^\sharp_{\mathbb{Q},v}$ is characterized by the properties that

(i) the action of the discrete group $\mathrm{GL}_V(\mathbb{Q})$ on $\bar{X}^\sharp_{\mathbb{Q},v}$ is continuous,
(ii) the subset $\bar{X}^\sharp_{\mathbb{Q},v}(B)$ is open, and
(iii) as a subspace, $\bar{X}^\sharp_{\mathbb{Q},v}(B)$ is homeomorphic to $\{x + yi \mid x \in \mathbb{R}, 0 < y \leqslant \infty\}$ as above.

3.2.5 The Borel–Serre and Satake topologies on $\bar{X}_{\mathbb{Q},v}$ (defined in §3.3 and §3.4) are characterized by the following properties:

(i) The subspace topology on $X_v \subset \bar{X}_{\mathbb{Q},v}$ coincides with the topology on $\mathfrak{H}$.
(ii) The action of the discrete group $\mathrm{GL}_V(\mathbb{Q})$ on $\bar{X}_{\mathbb{Q},v}$ is continuous.
(iii) The following sets (a) (resp., (b)) form a base of neighborhoods of ∞ for the Borel–Serre (resp., Satake) topology:

(a) the sets $U_f = \{x + yi \in \mathfrak{H} \mid y \geqslant f(x)\} \cup \{\infty\}$ for continuous $f: \mathbb{R} \to \mathbb{R}$,
(b) the sets $U_c = \{x + yi \in \mathfrak{H} \mid y \geqslant c\} \cup \{\infty\}$ with $c \in \mathbb{R}_{>0}$.

The Borel–Serre topology on $\bar{X}_{\mathbb{Q},v}$ is the image of the Borel–Serre topology on $\bar{X}^{\sharp}_{\mathbb{Q},v}$.

3.2.6 For example, the set $\{x + yi \in \mathfrak{H} \mid y > x\} \cup \{\infty\}$ is a neighborhood of ∞ for the Borel–Serre topology, but it is not a neighborhood of ∞ for the Satake topology.

3.2.7 For any subgroup Γ of $\mathrm{PGL}_2(\mathbb{Z})$ of finite index, the Borel–Serre and Satake topologies induce the same topology on the quotient space $X(\Gamma) = \Gamma\backslash\bar{X}_{\mathbb{Q},v}$. Under this quotient topology, $X(\Gamma)$ is compact Hausdorff. If Γ is the image of a congruence subgroup of $\mathrm{SL}_2(\mathbb{Z})$, then this is the usual topology on the modular curve $X(\Gamma)$.

3.3 Borel–Serre Topology

3.3.1 For a parabolic subgroup P of PGL_V, let $\bar{X}_{F,v}(P)$ (resp., $\bar{X}^{\sharp}_{F,v}(P)$) be the subset of $\bar{X}_{F,v}$ (resp., $\bar{X}^{\sharp}_{F,v}$) consisting of all elements (Q, μ) (resp., (Q, μ, s)) such that $Q \supset P$.

The action of $\mathrm{PGL}_V(F_v)$ on $\bar{X}_v$ induces an action of $P(F_v)$ on $\bar{X}_{F,v}(P)$. We have also an action of $P(F_v)$ on $\bar{X}^{\sharp}_{F,v}(P)$ given by

$$g(\alpha, s) = (g\alpha, g \circ s \circ g^{-1})$$

for $g \in P(F_v)$, $\alpha \in \bar{X}_{F,v}(P)$, and s a splitting of the filtration.

3.3.2 Fix a basis $(e_i)_i$ of V. Let P be a parabolic subgroup of PGL_V such that

- if $0 = V_{-1} \subsetneq V_0 \subsetneq \cdots \subsetneq V_m = V$ denotes the flag of F-subspaces corresponding to P, then each V_i is generated by the e_j with $1 \leqslant j \leqslant c(i)$, where $c(i) = \dim(V_i)$.

This condition on P is equivalent to the condition that P contains the Borel subgroup B of PGL_V consisting of all upper triangular matrices with respect to $(e_i)_i$. Where useful, we will identify PGL_V over F with PGL_d over F via the basis $(e_i)_i$.

Let

$$\Delta(P) = \{\dim(V_j) \mid 0 \leqslant j \leqslant m-1\} \subset \{1, \ldots, d-1\},$$

and let $\Delta'(P)$ be the complement of $\Delta(P)$ in $\{1, \ldots, d-1\}$. Let $\mathbb{R}^{d-1}_{\geqslant 0}(P)$ be the open subset of $\mathbb{R}^{d-1}_{\geqslant 0}$ given by

$$\mathbb{R}^{d-1}_{\geqslant 0}(P) = \{(t_i)_{1 \leqslant i \leqslant d-1} \in \mathbb{R}^{d-1}_{\geqslant 0} \mid t_i > 0 \text{ for all } i \in \Delta'(P)\}.$$

In particular, we have

$$\mathbb{R}^{d-1}_{\geqslant 0}(P) \cong \mathbb{R}^{\Delta'(P)}_{>0} \times \mathbb{R}^{\Delta(P)}_{\geqslant 0}.$$

3.3.3 With P as in 3.3.2, the map $\mathrm{PGL}_V(F_v) \times \mathbb{R}^{d-1}_{\geqslant 0} \to \bar{X}_v$ in 2.3.1 induces a map

$$\bar{\pi}_{P,v}\colon P(F_v) \times \mathbb{R}^{d-1}_{\geqslant 0}(P) \to \bar{X}_{F,v}(P),$$

which restricts to a map

$$\pi_{P,v}\colon P(F_v) \times \mathbb{R}^{d-1}_{>0} \to X_{F,v}.$$

The map $\bar{\pi}_{P,v}$ is induced by a map

$$\bar{\pi}^{\sharp}_{P,v}\colon P(F_v) \times \mathbb{R}^{d-1}_{\geqslant 0}(P) \to \bar{X}^{\sharp}_{F,v}(P)$$

defined as $(g, t) \mapsto g(P, \mu, s)$ where (P, μ) is as in 2.3.1 and s is the splitting of the filtration $(V_i)_{-1 \leqslant i \leqslant m}$ defined by the basis $(e_i)_i$. For this splitting s, we set

$$V^{(i)} = s(V_i/V_{i-1}) = \sum_{c(i-1)<j\leqslant c(i)} Fe_j$$

for $0 \leqslant i \leqslant m$ so that $V_i = V_{i-1} \oplus V^{(i)}$ and $V = \bigoplus_{i=0}^{m} V^{(i)}$. If $P = B$, then we will often omit the subscript B from our notation for these maps.

3.3.4 We review the Iwasawa decomposition. For v archimedean (resp., non-archimedean), let $A_v \leqslant \mathrm{PGL}_d(F_v)$ be the subgroup of elements of that lift to diagonal matrices in $\mathrm{GL}_d(F_v)$ with positive real entries (resp., with entries that are powers of ϖ_v). Let K_v denote the standard maximal compact subgroup of $\mathrm{PGL}_d(F_v)$ given by

$$K_v = \begin{cases} \mathrm{PO}_d(\mathbb{R}) & \text{if } v \text{ is real,} \\ \mathrm{PU}_d & \text{if } v \text{ is complex,} \\ \mathrm{PGL}_d(O_v) & \text{otherwise.} \end{cases}$$

Let B_u denote the upper-triangular unipotent matrices in the standard Borel B. The Iwasawa decomposition is given by the equality

$$\mathrm{PGL}_d(F_v) = B_u(F_v)A_vK_v.$$

3.3.5 If v is archimedean, then the expression of a matrix in $\mathrm{PGL}_d(F_v)$ as a product in the Iwasawa decomposition is unique.

3.3.6 If v is non-archimedean, then the Bruhat decomposition is $\mathrm{PGL}_d(k_v) = B(k_v)S_dB(k_v)$, where the symmetric group S_d of degree d is viewed as a subgroup of PGL_d over any field via the permutation representation on the standard basis. This implies that $\mathrm{PGL}_d(O_v) = B(O_v)S_d\,\mathrm{Iw}(O_v)$, where $\mathrm{Iw}(O_v)$ is the Iwahori subgroup consisting of those matrices in with upper triangular image in $\mathrm{PGL}_d(k_v)$. Combining this with the Iwasawa decomposition (in the notation of 3.3.4), we have

$$\mathrm{PGL}_d(F_v) = B_u(F_v)A_vS_d\,\mathrm{Iw}(O_v).$$

This decomposition is not unique, since $B_u(F_v) \cap \mathrm{Iw}(O_v) = B_u(O_v)$.

3.3.7 If v is archimedean, then there is a bijection $\mathbb{R}^{d-1}_{>0} \xrightarrow{\sim} A_v$ given by

$$t = (t_k)_{1 \leqslant k \leqslant d-1} \mapsto a = \begin{cases} \mathrm{diag}(r_1, \ldots, r_d)^{-1} & \text{if } v \text{ is real,} \\ \mathrm{diag}(r_1^{1/2}, \ldots, r_d^{1/2})^{-1} & \text{if } v \text{ is complex,} \end{cases}$$

where $r_i = \prod_{k=1}^{i-1} t_k^{-1}$ as in 2.3.1.

Proposition 3.3.8

(1) Let P be a parabolic subgroup of PGL_V *as in 3.3.2. Then the maps*

$$\bar{\pi}_{P,v}\colon P(F_v) \times \mathbb{R}^{d-1}_{\geqslant 0}(P) \to \bar{X}_{F,v}(P) \text{ and } \bar{\pi}^{\sharp}_{P,v}\colon P(F_v) \times \mathbb{R}^{d-1}_{\geqslant 0}(P) \to \bar{X}^{\sharp}_{F,v}(P)$$

of 3.3.3 are surjective.

(2) For the Borel subgroup B of 3.3.2, the maps

$$\pi_v\colon B_u(F_v) \times \mathbb{R}^{d-1}_{>0} \to X_v, \qquad \bar{\pi}_v\colon B_u(F_v) \times \mathbb{R}^{d-1}_{\geqslant 0} \to \bar{X}_{F,v}(B),$$
$$\text{and } \bar{\pi}^{\sharp}_v\colon B_u(F_v) \times \mathbb{R}^{d-1}_{\geqslant 0} \to \bar{X}^{\sharp}_{F,v}(B).$$

of 3.3.3 are all surjective.

(3) If v is archimedean, then π_v and $\bar{\pi}^{\sharp}_v$ are bijective.

(4) If v is non-archimedean, then $\bar{\pi}_v$ induces a bijection

$$(B_u(F_v) \times \mathbb{R}^{d-1}_{\geqslant 0})/\sim \; \to \bar{X}_{F,v}(B)$$

where $(g, (t_i)_i) \sim (g', (t'_i)_i)$ if and only if

(i) $t_i = t'_i$ for all i and

(ii) $|(g^{-1}g')_{ij}| \leqslant (\prod_{i \leqslant k < j} t_k)^{-1}$ for all $1 \leqslant i < j \leqslant d$, considering any $c \in \mathbb{R}$ to be less than $0^{-1} = \infty$.

Proof If $\bar{\pi}^{\sharp}_v$ is surjective, then for any parabolic P containing B, the restriction of $\bar{\pi}^{\sharp}_v$ to $B_u(F_v) \times \mathbb{R}^{d-1}_{\geqslant 0}(P)$ has image $\bar{X}^{\sharp}_{F,v}(P)$. Since $B_u(F_v) \subset P(F_v)$, this forces the subjectivity of $\bar{\pi}^{\sharp}_{P,v}$, hence of $\bar{\pi}_{P,v}$ as well. So, we turn our attention to (2)–(4). If $r \in \mathbb{R}^d_{>0}$, we let $\mu^{(r)} \in X_v$ denote the class of the norm attached to the basis $(e_i)_i$ and r.

Suppose first that v is archimedean. By the Iwasawa decomposition 3.3.4, and noting 3.3.5 and 2.2.2, we see that $B_u(F_v)A_v \to X_v$ given by $g \mapsto g\mu^{(1)}$ is bijective, where $\mu^{(1)}$ denotes the class of the norm attached to $(e_i)_i$ and $1 = (1, \ldots, 1) \in \mathbb{R}^d_{>0}$. For $t \in \mathbb{R}^{d-1}_{>0}$, let $a \in A_v$ be its image under the bijection in 3.3.7. Since $pa\mu^{(1)} = p\mu^{(r)}$, for $p \in B_u(F_v)$ and r as in 2.3.1, we have the bijectivity of π_v.

Consider the third map in (2). For $t \in \mathbb{R}_{\geqslant 0}^{d-1}$, let P be the parabolic that contains B and is determined by the set $\Delta(P)$ of $k \in \{1, \ldots, d-1\}$ for which $t_k = 0$. Let $(V_i)_{-1 \leqslant i \leqslant m}$ be the corresponding flag. Let M denote the Levi subgroup of P. (It is the quotient of $\prod_{i=0}^{m} \mathrm{GL}_{V^{(i)}} < \mathrm{GL}_V$ by scalars, where $V = \bigoplus_{i=0}^{m} V^{(i)}$ as in 3.3.3, and $M \cap B_u$ is isomorphic to the product of the upper-triangular unipotent matrices in each $\mathrm{PGL}_{V^{(i)}}$.) The product of the first maps in (2) for the blocks of M is a bijection

$$(M \cap B_u)(F_v) \times \mathbb{R}_{>0}^{\Delta'(P)} \xrightarrow{\sim} \prod_{i=0}^{m} X_{V_v^{(i)}} \subset \bar{X}_{F,v}(P),$$

such that (g, t') is sent to $(P, g\mu)$ in $\bar{X}_{F,v}$, where μ is the sequence of classes of norms determined by t' and the standard basis. The stabilizer of μ in B_u is the unipotent radical P_u of P, and this P_u acts simply transitively on the set of splittings for the graded quotients $(V_i/V_{i-1})_v$. Since $B_u = (M \cap B_u)P_u$, and this decomposition is unique, we have the desired bijectivity of $\bar{\pi}_v^{\sharp}$, proving (3).

Suppose next that v is non-archimedean. We prove the surjectivity of the first map in (2). Using the natural actions of A_v and the symmetric group S_d on $\mathbb{R}_{>0}^d$, we see that any norm on V_v can be written as $g\mu^{(r)}$, where $g \in \mathrm{PGL}_d(F_v)$ and $r = (r_i)_i \in \mathbb{R}_{>0}^d$, with r satisfying

$$r_1 \leqslant r_2 \leqslant \cdots \leqslant r_d \leqslant q_v r_1.$$

For such an r, the class $\mu^{(r)}$ is invariant under the action of $\mathrm{Iw}(O_v)$. Hence for such an r, any element of $S_d \,\mathrm{Iw}(O_v)\mu^{(r)} = S_d \mu^{(r)}$ is of the form $\mu^{(r')}$, where $r' = (r_{\sigma(i)})_i$ for some $\sigma \in S_d$. Hence, any element of $A_v S_d \,\mathrm{Iw}(O_v)\mu^{(r)}$ for such an r is of the form $\mu^{(r')}$ for some $r' = (r'_i)_i \in \mathbb{R}_{>0}^d$. This proves the surjecivity of the first map of (2). The surjectivity of the other maps in (2) is then shown using this, similarly to the archimedean case.

Finally, we prove (4). It is easy to see that the map $\bar{\pi}_v$ factors through the quotient by the equivalence relation. We can deduce the bijectivity in question from the bijectivity of $(B_u(F_v) \times \mathbb{R}_{>0}^{d-1})/\sim \to X_v$, replacing V by V_i/V_{i-1} as in the above arguments for the archimedean case. Suppose that $\pi_v(g, t) = \pi_v(1, t')$ for $g \in B_u(F_v)$ and $t, t' \in \mathbb{R}_{>0}^{d-1}$. We must show that $(g, t) \sim (1, t')$. Write $\pi_v(g, t) = g\mu^{(r)}$ and $\pi_v(1, t') = \mu^{(r')}$ with $r = (r_i)_i$ and $r' = (r'_i)_i \in \mathbb{R}_{>0}^d$ such that $r_1 = 1$ and $r_j/r_i = (\prod_{i \leqslant k < j} t_k)^{-1}$ for all $1 \leqslant i < j \leqslant d$, and similarly for r' and t'. It then suffices to check that $r' = r$ and $r_i|g_{ij}| \leqslant r_j$ for all $i < j$. Since $\mu^{(r)} = g^{-1}\mu^{(r')}$, there exists $c \in \mathbb{R}_{>0}$ such that

$$\max\{r_i|x_i| \mid 1 \leqslant i \leqslant d\} = c \max\{r'_i|(gx)_i| \mid 1 \leqslant i \leqslant d\}$$

for all $x = (x_i)_i \in F_v^d$. Taking $x = e_1$, we have $gx = e_1$ as well, so $c = 1$. Taking $x = e_i$, we obtain $r_i \geqslant r'_i$, and taking $x = g^{-1}e_i$, we obtain $r_i \leqslant r'_i$. Thus $r = r'$, and taking $x = e_j$ yields $r_j = \max\{r_i|g_{ij}| \mid 1 \leqslant i \leqslant j\}$, which tells us that $r_j \geqslant r_i|g_{ij}|$ for $i < j$. □

Proposition 3.3.9 *There is a unique topology on $\bar{X}_{F,v}$ (resp., $\bar{X}^{\sharp}_{F,v}$) satisfying the following conditions (i) and (ii).*

(i) For every parabolic subgroup P of PGL_V, the set $\bar{X}_{F,v}(P)$ (resp., $\bar{X}^{\sharp}_{F,v}(P)$) is open in $\bar{X}_{F,v}$ (resp., $\bar{X}^{\sharp}_{F,v}$).

(ii) For every parabolic subgroup P of PGL_V and basis $(e_i)_i$ of V such that P contains the Borel subgroup with respect to $(e_i)_i$, the topology on $\bar{X}_{F,v}(P)$ (resp., $\bar{X}^{\sharp}_{F,v}(P)$) is the topology as a quotient of $P(F_v) \times \mathbb{R}^{d-1}_{\geqslant 0}(P)$ under the surjection of 3.3.8(1).

This topology is also characterized by (i) and the following (ii)′.

(ii′) If B is a Borel subgroup of PGL_V consisting of upper triangular matrices with respect to a basis $(e_i)_i$ of V, then the topology on $\bar{X}_{F,v}(B)$ (resp., $\bar{X}^{\sharp}_{F,v}(B)$) is the topology as a quotient of $B_u(F_v) \times \mathbb{R}^{d-1}_{\geqslant 0}$ under the surjection of 3.3.8(2).

Proof The uniqueness is clear if we have existence of a topology satisfying (i) and (ii). Let $(e_i)_i$ be a basis of V, let B be the Borel subgroup of PGL_V with respect to this basis, and let P be a parabolic subgroup of PGL_V containing B. It suffices to prove that for the topology on $\bar{X}_{F,v}(B)$ (resp., $\bar{X}^{\sharp}_{F,v}(B)$) as a quotient of $B_u(F_v) \times \mathbb{R}^{d-1}_{\geqslant 0}(B)$, the subspace topology on $\bar{X}_{F,v}(P)$ (resp., $\bar{X}^{\sharp}_{F,v}(P)$) coincides with the quotient topology from $P(F_v) \times \mathbb{R}^{d-1}_{\geqslant 0}(P)$. For this, it is enough to show that the action of the topological group $P(F_v)$ on $\bar{X}_{F,v}(P)$ (resp., $\bar{X}^{\sharp}_{F,v}(P)$) is continuous with respect to the topology on $\bar{X}_{F,v}(P)$ (resp., $\bar{X}^{\sharp}_{F,v}(P)$) as a quotient of $B_u(F_v) \times \mathbb{R}^{d-1}_{\geqslant 0}(P)$. We must demonstrate this continuity.

Let $(V_i)_{-1\leqslant i\leqslant m}$ be the flag corresponding to P, and let $c(i) = \dim(V_i)$. For $0 \leqslant i \leqslant m$, we regard $\mathrm{GL}_{V^{(i)}}$ as a subgroup of GL_V via the decomposition $V = \bigoplus_{i=0}^{m} V^{(i)}$ of 3.3.3.

Suppose first that v is archimedean. For $0 \leqslant i \leqslant m$, let K_i be the compact subgroup of $\mathrm{GL}_{V^{(i)}}(F_v)$ that is the isotropy group of the norm on $V^{(i)}$ given by the basis $(e_j)_{c(i-1)<j\leqslant c(i)}$ and $(1,\ldots,1) \in \prod_{c(i-1)<j\leqslant c(i)} \mathbb{R}_{>0}$. We identify $\mathbb{R}^{d-1}_{>0}$ with A_v as in 3.3.7. By the Iwasawa decomposition 3.3.4 and its uniqueness in 3.3.5, the product on $P(F_v)$ induces a homeomorphism

$$(a,b,c)\colon P(F_v) \xrightarrow{\sim} B_u(F_v) \times \mathbb{R}^{d-1}_{>0} \times \left(\prod_{i=0}^{m} K_i\right)/\{z \in F_v^{\times} \mid |z| = 1\}.$$

We also have a product map $\phi\colon P(F_v) \times B_u(F_v) \times \mathbb{R}^{\Delta'(P)}_{>0} \to P(F_v)$, where we identify $t' \in \mathbb{R}^{\Delta'(P)}_{>0}$ with the diagonal matrix $\mathrm{diag}(r_1,\ldots,r_d)^{-1}$ if v is real and $\mathrm{diag}(r_1^{1/2},\ldots,r_d^{1/2})^{-1}$ if v is complex, with $r_j^{-1} = \prod_{c(i-1)<k<j} t'_k$ for $c(i-1) < j \leqslant c(i)$ as in 2.3.1. These maps fit in a commutative diagram

$$
\begin{array}{ccccc}
P(F_v) \times \mathbb{R}_{\geqslant 0}^{\Delta(P)} & \overset{(\phi,\mathrm{id})}{\twoheadleftarrow} & P(F_v) \times B_u(F_v) \times \mathbb{R}_{>0}^{\Delta'(P)} \times \mathbb{R}_{\geqslant 0}^{\Delta(P)} & \overset{(\mathrm{id},\bar{\pi}_{P,v})}{\twoheadrightarrow} & P(F_v) \times \bar{X}_{F,v}(P) \\
\downarrow & & & & \downarrow \\
B_u(F_v) \times \mathbb{R}_{\geqslant 0}^{d-1}(P) & & \xrightarrow{\quad\bar{\pi}_{P,v}\quad} & & \bar{X}_{F,v}(P)
\end{array}
$$

in which the right vertical arrow is the action of $P(F_v)$ on $\bar{X}_{F,v}(P)$, and the left vertical arrow is the continuous map

$$(u,t) \mapsto (a(u), b(u) \cdot (1,t)), \quad (u,t) \in P(F_v) \times \mathbb{R}_{\geqslant 0}^{\Delta(P)}$$

for $(1,t)$ the element of $\mathbb{R}_{\geqslant 0}^{d-1}(P)$ with $\mathbb{R}_{>0}^{\Delta'(P)}$-component 1 and $\mathbb{R}_{\geqslant 0}^{\Delta(P)}$-component t. (To see the commutativity, note that $c(u)$ commutes with the block-scalar matrix determined by $(1,t)$.) We also have a commutative diagram of the same form for $\bar{X}_{F,v}^{\sharp}$. Since the surjective horizontal arrows are quotient maps, we have the continuity of the action of $P(F_v)$.

Next, we consider the case that v is non-archimedean. For $0 \leqslant i \leqslant m$, let $S^{(i)}$ be the group of permutations of the set

$$I_i = \{j \in \mathbb{Z} \mid c(i-1) < j \leqslant c(i)\},$$

and regard it as a subgroup of $\mathrm{GL}_{V^{(i)}}(F)$. Let A_v be the subgroup of the diagonal torus of $\mathrm{PGL}_V(F_v)$ with respect to the basis $(e_i)_i$ with entries powers of a fixed uniformizer, as in 3.3.4.

Consider the action of $A_v \prod_{i=0}^m S^{(i)} \subset P(F_v)$ on $\mathbb{R}_{\geqslant 0}^{d-1}(P)$ that is compatible with the action of $P(F_v)$ on $\bar{X}_{F,v}(P)$ via the embedding $\mathbb{R}_{\geqslant 0}^{d-1}(P) \to \bar{X}_{F,v}(P)$. This action is described as follows. Any matrix $a = \mathrm{diag}(a_1, \ldots, a_d) \in A_v$ sends $t \in \mathbb{R}_{\geqslant 0}^{d-1}(P)$ to $(t_j|a_{j+1}||a_j|^{-1})_j \in \mathbb{R}_{\geqslant 0}^{d-1}(P)$. The action of $\prod_{i=0}^m S^{(i)}$ on $\mathbb{R}_{\geqslant 0}^{d-1}(P)$ is the unique continuous action which is compatible with the evident action of $\prod_{i=0}^m S^{(i)}$ on $\mathbb{R}_{>0}^d$ via the map $\mathbb{R}_{>0}^d \to \mathbb{R}_{\geqslant 0}^{d-1}(P)$ that sends $(r_i)_i$ to $(t_j)_j$, where $t_j = r_j/r_{j+1}$. That is, for

$$\sigma = (\sigma_i)_{0 \leqslant i \leqslant m} \in \prod_{i=0}^m S^{(i)},$$

let $f \in S_d$ be the unique permutation with $f|_{I_i} = \sigma_i^{-1}$ for all i. Then σ sends $t \in \mathbb{R}_{\geqslant 0}^{d-1}(P)$ to the element $t' = (t'_j)_j$ given by

$$t'_j = \begin{cases} \prod_{f(j) \leqslant k < f(j+1)} t_k & \text{if } f(j) < f(j+1), \\ \prod_{f(j+1) \leqslant k < f(j)} t_k^{-1} & \text{if } f(j+1) < f(j). \end{cases}$$

Let C be the compact subset of $\mathbb{R}^{d-1}_{\geqslant 0}(P)$ given by

$$C = \left\{ t = (t_j)_j \in \mathbb{R}^{d-1}_{\geqslant 0}(P) \cap [0,1]^{d-1} \mid \prod_{c(i-1)<j<c(i)} t_j \geqslant q_v^{-1} \text{ for all } 0 \leqslant i \leqslant m \right\}.$$

We claim that for each $x \in \mathbb{R}^{d-1}_{\geqslant 0}(P)$, there is a finite family $(h_k)_k$ of elements of $A_v \prod_{i=0}^m S^{(i)}$ such that the union $\bigcup_k h_k C$ is a neighborhood of x. This is quickly reduced to the following claim.

Claim Consider the natural action of $H = A_v S_d \subset \mathrm{PGL}_V$ on the quotient space $\mathbb{R}^d_{>0}/\mathbb{R}_{>0}$, with the class of $(a_j)_j$ in A_v acting as multiplication by $(|a_j|)_j$. Let C be the image of

$$\{r \in \mathbb{R}^d_{>0} \mid r_1 \leqslant r_2 \leqslant \cdots \leqslant r_d \leqslant q_v r_1\}$$

in $\mathbb{R}^d_{>0}/\mathbb{R}_{>0}$. Then for each $x \in \mathbb{R}^d_{>0}/\mathbb{R}_{>0}$, there is a finite family $(h_k)_k$ of elements of H such that $\bigcup_k h_k C$ is a neighborhood of x.

Proof of Claim This is a well-known statement in the theory of Bruhat-Tits buildings: the quotient $\mathbb{R}^d_{>0}/\mathbb{R}_{>0}$ is called the apartment of the Bruhat-Tits building X_v of PGL_V, and the set C is a $(d-1)$-simplex in this apartment. Any $(d-1)$-simplex in this apartment has the form hC for some $h \in H$, for any $x \in \mathbb{R}^d_{>0}/\mathbb{R}_{>0}$ there are only finitely many $(d-1)$-simplices in this apartment which contain x, and the union of these is a neighborhood of x in $\mathbb{R}^d_{>0}/\mathbb{R}_{>0}$.

By compactness of C, the topology on the neighborhood $\bigcup_k h_k C$ of x is the quotient topology from $\coprod_k h_k C$. Thus, it is enough to show that for each $h \in A_v \prod_{i=0}^m S^{(i)}$, the composition

$$P(F_v) \times B_u(F_v) \times hC \xrightarrow{(\mathrm{id}, \pi_{P,v})} P(F_v) \times \bar{X}_{F,v}(P) \to \bar{X}_{F,v}(P)$$

(where the second map is the action) and its analogue for $\bar{X}^{\sharp}_{F,v}$ are continuous.

For $0 \leqslant i \leqslant m$, let Iw_i be the Iwahori subgroup of $\mathrm{GL}_{V^{(i)}}(F_v)$ for the basis $(e_j)_{c(i-1)<j\leqslant c(i)}$. By the the Iwasawa and Bruhat decompositions as in 3.3.6, the product on $P(F_v)$ induces a continuous surjection

$$B_u(F_v) \times A_v \prod_{i=0}^m S^{(i)} \times \prod_{i=0}^m \mathrm{Iw}_i \to P(F_v),$$

and it admits continuous sections locally on $P(F_v)$. (Here, the middle group $A_v \prod_{i=0}^m S^{(i)}$ has the discrete topology.) Therefore, there exist an open covering $(U_\lambda)_\lambda$ of $P(F_v)$ and, for each λ, a subset $\mathcal{U}_\lambda$ of the above product mapping homeomorphically to U_λ, together with a continuous map

$$(a_\lambda, b_\lambda, c_\lambda)\colon \mathcal{U}_\lambda \to B_u(F_v) \times A_v \prod_{i=0}^{m} S^{(i)} \times \prod_{i=0}^{m} \mathrm{Iw}_i$$

such that its composition with the above product map is the map $\mathcal{U}_\lambda \xrightarrow{\sim} U_\lambda$. Let U'_λ denote the inverse image of U_λ under

$$P(F_v) \times B_u(F_v) \to P(F_v), \quad (g, g') \mapsto gg'h,$$

so that $(U'_\lambda)_\lambda$ is an open covering of $P(F_v) \times B_u(F_v)$. For any γ in the indexing set of the cover, let $\mathcal{U}'_{\lambda,\gamma}$ be the inverse image of U'_λ in $\mathcal{U}_\gamma \times B_u(F_v)$. Then the images of the $\mathcal{U}'_{\lambda,\gamma}$ form an open cover of $P(F_v) \times B_u(F_v)$ as well. Let $(a'_{\lambda,\gamma}, b'_{\lambda,\gamma})$ be the composition

$$\mathcal{U}'_{\lambda,\gamma} \to \mathcal{U}_\lambda \xrightarrow{(a_\lambda, b_\lambda)} B_u(F_v) \times A_v \prod_{i=0}^{m} S^{(i)}.$$

As $\prod_{i=0}^{m} \mathrm{Iw}_i$ fixes every element of C under its embedding in $\bar{X}_{F,v}(P)$, we have a commutative diagram

$$\begin{array}{ccc}
\mathcal{U}'_{\lambda,\gamma} \times hC & \hookrightarrow & P(F_v) \times B_u(F_v) \times hC \\
\downarrow & & \downarrow \\
B_u(F_v) \times \mathbb{R}^{d-1}_{\geqslant 0}(P) & \twoheadrightarrow & \bar{X}_{F,v}(P)
\end{array}$$

in which the left vertical arrow is

$$(u, hx) \mapsto (a'_{\lambda,\gamma}(u), b'_{\lambda,\gamma}(u)x)$$

for $x \in C$. We also have a commutative diagram of the same form for $\bar{X}^\sharp_{F,v}$. This proves the continuity of the action of $P(F_v)$. □

3.3.10 We call the topology on $\bar{X}_{F,v}$ (resp., $\bar{X}^\sharp_{F,v}$) in 3.3.9 the Borel–Serre topology. The Borel–Serre topology on $\bar{X}_{F,v}$ coincides with the quotient topology of the Borel–Serre topology on $\bar{X}^\sharp_{F,v}$. This topology on $\bar{X}_{F,v}$ is finer than the subspace topology from $\bar{X}_v$.

We define the Borel–Serre topology on $\bar{X}^\flat_{F,v}$ as the quotient topology of the Borel–Serre topology of $\bar{X}_{F,v}$. This topology on $\bar{X}^\flat_{F,v}$ is finer than the subspace topology from $\bar{X}^\flat_v$.

For a nonempty finite set S of places of F, we define the Borel–Serre topology on $\bar{X}_{F,S}$ (resp., $\bar{X}^\flat_{F,S}$) as the subspace topology for the product topology on $\prod_{v\in S} \bar{X}_{F,v}$ (resp., $\prod_{v\in S} \bar{X}^\flat_{F,v}$) for the Borel–Serre topology on each $\bar{X}_{F,v}$ (resp., $\bar{X}^\flat_{F,v}$).

3.4 Satake Topology

3.4.1 For a nonempty finite set of places S of F, we define the Satake topology on $\bar{X}_{F,S}$ and, under the assumption S contains all archimedean places, on $\bar{X}^{\flat}_{F,S}$.

The Satake topology is coarser than the Borel–Serre topology of 3.3.10. On the other hand, the Satake topology and the Borel–Serre topology induce the same topology on the quotient space by an arithmetic group (4.1.8). Thus, the Hausdorff compactness of this quotient space can be formulated without using the Satake topology (i.e., using only the Borel–Serre topology). However, arguments involving the Satake topology appear naturally in the proof of this property. One nice aspect of the Satake topology is that each point has an explicit base of neighborhoods (3.2.5, 3.4.9, 4.4.9).

3.4.2 Let H be a finite-dimensional vector space over a local field E. Let H' and H'' be E-subspaces of H such that $H' \supset H''$. Then a norm μ on H induces a norm ν on H'/H'' as follows. Let μ' be the restriction of μ to H'. Let $(\mu')^*$ be the norm on $(H')^*$ dual to μ'. Let ν^* be the restriction of $(\mu')^*$ to the subspace $(H'/H'')^*$ of $(H')^*$. Let ν be the dual of ν^*. This norm ν is given on $x \in H'/H''$ by

$$\nu(x) = \inf\{\mu(\tilde{x}) \mid \tilde{x} \in H' \text{ such that } \tilde{x} + H'' = x\}.$$

3.4.3 For a parabolic subgroup P of PGL_V, let $(V_i)_{-1 \leqslant i \leqslant m}$ be the corresponding flag. Set

$$\bar{X}_{F,S}(P) = \{(P', \mu) \in \bar{X}_{F,S} \mid P' \supset P\}.$$

For a place v of F, let us set

$$\mathfrak{Z}_{F,v}(P) = \prod_{i=0}^{m} X_{(V_i/V_{i-1})_v} \text{ and } \mathfrak{Z}_{F,S}(P) = \prod_{v \in S} \mathfrak{Z}_{F,v}(P).$$

We let $P(F_v)$ act on $\mathfrak{Z}_{F,v}(P)$ through $P(F_v)/P_u(F_v)$, using the $\mathrm{PGL}_{(V_i/V_{i-1})_v}(F_v)$-action on $X_{(V_i/V_{i-1})_v}$ for $0 \leqslant i \leqslant m$. We define a $P(F_v)$-equivariant map

$$\phi_{P,v} \colon \bar{X}_{F,v}(P) \to \mathfrak{Z}_{F,v}(P)$$

with the product of these over $v \in S$ giving rise to a map $\phi_{P,S} \colon \bar{X}_{F,S}(P) \to \mathfrak{Z}_{F,S}(P)$.

Let $(P', \mu) \in \bar{X}_{F,v}(P)$. Then the spaces in the flag $0 = V'_{-1} \subsetneq V'_0 \subsetneq \cdots \subsetneq V'_{m'} = V$ corresponding to P' form a subset of $\{V_i \mid -1 \leqslant i \leqslant m\}$. The image $\nu = (\nu_i)_{0 \leqslant i \leqslant m}$ of (P', μ) under $\phi_{P,v}$ is as follows: there is a unique j with $0 \leqslant j \leqslant m'$ such that

$$V'_j \supset V_i \supsetneq V_{i-1} \supset V'_{j-1},$$

and ν_i is the norm induced from μ_j on the subquotient $(V_i/V_{i-1})_v$ of $(V'_j/V'_{j-1})_v$, in the sense of 3.4.2. The $P(F_v)$-equivariance of $\phi_{P,v}$ is easily seen using the actions on norms of 2.1.5 and 2.1.7.

Though the following map is not used in this subsection, we introduce it here by way of comparison between $\bar{X}_{F,S}$ and $\bar{X}^{\flat}_{F,S}$.

3.4.4 Let W be a nonzero F-subspace of V, and set

$$\bar{X}^{\flat}_{F,S}(W) = \{(W', \mu) \in \bar{X}^{\flat}_{F,S} \mid W' \supset W\}.$$

For a place v of F, we have a map

$$\phi^{\flat}_{W,v}\colon \bar{X}^{\flat}_{F,v}(W) \to X_{W_v}$$

which sends $(W', \mu) \in \bar{X}^{\flat}_{F,v}(W)$ to the restriction of μ to W_v. The map $\phi^{\flat}_{W,v}$ is $P(F_v)$-equivariant, for P the parabolic subgroup of PGL_V consisting of all elements that preserve W. Setting $\mathfrak{Z}^{\flat}_{F,S}(W) = \prod_{v\in S} X_{W_v}$, the product of these maps over $v \in S$ provides a map $\phi^{\flat}_{W,S}\colon \bar{X}^{\flat}_{F,S}(W) \to \mathfrak{Z}^{\flat}_{F,S}(W)$.

3.4.5 For a finite-dimensional vector space H over a local field E, a basis $e = (e_i)_{1\leqslant i\leqslant d}$ of H, and a norm μ on H, we define the absolute value $|\mu : e| \in \mathbb{R}_{>0}$ of μ relative to e as follows. Suppose that μ is defined by a basis $e' = (e'_i)_{1\leqslant i\leqslant d}$ and a tuple $(r_i)_{1\leqslant i\leqslant d} \in \mathbb{R}^d_{>0}$. Let $h \in \mathrm{GL}_H(E)$ be the element such that $e' = he$. We then define

$$|\mu : e| = |\det(h)|^{-1} \prod_{i=1}^{d} r_i.$$

This is independent of the choice of e' and $(r_i)_i$. Note that we have

$$|g\mu : e| = |\det(g)|^{-1} |\mu : e|$$

for all $g \in \mathrm{GL}_H(E)$.

3.4.6 Let P and $(V_i)_i$ be as in 3.4.3, and let v be a place of F. Fix a basis $e^{(i)}$ of $(V_i/V_{i-1})_v$ for each $0 \leqslant i \leqslant m$. Then we have a map

$$\phi'_{P,v}\colon \bar{X}_{F,v}(P) \to \mathbb{R}^m_{\geqslant 0}, \qquad (P', \mu) \mapsto (t_i)_{1\leqslant i\leqslant m}$$

where $(t_i)_{1\leqslant i\leqslant m}$ is defined as follows. Let $(V'_j)_{-1\leqslant j\leqslant m'}$ be the flag associated to P'. Let $1 \leqslant i \leqslant m$. If V_{i-1} belongs to $(V'_j)_j$, let $t_i = 0$. If V_{i-1} does not belong to the last flag, then there is a unique j such that $V'_j \supset V_i \supset V_{i-2} \supset V'_{j-1}$. Let $\tilde{\mu}_j$ be a norm on $(V'_j/V'_{j-1})_v$ which belongs to the class μ_j, and let $\tilde{\mu}_{j,i}$ and $\tilde{\mu}_{j,i-1}$ be the norms induced by μ_j on the subquotients $(V_i/V_{i-1})_v$ and $(V_{i-1}/V_{i-2})_v$, respectively. We then let

$$t_i = |\tilde{\mu}_{j,i-1} : e^{(i-1)}|^{1/d_{i-1}} \cdot |\tilde{\mu}_{j,i} : e^{(i)}|^{-1/d_i},$$

where $d_i := \dim(V_i/V_{i-1})$.

The map $\phi'_{P,v}$ is $P(F_v)$-equivariant for the following action of $P(F_v)$ on $\mathbb{R}^m_{\geqslant 0}$. For $g \in P(F_v)$, let $\tilde{g} \in \mathrm{GL}_V(F_v)$ be a lift of g, and for $0 \leqslant i \leqslant m$, let $g_i \in \mathrm{GL}_{V_i/V_{i-1}}(F_v)$ be the element induced by $\tilde{g}$. Then $g \in P(F_v)$ sends $t \in \mathbb{R}^m_{\geqslant 0}$ to $t' \in \mathbb{R}^m_{\geqslant 0}$ where

$$t'_i = |\det(g_i)|^{1/d_i} \cdot |\det(g_{i-1})|^{-1/d_{i-1}} \cdot t_i.$$

If we have two families $e = (e^{(i)})_i$ and $f = (f^{(i)})_i$ of bases $e^{(i)}$ and $f^{(i)}$ of $(V_i/V_{i-1})_v$, and if the map $\phi'_{P,v}$ defined by e (resp., f) sends an element to t (resp., t'), then the same formula also describes the relationship between t and t', in this case taking g_i to be the element of $\mathrm{GL}_{V_i/V_{i-1}}$ such that $e^{(i)} = g_i f^{(i)}$.

3.4.7 Fix a basis $e^{(i)}$ of V_i/V_{i-1} for each $0 \leqslant i \leqslant m$. Then we have a map

$$\phi'_{P,S}\colon \bar{X}_{F,S}(P) \to \mathbb{R}^m_{\geqslant 0}, \qquad (P', \mu) \mapsto (t_i)_{1 \leqslant i \leqslant m}$$

where $t_i = \prod_{v \in S} t_{v,i}$, with $(t_{v,i})_i$ the image of (P', μ_v) under the map $\phi'_{P,v}$ of 3.4.6.

3.4.8 We define the Satake topology on $\bar{X}_{F,S}$ as follows.

For a parabolic subgroup P of PGL_V, consider the map

$$\psi_{P,S} := (\phi_{P,S}, \phi'_{P,S})\colon \bar{X}_{F,S}(P) \to \mathfrak{Z}_{F,S}(P) \times \mathbb{R}^m_{\geqslant 0}$$

from 3.4.3 and 3.4.7, which we recall depends on a choice of bases of the V_i/V_{i-1}. We say that a subset of $\bar{X}_{F,S}(P)$ is P-open if it is the inverse image of an open subset of $\mathfrak{Z}_{F,S}(P) \times \mathbb{R}^m_{\geqslant 0}$. By 3.4.6, the property of being P-open is independent of the choice of bases.

We define the Satake topology on $\bar{X}_{F,S}$ to be the weakest topology for which every P-open set for each parabolic subgroup P of PGL_V is open.

By this definition, we have:

3.4.9 Let $a \in \bar{X}_{F,S}$ be of the form (P, μ) for some μ. As U ranges over neighborhoods of the image $(\mu, 0)$ of a in $\mathfrak{Z}_{F,S}(P) \times \mathbb{R}^m_{\geqslant 0}$, the inverse images of the U in $\bar{X}_{F,S}(P)$ under $\psi_{P,S}$ form a base of neighborhoods of a in $\bar{X}_{F,S}$.

3.4.10 In §3.5 and §3.6, we explain that Satake topology on $\bar{X}_{F,S}$ is strictly coarser than the Borel–Serre topology for $d \geqslant 2$.

3.4.11 The Satake topology on $\bar{X}_{F,S}$ can differ from the subspace topology of the product topology for the Satake topology on each $\bar{X}_{F,v}$ with $v \in S$.

Example Let F be a real quadratic field, let $V = F^2$, and let $S = \{v_1, v_2\}$ be the set of real places of F. Consider the point $(\infty, \infty) \in (\mathfrak{H} \cup \{\infty\}) \times (\mathfrak{H} \cup \{\infty\}) \subset \bar{X}_{v_1} \times \bar{X}_{v_2}$ (see §3.2), which we regard as an element of $\bar{X}_{F,S}$. Then the sets

$$U_c := \{(x_1 + y_1 i, x_2 + y_2 i) \in \mathfrak{H} \times \mathfrak{H} \mid y_1 y_2 \geqslant c\} \cup \{(\infty, \infty)\}$$

with $c \in \mathbb{R}_{>0}$ form a base of neighborhoods of (∞, ∞) in $\bar{X}_{F,S}$ for the Satake topology, whereas the sets

$$U'_c := \{(x_1 + y_1 i, x_2 + y_2 i) \in \mathfrak{H} \times \mathfrak{H} \mid y_1 \geqslant c, y_2 \geqslant c\} \cup \{(\infty, \infty)\}$$

for $c \in \mathbb{R}_{>0}$ form a base of neighborhoods of (∞, ∞) in $\bar{X}_{F,S}$ for the topology induced by the product of Satake topologies on $\bar{X}_{F,v_1}$ and $\bar{X}_{F,v_2}$.

3.4.12 Let $G = \mathrm{PGL}_V$, and let Γ be a subgroup of $G(F)$.

- For a parabolic subgroup P of G, let $\Gamma_{(P)}$ be the subgroup of $\Gamma \cap P(F)$ consisting of all elements with image in the center of $(P/P_u)(F)$.
- For a nonzero F-subspace W of V, let $\Gamma_{(W)}$ denote the subgroup of elements of Γ that can be lifted to elements of $\mathrm{GL}_V(F)$ which fix every element of W.

3.4.13 We let $\mathbb{A}_F$ denote the adeles of F, let $\mathbb{A}_F^S$ denote the adeles of F outside of S, and let $\mathbb{A}_{F,S} = \prod_{v \in S} F_v$ so that $\mathbb{A}_F = \mathbb{A}_F^S \times \mathbb{A}_{F,S}$. Assume that S contains all archimedean places of F. Let $G = \mathrm{PGL}_V$, let K be a compact open subgroup of $G(\mathbb{A}_F^S)$, and let $\Gamma_K < G(F)$ be the inverse image of K under $G(F) \to G(\mathbb{A}_F^S)$.

The following proposition will be proved in 3.5.15.

Proposition 3.4.14 *For S, G, K and Γ_K as in 3.4.13, the Satake topology on $\bar{X}_{F,S}$ is the weakest topology such that for every parabolic subgroup P of G, a subset U of $\bar{X}_{F,S}(P)$ is open if*

(i) it is open for Borel–Serre topology, and
(ii) it is stable under the action of $\Gamma_{K,(P)}$ (see 3.4.12).

The following proposition follows easily from the fact that for any two compact open subgroups K and K' of $G(\mathbb{A}_F^S)$, the intersection $\Gamma_K \cap \Gamma_{K'}$ is of finite index in both Γ_K and $\Gamma_{K'}$.

Proposition 3.4.15 *For S, K and Γ_K as in 3.4.13, consider the weakest topology on $\bar{X}^\flat_{F,S}$ such that for every nonzero F-subspace W, a subset U of $\bar{X}^\flat_{F,S}(W)$ is open if*

(i) it is open for Borel–Serre topology, and
(ii) it is stable under the action of $\Gamma_{K,(W)}$ (see 3.4.12).

Then this topology is independent of the choice of K.

3.4.16 We call the topology in 3.4.15 the Satake topology on $\bar{X}^\flat_{F,S}$.

Proposition 3.4.17

(1) Let P be a parabolic subgroup of PGL_V. For both the Borel–Serre and Satake topologies on $\bar{X}_{F,S}$, the set $\bar{X}_{F,S}(P)$ is open in $\bar{X}_{F,S}$, and the action of the topological group $P(\mathbb{A}_{F,S})$ on $\bar{X}_{F,S}(P)$ is continuous.

(2) *The actions of the discrete group* $\mathrm{PGL}_V(F)$ *on the following spaces are continuous:* $\bar{X}_{F,S}$ *and* $\bar{X}^{\flat}_{F,S}$ *with their Borel–Serre topologies,* $\bar{X}_{F,S}$ *with the Satake topology, and assuming* S *contains all archimedean places,* $\bar{X}^{\flat}_{F,S}$ *with the Satake topology.*

Proposition 3.4.18 *Let* W *be a nonzero* F*-subspace of* V*. Then for the Borel–Serre topology, and for the Satake topology if* S *contains all archimedean places of* F*, the subset* $\bar{X}^{\flat}_{F,S}(W)$ *is open in* $\bar{X}^{\flat}_{F,S}$.

Part (1) of 3.4.17 was shown in §3.3 for the Borel–Serre topology, and the result for the Satake topology on $\bar{X}_{F,S}$ follows from it. The rest of 3.4.17 and 3.4.18 is easily proven.

3.5 Properties of $\bar{X}_{F,S}$

Let S be a nonempty finite set of places of F.

3.5.1 Let P and $(V_i)_{-1\leqslant i\leqslant m}$ be as before. Fix a basis $e^{(i)}$ of V_i/V_{i-1} for each i. Set

$$Y_0 = (\mathbb{R}^S_{>0} \cup \{(0)_{v\in S}\})^m \subset (\mathbb{R}^S_{\geqslant 0})^m.$$

The maps $\psi_{P,v} := (\phi_{P,v}, \phi'_{P,v})\colon \bar{X}_{F,v}(P) \to \mathfrak{Z}_{F,v}(P) \times \mathbb{R}^m_{\geqslant 0}$ of 3.4.3 and 3.4.6 for $v \in S$ combine to give the map

$$\psi_{P,S}\colon \bar{X}_{F,S}(P) \to \mathfrak{Z}_{F,S}(P) \times Y_0.$$

3.5.2 In addition to the usual topology on Y_0, we consider the weak topology on Y_0 that is the product topology for the topology on $\mathbb{R}^S_{>0} \cup \{(0)_{v\in S}\}$ which extends the usual topology on $\mathbb{R}^S_{>0}$ by taking the sets

$$\left\{(t_v)_{v\in S} \in \mathbb{R}^S_{>0} \mid \prod_{v\in S} t_v \leqslant c\right\} \cup \{(0)_{v\in S}\}$$

for $c \in \mathbb{R}_{>0}$ as a base of neighborhoods of $(0)_{v\in S}$. In the case that S consists of a single place, we have $Y_0 = \mathbb{R}^m_{\geqslant 0}$, and the natural topology and the weak topology on Y_0 coincide.

Proposition 3.5.3 *The map* $\psi_{P,S}$ *of 3.5.1 induces a homeomorphism*

$$P_u(\mathbb{A}_{F,S})\backslash \bar{X}_{F,S}(P) \xrightarrow{\sim} \mathfrak{Z}_{F,S}(P) \times Y_0$$

for the Borel–Serre topology (resp., Satake topology) on $\bar{X}_{F,S}$ *and the usual (resp., weak) topology on* Y_0*. This homeomorphism is equivariant for the action of* $P(\mathbb{A}_{F,S})$*, with the action of* $P(\mathbb{A}_{F,S})$ *on* Y_0 *being that of 3.4.6.*

This has the following corollary, which is also the main step in the proof.

Corollary 3.5.4 *For any place v of F, the map*

$$P_u(F_v)\backslash \bar{X}_{F,v}(P) \to \mathfrak{Z}_{F,v}(P) \times \mathbb{R}^m_{\geqslant 0}$$

is a homeomorphism for both the Borel–Serre and Satake topologies on $\bar{X}_{F,v}$.

We state and prove preliminary results towards the proof of 3.5.3.

3.5.5 Fix a basis $(e_i)_i$ of V and a parabolic subgroup P of PGL_V which satisfies the condition in 3.3.2 for this basis. Let $(V_i)_{-1\leqslant i\leqslant m}$ be the flag corresponding to P, and for each i, set $c(i) = \dim(V_i)$. We define two maps

$$\xi, \xi^{\star}\colon P_u(F_v) \times \mathfrak{Z}_{F,v}(P) \times \mathbb{R}^m_{\geqslant 0} \to \bar{X}_{F,v}(P).$$

3.5.6 First, we define the map ξ.

Set $\Delta(P) = \{c(0), \ldots, c(m-1)\}$. Let $\Delta_i = \{j \in \mathbb{Z} \mid c(i-1) < j < c(i)\}$ for $0 \leqslant i \leqslant m$. We then clearly have

$$\{1, \ldots, d-1\} = \Delta(P) \amalg \left(\coprod_{i=0}^{m} \Delta_i\right).$$

For $0 \leqslant i \leqslant m$, let $V^{(i)} = \sum_{c(i-1)<j\leqslant c(i)} Fe_j$, so $V_i = V_{i-1} \oplus V^{(i)}$. We have

$$\mathbb{R}^{d-1}_{\geqslant 0}(P) = \mathbb{R}^{\Delta(P)}_{\geqslant 0} \times \prod_{i=0}^{m} \mathbb{R}^{\Delta_i}_{>0} \cong \mathbb{R}^m_{\geqslant 0} \times \prod_{i=0}^{m} \mathbb{R}^{\Delta_i}_{>0}.$$

Let B be the Borel subgroup of PGL_V consisting of all upper triangular matrices for the basis $(e_i)_i$. Fix a place v of F. We consider two surjections

$$B_u(F_v) \times \mathbb{R}^{d-1}_{\geqslant 0}(P) \twoheadrightarrow \bar{X}_{F,v}(P),$$
$$B_u(F_v) \times \mathbb{R}^{d-1}_{\geqslant 0}(P) \twoheadrightarrow P_u(F_v) \times \mathfrak{Z}_{F,v}(P) \times \mathbb{R}^m_{\geqslant 0}.$$

The first is induced by the surjection $\bar{\pi}_v$ of 3.3.8.

The second map is obtained as follows. For $0 \leqslant i \leqslant m$, let B_i be the image of B in $\mathrm{PGL}_{V^{(i)}}$ under $P \to \mathrm{PGL}_{V_i/V_{i-1}} \cong \mathrm{PGL}_{V^{(i)}}$. Then B_i is a Borel subgroup of $\mathrm{PGL}_{V^{(i)}}$, and we have a canonical bijection

$$P_u(F_v) \times \prod_{i=0}^{m} B_{i,u}(F_v) \xrightarrow{\sim} B_u(F_v).$$

By 3.3.8, we have surjections $B_{i,u}(F_v) \times \mathbb{R}_{>0}^{\Delta_i} \twoheadrightarrow X_{(V_i/V_{i-1})_v}$ for $0 \leqslant i \leqslant m$. The second (continuous) surjection is then the composite

$$\begin{aligned} B_u(F_v) \times \mathbb{R}_{\geqslant 0}^{d-1}(P) &\xrightarrow{\sim} \left(P_u(F_v) \times \prod_{i=0}^{m} B_{i,u}(F_v)\right) \times \left(\mathbb{R}_{\geqslant 0}^{m} \times \prod_{i=0}^{m} \mathbb{R}_{>0}^{\Delta_i}\right) \\ &\twoheadrightarrow P_u(F_v) \times \left(\prod_{i=0}^{m} X_{(V_i/V_{i-1})_v}\right) \times \mathbb{R}_{\geqslant 0}^{m} \\ &= P_u(F_v) \times \mathfrak{Z}_{F,v}(P) \times \mathbb{R}_{\geqslant 0}^{m}. \end{aligned}$$

Proposition 3.5.7 *There is a unique surjective continuous map*

$$\xi\colon P_u(F_v) \times \mathfrak{Z}_{F,v}(P) \times \mathbb{R}_{\geqslant 0}^{m} \twoheadrightarrow \bar{X}_{F,v}(P)$$

for the Borel–Serre topology on $\bar{X}_{F,v}(P)$ that is compatible with the surjections from $B_u(F_v) \times \mathbb{R}_{\geqslant 0}^{d-1}(P)$ to these sets. This map induces a homeomorphism

$$\mathfrak{Z}_{F,v}(P) \times \mathbb{R}_{\geqslant 0}^{m} \xrightarrow{\sim} P_u(F_v)\backslash \bar{X}_{F,v}(P)$$

that restricts to a homeomorphism of $\mathfrak{Z}_{F,v}(P) \times \mathbb{R}_{>0}^{m}$ with $P_u(F_v)\backslash X_v$.

This follows from 3.3.8.

3.5.8 Next, we define the map $\xi^\star$.

For $g \in P_u(F_v)$, $(\mu_i)_i \in (X_{(V_i/V_{i-1})_v})_{0\leqslant i\leqslant m}$, and $(t_i)_{1\leqslant i\leqslant m} \in \mathbb{R}_{\geqslant 0}^{m}$, we let

$$\xi^\star(g, (\mu_i)_i, (t_i)_i) = g(P', \nu),$$

where P' and ν are as in (1) and (2) below, respectively.

(1) Let $J = \{c(i-1) \mid 1 \leqslant i \leqslant m,\ t_i = 0\}$. Write $J = \{c'(0), \dots, c'(m'-1)\}$ with $c'(0) < \cdots < c'(m'-1)$. Let $c'(-1) = 0$ and $c'(m') = d$. For $-1 \leqslant i \leqslant m'$, let

$$V_i' = \sum_{j=1}^{c'(i)} Fe_j \subset V.$$

Let $P' \supset P$ be the parabolic subgroup of PGL_V corresponding to the flag $(V_i')_i$.

(2) For $0 \leqslant i \leqslant m'$, set

$$J_i = \{j \mid c'(i-1) < c(j) \leqslant c'(i)\} \subset \{1, \dots, m\}.$$

We identify V_i'/V_{i-1}' with $\bigoplus_{j\in J_i} V^{(j)}$ via the basis $(e_k)_{c'(i-1)<k\leqslant c'(i)}$. We define a norm $\tilde{\nu}_i$ on V_i'/V_{i-1}' as follows. Let $\tilde{\mu}_j$ be the unique norm on $V^{(j)}$ which

belongs to μ_j and satisfies $|\tilde{\mu}_j : (e_k)_{c(j-1)<k\leqslant c(j)}| = 1$. For $x = \sum_{j\in J_i} x_j$ with $x_j \in V^{(j)}$, set

$$\tilde{\nu}_i(x) = \begin{cases} \sum_{j\in J_i}(r_j^2\tilde{\mu}_j(x_j)^2)^{1/2} & \text{if } v \text{ is real,} \\ \sum_{j\in J_i} r_j\tilde{\mu}_j(x_j) & \text{if } v \text{ is complex,} \\ \max_{j\in J_i}(r_j\tilde{\mu}_j(x_j)) & \text{if } v \text{ is non-archimedean,} \end{cases}$$

where for $j \in J_i$, we set

$$r_j = \prod_{\substack{\ell\in J_i \\ \ell<j}} t_\ell^{-1}.$$

Let $\nu_i \in X_{(V_i'/V_{i-1}')_v}$ be the class of the norm $\tilde{\nu}_i$.

We omit the proofs of the following two lemmas.

Lemma 3.5.9 *The composition*

$$P_u(F_v) \times \mathfrak{Z}_{F,v}(P) \times \mathbb{R}^m_{\geqslant 0} \xrightarrow{\xi^\star} \bar{X}_{F,v}(P) \xrightarrow{\psi_{P,v}} \mathfrak{Z}_{F,v}(P) \times \mathbb{R}^m_{\geqslant 0}$$

coincides with the canonical projection. Here, the definition of the second arrow uses the basis $(e_j \bmod V_{i-1})_{c(i-1)<j\leqslant c(i)}$ of V_i/V_{i-1}.

Lemma 3.5.10 *We have a commutative diagram*

$$\begin{array}{ccc} P_u(F_v) \times \mathfrak{Z}_{F,v}(P) \times \mathbb{R}^m_{\geqslant 0} & \xrightarrow{\xi} & \bar{X}_{F,v}(P) \\ \downarrow & & \| \\ P_u(F_v) \times \mathfrak{Z}_{F,v}(P) \times \mathbb{R}^m_{\geqslant 0} & \xrightarrow{\xi^\star} & \bar{X}_{F,v}(P) \end{array}$$

in which the left vertical arrow is $(u, \mu, t) \mapsto (u, \mu, t')$, for t' defined as follows. Let $I_i\colon X_{(V_i/V_{i-1})_v} \to \mathbb{R}^{\Delta_i}_{>0}$ be the unique continuous map for which the composition

$$B_{i,u}(F_v) \times \mathbb{R}^{\Delta_i}_{>0} \to X_{(V_i/V_{i-1})_v} \xrightarrow{I_i} \mathbb{R}^{\Delta_i}_{>0}$$

is projection onto the second factor, and for $j \in \Delta_i$, let $I_{i,j}\colon X_{(V_i/V_{i-1})_v} \to \mathbb{R}_{>0}$ denote the composition of I_i with projection onto the factor of $\mathbb{R}^{\Delta_i}_{>0}$ corresponding to j. Then

$$t_i' = t_i \cdot \prod_{j\in\Delta_{i-1}} I_{i-1,j}(\mu_i)^{\frac{j-c(i-2)}{c(i-1)-c(i-2)}} \cdot \prod_{j\in\Delta_i} I_{i,j}(\mu_i)^{\frac{c(i)-j}{c(i)-c(i-1)}}$$

for $1 \leqslant i \leqslant m$.

3.5.11 Proposition 3.5.3 is quickly reduced to Corollary 3.5.4, which now follows from 3.5.7, 3.5.9 and 3.5.10.

3.5.12 For two topologies T_1, T_2 on a set Z, we use $T_1 \geqslant T_2$ to denote that the identity map of Z is a continuous map from Z with T_1 to Z with T_2, and $T_1 > T_2$ to denote that $T_1 \geqslant T_2$ and $T_1 \neq T_2$. In other words, $T_1 \geqslant T_2$ if T_1 is finer than T_2 and $T_1 > T_2$ if T_1 is strictly finer than T_2.

By 3.5.3, the map $\psi_{P,S}\colon \bar{X}_{F,S}(P) \to \mathfrak{Z}_{F,S}(P) \times Y_0$ is continuous for the Borel–Serre topology on $\bar{X}_{F,S}$ and usual topology on Y_0. On $\bar{X}_{F,S}$, we therefore have

$$\text{Borel–Serre topology} \geqslant \text{Satake topology}.$$

Corollary 3.5.13 *For any nonempty finite set S of places of F, the map $\phi^\flat_{W,S}\colon \bar{X}^\flat_{F,S}(W) \to \mathfrak{Z}^\flat_{F,S}(W)$ of 3.4.4 is continuous for the Borel–Serre topology on $\bar{X}^\flat_{F,S}$. If S contains all archimedean places of F, it is continuous for the Satake topology.*

Proof The continuity for the Borel–Serre topology follows from the continuity of $\psi_{P,S}$, noting that the Borel–Serre topology on $\bar{X}^\flat_{F,S}$ is the quotient topology of the Borel–Serre topology on $\bar{X}_{F,S}$. Suppose that S contains all archimedean places. As $\phi^\flat_{W,S}$ is $\Gamma_{K,(W)}$-equivariant, and $\Gamma_{K,(W)}$ acts trivially on $\mathfrak{Z}^\flat_{F,S}(W)$, the continuity for the Satake topology is reduced to the continuity for the Borel–Serre topology. □

Remark 3.5.14 We remark that the map $\phi_{P,v}\colon \bar{X}_{F,v}(P) \to \mathfrak{Z}_{F,v}(P)$ of 3.4.3 need not be continuous for the topology on $\bar{X}_{F,v}$ as a subspace of $\bar{X}_v$. Similarly, the map $\phi^\flat_{W,v}\colon \bar{X}^\flat_{F,v}(W) \to X_{W_v}$ of 3.4.4 need not be continuous for the subspace topology on $\bar{X}^\flat_{F,v} \subset \bar{X}^\flat_v$. See 3.6.6 and 3.6.7.

3.5.15 We prove Proposition 3.4.14.

Proof Let $\alpha = (P, \mu) \in \bar{X}_{F,S}$. Let U be a neighborhood of α for the Borel–Serre topology which is stable under the action of $\Gamma_{K,(P)}$. By 3.5.12, it is sufficient to prove that there is a neighborhood W of α for the Satake topology such that $W \subset U$.

Let $(V_i)_{-1\leqslant i\leqslant m}$ be the flag corresponding to P, and let $V^{(i)}$ be as before. Let $\Gamma_1 = \Gamma_K \cap P_u(F)$, and let Γ_0 be the subgroup of Γ_K consisting of the elements that preserve $V^{(i)}$ and act on $V^{(i)}$ as a scalar for all i. Then Γ_1 is a normal subgroup of $\Gamma_{K,(P)}$ and $\Gamma_1\Gamma_0$ is a subgroup of $\Gamma_{K,(P)}$ of finite index.

Let

$$Y_1 = \left\{ (a_v)_{v\in S} \in \mathbb{R}^S_{>0} \mid \prod_{v\in S} a_v = 1 \right\}^m,$$

and set $s = \sharp S$. We have a surjective continuous map

$$\mathbb{R}^m_{\geqslant 0} \times Y_1 \twoheadrightarrow Y_0, \qquad (t, t') \mapsto (t_i^{1/s} t'_{v,i})_{v,i}.$$

The composition $\mathbb{R}^m_{\geqslant 0} \times Y_1 \to Y_0 \to \mathbb{R}^m_{\geqslant 0}$, where the second arrow is $(t_{v,i})_{v,i} \mapsto (\prod_{v\in S} t_{v,i})_i$, coincides with projection onto the first coordinate.

Let

$$\Phi = P_u(\mathbb{A}_{F,S}) \times Y_1 \text{ and } \Psi = \mathfrak{Z}_{F,S}(P) \times \mathbb{R}^m_{\geqslant 0}.$$

Consider the composite map

$$f\colon \Phi \times \Psi \to P_u(\mathbb{A}_{F,S}) \times \mathfrak{Z}_{F,S}(P) \times Y_0 \xrightarrow{(\xi_v^\star)_{v\in S}} \bar{X}_{F,S}(P).$$

The map f is $\Gamma_1\Gamma_0$-equivariant for the trivial action on Ψ and the following action on Φ: for $(g,t) \in \Phi$, $\gamma_1 \in \Gamma_1$ and $\gamma_0 \in \Gamma_0$, we have

$$\gamma_1\gamma_0 \cdot (g,t) = (\gamma_1\gamma_0 g \gamma_0^{-1}, \gamma_0 t),$$

where γ_0 acts on Y_1 via the embedding $\Gamma_K \to P(\mathbb{A}_{F,S})$ and the actions of the $P(F_v)$ described in 3.4.6. The composition

$$\Phi \times \Psi \xrightarrow{f} \bar{X}_{F,S}(P) \xrightarrow{\psi_{P,S}} \Psi$$

coincides with the canonical projection.

There exists a compact subset C of Φ such that $\Phi = \Gamma_1\Gamma_0 C$ for the above action of $\Gamma_1\Gamma_0$ on Φ. Let $\beta = (\mu, 0) \in \Psi$ be the image of α under $\psi_{P,S}$. For $x \in \Phi$, we have $f(x,\beta) = \alpha$. Hence, there is an open neighborhood $U'(x)$ of x in Φ and an open neighborhood $U''(x)$ of β in Ψ such that $U'(x) \times U''(x) \subset f^{-1}(U)$. Since C is compact, there is a finite subset R of C such that $C \subset \bigcup_{x\in R} U'(x)$. Let U'' be the open subset $\bigcap_{x\in R} U''(x)$ of Ψ, which contains β. The P-open set $W = \psi_{P,S}^{-1}(U'') \subset \bar{X}_{F,S}(P)$ is by definition an open neighborhood of α in the Satake topology on $\bar{X}_{F,S}$. We show that $W \subset f^{-1}(U)$. Since the map $\Phi \times \Psi \to \bar{X}_{F,S}(P)$ is surjective, it is sufficient to prove that the inverse image $\Phi \times U''$ of W in $\Phi \times \Psi$ is contained $f^{-1}(U)$. For this, we note that

$$\Phi \times U'' = \Gamma_1\Gamma_0 C \times U'' = \Gamma_1\Gamma_0 \left(\bigcup_{x\in R} U'(x) \times U'' \right) \subset \Gamma_1\Gamma_0 f^{-1}(U) = f^{-1}(U),$$

the last equality by the stability of U under the action of $\Gamma_{K,(P)} \supset \Gamma_1\Gamma_0$ and the $\Gamma_1\Gamma_0$-equivariance of f. □

3.5.16 In the case $d = 2$, the canonical surjection $\bar{X}_{F,S} \to \bar{X}^\flat_{F,S}$ is bijective. It is a homeomorphism for the Borel–Serre topology. If S contains all archimedean places of F, it is a homeomorphism for the Satake topology by 3.4.14.

3.6 Comparison of the Topologies

When considering $\bar{X}^{\flat}_{F,v}$, we assume that all places of F other than v are non-archimedean.

3.6.1 For $\bar{X}_{F,v}$ (resp., $\bar{X}^{\flat}_{F,v}$), we have introduced several topologies: the Borel–Serre topology, the Satake topology, and the subspace topology from $\bar{X}_v$ (resp., $\bar{X}^{\flat}_v$), which we call the weak topology. We compare these topologies below; note that we clearly have Borel–Serre topology $\geqslant$ Satake topology and Borel–Serre topology $\geqslant$ weak topology.

3.6.2 For both $\bar{X}_{F,v}$ and $\bar{X}^{\flat}_{F,v}$, the following hold:

(1) Borel–Serre topology $>$ Satake topology if $d \geqslant 2$,
(2) Satake topology $>$ weak topology if $d = 2$,
(3) Satake topology $\ngeqslant$ weak topology if $d > 2$.

We do not give full proofs of these statements. Instead, we describe some special cases that give clear pictures of the differences between these topologies. The general cases can be proven in a similar manner to these special cases.

Recall from 3.5.16 that in the case $d = 2$, the sets $\bar{X}_{F,v}$ and $\bar{X}^{\flat}_{F,v}$ are equal, their Borel–Serre topologies coincide, and their Satake topologies coincide.

3.6.3 We describe the case $d = 2$ of 3.6.2(1).

Take a basis $(e_i)_{i=1,2}$ of V. Consider the point $\alpha = (B, \mu)$ of $\bar{X}_{F,v}$, where B is the Borel subgroup of upper triangular matrices with respect to $(e_i)_i$, and μ is the unique element of $\mathfrak{Z}_{F,v}(B) = X_{F_v e_1} \times X_{V_v/F_v e_1}$.

Let $\bar{\pi}_v$ be the surjection of 3.3.8(2), and identify $B_u(F_v)$ with F_v in the canonical manner. The images of the sets

$$\{(x, t) \in F_v \times \mathbb{R}_{\geqslant 0} \mid t \leqslant c\} \subset B_u(F_v) \times \mathbb{R}_{\geqslant 0}$$

in $\bar{X}_{F,v}(B)$ for $c \in \mathbb{R}_{>0}$ form a base of neighborhoods of α for the Satake topology. Thus, while the image of the set

$$\{(x, t) \in F_v \times \mathbb{R}_{\geqslant 0} \mid t < |x|^{-1}\}$$

is a neighborhood of α for Borel–Serre topology, it is not a neighborhood of α for the Satake topology.

3.6.4 We prove 3.6.2(2) in the case that v is non-archimedean. The proof in the archimedean case is similar. Since all boundary points of $\bar{X}_{F,v} = \bar{X}^{\flat}_{F,v}$ are $\mathrm{PGL}_V(F)$-conjugate, to show 3.6.2(2), it is sufficient to consider any one boundary point. We consider α of 3.6.3 for a fixed basis $(e_i)_{i=1,2}$ of V.

For $x \in F_v$ and $y \in \mathbb{R}_{>0}$, let $\mu_{y,x}$ be the norm on V_v defined by

$$\mu_{y,x}(ae_1 + be_2) = \max(|a - xb|, y|b|).$$

The class of $\mu_{y,x}$ is the image of $(x, y^{-1}) \in B_u(F_v) \times \mathbb{R}_{>0}$. Any element of X_v is the class of the norm $\mu_{y,x}$ for some x, y. If we vary $x \in F_\infty$ and $y \in \mathbb{R}_{>0}$, the classes of $\mu_{y,x}$ in $\bar{X}_{F,v}$ converge under the Satake topology to the point α if and only if y approaches ∞. In $\bar{X}_v$, the point α is the class of the semi-norm ν on V_v^* defined by $\nu(ae_1^* + be_2^*) = |a|$. By 2.1.7,

$$\mu_{y,x}^* = \left(\mu_{y,0} \circ \begin{pmatrix} 1 & -x \\ 0 & 1 \end{pmatrix}\right)^* = \mu_{y,0}^* \circ \begin{pmatrix} 1 & 0 \\ x & 1 \end{pmatrix},$$

from which we see that

$$\mu_{y,x}^*(ae_1^* + be_2^*) = \max(|a|, y^{-1}|xa + b|).$$

Then $\mu_{y,x}^*$ is equivalent to the norm $\nu_{y,x}$ on V_v^* defined by

$$\nu_{y,x}(ae_1^* + be_2^*) = \min(1, y|x|^{-1}) \max(|a|, y^{-1}|xa + b|),$$

and the classes of the $\nu_{y,x}$ converge in $\bar{X}_v$ to the class of the semi-norm ν as $y \to \infty$. Therefore, the Satake topology is finer than the weak topology.

Now, the norm $\mu_{1,x}^*$ is equivalent to the norm $\nu_{1,x}$ on V_v^* defined above, which for sufficiently large x satisfies

$$\nu_{1,x}(ae_1 + be_2) = \max(|a/x|, |a + (b/x)|).$$

Thus, as $|x| \to \infty$, the sequence $\mu_{1,x}$ converges in $\bar{X}_v = \bar{X}_v^\flat$ to the class of the semi-norm ν. However, the sequence of classes of the norms $\mu_{1,x}$ does not converge to α in $\bar{X}_{F,v} = \bar{X}_{F,v}^\flat$ for the Satake topology, so the Satake topology is strictly finer than the weak topology.

3.6.5 We explain the case $d = 3$ of 3.6.2(3) in the non-archimedean case.

Take a basis $(e_i)_{1 \leqslant i \leqslant 3}$ of V. For $y \in \mathbb{R}_{>0}$, let μ_y be the norm on V_v defined by

$$\mu_y(ae_1 + be_2 + ce_3) = \max(|a|, y|b|, y^2|c|).$$

For $x \in F_v$, consider the norm $\mu_y \circ g_x$, where

$$g_x = \begin{pmatrix} 1 & 0 & 0 \\ 0 & 1 & x \\ 0 & 0 & 1 \end{pmatrix}.$$

If we vary $x \in F_\infty$ and let $y \in \mathbb{R}_{>0}$ approach ∞, then the class of $\mu_y \circ g_x$ in X_v converges under the Satake topology to the class $\alpha \in \bar{X}_{F,v}$ of the pair that is the Borel subgroup of upper triangular matrices and the unique element of $\prod_{i=0}^{2} X_{(V_i/V_{i-1})_v}$, where $(V_i)_{-1 \leqslant i \leqslant 2}$ is the corresponding flag. The quotient topology on $\bar{X}_{F,v}^\flat$ of the Satake topology on $\bar{X}_{F,v}$ is finer than the Satake topology on $\bar{X}_{F,v}^\flat$ by 3.4.14 and

3.4.15. Thus, if the Satake topology is finer than the weak topology on $\bar{X}_{F,v}$ or $\bar{X}^{\flat}_{F,v}$, then the composite $\mu_y \circ g_x$ should converge in $\bar{X}^{\flat}_v$ to the class of the semi-norm ν on V_v^* that satisfies $\nu(ae_1^* + be_2^* + ce_3^*) = |a|$. However, if $y \to \infty$ and $y^{-2}|x| \to \infty$, then the class of $\mu_y \circ g_x$ in X_v converges in $\bar{X}^{\flat}_v$ to the class of the semi-norm $ae_1^* + be_2^* + ce_3^* \mapsto |b|$. In fact, by 2.1.7 we have

$$\begin{aligned}(\mu_y \circ g_x)^*(ae_1^* + be_2^* + ce_3^*) &= \mu_y^* \circ (g_x^*)^{-1}(ae_1^* + be_2^* + ce_3^*)\\ &= \max(|a|,\ y^{-1}|b|,\ y^{-2}|-bx + c|) = y^{-2}|x|\nu_{y,x}\end{aligned}$$

where $\nu_{y,x}$ is the norm

$$ae_1^* + be_2^* + ce_3^* \mapsto \max(y^2|x|^{-1}|a|,\ y|x|^{-1}|b|,\ |-b + x^{-1}c|)$$

on V_v^*. The norms $\nu_{y,x}$ converge to the semi-norm $ae_1^* + be_2^* + ce_3^* \mapsto |b|$.

3.6.6 Let W be a nonzero subspace of V. We demonstrate that the map $\phi^{\flat}_{W,v}$: $\bar{X}^{\flat}_{F,v}(W) \to X_{W_v}$ of 3.4.4 given by restriction to W_v need not be continuous for the weak topology, even though by 3.5.13, it is continuous for the Borel–Serre topology and (if all places other than v are non-archimedean) for the Satake topology.

For example, suppose that v is non-archimedean and $d = 3$. Fix a basis $(e_i)_{1 \leqslant i \leqslant 3}$ of V, and let $W = Fe_1 + Fe_2$. Let μ be the class of the norm

$$ae_1 + be_2 \mapsto \max(|a|, |b|)$$

on W_v, and consider the element $(W, \mu) \in \bar{X}^{\flat}_{F,v}$. For $x \in F_v$ and $\epsilon \in \mathbb{R}_{>0}$, let $\mu_{x,\epsilon} \in X_v$ be the class of the norm

$$ae_1 + be_2 + ce_3 \mapsto \max(|a|, |b|, \epsilon^{-1}|c + bx|)$$

on V_v. Then $\mu^*_{x,\epsilon}$ is the class of the norm

$$ae_1^* + be_2^* + ce_3^* \mapsto \max(|a|, |b - xc|, \epsilon|c|)$$

on V_v^*. When $x \to 0$ and $\epsilon \to 0$, the last norm converges to the semi-norm

$$ae_1^* + be_2^* + ce_3^* \mapsto \max(|a|, |b|)$$

on V_v^*, and this implies that $\mu_{x,\epsilon}$ converges to (W, μ) for the weak topology. However, the restriction of $\mu_{x,\epsilon}$ to W_v is the class of the norm

$$ae_1 + be_2 \mapsto \max(|a|, |b|, \epsilon^{-1}|x||b|).$$

If $x \to 0$ and $\epsilon = r^{-1}|x| \to 0$ for a fixed $r > 1$, then the latter norms converge to the norm $ae_1 + be_2 \mapsto \max(|a|, r|b|)$, the class of which does not coincide with μ.

3.6.7 Let P be a parabolic subgroup of $\mathrm{PGL}_V(F)$. We demonstrate that the map $\phi_{P,v}\colon \bar{X}_{F,v}(P) \to \mathfrak{Z}_{F,v}(P)$ of 3.4.3 is not necessarily continuous for the weak topology, though by 3.5.4, it is continuous for the Borel–Serre topology and for the Satake topology.

Let $d = 3$ and W be as in 3.6.6, and let P be the parabolic subgroup of PGL_V corresponding to the flag

$$0 = V_{-1} \subset V_0 = W \subset V_1 = V.$$

In this case, the canonical map $\bar{X}_{F,v}(P) \to \bar{X}^{\flat}_{F,v}(W)$ is a homeomorphism for the weak topology on both spaces. It is also a homeomorphism for the Borel–Serre topology, and for the Satake topology if all places other than v are non-archimedean. Since $\mathfrak{Z}_{F,v}(P) = X_{(V_0)_v} \times X_{(V/V_0)_v} \cong X_{W_v}$, the argument of 3.6.6 shows that $\phi_{P,v}$ is not continuous for the weak topology.

3.6.8 For $d \geqslant 3$, the Satake topology on $\bar{X}^{\flat}_{F,v}$ does not coincide with the quotient topology for the Satake topology on $\bar{X}_{F,v}$, which is strictly finer. This is explained in 4.4.12.

3.7 Relations with Borel–Serre Spaces and Reductive Borel–Serre Spaces

3.7.1 In this subsection, we describe the relationship between our work and the theory of Borel–Serre and reductive Borel–Serre spaces (see Proposition 3.7.4). We also show that $\bar{X}^{\sharp}_{F,v}$ is not Hausdorff if v is a non-archimedean place.

3.7.2 Let G be a semisimple algebraic group over $\mathbb{Q}$. We recall the definitions of the Borel–Serre and reductive Borel–Serre spaces associated to G from [3] and [26, p. 190], respectively.

Let $\mathcal{Y}$ be the space of all maximal compact subgroups of $G(\mathbb{R})$. Recall from [3, Proposition 1.6] that for $K \in \mathcal{Y}$, the Cartan involution θ_K of $G_{\mathbb{R}} := \mathbb{R} \otimes_{\mathbb{Q}} G$ corresponding to K is the unique homomorphism $G_{\mathbb{R}} \to G_{\mathbb{R}}$ such that

$$K = \{g \in G(\mathbb{R}) \mid \theta_K(g) = g\}.$$

Let P be a parabolic subgroup of G, let S_P be the largest $\mathbb{Q}$-split torus in the center of P/P_u, and let A_P be the connected component of the topological group $S_P(\mathbb{R})$ containing the origin. We have

$$A_P \cong \mathbb{R}^r_{>0} \subset S_P(\mathbb{R}) \cong (\mathbb{R}^{\times})^r$$

for some integer r. We define an action of A_P on $\mathcal{Y}$ as follows (see [3, Sect. 3]). For $K \in \mathcal{Y}$, we have a unique subtorus $S_{P,K}$ of $P_{\mathbb{R}} = \mathbb{R} \otimes_{\mathbb{Q}} P$ over $\mathbb{R}$ such that the projection $P \to P/P_u$ induces an isomorphism

$$S_{P,K} \xrightarrow{\sim} (S_P)_{\mathbb{R}} := \mathbb{R} \otimes_{\mathbb{Q}} S_P$$

and such that the Cartan involution $\theta_K : G_{\mathbb{R}} \to G_{\mathbb{R}}$ of K satisfies $\theta_K(t) = t^{-1}$ for all $t \in S_{P,K}(\mathbb{R})$. For $t \in A_P$, let $t_K \in S_{P,K}(\mathbb{R})$ be the inverse image of t. Then A_P acts on $\mathcal{Y}$ by

$$A_P \times \mathcal{Y} \to \mathcal{Y}, \qquad (t, K) \mapsto t_K K t_K^{-1}.$$

The Borel–Serre space is the set of pairs (P, Z) such that P is a parabolic subgroup of G and Z is an A_P-orbit in $\mathcal{Y}$. The reductive Borel–Serre space is the quotient of the Borel–Serre space by the equivalence relation under which two elements (P, Z) and (P', Z') are equivalent if $(P', Z') = g(P, Z)$ (that is, $P = P'$ and $Z' = gZ$) for some $g \in P_u(\mathbb{R})$.

3.7.3 Now assume that $F = \mathbb{Q}$ and $G = \mathrm{PGL}_V$. Let v be the archimedean place of $\mathbb{Q}$.

We have a bijection between X_v and the set $\mathcal{Y}$ of all maximal compact subgroups of $G(\mathbb{R})$, whereby an element of X_v corresponds to its isotropy group in $G(\mathbb{R})$, which is a maximal compact subgroup.

Suppose that $K \in \mathcal{Y}$ corresponds to $\mu \in X_v$, with μ the class of a norm that in turn corresponds to a positive definite symmetric bilinear form $(\ ,\ )$ on V_v. The Cartan involution $\theta_K : G_{\mathbb{R}} \to G_{\mathbb{R}}$ is induced by the unique homomorphism $\theta_K : \mathrm{GL}_{V_v} \to \mathrm{GL}_{V_v}$ satisfying

$$(gx, \theta_K(g)y) = (x, y) \quad \text{for all } g \in \mathrm{GL}_V(\mathbb{R}) \text{ and } x, y \in V_v.$$

For a parabolic subgroup P of G corresponding to a flag $(V_i)_{-1 \leqslant i \leqslant m}$, we have

$$S_P = \left(\prod_{i=0}^{m} \mathbb{G}_{\mathrm{m},\mathbb{Q}}\right) / \mathbb{G}_{\mathrm{m},\mathbb{Q}},$$

where the ith term in the product is the group of scalars in $\mathrm{GL}_{V_i/V_{i-1}}$, and where the last $\mathbb{G}_{\mathrm{m},\mathbb{Q}}$ is embedded diagonally in the product. The above description of θ_K shows that $S_{P,K}$ is the lifting of $(S_P)_{\mathbb{R}}$ to $P_{\mathbb{R}}$ obtained through the orthogonal direct sum decomposition

$$V_v \cong \bigoplus_{i=0}^{m} (V_i/V_{i-1})_v$$

with respect to $(\ ,\ )$.

Proposition 3.7.4 *If v is the archimedean place of $\mathbb{Q}$, then $\bar{X}^{\sharp}_{\mathbb{Q},v}$ (resp., $\bar{X}_{\mathbb{Q},v}$) is the Borel–Serre space (resp., reductive Borel–Serre space) associated to* PGL_V.

Proof Denote the Borel–Serre space by $(\bar{X}^{\sharp}_{\mathbb{Q},v})'$ in this proof. We define a canonical map

$$\bar{X}^{\sharp}_{\mathbb{Q},v} \to (\bar{X}^{\sharp}_{\mathbb{Q},v})', \qquad (P, \mu, s) \mapsto (P, Z),$$

where Z is the subset of $\mathcal{Y}$ corresponding to the following subset Z' of X_v. Let $(V_i)_{-1 \leqslant i \leqslant m}$ be the flag corresponding to P. Recall that s is an isomorphism

$$s \colon \bigoplus_{i=0}^{m} (V_i/V_{i-1})_v \xrightarrow{\sim} V_v.$$

Then Z' is the subset of X_v consisting of classes of the norms

$$\tilde{\mu}^{(s)} \colon x \mapsto \left(\sum_{i=0}^{m} \tilde{\mu}_i (s^{-1}(x)_i)^2 \right)^{1/2}$$

on V_v, where $s^{-1}(x)_i \in (V_i/V_{i-1})_v$ denotes the ith component of $s^{-1}(x)$ for $x \in V_v$, and $\tilde{\mu} = (\tilde{\mu}_i)_{0 \leqslant i \leqslant m}$ ranges over all families of norms $\tilde{\mu}_i$ on $(V_i/V_{i-1})_v$ with class equal to μ_i. It follows from the description of $S_{P,K}$ in 3.7.3 that Z is an A_P-orbit.

For a parabolic subgroup P of G, let

$$(\bar{X}^{\sharp}_{\mathbb{Q},v})'(P) = \{(Q, Z) \in (\bar{X}^{\sharp}_{\mathbb{Q},v})' \mid Q \supset P\}.$$

By [3, 7.1], the subset $(\bar{X}^{\sharp}_{\mathbb{Q},v})'(P)$ is open in $(X^{\sharp}_{\mathbb{Q},v})'$.

Take a basis of V, and let B denote the Borel subgroup of PGL_V of upper-triangular matrices for this basis. By 3.3.8(3), we have a homeomorphism

$$B_u(\mathbb{R}) \times \mathbb{R}^{d-1}_{\geqslant 0} \xrightarrow{\sim} \bar{X}^{\sharp}_{\mathbb{Q},v}(B).$$

It follows from [3, 5.4] that the composition

$$B_u(\mathbb{R}) \times \mathbb{R}^{d-1}_{\geqslant 0} \to (\bar{X}^{\sharp}_{\mathbb{Q},v})'(B)$$

induced by the above map

$$\bar{X}^{\sharp}_{\mathbb{Q},v}(B) \to (\bar{X}^{\sharp}_{\mathbb{Q},v})'(B), \qquad (P, \mu, s) \mapsto (P, Z)$$

is also a homeomorphism. This proves that the map $\bar{X}^{\sharp}_{\mathbb{Q},v} \to (\bar{X}^{\sharp}_{\mathbb{Q},v})'$ restricts to a homeomorphism $\bar{X}^{\sharp}_{\mathbb{Q},v}(B) \xrightarrow{\sim} (\bar{X}^{\sharp}_{\mathbb{Q},v})'(B)$. Therefore, $\bar{X}^{\sharp}_{\mathbb{Q},v} \to (\bar{X}^{\sharp}_{\mathbb{Q},v})'$ is a homeomorphism as well. It then follows directly from the definitions that the reductive Borel–Serre space is identified with $\bar{X}_{\mathbb{Q},v}$. □

3.7.5 Suppose that F is a number field, let S be the set of all archimedean places of F, and let G be the Weil restriction $\mathrm{Res}_{F/\mathbb{Q}}\,\mathrm{PGL}_V$, which is a semisimple algebraic group over $\mathbb{Q}$. Then $\mathcal{Y}$ is identified with $X_{F,S}$, and $\bar{X}_{F,S}$ is related to the reductive Borel–Serre space associated to G but does not always coincide with it. We explain this below.

Let $(\bar{X}^{\sharp}_{F,S})'$ and $\bar{X}'_{F,S}$ be the Borel–Serre space and the reductive Borel–Serre space associated to G, respectively. Let $\bar{X}^{\sharp}_{F,S}$ be the subspace of $\prod_{v\in S}\bar{X}^{\sharp}_{F,v}$ consisting of all elements $(x_v)_{v\in S}$ such that the parabolic subgroup of G associated to x_v is independent of v. Then by similar arguments to the case $F=\mathbb{Q}$, we see that $\mathcal{Y}$ is canonically homeomorphic to $X_{F,S}$ and this homeomorphism extends uniquely to surjective continuous maps

$$(\bar{X}^{\sharp}_{F,S})' \to \bar{X}^{\sharp}_{F,S}, \qquad \bar{X}'_{F,S} \to \bar{X}_{F,S}.$$

However, these maps are not bijective unless F is $\mathbb{Q}$ or imaginary quadratic. We illustrate the differences between the spaces in the case that F is a real quadratic field and $d=2$.

Fix a basis $(e_i)_{i=1,2}$ of V. Let $\tilde{P}$ be the Borel subgroup of upper triangular matrices in PGL_V for this basis, and let P be the Borel subgroup $\mathrm{Res}_{F/\mathbb{Q}}\tilde{P}$ of G. Then $P/P_u \cong \mathrm{Res}_{F/\mathbb{Q}}\mathbb{G}_{m,F}$ and $S_P=\mathbb{G}_{m,\mathbb{Q}}\subset P/P_u$. We have the natural identifications $\mathcal{Y}=X_{F,S}=\mathfrak{H}\times\mathfrak{H}$. For $a\in\mathbb{R}_{>0}$, the set

$$Z_a := \{(yi, ayi)\in\mathfrak{H}\times\mathfrak{H} \mid y\in\mathbb{R}_{>0}\}$$

is an A_P-orbit. If $a\neq b$, the images of (P,Z_a) and (P,Z_b) in $(\bar{X}_{F,S})'$ do not coincide. On the other hand, both the images of (P,Z_a) and (P,Z_b) in $\bar{X}^{\sharp}_{F,S}$ coincide with $(x_v)_{v\in S}$, where $x_v=(P,\mu_v,s_v)$ with μ_v the unique element of $X_{F_ve_1}\times X_{V_v/F_ve_1}$ and s_v the splitting given by e_2.

Proposition 3.7.6 *If v is non-archimedean, then $\bar{X}^{\sharp}_{F,v}$ is not Hausdorff.*

Proof Fix $a,b\in B_u(F_v)$ with $a\neq b$, for a Borel subgroup B of PGL_V. When $t\in\mathbb{R}^{d-1}_{>0}$ is sufficiently near to $0=(0,\dots,0)$, the images of (a,t) and (b,t) in X_v coincide by 3.3.8(4) applied to $B_u(F_v)\times\mathbb{R}^{d-1}_{>0}\to X_v$. We denote this element of X_v by $c(t)$. The images $f(a)$ of $(a,0)$ and $f(b)$ of $(b,0)$ in $\bar{X}^{\sharp}_{F,v}$ are different. However, $c(t)$ converges to both $f(a)$ and $f(b)$ as t tends to 0. Thus, $\bar{X}^{\sharp}_{F,v}$ is not Hausdorff. □

3.7.7 Let F be a number field, S its set of archimedean places, and $G=\mathrm{Res}_{F/\mathbb{Q}}\,\mathrm{PGL}_V$, as in 3.7.5. Then $\bar{X}_{F,S}$ may be identified with the maximal Satake space for G of [23]. Its Satake topology was considered by Satake (see also [2, III.3]), and its Borel–Serre topology was considered by Zucker [27] (see also [14, 2.5]). The space $\bar{X}^{\flat}_{F,S}$ is also a Satake space corresponding to the standard projective representation of G on V viewed as a $\mathbb{Q}$-vector space.

4 Quotients by S-arithmetic Groups

As in §3, fix a global field F and a finite-dimensional vector space V over F.

4.1 Results on S-arithmetic Quotients

4.1.1 Fix a nonempty finite set S_1 of places of F which contains all archimedean places of F, fix a finite set S_2 of places of F which is disjoint from S_1, and let $S = S_1 \cup S_2$.

4.1.2 In the following, we take $\bar{X}$ to be one of the following two spaces:

(i) $\bar{X} := \bar{X}_{F,S_1}$,
(ii) $\bar{X} := \bar{X}^{\flat}_{F,S_1}$.

We endow $\bar{X}$ with either the Borel–Serre or the Satake topology.

4.1.3 Let $G = \mathrm{PGL}_V$, and let K be a compact open subgroup of $G(\mathbb{A}_F^S)$, with $\mathbb{A}_F^S$ as in 3.4.13.

We consider the two situations in which $(\mathfrak{X}, \bar{\mathfrak{X}})$ is taken to be one of the following pairs of spaces (for either choice of $\bar{X}$):

(I) $\mathfrak{X} := X_S \times G(\mathbb{A}_F^S)/K \ \subset \ \bar{\mathfrak{X}} := \bar{X} \times X_{S_2} \times G(\mathbb{A}_F^S)/K$,
(II) $\mathfrak{X} := X_S \ \subset \ \bar{\mathfrak{X}} := \bar{X} \times X_{S_2}$.

We now come to the main result of this paper.

Theorem 4.1.4 *Let the situations and notation be as in 4.1.1–4.1.3.*

(1) Assume we are in situation (I). Let Γ be a subgroup of $G(F)$. Then the quotient space $\Gamma\backslash\bar{\mathfrak{X}}$ is Hausdorff. It is compact if $\Gamma = G(F)$.
(2) Assume we are in situation (II). Let $\Gamma_K \subset G(F)$ be the inverse image of K under the canonical map $G(F) \to G(\mathbb{A}_F^S)$, and let Γ be a subgroup of Γ_K. Then the quotient space $\Gamma\backslash\bar{\mathfrak{X}}$ is Hausdorff. It is compact if Γ is of finite index in Γ_K.

4.1.5 The case $\Gamma = \{1\}$ of Theorem 4.1.4 shows that $\bar{X}_{F,S}$ and $\bar{X}^{\flat}_{F,S}$ are Hausdorff.

4.1.6 Let O_S be the subring of F consisting of all elements which are integral outside S. Take an O_S-lattice L in V. Then $\mathrm{PGL}_L(O_S)$ coincides with Γ_K for the compact open subgroup $K = \prod_{v \notin S} \mathrm{PGL}_L(O_v)$ of $G(\mathbb{A}_F^S)$. Hence Theorem 1.6 of the introduction follows from Theorem 4.1.4.

4.1.7 In the case that F is a number field and S (resp., S_1) is the set of all archimedean places of F, Theorem 4.1.4 in situation (II) is a special case of results of Satake [23] (resp., of Ji, Murty, Saper, and Scherk [14, Proposition 4.2]).

4.1.8 If in Theorem 4.1.4 we take $\Gamma = G(F)$ in part (1), or Γ of finite index in Γ_K in part (2), then the Borel–Serre and Satake topologies on $\bar{X}$ induce the same topology on the quotient space $\Gamma\backslash\bar{\mathfrak{X}}$. This can be proved directly, but it also follows from the compact Hausdorff property.

4.1.9 We show that some modifications of Theorem 4.1.4 are not good.

Consider the case $F = \mathbb{Q}$, $S = \{p, \infty\}$ for a prime number p, and $V = \mathbb{Q}^2$, and consider the S-arithmetic group $\mathrm{PGL}_2(\mathbb{Z}[\frac{1}{p}])$. Note that $\mathrm{PGL}_2(\mathbb{Z}[\frac{1}{p}])\backslash(\bar{X}_{\mathbb{Q},\infty} \times X_p)$ is compact Hausdorff, as is well known (and follows from Theorem 4.1.4). We show that some similar spaces are not Hausdorff. That is, we prove the following statements:

(1) $\mathrm{PGL}_2(\mathbb{Z}[\frac{1}{p}])\backslash(\bar{X}_{\mathbb{Q},p} \times X_\infty)$ is not Hausdorff.
(2) $\mathrm{PGL}_2(\mathbb{Z}[\frac{1}{p}])\backslash(\bar{X}_{\mathbb{Q},\infty} \times \mathrm{PGL}_2(\mathbb{Q}_p))$ is not Hausdorff.
(3) $\mathrm{PGL}_2(\mathbb{Q})\backslash(\bar{X}_{\mathbb{Q},\infty} \times \mathrm{PGL}_2(\mathbb{A}_{\mathbb{Q}}^\infty))$ is not Hausdorff.

Statement (1) shows that it is important to assume in 4.1.4 that S_1, not only S, contains all archimedean places. Statement (3) shows that it is important to take the quotient $G(\mathbb{A}_F^S)/K$ in situation (I) of 4.1.3.

Our proofs of these statements rely on the facts that the quotient spaces $\mathbb{Z}[\frac{1}{p}]\backslash\mathbb{R}$, $\mathbb{Z}[\frac{1}{p}]\backslash\mathbb{Q}_p$, and $\mathbb{Q}\backslash\mathbb{A}_{\mathbb{Q}}^\infty$ are not Hausdorff.

Proof of statements (1)–(3). For an element x of a ring R, let

$$g_x = \begin{pmatrix} 1 & x \\ 0 & 1 \end{pmatrix} \in \mathrm{PGL}_2(R).$$

In (1), for $b \in \mathbb{R}$, let h_b be the point $i + b$ of the upper half plane $\mathfrak{H} = X_\infty$. In (2), for $b \in \mathbb{Q}_p$, let $h_b = g_b \in \mathrm{PGL}_2(\mathbb{Q}_p)$. In (3), for $b \in \mathbb{A}_{\mathbb{Q}}^\infty$, let $h_b = g_b \in \mathrm{PGL}_2(\mathbb{A}_{\mathbb{Q}}^\infty)$. In (1) (resp., (2) and (3)), let $\infty \in \bar{X}_{\mathbb{Q},p}$ (resp., $\bar{X}_{\mathbb{Q},\infty}$) be the boundary point corresponding to the the Borel subgroup of upper triangular matrices.

In (1) (resp., (2), resp., (3)), take an element b of $\mathbb{R}$ (resp., $\mathbb{Q}_p$, resp., $\mathbb{A}_{\mathbb{Q}}^\infty$) which does not belong to $\mathbb{Z}[\frac{1}{p}]$ (resp., $\mathbb{Z}[\frac{1}{p}]$, resp., $\mathbb{Q}$). Then the images of (∞, h_0) and (∞, h_b) in the quotient space are different, but they are not separated. Indeed, in (1) and (2) (resp., (3)), some sequence of elements x of $\mathbb{Z}[\frac{1}{p}]$ (resp., $\mathbb{Q}$) will converge to b, in which case $g_x(\infty, h_0)$ converges to (∞, h_b) since $g_x\infty = \infty$. □

4.2 Review of Reduction Theory

We review important results in the reduction theory of algebraic groups: 4.2.2, 4.2.4, and a variant 4.2.6 of 4.2.2. More details may be found in the work of Borel [1] and Godement [8] in the number field case and Harder [12, 13] in the function field case.

Fix a basis $(e_i)_{1\leqslant i\leqslant d}$ of V. Let B be the Borel subgroup of $G = \mathrm{PGL}_V$ consisting of all upper triangular matrices for this basis. Let S be a nonempty finite set of places of F containing all archimedean places.

4.2.1 For $b = (b_v) \in \mathbb{A}_F^\times$, set $|b| = \prod_v |b_v|$. Let A_v be as in 3.3.4. We let $a \in \prod_v A_v$ denote the image of a diagonal matrix $\mathrm{diag}(a_1, \ldots, a_d)$ in $\mathrm{GL}_d(\mathbb{A}_F)$. The ratios $a_i a_{i+1}^{-1}$ are independent of the choice. For $c \in \mathbb{R}_{>0}$, we let $B(c) = B_u(\mathbb{A}_F)A(c)$, where

$$A(c) = \left\{ a \in \prod_v A_v \cap \mathrm{PGL}_d(\mathbb{A}_F) \mid |a_i a_{i+1}^{-1}| \geqslant c \text{ for all } 1 \leqslant i \leqslant d-1 \right\}.$$

Let $K^0 = \prod_v K_v^0 < G(\mathbb{A}_F)$, where K_v^0 is identified via $(e_i)_i$ with the standard maximal compact subgroup of $\mathrm{PGL}_d(F_v)$ of 3.3.4. Note that $B_u(F_v)A_v K_v^0 = B(F_v)K_v^0 = G(F_v)$ for all v.

We recall the following known result in reduction theory: see [8, Theorem 7] and [12, Satz 2.1.1].

Lemma 4.2.2 *For sufficiently small $c \in \mathbb{R}_{>0}$, one has $G(\mathbb{A}_F) = G(F)B(c)K^0$.*

4.2.3 Let the notation be as in 4.2.1. For a subset I of $\{1, \ldots, d-1\}$, let P_I be the parabolic subgroup of G corresponding to the flag consisting of 0, the F-subspaces $\sum_{1\leqslant j\leqslant i} Fe_j$ for $i \in I$, and V. Hence $P_I \supset B$ for all I, with $P_\varnothing = G$ and $P_{\{1,\ldots,d-1\}} = B$.

For $c' \in \mathbb{R}_{>0}$, let $B_I(c, c') = B_u(\mathbb{A}_F)A_I(c, c')$, where

$$A_I(c, c') = \{a \in A(c) \mid |a_i a_{i+1}^{-1}| \geqslant c' \text{ for all } i \in I\}.$$

Note that $B_I(c, c') = B(c)$ if $c \geqslant c'$.

The following is also known [12, Satz 2.1.2] (see also [8, Lemma 3]):

Lemma 4.2.4 *Fix $c \in \mathbb{R}_{>0}$ and a subset I of $\{1, \ldots, d-1\}$. Then there exists $c' \in \mathbb{R}_{>0}$ such that*

$$\{\gamma \in G(F) \mid B_I(c, c')K^0 \cap \gamma^{-1}B(c)K^0 \neq \varnothing\} \subset P_I(F).$$

4.2.5 We will use the following variant of 4.2.3.

Let $A_S = \prod_{v\in S} A_v$. For $c \in \mathbb{R}_{>0}$, let

$$A(c)_S = A_S \cap A(c) \text{ and } B(c)_S = B_u(\mathbb{A}_{F,S})A(c)_S.$$

For $c_1, c_2 \in \mathbb{R}_{>0}$, set

$$\begin{aligned} A(c_1, c_2)_S = \{&a \in A_S \mid \text{for all } v \in S \text{ and } 1 \leqslant i \leqslant d-1, \\ &|a_{v,i}a_{v,i+1}^{-1}| \geqslant c_1 \text{ and } |a_{v,i}a_{v,i+1}^{-1}| \geqslant c_2 |a_{w,i}a_{w,i+1}^{-1}| \text{ for all } w \in S\}. \end{aligned}$$

Note that $A(c_1, c_2)_S$ is empty if $c_2 > 1$. For a compact subset C of $B_u(\mathbb{A}_{F,S})$, we then set

$$B(C; c_1, c_2)_S = C \cdot A(c_1, c_2)_S.$$

Let $D_S = \prod_{v \in S} D_v$, where $D_v = K_v^0 < G(F_v)$ if v is archimedean, and $D_v < G(F_v)$ is identified with $S_d \operatorname{Iw}(O_v) < \mathrm{PGL}_d(F_v)$ using the basis $(e_i)_i$ otherwise. Here, S_d is the symmetric group of degree d and $\operatorname{Iw}(O_v)$ is the Iwahori subgroup of $\mathrm{PGL}_d(O_v)$, as in 3.3.6.

Lemma 4.2.6 *Let K be a compact open subgroup of $G(\mathbb{A}_F^S)$, let Γ_K be the inverse image of K under $G(F) \to G(\mathbb{A}_F^S)$, and let Γ be a subgroup of Γ_K of finite index. Then there exist c_1, c_2, C as above and a finite subset R of $G(F)$ such that*

$$G(\mathbb{A}_{F,S}) = \Gamma R \cdot B(C; c_1, c_2)_S D_S.$$

Proof This can be deduced from 4.2.2 by standard arguments in the following manner. By the Iwasawa decomposition 3.3.4, we have $G(\mathbb{A}_F^S) = B(\mathbb{A}_F^S) K^{0,S}$ where $K^{0,S}$ is the non-S-component of K^0. Choose a set E of representatives in $B(\mathbb{A}_F^S)$ of the finite set

$$B(F) \backslash B(\mathbb{A}_F^S) / (B(\mathbb{A}_F^S) \cap K^{0,S}).$$

Let $D^0 = D_S \times K^{0,S}$, and note that since $A_S \cap D_S = 1$, we can (by the Bruhat decomposition 3.3.6) replace K^0 by D^0 in Lemma 4.2.2. Using the facts that E is finite, $|a| = 1$ for all $a \in F^\times$, and D^0 is compact, we then have that there exists $c \in \mathbb{R}_{>0}$ such that

$$G(\mathbb{A}_F) = G(F)(B(c)_S \times E) D^0.$$

For any finite subset R of $G(F)$ consisting of one element from each of those sets $G(F) \cap K^{0,S} e^{-1}$ with $e \in E$ that are nonempty, we obtain from this that

$$G(\mathbb{A}_{F,S}) = \Gamma_K R \cdot B(c)_S D_S.$$

As Γ_K is a finite union of right Γ-cosets, we may enlarge R and replace Γ_K by Γ. Finally, we can replace $B(c)_S$ by $C \cdot A(c)_S$ for some C by the compactness of the image of

$$B_u(\mathbb{A}_{F,S}) \to \Gamma \backslash G(\mathbb{A}_{F,S}) / D_S$$

and then by $B(C; c_1, c_2)_S$ for some $c_1, c_2 \in \mathbb{R}_{>0}$ by the compactness of the cokernel of

$$\Gamma \cap B(\mathbb{A}_{F,S}) \to (B/B_u)(\mathbb{A}_{F,S})_1,$$

where $(B/B_u)(\mathbb{A}_{F,S})_1$ denotes the kernel of the homomorphism

$$(B/B_u)(\mathbb{A}_{F,S}) \to \mathbb{R}_{>0}^{d-1}, \qquad a B_u(\mathbb{A}_{F,S}) \mapsto \left(\prod_{v \in S} \left| \frac{a_{v,i}}{a_{v,i+1}} \right| \right)_{1 \leqslant i \leqslant d-1}. \square$$

4.3 $\bar{X}_{F,S}$ and Reduction Theory

4.3.1 Let S be a nonempty finite set of places of F containing all archimedean places. We consider $\bar{X}_{F,S}$. From the results 4.2.6 and 4.2.4 of reduction theory, we will deduce results 4.3.4 and 4.3.10 on $\bar{X}_{F,S}$, respectively. We will also discuss other properties of $\bar{X}_{F,S}$ related to reduction theory. Let G, $(e_i)_i$, and B be as in §4.2.

For $c_1, c_2 \in \mathbb{R}_{>0}$ with $c_2 \geqslant 1$, we define a subset $\mathfrak{T}(c_1, c_2)$ of $(\mathbb{R}^S_{\geqslant 0})^{d-1}$ by

$$\mathfrak{T}(c_1, c_2) = \left\{ t \in \left(\mathbb{R}^S_{\geqslant 0}\right)^{d-1} \,\middle|\, t_{v,i} \leqslant c_1,\ t_{v,i} \leqslant c_2 t_{w,i} \text{ for all } v, w \in S \text{ and } 1 \leqslant i \leqslant d-1 \right\}.$$

Let $Y_0 = (\mathbb{R}^S_{>0} \cup \{(0)_{v\in S}\})^{d-1}$ as in 3.5.1 (for the parabolic B), and note that $\mathfrak{T}(c_1, c_2) \subset Y_0$. Define the subset $\mathfrak{S}(c_1, c_2)$ of $\bar{X}_{F,S}(B)$ as the image of $B_u(\mathbb{A}_{F,S}) \times \mathfrak{T}(c_1, c_2)$ under the map

$$\pi_S = (\pi_v)_{v\in S} \colon B_u(\mathbb{A}_{F,S}) \times Y_0 \to \bar{X}_{F,S}(B),$$

with π_v as in 3.3.3. For a compact subset C of $B_u(\mathbb{A}_{F,S})$, we let $\mathfrak{S}(C; c_1, c_2) \subset \mathfrak{S}(c_1, c_2)$ denote the image of $C \times \mathfrak{T}(c_1, c_2)$ under π_S.

4.3.2 We give an example of the sets of 4.3.1.

Example Consider the case that $F = \mathbb{Q}$, the set S contains only the real place, and $d = 2$, as in §3.2. Fix a basis $(e_i)_{1\leqslant i\leqslant 2}$ of V. Identify $B_u(\mathbb{R})$ with $\mathbb{R}$ in the natural manner. We have

$$\mathfrak{S}(C; c_1, c_2) = \{x + yi \in \mathfrak{H} \mid x \in C, y \geqslant c_1^{-1}\} \cup \{\infty\},$$

which is contained in

$$\mathfrak{S}(c_1, c_2) = \{x + yi \in \mathfrak{H} \mid x \in \mathbb{R}, y \geqslant c_1^{-1}\} \cup \{\infty\}.$$

4.3.3 Fix a compact open subgroup K of $G(\mathbb{A}^S_F)$, and let $\Gamma_K \subset G(F)$ be the inverse image of K under $G(F) \to G(\mathbb{A}^S_F)$.

Proposition 4.3.4 *Let Γ be a subgroup of Γ_K of finite index. Then there exist c_1, c_2, C as in 4.3.1 and a finite subset R of $G(F)$ such that*

$$\bar{X}_{F,S} = \Gamma R \cdot \mathfrak{S}(C; c_1, c_2).$$

Proof It suffices to prove the weaker statement that there are c_1, c_2, C and R such that

$$X_S = \Gamma R \cdot (X_S \cap \mathfrak{S}(C; c_1, c_2)).$$

Indeed, we claim that the proposition follows from this weaker statement for the spaces in the product $\prod_{v\in S} X_{(V_i/V_{i-1})_v}$, where P_I is as in 4.2.3 for a subset I of $\{1,\dots,d-1\}$ and $(V_i)_{-1\leqslant i\leqslant m}$ is the corresponding flag. To see this, first note that there is a finite subset R' of $G(F)$ such that every parabolic subgroup of G has the form $\gamma P_I \gamma^{-1}$ for some I and $\gamma \in \Gamma R'$. It then suffices to consider $a = (P,\mu) \in \bar{X}_{F,S}$, where $P = P_I$ for some I, and $\mu \in \mathfrak{Z}_{F,S}(P)$. We use the notation of 3.5.6 and 3.5.1. By Proposition 3.5.7, the set $\bar{X}_{F,S}(P) \cap \mathfrak{S}(C; c_1, c_2)$ is the image under ξ of the image of $C \times \mathfrak{T}(c_1, c_2)$ in $P_u(\mathbb{A}_{F,S}) \times \mathfrak{Z}_{F,S}(P) \times Y_0$. Note that a has image $(1,\mu,0)$ in the latter set (for 1 the identity matrix of $P_u(\mathbb{A}_{F,S})$), and $\xi(1,\mu,0) = a$. Since the projection of $\mathfrak{T}(c_1,c_2)$ (resp., C) to $(\mathbb{R}^S_{>0})^{\Delta_i}$ (resp., $B_{i,u}(\mathbb{A}_{F,S})$) is the analogous set for c_1 and c_2 (resp., a compact subset), the claim follows.

For $v \in S$, we define subsets Q_v and Q'_v of X_v as follows. If v is archimedean, let $Q_v = Q'_v$ be the one point set consisting of the element of X_v given by the basis $(e_i)_i$ and $(r_i)_i$ with $r_i = 1$ for all i. If v is non-archimedean, let Q_v (resp., Q'_v) be the subset of X_v consisting of elements given by $(e_i)_i$ and $(r_i)_i$ such that $1 = r_1 \leqslant \cdots \leqslant r_d \leqslant q_v$ (resp., $r_1 = 1$ and $1 \leqslant r_i \leqslant q_v$ for $1 \leqslant i \leqslant d$). Then $X_v = G(F_v)Q_v$ for each $v \in S$. Hence by 4.2.6, there exist c'_1, c'_2, C as in 4.3.1 and a finite subset R of $G(F)$ such that

$$X_S = \Gamma R \cdot B(C; c'_1, c'_2)_S \cdot D_S Q_S,$$

where $Q_S = \prod_{v\in S} Q_v$.

We have $D_S Q_S = Q'_S$ for $Q'_S = \prod_{v\in S} Q'_v$, noting for archimedean (resp., non-archimedean) v that K^0_v (resp., $\mathrm{Iw}(O_v)$) stabilizes all elements of Q_v. We have $B(C; c'_1, c'_2)_S Q'_S \subset \mathfrak{S}(C; c_1, c_2)$, where

$$c_1 = \max\{q_v \mid v \in S_f\}(c'_1)^{-1} \text{ and } c_2 = \max\{q_v^2 \mid v \in S_f\}(c'_2)^{-1},$$

with S_f the set of all non-archimedean places in S (and taking the maxima to be 1 if $S_f = \varnothing$). □

4.3.5 For $v \in S$ and $1 \leqslant i \leqslant d-1$, let $t_{v,i} \colon \mathfrak{S}(c_1, c_2) \to \mathbb{R}_{\geqslant 0}$ be the map induced by $\phi'_{B,v} \colon \bar{X}_{F,v}(B) \to \mathbb{R}^{d-1}_{\geqslant 0}$ (see 3.4.6) and the ith projection $\mathbb{R}^{d-1}_{\geqslant 0} \to \mathbb{R}_{\geqslant 0}$. Note that $t_{v,i}$ is continuous.

4.3.6 Fix a subset I of $\{1,\dots,d-1\}$, and let P_I be the parabolic subgroup of G defined in 4.2.3. For $c_1, c_2, c_3 \in \mathbb{R}_{>0}$, let

$$\mathfrak{S}_I(c_1,c_2,c_3) = \{x \in \mathfrak{S}(c_1,c_2) \mid \min\{t_{v,i}(x) \mid v \in S\} \leqslant c_3 \text{ for each } i \in I\}.$$

4.3.7 For an element $a \in \bar{X}_{F,S}$, we define the parabolic type of a to be the subset

$$\{\dim(V_i) \mid 0 \leqslant i \leqslant m-1\}$$

of $\{1,\dots,d-1\}$, where $(V_i)_{-1\leqslant i\leqslant m}$ is the flag corresponding to the parabolic subgroup of G associated to a.

Lemma 4.3.8 *Let $a \in \bar{X}_{F,S}(B)$, and let J be the parabolic type of a. Then the parabolic subgroup of G associated to a is P_J.*

This is easily proved.

4.3.9 In the following, we will often consider subsets of $G(F)$ of the form $R_1 \Gamma_K R_2$, $\Gamma_K R$, or $R\Gamma_K$, where R_1, R_2, R are finite subsets of $G(F)$. These three types of cosets are essentially the same thing when we vary K. For finite subsets R_1, R_2 of $G(F)$, we have $R_1 \Gamma_K R_2 = R' \Gamma_{K'} = \Gamma_{K''} R''$ for some compact open subgroups K' and K'' of $G(\mathbb{A}_F^S)$ contained in K and finite subsets R' and R'' of $G(F)$.

Proposition 4.3.10 *Given $c_1 \in \mathbb{R}_{>0}$ and finite subsets R_1, R_2 of $G(F)$, there exists $c_3 \in \mathbb{R}_{>0}$ such that for all $c_2 \in \mathbb{R}_{>0}$ we have*

$$\{\gamma \in R_1 \Gamma_K R_2 \mid \gamma \mathfrak{S}_I(c_1, c_2, c_3) \cap \mathfrak{S}(c_1, c_2) \neq \varnothing\} \subset P_I(F).$$

Proof First we prove the weaker version that c_3 exists if the condition on $\gamma \in R_1 \Gamma_K R_2$ is replaced by $\gamma \mathfrak{S}_I(c_1, c_2, c_3) \cap \mathfrak{S}(c_1, c_2) \cap X_S \neq \varnothing$.

Let Q'_v for $v \in S$ and Q'_S be as in the proof of 4.3.4.

Claim 1 If $c'_1 \in \mathbb{R}_{>0}$ is sufficiently small (independent of c_2), then we have

$$X_S \cap \mathfrak{S}(c_1, c_2) \subset B(c'_1)_S Q'_S.$$

Proof of Claim 1 Any $x \in X_S \cap \mathfrak{S}(c_1, c_2)$ satisfies $t_{v,i}(x) \leqslant c_1$ for $1 \leqslant i \leqslant d-1$. Moreover, if $\prod_{v \in S} t_{v,i}(x)$ is sufficiently small relative to $(c'_1)^{-1}$ for all such i, then $x \in B(c'_1)_S Q'_S$. The claim follows.

Let C_v denote the compact set

$$C_v = \{g \in G(F_v) \mid g Q'_v \cap Q'_v \neq \varnothing\}.$$

If v is archimedean, then C_v is the maximal compact open subgroup K_v^0 of 4.2.1. Set $C_S = \prod_{v \in S} C_v$. We use the decomposition $G(\mathbb{A}_F) = G(\mathbb{A}_{F,S}) \times G(\mathbb{A}_F^S)$ to write elements of $G(\mathbb{A}_F)$ as pairs.

Claim 2 Fix $c'_1 \in \mathbb{R}_{>0}$. The subset $B(c'_1)_S C_S \times R_1 K R_2$ of $G(\mathbb{A}_F)$ is contained in $B(c) K^0$ for sufficiently small $c \in \mathbb{R}_{>0}$.

Proof of Claim 2 This follows from the compactness of the C_v for $v \in S$ and the Iwasawa decomposition $G(\mathbb{A}_F) = B(\mathbb{A}_F) K^0$.

Claim 3 Let c'_1 be as in Claim 1, and let $c \leqslant c'_1$. Let $c' \in \mathbb{R}_{>0}$. If $c_3 \in \mathbb{R}_{>0}$ is sufficiently small (independent of c_2), we have

$$X_S \cap \mathfrak{S}_I(c_1, c_2, c_3) \subset B_I(c, c')_S Q'_S,$$

where $B_I(c, c')_S = B(\mathbb{A}_{F,S}) \cap B_I(c, c')$.

Proof of Claim 3 An element $x \in B(c)_S Q'_S$ lies in $B_I(c, c')_S Q'_S$ if $\prod_{v \in S} t_{v,i}(x) \leqslant (c')^{-1}$ for all $i \in I$. An element $x \in X_S \cap \mathfrak{S}(c_1, c_2)$ lies in $X_S \cap \mathfrak{S}_I(c_1, c_2, c_3)$ if $\min\{t_{v,i}(x) \mid v \in S\} \leqslant c_3$ for all $i \in I$. In this case, x will lie in $B_I(c, c')_S Q'_S$ if $c_3 \leqslant (c')^{-1} c_1^{1-s}$, with $s = \sharp S$.

Let c'_1 be as in Claim 1, take c of Claim 2 for this c'_1 such that $c \leqslant c'_1$, and let $c' \in \mathbb{R}_{>0}$. Take c_3 satisfying the condition of Claim 3 for these c'_1, c, and c'.

Claim 4 If $X_S \cap \mathfrak{S}_I(c_1, c_2, c_3) \cap \gamma^{-1}\mathfrak{S}(c_1, c_2)$ is nonempty for some $\gamma \in R_1 \Gamma R_2 \subset G(F)$, then $B_I(c, c') \cap \gamma^{-1} B(c) K^0$ contains an element of $G(\mathbb{A}_{F,S}) \times \{1\}$.

Proof of Claim 4 By Claim 3, any $x \in X_S \cap \mathfrak{S}_I(c_1, c_2, c_3) \cap \gamma^{-1}\mathfrak{S}(c_1, c_2)$ lies in $g Q'_S$ for some $g \in B_I(c, c')_S$. By Claim 1, we have $\gamma x \in g' Q'_S$ for some $g' \in B(c'_1)_S$. Since $\gamma x \in \gamma g Q'_S \cap g' Q'_S$, we have $(g')^{-1} \gamma g \in C_S$. Hence $\gamma g \in B(c'_1)_S C_S$, and therefore $\gamma(g, 1) = (\gamma g, \gamma) \in B(c) K^0$ by Claim 2.

We prove the weaker version of 4.3.10: let $x \in X_S \cap \mathfrak{S}_I(c_1, c_2, c_3) \cap \gamma^{-1} \mathfrak{S}(c_1, c_2)$ for some $\gamma \in R_1 \Gamma R_2$. Then by Claim 4 and Lemma 4.2.4, with c' satisfying the condition of 4.2.4 for the given c, we have $\gamma \in P_I(F)$.

We next reduce the proposition to the weaker version, beginning with the following.

Claim 5 Let $c_1, c_2 \in \mathbb{R}_{>0}$. If $\gamma \in G(F)$ and $x \in \mathfrak{S}(c_1, c_2) \cap \gamma^{-1}\mathfrak{S}(c_1, c_2)$, then $\gamma \in P_J(F)$, where J is the parabolic type of x.

Proof of Claim 5 By Lemma 4.3.8, the parabolic subgroup associated to x is P_J and that associated to γx is P_J. Hence $\gamma P_J \gamma^{-1} = P_J$. Since a parabolic subgroup coincides with its normalizer, we have $\gamma \in P_J(F)$.

Fix $J \subset \{1, \ldots, d-1\}$, $\xi \in R_1$, and $\eta \in R_2$.

Claim 6 There exists $c_3 \in \mathbb{R}_{>0}$ such that if $\gamma \in \Gamma_K$ and $x \in \mathfrak{S}_I(c_1, c_2, c_3) \cap (\xi \gamma \eta)^{-1}\mathfrak{S}(c_1, c_2)$ is of parabolic type J, then $\xi \gamma \eta \in P_I(F)$.

Proof of Claim 6 Let $(V_i)_{-1 \leqslant i \leqslant m}$ be the flag corresponding to P_J. Suppose that we have $\gamma_0 \in \Gamma_K$ and $x_0 \in \mathfrak{S}(c_1, c_2) \cap (\xi \gamma_0 \eta)^{-1}\mathfrak{S}(c_1, c_2)$ of parabolic type J. By Claim 5, we have $\xi \gamma_0 \eta, \xi \gamma \eta \in P_J(F)$. Hence

$$\xi \gamma \eta = \xi (\gamma \gamma_0^{-1}) \xi^{-1} \xi \gamma_0 \eta \in \Gamma_{K'} \eta',$$

where K' is the compact open subgroup $\xi K \xi^{-1} \cap P_J(\mathbb{A}_F^S)$ of $P_J(\mathbb{A}_F^S)$, and $\eta' = \xi \gamma_0 \eta \in P_J(F)$. The claim follows from the weaker version of the proposition in which V is replaced by V_i/V_{i-1} (for $0 \leqslant i \leqslant m$), the group G is replaced by $\mathrm{PGL}_{V_i/V_{i-1}}$, the compact open subgroup K is replaced by the image of $\xi K \xi^{-1} \cap P_J(\mathbb{A}_F^S)$ in $\mathrm{PGL}_{V_i/V_{i-1}}(\mathbb{A}_F^S)$, the set R_1 is replaced by $\{1\}$, the set R_2 is replaced by the image of $\{\eta'\}$ in $\mathrm{PGL}_{V_i/V_{i-1}}(F)$, and $P_I(F)$ is replaced by the image of $P_I(F) \cap P_J(F)$ in $\mathrm{PGL}_{V_i/V_{i-1}}(F)$.

By Claim 6 for all J, ξ, and η, the result is proven. □

Lemma 4.3.11 *Let $1 \leqslant i \leqslant d-1$, and let $V' = \sum_{j=1}^{i} Fe_j$. Let $x \in X_v$ for some $v \in S$, and let $g \in \mathrm{GL}_V(F_v)$ be such that $gV' = V'$. For $1 \leqslant i \leqslant d-1$, we have*

$$\prod_{j=1}^{d-1} \left(\frac{t_{v,j}(gx)}{t_{v,j}(x)}\right)^{e(i,j)} = \frac{|\det(g_v\colon V'_v \to V'_v)|^{(d-i)/i}}{|\det(g_v\colon V_v/V'_v \to V_v/V'_v)|},$$

where

$$e(i,j) = \begin{cases} \frac{j(d-i)}{i} & \text{if } j \leqslant i, \\ d-j & \text{if } j \geqslant i. \end{cases}$$

Proof By the Iwasawa decomposition 3.3.4 and 3.4.5, it suffices to check this in the case that g is represented by a diagonal matrix $\mathrm{diag}(a_1, \dots, a_d)$. It follows from the definitions that $t_{v,j}(gx)t_{v,j}(x)^{-1} = |a_j a_{j+1}^{-1}|$, and the rest of the verification is a simple computation. □

Lemma 4.3.12 *Let i and V' be as in 4.3.11. Let R_1 and R_2 be finite subsets of $G(F)$. Then there exist $A, B \in \mathbb{R}_{>0}$ such that for all $\gamma \in \mathrm{GL}_V(F)$ with image in $R_1 \Gamma_K R_2 \subset G(F)$ and for which $\gamma V' = V'$, we have*

$$A \leqslant \prod_{v \in S} \frac{|\det(\gamma\colon V'_v \to V'_v)|^{(d-i)/i}}{|\det(\gamma\colon V_v/V'_v \to V_v/V'_v)|} \leqslant B.$$

Proof We may assume that R_1 and R_2 are one point sets $\{\xi\}$ and $\{\eta\}$, respectively. Suppose that an element γ_0 with the stated properties of γ exists. Then for any such γ, the image of $\gamma\gamma_0^{-1}$ in $G(F)$ belongs to $\xi\Gamma_K\xi^{-1} \cap P_{\{i\}}(F)$, and hence the image of $\gamma\gamma_0^{-1}$ in $G(\mathbb{A}_F^S)$ belongs to the compact subgroup $\xi K \xi^{-1} \cap P_{\{i\}}(\mathbb{A}_F^S)$ of $P_{\{i\}}(\mathbb{A}_F^S)$. Hence

$$|\det(\gamma\gamma_0^{-1} : V'_v \to V'_v)| = |\det(\gamma\gamma_0^{-1} : V_v/V'_v \to V_v/V'_v)| = 1$$

for every place v of F which does not belong to S. By Lemma 4.3.11 and the product formula, we have

$$\prod_{v \in S} (|\det(\gamma\gamma_0^{-1}\colon V'_v \to V'_v)|^{(d-i)/i} \cdot |\det(\gamma\gamma_0^{-1}\colon V_v/V'_v \to V_v/V'_v)|^{-1}) = 1,$$

so the value of the product in the statement is constant under our assumptions, proving the result. □

Proposition 4.3.13 *Fix $c_1, c_2 \in \mathbb{R}_{>0}$ and finite subsets R_1, R_2 of $G(F)$. Then there exists $A > 1$ such that if $x \in \mathfrak{S}(c_1, c_2) \cap \gamma^{-1}\mathfrak{S}(c_1, c_2)$ for some $\gamma \in R_1 \Gamma_K R_2$, then*

$$A^{-1} t_{v,i}(x) \leqslant t_{v,i}(\gamma x) \leqslant A t_{v,i}(x)$$

for all $v \in S$ and $1 \leqslant i \leqslant d-1$.

Proof By a limit argument, it is enough to consider $x \in X_S \cap \mathfrak{S}(c_1, c_2)$. Fix $v \in S$. For $x, x' \in X_S \cap \mathfrak{S}(c_1, c_2)$ and $1 \leqslant i \leqslant d-1$, let $s_i(x, x') = t_{v,i}(x')t_{v,i}(x)^{-1}$.

For each $1 \leqslant i \leqslant d-1$, take $c_3(i) \in \mathbb{R}_{>0}$ satisfying the condition in 4.3.10 for the set $I = \{i\}$ and both pairs of finite subsets R_1, R_2 and R_2^{-1}, R_1^{-1} of $G(F)$. Let

$$c_3 = \min\{c_3(i) \mid 1 \leqslant i \leqslant d-1\}.$$

For a subset I of $\{1, \dots, d-1\}$, let $Y(I)$ be the subset of $(X_S \cap \mathfrak{S}(c_1, c_2))^2$ consisting of all pairs (x, x') such that $x' = \gamma x$ for some $\gamma \in R_1 \Gamma_K R_2$ and such that

$$I = \{1 \leqslant i \leqslant d-1 \mid \min(t_{v,i}(x), t_{v,i}(x')) \leqslant c_3\}.$$

For the proof of 4.3.13, it is sufficient to prove the following statement (S_d), fixing I.

(S_d) There exists $A > 1$ such that $A^{-1} \leqslant s_i(x, x') \leqslant A$ for all $(x, x') \in Y(I)$ and $1 \leqslant i \leqslant d-1$.

By Proposition 4.3.10, if $\gamma \in R_1 \Gamma_K R_2$ is such that there exists $x \in X_S$ with $(x, \gamma x) \in Y(I)$, then $\gamma \in P_{\{i\}}(F)$ for all $i \in I$. Lemmas 4.3.11 and 4.3.12 then imply the following for all $i \in I$, noting that $c_2^{-1} t_{w,i}(y) \leqslant t_{v,i}(y) \leqslant c_2 t_{w,i}(y)$ for all $w \in S$ and $y \in X_S \cap \mathfrak{S}(c_1, c_2)$.

(T_i) There exists $B_i > 1$ such that for all $(x, x') \in Y(I)$, we have

$$B_i^{-1} \leqslant \prod_{j=1}^{d-1} s_j(x, x')^{e(i,j)} \leqslant B_i,$$

where $e(i, j)$ is as in 4.3.11.

We prove the following statement (S_i) for $0 \leqslant i \leqslant d-1$ by induction on i.

(S_i) There exists $A_i > 1$ such that $A_i^{-1} \leqslant s_j(x, x') \leqslant A_i$ for all $(x, x') \in Y(I)$ and all j such that $1 \leqslant j \leqslant i$ and j is not the largest element of $I \cap \{1, \dots, i\}$ (if it is nonempty).

That (S_0) holds is clear. Assume that (S_{i-1}) holds for some $i \geqslant 1$. If $i \notin I$, then since $c_3 \leqslant t_{v,i}(x) \leqslant c_1$ and $c_3 \leqslant t_{v,i}(x') \leqslant c_1$, we have

$$\frac{c_3}{c_1} \leqslant s_i(x, x') \leqslant \frac{c_1}{c_3},$$

and hence (S_i) holds with $A_i := \max(A_{i-1}, c_1 c_3^{-1})$.

Assume that $i \in I$. If $I \cap \{1, \dots, i-1\} = \varnothing$, then (S_i) is evidently true with $A_i := A_{i-1}$. If $I \cap \{1, \dots, i-1\} \neq \varnothing$, then let i' be the largest element of this intersection. We compare (T_i) and $(T_{i'})$. We have $e(i, j) = e(i', j)$ if $j \geqslant i$ and $e(i, j) < e(i', j)$ if $j < i$, so taking the quotient of the equations in $(T_{i'})$ and (T_i), we have

$$(B_i B_{i'})^{-1} \leqslant \prod_{j=1}^{i-1} s_j(x, x')^{e(i',j)-e(i,j)} \leqslant B_i B_{i'}.$$

Since (S_{i-1}) is assumed to hold, there then exists $a \in \mathbb{R}_{>0}$ such that

$$(B_i B_{i'})^{-1} A_{i-1}^{-a} \leqslant s_{i'}(x, x')^{e(i',i')-e(i,i')} \leqslant B_i B_{i'} A_{i-1}^{a}$$

As the exponent $e(i', i') - e(i, i')$ is nonzero, this implies that (S_i) holds.

By induction, we have (S_{d-1}). To deduce (S_d) from it, we may assume that I is nonempty, and let i be the largest element of I. Then (S_{d-1}) and (T_i) imply (S_d). □

Proposition 4.3.14 *Let $c_1, c_2 \in \mathbb{R}_{>0}$ and $a \in \bar{X}_{F,S}$. Let I be the parabolic type (4.3.7) of a. Fix a finite subset R of $G(F)$ and $1 \leqslant i \leqslant d-1$.*

(1) If $i \in I$, then for any $\epsilon > 0$, there exists a neighborhood U of a in $\bar{X}_{F,S}$ for the Satake topology such that $\max\{t_{v,i}(x) \mid v \in S\} < \epsilon$ *for all $x \in (\Gamma_K R)^{-1} U \cap \mathfrak{S}(c_1, c_2)$.*
(2) If $i \notin I$, then there exist a neighborhood U of a in $\bar{X}_{F,S}$ for the Satake topology and $c \in \mathbb{R}_{>0}$ such that $\min\{t_{v,i}(x) \mid v \in S\} \geqslant c$ *for all $x \in (\Gamma_K R)^{-1} U \cap \mathfrak{S}(c_1, c_2)$.*

Proof The first statement is clear by continuity of $t_{v,i}$ and the fact that $t_{v,i}(\gamma^{-1}a) = 0$ for all $\gamma \in G(F)$, and the second follows from 4.3.13, noting 4.3.4. □

Proposition 4.3.15 *Let $a \in \bar{X}_{F,S}$, and let P be the parabolic subgroup of* PGL_V *associated to a. Let $\Gamma_{K,(P)} \subset \Gamma_K$ be as in 3.4.12. Then there are $c_1, c_2 \in \mathbb{R}_{>0}$ and $\varphi \in G(F)$ such that $\Gamma_{K,(P)}\varphi\mathfrak{S}(c_1, c_2)$ is a neighborhood of a in $\bar{X}_{F,S}$ for the Satake topology.*

Proof This holds by definition of the Satake topology with $\varphi = 1$ if $a \in \bar{X}_{F,S}(B)$. In general, let I be the parabolic type of a. Then the parabolic subgroup associated to a has the form $\varphi P_I \varphi^{-1}$ for some $\varphi \in G(F)$. We have $\varphi^{-1}a \in \bar{X}_{F,S}(P_I) \subset \bar{X}_{F,S}(B)$. By that already proven case, there exists $\gamma \in \Gamma_{K,(P_I)}$ such that $\Gamma_{K,(P)}\varphi\gamma\mathfrak{S}(c_1, c_2)$ is a neighborhood of a for the Satake topoology. □

The following result can be proved in the manner of 4.4.8 for $\bar{X}^{\flat}_{F,S}$ below, replacing R by $\{\varphi\}$, and $\Gamma_{K,(W)}$ by $\Gamma_{K,(P)}$.

Lemma 4.3.16 *Let the notation be as in 4.3.15. Let U' be a neighborhood of $\varphi^{-1}a$ in $\bar{X}_{F,S}$ for the Satake topology. Then there is a neighborhood U of a in $\bar{X}_{F,S}$ for the Satake topology such that*

$$U \subset \Gamma_{K,(P)}\varphi(\mathfrak{S}(c_1, c_2) \cap U').$$

4.4 $\bar{X}^{\flat}_{F,S}$ and Reduction Theory

4.4.1 Let S be a finite set of places of F containing the archimedean places. In this subsection, we consider $\bar{X}^{\flat}_{F,S}$. Fix a basis $(e_i)_{1 \leqslant i \leqslant d}$ of V. Let $B \subset G = \mathrm{PGL}_V$ be the Borel subgroup of upper triangular matrices for $(e_i)_i$. Let K be a compact open subgroup of $G(\mathbb{A}_F^S)$.

4.4.2 Let $c_1, c_2 \in \mathbb{R}_{>0}$. We let $\mathfrak{S}^{\flat}(c_1, c_2)$ denote the image of $\mathfrak{S}(c_1, c_2)$ under $\bar{X}_{F,S} \to \bar{X}^{\flat}_{F,S}$. For $r \in \{1, \dots, d-1\}$, we then define

$$\mathfrak{S}^{\flat}_r(c_1, c_2) = \{(W, \mu) \in \mathfrak{S}^{\flat}(c_1, c_2) \mid \dim(W) \geqslant r\}.$$

Then the maps $t_{v,i}$ of 4.3.5 for $v \in S$ and $1 \leqslant i \leqslant r$ induce maps

$$t_{v,i} \colon \mathfrak{S}^{\flat}_r(c_1, c_2) \to \mathbb{R}_{>0} \ \ (1 \leqslant i \leqslant r-1) \text{ and } t_{v,r} \colon \mathfrak{S}^{\flat}_r(c_1, c_2) \to \mathbb{R}_{\geqslant 0}.$$

For $c_3 \in \mathbb{R}_{>0}$, we also set

$$\mathfrak{S}^{\flat}_r(c_1, c_2, c_3) = \{x \in \mathfrak{S}^{\flat}_r(c_1, c_2) \mid \min\{t_{v,r}(x) \mid v \in S\} \leqslant c_3\}.$$

Proposition 4.4.3 *Fix $c_1 \in \mathbb{R}_{>0}$ and finite subsets R_1, R_2 of $G(F)$. Then there exists $c_3 \in \mathbb{R}_{>0}$ such that for all $c_2 \in \mathbb{R}_{>0}$, we have*

$$\{\gamma \in R_1 \Gamma_K R_2 \mid \gamma \mathfrak{S}^{\flat}_r(c_1, c_2, c_3) \cap \mathfrak{S}^{\flat}_r(c_1, c_2) \neq \varnothing\} \subset P_{\{r\}}.$$

Proof Take $(W, \mu) \in \gamma \mathfrak{S}^{\flat}_r(c_1, c_2, c_3) \cap \mathfrak{S}^{\flat}_r(c_1, c_2)$, and let $r' = \dim W$. Let P be the parabolic subgroup of V corresponding to the flag $(V_i)_{-1 \leqslant i \leqslant d-r'}$ with $V_i = W + \sum_{j=r'+1}^{r'+i} Fe_i$ for $0 \leqslant i \leqslant d - r'$. Let $\mu' \in \mathfrak{Z}_{F,S}(P)$ be the unique element such that $a = (P, \mu') \in \bar{X}_{F,S}$ maps to (W, μ). Then $a \in \gamma \mathfrak{S}_{\{r\}}(c_1, c_2, c_3) \cap \mathfrak{S}(c_1, c_2)$, so we can apply 4.3.10. □

Proposition 4.4.4 *Fix $c_1, c_2 \in \mathbb{R}_{>0}$ and finite subsets R_1, R_2 of $G(F)$. Then there exists $A > 1$ such that if $x \in \mathfrak{S}^{\flat}_r(c_1, c_2) \cap \gamma^{-1} \mathfrak{S}^{\flat}_r(c_1, c_2)$ for some $\gamma \in R_1 \Gamma_K R_2$, then*

$$A^{-1} t_{v,i}(x) \leqslant t_{v,i}(\gamma x) \leqslant A t_{v,i}(x)$$

for all $v \in S$ and $1 \leqslant i \leqslant r$.

Proof This follows from 4.3.13. □

We also have the following easy consequence of Lemma 4.3.8.

Lemma 4.4.5 *Let a be in the image of $\bar{X}_{F,S}(B) \to \bar{X}^{\flat}_{F,S}$, and let r be the dimension of the F-subspace of V associated to a. Then the F-subspace of V associated to a is $\sum_{i=1}^{r} Fe_i$.*

Proposition 4.4.6 *Let $a \in \bar{X}^{\flat}_{F,S}$ and let r be the dimension of the F-subspace of V associated to a. Let $c_1, c_2 \in \mathbb{R}_{>0}$. Fix a finite subset R of $G(F)$.*

(1) For any $\epsilon > 0$, there exists a neighborhood U of a in $\bar{X}^{\flat}_{F,S}$ for the Satake topology such that $\max\{t_{v,r}(x) \mid v \in S\} < \epsilon$ for all $x \in (\Gamma_K R)^{-1}U \cap \mathfrak{S}^{\flat}_r(c_1, c_2)$.

(2) If $1 \leqslant i < r$, then there exist a neighborhood U of a in $\bar{X}^{\flat}_{F,S}$ for the Satake topology and $c \in \mathbb{R}_{>0}$ such that $\min\{t_{v,i}(x) \mid v \in S\} \geqslant c$ for all $x \in (\Gamma_K R)^{-1}U \cap \mathfrak{S}^{\flat}_r(c_1, c_2)$.

Proof This follows from 4.4.4, as in the proof of 4.3.14. □

Proposition 4.4.7 *Let W be an F-subspace of V of dimension $r \geqslant 1$. Let Φ be set of $\varphi \in G(F)$ such that $\varphi(\sum_{i=1}^{r} Fe_i) = W$.*

(1) There exists a finite subset R of Φ such that for any $a \in \bar{X}^{\flat}_{F,S}(W)$, there exist $c_1, c_2 \in \mathbb{R}_{>0}$ for which the set $\Gamma_{K,(W)} R\mathfrak{S}^{\flat}_r(c_1, c_2)$ is a neighborhood of a in the Satake topology.

(2) For any $\varphi \in \Phi$ and $a \in \bar{X}^{\flat}_{F,S}$ with associated subspace W, there exist $c_1, c_2 \in \mathbb{R}_{>0}$ such that $a \in \varphi\mathfrak{S}^{\flat}_r(c_1, c_2)$ and $\Gamma_{K,(W)}\varphi\mathfrak{S}^{\flat}_r(c_1, c_2)$ is a neighborhood of a in the Satake topology.

Proof We may suppose without loss of generality that $W = \sum_{j=1}^{r} Fe_j$, in which case $\Phi = G(F)_{(W)}$ (see 3.4.12). Consider the set $\mathcal{Q}$ of all parabolic subgroups Q of G such that W is contained in the smallest nonzero subspace of V preserved by Q. Any $Q \in \mathcal{Q}$ has the form $Q = \varphi P_I \varphi^{-1}$ for some $\varphi \in G(F)_{(W)}$ and subset I of $J := \{i \in \mathbb{Z} \mid r \leqslant i \leqslant d-1\}$. There exists a finite subset R of $G(F)_{(W)}$ such that we may always choose $\varphi \in \Gamma_{K,(W)} R$.

By 4.4.5, an element of $\bar{X}_{F,S}(B)$ has image in $\bar{X}^{\flat}_{F,S}(W)$ if and only if the parabolic subgroup associated to it has the form P_I for some $I \subset J$. The intersection of the image of $\bar{X}_{F,S}(B) \to \bar{X}^{\flat}_{F,S}$ with $\bar{X}^{\flat}_{F,S}(W)$ is the union of the $\mathfrak{S}^{\flat}_r(c_1, c_2)$ with $c_1, c_2 \in \mathbb{R}_{>0}$. By the above, for any $a \in \bar{X}^{\flat}_{F,S}(W)$, we may choose $\xi \in \Gamma_{K,(W)} R$ such that $\xi^{-1}a$ is in this intersection, and part (1) follows. Moreover, if W is the subspace associated to a, then $\varphi^{-1}a \in \bar{X}^{\flat}_{F,S}(W)$ is in the image of $\bar{X}_{F,S}(B)$ for all $\varphi \in G(F)_{(W)}$, from which (2) follows. □

Lemma 4.4.8 *Let W, Φ, R be as in 4.4.7, fix $a \in \bar{X}^{\flat}_{F,S}(W)$, and let $c_1, c_2 \in \mathbb{R}_{>0}$ be as in 4.4.7(1) for this a. For each $\varphi \in R$, let U_φ be a neighborhood of $\varphi^{-1}a$ in $\bar{X}^{\flat}_{F,S}$ for the Satake topology. Then there is a neighborhood U of a in $\bar{X}^{\flat}_{F,S}$ for the Satake topology such that*

$$U \subset \bigcup_{\varphi \in R} \Gamma_{K,(W)}\varphi(\mathfrak{S}^{\flat}_r(c_1, c_2) \cap U_\varphi).$$

Proof We may assume that each $\varphi(U_\varphi)$ is stable under the action of $\Gamma_{K,(W)}$. Let

$$U = \Gamma_{K,(W)} R\mathfrak{S}^{\flat}_r(c_1, c_2) \cap \bigcap_{\varphi \in R} \varphi(U_\varphi).$$

Then U is a neighborhood of a by 4.4.7(1). Let $x \in U$. Take $\gamma \in \Gamma_{K,(W)}$ and $\varphi \in R$ such that $x \in \gamma\varphi\mathfrak{S}_r^\flat(c_1, c_2)$. Since $\varphi(U_\varphi)$ is $\Gamma_{K,(W)}$-stable, $\gamma^{-1}x \in \varphi(U_\varphi)$ and hence $\varphi^{-1}\gamma^{-1}x \in \mathfrak{S}_r^\flat(c_1, c_2) \cap U_\varphi$. □

Proposition 4.4.9 *Let $a = (W, \mu) \in \bar{X}^\flat_{F,S}$, and let $r = \dim(W)$. Take $\varphi \in G(F)$ and $c_1, c_2 \in \mathbb{R}_{>0}$ as in 4.4.7(2) such that $\Gamma_{K,(W)}\varphi\mathfrak{S}_r^\flat(c_1, c_2)$ is a neighborhood of a. Let $\phi^\flat_{W,S} \colon \bar{X}^\flat_{F,S}(W) \to \mathfrak{Z}^\flat_{F,S}(W)$ be as in 3.4.4. For any neighborhood U of $\mu = \phi^\flat_{W,S}(a)$ in $\mathfrak{Z}^\flat_{F,S}(W)$ and any $\epsilon \in \mathbb{R}_{>0}$, set*

$$\Phi(U, \epsilon) = (\phi^\flat_{W,S})^{-1}(U) \cap \Gamma_{K,(W)}\varphi\{x \in \mathfrak{S}_r^\flat(c_1, c_2) \mid t_{v,r}(x) < \epsilon \text{ for all } v \in S\}.$$

Then the set of all $\Phi(U, \epsilon)$ forms a base of neighborhoods of a in $\bar{X}^\flat_{F,S}$ under the Satake topology. □

Proof We may suppose that $W = \sum_{i=1}^r Fe_i$ without loss of generality, in which case $\varphi \in G(F)_{(W)}$. Let P be the smallest parabolic subgroup containing B with flag $(V_i)_{-1\leqslant i\leqslant m}$ such that $V_0 = W$ and $m = d - r$. Let Q be the parabolic of all elements that preserve W. We then have $G \supset Q \supset P \supset B$. Let B' be the Borel subgroup of $\mathrm{PGL}_{V/W}$ that is the image of P and which we regard as a subgroup of G using $(e_{r+i})_{1\leqslant i\leqslant m}$ to split $V \to V/W$.

Let

$$f_v \colon Q_u(F_v) \times \bar{X}_{V/W,F,v}(B') \times X_{W_v} \times \mathbb{R}_{\geqslant 0} \to \bar{X}_{F,v}(P)$$

be the unique surjective continuous map such that $\xi = f_v \circ h$, where ξ is as in 3.5.6 and h is defined as the composition

$$\begin{aligned} P_u(F_v) \times X_{W_v} \times \mathbb{R}^m_{\geqslant 0} &\xrightarrow{\sim} Q_u(F_v)\times B'_u(F_v) \times X_{W_v} \times \mathbb{R}_{\geqslant 0} \times \mathbb{R}^{m-1}_{\geqslant 0} \\ &\to Q_u(F_v) \times \bar{X}_{V/W,F,v}(B') \times X_{W_v} \times \mathbb{R}_{\geqslant 0} \end{aligned}$$

of the map induced by the isomorphism $P_u(F_v) \xrightarrow{\sim} Q_u(F_v) \times B'_u(F_v)$ and the map induced by the surjection $\bar{\pi}_{B',v} \colon B'_u(F_v) \times \mathbb{R}^{m-1}_{\geqslant 0} \to \bar{X}_{V/W,F,v}(B')$ of 3.3.8(2). The existence of f_v follows from 3.3.8(4).

Set $Y_0 = \mathbb{R}^S_{>0} \cup \{(0)_{v\in S}\}$, and let

$$f_S \colon Q_u(\mathbb{A}_{F,S}) \times \bar{X}_{V/W,F,S}(B') \times \mathfrak{Z}^\flat_{F,S}(W) \times Y_0 \to \bar{X}_{F,S}(P)$$

be the product of the maps f_v. Let $t_{v,r} \colon \bar{X}_{F,v}(P) \to \mathbb{R}_{\geqslant 0}$ denote the composition

$$\bar{X}_{F,v}(P) \to \bar{X}_{F,v}(B) \xrightarrow{\phi'_{B,v}} \mathbb{R}^{d-1}_{\geqslant 0} \to \mathbb{R}_{\geqslant 0},$$

where the last arrow is the rth projection. The composition of f_S with $(t_{v,r})_{v\in S}$ is projection onto Y_0 by 3.5.7 and 3.4.6.

Let $\bar{X}_{F,S}(W)$ denote the inverse image of $\bar{X}^{\flat}_{F,S}(W)$ under the canonical surjection $\Pi_S : \bar{X}_{F,S} \to \bar{X}^{\flat}_{F,S}$. Combining f_S with the action of $G(F)_{(W)}$, we obtain a surjective map

$$f'_S : G(F)_{(W)} \times (Q_u(\mathbb{A}_{F,S}) \times \bar{X}_{V/W,F,S}(B') \times \mathfrak{Z}^{\flat}_{F,S}(W) \times Y_0) \to \bar{X}_{F,S}(W),$$
$$f'_S(g, z) = gf_S(z).$$

The composition of f'_S with $\phi^{\flat}_{W,S} \circ \Pi_S$ is projection onto $\mathfrak{Z}^{\flat}_{F,S}(W)$ by 3.5.9 and 3.5.10.

Applying 4.3.4 with V/W in place of V, there exists a compact subset C of $Q_u(\mathbb{A}_{F,S}) \times \bar{X}_{V/W,F,S}(B')$ and a finite subset R of $G(F)_{(W)}$ such that $f'_S(\Gamma_{K,(W)} R \times C \times \mathfrak{Z}^{\flat}_{F,S}(W) \times Y_0) = \bar{X}_{F,S}(W)$. Consider the restriction of $\Pi_S \circ f'_S$ to a surjective map

$$\lambda_S : \Gamma_{K,(W)} R \times C \times \mathfrak{Z}^{\flat}_{F,S}(W) \times Y_0 \to \bar{X}^{\flat}_{F,S}(W).$$

We may suppose that R contains φ, since it lies in $G(F)_{(W)}$.

Now, let U' be a neighborhood of a in $\bar{X}^{\flat}_{F,S}(W)$ for the Satake topology. It is sufficient to prove that there exist an open neighborhood U of μ in $\mathfrak{Z}^{\flat}_{F,S}(W)$ and $\epsilon \in \mathbb{R}_{>0}$ such that $\Phi(U, \epsilon) \subset U'$. For $\epsilon \in \mathbb{R}_{>0}$, set $Y_\epsilon = \{(t_v)_{v \in S} \in Y_0 \mid t_v < \epsilon \text{ for all } v \in S\}$.

For any $x \in C$, we have $\lambda_S(\alpha, x, \mu, 0) = (W, \mu) \in U'$ for all $\alpha \in R$. By the continuity of λ_S, there exist a neighborhood $D(x) \subset Q_u(\mathbb{A}_{F,S}) \times \bar{X}_{V/W,F,S}(B')$ of x, a neighborhood $U(x) \subset \mathfrak{Z}^{\flat}_{F,v}(W)$ of μ, and $\epsilon(x) \in \mathbb{R}_{>0}$ such that

$$\lambda_S(R \times D(x) \times U(x) \times Y_{\epsilon(x)}) \subset U'.$$

Since C is compact, some finite collection of the sets $D(x)$ cover C. Thus, there exist a neighborhood U of μ in $\mathfrak{Z}^{\flat}_{F,v}(W)$ and $\epsilon \in \mathbb{R}_{>0}$ such that $\lambda_S(R \times C \times U \times Y_\epsilon) \subset U'$. Since U' is $\Gamma_{K,(W)}$-stable by 3.4.15, we have $\lambda_S(\Gamma_{K,(W)} R \times C \times U \times Y_\epsilon) \subset U'$.

Let $y \in \Phi(U, \epsilon)$, and write $y = gx$ with $g \in \Gamma_{K,(W)}\varphi$ and $x \in \mathfrak{S}^{\flat}_r(c_1, c_2)$ such that $t_{v,r}(x) < \epsilon$ for all $v \in S$. Since $\Phi(U, \epsilon) \subset \bar{X}^{\flat}_{F,S}(W)$, we may by our above remarks write $y = \lambda_S(g, c, \nu, t) = g\Pi_S(f_S(c, \nu, t))$, where $c \in C$, $\nu = \phi^{\flat}_{W,S}(y)$, and $t = (t_{v,r}(x))_{v \in S}$. Since $\nu \in U$ and $t \in Y_\epsilon$ by definition, y is contained in U'. Therefore, we have $\Phi(U, \epsilon) \subset U'$. □

Example 4.4.10 Consider the case $F = \mathbb{Q}$, $S = \{v\}$ with v the archimedean place, and $d = 3$. We construct a base of neighborhoods of a point in $\bar{X}^{\flat}_{\mathbb{Q},v}$ for the Satake topology.

Fix a basis $(e_i)_{1 \leqslant i \leqslant 3}$ of V. Let $a = (W, \mu) \in \bar{X}^{\flat}_{\mathbb{Q},v}$, where $W = \mathbb{Q}e_1$, and μ is the unique element of X_{W_v}.

For $c \in \mathbb{R}_{>0}$, let U_c be the subset of $X_v = \mathrm{PGL}_3(\mathbb{R})/\mathrm{PO}_3(\mathbb{R})$ consisting of the elements

$$\begin{pmatrix} 1 & 0 \\ 0 & \gamma \end{pmatrix} \begin{pmatrix} 1 & x_{12} & x_{13} \\ 0 & 1 & x_{23} \\ 0 & 0 & 1 \end{pmatrix} \begin{pmatrix} y_1 y_2 & 0 & 0 \\ 0 & y_2 & 0 \\ 0 & 0 & 1 \end{pmatrix}$$

such that $\gamma \in \mathrm{PGL}_2(\mathbb{Z})$, $x_{ij} \in \mathbb{R}$, $y_1 \geqslant c$, and $y_2 \geqslant \frac{\sqrt{3}}{2}$. When γ, x_{ij} and y_2 are fixed and $y_1 \to \infty$, these elements converge to a in $\bar{X}^{\flat}_{\mathbb{Q},v}$ under the Satake topology. When γ, x_{ij}, and y_1 are fixed and $y_2 \to \infty$, they converge in the Satake topology to

$$\mu(\gamma, x_{12}, y_1) := \begin{pmatrix} 1 & 0 \\ 0 & \gamma \end{pmatrix} \begin{pmatrix} 1 & x_{12} & 0 \\ 0 & 1 & 0 \\ 0 & 0 & 1 \end{pmatrix} \mu(y_1),$$

where $\mu(y_1)$ is the class in $\bar{X}^{\flat}_{\mathbb{Q},v}$ of the semi-norm $a_1 e_1^* + a_2 e_2^* + a_3 e_3^* \mapsto (a_1^2 y_1^2 + a_2^2)^{1/2}$ on V_v^*.

The set of

$$\bar{U}_c := \{a\} \cup \{\mu(\gamma, x, y) \mid \gamma \in \mathrm{PGL}_2(\mathbb{Z}), x \in \mathbb{R}, y \geqslant c\} \cup U_c.$$

is a base of neighborhoods for a in $\bar{X}^{\flat}_{\mathbb{Q},v}$ under the Satake topology. Note that $\mathfrak{H} = \mathrm{SL}_2(\mathbb{Z})\{z \in \mathfrak{H} \mid \mathrm{Im}(z) \geqslant \frac{\sqrt{3}}{2}\}$, which is the reason for the appearance of $\frac{\sqrt{3}}{2}$. It can of course be replaced by any $b \in \mathbb{R}_{>0}$ such that $b \leqslant \frac{\sqrt{3}}{2}$.

4.4.11 We continue with Example 4.4.10. Under the canonical surjection $\bar{X}_{\mathbb{Q},v} \to \bar{X}^{\flat}_{\mathbb{Q},v}$, the inverse image of $a = (W, \mu)$ in $\bar{X}_{\mathbb{Q},v}$ is canonically homeomorphic to $\bar{X}_{(V/W)_v} = \mathfrak{H} \cup \mathbb{P}^1(\mathbb{Q})$ under the Satake topology on both spaces. This homeomorphism sends $x + y_2 i \in \mathfrak{H}$ ($x \in \mathbb{R}$, $y_2 \in \mathbb{R}_{>0}$) to the limit for the Satake topology of

$$\begin{pmatrix} 1 & 0 & 0 \\ 0 & 1 & x \\ 0 & 0 & 1 \end{pmatrix} \begin{pmatrix} y_1 y_2 & 0 & 0 \\ 0 & y_2 & 0 \\ 0 & 0 & 1 \end{pmatrix} \in \mathrm{PGL}_3(\mathbb{R})/\mathrm{PO}_3(\mathbb{R})$$

as $y_1 \to \infty$. (This limit in $\bar{X}_{\mathbb{Q},v}$ depends on x and y_2, but the limit in $\bar{X}^{\flat}_{\mathbb{Q},v}$ is a.)

4.4.12 In the example of 4.4.10, we explain that the quotient topology on $\bar{X}^{\flat}_{\mathbb{Q},v}$ of the Satake topology on $\bar{X}_{\mathbb{Q},v}$ is different from the Satake topology on $\bar{X}^{\flat}_{\mathbb{Q},v}$.

For a map

$$f\colon \mathrm{PGL}_2(\mathbb{Z})/\begin{pmatrix} 1 & \mathbb{Z} \\ 0 & 1 \end{pmatrix} \to \mathbb{R}_{>0},$$

define a subset U_f of X_v as in the definition of U_c but replacing the condition on γ, x_{ij}, y_i by $\gamma \in \mathrm{PGL}_2(\mathbb{Z})$, $x_{ij} \in \mathbb{R}$, $y_1 \geqslant f(\gamma)$, and $y_2 \geqslant \frac{\sqrt{3}}{2}$. Let

$$\bar{U}_f = \{a\} \cup \{\mu(\gamma, x, y) \mid \gamma \in \mathrm{PGL}_2(\mathbb{Z}), x \in \mathbb{R}, y \geqslant f(\gamma)\} \cup U_f.$$

When f varies, the $\bar{U}_f$ form a base of neighborhoods of a in $\bar{X}^{\flat}_{\mathbb{Q},v}$ for the quotient topology of the Satake topology on $\bar{X}_{\mathbb{Q},v}$. On the other hand, if $\inf\{f(\gamma) \mid \gamma \in \mathrm{PGL}_2(\mathbb{Z})\} = 0$, then $\bar{U}_f$ is not a neighborhood of a for the Satake topology on $\bar{X}^{\flat}_{\mathbb{Q},v}$.

4.5 Proof of the Main Theorem

In this subsection, we prove Theorem 4.1.4. We begin with the quasi-compactness asserted therein. Throughout this subsection, we set $Z = X_{S_2} \times G(\mathbb{A}^S_F)/K$ in situation (I) and $Z = X_{S_2}$ in situation (II), so $\bar{\mathfrak{X}} = \bar{X} \times Z$.

Proposition 4.5.1 *In situation (I) of 4.1.3, the quotient $G(F)\backslash\bar{\mathfrak{X}}$ is quasi-compact. In situation (II), the quotient $\Gamma\backslash\bar{\mathfrak{X}}$ is quasi-compact for any subgroup Γ of Γ_K of finite index.*

Proof We may restrict to case (i) of 4.1.2 that $\bar{X} = \bar{X}_{F,S_1}$, as $\bar{X}^{\flat}_{F,S_1}$ of case (ii) is a quotient of $\bar{X}_{F,S_1}$ (under the Borel–Serre topology). In situation (I), we claim that there exist $c_1, c_2 \in \mathbb{R}_{>0}$, a compact subset C of $B_u(\mathbb{A}_{F,S})$, and a compact subset C' of Z such that $\bar{\mathfrak{X}} = G(F)(\mathfrak{S}(C; c_1, c_2) \times C')$. In situation (II), we claim that there exist c_1, c_2, C, C' as above and a finite subset R of $G(F)$ such that $\bar{\mathfrak{X}} = \Gamma R(\mathfrak{S}(C; c_1, c_2) \times C')$. It follows that in situation (I) (resp., (II)), there is a surjective continuous map from the compact space $C \times \mathfrak{T}(c_1, c_2) \times C'$ (resp., $R \times C \times \mathfrak{T}(c_1, c_2) \times C'$) onto the quotient space under consideration, which yields the proposition.

For any compact open subgroup K' of $G(\mathbb{A}^{S_1}_F)$, the set $G(F)\backslash G(\mathbb{A}^{S_1}_F)/K'$ is finite. Each X_v for $v \in S_2$ may be identified with the geometric realization of the Bruhat-Tits building for PGL_{V_v}, the set of i-simplices of which for a fixed i can be identified with $G(F_v)/K'_v$ for some K'. So, we see that in situation (I) (resp., (II)), there is a compact subset D of Z such that $Z = G(F)D$ (resp., $Z = \Gamma D$).

Now fix such a compact open subgroup K' of $G(\mathbb{A}^{S_1}_F)$. By 4.3.4, there are $c_1, c_2 \in \mathbb{R}_{>0}$, a compact subset C of $P_u(\mathbb{A}_{F,S_1})$, and a finite subset R' of $G(F)$ such that $\bar{X}_{F,S} = \Gamma_{K'} R' \mathfrak{S}(C; c_1, c_2)$. We consider the compact subset $C' := (R')^{-1} K' D$ of Z.

Let $(x, y) \in \bar{\mathfrak{X}}$, where $x \in \bar{X}_{F,S}$ and $y \in Z$. Write $y = \gamma z$ for some $z \in D$ and $\gamma \in G(F)$ (resp., $\gamma \in \Gamma$) in situation (I) (resp., (II)). In situation (II), we write $\Gamma\Gamma_{K'} R' = \Gamma R$ for some finite subset R of $G(F)$. Write $\gamma^{-1} x = \gamma' \varphi s$ where $\gamma' \in \Gamma_{K'}$, $\varphi \in R'$, $s \in \mathfrak{S}(C; c_1, c_2)$. We have

$$(x, y) = \gamma(\gamma^{-1} x, z) = \gamma(\gamma' \varphi s, z) = (\gamma\gamma'\varphi)(s, \varphi^{-1}(\gamma')^{-1} z).$$

As $\gamma\gamma'\varphi$ lies in $G(F)$ in situation (I) and in ΓR in situation (II), we have the claim. □

4.5.2 To prove Theorem 4.1.4, it remains only to verify the Hausdorff property. For this, it is sufficient to prove the following.

Proposition 4.5.3 *Let $\Gamma = G(F)$ in situation (I) of 4.1.3, and let $\Gamma = \Gamma_K$ in situation (II). For every $a, a' \in \bar{\mathfrak{X}}$, there exist neighborhoods U of a and U' of a' such that if $\gamma \in \Gamma$ and $\gamma U \cap U' \neq \varnothing$, then $\gamma a = a'$.*

In the rest of this subsection, let the notation be as in 4.5.3. It is sufficient to prove 4.5.3 for the Satake topology on $\bar{\mathfrak{X}}$. In 4.5.4–4.5.8, we prove 4.5.3 in situation (II) for $S = S_1$. That is, we suppose that $\bar{\mathfrak{X}} = \bar{X}$. In 4.5.9 and 4.5.10, we deduce 4.5.3 in general from this case.

Lemma 4.5.4 *Assume that $\bar{\mathfrak{X}} = \bar{X}_{F,S}$. Suppose that $a, a' \in \bar{\mathfrak{X}}$ have distinct parabolic types (4.3.7). Then there exist neighborhoods U of a and U' of a' such that $\gamma U \cap U' = \varnothing$ for all $\gamma \in \Gamma$.*

Proof Let I (resp., I') be the parabolic type of a (resp., a'). We may assume that there exists an $i \in I$ with $i \notin I'$.

By 4.3.15, there exist $\varphi, \psi \in G(F)$ and $c_1, c_2 \in \mathbb{R}_{>0}$ such that $\Gamma_K \varphi \mathfrak{S}(c_1, c_2)$ is a neighborhood of a and $\Gamma_K \psi \mathfrak{S}(c_1, c_2)$ is a neighborhood of a'. By 4.3.14(2), there exist a neighborhood $U' \subset \Gamma_K \psi \mathfrak{S}(c_1, c_2)$ of a' and $c \in \mathbb{R}_{>0}$ with the property that $\min\{t_{v,i}(x) \mid v \in S\} \geqslant c$ for all $x \in (\Gamma_K \psi)^{-1} U' \cap \mathfrak{S}(c_1, c_2)$. Let $A \in \mathbb{R}_{>1}$ be as in 4.3.13 for these c_1, c_2 for $R_1 = \{\varphi^{-1}\}$ and $R_2 = \{\psi\}$. Take $\epsilon \in \mathbb{R}_{>0}$ such that $A\epsilon \leqslant c$. By 4.3.14(1), there exists a neighborhood $U \subset \Gamma_K \varphi \mathfrak{S}(c_1, c_2)$ of a such that $\max\{t_{v,i}(x) \mid v \in S\} < \epsilon$ for all $x \in (\Gamma_K \varphi)^{-1} U \cap \mathfrak{S}(c_1, c_2)$.

We prove that $\gamma U \cap U' = \varnothing$ for all $\gamma \in \Gamma_K$. If $x \in \gamma U \cap U'$, then we may take $\delta, \delta' \in \Gamma_K$ such that $(\delta\varphi)^{-1}\gamma^{-1}x \in \mathfrak{S}(c_1, c_2)$ and $(\delta'\psi)^{-1}x \in \mathfrak{S}(c_1, c_2)$. Since

$$(\delta\varphi)^{-1}\gamma^{-1}x = \varphi^{-1}(\delta^{-1}\gamma^{-1}\delta')\psi(\delta'\psi)^{-1}x \in \varphi^{-1}\Gamma_K\psi \cdot (\delta'\psi)^{-1}x,$$

we have by 4.3.13 that

$$c \leqslant t_{v,i}((\delta'\psi)^{-1}x) \leqslant A t_{v,i}((\delta\varphi)^{-1}\gamma^{-1}x) < A\epsilon,$$

for all $v \in S$ and hence $c < A\epsilon$, a contradiction. □

Lemma 4.5.5 *Assume that $\bar{\mathfrak{X}} = \bar{X}^{\flat}_{F,S}$. Let $a, a' \in \bar{\mathfrak{X}}$ and assume that the dimension of the F-subspace associated to a is different from that of a'. Then there exist neighborhoods U of a and U' of a' such that $\gamma U \cap U' = \varnothing$ for all $\gamma \in \Gamma$.*

Proof The proof is similar to that of 4.5.4. In place of 4.3.13, 4.3.14, and 4.3.15, we use 4.4.4, 4.4.6, and 4.4.7, respectively. □

Lemma 4.5.6 *Let P be a parabolic subgroup of G. Let $a, a' \in \mathfrak{Z}_{F,S}(P)$ (see 3.4.3), and let R_1 and R_2 be finite subsets of $G(F)$. Then there exist neighborhoods U of a and U' of a' in $\mathfrak{Z}_{F,S}(P)$ such that $\gamma a = a'$ for every $\gamma \in R_1 \Gamma_K R_2 \cap P(F)$ for which $\gamma U \cap U' \neq \varnothing$.*

Proof For each $\xi \in R_1$ and $\eta \in R_2$, the set $\xi\Gamma_K\eta \cap P(F)$ is a $\xi\Gamma_K\xi^{-1} \cap P(F)$-orbit for the left action of $\xi\Gamma_K\xi^{-1}$. Hence its image in $\prod_{i=0}^{m} \mathrm{PGL}_{V_i/V_{i-1}}(\mathbb{A}_{F,S})$ is discrete, for $(V_i)_{-1\leqslant i\leqslant m}$ the flag corresponding to P, and thus the image of $R_1\Gamma_K R_2 \cap P(F)$ in $\prod_{i=0}^m \mathrm{PGL}_{V_i/V_{i-1}}(\mathbb{A}_{F,S})$ is discrete as well. On the other hand, for any compact neighborhoods U of a and U' of a', the set

$$\left\{ g \in \prod_{i=0}^{m} \mathrm{PGL}_{V_i/V_{i-1}}(\mathbb{A}_{F,S}) \mid gU \cap U' \neq \varnothing \right\}$$

is compact. Hence the intersection $M := \{\gamma \in R_1\Gamma_K R_2 \cap P(F) \mid \gamma U \cap U' \neq \varnothing\}$ is finite. If $\gamma \in M$ and $\gamma a \neq a'$, then replacing U and U' by smaller neighborhoods of a and a', respectively, we have $\gamma U \cap U' = \varnothing$. Hence for sufficiently small neighborhoods U and U' of a and a', respectively, we have that if $\gamma \in M$, then $\gamma a = a'$. □

Lemma 4.5.7 *Let W be an F-subspace of V. Let $a, a' \in \mathfrak{Z}^{\flat}_{F,S}(W)$ (see 3.4.4), and let R_1 and R_2 be finite subsets of $G(F)$. Let P be the parabolic subgroup of G consisting of all elements which preserve W. Then there exist neighborhoods U of a and U' of a' in $\mathfrak{Z}^{\flat}_{F,S}(W)$ such that $\gamma a = a'$ for every $\gamma \in R_1\Gamma_K R_2 \cap P(F)$ for which $\gamma U \cap U' \neq \varnothing$.*

Proof This is proven in the same way as 4.5.6. □

4.5.8 We prove 4.5.3 in situation (II), supposing that $S = S_1$.

In case (i) (that is, $\bar{X} = \bar{\mathfrak{X}} = \bar{X}_{F,S}$), we may assume by 4.5.4 that a and a' have the same parabolic type I. In case (ii) (that is, $\bar{X} = \bar{\mathfrak{X}} = \bar{X}^{\flat}_{F,S}$), we may assume by 4.5.5 that the dimension r of the F-subspace of V associated to a coincides with that of a'. In case (i) (resp., (ii)), take $c_1, c_2 \in \mathbb{R}_{>0}$ and elements φ and ψ (resp., finite subsets R and R') of $G(F)$ such that c_1, c_2, φ (resp., c_1, c_2, R) satisfy the condition in 4.3.15 (resp., 4.4.7) for a and c_1, c_2, ψ (resp., c_1, c_2, R') satisfy the condition in 4.3.15 (resp., 4.4.7) for a'. In case (i), we set $R = \{\varphi\}$ and $R' = \{\psi\}$.

Fix a basis $(e_i)_{1\leqslant i\leqslant d}$ of V. In case (i) (resp., (ii)), denote $\mathfrak{S}(c_1, c_2)$ (resp., $\mathfrak{S}^{\flat}_r(c_1, c_2)$) by $\mathfrak{S}$. In case (i), let $P = P_I$, and let $(V_i)_{-1\leqslant i\leqslant m}$ be the associated flag. In case (ii), let $W = \sum_{i=1}^r Fe_i$, and let P be the parabolic subgroup of G consisting of all elements which preserve W.

Note that in case (i) (resp., (ii)), for all $\varphi \in R$ and $\psi \in R'$, the parabolic subgroup P is associated to $\varphi^{-1}a$ and to $\psi^{-1}a'$ (resp., W is associated to $\varphi^{-1}a$ and to $\psi^{-1}a'$) and hence these elements are determined by their images in $\mathfrak{Z}_{F,S}(P)$ (resp. $\mathfrak{Z}^{\flat}_{F,S}(W)$).

In case (i) (resp., case (ii)), apply 4.5.6 (resp., 4.5.7) to the images of $\varphi^{-1}a$ and $\psi^{-1}a'$ for $\varphi \in R$, $\psi \in R'$ in $\mathfrak{Z}_{F,S}(P)$ (resp., $\mathfrak{Z}^{\flat}_{F,S}(W)$). By this, and by 4.3.10 for case (i) and 4.4.3 for case (ii), we see that there exist neighborhoods U_φ of $\varphi^{-1}a$ for each $\varphi \in R$ and U'_ψ of $\psi^{-1}a'$ for each $\psi \in R'$ for the Satake topology with the following two properties:

(A) $\{\gamma \in (R')^{-1}\Gamma_K R \mid \gamma(\mathfrak{S} \cap U_\varphi) \cap (\mathfrak{S} \cap U'_\psi) \neq \varnothing$ for some $\varphi \in R, \psi \in R'\}$ $\subset P(F)$,

(B) if $\gamma \in (R')^{-1}\Gamma_K R \cap P(F)$ and $\gamma U_\varphi \cap U'_\psi \neq \varnothing$ for $\varphi \in R$ and $\psi \in R'$, then $\gamma\varphi^{-1}a = \psi^{-1}a'$.

In case (i) (resp., (ii)), take a neighborhood U of a satisfying the condition in 4.3.16 (resp., 4.4.8) for $(U_\varphi)_{\varphi\in R}$, and take a neighborhood U' of a' satisfying the condition in 4.3.16 (resp., 4.4.8) for $(U'_\psi)_{\psi\in R'}$. Let $\gamma \in \Gamma_K$ and assume $\gamma U \cap U' \neq \varnothing$. We prove $\gamma a = a'$. Take $x \in U$ and $x' \in U'$ such that $\gamma x = x'$. By 4.3.16 (resp., 4.4.8), there are $\varphi \in R$, $\psi \in R'$, and $\epsilon \in \Gamma_{K,(\varphi P\varphi^{-1})}$ and $\delta \in \Gamma_{K,(\psi P\psi^{-1})}$ in case (i) (resp., $\epsilon \in \Gamma_{K,(\varphi W)}$ and $\delta \in \Gamma_{K,(\psi W)}$ in case (ii)) such that $\varphi^{-1}\epsilon^{-1}x \in \mathfrak{S} \cap U_\varphi$ and $\psi^{-1}\delta^{-1}x' \in \mathfrak{S} \cap U'_\psi$. Since

$$(\psi^{-1}\delta^{-1}\gamma\epsilon\varphi)\varphi^{-1}\epsilon^{-1}x = \psi^{-1}\delta^{-1}x',$$

we have $\psi^{-1}\delta^{-1}\gamma\epsilon\varphi \in P(F)$ by property (A). By property (B), we have

$$(\psi^{-1}\delta^{-1}\gamma\epsilon\varphi)\varphi^{-1}a = \psi^{-1}a'.$$

Since $\epsilon a = a$ and $\delta a' = a'$, this proves $\gamma a = a'$.

We have proved 4.5.3 in situation (II) under the assumption $S = S_1$. In the following 4.5.9 and 4.5.10, we reduce the general case to that case.

Lemma 4.5.9 *Let $a, a' \in Z$. In situation (I) (resp., (II)), let $H = G(\mathbb{A}_F^{S_1})$ (resp., $H = G(\mathbb{A}_{F,S_2})$). Then there exist neighborhoods U of a and U' of a' in Z such that $ga = a'$ for all $g \in H$ for which $gU \cap U' \neq \varnothing$.*

Proof For any compact neighborhoods U of a and U' of a', the set $M := \{g \in H \mid gU \cap U' \neq \varnothing\}$ is compact. By definition of Z, there exist a compact open subgroup N of H and a compact neighborhood U of a such that $gx = x$ for all $g \in N$ and $x \in U$. For such a choice of U, the set M is stable under the right translation by N, and M/N is finite because M is compact and N is an open subgroup of H. If $g \in M$ and if $ga \neq a'$, then by shrinking the neighborhoods U and U', we have that $gU \cap U' = \varnothing$. As M/N is finite, we have sufficiently small neighborhoods U and U' such that if $g \in M$ and $gU \cap U' \neq \varnothing$, then $ga = a'$. □

4.5.10 We prove Proposition 4.5.3.

Let H be as in Lemma 4.5.9. Write $a = (a_{S_1}, a_Z)$ and $a' = (a'_{S_1}, a'_Z)$ as elements of $\bar{X} \times Z$. By 4.5.9, there exist neighborhoods U_Z of a_Z and U'_Z of a'_Z in Z such that if $g \in H$ and $gU_Z \cap U'_Z \neq \varnothing$, then $ga = a'$. The set $K' := \{g \in H \mid ga_Z = a_Z\}$ is a compact open subgroup of H. Let Γ' be the inverse image of K' under $\Gamma \to H$, where $\Gamma = G(F)$ in situation (I). In situation (II), the group Γ' is of finite index in the inverse image of the compact open subgroup $K' \times K$ under $G(F) \to G(\mathbb{A}_F^{S_1})$. In both situations, the set $M := \{\gamma \in \Gamma \mid \gamma a_Z = a'_Z\}$ is either empty or a Γ'-torsor for the right action of Γ'.

Assume first that $M \neq \varnothing$, in which case we may choose $\theta \in \Gamma$ such that $M = \theta\Gamma'$. Since we have proven 4.5.3 in situation (II) for $S_1 = S$, there exist neighborhoods U_{S_1} of a_{S_1} and U'_{S_1} of $\theta^{-1}a'_{S_1}$ such that if $\gamma \in \Gamma'$ satisfies $\gamma U_{S_1} \cap U'_{S_1} \neq \varnothing$, then $\gamma a_{S_1} = \theta^{-1}a'_{S_1}$. Let $U = U_{S_1} \times U_Z$ and $U' = \theta U'_{S_1} \times U'_Z$, which are neighborhoods of a and a' in $\bar{\mathfrak{X}}$, respectively. Suppose that $\gamma \in \Gamma$ satisfies $\gamma U \cap U' \neq \varnothing$. Then, since $\gamma U_Z \cap U'_Z \neq \varnothing$, we have $\gamma a_Z = a'_Z$ and hence $\gamma = \theta\gamma'$ for some $\gamma' \in \Gamma'$.

Since $\theta\gamma' U_{S_1} \cap \theta U'_{S_1} \neq \varnothing$, we have $\gamma' U_{S_1} \cap U'_{S_1} \neq \varnothing$, and hence $\gamma' a_{S_1} = \theta^{-1} a'_{S_1}$. That is, we have $\gamma a_{S_1} = a'_{S_1}$, so $\gamma a = a'$.

In the case that $M = \varnothing$, take any neighborhoods U_{S_1} of a_{S_1} and U'_{S_1} of a'_{S_1}, and set $U = U_{S_1} \times U_Z$ and $U' = U'_{S_1} \times U'_Z$. Any $\gamma \in \Gamma$ such that $\gamma U \cap U' \neq \varnothing$ is contained in M, so no such γ exists.

4.6 Supplements to the Main Theorem

We use the notation of §4.1 throughout this subsection, and in particular, we let Γ be as in Theorem 4.1.4. For $a \in \bar{\mathfrak{X}}$, let $\Gamma_a < \Gamma$ denote the stabilizer of a.

Theorem 4.6.1 *Let $\Gamma = G(F)$ in situation (I), and let Γ be a subgroup of Γ_K of finite index in situation (II). For $a \in \bar{\mathfrak{X}}$ (with either the Borel–Serre or the Satake topology), there is an open neighborhood U of the image of a in $\Gamma_a \backslash \bar{\mathfrak{X}}$ such that the image U' of U under the quotient map $\Gamma_a \backslash \bar{\mathfrak{X}} \to \Gamma \backslash \bar{\mathfrak{X}}$ is open and the map $U \to U'$ is a homeomorphism.*

Proof By the case $a = a'$ of Proposition 4.5.3, there is an open neighborhood $U'' \subset \bar{\mathfrak{X}}$ of a such that if $\gamma \in \Gamma_K$ and $\gamma U'' \cap U'' \neq \varnothing$, then $\gamma a = a$. Then the subset $U := \Gamma_a \backslash \Gamma_a U''$ of $\Gamma_a \backslash \bar{\mathfrak{X}}$ is open and has the desired property. □

Proposition 4.6.2 *Suppose that $S = S_1$, and let $a \in \bar{\mathfrak{X}}$. Let $\Gamma = G(F)$ in situation (I), and let Γ be a subgroup of Γ_K of finite index in situation (II).*

(1) Take $\bar{X} = \bar{X}_{F,S}$, and let P be the parabolic subgroup associated to a. Then $\Gamma_{(P)}$ (as in 3.4.12) is a normal subgroup of Γ_a of finite index.
(2) Take $\bar{X} = \bar{X}^{\flat}_{F,S}$, and let W be the F-subspace of V associated to a. Then $\Gamma_{(W)}$ (as in 3.4.12) is a normal subgroup of Γ_a of finite index.

Proof We prove (1), the proof of (2) being similar. Let $(V_i)_{-1 \leqslant i \leqslant m}$ be the flag corresponding to P. The image of $\Gamma \cap P(F)$ in $\prod_{i=0}^m \mathrm{PGL}_{V_i/V_{i-1}}(\mathbb{A}_{F,S})$ is discrete. On the other hand, the stabilizer in $\prod_{i=0}^m \mathrm{PGL}_{V_i/V_{i-1}}(\mathbb{A}_{F,S})$ of the image of a in $\mathfrak{Z}_{F,S}(P)$ is compact. Hence the image of Γ_a in $\prod_{i=0}^m \mathrm{PGL}_{V_i/V_{i-1}}(F)$, which is isomorphic to $\Gamma_a/\Gamma_{(P)}$, is finite. □

Theorem 4.6.3 *Assume that F is a function field and $\bar{X} = \bar{X}_{F,S_1}$, where S_1 consists of a single place v. Let Γ be as in Theorem 4.6.1. Then the inclusion map $\Gamma \backslash \mathfrak{X} \hookrightarrow \Gamma \backslash \bar{\mathfrak{X}}$ is a homotopy equivalence.*

Proof Let $a \in \bar{\mathfrak{X}}$. In situation (I) (resp., (II)), write $a = (a_v, a^v)$ with $a_v \in \bar{X}_{F,v}$ and $a^v \in X_{S_2} \times G(\mathbb{A}_F^S)/K$ (resp., X_{S_2}). Let K' be the isotropy subgroup of a^v in $G(\mathbb{A}_F^v)$ (resp., $\prod_{w \in S_2} G(F_w)$), and let $\Gamma' < \Gamma$ be the inverse image of K' under the map $\Gamma \to G(\mathbb{A}_F^v)$ (resp., $\Gamma \to \prod_{w \in S_2} G(F_w)$).

Let P be the parabolic subgroup associated to a. Let Γ_a be the isotropy subgroup of a in Γ, which is contained in $P(F)$ and equal to the isotropy subgroup Γ'_{a_v} of a_v

in Γ'. In situation (I) (resp., (II)), take a Γ_a-stable open neighborhood D of a^v in $X_{S_2} \times G(\mathbb{A}_F^S)/K$ (resp., X_{S_2}) that has compact closure.

Claim 1 The subgroup $\Gamma_D := \{\gamma \in \Gamma_a \mid \gamma x = x \text{ for all } x \in D\}$ of Γ_a is normal of finite index.

Proof of Claim 1 Normality follows from the Γ_a-stability of D. For any x in the closure $\bar{D}$ of D, there exists an open neighborhood V_x of x and a compact open subgroup N_x of $G(\mathbb{A}_F^v)$ (resp., $\prod_{w \in S_2} G(F_w)$) in situation (I) (resp., (II)) such that $gy = y$ for all $g \in N_x$ and $y \in V_x$. For a finite subcover $\{V_{x_1}, \ldots, V_{x_n}\}$ of $\bar{D}$, the group Γ_D is the inverse image in Γ_a of $\bigcap_{i=1}^n N_{x_i}$, so is of finite index.

Claim 2 The subgroup $H := \Gamma_D \cap P_u(F)$ of Γ_a is normal of finite index.

Proof of Claim 2 Normality is immediate from Claim 1 as $P_u(F)$ is normal in $P(F)$. Let $H' = \Gamma'_{(P)} \cap P_u(F)$, which has finite index in $\Gamma'_{(P)}$ and equals $\Gamma' \cap P_u(F)$ by definition of $\Gamma'_{(P)}$. Since $\Gamma'_{(P)} \subset \Gamma'_{a_v} \subset \Gamma'$ and $\Gamma'_{a_v} = \Gamma_a$, we have $H' = \Gamma_a \cap P_u(F)$ as well. By Claim 1, we then have that H' contains H with finite index, so H has finite index in $\Gamma'_{(P)}$. Proposition 4.6.2(1) tells us that $\Gamma'_{(P)}$ is of finite index in $\Gamma'_{a_v} = \Gamma_a$.

Let $(V_i)_{-1 \leqslant i \leqslant m}$ be the flag corresponding to P. By Corollary 3.5.4, we have a homeomorphism

$$\chi \colon P_u(F_v)\backslash \bar{X}_{F,v}(P) \xrightarrow{\sim} \mathfrak{Z}_{F,v}(P) \times \mathbb{R}^m_{\geqslant 0}$$

on quotient spaces arising from the $P(F_v)$-equivariant homeomorphism $\psi_{P,v} = (\phi_{P,v}, \phi'_{P,v})$ of 3.5.1 (see 3.4.3 and 3.4.6).

Claim 3 For a sufficiently small open neighborhood U of $0 = (0, \ldots, 0)$ in $\mathbb{R}^m_{\geqslant 0}$, the map χ induces a homeomorphism

$$\chi_U \colon H\backslash \bar{X}_{F,v}(P)_U \xrightarrow{\sim} \mathfrak{Z}_{F,v}(P) \times U,$$

where $\bar{X}_{F,v}(P)_U$ denotes the inverse image of U under $\phi'_{P,v} \colon \bar{X}_{F,v}(P) \to \mathbb{R}^m_{\geqslant 0}$.

Proof of Claim 3 By definition, χ restricts to a homeomorphism

$$P_u(F_v)\backslash \bar{X}_{F,v}(P)_U \xrightarrow{\sim} \mathfrak{Z}_{F,v}(P) \times U$$

for any open neighborhood U of 0. For a sufficiently large compact open subset C of $P_u(F_v)$, we have $P_u(F_v) = HC$. For U sufficiently small, every $g \in C$ fixes all $x \in \bar{X}_{F,v}(P)_U$, which yields the claim.

Claim 4 The map χ_U and the identity map on D induce a homeomorphism

$$\chi_{U,a} \colon \Gamma_a\backslash(\bar{X}_{F,v}(P)_U \times D) \xrightarrow{\sim} (\Gamma_a\backslash(\mathfrak{Z}_{F,v}(P) \times D)) \times U.$$

Proof of Claim 4 The quotient group Γ_a/H is finite by Claim 2. Since the determinant of an automorphism of V_i/V_{i-1} of finite order has trivial absolute value at v, the Γ_a-action on $\mathbb{R}^m_{\geqslant 0}$ is trivial. Since H acts trivially on D, the claim follows from Claim 3.

Now let $c \in \mathbb{R}^m_{>0}$, and set $U = \{t \in \mathbb{R}^m_{\geqslant 0} \mid t_i < c \text{ for all } 1 \leqslant i \leqslant m\}$. Set $(X_v)_U = X_v \cap \bar{X}_{F,v}(P)_U$. If c is sufficiently small, then

$$(\Gamma_a \backslash (\mathfrak{Z}_{F,v}(P) \times D)) \times (U \cap \mathbb{R}^m_{>0}) \hookrightarrow (\Gamma_a \backslash (\mathfrak{Z}_{F,v}(P) \times D)) \times U$$

is a homotopy equivalence, and we can apply $\chi_{U,a}^{-1}$ to both sides to see that the inclusion map

$$\Gamma_a \backslash ((X_v)_U \times D) \hookrightarrow \Gamma_a \backslash (\bar{X}_{F,v}(P)_U \times D)$$

is also a homotopy equivalence. By Theorem 4.6.1, this proves Theorem 4.6.3. □

Remark 4.6.4 Theorem 4.6.3 is well-viewed as a function field analogue of the homotopy equivalence for Borel–Serre spaces of [3].

4.6.5 Theorem 4.1.4 remains true if we replace $G = \mathrm{PGL}_V$ by $G = \mathrm{SL}_V$ in 4.1.3 and 4.1.4. It also remains true if we replace $G = \mathrm{PGL}_V$ by $G = \mathrm{GL}_V$ and we replace $\bar{\mathfrak{X}}$ in 4.1.4 in situation (I) (resp., (II)) by

$$\bar{X} \times X_{S_2} \times (\mathbb{R}^S_{>0} \times G(\mathbb{A}^S_F)/K)_1 \quad (\text{resp., } \bar{X} \times X_{S_2} \times (\mathbb{R}^S_{>0})_1),$$

where $(\ )_1$ denotes the kernel of

$$((a_v)_{v \in S}, g) \mapsto |\det(g)| \prod_{v \in S} a_v \quad (\text{resp., } (a_v)_{v \in S} \mapsto \prod_{v \in S} a_v),$$

and $\gamma \in \mathrm{GL}_V(F)$ (resp., $\gamma \in \Gamma_K$) acts on this kernel by multiplication by $((|\det(\gamma)|_v)_{v \in S}, \gamma)$ (resp., $(|\det(\gamma)|_v)_{v \in S}$).

Theorems 4.6.1 and 4.6.3 also remain true under these modifications. These modified versions of the results are easily reduced to the original case $G = \mathrm{PGL}_V$.

4.7 Subjects Related to This Paper

4.7.1 In this subsection, as possibilities of future applications of this paper, we describe connections with the study of

- toroidal compactifications of moduli spaces of Drinfeld modules (4.7.2–4.7.5)
- the asymptotic behavior of Hodge structures and p-adic Hodge structures associated to a degenerating family of motives over a number field (4.7.6, 4.7.7), and
- modular symbols over function fields (4.7.8, 4.7.9).

4.7.2 In [21], Pink constructed a compactification of the moduli space of Drinfeld modules that is similar to the Satake-Baily-Borel compactification of the moduli space of polarized abalian varieties. In a special case, it had been previously constructed by Kapranov [15].

In [20], Pink, sketched a method for constructing a compactification of the moduli space of Drinfeld modules that is similar to the toroidal compactification of the moduli space of polarized abelian varieties (in part, using ideas of K. Fujiwara). However, the details of the construction have not been published. Our plan for constructing toroidal compactifications seems to be different from that of [20].

4.7.3 We give a rough outline of the relationship that we envision between this paper and the analytic theory of toroidal compactifications. Suppose that F is a function field, and fix a place v of F. Let O be the ring of all elements of F which are integral outside v. In [6], the notion of a Drinfeld O-module of rank d is defined, and the moduli space of such Drinfeld modules is constructed.

Let $\mathbb{C}_v$ be the completion of an algebraic closure of F_v and let $|\ |: \mathbb{C}_v \to \mathbb{R}_{\geqslant 0}$ be the absolute value which extends the normalized absolute value of F_v. Let $\Omega \subset \mathbb{P}^{d-1}(\mathbb{C}_v)$ be the $(d-1)$-dimensional Drinfeld upper half-space consisting of all points $(z_1 : \ldots : z_d) \in \mathbb{P}^{d-1}(\mathbb{C}_v)$ such that $(z_i)_{1 \leqslant i \leqslant d}$ is linearly independent over F_v.

For a compact open subgroup K of $\mathrm{GL}_d(\mathbb{A}_F^v)$, the set of $\mathbb{C}_v$-points of the moduli space M_K of Drinfeld O-modules of rank d with K-level structure is expressed as

$$M_K(\mathbb{C}_v) = \mathrm{GL}_d(F)\backslash(\Omega \times \mathrm{GL}_d(\mathbb{A}_F^v)/K)$$

(see [6]).

Consider the case $V = F^d$ in §3 and §4. We have a map $\Omega \to X_v$ which sends $(z_1 : \cdots : z_d) \in \Omega$ to the class of the the norm $V_v = F_v^d \to \mathbb{R}_{\geqslant 0}$ given by $(a_1, \ldots, a_d) \mapsto |\sum_{i=1}^d a_i z_i|$ for $a_i \in F_v$. This map induces a canonical continuous map

(1) $M_K(\mathbb{C}_v) = \mathrm{GL}_d(F)\backslash(\Omega \times \mathrm{GL}_d(\mathbb{A}_F^v)/K) \to \mathrm{GL}_d(F)\backslash(X_v \times \mathrm{GL}_d(\mathbb{A}_F^v)/K)$.

The map (1) extends to a canonical continuous map

(2) $\bar{M}_K^{\mathrm{KP}}(\mathbb{C}_v) \to \mathrm{GL}_d(F)\backslash(\bar{X}_{F,v}^{\flat} \times \mathrm{GL}_d(\mathbb{A}_F^v)/K)$,

where $\bar{M}_K^{\mathrm{KP}}$ denotes the compactification of Kapranov and Pink of M_K. In particular, $\bar{M}_K^{\mathrm{KP}}$ is related to $\bar{X}_{F,v}^{\flat}$. On the other hand, the toroidal compactifications of M_K should be related to $\bar{X}_{F,v}$. If we denote by $\bar{M}_K^{\mathrm{tor}}$ the projective limit of all toroidal compactifications of M_K, then the map of (1) should extend to a canonical continuous map

(3) $\bar{M}_K^{\mathrm{tor}}(\mathbb{C}_v) \to \mathrm{GL}_d(F)\backslash(\bar{X}_{F,v} \times \mathrm{GL}_d(\mathbb{A}_F^v)/K)$.

that fits in a commutative diagram

$$\begin{array}{ccc} \bar{M}_K^{\mathrm{tor}}(\mathbb{C}_v) & \longrightarrow & \mathrm{GL}_d(F)\backslash(\bar{X}_{F,v} \times \mathrm{GL}(\mathbb{A}_F^v)/K) \\ \downarrow & & \downarrow \\ M_K^{\mathrm{KP}}(\mathbb{C}_v) & \longrightarrow & \mathrm{GL}_d(F)\backslash(\bar{X}^{\flat}_{F,v} \times \mathrm{GL}_d(\mathbb{A}_F^v)/K). \end{array}$$

4.7.4 The expected map of 4.7.3(3) is the analogue of the canonical continuous map from the projective limit of all toroidal compactifications of the moduli space of polarized abelian varieties to the reductive Borel–Serre compactification (see [10, 16]).

From the point of view of our study, the reductive Borel–Serre compactification and $\bar{X}_{F,v}$ are enlargements of spaces of norms. A polarized abelian variety A gives a norm on the polarized Hodge structure associated to A (the Hodge metric). This relationship between a polarized abelian variety and a norm forms the foundation of the relationship between the toroidal compactifications of a moduli space of polarized abelian varieties and the reductive Borel–Serre compactification. This is similar to the relationship between M_K and the space of norms X_v given by the map of 4.7.3(1), as well as the relationship between $\bar{M}_K^{\mathrm{tor}}$ and $\bar{X}_{F,v}$ given by 4.7.3(3).

4.7.5 In the usual theory of toroidal compactifications, cone decompositions play an important role. In the toroidal compactifications of 4.7.3, the simplices of Bruhat-Tits buildings (more precisely, the simplices contained in the fibers of $\bar{X}_{F,v} \to \bar{X}^{\flat}_{F,v}$) should play the role of the cones in cone decompositions.

4.7.6 We are preparing a paper in which our space $\bar{X}_{F,S}$ with F a number field and with S containing a non-archimedean place appears in the following way.

Let F be a number field, and let Y be a polarized projective smooth variety over F. Let $m \geqslant 0$, and let $V = H^m_{\mathrm{dR}}(Y)$ be the de Rham cohomology. For a place v of F, let $V_v = F_v \otimes_F V$.

For an archimedean place v of F, it is well known that V_v has a Hodge metric. For v non-archimedean, we can under certain assumptions define a Hodge metric on V_v by the method illustrated in the example of 4.7.7 below. The $[F_v : \mathbb{Q}_v]$-powers of these Hodge metrics for $v \in S$ are norms and therefore provide an element of $\prod_{v\in S} X_{V_v}$. When Y degenerates, this element of $\prod_{v\in S} X_{V_v}$ can converge to a boundary point of $\bar{X}_{F,S}$.

4.7.7 Let Y be an elliptic curve over a number field F, and take $m = 1$.

Let v be a non-archimedean place of F, and assume that $F_v \otimes_F Y$ is a Tate elliptic curve of q-invariant $q_v \in F_v^\times$ with $|q_v| < 1$. Then the first log-crystalline cohomology group D of the special fiber of this elliptic curve is a free module of rank 2 over the Witt vectors $W(k_v)$ with a basis (e_1, e_2) on which the Frobenius φ acts as $\varphi(e_1) = e_1$ and $\varphi(e_2) = pe_2$, where p is the characteristic of k_v. The first ℓ-adic étale cohomology group of this elliptic curve is a free module of rank 2 over $\mathbb{Z}_\ell$ with a basis $(e_{1,\ell}, e_{2,\ell})$ such that the inertia subgroup of $\mathrm{Gal}(\bar{F}_v/F_v)$ fixes e_1. The monodromy operator N satisfies

$$Ne_2 = \xi_v' e_1, \quad Ne_1 = 0, \quad Ne_{2,\ell} = \xi_v' e_{1,\ell}, \quad Ne_{1,\ell} = 0$$

where $\xi'_v = \mathrm{ord}_{F_v}(q_v) > 0$. The standard polarization $\langle\ ,\ \rangle$ of the elliptic curve satisfies $\langle e_1, e_2\rangle = 1$ and hence

$$\langle Ne_2, e_2\rangle = \xi'_v, \quad \langle e_1, N^{-1}e_1\rangle = 1/\xi'_v \quad \langle Ne_{2,\ell}, e_{2,\ell}\rangle = \xi'_v, \quad \langle e_{1,\ell}, N^{-1}e_{1,\ell}\rangle = 1/\xi'_v.$$

For $V = H^1_{\mathrm{dR}}(Y)$, we have an isomorphism $V_v \cong F_v \otimes_{W(k_v)} D$. The Hodge metric $|\ |_v$ on V_v is defined by

$$|a_1e_1 + a_2e_2|_v = \max(\xi_v^{-1/2}|a_1|_p, \xi_v^{1/2}|a_2|_p)$$

for $a_1, a_2 \in F_v$, where $|\ |_p$ denotes the absolute value on F_v satisfying $|p|_p = p^{-1}$ and

$$\xi_v := -\xi'_v \log(|\varpi_v|_p) = -\log(|q_v|_p) > 0,$$

where ϖ_v is a prime element of F_v. That is, to define the Hodge metric on V_v, we modify the naive metric (coming from the integral structure of the log-crystalline cohomology) by using ξ_v which is determined by the polarization $\langle\ ,\ \rangle$, the monodromy operator N, and the integral structures of the log-crystalline and ℓ-adic cohomology groups (for $\ell \neq p$).

This is similar to what happens at an archimedean place v. We have $Y(\mathbb{C}) \cong \mathbb{C}^\times/q_v^{\mathbb{Z}}$ with $q_v \in F_v^\times$. Assume for simplicity that we can take $|q_v| < e^{-2\pi}$ where $|\ |$ denotes the usual absolute value. Then q_v is determined by $F_v \otimes_F Y$ uniquely. Let $\xi := -\log(|q_v|) > 2\pi$. If v is real, we further assume that $q_v > 0$ and that we have an isomorphism $Y(F_v) \cong F_v^\times/q_v^{\mathbb{Z}}$ which is compatible with $Y(\mathbb{C}) \cong \mathbb{C}^\times/q_v^{\mathbb{Z}}$. Then in the case v is real (resp., complex), there is a basis (e_1, e_2) of V_v such that $(e_1, (2\pi i)^{-1}e_2)$ is a $\mathbb{Z}$-basis of $H^1(Y(\mathbb{C}), \mathbb{Z})$ and such that the Hodge metric $|\ |_v$ on V_v satisfies $|e_1|_v = \xi_v^{-1/2}$ and $|e_2|_v = \xi_v^{1/2}$ (resp., $||e_2|_v - \xi_v^{1/2}| \leqslant \pi\xi_v^{-1/2}$).

Consider for example the elliptic curves $y^2 = x(x-1)(x-t)$ with $t \in F = \mathbb{Q}$, $t \neq 0, 1$. As t approaches $1 \in \mathbb{Q}_v$ for all $v \in S$, the elliptic curves $F_v \otimes_F Y$ satisfy the above assumptions for all $v \in S$, and each q_v approaches 0, so ξ_v tends to ∞. The corresponding elements of $\prod_{v\in S} X_{V_v}$ defined by the classes of the $|\ |_v$ for $v \in S$ converge to the unique boundary point of $\bar{X}_{F,S}$ with associated parabolic equal to the Borel subgroup of upper triangular matrices in PGL_V for the basis (e_1, e_2).

We hope that this subject about $\bar{X}_{F,S}$ is an interesting direction to be studied. It may be related to the asymptotic behaviors of heights of motives in degeneration studied in [18].

4.7.8 Suppose that F is a function field and let v be a place of F. Let Γ be as in 1.3.

Kondo and Yasuda [17] proved that the image of $H_{d-1}(\Gamma\backslash X_v, \mathbb{Q}) \to H^{\mathrm{BM}}_{d-1}(\Gamma\backslash X_v, \mathbb{Q})$ is generated by modular symbols, where H^{BM}_* denotes Borel-Moore homology. Our hope is that the compactification $\Gamma\backslash\bar{X}_{F,v}$ of $\Gamma\backslash X_v$ is useful in further studies of modular symbols.

Let $\partial := \bar{X}_{F,v} \setminus X_v$. Then we have an isomorphism

$$H^{\mathrm{BM}}_*(\Gamma\backslash X_v, \mathbb{Q}) \cong H_*(\Gamma\backslash\bar{X}_{F,v}, \Gamma\backslash\partial, \mathbb{Q})$$

and an exact sequence

$$\cdots \to H_i(\Gamma\backslash\bar{X}_{F,v}, \mathbb{Q}) \to H_i(\Gamma\backslash\bar{X}_{F,v}, \Gamma\backslash\partial, \mathbb{Q}) \\ \to H_{i-1}(\Gamma\backslash\partial, \mathbb{Q}) \to H_{i-1}(\Gamma\backslash\bar{X}_{F,v}, \mathbb{Q}) \to \cdots.$$

Since $\Gamma\backslash X_v \to \Gamma\backslash\bar{X}_{F,v}$ is a homotopy equivalence by Theorem 4.6.3, we have

$$H_*(\Gamma\backslash X_v, \mathbb{Q}) \xrightarrow{\sim} H_*(\Gamma\backslash\bar{X}_{F,v}, \mathbb{Q}).$$

Hence the result of Kondo and Yasuda shows that the kernel of

$$H_{d-1}^{\mathrm{BM}}(\Gamma\backslash X_v, \mathbb{Q}) \cong H_{d-1}(\Gamma\backslash\bar{X}_{F,v}, \Gamma\backslash\partial, \mathbb{Q}) \to H_{d-2}(\Gamma\backslash\partial, \mathbb{Q})$$

is generated by modular symbols.

If we want to prove that $H_{d-1}^{\mathrm{BM}}(\Gamma\backslash X_v, \mathbb{Q})$ is generated by modular symbols, it is now sufficient to prove that the kernel of $H_{d-2}(\Gamma\backslash\partial, \mathbb{Q}) \to H_{d-2}(\Gamma\backslash\bar{X}_{F,v}, \mathbb{Q})$ is generated by the images (i.e., boundaries) of modular symbols.

4.7.9 In 4.7.8, assume $d = 2$. Then we can prove that $H_1^{\mathrm{BM}}(\Gamma\backslash X_v, \mathbb{Q})$ is generated by modular symbols. In this case $H_0(\Gamma\backslash\partial, \mathbb{Q}) = \mathrm{Map}(\Gamma\backslash\partial, \mathbb{Q}) \to H_0(\bar{X}_{F,v}, \mathbb{Q}) = \mathbb{Q}$ is just the summation map and it is clear that the kernel of it is generated by the boundaries of modular symbols.

Acknowledgements The work of the first two authors was supported in part by the National Science Foundation under Grant No. 1001729. The work of the third author was partially supported by the National Science Foundation under Grant Nos. 1401122/1661568 and 1360583, and by a grant from the Simons Foundation (304824 to R.S.).

References

1. Borel, A.: Some finiteness properties of adele groups over number fields. Publ. Math. Inst. Hautes Études Sci. **16**, 5–30 (1963)
2. Borel, A., Ji, L.: Compactifications of Symmetric and Locally Symmetric Spaces, Mathematics: Theory & Applications. Birkhäuser, Boston, MA (2006)
3. Borel, A., Serre, J.P.: Corners and arithmetic groups. Comment. Math. Helv. **4**, 436–491 (1973)
4. Bruhat, F., Tits, J.: Groupes réductifs sur un corps local: I. Données radicielles valuées, Publ. Math. Inst. Hautes Études Sci. **41**, 5–251 (1972)
5. Deligne, P., Husemöller, D.: Survey of Drinfel'd modules, Current trends in arithmetical algebraic geometry (Arcata, Calif., 1985) Contemp. Math. **67**, Amer. Math. Soc. Providence, RI, 25–91 (1987)
6. Drinfeld, V.G.: *Elliptic modules*, Mat. Sb. (N.S.) **94**(136), 594–627, 656 (1974) (Russian), English translation: Math. USSR-Sb. **23**, 561–592 (1974)
7. Gérardin, P.: *Harmonic functions on buildings of reductive split groups*, Operator algebras and group representations, Monogr. Stud. Math. **17**, Pitman, Boston, MA, 208–221 (1984)
8. Godement, R.: *Domaines fondamentaux des groupes arithmétiques*, Séminaire Bourbaki **8**(257), 201–205 (1962–1964)

9. Goldman, O., Iwahori, N.: The space of p-adic norms. Acta Math. **109**, 137–177 (1963)
10. Goresky, M., Tai, Y.: Toroidal and reductive Borel-Serre compactifications of locally symmetric spaces. Amer. J. Math. **121**, 1095–1151 (1999)
11. Guivarc'h, Y., Rémy, B.: Group-theoretic compactifcation of Bruhat-Tits buildings. Ann. Sci. Éc. Norm. Supér. **39**, 871–920 (2006)
12. Harder, G.: Minkowskische Reduktionstheorie über Funktionenkörpern. Invent. Math. **7**, 33–54 (1969)
13. Harder, G.: Chevalley groups over function fields and automorphic forms. Ann. Math. **100**, 249–306 (1974)
14. Ji, L., Murty, V.K., Saper, L., Scherk, J.: The fundamental group of reductive Borel-Serre and Satake compactifications. Asian J. Math. **19**, 465–485 (2015)
15. Kapranov, M.M.: Cuspidal divisors on the modular varieties of elliptic modules. Izv. Akad. Nauk SSSR Ser. Mat. **51**, 568–583, 688 (1987), English translation: Math. USSR-Izv. **30**, 533–547 (1988)
16. Kato, K., Usui, S.: Classifying spaces of degenerating polarized Hodge structures. Ann. Math. Stud., Princeton Univ. Press (2009)
17. Kondo, S., Yasuda, S.: The Borel-Moore homology of an arithmetic quotient of the Bruhat-Tits building of PGL of a non-archimedean local field in positive characteristic and modular symbols, preprint, arXiv:1406.7047
18. Koshikawa, T.: On heights of motives with semistable reduction, preprint, arXiv:1505.01873
19. Landvogt, E.: A compactification of the Bruhat-Tits building. In: Lecture Notes in Mathematics, vol. 1619. Springer, Berlin (1996)
20. Pink, R.: On compactification of Drinfeld moduli schemes. Sûrikaisekikenkyûsho Kôkyûroku **884**, 178–183 (1994)
21. Pink, R.: Compactification of Drinfeld modular varieties and Drinfeld modular forms of arbitrary rank. Manuscripta Math. **140**, 333–361 (2013)
22. Satake, I.: On representations and compactifications of symmetric Riemannian spaces. Ann. Math. **71**, 77–110 (1960)
23. Satake, I.: On compactifications of the quotient spaces for arithmetically defined discontinuous groups. Ann. Math. **72**, 555–580 (1960)
24. Werner, A.: Compactification of the Bruhat-Tits building of PGL by lattices of smaller rank. Doc. Math. **6**, 315–341 (2001)
25. Werner, A.: Compactification of the Bruhat-Tits building of PGL by semi-norms. Math. Z. **248**, 511–526 (2004)
26. Zucker, S.: L^2-cohomology of warped products and arithmetic groups. Invent. Math. **70**, 169–218 (1982)
27. Zucker, S.: Satake compactifications. Comment. Math. Helv. **58**, 312–343 (1983)

On the Structure of Selmer Groups

Ralph Greenberg

1 Introduction

Our objective in this paper is to prove a rather broad generalization of some classical theorems in Iwasawa theory. We begin by recalling two of those old results. The first is a theorem of Iwasawa, which we state in terms of Galois cohomology. Suppose that K is a totally real number field and that ψ is a totally odd Hecke character for K of finite order. We can view ψ as a character of the absolute Galois group G_K. Let K_ψ be the corresponding cyclic extension of K and let $\Delta = \mathrm{Gal}(K_\psi/K)$. Then ψ becomes a faithful character of Δ and K_ψ is a CM field. Now let p be an odd prime. For simplicity, we will assume that the order of ψ divides $p-1$. We can then view ψ as a character with values in $\mathbf{Z}_p^\times$. Let D be the Galois module which is isomorphic to $\mathbf{Q}_p/\mathbf{Z}_p$ as a group and on which G_K acts by ψ. Let K_∞ denote the cyclotomic $\mathbf{Z}_p$-extension of K. Thus $\Gamma = \mathrm{Gal}(K_\infty/K)$ is isomorphic to $\mathbf{Z}_p$. Define $S(K_\infty, D)$ to be the subgroup of $H^1(K_\infty, D)$ consisting of everywhere unramified cocycle classes. As is usual in Iwasawa theory, we can view $S(K_\infty, D)$ as a discrete Λ-module, where $\Lambda = \mathbf{Z}_p[[\Gamma]]$ is the completed group algebra for Γ over $\mathbf{Z}_p$. Iwasawa's theorem asserts that the Pontryagin dual of $S(K_\infty, D)$ has no nonzero, finite Λ-submodules.

The Selmer group for the above Galois module D over K_∞, as defined in [3], is precisely $S(K_\infty, D)$. Let $K_{\infty,\psi} = K_\psi K_\infty$, the cyclotomic $\mathbf{Z}_p$-extension of K_ψ. Under the restriction map $H^1(K_\infty, D) \to H^1(K_{\infty,\psi}, D)^\Delta$, one can identify $S(K_\infty, D)$ with $\mathrm{Hom}(X^{(\psi)}, D)$, where X is a certain Galois group on which Δ acts.

Dedicated to John Coates

Research supported in part by National Science Foundation grant DMS-0200785.

R. Greenberg (✉)
Department of Mathematics, University of Washington, Box 354350, Seattle, WA 98195-4350, USA
e-mail: greenber@math.washington.edu

D. Loeffler and S.L. Zerbes (eds.), *Elliptic Curves, Modular Forms and Iwasawa Theory*, Springer Proceedings in Mathematics & Statistics 188, DOI 10.1007/978-3-319-45032-2_6

To be precise, one takes $X = \mathrm{Gal}(L_\infty/K_{\infty,\psi})$, where L_∞ denotes the maximal, abelian pro-p extension of $K_{\infty,\psi}$ which is unramified at all primes of $K_{\infty,\psi}$. There is a canonical action of $G = \mathrm{Gal}(K_{\infty,\psi}/K)$ on X (defined by conjugation). Furthermore, we can identify Δ and Γ with $\mathrm{Gal}(K_{\infty,\psi}/K_\infty)$ and $\mathrm{Gal}(K_{\infty,\psi}/K_\psi)$, respectively, so that $G \cong \Delta \times \Gamma$. We define $X^{(\psi)}$ to be $e_\psi X$, where $e_\psi \in \mathbf{Z}_p[\Delta]$ is the idempotent for ψ. Iwasawa proved that $X^{(\psi)}$ has no nonzero, finite Λ-submodules. The theorem stated above is equivalent to that result.

To state the second classical result, suppose that K is any number field and that E is an elliptic curve defined over K with good, ordinary reduction at the primes of K lying above p. The p-primary subgroup $\mathrm{Sel}_E(K_\infty)_p$ of the Selmer group for E over K_∞ is again a discrete Λ-module. If $D = E[p^\infty]$, then $\mathrm{Sel}_E(K_\infty)_p$ can again be identified with the Selmer group for the Galois module D over K_∞ as defined in [3]. Its Pontryagin dual $X_E(K_\infty)$ is a finitely-generated Λ-module. Mazur conjectured that $X_E(K_\infty)$ is a torsion Λ-module. If this is so, and if one makes the additional assumption that $E(K)$ has no element of order p, then one can show that $X_E(K_\infty)$ has no nonzero, finite Λ-submodule.

The above results take the following form: $\mathcal{S}$ is a certain discrete Λ-module. The Pontryagin dual $\mathcal{X} = \mathrm{Hom}(\mathcal{S}, \mathbf{Q}_p/\mathbf{Z}_p)$ is finitely generated as a Λ-module. The results assert that $\mathcal{X}$ has no nonzero finite Λ-submodule. An equivalent statement about $\mathcal{S}$ is the following: *There exists a nonzero element $\theta \in \Lambda$ such that $\pi\mathcal{S} = \mathcal{S}$ for all irreducible elements $\pi \in \Lambda$ not dividing θ.* We then say that $\mathcal{S}$ is an "*almost divisible*" Λ-module. Note that Λ is isomorphic to $\mathbf{Z}_p[[T]]$, a formal power series ring over $\mathbf{Z}_p$ in one variable, and hence is a unique factorization domain. Thus, one can equivalently say that $\lambda\mathcal{S} = \mathcal{S}$ for all $\lambda \in \Lambda$ which are relatively prime to θ. This definition makes sense in a much more general setting, as we now describe.

Suppose that Λ is isomorphic to a formal power series ring over $\mathbf{Z}_p$, or over $\mathbf{F}_p$, in a finite number of variables. Suppose that $\mathcal{S}$ is a discrete Λ-module and that its Pontryagin dual $\mathcal{X}$ is finitely generated. We then say that $\mathcal{S}$ is a cofinitely generated Λ-module. We say that $\mathcal{S}$ is an *almost divisible* Λ-module if any one of the five equivalent statements given below is satisfied. In the statements, the set of prime ideals of Λ of height 1 is denoted by $\mathrm{Spec}_{ht=1}(\Lambda)$. Note that since Λ is a UFD, all such prime ideals Π are principal. Also, if we say *almost all*, we mean *all but finitely many*. The notation $\mathcal{X}[\Pi]$ denotes the Λ-submodule of $\mathcal{X}$ consisting of elements annihilated by Π. This is also denoted by $\mathcal{X}[\pi]$, where π is a generator of Π. In the fifth statement, recall that a finitely-generated Λ-module $\mathcal{Z}$ is said to be pseudo-null if there exist two relatively prime elements of Λ which annihilate $\mathcal{Z}$.

- *We have $\Pi\mathcal{S} = \mathcal{S}$ for almost all $\Pi \in \mathrm{Spec}_{ht=1}(\Lambda)$.*
- *There exists a nonzero element θ in Λ such that $\pi\mathcal{S} = \mathcal{S}$ for all irreducible elements π of Λ not dividing θ.*
- *We have $\mathcal{X}[\Pi] = 0$ for almost all $\Pi \in \mathrm{Spec}_{ht=1}(\Lambda)$.*
- *The set $\mathrm{Ass}_\Lambda(\mathcal{Y})$ of associated prime ideals for the torsion Λ-submodule $\mathcal{Y}$ of $\mathcal{X}$ contains only prime ideals of height 1.*
- *The Λ-module $\mathcal{X}$ has no nonzero, pseudo-null submodules.*

We refer the reader to [6] (and Proposition 2.4, in particular) for further discussion, including an explanation of the equivalence of all of the above statements.

We will consider *Selmer groups* that arise in the following very general context. Suppose that K is a finite extension of $\mathbf{Q}$ and that Σ is a finite set of primes of K. Let K_Σ denote the maximal extension of K unramified outside of Σ. We assume that Σ contains all archimedean primes and all primes lying over some fixed rational prime p. The Selmer groups that we consider in this article are associated to a continuous representation

$$\rho : \mathrm{Gal}(K_\Sigma/K) \longrightarrow GL_n(R)$$

where R is a complete Noetherian local ring. Let $\mathfrak{M}$ denote the maximal ideal of R. We assume that the residue field $R/\mathfrak{M}$ is finite and has characteristic p. Hence R is compact in its $\mathfrak{M}$-adic topology. Let $\mathcal{T}$ be the underlying free R-module on which $\mathrm{Gal}(K_\Sigma/K)$ acts via ρ. We define $\mathcal{D} = \mathcal{T} \otimes_R \widehat{R}$, where $\widehat{R} = \mathrm{Hom}(R, \mathbf{Q}_p/\mathbf{Z}_p)$ is the Pontryagin dual of R with a trivial action of $\mathrm{Gal}(K_\Sigma/K)$. That Galois group acts on $\mathcal{D}$ through its action on the first factor $\mathcal{T}$. Thus, $\mathcal{D}$ is a discrete abelian group which is isomorphic to $\widehat{R}^n$ as an R-module and which has a continuous R-linear action of $\mathrm{Gal}(K_\Sigma/K)$.

The Galois cohomology group $H^1(K_\Sigma/K, \mathcal{D})$ can be considered as a discrete R-module too. It is a cofinitely generated R-module in the sense that its Pontryagin dual is finitely generated as an R-module. (See Proposition 3.2 in [6].) Suppose that one specifies an R-submodule $L(K_v, \mathcal{D})$ of $H^1(K_v, \mathcal{D})$ for each $v \in \Sigma$. We will denote such a specification by $\mathcal{L}$ for brevity. Let

$$P(K, \mathcal{D}) = \prod_{v \in \Sigma} H^1(K_v, \mathcal{D}) \quad \text{and} \quad L(K, \mathcal{D}) = \prod_{v \in \Sigma} L(K_v, \mathcal{D}).$$

Thus, $L(K, \mathcal{D})$ is an R-submodule of $P(K, \mathcal{D})$. Let $\mathcal{Q}_{\mathcal{L}}(K, \mathcal{D})$ denote the quotient $P(K, \mathcal{D})\big/L(K, \mathcal{D})$. Thus,

$$\mathcal{Q}_{\mathcal{L}}(K, \mathcal{D}) = \prod_{v \in \Sigma} \mathcal{Q}_{\mathcal{L}}(K_v, \mathcal{D}), \quad \text{where} \quad \mathcal{Q}_{\mathcal{L}}(K_v, \mathcal{D}) = H^1(K_v, \mathcal{D})\big/L(K_v, \mathcal{D}).$$

The natural global-to-local restriction maps for $H^1(\,\cdot\,, \mathcal{D})$ induce a map

$$\phi_{\mathcal{L}} : H^1(K_\Sigma/K, \mathcal{D}) \longrightarrow \mathcal{Q}_{\mathcal{L}}(K, \mathcal{D}). \tag{1}$$

The kernel of $\phi_{\mathcal{L}}$ will be denoted by $S_{\mathcal{L}}(K, \mathcal{D})$. It is the "*Selmer group*" for $\mathcal{D}$ over K corresponding to the specification $\mathcal{L}$.

It is clear that $S_{\mathcal{L}}(K, \mathcal{D})$ is an R-submodule of $H^1(K_\Sigma/K, \mathcal{D})$ and so is also a discrete, cofinitely generated R-module. For a fixed set Σ, the smallest possible Selmer group occurs when we take $L(K_v, \mathcal{D}) = 0$ for all $v \in \Sigma$. The Selmer group corresponding to that choice will be denoted by $\text{Ш}^1(K, \Sigma, \mathcal{D})$. That is,

$$\text{Ш}^1(K, \Sigma, \mathcal{D}) = \ker\Big(H^1(K_\Sigma/K, \mathcal{D}) \longrightarrow \prod_{v \in \Sigma} H^1(K_v, \mathcal{D})\Big)$$

Obviously, we have $\text{Ш}^1(K, \Sigma, \mathcal{D}) \subseteq S_{\mathcal{L}}(K, \mathcal{D})$ for any choice of the specification $\mathcal{L}$.

In addition to the above assumptions about R, suppose that R is a domain. Let $d = m + 1$ denote the Krull dimension of R, where $m \geq 0$. (We will assume that R is not a field. Our results are all trivial in that case.) A theorem of Cohen [2] implies that R is a finite, integral extension of a subring Λ which is isomorphic to one of the formal power series rings $\mathbf{Z}_p[[T_1, ..., T_m]]$ or $\mathbf{F}_p[[T_1, ..., T_{m+1}]]$, depending on whether R has characteristic 0 or p. Although such a subring is far from unique, it will be convenient to just fix a choice. A cofinitely generated R-module $\mathcal{S}$ will also be a cofinitely generated Λ-module. All the results that we will prove in this paper could be viewed as statements about the structure of the Selmer groups as R-modules. But they are equivalent to the corresponding statements about their structure as Λ-modules and that is how we will formulate and prove them. Those equivalences are discussed in some detail in [6], Sect. 2. In particular, if $\mathcal{S}$ is a discrete, cofinitely generated R-module, then we say that $\mathcal{S}$ is divisible (resp., almost divisible) as an R-module if $P\mathcal{S} = \mathcal{S}$ for all (resp., almost all) $P \in \mathrm{Spec}_{ht=1}(R)$. One result is that $\mathcal{S}$ is almost divisible as an R-module if and only if $\mathcal{S}$ is almost divisible as a Λ-module. (See statement **1** on page 350 of [6].) A similar equivalence is true for divisibility, but quite easy to prove.

One basic assumption that we will make about R is that it contain a subring Λ of the form described in the previous paragraph, that R is finitely-generated as a Λ-module, and that R is also reflexive as a Λ-module. If these assumptions are satisfied, we say that R is a "*reflexive ring*". In the case where R is also assumed to be a domain, one can equivalently require that R is the intersection of all its localizations at prime ideals of height 1. See part D, Sect. 2 in [6] for the explanation of the equivalence. In the literature, one sometimes finds the term "*weakly Krull domain*" for such a domain. The class of reflexive domains is rather large. For example, if R is integrally closed or Cohen–Macaulay, then it turns out that R is reflexive. There are important examples (from Hida theory), where R is not necessarily a domain, but is still a free (and hence reflexive) module over a suitable subring Λ.

The main results of this paper assert that if we make certain hypotheses about $\mathcal{D}$ and $\mathcal{L}$, then $S_{\mathcal{L}}(K, \mathcal{D})$ will be almost divisible. Some of the hypotheses are those needed for Theorem 1 in [6] which gives sufficient conditions for $H^1(K_\Sigma/K, \mathcal{D})$ itself to be almost divisible. That theorem will be stated later (as Proposition 2.6.1.) and is our starting point. The basic approach for deducing the almost divisibility of a Λ-submodule of $H^1(K_\Sigma/K, \mathcal{D})$, defined by imposing local conditions corresponding to a specification $\mathcal{L}$, will be described in Sect. 3. Some of the needed hypotheses will be discussed in Sect. 2. We also state there some results from [7] concerning the surjectivity of the map $\phi_{\mathcal{L}}$. We will apply those results not just to $\mathcal{D}$, but also to the corresponding map for $\mathcal{D}[\Pi]$, where $\Pi \in \mathrm{Spec}_{ht=1}(\Lambda)$. Our main results concerning the almost divisibility of $S_{\mathcal{L}}(K, \mathcal{D})$ will be proved in Sect. 4.1.

We show in Sect. 4.2 how to prove the classical theorems mentioned above from the point of view of this paper.

This paper is part of a series of papers concerning foundational questions in Iwasawa theory. The results discussed above depend on the results proved in [6, 7], the first papers in this series. A subsequent paper will use the results we prove here to study the behavior of Selmer groups under specialization. In particular, one would like to understand how the "*characteristic ideal*" or "*characteristic divisor*" for a Selmer group associated to the representation $\rho_P : \mathrm{Gal}(K_\Sigma/K) \longrightarrow GL_n(R/P)$, the reduction of ρ modulo a prime ideal P of R, is related to the characteristic ideal or divisor associated to a Selmer group for ρ itself. Such a question has arisen many times in the past. Consequently, for the purpose of studying exactly that question, one can find numerous special cases of the results of this paper in the literature on Iwasawa theory.

2 Various Hypotheses

The R-module $\mathcal{T}$ is a free R-module and so we say that $\mathcal{D}$ is a cofree R-module. We also define $\mathcal{T}^* = \mathrm{Hom}(\mathcal{D}, \mu_{p^\infty})$. We can consider $\mathcal{T}^*$ as a module over the ring R^{op}, which is just R since that ring is commutative. It is clear that $\mathcal{T}^*$ is also a free R-module and that the discrete R-module $\mathcal{D}^* = \mathcal{T}^* \otimes_R \widehat{R}$ is cofree. It will be simpler and more useful to formulate the hypotheses in terms of their structure as Λ-modules rather than R-modules.

2.1 *Hypotheses Involving Reflexivity*

Recall that Λ is isomorphic to a formal power series ring in a finite number of variables over either $\mathbf{Z}_p$ or $\mathbf{F}_p$. Reflexive Λ-modules play an important role here. A detailed discussion of the definition can be found in Sect. 2, part C, of [6]. We almost always will assume the following hypothesis.

- RFX($\mathcal{D}$): *The Λ-module $\mathcal{T}$ is reflexive.*

Equivalently, since $\mathcal{T}$ is free as an R-module, RFX($\mathcal{D}$) means that the ring R is reflexive as a Λ-module. That is, R is a reflexive ring in the sense defined in the introduction. We are still always assuming that R is a complete Noetherian local ring with finite residue field of characteristic p.

We will say that $\mathcal{D}$ is a coreflexive Λ-module if RFX($\mathcal{D}$) holds. This terminology is appropriate because $\mathcal{D}$ is isomorphic to the Λ-module $\widehat{R}^n$ (ignoring the Galois action) and its Pontryagin dual is the reflexive Λ-module R^n. One important role of RFX($\mathcal{D}$) is to guarantee that $\mathcal{D}[\pi]$ is a divisible (Λ/Π)-module for all prime ideals $\Pi = (\pi)$ in Λ of height 1. That property is equivalent to requiring that $\mathcal{D}$ be coreflexive as a Λ-module. See Corollary 2.6.1 in [6] for the proof.

The next two hypotheses involve $\mathcal{T}^*$ and are of a local nature. They could be formulated just in terms of $\mathcal{D}$, but the statements would become more complicated. Note that if RFX($\mathcal{D}$) holds, then $\mathcal{T}^*$ is also a reflexive Λ-module. We suppose that v is a prime of K and that K_v is the completion of K at v. We usually consider just the primes $v \in \Sigma$.

- $\mathrm{LOC}_v^{(1)}(\mathcal{D})$: *We have* $(\mathcal{T}^*)^{G_{K_v}} = 0$.
- $\mathrm{LOC}_v^{(2)}(\mathcal{D})$: *The* Λ*-module* $\mathcal{T}^*/(\mathcal{T}^*)^{G_{K_v}}$ *is reflexive.*

Assumptions $\mathrm{LOC}_v^{(1)}(\mathcal{D})$ and $\mathrm{LOC}_v^{(2)}(\mathcal{D})$ play a crucial role in proving Theorem 1 in [6]. Just as in that result, we will usually assume $\mathrm{LOC}_v^{(1)}(\mathcal{D})$ for at least one non-archimedean prime $v \in \Sigma$ and $\mathrm{LOC}_v^{(2)}(\mathcal{D})$ for all $v \in \Sigma$. One can find a general discussion of when those hypotheses are satisfied in part F, Sect. 5 of [6]. One obvious remark is that since $\mathcal{T}^*$ is a torsion-free Λ-module, $\mathrm{LOC}_v^{(1)}$ is satisfied if and only if $\mathrm{rank}_\Lambda\big((\mathcal{T}^*)^{G_{K_v}}\big) = 0$. It is also obvious that $\mathcal{T}^*/(\mathcal{T}^*)^{G_{K_v}}$ is at least torsion-free as a Λ-module. Furthermore, note that if RFX($\mathcal{D}$) is true, then $\mathrm{LOC}_v^{(2)}(\mathcal{D})$ follows from $\mathrm{LOC}_v^{(1)}(\mathcal{D})$. Notice also that if $\mathrm{LOC}_v^{(1)}(\mathcal{D})$ and $\mathrm{LOC}_v^{(2)}(\mathcal{D})$ are both true for some prime v, then RFX($\mathcal{D}$) is also true. Nevertheless, our propositions will often include RFX($\mathcal{D}$) as a hypothesis even though it may actually be implied by other hypotheses.

2.2 Locally Trivial Cocycle Classes

The following much more subtle hypothesis is also needed in the proof of Theorem 1 in [6], where it is referred to as Hypothesis L. As we explain there, it can be viewed as a generalization of Leopoldt's Conjecture for number fields. That special case occurs when $\Lambda = \mathbf{Z}_p$, $\mathcal{D} = \mathbf{Q}_p/\mathbf{Z}_p$, and G_K acts trivially on $\mathcal{D}$. For the formulation, we define

$$Ш^2(K, \Sigma, \mathcal{D}) \;=\; \ker\Big(H^2(K_\Sigma/K, \mathcal{D}) \longrightarrow \prod_{v\in\Sigma} H^2(K_v, \mathcal{D})\Big),$$

which is a discrete, cofinitely-generated Λ-module.

- LEO($\mathcal{D}$): *The* Λ*-module* $Ш^2(K, \Sigma, \mathcal{D})$ *is cotorsion.*

A long discussion about the validity of the above hypothesis can be found in the last few pages of Sect. 6, part D, in [6]. There are situations where it fails to be satisfied. Also, Sect. 4 of that paper derives a natural lower bound on the Λ-corank of $H^1(K_\Sigma/K, \mathcal{D})$ from the duality theorems of Poitou and Tate. Hypothesis LEO($\mathcal{D}$) is equivalent to the statement that $\mathrm{corank}_\Lambda\big(H^1(K_\Sigma/K, \mathcal{D})\big)$ is equal to that lower bound. That equivalence is the content of Propositions 4.3 and 4.4 in [6]. Furthermore, one part of Theorem 1 in that paper asserts that if RFX($\mathcal{D}$) is satisfied, and if we assume $\mathrm{LOC}_v^{(1)}(\mathcal{D})$ for at least one non-archimedean prime $v \in \Sigma$ and $\mathrm{LOC}_v^{(2)}(\mathcal{D})$ for all $v \in \Sigma$, then $\mathrm{LEO}(\mathcal{D})$ means that $\text{Ш}^2(K, \Sigma, \mathcal{D})$ actually vanishes.

2.3 Hypotheses Involving $\mathcal{L}$

None of the hypotheses stated above involves the specification $\mathcal{L}$. We now mention two hypotheses which do involve $\mathcal{L}$, one of which implies the other. They are statements about the cokernel of the map $\phi_{\mathcal{L}}$ defined in the introduction. The first plays an important role in studying Selmer groups. The second appears weaker, but often is sufficient to imply the first.

- $\mathrm{SUR}(\mathcal{D}, \mathcal{L})$: *The map $\phi_{\mathcal{L}}$ defining $S_{\mathcal{L}}(K, \mathcal{D})$ is surjective.*

An obvious necessary condition for this to be satisfied is the following equality for the coranks:

- $\mathrm{CRK}(\mathcal{D}, \mathcal{L})$: *We have*

$$\mathrm{corank}_{\Lambda}\left(H^1(K_{\Sigma}/K, \mathcal{D})\right) = \mathrm{corank}_{\Lambda}\left(S_{\mathcal{L}}(K, \mathcal{D})\right) + \mathrm{corank}_{\Lambda}\left(Q_{\mathcal{L}}(K, \mathcal{D})\right).$$

This just means that $\mathrm{coker}(\phi_{\mathcal{L}})$ is a cotorsion Λ-module. Proposition 3.2.1 in [7] shows that $\mathrm{CRK}(\mathcal{D}, \mathcal{L})$, together with various additional assumptions, actually implies $\mathrm{SUR}(\mathcal{D}, \mathcal{L})$. One has the following obvious inequality:

$$\mathrm{corank}_{\Lambda}\left(S_{\mathcal{L}}(K, \mathcal{D})\right) \geqslant \mathrm{corank}_{\Lambda}\left(H^1(K_{\Sigma}/K, \mathcal{D})\right) - \mathrm{corank}_{\Lambda}\left(Q_{\mathcal{L}}(K, \mathcal{D})\right) \quad (2)$$

Thus, $\mathrm{CRK}(\mathcal{D}, \mathcal{L})$ is equivalent to having equality here. Of course, $\mathrm{CRK}(\mathcal{D}, \mathcal{L})$, and hence $\mathrm{SUR}(\mathcal{D}, \mathcal{L})$, can fail simply because the quantity on the right side is negative. Verifying $\mathrm{CRK}(\mathcal{D}, \mathcal{L})$ is quite a difficult problem in many interesting cases.

It is worth recalling what the formulas for global and local Euler–Poincaré characteristics tell us about the coranks on the right side of (2). One can find proofs in Sect. 4 of [6]. For any prime v of K, we use the notation $\mathcal{D}(K_v)$ as an abbreviation for $H^0(K_v, \mathcal{D})$, a Λ-submodule of $\mathcal{D}$. Similarly, $\mathcal{D}(K)$ will denote $H^0(K, \mathcal{D})$. Let $r_1(K)$ and $r_2(K)$ denote the number of real primes and complex primes of K, respectively. We give formulas for the Λ-coranks of the global and local H^1's. For the global H^1, we have

$$\begin{aligned}&\mathrm{corank}_{\Lambda}\left(H^1(K_{\Sigma}/K, \mathcal{D})\right)\\ &\quad = \mathrm{corank}_{\Lambda}\left(\mathcal{D}(K)\right) + \mathrm{corank}_{\Lambda}\left(H^2(K_{\Sigma}/K, \mathcal{D})\right) + \delta_{\Lambda}(K, \mathcal{D}),\end{aligned}$$

where $\delta_{\Lambda}(K, \mathcal{D}) = \left(r_1(K) + r_2(K)\right)\mathrm{corank}_{\Lambda}\left(\mathcal{D}\right) - \sum_{v \text{ real}} \mathrm{corank}_{\Lambda}\left(\mathcal{D}(K_v)\right)$.

Now assume that v is a non-archimedean prime. Recall that $\mathcal{D}^*$ denotes the R-module $\mathcal{T}^* \otimes_R \widehat{R}$. If v does not lie over p, then the local Euler–Poincaré characteristic is 0 and we have

$$\operatorname{corank}_\Lambda \left(H^1(K_v, \mathcal{D})\right) = \operatorname{corank}_\Lambda \left(\mathcal{D}(K_v)\right) + \operatorname{corank}_\Lambda \left(\mathcal{D}^*(K_v)\right).$$

To justify replacing the Λ-corank of $H^2(K_v, \mathcal{D})$ by that of $\mathcal{D}^*(K_v)$ in the above formula as well as the formula below, one uses the fact that the Pontryagin dual of $H^2(K_v, \mathcal{D})$ is isomorphic to $H^0(K_v, \mathcal{T}^*)$. Proposition 3.10 in [6] implies that the Λ-rank of $H^0(K_v, \mathcal{T}^*)$ is equal to the Λ-corank of $H^0(K_v, \mathcal{D}^*)$. If v lies over p, then we have

$$\begin{aligned} &\operatorname{corank}_\Lambda \left(H^1(K_v, \mathcal{D})\right) \\ &\quad = \operatorname{corank}_\Lambda \left(\mathcal{D}(K_v)\right) + \operatorname{corank}_\Lambda \left(\mathcal{D}^*(K_v)\right) + [K_v : \mathbf{Q}_p] \operatorname{corank}_\Lambda \left(\mathcal{D}\right). \end{aligned}$$

If v is archimedean, then $H^1(K_v, \mathcal{D})$ vanishes unless $p = 2$ and v is real. Even for $p = 2$, its Λ-corank is 0 unless Λ is a power series ring over $\mathbf{F}_2$. In that case, one has the following formula when v is real:

$$\operatorname{corank}_\Lambda \left(H^1(K_v, \mathcal{D})\right) = 2 \operatorname{corank}_\Lambda \left(\mathcal{D}(K_v)\right) - n.$$

Here $n = \operatorname{corank}_\Lambda \left(\mathcal{D}\right)$. See page 380 of [6] for the simple justification.

Finally, we have the obvious formula

$$\operatorname{corank}_\Lambda \left(\mathcal{Q}_{\mathcal{L}}(K_v, \mathcal{D})\right) = \operatorname{corank}_\Lambda \left(H^1(K_v, \mathcal{D})\right) - \operatorname{corank}_\Lambda \left(L(K_v, \mathcal{D})\right)$$

and so the above formulas for the Λ-coranks of $H^1(K_v, \mathcal{D})$ for $v \in \Sigma$, and the specification $\mathcal{L}$, determine the Λ-corank of $\mathcal{Q}_{\mathcal{L}}(K, D)$.

2.4 Behavior Under Specialization

In some proofs, Selmer groups for $\mathcal{D}[\Pi]$, as well as for $\mathcal{D}$, will occur. Here Π is a prime ideal of Λ and $\mathcal{D}[\Pi]$ is a discrete, cofinitely-generated module over the ring Λ/Π. Various other modules over Λ/Π will arise. Now Λ/Π is a complete, Noetherian, local ring, and therefore (just as for R in the introduction), it is a finite, integral extension of a subring Λ' which is isomorphic to a formal power series ring over $\mathbf{Z}_p$ or $\mathbf{F}_p$. We fix such a choice for each Π and denote Λ' by Λ_Π. If Λ has Krull dimension d, then Λ_Π has Krull dimension $d - 1$. Of course, some results could be easily stated or proved just in terms of Λ/Π itself.

Many of the above hypotheses are not preserved when the Λ-module $\mathcal{D}$ is replaced by the Λ_Π-module $\mathcal{D}[\Pi]$. For example, even if RFX($\mathcal{D}$) is satisfied, $\mathcal{D}[\Pi]$ may fail to be reflexive as a Λ_Π-module and so RFX($\mathcal{D}[\Pi]$) may fail to be satisfied. In general, all one can say is that RFX($\mathcal{D}$) implies that $\mathcal{D}[\Pi]$ is a divisible Λ_Π-module for all $\Pi \in \mathrm{Spec}_{ht=1}(\Lambda)$. The situation is better for $\mathrm{LOC}_v^{(1)}(\mathcal{D})$ and LEO($\mathcal{D}$). We have the following equivalences.

- *Assume that* RFX($\mathcal{D}$) *is satisfied. Then* $\mathrm{LOC}_v^{(1)}(\mathcal{D})$ *is true if and only if* $\mathrm{LOC}_v^{(1)}(\mathcal{D}[\Pi])$ *is true for almost all* $\Pi \in \mathrm{Spec}_{ht=1}(\Lambda)$.
- LEO($\mathcal{D}$) *is true if and only if* LEO($\mathcal{D}[\Pi]$) *is true for almost all* $\Pi \in \mathrm{Spec}_{ht=1}(\Lambda)$.

These assertions follow easily from results in [6]. For the first statement, one should see Remarks 3.5.1 or 3.10.2 there. Note that $\mathrm{LOC}_v^{(1)}(\mathcal{D}[\Pi])$ and $\mathrm{LOC}_v^{(2)}(\mathcal{D}[\Pi])$ are statements about the (Λ/Π)-module $\mathrm{Hom}(\mathcal{D}[\Pi], \mu_{p^\infty})$, which is isomorphic to $\mathcal{T}^*/\Pi\mathcal{T}^*$. The second of the above equivalences follows from Lemma 4.4.1 and Remark 2.1.3 in [6]. We will prove a similar equivalence for CRK($\mathcal{D}, \mathcal{L}$) in Sect. 3.4.

2.5 A Result About Almost Divisibility

In addition to SUR($\mathcal{D}, \mathcal{L}$) and CRK($\mathcal{D}, \mathcal{L}$), there will be various other hypotheses concerning the specification $\mathcal{L}$. If $L(K_v, \mathcal{D})$ is Λ-divisible (resp., almost Λ-divisible) for all $v \in \Sigma$, then we will say that $\mathcal{L}$ is Λ-divisible (resp, almost Λ-divisible). Consider another specification $\mathcal{L}'$ for $\mathcal{D}$ and let $L'(K_v, \mathcal{D})$ denote the corresponding subgroup of $H^1(K_v, \mathcal{D})$ for each $v \in \Sigma$. We write $\mathcal{L}' \subseteq \mathcal{L}$ if $L'(K_v, \mathcal{D}) \subseteq L(K_v, \mathcal{D})$ for all $v \in \Sigma$. In particular, if $L'(K_v, \mathcal{D}) = L(K_v, \mathcal{D})_{\Lambda\text{-}div}$ for each $v \in \Sigma$, then we will refer to the specification $\mathcal{L}'$ as the maximal Λ-divisible subspecification of $\mathcal{L}$, which we denote simply by $\mathcal{L}_{div}$. One assumption that we will usually make is that $\mathcal{L}$ is almost divisible. Its importance is clear from the following proposition.

Proposition 2.5.1 *Assume that $\mathcal{L}'$ and $\mathcal{L}$ are specifications for $\mathcal{D}$ and that $\mathcal{L}' \subseteq \mathcal{L}$. Assume also that SUR($\mathcal{D}, \mathcal{L}'$) is true. Then SUR($\mathcal{D}, \mathcal{L}$) is also true, we have the inclusion $S_{\mathcal{L}'}(K, \mathcal{D}) \subseteq S_{\mathcal{L}}(K, \mathcal{D})$, and*

$$S_{\mathcal{L}}(K, \mathcal{D})/S_{\mathcal{L}'}(K, \mathcal{D}) \cong \prod_{v \in \Sigma} L(K_v, \mathcal{D})/L'(K_v, \mathcal{D})$$

as Λ-modules. In particular, if SUR($\mathcal{D}, \mathcal{L}_{div}$) is true and $S_{\mathcal{L}_{div}}(K, \mathcal{D})$ is almost Λ-divisible, then $S_{\mathcal{L}}(K, \mathcal{D})$ is almost Λ-divisible if and only if $\mathcal{L}$ is almost Λ-divisible. If SUR($\mathcal{D}, \mathcal{L}_{div}$) is true and $S_{\mathcal{L}}(K, \mathcal{D})$ is almost Λ-divisible, then $\mathcal{L}$ must be almost Λ-divisible.

Thus, under certain assumptions, the structure of $S_{\mathcal{L}}(K, \mathcal{D})$ can be related to that of $S_{\mathcal{L}_{div}}(K, \mathcal{D})$ and the quotient Λ-modules $L(K_v, \mathcal{D})/L(K_v, \mathcal{D})_{div}$ for $v \in \Sigma$. Since all of those quotients are cofinitely-generated, cotorsion Λ-module for all $v \in \Sigma$, it follows that CRK($\mathcal{D}, \mathcal{L}$) is true if and only if CRK($\mathcal{D}, \mathcal{L}_{div}$) is true.

Proof Most of the statements are clear from the definitions. For the isomorphism, consider the following maps:

$$H^1(K_\Sigma/K, \mathcal{D}) \xrightarrow{\phi_{\mathcal{L}'}} \mathcal{Q}_{\mathcal{L}'}(K, \mathcal{D}) \xrightarrow{\psi} \mathcal{Q}_{\mathcal{L}}(K, \mathcal{D})$$

where ψ is the natural map, the canonical homomorphism whose kernel is the direct product in the proposition. The map ψ is surjective and the composition is $\phi_{\mathcal{L}}$. If $\phi_{\mathcal{L}'}$ is surjective, then it follows that $\phi_{\mathcal{L}}$ is also surjective and that $S_{\mathcal{L}}(K, \mathcal{D})\big/S_{\mathcal{L}'}(K, \mathcal{D})$ is isomorphic to $\ker(\psi)$. The stated isomorphism follows immediately. For the final statements, one takes $\mathcal{L}' = \mathcal{L}_{div}$. Note that if $S_{\mathcal{L}}(K, \mathcal{D})$ is almost divisible, and if one assumes $\mathrm{SUR}(\mathcal{D}, \mathcal{L}_{div})$, then there is a surjective homomorphism from $S_{\mathcal{L}}(K, \mathcal{D})$ to $L(K_v, \mathcal{D})\big/L(K_v, \mathcal{D})_{\Lambda\text{-}div}$, which must therefore be almost divisible too. This implies that $L(K_v, \mathcal{D})$ is then almost divisible for all $v \in \Sigma$. Thus, $\mathcal{L}$ is almost divisible. Moreover, if a discrete, cofinitely-generated Λ-module $\mathcal{S}$ contains an almost divisible Λ-submodule $\mathcal{S}'$, then it is clear that $\mathcal{S}$ is almost divisible if and only if $\mathcal{S}/\mathcal{S}'$ is almost divisible. □

2.6 The Main Results in [6] and [7]

The following result is proved in [6]. It is part of the Theorem 1 which we alluded to before. It plays a crucial role in this paper because we will study when $S_{\mathcal{L}}(K, \mathcal{D})$ is almost divisible as a Λ-module under the assumption that $H^1(K_\Sigma/K, \mathcal{D})$ is almost divisible, as outlined in the next section.

Proposition 2.6.1 *Suppose that* $\mathrm{RFX}(\mathcal{D})$ *and* $\mathrm{LEO}(\mathcal{D})$ *are both satisfied, that* $\mathrm{LOC}_v^{(2)}(\mathcal{D})$ *is satisfied for all* v *in* Σ*, and that there exists a non-archimedean prime* $\eta \in \Sigma$ *such that* $\mathrm{LOC}_\eta^{(1)}(\mathcal{D})$ *is satisfied. Then* $H^1(K_\Sigma/K, \mathcal{D})$ *is an almost divisible* Λ*-module.*

Another part of Theorem 1 is the following.

Proposition 2.6.2 *Suppose that* $\mathrm{RFX}(\mathcal{D})$ *is satisfied, that* $\mathrm{LOC}_v^{(2)}(\mathcal{D})$ *is both satisfied for all* v *in* Σ*, and that there exists a non-archimedean prime* $\eta \in \Sigma$ *such that* $\mathrm{LOC}_\eta^{(1)}(\mathcal{D})$ *is satisfied. Then* $Ш^2(K, \Sigma, \mathcal{D})$ *is a coreflexive* Λ*-module.*

The conclusion in this result has the interesting consequence that the Pontryagin dual of $Ш^2(K, \Sigma, \mathcal{D})$ is torsion-free as a Λ-module. It follows that $Ш^2(K, \Sigma, \mathcal{D})$ is Λ-divisible. Hence either $Ш^2(K, \Sigma, \mathcal{D})$ has positive Λ-corank or $Ш^2(K, \Sigma, \mathcal{D})$ vanishes under the assumptions in Proposition 2.6.2.

We now state the main result that we need from [7]. It is Proposition 3.2.1 there.

Proposition 2.6.3 *Suppose that* $\mathcal{D}$ *is divisible as a* Λ*-module. Assume that* $\mathrm{LEO}(\mathcal{D})$, $\mathrm{CRK}(\mathcal{D}, \mathcal{L})$*, and also at least one of the following additional assumptions is satisfied.*

(a) $\mathcal{D}[\mathfrak{m}]$ *has no subquotient isomorphic to* μ_p *for the action of* G_K,
(b) $\mathcal{D}$ *is a cofree* Λ*-module and* $\mathcal{D}[\mathfrak{m}]$ *has no quotient isomorphic to* μ_p *for the action of* G_K,
(c) There is a prime $\eta \in \Sigma$ *satisfying the following properties: (i)* $H^0(K_\eta, \mathcal{T}^*) = 0$, *and (ii)* $Q_{\mathcal{L}}(K_\eta, \mathcal{D})$ *is divisible as a* Λ*-module.*

Then $\phi_{\mathcal{L}}$ *is surjective.*

As mentioned in the introduction, we will apply the above result not just to $\mathcal{D}$, but also to $\mathcal{D}[\Pi]$ for prime ideals Π of Λ of height 1. Fortunately, if $\mathcal{D}$ is itself coreflexive as a Λ-module, then $\mathcal{D}[\Pi]$ is divisible as a (Λ/Π)-module, and hence satisfies the first hypothesis in the above proposition.

3 An Outline

3.1 An Exact Sequence

Assume that SUR$(\mathcal{D}, \mathcal{L})$ is satisfied. We will denote $\phi_{\mathcal{L}}$ just by ϕ, although we will continue to indicate the $\mathcal{L}$ for other objects. We have an exact sequence

$$0 \longrightarrow S_{\mathcal{L}}(K, \mathcal{D}) \longrightarrow H^1(K_\Sigma/K, \mathcal{D}) \stackrel{\phi}{\longrightarrow} Q_{\mathcal{L}}(K, \mathcal{D}) \longrightarrow 0 \qquad (3)$$

of discrete Λ-modules. Suppose that $\Pi \in \mathrm{Spec}_{ht=1}(\Lambda)$ and that π is a generator of Π. Applying the snake lemma to the exact sequence (3) and to the endomorphisms of each of the above modules induced by multiplication by π, we obtain the following important exact sequence. We refer to it as the *snake lemma sequence for* Π.

$$H^1(K_\Sigma/K, \mathcal{D})[\Pi] \xrightarrow{\alpha_\Pi} Q_{\mathcal{L}}(K, \mathcal{D})[\Pi] \longrightarrow S_{\mathcal{L}}(K, \mathcal{D})/\Pi S_{\mathcal{L}}(K, \mathcal{D}) \longrightarrow H^1(K_\Sigma/K, \mathcal{D})/\Pi H^1(K_\Sigma/K, \mathcal{D})$$

Now assume additionally that $H^1(K_\Sigma/K, \mathcal{D})$ is an almost divisible Λ-module. The last term in the above exact sequence is then trivial for almost all $\Pi \in \mathrm{Spec}_{ht=1}(\Lambda)$. Therefore, under these assumptions, the assertion that $S_{\mathcal{L}}(K, \mathcal{D})$ is almost divisible is equivalent to the assertion that α_Π is surjective for almost all $\Pi \in \mathrm{Spec}_{ht=1}(\Lambda)$. We study the surjectivity of α_Π by considering the (Λ/Π)-module $\mathcal{D}[\pi]$.

3.2 The Cokernel of α_Π

If $\mathcal{D}$ arises from a representation ρ as described in the introduction, and if R is a domain, then $\mathcal{D}$ will be Λ-divisible. However, here we will just assume that $\mathcal{D}$ is divisible by π and cofinitely generated as a Λ-module. We then have an exact sequence

$$0 \longrightarrow \mathcal{D}[\Pi] \longrightarrow \mathcal{D} \longrightarrow \mathcal{D} \longrightarrow 0$$

induced by multiplication by π. As a consequence, the following global and local "*specialization*" maps are surjective:

$$h_\Pi : H^1(K_\Sigma/K, \mathcal{D}[\Pi]) \longrightarrow H^1(K_\Sigma/K, \mathcal{D})[\Pi],$$
$$h_{\Pi,v} : H^1(K_v, \mathcal{D}[\Pi]) \longrightarrow H^1(K_v, \mathcal{D})[\Pi]$$

We can compare the exact sequence (3) with an analogous sequence for $\mathcal{D}[\Pi]$, viewed as a (Λ/Π)-module. For this purpose, we define a specification $\mathcal{L}_\Pi$ for $\mathcal{D}[\Pi]$ as follows: For each $v \in \Sigma$, let us take

$$L(K_v, \mathcal{D}[\Pi]) = h_{\Pi,v}^{-1}\big(L(K_v, \mathcal{D})[\Pi] \big)$$

which is a (Λ/Π)-submodule of $H^1(K_v, \mathcal{D}[\Pi])$. If we think of $\mathcal{L}$ as fixed, we will refer to the specification $\mathcal{L}_\Pi$ just defined as the "*$\mathcal{L}$-maximal specification for $\mathcal{D}[\Pi]$*". Using the analogous notation to that for $\mathcal{D}$, we define

$$P(K, \mathcal{D}[\Pi]) = \prod_{v\in\Sigma} H^1(K_v, \mathcal{D}[\Pi]),$$
$$Q_{\mathcal{L}_\Pi}(K, \mathcal{D}[\Pi]) = P(K, \mathcal{D}[\Pi])/L(K, \mathcal{D}[\Pi])$$

where $L(K, \mathcal{D}[\Pi]) = \prod_{v\in\Sigma} L(K_v, \mathcal{D}[\Pi])$. We can then define the corresponding global-to-local map

$$\phi_{\mathcal{L}_\Pi} : H^1(K_\Sigma/K, \mathcal{D}[\Pi]) \longrightarrow Q_{\mathcal{L}_\Pi}(K, \mathcal{D}[\Pi])$$

We will usually denote the map $\phi_{\mathcal{L}_\Pi}$ simply by ϕ_Π. The product of the $h_{\Pi,v}$'s for $v \in \Sigma$ defines a map $b_\Pi : P(K, \mathcal{D}[\Pi]) \to P(K, \mathcal{D})[\Pi]$. Note that the image of $L(K, \mathcal{D}[\Pi])$ under b_Π is contained in $L(K, \mathcal{D})$ and so we get a well-defined map

$$q_\Pi : Q_{\mathcal{L}_\Pi}(K, \mathcal{D}[\Pi]) \longrightarrow Q_{\mathcal{L}}(K, \mathcal{D})[\Pi].$$

Lemma 3.2.1 *Assume that $\mathcal{D}$ is divisible by π and that $L(K_v, \mathcal{D})$ is divisible by π for all $v \in \Sigma$. Then q_Π is an isomorphism.*

Proof The definition of $\mathcal{L}_\Pi$ implies that q_Π is injective without any assumptions. Furthermore, a snake lemma argument shows that if $L(K, \mathcal{D})$ is divisible by π, then the map $c_\Pi : P(K, \mathcal{D})[\Pi] \to Q_\mathcal{L}(K, \mathcal{D})[\Pi]$ will be surjective. Since the map b_Π is also surjective, it would then follow that $c_\Pi \circ b_\Pi$ is also surjective. This would imply that q_Π is surjective. □

Consequently, if we assume that $\mathcal{D}$ is almost Λ-divisible and that $\mathcal{L}$ is almost Λ-divisible, we see that q_Π is then an isomorphism for almost all $\Pi \in \mathrm{Spec}_{ht=1}(\Lambda)$.

The map α_Π is induced by the map ϕ and is defined without making any assumptions. We have the following commutative diagram whose rows are exact:

$$\begin{array}{ccccccc}
0 \longrightarrow & S_{\mathcal{L}_\Pi}(K, \mathcal{D}[\Pi]) & \longrightarrow & H^1(K_\Sigma/K, \mathcal{D}[\Pi]) & \xrightarrow{\phi_\Pi} & Q_{\mathcal{L}_\Pi}(K, \mathcal{D}[\Pi]) \\
 & \downarrow{\scriptstyle s_\Pi} & & \downarrow{\scriptstyle h_\Pi} & & \downarrow{\scriptstyle q_\Pi} \\
0 \longrightarrow & S_\mathcal{L}(K, \mathcal{D})[\Pi] & \longrightarrow & H^1(K_\Sigma/K, \mathcal{D})[\Pi] & \xrightarrow{\alpha_\Pi} & Q_\mathcal{L}(K, \mathcal{D})[\Pi]
\end{array}$$

The second and third vertical maps have been defined and make that part of the diagram commutative, and so the map s_Π is induced from h_Π. Although it is not needed now, we remark in passing that the injectivity of the map q_Π and the surjectivity of the map h_Π imply that s_Π is also surjective. But the important consequence for us is that q_Π maps $\mathrm{im}(\phi_\Pi)$ isomorphically to $\mathrm{im}(\alpha_\Pi)$ and therefore induces an isomorphism

$$\mathrm{coker}(\alpha_\Pi) \cong \mathrm{coker}(\phi_\Pi) \tag{4}$$

under the assumptions in Lemma 3.2.1. In particular, the surjectivity of α_Π and ϕ_Π would then be equivalent.

To summarize, if we assume that $\mathcal{D}$, $H^1(K_\Sigma/K, \mathcal{D})$, and the specification $\mathcal{L}$ are almost Λ-divisible, and that $\mathrm{SUR}(\mathcal{D}, \mathcal{L})$ holds, then $S_\mathcal{L}(K, \mathcal{D})$ is almost Λ-divisible if and only if ϕ_Π is surjective for almost all $\Pi \in \mathrm{Spec}_{ht=1}(\Lambda)$.

Remark 3.2.2 (Divisibility by $\mathfrak{m}$*)* One can ask if $\mathfrak{m}S_\mathcal{L}(K, \mathcal{D}) = S_\mathcal{L}(K, \mathcal{D})$, where $\mathfrak{m}$ denotes the maximal ideal of Λ. This would mean that the Pontryagin dual $\mathcal{X}$ of $S_\mathcal{L}(K, \mathcal{D})$ has no nonzero, finite Λ-submodules. For if $\mathcal{Z}$ is the maximal finite Λ-submodule of $\mathcal{X}$, then $\mathcal{Z}[\mathfrak{m}] = \mathcal{X}[\mathfrak{m}]$ is the Pontryagin dual of the quotient Λ-module $S_\mathcal{L}(K, \mathcal{D})/\mathfrak{m}S_\mathcal{L}(K, \mathcal{D})$. This is trivial if and only if $\mathcal{Z}$ itself is trivial.

Assume that $\mathrm{SUR}(\mathcal{D}, \mathcal{L})$ is satisfied and that $H^1(K_\Sigma/K, \mathcal{D})$ is almost Λ-divisible. One sees easily that if $\mathcal{Z} \neq 0$, then $\mathcal{Z}[\Pi] \neq 0$ for all $\Pi \in \mathrm{Spec}_{ht=1}(\Lambda)$. As a consequence of the snake lemma sequence, if one can show that α_Π is surjective for infinitely many Π's in $\mathrm{Spec}_{ht=1}(\Lambda)$, then it would follow as a consequence that $\mathfrak{m}S_\mathcal{L}(K, \mathcal{D}) = S_\mathcal{L}(K, \mathcal{D})$. This observation is especially useful if Λ has Krull dimension 2. In that case, it follows that $S_\mathcal{L}(K, \mathcal{D})$ is almost divisible if and only if α_Π is surjective for infinitely many Π's in $\mathrm{Spec}_{ht=1}(\Lambda)$. ♢

3.3 Behavior of the Corank Hypothesis Under Specialization

We can now complete the discussion in Sect. 2.4. We want to justify the following equivalence.

- CRK($\mathcal{D}$, $\mathcal{L}$) *is true if and only if CRK*($\mathcal{D}[\Pi], \mathcal{L}_\Pi$) *is true for almost all* $\Pi \in \mathrm{Spec}_{ht=1}(\Lambda)$.

According to Remark 2.1.3 in [6], coker(ϕ) is Λ-cotorsion if and only if $\mathrm{coker}(\alpha_\Pi)$ is (Λ/Π)-cotorsion for almost all $\Pi \in \mathrm{Spec}_{ht=1}(\Lambda)$. If we assume that $\mathcal{L}$ is almost Λ-divisible, then we have the isomorphism (4) for almost all Π's. Since Λ/Π is a finitely-generated Λ_Π-module, it follows that coker(ϕ) is Λ-cotorsion if and only if $\mathrm{coker}(\phi_\Pi)$ is Λ_Π-cotorsion for almost all $\Pi \in \mathrm{Spec}_{ht=1}(\Lambda)$, which is the stated equivalence.

The assumption that $\mathcal{L}$ is almost Λ-divisible is not needed. Suppose that $\Pi = (\pi)$ is an arbitrary element of $\mathrm{Spec}_{ht=1}(\Lambda)$. Referring to the discussion in Sect. 3.2, we have an injective map

$$\mathrm{coker}(\phi_\Pi) \longrightarrow \mathrm{coker}(\alpha_\Pi). \tag{5}$$

induced by q_Π. Furthermore, the cokernel of (5) is isomorphic to $\mathrm{coker}(q_\Pi)$. The stated equivalence will follow if we show that $\mathrm{coker}(q_\Pi)$ is (Λ/Π)-cotorsion for almost all Π's. Using the notation from Sect. 3.2, we have $\mathrm{coker}(q_\Pi) = \mathrm{coker}(c_\Pi)$. We then obtain another injective map

$$\mathrm{coker}(q_\Pi) \longrightarrow L(K, \mathcal{D})/\pi L(K, \mathcal{D}).$$

Thus, it suffices to show that $L(K, \mathcal{D})/\Pi L(K, \mathcal{D})$ is (Λ/Π)-cotorsion for almost all $\Pi \in \mathrm{Spec}_{ht=1}(\Lambda)$.

In general, suppose that $\mathcal{A}$ is a discrete, cofinitely-generated Λ-module and that $\mathcal{X}$ is the Pontryagin dual of $\mathcal{A}$. Thus, we have a perfect pairing $\mathcal{A} \times \mathcal{X} \to \mathbf{Q}_p/\mathbf{Z}_p$. Let $\mathcal{Z}$ denote the maximal pseudo-null Λ-submodule of $\mathcal{X}$ and let $\mathcal{B}$ denote the orthogonal complement of $\mathcal{Z}$ under that pairing. Thus, $\mathcal{B} \subseteq \mathcal{A}$. Let $C = \mathcal{A}/\mathcal{B}$. The Pontryagin duals of $\mathcal{B}$ and C are $\mathcal{X}/\mathcal{Z}$ and $\mathcal{Z}$, respectively. It follows that $\mathcal{B}$ is the maximal almost Λ-divisible Λ-submodule of $\mathcal{A}$. Furthermore, by definition, $\mathcal{Z}$ is annihilated by a nonzero element of Λ relatively prime to π, and so $\mathcal{Z}[\Pi]$ is a torsion (Λ/Π)-module. Thus, $C/\Pi C$ is (Λ/Π)-cotorsion. If we choose Π so that $\Pi\mathcal{B} = \mathcal{B}$, it follows that $\mathcal{A}/\Pi\mathcal{A} \cong C/\Pi C$. Applying these considerations to $\mathcal{A} = L(K, \mathcal{D})$, we see that $L(K, \mathcal{D})/\Pi L(K, \mathcal{D})$ is indeed (Λ/Π)-cotorsion for almost all $\Pi \in \mathrm{Spec}_{ht=1}(\Lambda)$.

3.4 The Case Where $\phi_{\mathcal{L}}$ is Not Surjective

We assume in this section that $\mathcal{D}$, $H^1(K_\Sigma/K, \mathcal{D})$, and the specification $\mathcal{L}$ are almost Λ-divisible, but not that SUR($\mathcal{D}$, $\mathcal{L}$) holds. In the exact sequence (3), we can simply replace $Q_{\mathcal{L}}(K, \mathcal{D})$ by the image of $\phi = \phi_{\mathcal{L}}$, which we will denote by $Q'_{\mathcal{L}}(K, \mathcal{D})$. We then can consider the map

$$\alpha'_{\Pi} : H^1(K_{\Sigma}/K, \mathcal{D})[\Pi] \longrightarrow Q'_{\mathcal{L}}(K, \mathcal{D})[\Pi].$$

Applying the snake lemma as before, and using the assumption that $H^1(K_{\Sigma}/K, \mathcal{D})$ is almost Λ-divisible, we see that $S_{\mathcal{L}}(K, \mathcal{D})$ is almost Λ-divisible if and only if α'_{Π} is surjective for almost all $\Pi \in \mathrm{Spec}_{ht=1}(\Lambda)$.

We have an exact sequence

$$0 \longrightarrow Q'_{\mathcal{L}}(K, \mathcal{D}) \longrightarrow Q_{\mathcal{L}}(K, \mathcal{D}) \longrightarrow \mathrm{coker}(\phi) \longrightarrow 0. \tag{6}$$

The kernel of the natural map $\xi_{\Pi} : Q_{\mathcal{L}}(K, \mathcal{D})[\Pi] \to \mathrm{coker}(\phi)[\Pi]$ is $Q'_{\mathcal{L}}(K, \mathcal{D})[\Pi]$ which clearly contains $\mathrm{im}(\alpha_{\Pi}) = \mathrm{im}(\alpha'_{\Pi})$. We then obtain a map

$$\tilde{\xi}_{\Pi} : \mathrm{coker}(\alpha_{\Pi}) \longrightarrow \mathrm{coker}(\phi)[\Pi]$$

whose kernel is $\mathrm{coker}(\alpha'_{\Pi})$. Thus, α'_{Π} is surjective if and only if $\ker(\tilde{\xi}_{\Pi})$ is trivial.

Choose $\Pi \in \mathrm{Spec}_{ht=1}(\Lambda)$ so that the assumptions in Lemma 3.2.1 are satisfied. Then q_{Π} induces the isomorphism (4) and hence determines an isomorphism from $\ker\big(\tilde{\xi}_{\Pi}\big)$ to a certain subgroup of $\mathrm{coker}(\phi_{\Pi})$, namely the subgroup

$$q_{\Pi}^{-1}\big(\ker(\xi_{\Pi})\big)/\mathrm{im}(\phi_{\Pi}) = \ker\big(\xi_{\Pi} \circ q_{\Pi}\big)/\mathrm{im}(\phi_{\Pi}). \tag{7}$$

Note that $\xi_{\Pi} \circ q_{\Pi}$ is a map from $Q_{\mathcal{L}_{\Pi}}(K, \mathcal{D}[\Pi])$ to $\mathrm{coker}(\phi)[\Pi]$ and induces a map from $\mathrm{coker}(\phi_{\Pi})$ to $\mathrm{coker}(\phi)[\Pi]$ whose kernel is (7). We can conclude that α'_{Π} is surjective if and only if the map

$$\mathrm{coker}(\phi_{\Pi}) \longrightarrow \mathrm{coker}(\phi)[\Pi] \tag{8}$$

is injective.

Consequently, $S_{\mathcal{L}}(K, \mathcal{D})$ is almost Λ-divisible if and only if (8) is injective for almost all $\Pi \in \mathrm{Spec}_{ht=1}(\Lambda)$. Note that (8) is surjective for almost all Π's. To see this, note that $Q'_{\mathcal{L}}(K, \mathcal{D})$ is a quotient of $H^1(K_{\Sigma}/K, \mathcal{D})$ and hence is almost Λ-divisible. Applying the snake Lemma to (6), it follows that ξ_{Π} is surjective for almost all $\Pi \in \mathrm{Spec}_{ht=1}(\Lambda)$. The same is true for the map q_{Π}, hence for $\xi_{\Pi} \circ q_{\Pi}$, and therefore for the map (8).

Examples exist where ϕ fails to be surjective, but α'_{Π} is surjective for almost all $\Pi \in \mathrm{Spec}_{ht=1}(\Lambda)$. This is discussed in the next section. The type of example considered there involves choosing a suitable $\mathbf{Z}_p$-extension K_{∞} of K and a suitable β in the maximal ideal of Λ to define a Galois module $\mathcal{T}^*$ of Λ-rank 1. The Galois module $\mathcal{D}$ is then $\mathrm{Hom}(\mathcal{T}^*, \mu_{p^{\infty}})$. When we take $\Lambda = \mathbf{Z}_p[[T]]$, both groups in (8) are finite for almost all $\Pi \in \mathrm{Spec}_{ht=1}(\Lambda)$. Hence injectivity follows from surjectivity by simply showing that their orders are equal.

3.5 A Special Case

We discuss a way to verify that (8) is injective in a very special situation. A specific example will be given at the end of Sect. 4.4. In general, suppose that $\mathcal{D}$ satisfies the assumptions in Proposition 2.6.2 and that LEO($\mathcal{D}$) and CRK($\mathcal{D}$, $\mathcal{L}$) are satisfied. Then $Ш^1(K, \Sigma, \mathcal{T}^*) = 0$. Suppose also that

$$L(K_v, \mathcal{D}) \subseteq H^1(K_v, \mathcal{D})_{\Lambda-div} \tag{9}$$

for all $v \in \Sigma$. Under these assumptions, we have

$$\widehat{\mathrm{coker}(\phi)} \cong H^1(K_\Sigma/K, \mathcal{T}^*)_{\Lambda-tors} \cong H^0(K, \mathcal{T}^*/\theta\mathcal{T}^*)$$

as Λ-modules, where $\theta \in \Lambda$ is any nonzero annihilator for $H^1(K_\Sigma/K, \mathcal{T}^*)_{\Lambda-tors}$. The first isomorphism follows from Propositions 2.3.1 and 3.1.1 in [7]. The second follows from Proposition 2.2.2 in that paper.

Now assume also that $\mathcal{D}$ is a cofree Λ-module of corank 1. Then $\mathcal{T}^*$ is a free Λ-module of rank 1. Suppose that the action of G_K on $\mathcal{T}^*$ factors through $\Gamma = \mathrm{Gal}(K_\infty/K)$, where K_∞ is a $\mathbf{Z}_p$-extension of K. Hence the image of G_K in $\Lambda^\times$ is generated topologically by an element $1+\beta$, where $\beta \in \mathfrak{m}$. We assume that $p \nmid \beta$. Note that $1+\beta$ has infinite order. We can choose θ (as above) so that $\beta|\theta$. We then have

$$\widehat{\mathrm{coker}(\phi)} \cong (\beta^{-1}\theta)/\theta \cong \Lambda/(\beta) \tag{10}$$

as Λ-modules. Let $B = (\beta)$. Therefore, $\mathrm{coker}(\phi)[\Pi]$ is isomorphic to the Pontryagin dual of $\Lambda/(B+\Pi)$ as discrete Λ-modules.

In addition to the above assumptions, let us now assume that $\Lambda \cong \mathbf{Z}_p[[T]]$. Then Λ has Krull dimension 2 and Λ/B is a free $\mathbf{Z}_p$-module of some rank. Furthermore, if $\Pi \neq (p)$, then Λ/Π is a finite integral extension of $\mathbf{Z}_p$ and is free as a $\mathbf{Z}_p$-module. Note that $\mathcal{D}[\Pi]$ is $\mathbf{Z}_p$-cofree and hence $\mathbf{Z}_p$-divisible. If $B \not\subset \Pi$, then $\Lambda/(B+\Pi)$ is finite. Since the map (8) is surjective, injectivity will follow if one can verify that $\mathrm{coker}(\phi_\Pi)$ has the same order as $\Lambda/(B+\Pi)$.

We add one more assumption. For each $v \in \Sigma$, let Γ_v be the decomposition subgroup of Γ for v. We will assume that Γ_v is nontrivial for all $v \in \Sigma$. Thus, $[\Gamma : \Gamma_v]$ is finite. A topological generator for Γ_v acts as multiplication by $1+\beta_v$, where $\beta_v \in \mathfrak{m}$. Note that $1+\beta_v = (1+\beta)^a$, where $a \in \mathbf{Z}_p$ and $a \neq 0$. Since $p \nmid \beta$, it follows that $p \nmid \beta_v$. Let $B_v = (\beta_v)$. Then $H^0(K_v, \mathcal{D}) = \mathcal{D}[B_v]$ is cofree as a Λ/B_v-module and hence is almost divisible as a Λ-module. It follows that $h_{\Pi,v}$ is an isomorphism for almost all $\Pi \in \mathrm{Spec}_{ht=1}(\Lambda)$ and all $v \in \Sigma$. Also, as we show in the lemma below, $H^1(K_v, \mathcal{D})_{\Lambda-div}$ is Λ-cofree. Consequently, $H^1(K_v, \mathcal{D})_{\Lambda-div}[\Pi]$ is $\mathbf{Z}_p$-cofree (e.g., $\mathbf{Z}_p$-divisible) if $p \notin \Pi$. Assuming that $h_{\Pi,v}$ is injective, the same is true for its inverse image under the map $h_{\Pi,v}$. Consequently, for almost all $\Pi \in \mathrm{Spec}_{ht=1}(\Lambda)$, the inclusion (9) holds when $\mathcal{D}$ is replaced by $\mathcal{D}[\Pi]$.

Note that if $p \notin \Pi$, then $\mathcal{D}[\Pi]$ is $\mathbf{Z}_p$-cofree of finite corank. It is just a divisible group. Furthermore, for almost all $\Pi \in \mathrm{Spec}_{ht=1}(\Lambda)$, both LEO($\mathcal{D}[\Pi]$) and CRK($\mathcal{D}[\Pi], \mathcal{L}_\Pi$) are satisfied. (See Sects. 2.4 and 3.3.) For such II, it follows that $Ш^1(K, \Sigma, \mathcal{T}^*/\Pi\mathcal{T}^*)$ vanishes, that $\mathrm{coker}(\phi_\Pi)$ is finite, and that we can determine its order when the analogue of the inclusion (9) holds for $\mathcal{D}[\Pi]$. In what follows, we will denote $\mathcal{T}^*/\Pi\mathcal{T}^*$ more simply by $\mathcal{T}^*_\Pi$.

For Π as above, Propositions 2.3.1 and 3.1.1 in [7] imply that $\mathrm{coker}(\phi_\Pi)$ is isomorphic to the Pontryagin dual of $H^1(K_\Sigma/K, \mathcal{T}^*_\Pi)_{\mathbf{Z}_p-tors}$. Since this last group is finite, we can choose m sufficiently large so that p^m annihilates that group. If we assume that $\beta \notin \Pi$, then Proposition 2.2.2 in [7] implies that this last group is isomorphic to $H^0(K, \mathcal{T}^*_\Pi/p^m\mathcal{T}^*_\Pi)$. Hence, $\mathrm{coker}(\phi_\Pi)$ has the same order as the kernel of multiplication by β on the finite group $\mathcal{T}^*_\Pi/p^m\mathcal{T}^*_\Pi$. This is the same as the order of the cokernel of multiplication by β on that group, which is $\mathcal{T}^*_\Pi/(\beta, p^m)\mathcal{T}^*_\Pi$. By taking $m >> 0$, one can conclude that $\mathrm{coker}(\phi_\Pi)$ has the same order as $\mathcal{T}^*_\Pi/\beta\mathcal{T}^*_\Pi$ since that group is finite. Since $\mathcal{T}^*$ is free of rank 1 over Λ, it follows that $\mathrm{coker}(\phi_\Pi)$ has the same order as $\Lambda/(\Pi + B)$ for almost all $\Pi \in \mathrm{Spec}_{ht=1}(\Lambda)$, which is indeed equal to the order of $\mathrm{coker}(\phi)[\Pi]$.

To complete this discussion, we need the following lemma.

Lemma 3.5.1 *With the above assumptions, $H^1(K_v, \mathcal{D})_{\Lambda-div}$ is Λ-cofree.*

Proof Let $c_v = \mathrm{corank}_\Lambda\left(H^1(K_v, \mathcal{D})\right) = \mathrm{corank}_\Lambda\left(H^1(K_v, \mathcal{D})_{\Lambda-div}\right)$. It suffices to show that $H^1(K_v, \mathcal{D})_{\Lambda-div}[\Pi]$ is (Λ/Π)-cofree of corank c_v for at least one prime ideal $\Pi \in \mathrm{Spec}_{ht=1}(\Lambda)$. For it would then follow by Nakayama's lemma that the Pontryagin dual of $H^1(K_v, \mathcal{D})_{\Lambda-div}$ can be generated by c_v elements as a Λ-module and hence must be a free Λ-module of rank c_v.

For almost all $\Pi \in \mathrm{Spec}_{ht=1}(\Lambda)$, the (Λ/Π)-coranks of $H^1(K_v, \mathcal{D})_{\Lambda-div}[\Pi]$ and $H^1(K_v, \mathcal{D})[\Pi]$ are both equal to c_v. This follows from Remark 2.1.3 in [6]. We will assume in this proof that Π is chosen in that way. To simplify the discussion, we will also assume that Π is chosen so that $\Lambda/\Pi \cong \mathbf{Z}_p$. Define

$$A_v = H^1(K_v, \mathcal{D})/H^1(K_v, \mathcal{D})_{\Lambda-div},$$
$$A_{\Pi,v} = H^1(K_v, \mathcal{D}[\Pi])/H^1(K_v, \mathcal{D}[\Pi])_{\mathbf{Z}_p-div}.$$

By Poitou–Tate duality, we find that the Pontryagin dual of A_v is isomorphic to $H^1(K_v, \mathcal{T}^*)_{\Lambda-tors}$. Just as argued above in the global case, it follows that $A_v[\Pi]$ is finite and is isomorphic to the Pontryagin dual of $\Lambda/(B_v + \Pi)$ for almost all $\Pi \in \mathrm{Spec}_{ht=1}(\Lambda)$. Furthermore, we have an isomorphism between the groups $H^1(K_v, \mathcal{T}^*/\Pi\mathcal{T}^*)_{\mathbf{Z}_p-tors}$ and $\Lambda/(B_v + \Pi)$ for almost all $\Pi \in \mathrm{Spec}_{ht=1}(\Lambda)$. By Poitou–Tate duality again, the Pontryagin dual of $H^1(K_v, \mathcal{T}^*/\Pi\mathcal{T}^*)_{\mathbf{Z}_p-tors}$ is in turn isomorphic to $A_{\Pi,v}$. Thus, it follows that $A_v[\Pi]$ is finite and isomorphic to $A_{\Pi,v}$ for almost all $\Pi \in \mathrm{Spec}_{ht=1}(\Lambda)$. We assume that Π is chosen in this way.

Since Π is principal, a snake lemma argument gives us the following exact sequence.

$$0 \longrightarrow H^1(K_v, \mathcal{D})_{\Lambda-div}[\Pi] \longrightarrow H^1(K_v, \mathcal{D})[\Pi] \longrightarrow A_v[\Pi] \longrightarrow 0.$$

By definition, we also have the exact sequence

$$0 \longrightarrow H^1(K_v, \mathcal{D}[\Pi])_{\mathbf{Z}_p-div} \longrightarrow H^1(K_v, \mathcal{D}[\Pi]) \longrightarrow A_{\Pi,v} \longrightarrow 0.$$

Now the natural map $H^1(K_v, \mathcal{D}[\Pi]) \to H^1(K_v, \mathcal{D})[\Pi]$ is an isomorphism for almost all Π because $H^0(K_v, \mathcal{D}) = \mathcal{D}[\beta_v]$ is an almost divisible Λ-module. For such Π, it is clear that the image of $H^1(K_v, \mathcal{D}[\Pi])_{\mathbf{Z}_p-div}$ under that natural map is precisely the maximal $\mathbf{Z}_p$-divisible submodule of $H^1(K_v, \mathcal{D})[\Pi]$ and hence is contained in $H^1(K_v, \mathcal{D})_{\Lambda-div}[\Pi]$. The fact that $A_v[\Pi]$ and $A_{\Pi,v}$ have the same order implies that the natural map induces an isomorphism

$$H^1(K_v, \mathcal{D}[\Pi])_{\mathbf{Z}_p-div} \longrightarrow H^1(K_v, \mathcal{D})_{\Lambda-div}[\Pi]$$

of (Λ/Π)-modules. Since the $\mathbf{Z}_p$-coranks of each is c_v, it then follows that $H^1(K_v, \mathcal{D})_{\Lambda-div}[\Pi]$ is indeed (Λ/Π)-cofree of corank c_v. □

Remark 3.5.2 Suppose that $\Pi \in \mathrm{Spec}_{ht=1}(\Lambda)$ and that $A_v[\Pi]$ has positive (Λ/Π)-corank. Since A_v is a cofinitely generated, cotorsion Λ-module, this means that $\widehat{A_v}[\Pi]$ has positive (Λ/Π)-rank, and so the same is true for $H^1(K_v, \mathcal{T}^*)_{\Lambda-tors}[\Pi]$. Consequently, $H^0(K_v, \mathcal{T}^*/\Pi\mathcal{T}^*)$ has positive (Λ/Π)-rank. Now

$$\mathcal{T}^*/\Pi\mathcal{T}^* \cong \mathrm{Hom}(\mathcal{D}[\Pi], \mu_{p^\infty}).$$

If $\Pi \neq (p)$, it follows that $A_v[\Pi]$ has positive (Λ/Π)-corank if and only if the group $\mathrm{Hom}_{G_{K_v}}(\mathcal{D}[\Pi], \mu_{p^\infty})$ is infinite.

Assume now that G_{K_v} acts on $\mathcal{D}[\Pi]$ through a finite quotient group. Since $p \nmid \beta_v$, one sees easily that $\Pi \neq (p)$. Note that $K_v(\mu_{p^\infty})/K_v$ is an infinite extension. Consequently, it follows that $A_v[\Pi]$ is finite. Furthermore, if J is a product of such prime ideals, then $A_v[J]$ is also finite. Therefore, if L_v is a Λ-submodule of $H^1(K_v, \mathcal{D})$ which is annihilated by such an ideal J and if L_v is divisible as a group, then we must have the inclusion $L_v \subseteq H^1(K_v, \mathcal{D})_{\Lambda-div}$.

4 Sufficient Conditions for Almost Divisibility

We will prove a rather general result in Sect. 4.1. Section 4.2 discusses the verification of various hypotheses in that result. Section 4.3 will concern a special case (although still quite general) where several of the hypotheses are automatically satisfied.

4.1 The Main Theorem

We prove the following result.

Proposition 4.1.1 *Suppose that* $\mathrm{RFX}(\mathcal{D})$ *and* $\mathrm{LEO}(\mathcal{D})$ *are both satisfied, that* $\mathrm{LOC}_v^{(2)}(\mathcal{D})$ *is satisfied for all* v *in* Σ, *and that there exists a non-archimedean prime* $\eta \in \Sigma$ *such that* $\mathrm{LOC}_\eta^{(1)}(\mathcal{D})$ *is satisfied. Suppose also that* $\mathcal{L}$ *is almost divisible, that* $\mathrm{CRK}(\mathcal{D}, \mathcal{L})$ *is satisfied, and also that at least one of the following additional assumptions is satisfied.*

(a) $\mathcal{D}[\mathfrak{m}]$ *has no subquotient isomorphic to* μ_p *for the action of* G_K,
(b) $\mathcal{D}$ *is a cofree* Λ*-module and* $\mathcal{D}[\mathfrak{m}]$ *has no quotient isomorphic to* μ_p *for the action of* G_K,
(c) There is a prime $\eta \in \Sigma$ *which satisfies* $\mathrm{LOC}_\eta^{(1)}(\mathcal{D})$ *and such that* $Q_{\mathcal{L}}(K_\eta, \mathcal{D})$ *is coreflexive as a* Λ*-module.*

Then $S_{\mathcal{L}}(K, \mathcal{D})$ *is an almost divisible* Λ*-module.*

Proof First of all, $\mathrm{RFX}(\mathcal{D})$, $\mathrm{LEO}(\mathcal{D})$, and the assumptions about $\mathrm{LOC}_v^{(1)}$ and $\mathrm{LOC}_v^{(2)}$ are sufficient to imply that $H^1(K_\Sigma/K, \mathcal{D})$ is an almost divisible Λ-module. This follows from Proposition 2.6.1. Secondly, since $\mathrm{RFX}(\mathcal{D})$ holds, $\mathcal{D}$ is certainly Λ-divisible. We can apply Proposition 2.6.3 to conclude that $\mathrm{SUR}(\mathcal{D}, \mathcal{L})$ is satisfied too.

Thus, as described in Sect. 3.1, it suffices to show that the map

$$\alpha_\Pi : H^1(K_\Sigma/K, \mathcal{D})[\Pi] \longrightarrow Q_{\mathcal{L}}(K, \mathcal{D})[\Pi]$$

is surjective for almost all $\Pi = (\pi)$ in $\mathrm{Spec}_{ht=1}(\Lambda)$. In the rest of this proof, we will exclude finitely many Π's in $\mathrm{Spec}_{ht=1}(\Lambda)$ in each step, and altogether just finitely many. We will follow the approach outlined in Sect. 3, reducing the question to studying $\mathrm{coker}(\phi_\Pi)$ and then applying Proposition 2.6.3. We want to apply that proposition to $\mathcal{D}[\Pi]$ and so must verify the appropriate hypotheses. At each step, we consider just the Π's which have not been already excluded. As described in Sect. 2, we regard various (Λ/Π)-modules as modules over a certain subring Λ_Π.

Since $\mathrm{RFX}(\mathcal{D})$ holds for $\mathcal{D}$, it follows that $\mathcal{D}[\Pi]$ is a divisible (Λ/Π)-module. Corollary 2.6.1 in [6] justifies that assertion. Therefore, $\mathcal{D}[\Pi]$ is also divisible as a Λ_Π-module. Furthermore, the assumption $\mathrm{LEO}(\mathcal{D})$ means that $\text{Ш}^2(K, \Sigma, \mathcal{D})$ is Λ-cotorsion. Consequently, $\text{Ш}^2(K, \Sigma, \mathcal{D})[\Pi]$ is a cotorsion (Λ/Π)-module for almost all $\Pi \in \mathrm{Spec}_{ht=1}(\Lambda)$. This follows from Remark 2.1.3 in [6]. The same is true for $\text{Ш}^2(K, \Sigma, \mathcal{D}[\Pi])$ according to Lemma 4.1.1 in [6]. Recall that Λ/Π is finitely-generated as a Λ_Π-module. It follows that $\mathrm{LEO}(\mathcal{D}[\Pi])$ holds for almost all $\Pi \in \mathrm{Spec}_{ht=1}(\Lambda)$.

The fact that $\mathrm{CRK}(\mathcal{D}, \mathcal{L})$ is satisfied implies that $\mathrm{CRK}(\mathcal{D}[\Pi], \mathcal{L}_\Pi)$ is satisfied for almost all $\Pi \in \mathrm{Spec}_{ht=1}(\Lambda)$. This follows from Sect. 3.4. Thus, we can assume from here on that $\mathrm{coker}(\phi_\Pi)$ is Λ_Π-cotorsion. Now we consider the additional assumptions. Each implies the corresponding assumption in Proposition 2.6.3.

Once we verify that assertion, it will then follow that ϕ_Π is surjective for almost all $\Pi \in \mathrm{Spec}_{ht=1}(\Lambda)$. Hence the same thing will be true for α_Π. This will prove that $S_{\mathcal{L}}(K, \mathcal{D})$ is indeed almost divisible as a Λ-module.

First assume that *(a)* is satisfied. Let $\mathfrak{m}_\Pi$ denote the maximal ideal of Λ_Π. Using Proposition 3.8 in [6], it follows that $\mathcal{D}[\Pi][\mathfrak{m}_\Pi]$ indeed has no subquotient isomorphic to μ_p. Now assume that *(b)* is satisfied. Then $\mathcal{D}[\Pi]$ is cofree as a (Λ/Π)-module. Since Π is principal, Λ/Π is a complete intersection. According to Proposition 3.1.20 in [1], it follows that Λ/Π is a Cohen–Macaulay domain. Proposition 2.2.11 in [1] then implies that Λ/Π is a free Λ_Π-module. Hence $\mathcal{D}[\Pi]$ is cofree as a Λ_Π-module. Furthermore, $\mathcal{D}[\mathfrak{m}] = \mathcal{D}[\Pi][\mathfrak{m}]$ has no quotient isomorphic to μ_p for the action of G_K. Remark 3.2.2 in [7] implies that the same thing is true for $\mathcal{D}[\Pi][\mathfrak{m}_\Pi]$. Thus, the assumption (b) in Proposition 2.6.3 for the Λ_Π-module $\mathcal{D}[\Pi]$ is indeed satisfied.

Now assume that (c) is satisfied. As pointed out in Sect. 2.4, $\mathrm{LOC}_\eta^{(1)}(\mathcal{D}[\Pi])$ is satisfied for almost all $\Pi \in \mathrm{Spec}_{ht=1}(\Lambda)$. Since $\mathcal{D}$ is Λ-divisible and $L(K_\eta, \mathcal{D})$ is almost Λ-divisible, we have

$$Q_{\mathcal{L}_\Pi}(K_\eta, \mathcal{D}[\Pi]) \cong Q_{\mathcal{L}}(K_\eta, \mathcal{D})[\Pi]$$

for almost all Π's. It suffices to have $L(K_\eta, \mathcal{D})$ divisible by π. The assumption that $Q_{\mathcal{L}}(K_\eta, \mathcal{D})$ is a coreflexive Λ-module then implies that $Q_{\mathcal{L}_\Pi}(K_\eta, \mathcal{D}[\Pi])$ is (Λ/Π)-divisible, and hence Λ_Π-divisible, which is the only assumption in Proposition 2.6.3(c) left to verify. □

4.2 Non-primitive Selmer Groups

Suppose that Σ_0 is a subset of Σ consisting of non-archimedean primes. Consider the map

$$\phi_{\mathcal{L},\Sigma_0} \; : \; H^1(K_\Sigma/K, \mathcal{D}) \longrightarrow \prod_{v \in \Sigma - \Sigma_0} Q_{\mathcal{L}}(K_v, \mathcal{D}).$$

We denote the kernel of $\phi_{\mathcal{L},\Sigma_0}$ by $S_{\mathcal{L}}^{\Sigma_0}(K, \mathcal{D})$. We refer to this group as the non-primitive Selmer group corresponding to the specification $\mathcal{L}$ and the set Σ_0. It is defined just as $S_{\mathcal{L}}(K, \mathcal{D})$, but one omits the local conditions for the specification $\mathcal{L}$ corresponding to the primes $v \in \Sigma_0$. Of course, we have the obvious inclusion $S_{\mathcal{L}}(K, \mathcal{D}) \subseteq S_{\mathcal{L}}^{\Sigma_0}(K, \mathcal{D})$ and the corresponding quotient $S_{\mathcal{L}}^{\Sigma_0}(K, \mathcal{D})/S_{\mathcal{L}}(K, \mathcal{D})$ is isomorphic to a Λ-submodule of $\prod_{v \in \Sigma_0} Q_{\mathcal{L}}(K_v, \mathcal{D})$. In effect, $S_{\mathcal{L}}^{\Sigma_0}(K, \mathcal{D})$ is the Selmer group corresponding to a new specification $\mathcal{L}'$, where we simply replace $L(K_v, \mathcal{D})$ by $L'(K_v, \mathcal{D}) = H^1(K_v, \mathcal{D})$ for all $v \in \Sigma_0$. Thus, we now have $Q_{\mathcal{L}'}(K_v, \mathcal{D}) = 0$ for $v \in \Sigma_0$.

If we assume that $\mathrm{SUR}(\mathcal{D}, \mathcal{L})$ is satisfied, then it obviously follows that $\mathrm{SUR}(\mathcal{D}, \mathcal{L}')$ is satisfied. Furthermore, we have

$$S_{\mathcal{L}}^{\Sigma_0}(K, \mathcal{D})/S_{\mathcal{L}}(K, \mathcal{D}) \cong \prod_{v \in \Sigma_0} \mathcal{Q}_{\mathcal{L}}(K_v, \mathcal{D}).$$

In general, $\text{coker}(\phi_{\mathcal{L}'})$ is clearly a quotient of $\text{coker}(\phi_{\mathcal{L}})$, and hence if we assume that CRK($\mathcal{D}$, $\mathcal{L}$) is satisfied, then so is CRK($\mathcal{D}$, $\mathcal{L}'$). The following proposition then follows immediately from Proposition 4.1.1(c).

Proposition 4.2.1 *Suppose that* RFX($\mathcal{D}$) *and* LEO($\mathcal{D}$) *are both satisfied, that* $\text{LOC}_v^{(2)}(\mathcal{D})$ *is satisfied for all* v *in* Σ, *and that there exists a non-archimedean prime* $\eta \in \Sigma_0$ *such that* $\text{LOC}_\eta^{(1)}(\mathcal{D})$ *is satisfied. Suppose also that* $\mathcal{L}$ *is almost divisible and that* CRK($\mathcal{D}$, $\mathcal{L}$) *is satisfied. Then* $S_{\mathcal{L}}^{\Sigma_0}(K, \mathcal{D})$ *is an almost divisible* Λ*-module.*

Remark 4.2.2 Suppose that η is a non-archimedean prime not dividing p. Regarding $\mathcal{D}[\mathfrak{m}]$ as an $\mathbf{F}_p$-representation space for G_{K_η}, suppose that it has no subquotients isomorphic to μ_p or to $\mathbf{Z}/p\mathbf{Z}$ (with trivial action of G_{K_η}). According to Proposition 3.1 in [6], the G_{K_η}-module $\mathcal{D}[\mathfrak{m}^t]$ has the same property for all $t \geq 1$. The local duality theorems imply that $H^0(K_\eta, \mathcal{D}[\mathfrak{m}^t])$ and $H^2(K_\eta, \mathcal{D}[\mathfrak{m}^t])$ both vanish, and therefore that $H^1(K_\eta, \mathcal{D}[\mathfrak{m}^t]) = 0$. It follows that $H^1(K_\eta, \mathcal{D}) = 0$. If we let $\Sigma_0 = \{\eta\}$, then we have $S_{\mathcal{L}}^{\Sigma_0}(K, \mathcal{D}) = S_{\mathcal{L}}(K, \mathcal{D})$. The hypothesis $\text{LOC}_\eta^{(1)}(\mathcal{D})$ is also satisfied. Consequently, if the other assumptions in Proposition 4.2.1 are satisfied, it follows that $S_{\mathcal{L}}(K, \mathcal{D})$ is almost divisible as a Λ-module. Alternatively, in this case, $\mathcal{Q}_{\mathcal{L}}(K_\eta, \mathcal{D})$ vanishes and so is certainly coreflexive, making assumption (c) in Proposition 4.1.1 satisfied.

4.3 Verifying the Hypotheses

We will discuss the various hypotheses in Proposition 4.1.1. Some of them are already needed for Propositions 2.6.1 and 2.6.3, and we may simply refer to discussions in [6, 7]. We have nothing additional to say about RFX($\mathcal{D}$). If $\mathcal{D}$ is R-cofree, then that hypothesis is just that R is a reflexive ring.

4.3.1 The Local Hypotheses

There is a discussion of the verification of $\text{LOC}_v^{(1)}(\mathcal{D})$ and $\text{LOC}_v^{(2)}(\mathcal{D})$ in Sect. 5, part F of [6]. Most commonly, $\text{LOC}_v^{(1)}(\mathcal{D})$ is satisfied for all non-archimedean primes $v \subset \Sigma$ simply because $H^0(K_v, \mathcal{T}^*) = 0$ for those v's. That is a rather mild condition, although we mention one kind of example in Sect. 4.4 where it may fail to be satisfied. Such examples were one motivation for introducing $\text{LOC}_v^{(2)}(\mathcal{D})$ as a hypothesis in [6]. Another motivation is that for archimedean primes, $H^0(K_v, \mathcal{T}^*)$ is often nontrivial, but $\text{LOC}_v^{(2)}(\mathcal{D})$ may still be satisfied. The archimedean primes are only an issue when $p = 2$.

4.3.2 The Hypotheses CRK($\mathcal{D}$, $\mathcal{L}$) and LEO($\mathcal{D}$)

As mentioned before, a discussion of LEO($\mathcal{D}$) can be found in Sect. 6, part D of [6]. It is called hypothesis L there. Of course, the validity of CRK($\mathcal{D}$, $\mathcal{L}$) is related to the choice of the specification $\mathcal{L}$. We will discuss one rather natural way of choosing a specification below. Let $c_{\mathcal{L}}(K, \mathcal{D})$ denote the Λ-corank of the cokernel of $\phi_{\mathcal{L}}$. Thus, CRK($\mathcal{D}$, $\mathcal{L}$) means that $c_{\mathcal{L}}(K, \mathcal{D}) = 0$. As discussed in the introduction to [7], one has an equation

$$s_{\mathcal{L}}(K, \mathcal{D}) = b_1(K, \mathcal{D}) - q_{\mathcal{L}}(K, \mathcal{D}) + c_{\mathcal{L}}(K, \mathcal{D}) + \mathrm{corank}_{\Lambda}\left(\text{Ш}^2(K, \Sigma, \mathcal{D})\right),$$

where $s_{\mathcal{L}}(K, \mathcal{D})$ and $q_{\mathcal{L}}(K, \mathcal{D})$ are the Λ-coranks of $S_{\mathcal{L}}(K, \mathcal{D})$ and $Q_{\mathcal{L}}(K, \mathcal{D})$, respectively. The integer $b_1(K, \mathcal{D})$ is defined just in terms of the Euler–Poincaré characteristic for $\mathcal{D}$ and the Λ-coranks of some local Galois cohomology groups, and does not depend on $\mathcal{L}$. It occurs in Proposition 4.3 in [6]. One then has a lower bound

$$s_{\mathcal{L}}(K, \mathcal{D}) \geq b_1(K, \mathcal{D}) - q_{\mathcal{L}}(K, \mathcal{D})$$

and equality means that both CRK($\mathcal{D}$, $\mathcal{L}$) and LEO($\mathcal{D}$) are satisfied. The simplest case is where $\mathcal{L}$ is chosen so that $q_{\mathcal{L}}(K, \mathcal{D}) = b_1(K, \mathcal{D})$. In this case, the equality means that $S_{\mathcal{L}}(K, \mathcal{D})$ is a cotorsion Λ-module.

4.3.3 The Additional Assumptions in Proposition 4.1.1

Remark 3.2.2 in [7] discusses assumptions (a) and (b). It includes some observations when $\mathcal{D}$ arises from an n-dimensional representation ρ of $\mathrm{Gal}(K_\Sigma/K)$ over a ring R, as in the introduction. One observation is that if $n \geq 2$ and if the residual representation $\widetilde{\rho}$ is irreducible over the finite field $R/\mathfrak{M}$, then hypothesis (a) is satisfied. The residual representation gives the action of $\mathrm{Gal}(K_\Sigma/K)$ on $\mathcal{D}[\mathfrak{M}]$. Another observation in that remark is that $\mathcal{D}[\mathfrak{m}]$ has a quotient isomorphic to μ_p if and only if $\mathcal{D}[\mathfrak{M}]$ has such a quotient.

We now discuss hypothesis (c). This will be useful if $\mathcal{D}[\mathfrak{m}]$ has a quotient or subquotient isomorphic to μ_p for the action of G_K. We will assume that η is a non-archimedean prime in Σ and that $\mathrm{LOC}^{(1)}_{\eta}(\mathcal{D})$ is satisfied. The issue is the coreflexivity of $Q_{\mathcal{L}}(K_\eta, \mathcal{D})$ as a Λ-module.

Let us now make the following two assumptions: *(i)* $H^1(K_\eta, \mathcal{D})$ is Λ-coreflexive, *(ii)* $L(K_\eta, \mathcal{D})$ is almost Λ-divisible. The coreflexivity of the discrete Λ-module $Q_{\mathcal{L}}(K_\eta, \mathcal{D})$ then follows easily. To see this, suppose that $\mathcal{A}$ is a cofinitely generated, coreflexive, discrete Λ-module and that $\mathcal{B}$ is an almost divisible Λ-submodule of $\mathcal{A}$. Let $\mathcal{X}$ be the Pontryagin dual of $\mathcal{A}$ and let $\mathcal{Y}$ be the orthogonal complement of $\mathcal{B}$ under the perfect pairing $\mathcal{A} \times \mathcal{X} \to \mathbf{Q}_p/\mathbf{Z}_p$. Then $\mathcal{X}$ is a finitely-generated, reflexive Λ-module. Furthermore, $\mathcal{X}/\mathcal{Y}$ is the Pontryagin dual of $\mathcal{B}$ and hence has no nonzero pseudo-null Λ-submodules. However, the reflexive hull $\widetilde{\mathcal{Y}}$ of $\mathcal{Y}$ must be contained in $\mathcal{X}$ and the quotient $\widetilde{\mathcal{Y}}/\mathcal{Y}$ is a pseudo-null Λ-module, and so must be zero. It follows that $\mathcal{Y}$ is reflexive as a Λ-module and hence that its Pontryagin dual $\mathcal{A}/\mathcal{B}$ is a coreflexive Λ-module.

Section 5, part D, of [6] gives some sufficient conditions for $H^1(K_\eta, \mathcal{D})$ to be coreflexive. One condition requires the assumption that μ_p is not a quotient of $\mathcal{D}[\mathfrak{m}]$ as a G_{K_η}-module. However, that assumption clearly implies assumption (*a*) in Proposition 4.1.1. Another more subtle sufficient condition is given in Proposition 5.9 in [6]. It involves $\mathcal{T}^* \otimes_\Lambda \widehat{\Lambda}$ which is denoted by $\mathcal{D}^*$ there. We are assuming that $H^0(K_\eta, \mathcal{T}^*) = 0$. Equivalently, that means that $\mathcal{D}^*(K_\eta) = H^0(K_\eta, \mathcal{D}^*)$ is Λ-cotorsion. Its Pontryagin dual $\widehat{\mathcal{D}^*(K_\eta)}$ is a torsion Λ-module. The result from [6] is that if $\mathcal{D}$ is Λ-cofree and if every associated prime ideal for the torsion Λ-module $\widehat{\mathcal{D}^*(K_\eta)}$ has height at least 3, then $H^1(K_\eta, \mathcal{D})$ is coreflexive as a Λ-module. Some interesting cases where this criterion is satisfied will be discussed in [8].

Even if $H^1(K_\eta, \mathcal{D})$ fails to be coreflexive, it is still possible for the quotient Λ-module $Q_{\mathcal{L}}(K_\eta, \mathcal{D})$ to be coreflexive. Consider the following natural way to specify a choice of $L(K_\eta, \mathcal{D})$. Suppose that C_η is a G_{K_η}-invariant Λ-submodule of $\mathcal{D}$ and that $H^2(K_\eta, C_\eta)$ vanishes. Then we can define

$$L(K_\eta, \mathcal{D}) = \operatorname{im}\big(H^1(K_\eta, C_\eta) \longrightarrow H^1(K_\eta, \mathcal{D}) \big).$$

Let $\mathcal{E}_\eta = \mathcal{D}/C_\eta$. The map $H^1(K_\eta, \mathcal{D}) \to H^1(K_\eta, \mathcal{E}_\eta)$ is surjective and its kernel is $L(K_\eta, \mathcal{D})$. If $\eta \nmid p$, then one can take $C_\eta = 0$ and hence $L(K_\eta, \mathcal{D}) = 0$. This is often a useful choice. If $\eta | p$, then one often will make a nontrivial choice of C_η. This kind of definition occurs in [4] for primes above p when a Galois representation ρ satisfies something we called a "Panchiskin condition." (See Sect. 4 in [4].) Under the stated assumptions, we have

$$Q_{\mathcal{L}}(K_\eta, \mathcal{D}) \;\cong\; H^1(K_\eta, \mathcal{E}_\eta)$$

as Λ-modules. Propositions 5.8 and 5.9 from [6] then give the following result.

Proposition 4.3.1 *In addition to the assumption that* $H^2(K_\eta, C_\eta) = 0$, *suppose that either one of the following assumptions is satisfied.*

(i) $\mathcal{E}_\eta$ *is* Λ*-coreflexive and* $\mathcal{E}_\eta[\mathfrak{m}]$ *has no subquotient isomorphic to* μ_p *as a* G_{K_η}*-module,*
(ii) $\mathcal{E}_\eta$ *is* Λ*-cofree and every associated prime ideal for the* Λ*-module* $\widehat{\mathcal{E}_\eta^*(K_\eta)}$ *has height at least* 3.

Then the Λ*-module* $Q_{\mathcal{L}}(K_\eta, \mathcal{D})$ *is coreflexive.*

Concerning (*i*), note that it may be satisfied even if assumption (*a*) in Proposition 4.1.1 fails to be satisfied. One such situation will be mentioned in Sect. 4.4.

We will also want $L(K_\eta, C_\eta)$ to be almost Λ-divisible. The following result follows immediately from Proposition 5.3 in [6].

Proposition 4.3.2 *Assume that C_η is Λ-coreflexive and that $H^2(K_\eta, C_\eta) = 0$. Then $H^1(K_\eta, C_\eta)$ is almost Λ-divisible. Hence the image of $H^1(K_\eta, C_\eta)$ in $H^1(K_\eta, \mathcal{D})$ is also almost Λ-divisible.*

4.4 The Two Classical Results

Let p be an odd prime. Suppose that T is a free $\mathbf{Z}_p$-module of rank n which has a continuous action of $\mathrm{Gal}(K_\Sigma/K)$. Thus, we have a continuous homomorphism $\mathrm{Gal}(K_\Sigma/K) \to \mathrm{Aut}_{\mathbf{Z}_p}(T)$. Suppose also that K_∞ is the cyclotomic $\mathbf{Z}_p$-extension of K and let $\Gamma = \mathrm{Gal}(K_\infty/K)$. Let $\Lambda = \mathbf{Z}_p[[\Gamma]]$ denote the completed group ring for Γ over $\mathbf{Z}_p$. Thus, Λ is isomorphic to a formal power series ring $\mathbf{Z}_p$ in one variable. In this situation, one can define a free Λ-module $\mathcal{T}$ of rank n together with a homomorphism $\rho : \mathrm{Gal}(K_\Sigma/K) \to \mathrm{Aut}_\Lambda(\mathcal{T})$. This is described in Sect. 5 of [7] in detail, where $\mathcal{T}$ is denoted by $T \otimes \kappa$, and also in [4] where it is called the cyclotomic deformation of T. Here κ is the natural embedding of Γ into $\Lambda^\times$ and one thinks of $\mathcal{T}$ as the twist of T by the $\Lambda^\times$-valued character κ.

Just as in the introduction, taking $R = \Lambda$, one can define $\mathcal{D} = \mathcal{T} \otimes_\Lambda \widehat{\Lambda}$. This discrete, Λ-cofree $\mathrm{Gal}(K_\Sigma/K)$-module $\mathcal{D}$ is denoted by $D \otimes \kappa$ in [7], where $D = T \otimes_{\mathbf{Z}_p} (\mathbf{Q}_p/\mathbf{Z}_p)$. We think of $\mathcal{D}$ as the $\mathrm{Gal}(K_\Sigma/K)$-module obtained from D by inducing from $\mathrm{Gal}(K_\Sigma/K_\infty)$ up to $\mathrm{Gal}(K_\Sigma/K)$. We have $D \cong \mathcal{D}[I]$ as $\mathrm{Gal}(K_\Sigma/K)$, where I denotes the augmentation ideal in Λ, Consequently, we have $D[p] \cong \mathcal{D}[\mathfrak{m}]$, where $\mathfrak{m}$ is the maximal ideal of Λ.

Many of our hypothesis are automatically satisfied. Obviously, RFX($\mathcal{D}$) is satisfied. Furthermore, Lemma 5.2.2 in [7] shows that $\mathrm{LOC}_\eta^{(1)}(\mathcal{D})$ is satisfied for all non-archimedean primes η in Σ. This is so because only the archimedean primes can split completely in K_∞/K. Since p is assumed to be odd, if η is archimedean, then $(\mathcal{T}^*)^{G_{K_\eta}}$ is a direct summand in $\mathcal{T}^*$ and hence $\mathrm{LOC}_\eta^{(2)}(\mathcal{D})$ is satisfied. It is reasonable to conjecture that LEO($\mathcal{D}$) is always satisfied. This is stated as Conjecture 5.2.1 in [7] and is equivalent to conjecture L stated in the introduction to [6]. Section 5.2 in [6] discusses its validity. It is proved in certain special cases. In the examples that we will discuss below, LEO($\mathcal{D}$) is indeed satisfied as well as CRK($\mathcal{D}$, $\mathcal{L}$).

Consider the case where $T = T_p(E)$, the p-adic Tate module for an elliptic curve defined over K. We then have $T/pT \cong E[p]$. Let Σ be a finite set of primes of K including the primes dividing p, the infinite primes, and the primes where E has bad reduction. The properties of the Weil pairing $E[p] \times E[p] \to \mu_p$ show that assumption (b) is satisfied if and only if $E(K)$ has no element of order p. Assume that E has good, ordinary reduction at the primes of K lying over p. There is a natural choice of a specification $\mathcal{L}$ in this case because the Panchiskin condition is satisfied. See the discussion in Sect. 4.3. One chooses $L(K_\eta, \mathcal{D}) = 0$ if $\eta \nmid p$. If $\eta | p$, let C_η denote the kernel of the reduction map $E[p^\infty] \to \overline{E}_\eta[p^\infty]$, where $\overline{E}_\eta$ is the reduction of E at η. Let $C_\eta = C_\eta \otimes \kappa$. Then $\mathcal{E}_\eta = \overline{E}_\eta[p^\infty] \otimes \kappa$. Note that $\mathcal{L}$ is almost divisible.

The formulas in Sect. 2.3 show that $\delta_\Lambda(K, \mathcal{D}) = [K : \mathbf{Q}]$, which is a lower bound on $\mathrm{corank}_\Lambda\left(H^1(K_\Sigma/K, \mathcal{D})\right)$. However, the local formulas show easily that $Q_{\mathcal{L}}(K_v, \mathcal{D})$ has Λ-corank 0 when $v \nmid p$ and Λ-corank $[K_v : \mathbf{Q}_p]$ when $v|p$. Therefore, $\mathrm{corank}_\Lambda\left(Q_{\mathcal{L}}(K, \mathcal{D})\right) = [K : \mathbf{Q}]$. Thus, if $S_{\mathcal{L}}(K, \mathcal{D})$ is Λ-cotorsion, then inequality (2) shows that both LEO($\mathcal{D}$) and CRK($\mathcal{D}$, $\mathcal{L}$) are satisfied.

The above discussion shows that if $E(K)$ has no element of order p and if $S_{\mathcal{L}}(K, \mathcal{D})$ is Λ-cotorsion, then Proposition 4.1.1 implies that $S_{\mathcal{L}}(K, \mathcal{D})$ is an almost divisible Λ-module. The second classical result stated in the introduction follows from this because $\mathrm{Sel}_E(K_\infty)$ can be identified with the Selmer group attached to D over K_∞. However, Proposition 3.2 in [4] gives an isomorphism between that Selmer group and $S_{\mathcal{L}}(K, \mathcal{D})$ (with a Λ-module structure modified by the involution of Λ induced from $\gamma \to \gamma^{-1}$ for $\gamma \in \Gamma$).

Now suppose that K is totally real, that $T \cong \mathbf{Z}_p$, and that G_K acts on T by a totally odd character ψ. Since p is odd, the order of ψ divides $p-1$. Let Σ be a finite set of primes of K including the primes dividing p, the infinite primes, and the primes dividing the conductor of ψ. Define D and $\mathcal{D}$ as described above. Thus, $\mathcal{D}$ is Λ-cofree and has Λ-corank 1. We take the following specification $\mathcal{L}$:

$$L(K_v, \mathcal{D}) = \ker\left(H^1(K_v, \mathcal{D}) \to H^1(K_v^{unr}, \mathcal{D})\right)$$

for all $v \in \Sigma$. Here K_v^{unr} denotes the maximal unramified extension of K_v. Thus, $S_{\mathcal{L}}(K, \mathcal{D})$ consists of locally unramified cocycle classes in $H^1(K_\Sigma/K, \mathcal{D})$ (or equivalently, cocycle classes in $H^1(K, \mathcal{D})$ which are unramified at all primes v of K.). Just as in the elliptic curve case, one can identify $S_{\mathcal{L}}(K, \mathcal{D})$ (slightly modifying the Λ-module structure) with $S(K_\infty.D)$ (as defined in the introduction) and hence the Pontryagin dual of $S_{\mathcal{L}}(K, \mathcal{D})$ can be identified with $X^{(\psi)}$, where $X = \mathrm{Gal}(L_\infty/K_{\infty,\psi})$. Iwasawa proved that X is a finitely generated, torsion Λ-module. Hence, $S_{\mathcal{L}}(K, \mathcal{D})$ is a cofinitely generated, cotorsion Λ-module. As we explain below, $L(K_v, \mathcal{D})$ is Λ-cotorsion for all v. Furthermore, the formulas in Sect. 2.3 show that the Λ-corank of $H^1(K_\Sigma/K, \mathcal{D})$ is at least $[K : \mathbf{Q}]$ and the Λ-corank of $Q_{\mathcal{L}}(K, \mathcal{D})$ is equal to $[K : \mathbf{Q}]$. The fact that $S_{\mathcal{L}}(K, \mathcal{D})$ has Λ-corank 0 implies that the Λ-corank of $H^1(K_\Sigma/K, \mathcal{D})$ is equal to $[K : \mathbf{Q}]$ and that CRK($\mathcal{D}$, $\mathcal{L}$) is satisfied. It also follows that $H^2(K_\Sigma/K, \mathcal{D})$ has Λ-corank 0, and hence the same is true for $\text{Ш}^2(K_\infty, \Sigma, D)$. Thus, LEO($\mathcal{D}$) is satisfied. We now show that $\mathcal{L}$ is almost divisible.

Let $\mathcal{D}(K_v^{unr})$ denote $H^0(K_v^{unr}, \mathcal{D})$. The inflation-restriction sequence shows that

$$L(K_v, \mathcal{D}) \;\cong\; H^1\left(K_v^{unr}/K_v,\; \mathcal{D}(K_v^{unr})\right)$$

as Λ-modules. Let ψ_v be the restriction of ψ to the decomposition subgroup Δ_v of $\Delta = \mathrm{Gal}(K_\psi/K)$. Then ψ_v is a faithful character of Δ_v and has order dividing $p-1$. We can regard ψ_v as a character of G_{K_v} and it defines a faithful character of $\mathrm{Gal}(K_{v,\psi_v}/K_v)$ for a certain cyclic extension K_{v,ψ_v} of K_v. Let $K_{v,\infty}$ be the cyclotomic $\mathbf{Z}_p$-extension of K_v and let $\Gamma_v = \mathrm{Gal}(K_{v,\infty}/K)$. The action of G_{K_v} on $\mathcal{D}$ factors through $\mathrm{Gal}(K_{v,\psi_v}K_{v,\infty}/K_v)$ which is isomorphic to $\Delta_v \times \Gamma_v$, where we have identified Δ_v and Γ_v with subgroups of $\mathrm{Gal}(K_{v,\psi_v}K_{v,\infty}/K_v)$ in an obvious way. Note that Δ_v is a cyclic group of order dividing $p-1$. The inertia subgroup of Δ_v is also cyclic and a generator will act on $\mathcal{D}$ as multiplication by a root of unity ε_v of order dividing $p-1$.

If the restriction of ψ_v to $G_{K_v^{unr}}$ is nontrivial, then $\varepsilon_v \neq 1$ and hence $\mathcal{D}(K_v^{unr}) = 0$. It follows that $L(K_v, \mathcal{D}) = 0$ for such v. We assume now that ψ_v is unramified at v and hence $\varepsilon_v = 1$. The restriction map

$$H^1\big(K_v^{unr}/K_v,\ \mathcal{D}(K_v^{unr})\big) \longrightarrow H^1\big(K_v^{unr}/K_{v,\psi_v},\ \mathcal{D}(K_v^{unr})\big)^{\Delta_v}$$

is injective. Also, we have an isomorphism

$$H^1\big(K_v^{unr}/K_{v,\psi_v},\ \mathcal{D}(K_v^{unr})\big)^{\Delta_v} \cong \mathrm{Hom}_{\Delta_v}\big(\Gamma_v,\ \mathcal{D}(K_v^{unr})/(\gamma_v - 1)\mathcal{D}(K_v^{unr})\big). \tag{11}$$

The action of Δ_v on Γ_v (by conjugation) is trivial. On the other hand, Δ_v is cyclic and a generator δ_v acts on $\mathcal{D}$ as multiplication by a root of unity ζ_v of order dividing $p-1$. Hence, if ψ_v is nontrivial, then $\zeta_v \neq 1$ and

$$H^0\big(\Delta_v, \mathcal{D}(K_v^{unr})/(\gamma_v - 1)\mathcal{D}(K_v^{unr})\big)$$

must vanish. It follows that the right side in (11) is trivial. Therefore, we must have $L(K_v, \mathcal{D}) = 0$ in this case too.

We assume now that ψ_v is trivial and hence the action of G_{K_v} on $\mathcal{D}$ factors through $\mathrm{Gal}(K_{v,\infty}/K_v)$. If $v \nmid p$, then v is unramified in K_∞/K and hence $K_{v,\infty} \subset K_v^{unr}$. Thus, $\mathcal{D}(K_v^{unr}) = \mathcal{D}$. Furthermore, $\mathrm{Gal}(K_v^{unr}/K_v)$ contains a unique subgroup P isomorphic to $\mathbf{Z}_p$ and the restriction map $P \to \Gamma_v$ is an isomorphism. The action of P on $\mathcal{D}$ is through this isomorphism. Let γ_v be a topological generator for Γ_v. The restriction map

$$H^1(K_v^{unr}/K_v, \mathcal{D}) \longrightarrow H^1(P, \mathcal{D})$$

is injective. Also, $H^1(P, \mathcal{D}) \cong \mathcal{D}/(\gamma_v - 1)\mathcal{D}$ vanishes because $\gamma_v - 1$ acts on $\mathcal{D}$ as multiplication by a nonzero element of Λ and $\mathcal{D}$ is Λ-divisible. The above remarks show that $L(K_v, \mathcal{D}) = 0$ for all $v \nmid p$.

Now consider primes v of K lying over p. If ψ_v is nontrivial, then $L(K_v, \mathcal{D}) = 0$, as shown above. Assuming that ψ_v is trivial, the action of G_{K_v} on $\mathcal{D}$ factors through Γ_v. By definition, $\Gamma_v \subseteq \Gamma$ is identified with a subgroup of $\Lambda^\times = \mathbf{Z}_p[[\Gamma]]^\times$ in a canonical way, and one sees that γ_v acts on $\mathcal{D}$ as multiplication by $1 + \beta_v$, where $\beta_v \in \Lambda$ and $p \nmid \beta_v$. The inertia subgroup of Γ_v is topologically generated by $\gamma_v^{p^a}$ for some $a \geq 0$. Also, $\gamma_v^{p^a} - 1$ acts as multiplication by $\beta_v^{p^a} - 1$, an element of Λ which is not divisible by p. It follows that $\mathcal{D}(K_v^{unr}) = \mathcal{D}[\beta_v^{p^a} - 1]$ is a divisible group. It also follows that $\mathcal{D}(K_v^{unr})$ is a cotorsion Λ-module. Furthermore, as before, $H^1\big(K_v^{unr}/K_v, \mathcal{D}(K_v^{unr})\big)$ is isomorphic to a certain quotient of $\mathcal{D}(K_v^{unr})$.

Hence $L(K_v, \mathcal{D})$ is cotorsion as a Λ-module and is a divisible group. Hence, its Pontryagin dual has no nonzero finite Λ-submodules. Thus, in this case, $L(K_v, \mathcal{D})$ may be nontrivial, but it is still almost divisible as a Λ-module.

Assume that $\psi \neq \omega$. Now $\mathcal{D}[\mathfrak{m}]$ is a 1-dimensional $\mathbf{F}_p$-vector space on which G_K acts by ψ. Hence, assumption *(a)* in Proposition 4.1.1 is satisfied. It then follows that the Λ-module $S_{\mathcal{L}}(K, \mathcal{D})$ is indeed almost divisible, and hence Iwasawa's theorem is proved when $\psi \neq \omega$.

In the case $\psi = \omega$, then we are in the setting of Sect. 3.5 and must use the results from Sects. 3.4 and 3.5. In this case, (10) shows that $\phi_{\mathcal{L}}$ is not surjective. We must show that $L(K_v, \mathcal{D}) \subseteq H^1(K_v, \mathcal{D})_{\Lambda-div}$ for all $v \in \Sigma$. Now, $L(K_v, \mathcal{D})$ is nontrivial only when $v|p$ and and ψ_v is trivial. But in that case, $L(K_v, \mathcal{D})$ is a quotient of $\mathcal{D}[\beta_v^{p^a} - 1]$ and is annihilated by $J = (\beta_v^{p^a} - 1)$. Now J is a product of prime ideals of height 1 which contain $\beta_v^{p^a} - 1$. Hence, G_{K_v} acts on $\mathcal{D}[\Pi]$ through a finite quotient group. Remark 3.5.2 then implies that $L(K_v, \mathcal{D}) \subseteq H^1(K_v, \mathcal{D})_{\Lambda-div}$. This is true for all $v \in \Sigma$. Consequently, $S_{\mathcal{L}}(K, \mathcal{D})$ is an almost divisible Λ-module.

4.5 Examples Where Almost Divisibility Fails

We consider two variants of the classical examples mentioned in Sect. 4.4. We will follow the notation described there and the Selmer groups will be defined in exactly the same way. In one example, all the hypotheses in Proposition 4.1.1 are satisfied, except that none of the additional assumptions (*a*), (*b*) or (*c*) hold. In another example, it is CRK($\mathcal{D}$, $\mathcal{L}$) which is not satisfied.

Let $p = 5$. Let E be the elliptic curve over $\mathbf{Q}$ of conductor 11 such that $E(\mathbf{Q}) = 0$. It is the second curve in Cremona's tables and has good, ordinary reduction at p. The curve E has an isogeny of degree p defined over $\mathbf{Q}$ whose kernel Φ is isomorphic to μ_p for the action of $G_{\mathbf{Q}}$. Also, the action of $G_{\mathbf{Q}}$ on $E[p]/\Phi \cong \mathbf{Z}/p\mathbf{Z}$ is trivial. Let $K = \mathbf{Q}(\mu_p)$ and let $T = T_p(E)$ as in Sect. 4.4. Note that $E(K)$ has a point of order p. A theorem of Kato, or a direct calculation, implies that $S_{\mathcal{L}}(K, \mathcal{D})$ is Λ-cotorsion, and hence CRK($\mathcal{D}$, $\mathcal{L}$) is satisfied. It is clear that $\mathcal{D}[\mathfrak{m}] = E[p]$ has a quotient $E[p]/\Phi$ isomorphic to μ_p for the action of G_K. Thus, assumptions (*a*) and (*b*) fail to hold. We take Σ to be the set of primes lying above ∞, p, or 11. Assumption (*c*) fails to hold too. For if η lies over 11, one finds that $Q_{\mathcal{L}}(K_\eta, \mathcal{D})$ is Λ-cotorsion, but nontrivial, and hence cannot be coreflexive. If η lies over p, one finds that $Q_{\mathcal{L}}(K_\eta, \mathcal{D})$ is not Λ-divisible and hence is not coreflexive. In this example, $S_{\mathcal{L}}(K, \mathcal{D})$ can be identified with $\mathrm{Sel}_E(K_\infty)_p$ as Λ-modules (up to an involution of Λ). It is shown in [5], pp. 127–128, that $\mathrm{Sel}_E(K_\infty)_p$ has a direct summand as a Λ-module which is of order p. Hence, the Pontryagin dual of $\mathrm{Sel}_E(K_\infty)_p$ has a submodule isomorphic to $\Lambda/\mathfrak{m}$. And so, in this example, $S_{\mathcal{L}}(K, \mathcal{D})$ fails to be almost divisible as a Λ-module.

Let p be any odd prime. Suppose that K is a totally real number field, that $T \cong \mathbf{Z}_p$, and that G_K acts on T by a totally even character ψ. In this case, K_ψ is totally real. It is conjectured that $X = \mathrm{Gal}(L_\infty/K_{\infty,\psi})$ is finite. We refer the reader to [9], pp. 350, 351, for more discussion and references concerning this conjecture. There are many examples when $K = \mathbf{Q}$, ψ has order 2, and $p = 3$, where $X^{(\psi)}$ turns out to be finite, but nonzero. It would then follow that $S_{\mathcal{L}}(K, \mathcal{D})$ is finite and nonzero, and hence fails to be almost divisible as a Λ-module. The weak Leopoldt conjecture holds for the $\mathbf{Z}_p$-extension $K_{\infty,\psi}/K_\psi$ and this implies that the Λ-corank of $H^1(K_\Sigma/K, \mathcal{D})$ is zero. We refer the reader to page 344 in [6] for an explanation. In contrast, the local formula in Sect. 2.3 implies that $Q_{\mathcal{L}}(K, \mathcal{D})$ has positive Λ-corank. Therefore, CRK($\mathcal{D}$, $\mathcal{L}$) cannot be satisfied.

References

1. Bruns, W., Herzog, J.: Cohen–Macaulay Rings, Cambridge Studies in Advanced Math, vol. 39. Cambridge University Press (1998)
2. Cohen, I.S.: On the structure and ideal theory of complete local rings. Trans. Am. Math. Soc. **59**, 54–106 (1946)
3. Greenberg, R.: Iwasawa theory for p-adic representations. Adv. Stud. Pure Math. **17**, 97–137 (1989)
4. Greenberg, R.: Iwasawa theory and p-adic deformations of motives. Proc. Symp. Pure Math. **55**(II), 193–223 (1994)
5. Greenberg, R.: Iwasawa theory for elliptic curves. Lect. Notes Math. **1716**, 51–144 (1999)
6. Greenberg, R.: On the structure of certain Galois cohomology groups, Documenta Math. Extra Volume Coates, 357–413 (2006)
7. Greenberg, R.: Surjectivity of the global-to-local map defining a Selmer group. Kyoto J. Math. **50**, 853–888 (2011)
8. Greenberg, R.: Iwasawa theory for Z_p^m-extensions, in preparation
9. Greenberg, R.: Iwasawa theory—past and present. Adv. Stud. Pure Math. **30**, 335–385 (2001)

Control of Λ-adic Mordell–Weil Groups

Haruzo Hida

Abstract The (pro) Λ-MW group is a projective limit of Mordell–Weil groups over a number field k (made out of modular Jacobians) with an action of the Iwasawa algebra and the "big" Hecke algebra. We prove a control theorem of the ordinary part of the Λ-MW groups under mild assumptions. We have proven a similar control theorem for the dual completed inductive limit in [21].

Keywords Modular curve · Hecke algebra · Modular deformation · Analytic family of modular forms · Mordell–Weil group · Modular Jacobian

MSCs primary: 11G40, 11F25, 11F32, 11G18, 14H40 · secondary: 11D45, 11G05, 11G10

1 Introduction

Fix a prime p. This article concerns weight 2 cusp forms of level Np^r for $r > 0$ and $p \nmid N$, and for small primes $p = 2, 3$, they exist only when $N > 2$; thus, we may assume $Np^r \geq 4$. Then the open curve $Y_1(Np^r)$ (obtained from $X_1(Np^r)$ removing all cusps) gives the fine smooth moduli scheme classifying elliptic curves E with an embedding $\mu_{Np^r} \hookrightarrow E$. We applied in [17, 20] the techniques of $U(p)$-isomorphisms to Barsotti–Tate groups of modular Jacobian varieties of high p-power level (with the fixed prime-to-p level N). In this article, we apply the same techniques of $U(p)$-isomorphisms to the projective limit of Mordell–Weil groups of

To John Coates

The author is partially supported by the NSF grants: DMS 0753991 and DMS 1464106.

H. Hida (✉)
Department of Mathematics, UCLA, Los Angeles, CA 90095-1555, USA
e-mail: hida@math.ucla.edu

D. Loeffler and S.L. Zerbes (eds.), *Elliptic Curves, Modular Forms and Iwasawa Theory*, Springer Proceedings in Mathematics & Statistics 188, DOI 10.1007/978-3-319-45032-2_7

the Jacobians and see what we can say (see Sect. 3 for $U(p)$-isomorphisms). We study the (inductive limit of) Tate–Shafarevich groups of the Jacobians in another article [22].

Let $X_r = X_1(Np^r)_{/\mathbb{Q}}$ be the compactified moduli of the classification problem of pairs (E, ϕ) of an elliptic curve E and an embedding $\phi : \mu_{Np^r} \hookrightarrow E[Np^r]$. Write $J_{r/\mathbb{Q}}$ for the Jacobian whose origin is given by the infinity cusp $\infty \in X_r(\mathbb{Q})$ of X_r. For a number field k, we consider the group of k-rational points $J_r(k)$. Put $\widehat{J}_r(k) := \varprojlim_n J_r(k)/p^n J_r(k)$ (as a compact p-profinite module). The Albanese functoriality of Jacobians (twisted by the Weil involutions) gives rise to a projective system $\{\widehat{J}_r(k)\}_r$ compatible with Hecke operators (see Sect. 6 for details of twisting), and we have

$$\widehat{J}_\infty(k) = \varprojlim_r \widehat{J}_r(k)$$

equipped with the projective limit compact topology. By Picard functoriality, we have an injective limit $J_\infty(k) = \varinjlim_r \widehat{J}_r(k)$ (with the injective limit of the compact topology of $\widehat{J}_r(k)$) and $J_\infty[p^\infty]_{/\mathbb{Q}} = \varinjlim_r J_r[p^\infty]_{/\mathbb{Q}}$ (the injective limit of the p-divisible Barsotti–Tate group). We define

$$\check{J}_\infty(k) = \varprojlim_n J_\infty(k)/p^n J_\infty(k).$$

An fppf sheaf $\mathcal{F}$ (over $\mathrm{Spec}(k)$) is a presheaf functor from the fppf site over $\mathrm{Spec}(k)$ to the category of abelian groups satisfying the sheaf condition for an fppf covering $\{U_i\}$ of $T_{/k}$, that is, the exactness of

$$0 \to \mathcal{F}(T) \xrightarrow{\mathrm{Res}_{U_i/T}} \prod_i \mathcal{F}(U_i) \xrightarrow{\mathrm{Res}_{U_{ij}/U_i} - \mathrm{Res}_{U_{ij}/U_j}} \prod_{i,j} \mathcal{F}(U_{ij}), \qquad \text{(L)}$$

where $\mathrm{Res}_{U/V}$ indicates the restriction map relative to $U \to V$ and $U_{ij} := U_i \times_T U_j$. Since the category of fppf sheaves over $\mathbb{Q}$ (e.g., [4, §4.3.7]) is an abelian category (cf. [10, II.2.15]), if we apply a left exact functor (of the category of abelian groups into itself) to the value of a sheaf, it preserves the sheaf condition given by the left exactness (L). Thus projective limits and injective limits exist inside the category of fppf sheaves. We may thus regard

$$R \mapsto \widehat{J}_\infty(R) := \varprojlim_r (J_r(R) \otimes_{\mathbb{Z}} \mathbb{Z}_p) \text{ and } R \mapsto J_\infty(R)$$

as fppf sheaves over the fppf site over $\mathbb{Q}$ for an fppf extension $R_{/k}$, though we do not use this fact much (as we compute $\widehat{J}_\infty(k)$ as a limit of $\widehat{J}_s(k)$ not using sheaf properties of $\widehat{J}_\infty$). If one extends $\widehat{J}_s$ to the ind-category of fppf extensions, we no longer have projective limit expression. We have given detailed description of the value $\widehat{J}_s(R)$ in [21, §2] and we will give a brief outline of this in Sect. 2 in the text.

We can think of the sheaf endomorphism algebra $\mathrm{End}(J_{\infty/\mathbb{Q}})$ (in which we have Hecke operators $T(n)$ and $U(l)$ for $l|Np$).

The Hecke operator $U(p)$ acts on $J_r(k)$, and the p-adic limit $e = \lim_{n\to\infty} U(p)^{n!}$ is well defined on $\widehat{J}_r(k)$. As is well known (cf. [17, 28]; see an exposition on this in Sect. 6), $T(n)$, $U(l)$ and diamond operators are endomorphisms of the injective (resp. projective) systems $\{J_s(k)\}_s$ (resp. $\{\widehat{J}_s(k)\}_s$). The projective system comes from w-twisted Albanese functoriality for the Weil involution w (as we need to twist in order to make the system compatible with $U(p)$; see Sect. 6 for the twisting). The image of e is called the *ordinary* part. We attach as the superscript or the subscript "ord" to indicate the ordinary part. Since these $\mathbb{Z}_p$-modules have natural action of the Iwasawa algebra Λ through diamond operators, we call in particular the group $\widehat{J}_\infty(k)^{\mathrm{ord}}$ the pro Λ-MW group ("MW" stands for Mordell–Weil). We define the Λ-BT group $\mathcal{G}_{/\mathbb{Q}}$ by the ordinary part $J_\infty[p^\infty]^{\mathrm{ord}}_{/\mathbb{Q}}$ of $J_\infty[p^\infty]_{/\mathbb{Q}}$ whose detailed study is made in [20, §4]. Though in [20], we made an assumption that $p \geq 5$, as for the results over $\mathbb{Q}$ in [20, §4], they are valid without any change for $p = 2, 3$ as verified in [15] for $p = 2$ (and the prime $p = 3$ can be treated in the same manner as in [16] or [20, §4]). Thus we use control result over $\mathbb{Q}$ of $\mathcal{G}$ in this paper without assuming $p \geq 5$. Its Tate module $T\mathcal{G} := \mathrm{Hom}_{\mathbb{Z}_p}(\Lambda^\vee, \mathcal{G})$ is a continuous $\Lambda[\mathrm{Gal}(\overline{\mathbb{Q}}/\mathbb{Q})]$-module under the profinite topology, where $M^\vee = \mathrm{Hom}_{\mathbb{Z}_p}(M, \mathbb{Q}_p/\mathbb{Z}_p)$ (Pontryagin dual) for $\mathbb{Z}_p$-modules M. We define the big Hecke algebra $\mathbf{h} = \mathbf{h}(N)$ to be the Λ-subalgebra of $\mathrm{End}_\Lambda(T\mathcal{G})$ generated by Hecke operators $T(n)$ $(n = 1, 2, \dots)$. Then $\widehat{J}_\infty(k)^{\mathrm{ord}}$ and $\check{J}_\infty(k)^{\mathrm{ord}}$ are naturally continuous $\mathbf{h}$-modules. Take a connected component $\mathrm{Spec}(\mathbb{T})$ of $\mathrm{Spec}(\mathbf{h})$ and define the direct factors

$$\widehat{J}_s(k)^{\mathrm{ord}}_{\mathbb{T}} := \widehat{J}_s(k)^{\mathrm{ord}} \otimes_{\mathbf{h}} \mathbb{T} \ (s = 1, 2, \dots, \infty) \ \text{ and } \ T\mathcal{G}_{\mathbb{T}} := T\mathcal{G} \otimes_{\mathbf{h}} \mathbb{T}$$

of $\widehat{J}_\infty(k)^{\mathrm{ord}}$ and $T\mathcal{G}$, respectively. In this introduction, for simplicity, we assume that the component $\mathbb{T}$ cuts out $\widehat{J}_\infty(k)^{\mathrm{ord}}_{\mathbb{T}}$ from $\widehat{J}_\infty(k)^{\mathrm{ord}}$ a part with potentially good reduction modulo p (meaning that $\mathcal{G}_{\mathbb{T}}[\gamma^{p^s} - 1]$ extends to Λ-BT group over $\mathbb{Z}_p[\mu_{p^s}]$ for all s). This is to avoid technicality coming from potentially multiplicative reduction of factors of J_s outside $\widehat{J}_s(k)^{\mathrm{ord}}_{\mathbb{T}}$.

The maximal torsion-free part Γ of $\mathbb{Z}_p^\times$ (which is a p-profinite cyclic group) acts on these modules by the diamond operators. In other words, for modular curves X_r and $X_0(Np^r)$, we identify $\mathrm{Gal}(X_r/X_0(Np^r))$ with $(\mathbb{Z}/Np^r\mathbb{Z})^\times$, and Γ acts on J_r through its image in $\mathrm{Gal}(X_r/X_0(Np^r))$. Therefore the Iwasawa algebra $\Lambda = \mathbb{Z}_p[[\Gamma]] = \varprojlim_r \mathbb{Z}_p[\Gamma/\Gamma^{p^r}]$ acts on the pro Λ-MW group, the ind Λ-MW group, the Λ-BT group and its Tate module. Then $T\mathcal{G}$ is known to be free of finite rank over Λ [15, 17] and [20, §4]. A prime $P \in \mathrm{Spec}(\mathbb{T})(\overline{\mathbb{Q}}_p)$ is called *arithmetic* of weight 2 if P factors through $\mathrm{Spec}(\mathbb{T} \otimes_\Lambda \mathbb{Z}_p[\Gamma/\Gamma^{p^r}])$ for some $r > 0$. Associated to P is a unique Hecke eigenform of weight 2 on $X_1(Np^r)$ for some $r > 0$. Write B_P for the Shimura's abelian quotient associated to f_P of the jacobian J_r. Let $\mathcal{A}_{\mathbb{T}}$ be the set of all principal arithmetic points of $\mathrm{Spec}(\mathbb{T})(\overline{\mathbb{Q}}_p)$ of weight 2 and put $\Omega_{\mathbb{T}} := \{P \in \mathcal{A}_{\mathbb{T}} | B_P$ has good reduction over $\mathbb{Z}_p[\mu_{p^\infty}]\}$. The word "principal" means, as a prime ideal of $\mathbb{T}$, it is generated by a single element, often written as α. In this article, we prove control results for the pro Λ-MW group $\widehat{J}_\infty(k)^{\mathrm{ord}}$ and study the control of the ind Λ-MW-groups $\check{J}_\infty(k)^{\mathrm{ord}}$ in the twin paper [21, Theorem 6.5]. Take a topological generator $\gamma = 1 + p^\epsilon$ of Γ, and regard γ as a group element of $\Lambda = \mathbb{Z}_p[[\Gamma]]$, where $\epsilon = 1$ if $p > 2$ and $\epsilon = 2$ if $p = 2$. We use this definition of ϵ throughout the paper (and we assume that $r \geq \epsilon$ if the exponent $r - \epsilon$ shows up in a formula). We fix a finite set S of places of $\mathbb{Q}$ containing all places $v|Np$ and the archimedean place. Here is a simplified statement of our finial result:

Theorem. *If $\mathbb{T}$ is an integral domain, for almost all principal arithmetic prime $P = (\alpha) \in \mathcal{A}_{\mathbb{T}}$, we have the following canonical exact sequence up to finite error of Hecke modules:*

$$0 \to \widehat{J}^{\mathrm{ord}}_\infty(k)_{\mathbb{T}} \xrightarrow{\alpha} \widehat{J}^{\mathrm{ord}}_\infty(k)_{\mathbb{T}} \xrightarrow{\rho_\infty} \widehat{B}^{\mathrm{ord}}_P(k)_{\mathbb{T}}. \tag{1}$$

This theorem will be proven as Theorem 9.2. The exact sequence in the theorem is a Mordell–Weil analogue of a result of Nekovář in [27, 12.7.13.4] for Selmer groups and implies that $\widehat{J}^{\mathrm{ord}}_\infty(k)$ is a Λ-module of finite type. In the text, we prove a stronger result showing finiteness of $\mathrm{Coker}(\rho_\infty)$ for almost all principal arithmetic primes P if the ordinary part of Selmer group of B_{P_0} is finite for one principal arithmetic prime P_0 (see Theorem 10.1).

Put $\check{J}_\infty(k)^*_{\mathrm{ord}} := \mathrm{Hom}_{\mathbb{Z}_p}(\check{J}_\infty(k)^{\mathrm{ord}}, \mathbb{Z}_p)$. In [21, Theorem 1.1], we proved the following exact sequence:

$$\check{J}_\infty(k)^*_{\mathrm{ord},P} \xrightarrow{\alpha} \check{J}_\infty(k)^*_{\mathrm{ord},P} \to \widehat{A}_P(k)^*_{\mathrm{ord},P} \to 0$$

for arithmetic P of weight 2, in addition to the finiteness of $\check{J}_\infty(k)^*_{\mathrm{ord}}$ as a Λ-module. This sequence is a localization at P of the natural one. The two sequences could be dual each other if we have a Λ-adic version of the Néron–Tate height pairing.

Here is some notation for Hecke algebras used throughout the paper. Let

$$h_r(\mathbb{Z}) = \mathbb{Z}[T(n), U(l) : l|Np, (n, Np) = 1] \subset \mathrm{End}(J_r),$$

and put $h_r(R) = h_r(\mathbb{Z}) \otimes_{\mathbb{Z}} R$ for any commutative ring R. Then we define $\mathbf{h}_r = e(h_r(\mathbb{Z}_p))$. The restriction morphism $h_s(\mathbb{Z}) \ni h \mapsto h|_{J_r} \in h_r(\mathbb{Z})$ for $s > r$ induces a projective system $\{\mathbf{h}_r\}_r$ whose limit gives rise to a big ordinary Hecke algebra

$$\mathbf{h} = \mathbf{h}(N) := \varprojlim_r \mathbf{h}_r.$$

Writing $\langle l \rangle$ (the diamond operator) for the action of l as an element of $(\mathbb{Z}/Np^r\mathbb{Z})^\times = \mathrm{Gal}(X_r/X_0(Np^r))$, we have an identity $l\langle l \rangle = T(l)^2 - T(l^2) \in h_r(\mathbb{Z}_p)$ for all primes $l \nmid Np$. Thus we have a canonical Λ-algebra structure $\Lambda = \mathbb{Z}_p[[\Gamma]] \hookrightarrow \mathbf{h}$. It is now well known that $\mathbf{h}$ is a free of finite rank over Λ and $\mathbf{h}_r = \mathbf{h} \otimes_\Lambda \Lambda/(\gamma^{p^{r-\epsilon}} - 1)$ (cf. [16]). Though the construction of the big Hecke algebra is intrinsic, to relate an algebra homomorphism $\lambda : \mathbf{h} \to \overline{\mathbb{Q}}_p$ killing $\gamma^{p^r} - 1$ for $r > 0$ to a classical Hecke eigenform, we need to fix (once and for all) an embedding $\overline{\mathbb{Q}} \xrightarrow{i_p} \overline{\mathbb{Q}}_p$ of the algebraic closure $\overline{\mathbb{Q}}$ in $\mathbb{C}$ into a fixed algebraic closure $\overline{\mathbb{Q}}_p$ of $\mathbb{Q}_p$. We write i_∞ for the inclusion $\overline{\mathbb{Q}} \subset \mathbb{C}$.

The following two Sects. 3 and 4 (after a description of sheaves associated to abelian varieties) about $U(p)$-isomorphisms are an expanded version of a conference talk at CRM (see http://www.crm.umontreal.ca/Representations05/indexen.html) in September of 2005 which was not posted in the author's web page, though the lecture notes of the two lectures [18] at CRM earlier than the conference have been posted. While converting [18] into a research article [20], the author found an application to Mordell–Weil groups of modular Jacobians. The author is grateful for CRM's invitation to speak. The author would like to thank the referee of this paper for careful reading (and the proof of (10.4) in the old version is incomplete as was pointed out by the referee). Heuristically, as explained just after Theorem 10.1, this point does not cause much trouble as we are dealing with the standard tower for which the root number for members of the family is not equal to -1 for most arithmetic point; so, presumably, the Mordell Weil group of B_P is finite for most P.

2 Sheaves Associated to Abelian Varieties

Here is a general fact proven in [21, §2] about sheaves associated to abelian varieties. Let $0 \to A \to B \to C \to 0$ be an exact sequence of algebraic groups proper over a field k. The field k is either a number field or a finite extension of the l-adic field $\mathbb{Q}_l$ for a prime l. We assume that B and C are abelian varieties. However A can be an extension of an abelian variety by a finite (étale) group.

If k is a number field, let S be a set of places including all archimedean places of k such that all members of the above exact sequence have good reduction outside S. We use the symbol K for k^S (the maximal extension unramified outside S) if k is a number field and for $\overline{k}$ (an algebraic closure of k) if k is a finite extension of $\mathbb{Q}_l$. A general field extension of k is denoted by κ. We consider the étale topology, the smooth topology and the fppf topology on the small site over Spec(k). Here under the smooth topology, covering families are made of faithfully flat smooth morphisms.

For the moment, assume that k is a number field. In this case, for an extension X of abelian variety defined over k by a finite étale group scheme, we define $\widehat{X}(\kappa) := X(\kappa) \otimes_{\mathbb{Z}} \mathbb{Z}_p$ for an fppf extension κ over k. By Mordell–Weil theorem (and its extension to fields of finite type over $\mathbb{Q}$; e.g., [2, IV]), we have $\widehat{X}(\kappa) = \varprojlim_n X(\kappa)/p^n X(\kappa)$ if κ is a field extension of k of finite type. We may regard the sequence $0 \to \widehat{A} \to \widehat{B} \to \widehat{C} \to 0$ as an exact sequence of fppf abelian sheaves over k (or over any subring of k over which B and C extends to abelian schemes). Since we find a complementary abelian subvariety C' of B such that C' is isogenous to C and $B = A + C'$ with finite $A \cap C'$, adding the primes dividing the order $|A \cap C'|$ to S, the intersection $A \cap C' \cong \mathrm{Ker}(C' \to C)$ extends to an étale finite group scheme outside S; so, $C'(K) \to C(K)$ is surjective. Thus we have an exact sequence of $\mathrm{Gal}(K/k)$-modules

$$0 \to A(K) \to B(K) \to C(K) \to 0.$$

Note that $\widehat{A}(K) = A(K) \otimes_{\mathbb{Z}} \mathbb{Z}_p := \bigcup_F \widehat{A}(F)$ for F running over all finite extensions of k inside K. Then we have an exact sequence

$$0 \to \widehat{A}(K) \to \widehat{B}(K) \to \widehat{C}(K) \to 0. \tag{2}$$

Now assume that k is a finite extension of $\mathbb{Q}_l$. Again we use F to denote a finite field extension of k. Then $A(F) \cong O_F^{\dim A} \oplus \Delta_F$ for a finite group Δ_F for the l-adic integer ring O_F of F (by [23] or [32]). Thus if $l \neq p$, $\widehat{A}(F) := \varprojlim_n A(F)/p^n A(F) = \Delta_F \otimes_{\mathbb{Z}} \mathbb{Z}_p = A[p^\infty](F)$. Recall $K = \overline{k}$. Then $\widehat{A}(K) = A[p^\infty](K)$ (for $A[p^\infty] = \varinjlim_n A[p^n]$ with $A[p^n] = \mathrm{Ker}(p^n : A \to A)$); so, defining $\widehat{A}$, $\widehat{B}$ and $\widehat{C}$ by $A[p^\infty]$, $B[p^\infty]$ and $C[p^\infty]$ as fppf abelian sheaves, we again have the exact sequence (2) of $\mathrm{Gal}(\overline{k}/k)$-modules:

$$0 \to \widehat{A}(K) \to \widehat{B}(K) \to \widehat{C}(K) \to 0$$

and an exact sequence of fppf abelian sheaves

$$0 \to \widehat{A} \to \widehat{B} \to \widehat{C} \to 0$$

whose value at a finite field extension $\kappa/\mathbb{Q}_l$ coincides with $\widehat{X}(\kappa) = \varprojlim_n X(\kappa)/p^n X(\kappa)$ for $X = A, B, C$.

Suppose $l = p$. For any module M, we define $M^{(p)}$ by the maximal prime-to-p torsion submodule of M. For $X = A, B, C$ and an fppf extension $R_{/k}$, the sheaf $R \mapsto X^{(p)}(R) = \varinjlim_{p \nmid N} X[N](R)$ is an fppf abelian sheaf. Then we define the fppf abelian sheaf $\widehat{X}$ by the sheaf quotient $X/X^{(p)}$. Since $X(F) = O_F^{\dim X} \oplus X[p^\infty](F) \oplus X^{(p)}(F)$ for a finite field extension $F_{/k}$, over the étale site on k, $\widehat{X}$ is the sheaf associated to a presheaf $R \mapsto O_F^{\dim X} \oplus X[p^\infty](R)$. If X has semi-stable reduction over O_F, we have $\widehat{X}(F) = X^\circ(O_F) + X[p^\infty](F) \subset X(F)$ for the formal group X° of the identity connected component of the Néron model of X over O_F [32]. Since X becomes semi-stable over a finite Galois extension F_0/k, in general $\widehat{X}(F) = H^0(\mathrm{Gal}(F_0F/F), X(F_0F))$ for any finite extension $F_{/K}$ (or more generally for each finite étale extension $F_{/k}$); so, $F \mapsto \widehat{X}(F)$ is a sheaf over the étale site on k. Thus by [10, II.1.5], the sheafication coincides over the étale site with the presheaf $F \mapsto \varprojlim_n X(F)/p^n X(F)$. Thus we conclude $\widehat{X}(F) = \varprojlim_n X(F)/p^n X(F)$ for any étale finite extensions $F_{/k}$. Moreover $\widehat{X}(K) = \bigcup_{K/F/k} \widehat{X}(F)$. Applying the snake lemma to the commutative diagram with exact rows (in the category of fppf abelian sheaves):

$$\begin{array}{ccccc} A^{(p)} & \xrightarrow{\hookrightarrow} & B^{(p)} & \xrightarrow{\twoheadrightarrow} & C^{(p)} \\ \cap\downarrow & & \cap\downarrow & & \cap\downarrow \\ A & \underset{\hookrightarrow}{\longrightarrow} & B & \underset{\twoheadrightarrow}{\longrightarrow} & C, \end{array}$$

the cokernel sequence gives rise to an exact sequence of fppf abelian sheaves over k:

$$0 \to \widehat{A} \to \widehat{B} \to \widehat{C} \to 0$$

and an exact sequence of $\mathrm{Gal}(\overline{k}/k)$-modules

$$0 \to \widehat{A}(K) \to \widehat{B}(K) \to \widehat{C}(K) \to 0.$$

In this way, we extended the sheaves $\widehat{A}$, $\widehat{B}$, $\widehat{C}$ to fppf abelian sheaves keeping the exact sequence $\widehat{A} \hookrightarrow \widehat{B} \twoheadrightarrow \widehat{C}$ intact. However note that our way of defining $\widehat{X}$ for $X = A, B, C$ depends on the base field $k = \mathbb{Q}, \mathbb{Q}_p, \mathbb{Q}_l$. Here is a summary for fppf algebras $R_{/k}$:

$$\widehat{X}(R) = \begin{cases} X(R) \otimes_{\mathbb{Z}} \mathbb{Z}_p & \text{if } [k : \mathbb{Q}] < \infty, \\ X[p^\infty](R) & \text{if } [k : \mathbb{Q}_l] < \infty (l \neq p), \\ (X/X^{(p)})(R) \text{ as a sheaf quotient} & \text{if } [k : \mathbb{Q}_p] < \infty. \end{cases} \tag{S}$$

Here is a sufficient condition when $\widehat{X}(\kappa)$ is given by the projective limit: $\varprojlim_n X(\kappa)/p^n X(\kappa)$ for $X = A, B$ or C:

$$\widehat{X}(\kappa) = \varprojlim_n \widehat{X}(\kappa)/p^n \widehat{X}(\kappa)$$
$$\text{if } \begin{cases} [k : \mathbb{Q}] < \infty \text{ and } \kappa \text{ is a field of finite type over } k \\ [k : \mathbb{Q}_l] < \infty \text{ with } l \neq p \text{ and } \kappa \text{ is a field of finite type over } k \\ [k : \mathbb{Q}_p] < \infty \text{ and } \kappa \text{ is a finite algebraic extension over } k. \end{cases} \tag{3}$$

A slightly weaker sufficient condition for $\widehat{X}(\kappa) = \varprojlim_n \widehat{X}(\kappa)/p^n \widehat{X}(\kappa)$ is proven in [21, Lemma 2.1].

For a sheaf X under the topology ?, we write $H^\bullet_?(X)$ for the cohomology group $H^1_?(\mathrm{Spec}(k), X)$ under the topology ?. If we have no subscript, $H^1(X)$ means the Galois cohomology $H^\bullet(\mathrm{Gal}(K/k), X)$ for the $\mathrm{Gal}(K/k)$-module X. For any $\mathbb{Z}_p$-module M, we put $T_p M = \varprojlim_n M[p^n] = \mathrm{Hom}_{\mathbb{Z}_p}(\mathbb{Q}_p/\mathbb{Z}_p, M)$.

The following fact is essentially proven in [21, Lemma 2.2] (where it was proven for finite S but the same proof works for infinite S as is obvious from the fact that it works under fppf topology):

Lemma 2.1 *Let X be an extension of an abelian variety over k by a finite étale group scheme of order prime to p. Then, we have a canonical injection*

$$\varprojlim_n \widehat{X}(k)/p^n\widehat{X}(k) \hookrightarrow \varprojlim_n H^1(X[p^n]).$$

Similarly, for any fppf or smooth extension κ/k of finite type which is an integral domain, we have an injection

$$\varprojlim_n \widehat{X}(\kappa)/p^n\widehat{X}(\kappa) \hookrightarrow \varprojlim_n H^1_?(\mathrm{Spec}(\kappa), X[p^n])$$

for $? = \mathrm{fppf}$ or sm according as κ/k is an fppf extension or a smooth extension of finite type. For Galois cohomology, we have an exact sequence for $j = 0, 1$:

$$0 \to \varprojlim_n H^j(\widehat{X}(k))/p^n H^j(\widehat{X}(k)) \to \varprojlim_n H^{j+1}(X[p^n]) \to T_p H^{j+1}(X).$$

The natural map: $\varprojlim_n H^{j+1}(X[p^n]) \xrightarrow{\pi} T_p H^{j+1}(X)$ is surjective if either $j = 0$ or k is local or S is finite. In particular, $H^1(T_p X)$ for $T_p X = \varprojlim_n X[p^n]$ is equal to $\varprojlim_n H^1(X[p^n])$, and

$$0 \to \widehat{X}(k) \to H^1(T_p X) \to T_p H^1(X)[p^n] \to 0$$

is exact.

We shall give a detailed proof of the surjectivity of π for Galois cohomology (which we will use) along with a sketch of the proof of the exactness.

Proof By p-divisibility, we have the sheaf exact sequence under the étale topology over $\mathrm{Spec}(\kappa)$

$$0 \to X[p^n] \to X \xrightarrow{p^n} X \to 0.$$

This implies, we have an exact sequence

$$0 \to X[p^n](K) \to X(K) \xrightarrow{p^n} X(K) \to 0.$$

By the long exact sequence associated to this sequence, for a finite intermediate extension $K/\kappa/k$, we have exactness of

$$0 \to H^j(X(\kappa))/p^n H^j(X(\kappa)) \to H^{j+1}(X[p^n]) \to H^{j+1}(X)[p^n] \to 0. \qquad (*)$$

Passing to the limit (with respect to n), we have the exactness of

$$0 \to \varprojlim_n H^j(X(\kappa))/p^n H^j(X(\kappa)) \to H^{j+1}(T_p X) \to T_p H^{j+1}(X).$$

as $\varprojlim_n H^{j+1}(X[p^n]) = H^{j+1}(\varprojlim_n X[p^n]) = H^j(T_pX)$ for $j = 0, 1$ without assumption if $j = 0$ and assuming S is finite if $j = 1$ (because of finiteness of $X[p^n](K)$ and p-divisibility of X; e.g., [12, Corollary 2.7.6] and [22, Lemma 7.1 (2)]).

Assume $\kappa = k$. If k is local or S is finite, by Tate duality, all the terms of $(*)$ is finite; so, the surjectivity of $(*)$ is kept after passing to the limit. If $j = 0$ and $\kappa = k$, $X(k)/p^nX(k)$ is a finite module; so, the sequences $(*)$ satisfied Mittag–Leffler condition. Thus again the surjectivity of $(*)$ is kept after passing to the limit. □

For finite S, the following module structure of $H^1(\widehat{A})$ is well known (see [9, Corollary I.4.15] or [21, Lemma 2.3]):

Lemma 2.2 *Let k be a finite extension of $\mathbb{Q}$ or $\mathbb{Q}_l$ for a prime l. Suppose that S is finite if k is a finite extension of $\mathbb{Q}$. Let $A_{/k}$ be an abelian variety. Then $H^1(A) \otimes_{\mathbb{Z}} \mathbb{Z}_p = H^1(\widehat{A})$ is isomorphic to the discrete module $(\mathbb{Q}_p/\mathbb{Z}_p)^r \oplus \Delta$ for a finite $r \geq 0$ and a finite p-torsion group Δ.*

Hereafter we assume that S is a finite set unless otherwise indicated.

3 $U(p)$-isomorphisms for Group Cohomology

For $\mathbb{Z}[U]$-modules X and Y, we call a $\mathbb{Z}[U]$-linear map $f : X \to Y$ a U-injection (resp. a U-surjection) if $\mathrm{Ker}(f)$ is killed by a power of U (resp. $\mathrm{Coker}(f)$ is killed by a power of U). If f is an U-injection and U-surjection, we call f is a U-isomorphism. If $X \to Y$ is a U-isomorphism, we write $X \cong_U Y$. In terms of U-isomorphisms (for $U = U(p), U^*(p)$, we describe briefly the facts we need in this article (and in later sections, we fill in more details in terms of the ordinary projector e and the co-ordinary projector $e^* := \lim_{n\to\infty} U^*(p)^{n!}$).

Let N be a positive integer prime to p. We consider the (open) modular curve $Y_1(Np^r)_{/\mathbb{Q}}$ which classifies elliptic curves E with an embedding $\phi : \mu_{p^r} \hookrightarrow E[p^r] = \mathrm{Ker}(p^r : E \to E)$ of finite flat groups. Let $R_i = \mathbb{Z}_{(p)}[\mu_{p^i}]$ and $K_i = \mathbb{Q}[\mu_{p^i}]$. For a valuation subring or a subfield R of K_∞ over $\mathbb{Z}_{(p)}$ with quotient field K, we write $X_{r/R}$ for the normalization of the j-line $\mathbf{P}(j)_{/R}$ in the function field of $Y_1(Np^r)_{/K}$. The group $z \in (\mathbb{Z}/p^r\mathbb{Z})^\times$ acts on X_r by $\phi \mapsto \phi \circ z$, as $\mathrm{Aut}(\mu_{Np^r}) \cong (\mathbb{Z}/Np^r\mathbb{Z})^\times$. Thus $\Gamma = 1 + p^\epsilon\mathbb{Z}_p = \gamma^{\mathbb{Z}_p}$ acts on X_r (and its Jacobian) through its image in $(\mathbb{Z}/Np^r\mathbb{Z})^\times$. Hereafter we take $U = U(p), U^*(p)$ for the Hecke–Atkin operator $U(p)$.

Let $J_{r/R} = \mathrm{Pic}^0_{X_{r/R}}$ be the connected component of the Picard scheme. We state a result comparing $J_{r/R}$ and the Néron model of $J_{r/K}$ over R. Thus we assume that R is a valuation ring. By [8, 5.5.1, 13.5.6, 13.11.4], $X_{r/R}$ is regular; the reduction $X_r \otimes_R \mathbb{F}_p$ is a union of irreducible components, and the component containing the ∞ cusp has geometric multiplicity 1. Then by [1, Theorem 9.5.4], $J_{r/R}$ gives the identity connected component of the Néron model of the Jacobian of $X_{r/R}$. In this paper, we do not use these fine integral structure of $X_{r/R}$ but work with $X_{r/\mathbb{Q}}$. We just wanted to note these facts for possible use in our future articles.

We write $X^s_{r/R}$ for the normalization of the j-line of the canonical $\mathbb{Q}$-curve associated to the modular curve for the congruence subgroup $\Gamma^r_s = \Gamma_1(Np^r) \cap \Gamma_0(p^s)$ for $0 < r \leq s$. We denote $\mathrm{Pic}^0_{X^r_{s/R}}$ by $J^r_{s/R}$. Similarly, as above, $J^r_{s/R}$ is the connected component of the Néron model of $X^r_{s/K}$. Note that, for $\alpha_m = \left(\begin{smallmatrix} 1 & 0 \\ 0 & p^m \end{smallmatrix}\right)$,

$$\begin{aligned}\Gamma^r_s \backslash \Gamma^r_s \alpha_{s-r} \Gamma_1(Np^r) &= \left\{ \left(\begin{smallmatrix} 1 & a \\ 0 & p^{s-r} \end{smallmatrix}\right) \middle|\, a \mod p^{s-r} \right\} \\ &= \Gamma_1(Np^r) \backslash \Gamma_1(Np^r) \alpha_{s-r} \Gamma_1(Np^r). \qquad (4)\end{aligned}$$

Write $U^s_r(p^{s-r}) : J^s_{r/R} \to J_{r/R}$ for the Hecke operator of $\Gamma^s_r \alpha_{s-r} \Gamma_1(Np^r)$. Strictly speaking, the Hecke operator induces a morphism of the generic fiber of the Jacobians and then extends to their connected components of the Néron models by the functoriality of the model (or by Picard functoriality). Then we have the following commutative diagram from the above identity, first over $\mathbb{C}$, then over K and by Picard functoriality over R:

$$\begin{array}{ccc} J_{r/R} & \xrightarrow{\pi^*} & J^r_{s/R} \\ {\scriptstyle u}\downarrow & \swarrow{\scriptstyle u'} & \downarrow{\scriptstyle u''} \\ J_{r/R} & \xrightarrow{\pi^*} & J^r_{s/R}, \end{array} \qquad (5)$$

where the middle u' is given by $U^s_r(p^{s-r})$ and u and u'' are $U(p^{s-r})$. Thus

- (u1) The map $\pi^* : J_{r/R} \to J^r_{s/R}$ is a $U(p)$-isomorphism (for the projection $\pi : X^r_s \to X_r$).

Taking the dual $U^*(p)$ of $U(p)$ with respect to the Rosati involution associated to the canonical polarization of the Jacobians, we have a dual version of the above diagram for $s > r > 0$:

$$\begin{array}{ccc} J_{r/R} & \xleftarrow{\pi_*} & J^r_{s/R} \\ {\scriptstyle u^*}\uparrow & \nearrow{\scriptstyle u'^*} & \uparrow{\scriptstyle u''^*} \\ J_{r/R} & \xleftarrow{\pi_*} & J^r_{s/R}. \end{array} \qquad (6)$$

Here the superscript "$*$" indicates the Rosati involution of the canonical divisor of the Jacobians, and $u^* = U^*(p)^{s-r}$ for the level $\Gamma_1(Np^r)$ and $u''^* = U^*(p)^{s-r}$ for Γ^r_s. Note that these morphisms come from the following coset decomposition, for $\beta_m := \left(\begin{smallmatrix} p^m & 0 \\ 0 & 1 \end{smallmatrix}\right) \Gamma_1(Np^r)$,

$$\Gamma^r_s\backslash\Gamma^r_s\beta_{s-r}\Gamma_1(Np^r) = \left\{ \left(\begin{smallmatrix} p^{s-r} & a \\ 0 & 1 \end{smallmatrix}\right) \middle| a \mod p^{s-r} \right\}$$
$$= \Gamma_1(Np^r)\backslash\Gamma_1(Np^r)\beta_{s-r}\Gamma_1(Np^r). \quad (7)$$

From this, we get

(u*1) The map $\pi_* : J_{r/R} \to J^r_{s/R}$ is a $U^*(p)$-isomorphism, where π_* is the dual of π^*.

In particular, if we take the ordinary and the co-ordinary projector $e = \lim_{n\to\infty} U(p)^{n!}$ and $e^* = \lim_{n\to\infty} U^*(p)^{n!}$ on $J[p^\infty]$ for $J = J_{r/R}, J_{s/R}, J^r_{s/R}$, noting $U(p^m) = U(p)^m$, we have

$$\pi^* : J^{\mathrm{ord}}_{r/R}[p^\infty] \cong J^{r,\mathrm{ord}}_{s/R}[p^\infty] \text{ and } \pi_* : J^{r,\text{co-ord}}_{s/R}[p^\infty] \cong J^{\text{co-ord}}_{r/R}[p^\infty]$$

where "ord" (resp. "co-ord") indicates the image of the projector e (resp. e^*). For simplicity, we write $\mathcal{G}_{r/R} := J^{\mathrm{ord}}_{r/R}[p^\infty]_{/R}$, and we set $\mathcal{G} := \varinjlim_r \mathcal{G}_r$.

Pick a congruence subgroup Γ defining the modular curve $X(\mathbb{C}) = \Gamma\backslash(\mathfrak{H} \sqcup \mathbf{P}^1(\mathbb{Q}))$, and write its Jacobian as J. We now identify $J(\mathbb{C})$ with a subgroup of $H^1(\Gamma, \mathbf{T})$ (for the trivial Γ-module $\mathbf{T} := \mathbb{R}/\mathbb{Z} \cong \{z \in \mathbb{C}^\times : |z| = 1\}$ with trivial Γ-action). Since $\Gamma^r_s \rhd \Gamma_1(Np^s)$, consider the finite cyclic quotient group $C := \frac{\Gamma^r_s}{\Gamma_1(Np^s)}$. By the inflation restriction sequence, we have the following commutative diagram with exact rows:

$$\begin{array}{ccccccc}
H^1(C,\mathbf{T}) & \xrightarrow{\hookrightarrow} & H^1(\Gamma^r_s,\mathbf{T}) & \longrightarrow & H^1(\Gamma_1(Np^s),\mathbf{T})^{\gamma^{p^r}=1} & \longrightarrow & H^2(C,\mathbf{T}) \\
\uparrow & & \cup\uparrow & & \uparrow\cup & & \uparrow \\
? & \longrightarrow & J^r_s(\mathbb{C}) & \longrightarrow & J_s(\mathbb{C})[\gamma^{p^{r-e}}-1] & \longrightarrow & ?.
\end{array} \quad (8)$$

Since C is a finite cyclic group of order p^{s-r} (with generator g) acting trivially on $\mathbf{T}$, we have $H^1(C, \mathbf{T}) = \mathrm{Hom}(C, \mathbf{T}) \cong C$ and

$$H^2(C, \mathbf{T}) = \mathbf{T}/(1 + g + \cdots + g^{p^{s-r}-1}) = \mathbf{T}/p^{s-r}\mathbf{T} = 0.$$

By the same token, replacing $\mathbf{T}$ by $\mathbb{T}_p := \mathbb{Q}_p/\mathbb{Z}_p$, we get $H^2(C, \mathbb{T}_p) = 0$. By a sheer computation (cf. [17, Lemma 6.1]), we confirm that $U(p)$ acts on $H^1(C, \mathbf{T})$ and $H^1(C, \mathbb{T}_p)$ via multiplication by its degree p, and hence $U(p)^{s-r}$ kill $H^1(C, \mathbf{T})$ and $H^1(C, \mathbb{T}_p)$. We record what we have proven:

$$U(p)^{s-r}(H^1(C, \mathbb{T}_p)) = H^2(C, \mathbf{T}) = H^2(C, \mathbb{T}_p) = 0. \quad (9)$$

This fact has been exploited by the author (for example, [17] and [20]) to study the modular Barsotti–Tate groups $J_s[p^\infty]$.

4 $U(p)$-isomorphisms for Arithmetic Cohomology

To good extent, we reproduce the results and proofs in [21, §3] as it is important in the sequel. Let $X \to Y \to S$ be proper morphisms of noetherian schemes. We now replace $H^1(\Gamma, \mathbf{T})$ in the above diagram (8) by

$$H^0_{\mathrm{fppf}}(T, R^1 f_* \mathbb{G}_m) = R^1 f_* O_X^\times(T) = \mathrm{Pic}_{X/S}(T)$$

for S-scheme T and the structure morphism $f : X \to S$, and do the same analysis as in Sect. 3 for arithmetic cohomology in place of group cohomology (via the moduli theory of Katz–Mazur and Drinfeld; cf., [8]). Write the morphisms as $X \xrightarrow{\pi} Y \xrightarrow{g} S$ with $f = g \circ \pi$. Assume that π is finite flat.

Suppose that f and g have compatible sections $S \xrightarrow{s_g} Y$ and $S \xrightarrow{s_f} X$ so that $\pi \circ s_f = s_g$. Then we get (e.g., [1, Sect. 8.1])

$$\mathrm{Pic}_{X/S}(T) = \mathrm{Ker}(s_f^1 : H^1_{\mathrm{fppf}}(X_T, O_X^\times) \to H^1_{\mathrm{fppf}}(T, O_T^\times))$$
$$\mathrm{Pic}_{Y/S}(T) = \mathrm{Ker}(s_g^1 : H^1_{\mathrm{fppf}}(Y_T, O_{Y_T}^\times) \to H^1_{\mathrm{fppf}}(T, O_T^\times))$$

for any S-scheme T, where $s_f^q : H^q(X_T, O_{X_T}^\times) \to H^q(T, O_T^\times)$ and $s_g^n : H^n(Y_T, O_{Y_T}^\times) \to H^n(T, O_T^\times)$ are morphisms induced by s_f and s_g, respectively. Here $X_T = X \times_S T$ and $Y_T = Y \times_S T$. We suppose that the functors $\mathrm{Pic}_{X/S}$ and $\mathrm{Pic}_{Y/S}$ are representable by group schemes whose connected components are smooth (for example, if X, Y are curves and $S = \mathrm{Spec}(k)$ for a field k; see [1, Theorem 8.2.3 and Proposition 8.4.2]). We then put $J_? = \mathrm{Pic}^0_{?/S}$ $(? = X, Y)$. Anyway we suppose hereafter also that X, Y, S are varieties (in the sense of [3, II.4]).

For an fppf covering $\mathcal{U} \to Y$ and a presheaf $P = P_Y$ on the fppf site over Y, we define via Čech cohomology theory an fppf presheaf $\mathcal{U} \mapsto \check{H}^q(\mathcal{U}, P)$ denoted by $\underline{\check{H}}^q(P_Y)$ (see [10, III.2.2 (b)]). The inclusion functor from the category of fppf sheaves over Y into the category of fppf presheaves over Y is left exact. The derived functor of this inclusion of an fppf sheaf $F = F_Y$ is denoted by $\underline{H}^\bullet(F_Y)$ (see [10, III.1.5 (c)]). Thus $\underline{H}^\bullet(\mathbb{G}_{m/Y})(\mathcal{U}) = H^\bullet_{\mathrm{fppf}}(\mathcal{U}, O_{\mathcal{U}}^\times)$ for a Y-scheme $\mathcal{U}$ as a presheaf (here $\mathcal{U}$ varies in the small fppf site over Y).

Instead of the Hochschild–Serre spectral sequence producing the top row of the diagram (8), assuming that f, g and π are all faithfully flat of finite presentation, we use the spectral sequence of Čech cohomology of the flat covering $\pi : X \twoheadrightarrow Y$ in the fppf site over Y [10, III.2.7]:

$$\check{H}^p(X_T/Y_T, \underline{H}^q(\mathbb{G}_{m/Y})) \Rightarrow H^n_{\mathrm{fppf}}(Y_T, O_{Y_T}^\times) \underset{\iota}{\xrightarrow{\sim}} H^n(Y_T, O_{Y_T}^\times) \tag{10}$$

for each S-scheme T. Here $F \mapsto H^n_{\mathrm{fppf}}(Y_T, F)$ (resp. $F \mapsto H^n(Y_T, F)$) is the right derived functor of the global section functor: $F \mapsto F(Y_T)$ from the category of fppf sheaves (resp. Zariski sheaves) over Y_T to the category of abelian groups. The canonical isomorphism ι is the one given in [10, III.4.9].

By the sections $s_?$, we have a splitting $H^q(X_T, O^{\times}_{X_T}) = \mathrm{Ker}(s^q_f) \oplus H^q(T, O^{\times}_T)$ and $H^n(Y_T, O^{\times}_{Y_T}) = \mathrm{Ker}(s^n_g) \oplus H^n(T, O^{\times}_T)$. Write $\underline{H}^{\bullet}_{Y_T}$ for $\underline{H}^{\bullet}(\mathbb{G}_{m/Y_T})$ and $\check{H}^{\bullet}(\underline{H}^0_{Y_T})$ for $\check{H}^{\bullet}(Y_T/X_T, \underline{H}^0_{Y_T})$. Since

$$\mathrm{Pic}_{X/S}(T) = \mathrm{Ker}(s^1_{f,T} : H^1(X_T, O^{\times}_{X_T}) \to H^1(T, O^{\times}_T))$$

for the morphism $f : X \to S$ with a section [1, Proposition 8.1.4], from the spectral sequence (10), we have the following commutative diagram with exact rows:

$$\begin{array}{ccccccc}
\check{H}^1(\underline{H}^0_{Y_T}) & \hookrightarrow & H^1(T, O^{\times}_T) \oplus \mathrm{Ker}(s^1_{g,T}) & \xrightarrow{a} & \check{H}^0(\frac{X_T}{Y_T}, \underline{H}^1(\mathbb{G}_{m,Y})) & \longrightarrow & H^2(\underline{H}^0_{Y_T}) \\
\uparrow \| & & \uparrow \wr & & \uparrow \| & & \uparrow \| \\
\check{H}^1(\underline{H}^0_{Y_T}) & \longrightarrow & \mathrm{Pic}_T \oplus \mathrm{Pic}_{Y/S}(T) & \xrightarrow{b} & \check{H}^0(\frac{X_T}{Y_T}, \mathrm{Pic}_Y(T)) & \longrightarrow & \check{H}^2(\underline{H}^0_{Y_T}) \\
\uparrow & & \uparrow \cup & & \uparrow \cup & & \uparrow \\
?_1 & \longrightarrow & \mathrm{Pic}_T \oplus J_Y(T) & \xrightarrow{c} & \mathrm{Pic}_T \oplus \check{H}^0(\frac{X_T}{Y_T}, J_X(T)) & \longrightarrow & ?_2,
\end{array} \tag{11}$$

where we have written $J_? = \mathrm{Pic}^0_{?/S}$ (the identity connected component of $\mathrm{Pic}_{?/S}$). Here the horizontal exactness at the top two rows follows from the spectral sequence (10) (see [10, Appendix B]).

Take a correspondence $U \subset Y \times_S Y$ given by two finite flat projections $\pi_1, \pi_2 : U \to Y$ of constant degree (i.e., $\pi_{j,*}O_U$ is locally free of finite rank $\deg(\pi_j)$ over O_Y). Consider the pullback $U_X \subset X \times_S X$ given by the Cartesian diagram:

$$\begin{array}{ccc}
U_X = U \times_{Y \times_S Y} (X \times_S X) & \longrightarrow & X \times_S X \\
\downarrow & & \downarrow \\
U & \hookrightarrow & Y \times_S Y
\end{array}$$

Let $\pi_{j,X} = \pi_j \times_S \pi : U_X \twoheadrightarrow X$ $(j = 1, 2)$ be the projections.

Consider a new correspondence $U^{(q)}_X = \overbrace{U_X \times_Y U_X \times_Y \cdots \times_Y U_X}^{q}$, whose projections are the iterated product

$$\pi_{j,X^{(q)}} = \pi_{j,X} \times_Y \cdots \times_Y \pi_{j,X} : U^{(q)}_X \to X^{(q)} \; (j = 1, 2).$$

Here is the first step to prove a result analogous to (9) for arithmetic cohomology.

Lemma 4.1 *Let the notation and the assumption be as above. In particular,* $\pi : X \to Y$ *is a finite flat morphism of geometrically reduced proper schemes over*

$S = \mathrm{Spec}(k)$ *for a field k. Suppose that X and U_X are proper schemes over a field k satisfying one of the following conditions:*

1. *U_X is geometrically reduced, and for each geometrically connected component X° of X, its pull back to U_X by $\pi_{2,X}$ is also connected; i.e., $\pi^0(X) \xrightarrow[\sim]{\pi_{2,X}^*} \pi^0(U_X)$;*
2. $(f \circ \pi_{2,X})_* \mathcal{O}_{U_X} = f_* \mathcal{O}_X$.

If $\pi_2 : U \to Y$ has constant degree $\deg(\pi_2)$, then, for each $q > 0$, the action of $U_X^{(q)}$ on $H^0(X^{(q)}, \mathcal{O}_{X^{(q)}}^\times)$ factors through the multiplication by $\deg(\pi_2) = \deg(\pi_{2,X})$.

This result is given as [21, Lemma 3.1, Corollary 3.2].

To describe the correspondence action of U on $H^0(X, \mathcal{O}_X^\times)$ in down-to-earth terms, let us first recall the Čech cohomology: for a general S-scheme T,

$$\check{H}^q(\frac{X_T}{Y_T}, \underline{H}^0(\mathbb{G}_{m/Y})) = \frac{\left\{ (c_{i_0,\dots,i_q}) \,\middle|\, \begin{array}{l} c_{i_0,\dots,i_q} \in H^0(X_T^{(q+1)}, \mathcal{O}_{X_T^{(q+1)}}^\times) \text{ and} \\ \prod_j (c_{i_0\dots\check{i}_j\dots i_{q+1}} \circ p_{i_0\dots\check{i}_j\dots i_{q+1}})^{(-1)^j} = 1 \end{array} \right\}}{\{db_{i_0\dots i_q} = \prod_j (b_{i_0\dots\check{i}_j\dots i_q} \circ p_{i_0\dots\check{i}_j\dots i_q})^{(-1)^j} | b_{i_0\dots\check{i}_j\dots i_q} \in H^0(X_T^{(q)}, \mathcal{O}_{X_T^{(q)}}^\times)\}} \tag{12}$$

where we agree to put $H^0(X_T^{(0)}, \mathcal{O}_{X_T}^{(0)}) = 0$ as a convention,

$$X_T^{(q)} = \overbrace{X \times_Y X \times_Y \cdots \times_Y X}^{q} \times_S T, \; \mathcal{O}_{X_T^{(q)}} = \overbrace{\mathcal{O}_X \times_{\mathcal{O}_Y} \mathcal{O}_X \times_{\mathcal{O}_Y} \cdots \times_{\mathcal{O}_Y} \mathcal{O}_X}^{q} \times_{\mathcal{O}_S} \mathcal{O}_T,$$

the identity $\prod_j (c \circ p_{i_0\dots\check{i}_j\dots i_{q+1}})^{(-1)^j} = 1$ takes place in $\mathcal{O}_{X_T^{(q+2)}}$ and $p_{i_0\dots\check{i}_j\dots i_{q+1}} : X_T^{(q+2)} \to X_T^{(q+1)}$ is the projection to the product of X the j-th factor removed. Since $T \times_T T \cong T$ canonically, we have $X_T^{(q)} \cong \overbrace{X_T \times_T \cdots \times_T X_T}^{q}$ by transitivity of fiber product.

Consider $\alpha \in H^0(X, \mathcal{O}_X)$. Then we lift $\pi_{1,X}^* \alpha = \alpha \circ \pi_{1,X} \in H^0(U_X, \mathcal{O}_{U_X})$. Put $\alpha_U := \pi_{1,X}^* \alpha$. Note that $\pi_{2,X,*}\mathcal{O}_{U_X}$ is locally free of rank $d = \deg(\pi_2)$ over $\mathcal{O}_X$, the multiplication by α_U has its characteristic polynomial $P(T)$ of degree d with coefficients in $\mathcal{O}_X$. We define the norm $N_U(\alpha_U)$ to be the constant term $P(0)$. Since α is a global section, $N_U(\alpha_U)$ is a global section, as it is defined everywhere locally. If $\alpha \in H^0(X, \mathcal{O}_X^\times)$, $N_U(\alpha_U) \in H^0(X, \mathcal{O}_X^\times)$. Then define $U(\alpha) = N_U(\alpha_U)$, and in this way, U acts on $H^0(X, \mathcal{O}_X^\times)$.

For a degree q Čech cohomology class $[c] \in \check{H}^q(X_{/Y}, \underline{H}^0(\mathbb{G}_{m/Y}))$ with a Čech q-cocycle $c = (c_{i_0,\dots,i_q})$, $U([c])$ is given by the cohomology class of the Čech cocycle $U(c) = (U(c_{i_0,\dots,i_q}))$, where $U(c_{i_0,\dots,i_q})$ is the image of the global section $c_{i_0,\dots,i_q}$ under U. Indeed, $(\pi_{1,X}^* c_{i_0,\dots,i_q})$ plainly satisfies the cocycle condition, and $(N_U(\pi_{1,X}^* c_{i_0,\dots,i_q}))$ is again a Čech cocycle as N_U is a multiplicative homomorphism. By the same token, this operation sends coboundaries to coboundaries, and define the action of U on the cohomology group. We get the following vanishing result (cf. (9)):

Proposition 4.2 *Suppose that $S = \mathrm{Spec}(k)$ for a field k. Let $\pi : X \to Y$ be a finite flat covering of (constant) degree d of geometrically reduced proper varieties over k, and let $Y \overset{\pi_1}{\leftarrow} U \overset{\pi_2}{\rightarrow} Y$ be two finite flat coverings (of constant degree) identifying the correspondence U with a closed subscheme $U \overset{\pi_1\times\pi_2}{\hookrightarrow} Y \times_S Y$. Write $\pi_{j,X} : U_X = U \times_Y X \to X$ for the base-change to X. Suppose one of the conditions* (1) *and* (2) *of Lemma 4.1 for (X, U). Then*

1. *The correspondence $U \subset Y \times_S Y$ sends $\check{H}^q(\underline{H}^0_Y)$ into* $\deg(\pi_2)(\check{H}^q(\underline{H}^0_Y))$ *for all $q > 0$.*
2. *If d is a p-power and* $\deg(\pi_2)$ *is divisible by p, $\check{H}^q(\underline{H}^0_Y)$ for $q > 0$ is killed by U^M if $p^M \geq d$.*
3. *The cohomology $\check{H}^q(\underline{H}^0_Y)$ with $q > 0$ is killed by d.*

This follows from Lemma 4.1, because on each Čech q-cocycle (whose value is a global section of iterated product $X_T^{(q+1)}$), the action of U is given by $U^{(q+1)}$ by (12). See [21, Proposition 3.3] for a detailed proof. We can apply the above proposition to $(U, X, Y) = (U(p), X_s, X_s^r)$ with U given by $U(p) \subset X_s^r \times X_s^r$ over $\mathbb{Q}$. Indeed, $U := U(p) \subset X_s^r \times X_s^r$ corresponds to $X(\Gamma)$ given by $\Gamma = \Gamma_1(Np^r) \cap \Gamma_0(p^{s+1})$ and U_X is given by $X(\Gamma')$ for $\Gamma' = \Gamma_1(Np^s) \cap \Gamma_0(p^{s+1})$ both geometrically irreducible curves. In this case π_1 is induced by $z \mapsto \frac{z}{p}$ on the upper complex plane and π_2 is the natural projection of degree p. Moreover, $\deg(X_s/X_s^r) = p^{s-r}$ and $\deg(\pi_2) = p$.

An easy criterion to see $\pi^0(U_X^{(q)}) = \pi^0(X^{(q)})$ (which will not be used in this paper), we can offer

Lemma 4.3 *For a finite flat covering $V \overset{\pi}{\to} X \overset{f}{\to} Y$ of geometrically irreducible varieties over a field k, if a fiber $f \circ \pi$ of a k-closed point $y \in Y$ of V is made of a single closed point $v \in V(k)$ (as a topological space), then $V^{(q)} := \overbrace{V \times_Y V \times_Y \cdots \times_Y V}^{q}$ and $X^{(q)}$ are geometrically connected.*

Proof The q-fold tensor product of the stalks at v given by

$$O_{V,v}^{(q)} := \overbrace{O_{V,v} \otimes_{O_{Y,y}} O_{V,v} \otimes_{O_{Y,y}} \cdots \otimes_{O_{Y,y}} O_{V,v}}^{q}$$

is a local ring whose residue field is that of y. This fact holds true for the base change $V_{/k'} \to X_{/k'} \to Y_{/k'}$ for any algebraic extension k'/k; so, $V^{(q)}$ and $X^{(q)}$ are geometrically connected □

Assume that a finite group G acts on $X_{/Y}$ faithfully. Then we have a natural morphism $\phi : X \times G \to X \times_Y X$ given by $\phi(x, \sigma) = (x, \sigma(x))$. In other words, we have a commutative diagram

$$\begin{array}{ccc} X \times G & \xrightarrow{(x,\sigma)\mapsto\sigma(x)} & X \\ {\scriptstyle (x,\sigma)\mapsto x}\downarrow & & \downarrow \\ X & \longrightarrow & Y, \end{array}$$

which induces $\phi : X \times G \to X \times_Y X$ by the universality of the fiber product. Suppose that ϕ is surjective; for example, if Y is a geometric quotient of X by G; see [5, §1.8.3]). Under this map, for any fppf abelian sheaf F, we have a natural map $\check{H}^0(X/Y, F) \to H^0(G, F(X))$ sending a Čech 0-cocycle $c \in H^0(X, F) = F(X)$ (with $p_1^* c = p_2^* c$) to $c \in H^0(G, F(X))$. Obviously, by the surjectivity of ϕ, the map $\check{H}^0(X/Y, F) \to H^0(G, F(X))$ is an isomorphism (e.g., [10, Example III.2.6, p. 100]). Thus we get

Lemma 4.4 *Let the notation be as above, and suppose that ϕ is surjective. For any scheme T fppf over S, we have a canonical isomorphism: $\check{H}^0(X_T/Y_T, F) \cong H^0(G, F(X_T))$.*

We now assume $S = \mathrm{Spec}(k)$ for a field k and that X and Y are proper reduced connected curves. Then we have from the diagram (11) with the exact middle two columns and exact horizontal rows:

$$\begin{array}{ccccccc} 0 & \longrightarrow & \mathbb{Z} & = & \mathbb{Z} & \longrightarrow & 0 \\ \uparrow & & {\scriptstyle \deg}\uparrow{\scriptstyle \text{onto}} & & {\scriptstyle \deg}\uparrow{\scriptstyle \text{onto}} & & \uparrow \\ \check{H}^1(\underline{H}^0_Y) & \longrightarrow & \mathrm{Pic}_{Y/S}(T) & \xrightarrow{b} & \check{H}^0(\frac{X_T}{Y_T}, \mathrm{Pic}_{Y/S}(T)) & \longrightarrow & \check{H}^2(\underline{H}^0_Y) \\ \uparrow & & \cup\uparrow & & \uparrow\cup & & \uparrow \\ ?_1 & \longrightarrow & J_Y(T) & \xrightarrow[c]{} & \check{H}^0(\frac{X_T}{Y_T}, J_X(T)) & \longrightarrow & ?_2, \end{array}$$

Thus we have $?_j = \check{H}^j(\underline{H}^0_Y)$ $(j = 1, 2)$.

By Proposition 4.2, if $q > 0$ and X/Y is of degree p-power and $p|\deg(\pi_2)$, $\check{H}^q(\underline{H}^0_Y)$ is a p-group, killed by U^M for $M \gg 0$. Taking $(X, Y, U)_{/S}$ to be $(X_{s/\mathbb{Q}}, X^r_{s/\mathbb{Q}}, U(p))_{/\mathbb{Q}}$ for $s > r \geq 1$, we get for the projection $\pi : X_s \to X^r_s$

Corollary 4.5 *Let F be a number field or a finite extension of $\mathbb{Q}_l$ for a prime l. Then we have*

(u) The map $\pi^* : J^r_{s/\mathbb{Q}}(F) \to \check{H}^0(X_s/X^r_s, J_{s/\mathbb{Q}}(F)) \overset{(*)}{=} J_{s/\mathbb{Q}}(F)[\gamma^{p^{r-\epsilon}} - 1]$ is a $U(p)$-isomorphism,
where $J_{s/\mathbb{Q}}(F)[\gamma^{p^{r-\epsilon}} - 1] = \mathrm{Ker}(\gamma^{p^{r-\epsilon}} - 1 : J_s(F) \to J_s(F))$.

Here the identity at $(*)$ follows from Lemma 4.4. The kernel $A \mapsto \mathrm{Ker}(\gamma^{p^{r-\epsilon}} - 1 : J_s(A) \to J_s(A))$ is an abelian fppf sheaf (as the category of abelian fppf sheaves is abelian and regarding a sheaf as a presheaf is a left exact functor), and it is represented by the scheme theoretic kernel $J_{s/\mathbb{Q}}[\gamma^{p^{r-\epsilon}} - 1]$ of the endomorphism $\gamma^{p^{r-\epsilon}} - 1$ of $J_{s/\mathbb{Q}}$. From the exact sequence $0 \to J_s[\gamma^{p^{r-\epsilon}} - 1] \to J_s \xrightarrow{\gamma^{p^{r-\epsilon}}-1} J_s$, we get another exact sequence

$$0 \to J_s[\gamma^{p^{r-\epsilon}} - 1](F) \to J_s(F) \xrightarrow{\gamma^{p^{r-\epsilon}}-1} J_s(F).$$

Thus

$$J_{s/\mathbb{Q}}(F)[\gamma^{p^{r-\epsilon}} - 1] = J_{s/\mathbb{Q}}[\gamma^{p^{r-\epsilon}} - 1](F).$$

The above (u) combined with (u1) implies (u2) below:

(u2) The map $\pi^* : J_{r/\mathbb{Q}} \to J_{s/\mathbb{Q}}[\gamma^{p^{r-\epsilon}} - 1] = \mathrm{Ker}(\gamma^{p^{r-\epsilon}} - 1 : J_{s/\mathbb{Q}} \to J_{s/\mathbb{Q}})$ is a $U(p)$-isomorphism.

Actually we can reformulate these facts as

Lemma 4.6 *Then we have morphisms*

$$\iota_s^r : J_{s/\mathbb{Q}}[\gamma^{p^{r-\epsilon}} - 1] \to J^r_{s/\mathbb{Q}} \text{ and } \iota_s^{r,*} : J^r_{s/\mathbb{Q}} \to J_{s/\mathbb{Q}}/(\gamma^{p^{r-\epsilon}} - 1)(J_{s/\mathbb{Q}})$$

satisfying the following commutative diagrams:

$$\begin{array}{ccc} J^r_{s/\mathbb{Q}} & \xrightarrow{\pi^*} & J_{s/\mathbb{Q}}[\gamma^{p^{r-\epsilon}} - 1] \\ {\scriptstyle u}\downarrow & \swarrow{\scriptstyle \iota_s^r} & \downarrow{\scriptstyle u''} \\ J^r_{s/\mathbb{Q}} & \xrightarrow{\pi^*} & J_{s/\mathbb{Q}}[\gamma^{p^{r-\epsilon}} - 1], \end{array} \qquad (13)$$

and

$$\begin{array}{ccc} J^r_{s/\mathbb{Q}} & \xleftarrow{\pi_*} & J_{s/\mathbb{Q}}/(\gamma^{p^{r-\epsilon}} - 1)(J_{s/\mathbb{Q}}) \\ {\scriptstyle u^*}\uparrow & \nearrow{\scriptstyle \iota_s^{r,*}} & \uparrow{\scriptstyle u''^*} \\ J^r_{s/\mathbb{Q}} & \xleftarrow{\pi_*} & J_{s/\mathbb{Q}}/(\gamma^{p^{r-\epsilon}} - 1)(J_{s/\mathbb{Q}}), \end{array} \qquad (14)$$

where u and u'' are $U(p^{s-r}) = U(p)^{s-r}$ and u^* and u''^* are $U^*(p^{s-r}) = U^*(p)^{s-r}$. In particular, for an fppf extension $T_{/\mathbb{Q}}$, the evaluated map at T: $(J_{s/\mathbb{Q}}/(\gamma^{p^{r-\epsilon}} - 1)(J_{s/\mathbb{Q}}))(T) \xrightarrow{\pi_*} J^r_s(T)$ (resp. $J^r_s(T) \xrightarrow{\pi^*} J_s[\gamma^{p^{r-\epsilon}} - 1](T)$) is a $U^*(p)$-isomorphism (resp. $U(p)$-isomorphism).

Note here that the natural homomorphism:

$$\frac{J_s(T)}{(\gamma^{p^{r-\epsilon}} - 1)(J_s(T))} \to (J_{s/\mathbb{Q}}/(\gamma^{p^{r-\epsilon}} - 1)(J_{s/\mathbb{Q}}))(T)$$

may have non-trivial kernel and cokernel which may not be killed by a power of $U^*(p)$. In other words, the left-hand-side is an fppf presheaf (of T) and the right-hand-side is its sheafication. On the other hand, $T \mapsto J_s[\gamma^{p^{r-\epsilon}} - 1](T)$ is already an fppf abelian sheaf; so, $J_s^r(T) \xrightarrow{\pi^*} J_s[\gamma^{p^{r-\epsilon}} - 1](T)$ is a $U(p)$-isomorphism without ambiguity.

Proof We first prove the assertion for π^*. We note that the category of groups schemes fppf over a base S is a full subcategory of the category of abelian fppf sheaves. We may regard $J^r_{s/\mathbb{Q}}$ and $J_s[\gamma^{p^{r-\epsilon}} - 1]_{/\mathbb{Q}}$ as abelian fppf sheaves over $\mathbb{Q}$ in this proof. Since these sheaves are represented by (reduced) algebraic groups over $\mathbb{Q}$, we can check being $U(p)$-isomorphism by evaluating the sheaf at a field k of characteristic 0 (e.g., [4, Lemma 4.18]). By Proposition 4.2 (2) applied to $X = X_{s/k} = X_s \times_{\mathbb{Q}} k$ and $Y = X^r_{s/k}$ (with $S = \mathrm{Spec}(k)$ and $s \geq r$),

$$\mathcal{K} := \mathrm{Ker}(J^r_{s/\mathbb{Q}} \to J_{s/\mathbb{Q}}[\gamma^{p^{r-\epsilon}} - 1])$$

is killed by $U(p)^{s-r}$ as $d = p^{s-r} = \deg(X_s/X_s^r)$. Thus we get

$$\mathcal{K} \subset \mathrm{Ker}(U(p)^{s-r} : J^r_{s/\mathbb{Q}} \to J^r_{s/\mathbb{Q}}).$$

Since the category of fppf abelian sheaves is an abelian category (because of the existence of the sheafication functor from presheaves to sheaves under fppf topology described in [10, §II.2]), the above inclusion implies the existence of ι^r_s with $\pi^* \circ \iota^r_s = U(p)^{s-r}$ as a morphism of abelian fppf sheaves. Since the category of group schemes fppf over a base S is a full subcategory of the category of abelian fppf sheaves, all morphisms appearing in the identity $\pi^* \circ \iota^r_s = U(p)^{s-r}$ are morphism of group schemes. This proves the assertion for π^*.

Note that the second assertion is the dual of the first; so, it can be proven reversing all the arrows and replacing $J_s[\gamma^{p^{r-\epsilon}} - 1]_{/\mathbb{Q}}$ (resp. π^*, $U(p)$) by the quotient $J_s/(\gamma^{p^{r-\epsilon}} - 1)J_s$ as fppf abelian sheaves (resp. π_*, $U^*(p)$). Since $J_s/(\gamma^{p^{r-\epsilon}} - 1)(J_s)$ and J^r_s are abelian schemes over $\mathbb{Q}$, the quotient abelian scheme $J_s/(\gamma^{p^{r-\epsilon}} - 1)(J_s)$ is the dual of $J_s[\gamma^{p^{r-\epsilon}} - 1]$ and $\iota^{r,*}_s$ is the dual of ι^r_s. □

By the second diagram of the above lemma, we get

(u*) The map $J_s/(\gamma^{p^{r-\epsilon}} - 1)(J_s)_{/\mathbb{Q}} \xrightarrow{\pi_*} J^r_{s/\mathbb{Q}}$ is a $U^*(p)$-isomorphism of abelian fppf sheaves.

As a summary, we have

Corollary 4.7 *Then the morphism $\pi : X_s \to X^r_s$ induces an isogeny*

$$\overline{\pi}_* : J_s/(\gamma^{p^{r-\epsilon}} - 1)(J_s)_{/\mathbb{Q}} \to J^r_{s/\mathbb{Q}}$$

whose kernel is killed by a sufficiently large power of $U^*(p)$, and the pull-back map π^* induces an isogeny $\underline{\pi}^* : J_s[\gamma^{p^{r-\epsilon}} - 1] \to J^r_s$ whose kernel is killed by a high *power of $U(p)$. Moreover, for a finite extension F of $\mathbb{Q}$ or $\mathbb{Q}_l$ (for a prime l not necessarily equal to p), $\underline{\pi}^* : J_s[\gamma^{p^{r-\epsilon}} - 1](F) \to J_s^r(F)$ is a $U(p)$-isomorphism.*

Proof Let $C \subset \mathrm{Aut}(X_s)$ be the cyclic group generated by the action of $\gamma^{p^{r-\epsilon}}$. Then $X_{s/\overline{\mathbb{Q}}}/X^r_{s/\overline{\mathbb{Q}}}$ is an étale covering with Galois group C (even unramified at cusps). Thus $Lie(J_s^r) = H^1(X_s^r, O_{X_s^r}) = H_0(C, H^1(X_s, O_{X_s})) = H_0(C, Lie(J_s))$. This shows that $\overline{\pi}_*$ is an isogeny over $\overline{\mathbb{Q}}$ and hence over $\mathbb{Q}$, which is a $U^*(p)$-isomorphism by Lemma 4.6. By taking dual, $\underline{\pi}^*$ is also an isogeny, which is a $U(p)$-isomorphism even after evaluating the fppf sheaves at F by Lemma 4.6 and the remark following the lemma. This proves the corollary. □

Then we get

(u*2) The map $J_s/(\gamma^{p^{r-\epsilon}} - 1)(J_s)_{/\mathbb{Q}} \to J_{r/\mathbb{Q}}$ is a $U^*(p)$-isomorphism of abelian fppf sheaves.

We can prove (u*2) in a more elementary way. We describe the easier proof. Identify $J_s(\mathbb{C}) = H^1(X_s, \mathbf{T})$ whose Pontryagin dual is given by $H_1(X_s, \mathbb{Z})$. If $k = \mathbb{Q}$, we have the Pontryagin dual version of (u2):

$$H_1(X_r, \mathbb{Z}) \overset{\pi_*}{\longleftarrow} H_1(X_s, \mathbb{Z})/(\gamma^{p^{r-\epsilon}} - 1)(H_1(X_s, \mathbb{Z})) \text{ is a } U^*(p)\text{-isomorphism.} \quad (15)$$

Since $J_{s,\mathbb{Q}}(\mathbb{C}) \cong H_1(X_s, \mathbb{R})/H_1(X_s, \mathbb{Z})$ as Lie groups, we get

$$J_r(\mathbb{C}) \overset{\pi_*}{\longleftarrow} J_s(\mathbb{C})/(\gamma^{p^{r-\epsilon}} - 1)(J_s(\mathbb{C})) \text{ is a } U^*(p)\text{-isomorphism.} \quad (16)$$

This implies (u*2). By (16), writing $\overline{\mathbb{Q}}$ for the algebraic closure of $\mathbb{Q}$ in $\mathbb{C}$ and taking algebraic points, we get that

$$J_r(\overline{\mathbb{Q}}) \overset{\pi_*}{\longleftarrow} (J_s/(\gamma^{p^{r-\epsilon}} - 1)(J_s))(\overline{\mathbb{Q}}) = J_s(\overline{\mathbb{Q}})/(\gamma^{p^{r-\epsilon}} - 1)(J_s(\overline{\mathbb{Q}})) \quad (17)$$

is a $U^*(p)$-isomorphism.

Remark 4.8 The $U(p)$-isomorphisms of Jacobians do not kill the part associated to finite slope Hecke eigenforms. Thus the above information includes not just the information of p-ordinary forms but also those of finite slope Hecke eigenforms.

5 Control of Λ-MW Groups as Fppf Sheaves

Let k be either a number field in $\overline{\mathbb{Q}}$ or a finite extension of $\mathbb{Q}_l$ in $\overline{\mathbb{Q}}_l$ for a prime l. Write O_k (resp. W) for the (resp. l-adic) integer ring of k if k is a number field (resp. a finite extension of $\mathbb{Q}_l$). For an abelian variety $A_{/k}$, we have $\widehat{A}(\kappa) := \varprojlim A(\kappa)/p^n A(\kappa)$ for a finite field extension κ/k as in (3). A down-to-earth description of the value of $\widehat{A}(\kappa)$ is given by (S) just above (3).

We study $J_r(k)$ equipped with the topology $J_k(k)$ induced from k (so, it is discrete if k is a number field and is l-adic if k is a finite extension of $\mathbb{Q}_l$). The p-adic limits $e = \lim_{n\to\infty} U(p)^{n!}$ and $e^* = \lim_{n\to\infty} U^*(p)^{n!}$ are well defined on $\widehat{J}_r(k)$. The Albanese functoriality gives rise to a projective system $\{\widehat{J}_s(k), \pi_{s,r,*}\}_s$ for the covering map $\pi_{s,r} : X_s \to X_r$ $(s > r)$, and we have

$$\widetilde{J}_\infty(k) = \varprojlim_r \widehat{J}_r(k) \ \text{(with projective limit of } p\text{ -profinite compact topology)}$$

on which the co-ordinary projector $e^* = \lim_{n\to\infty} U^*(p)^{n!}$ acts. As before, adding superscript or subscript "ord" (resp. "co-ord"), we indicate the image of e (resp. e^*) depending on the situation.

We study mainly in this paper the control theorems of the w-twisted version $\widehat{J}_\infty(k)^{\text{ord}}$ (which we introduce in Sect. 6) of $\widetilde{J}_\infty(k)^{\text{co-ord}}$ under the action of Γ and Hecke operators, and we have studied $\check{J}^{\text{ord}}_\infty(k)$ in [21] in a similar way. Here the word "w-twisting" means modifying the transition maps by the Weil involution at each step. As fppf sheaves, we have an isomorphism $i : \widehat{J}_\infty(k)^{\text{ord}} \cong \widetilde{J}_\infty(k)^{\text{co-ord}}$ but $i \circ T(n) = T^*(n) \circ i$ for all n. Hereafter, unless otherwise mentioned, once our fppf abelian sheaf is evaluated at k, all morphisms are continuous with respect to the topology defined above (and we do not mention continuity often).

From (u1), we get

$$J_r(k) \xrightarrow{\pi^*} J^r_s(k) \text{ is a } U(p)\text{-isomorphism (for the projection } \pi : X^r_s \to X_r). \quad (18)$$

The dual version (following from (u*1)) is

$$J^r_s(k) \xrightarrow{\pi_*} J_r(k) \text{ is a } U^*(p)\text{-isomorphism, where } \pi_* \text{ is the dual of } \pi^*. \quad (19)$$

From (18) and (19), we get

Lemma 5.1 *For a field k as above, we have*

$$\pi_* : \widehat{J}^r_s(k)^{\text{co-ord}} \cong \widehat{J}_r(k)^{\text{co-ord}} \ \textit{and}\ \pi^* : \widehat{J}_r(k)^{\text{ord}} \cong \widehat{J}^r_s(k)^{\text{ord}}$$

for all $0 < r < s$ *with the projection* $\pi : X^r_s \twoheadrightarrow X_r$.

From Corollary 4.7 (or Lemma 4.6 combined with (u*2) and (u2)), for any field k, we get

(I) $\pi^* : J_r(k) \to J_s[\gamma^{p^{r-\epsilon}} - 1](k)$ is a $U(p)$-isomorphism, and obviously, $\pi^* : J_r \to J_s[\gamma^{p^{r-\epsilon}} - 1]$ is a $U(p)$-isomorphism of abelian fppf sheaves.
(P) $\pi_* : J_r \to J_s/(\gamma^{p^{r-\epsilon}} - 1)J_s$ is a $U^*(p)$-isomorphism of fppf abelian sheaves.

Note that (P) does not mean that $\frac{\widehat{J}_s(k)}{(\gamma^{p^{r-\epsilon}}-1)(\widehat{J}_s(k))} \to \widehat{J}_r(k)$ is a $U^*(p)$-isomorphism (as the sheaf quotient $J_s/(\gamma^{p^{r-\epsilon}} - 1)J_s$ and the corresponding presheaf quotient could be different).

We now claim

Lemma 5.2 *For integers $0 < r < s$, we have isomorphisms of fppf abelian sheaves*

$$\pi^* : \widehat{J}_r^{\mathrm{ord}} \cong \widehat{J}_s[\gamma^{p^{r-\epsilon}} - 1]^{\mathrm{ord}} \text{ and } \pi_* : (\widehat{\frac{J_s}{(\gamma^{p^{r-\epsilon}} - 1)J_s}})^{\text{co-ord}} \cong \widehat{J}_r^{\text{co-ord}}.$$

The first isomorphism π^* induces an isomorphism: $\widehat{J}_r^{\mathrm{ord}}(T) \cong \widehat{J}_s[\gamma^{p^{r-\epsilon}} - 1]^{\mathrm{ord}}(T)$ for any fppf extension $T_{/k}$, but the morphism $\frac{\widehat{J}_s(T)^{\text{co-ord}}}{(\gamma^{p^{r-\epsilon}}-1)(\widehat{J}_s(T))^{\text{co-ord}}} \to \widehat{J}_r(T)^{\text{co-ord}}$ induced by the second one may not be an isomorphism.

Proof By (I) above, $\widehat{J}_r^{\mathrm{ord}} \cong \widehat{A}^{\mathrm{ord}}$ for the abelian variety $A = J_s[\gamma^{p^{r-\epsilon}} - 1]$ and $\widehat{A}$ as in (S) above (3). We consider the following exact sequence

$$0 \to A \to J_s \xrightarrow{\gamma^{p^{r-\epsilon}}-1} J_s.$$

This produces another exact sequence $0 \to \widehat{A} \to \widehat{J}_s \xrightarrow{\gamma^{p^{r-\epsilon}}-1} \widehat{J}_s$; so, we get $\widehat{A} \cong \widehat{J}_s[\gamma^{p^{r-\epsilon}} - 1]$. Taking ordinary part and combining with the identity: $\widehat{J}_r^{\mathrm{ord}} \cong \widehat{A}^{\mathrm{ord}}$, we conclude $\widehat{J}_r^{\mathrm{ord}} \cong \widehat{J}_s[\gamma^{p^{r-\epsilon}} - 1]^{\mathrm{ord}}$. This holds true after evaluation at T as the presheaf-kernel of a sheaf morphism is still a sheaf. The second assertion is the dual of the first. □

Passing to the limit, Lemmas 5.1 and 5.2 tells us

Theorem 5.3 *Let k be either a number field or a finite extension of $\mathbb{Q}_l$. Then we have isomorphisms of fppf abelian sheaves over k:*

(a) $J_\infty^{\mathrm{ord}}[\gamma^{p^{r-\epsilon}} - 1] \cong \widehat{J}_r^{\mathrm{ord}}$;
(b) $(\widetilde{J}_\infty/(\gamma^{p^{r-\epsilon}} - 1)(\widetilde{J}_\infty))^{\text{co-ord}} \cong \widehat{J}_r^{\text{co-ord}}$

where we put $\widetilde{J}_\infty/(\gamma^{p^{r-\epsilon}} - 1)(\widetilde{J}_\infty)^{\text{co-ord}} := \varprojlim_s \widehat{J_s/(\gamma^{p^{r-\epsilon}} - 1)(J_s)}^{\text{co-ord}}$ *as an fppf sheaf.*

Proof The assertion (b) is just the projective limit of the corresponding statement in Lemma 5.2.

We prove (a). Since injective limit always preserves exact sequences, we have

$$0 \to \widehat{J}_r(k)^{\mathrm{ord}} \to \varinjlim_s \widehat{J}_s(k)^{\mathrm{ord}} \xrightarrow{\gamma^{p^{r-\epsilon}}-1} \varinjlim_s \widehat{J}_s(k)^{\mathrm{ord}}$$

is exact, showing (a). □

See [21, Proposition 6.4] for a control result similar to (a) for $\check{J}_\infty^{\mathrm{ord}}$.

Remark 5.4 As is clear from the warning after *(P)*, the isomorphism *(b)* does **not** mean that

$$\varprojlim_s \left\{ \frac{\widehat{J}_s(T)}{(\gamma^{p^{r-\epsilon}} - 1)(\widehat{J}_s(T))} \right\}^{\text{co-ord}} \to \widehat{J}_r(T)^{\text{co-ord}}$$

for each fppf extension $T_{/k}$ is an isomorphism. The kernel and the cokernel of this map will be studied in Sect. 9.

6 Sheaves Associated to Modular Jacobians

We fix an element $\zeta \in \mathbb{Z}_p(1) = \varprojlim_n \mu_{p^n}(\overline{\mathbb{Q}})$; so, ζ is a coherent sequence of generators ζ_{p^n} of $\mu_{p^n}(\overline{\mathbb{Q}})$ (i.e., $\zeta_{p^{n+1}}^p = \zeta_{p^n}$). We also fix a generator ζ_N of $\mu_N(\overline{\mathbb{Q}})$, and put $\zeta_{Np^r} := \zeta_N \zeta_{p^r}$. Identify the étale group scheme $\mathbb{Z}/Np^n\mathbb{Z}_{/\mathbb{Q}[\zeta_N,\zeta_{p^n}]}$ with μ_{Np^n} by sending $m \in \mathbb{Z}$ to $\zeta_{Np^n}^m$. Then for a couple $(E, \phi_{Np^r} : \mu_{Np^r} \hookrightarrow E)_{/K}$ over a $\mathbb{Q}[\mu_{p^r}]$-algebra K, let $\phi^* : E[Np^r] \twoheadrightarrow \mathbb{Z}/Np^r\mathbb{Z}$ be the Cartier dual of ϕ_{Np^r}. Then ϕ^* induces $E[Np^r]/\operatorname{Im}(\phi_{Np^r}) \cong \mathbb{Z}/Np^r\mathbb{Z}$. Define $i : \mathbb{Z}/p^r\mathbb{Z} \cong (E/\operatorname{Im}(\phi_{Np^r}))[Np^r]$ by the inverse of ϕ^*. Then we define $\varphi_{Np^r} : \mu_{Np^r} \hookrightarrow E/\operatorname{Im}(\phi_{Np^r})$ by $\varphi_{Np^r} : \mu_{Np^r} \cong \mathbb{Z}/Np^r\mathbb{Z} \xrightarrow{i} (E/\operatorname{Im}(\phi_{Np^r}])[p^r] \subset E/\operatorname{Im}(\phi_{Np^r})$. This induces an automorphism w_r of X_r defined over $\mathbb{Q}[\mu_{Np^r}]$, which in turn induces an automorphism w_r of $J_{r/\mathbb{Q}[\zeta_{Np^r}]})$. We have the following well known commutative diagram (e.g., [11, Sect. 4.6]):

$$\begin{array}{ccc} J_r & \xrightarrow{T(n)} & J_r \\ {\scriptstyle w_r}\downarrow\wr & & {\scriptstyle w_r}\downarrow\wr \\ J_r & \xrightarrow[T^*(n)]{} & J_r. \end{array}$$

Let $P \in \operatorname{Spec}(\mathbf{h})(\overline{\mathbb{Q}}_p)$ be an arithmetic point of weight 2. Then we have a p-stabilized Hecke eigenform form f_P associated to P; i.e., $f_P|T(n) = P(T(n))f_P$ for all n. Then $f_P^* = w_r(f_P)$ is the dual common eigenform of $T^*(n)$. If f_P is new at every prime $l|pN$, f_P^* is a constant multiple of the complex conjugate f_P^c of f_P (but otherwise, it is different).

We then define as described in (S) in Sect. 2 an fppf abelian sheaf $\widehat{X}$ for any abelian variety quotient or subgroup variety X of $J_{s/k}$ over the fppf site over $k = \mathbb{Q}$ and $\mathbb{Q}_l$ (note here the explicit value of $\widehat{J}_s$ depends on k as in (S)).

Pick an automorphism $\sigma \in \operatorname{Gal}(\mathbb{Q}(\mu_{Np^r})/\mathbb{Q})$ with $\zeta_{Np^r}^\sigma = \zeta_{Np^r}^z$ for $z \in (\mathbb{Z}/Np^r\mathbb{Z})^\times$. Since w_r^σ is defined with respect to $\zeta_{Np^r}^\sigma = \zeta_{Np^r}^z$, we find $w_r^\sigma = \langle z\rangle \circ w_r = w_r \circ \langle z\rangle^{-1}$ (see [25, p. 237] and [24, 2.5.6]). Here $\langle z\rangle$ is the image of z in $(\mathbb{Z}/Np^r\mathbb{Z})^\times = \operatorname{Gal}(X_r/X_0(Np^r))$. Let $\pi_{s,r,*} : J_s \to J_r$ for $s > r$ be the morphism induced by the covering map $X_s \twoheadrightarrow X_r$ through Albanese functoriality. Then we define $\pi_s^r = w_r \circ \pi_{s,r,*} \circ w_s$. Then $(\pi_s^r)^\sigma = w_r\langle z^{-1}\rangle\pi_{s,r,*}\langle z\rangle w_s = \pi_s^r$ for all $\sigma \in \operatorname{Gal}(\mathbb{Q}(\mu_{Np^s})/\mathbb{Q})$; thus, π_s^r is well defined over $\mathbb{Q}$, and satisfies $T(n) \circ \pi_s^r = \pi_s^r \circ T(n)$ for all n prime to Np and $U(q) \circ \pi_s^r = \pi_s^r \circ U(q)$ for all $q|Np$. Since $w_r^2 = 1$, by this w-twisting, the projective system $\{J_s, \pi_{s,r,*}\}$ equivariant under $T^*(n)$ is transformed into the isomorphic projective system $\{J_s, \pi_s^r\}_{s>r}$ (of abelian varieties defined over $\mathbb{Q}$) which is Hecke equivariant (i.e., $T(n)$ and $U(l)$-equivariant). Thus what we proved for the co-ordinary part of the projective system $\{\widehat{J}_s, \pi_{s,r,*}\}$ is valid for the ordinary part of the projective system $\{\widehat{J}_s, \pi_s^r\}_{s>r}$. If X_s is either an algebraic subgroup or an abelian variety quotient of J_s and π_s^r produces a projective system $\{X_s\}_s$ we define $\widehat{X}_\infty := \varprojlim_s \widehat{X}_s(R)$ for an fppf extension R of $k = \mathbb{Q}, \mathbb{Q}_l$ (again the definition of $\widehat{X}_s$ and hence $\widehat{X}_\infty$ depends on k). For each ind-object $R = \varinjlim_i R_i$ of fppf, smooth or étale algebras R_i/k, we define $\widehat{X}_\infty(R) = \varinjlim_i \widehat{X}_\infty(R_i)$.

Lemma 6.1 *Let K/k be the Galois extension as in Sect. 2. Then the* $\mathrm{Gal}(K/k)$*-action on $\widehat{X}_\infty(K)$ is continuous under the discrete topology on $\widehat{X}_\infty(K)$. In particular, the Galois cohomology group $H^q(\widehat{X}_\infty(K)) := H^q(\mathrm{Gal}(K/\kappa), \widehat{X}_\infty(K))$ for $q > 0$ is a torsion $\mathbb{Z}_p$-module for any intermediate extension $K/\kappa/k$.*

Proof By definition, $\widehat{X}_\infty(K) = \bigcup_{K/F/k} \widehat{X}_\infty(F)$, and for all finite intermediate extensions $K/F/k$ we have $\widehat{X}_\infty(F) \subset H^0(\mathrm{Gal}(K/F), \widehat{X}_\infty(K))$. Thus $\widehat{X}_\infty(K) = \varinjlim_F H^0(\mathrm{Gal}(K/F), \widehat{X}_\infty(K))$, which implies the continuity of the action under the discrete topology. Then the torsion property follows from [7, Corollary 4.26]. □

Let $\iota : C_{r/\mathbb{Q}} \subset J_{r/\mathbb{Q}}$ be an abelian subvariety stable under Hecke operators (including $U(l)$ for $l|Np$) and w_r and ${}^t\iota : J_{r/\mathbb{Q}} \twoheadrightarrow {}^tC_{r/\mathbb{Q}}$ be the dual abelian quotient. We then define $\pi : J_r \twoheadrightarrow D_r$ by $D_r := {}^tC_r$ and $\pi = {}^tw_r \circ {}^t\iota_r \circ w_r$ for the map ${}^tw_r \in \mathrm{Aut}({}^tC_{r/\mathbb{Q}[\mu_{p^r}]})$ dual to $w_r \in \mathrm{Aut}(C_{r/\mathbb{Q}[\mu_{p^r}]})$. Again π is defined over $\mathbb{Q}$. Then ι and π are Hecke equivariant. Let $\iota_s : C_s := \pi_{s,r}^*(C) \subset J_s$ for $s > r$ and D_s be the quotient abelian variety of J_s defined in the same way taking C_s in place of C_r (and replacing r by s). Put $\pi_s : J_s \twoheadrightarrow D_s$ which is Hecke equivariant.

Since the two morphisms $J_r \to J_s^r$ and $J_s^r \to J_s[\gamma^{p^{r-\epsilon}} - 1]$ (Picard functoriality) are $U(p)$-isomorphism of fppf abelian sheaves by (u1) and Corollary 4.5, we get the following two isomorphisms of fppf abelian sheaves for $s > r > 0$:

$$C_r[p^\infty]^{\mathrm{ord}} \xrightarrow[\pi_{r,s}^*]{\sim} C_s[p^\infty]^{\mathrm{ord}} \text{ and } \widehat{C}_r^{\mathrm{ord}} \xrightarrow[\pi_{r,s}^*]{\sim} \widehat{C}_s^{\mathrm{ord}}, \tag{20}$$

since $\widehat{C}_s^{\mathrm{ord}}$ is the isomorphic image of $\widehat{C}_r^{\mathrm{ord}} \subset \widehat{J}_r$ in $\widehat{J}_s[\gamma^{p^{r-\epsilon}} - 1]$. By w-twisted Cartier duality [20, §4], we have

$$D_s[p^\infty]^{\mathrm{ord}} \xrightarrow[\pi_s^r]{\sim} D_r[p^\infty]^{\mathrm{ord}}. \tag{21}$$

Thus by Kummer sequence in Lemma 2.1, we have the following commutative diagram

$$\begin{array}{ccc}
\widehat{D}_s^{\mathrm{ord}}(\kappa) \otimes \mathbb{Z}/p^m\mathbb{Z} = (D_s(\kappa) \otimes \mathbb{Z}/p^m\mathbb{Z})^{\mathrm{ord}} & \xrightarrow[\hookrightarrow]{} & H^1(D_s[p^m]^{\mathrm{ord}}) \\
\pi_s^r \downarrow & & \wr \downarrow (21) \\
\widehat{D}_r^{\mathrm{ord}}(\kappa) \otimes \mathbb{Z}/p^m\mathbb{Z} = (D_r(\kappa) \otimes \mathbb{Z}/p^m\mathbb{Z})^{\mathrm{ord}} & \xrightarrow[\hookrightarrow]{} & H^1(D_r[p^m]^{\mathrm{ord}})
\end{array}$$

This shows

$$\widehat{D}_s^{\mathrm{ord}}(\kappa) \otimes \mathbb{Z}/p^m\mathbb{Z} \cong \widehat{D}_r^{\mathrm{ord}}(\kappa) \otimes \mathbb{Z}/p^m\mathbb{Z}.$$

Passing to the limit, we get

$$\widehat{D}_s^{\mathrm{ord}} \underset{\pi_s^r}{\xrightarrow{\sim}} \widehat{D}_r^{\mathrm{ord}} \text{ and } (D_s \otimes_{\mathbb{Z}} \mathbb{T}_p)^{\mathrm{ord}} \underset{\pi_s^r}{\xrightarrow{\sim}} (D_r \otimes_{\mathbb{Z}} \mathbb{T}_p)^{\mathrm{ord}} \tag{22}$$

as fppf abelian sheaves. In short, we get

Lemma 6.2 *Suppose that κ is a field extension of finite type of either a number field or a finite extension of $\mathbb{Q}_l$. Then we have the following isomorphism*

$$\widehat{C}_r(\kappa)^{\mathrm{ord}} \underset{\pi_{s,r}^*}{\xrightarrow{\sim}} \widehat{C}_s(\kappa)^{\mathrm{ord}} \text{ and } \widehat{D}_s(\kappa)^{\mathrm{ord}} \underset{\pi_s^r}{\xrightarrow{\sim}} \widehat{D}_r(\kappa)^{\mathrm{ord}}$$

for all $s > r$ including $s = \infty$.

By computation, $\pi_s^r \circ \pi_{r,s}^* = p^{s-r}U(p^{s-r})$. To see this, as Hecke operators coming from Γ_s-coset operations, $\pi_{r,s}^* = [\Gamma_s]$ (restriction map) and $\pi_{r,s,*} = [\Gamma_r]$ (trace operator for Γ_r/Γ_s). Thus we have

$$\begin{aligned}\pi_s^r \circ \pi_{r,s}^*(x) &= x|[\Gamma_s] \cdot w_s \cdot [\Gamma_r] \cdot w_r = x|[\Gamma_s] \cdot [w_s w_r] \cdot [\Gamma_r] \\ &= x|[\Gamma_s^r : \Gamma_s][\Gamma_s^r \left(\begin{smallmatrix} 1 & 0 \\ 0 & p^{s-r}\end{smallmatrix}\right) \Gamma_r] = p^{s-r}(x|U(p^{s-r})).\end{aligned} \tag{23}$$

Corollary 6.3 *We have the following two commutative diagrams for $s' > s$*

$$\begin{array}{ccc} \widehat{C}_{s'}^{\mathrm{ord}} & \underset{\pi_{s,s'}^*}{\xleftarrow{\sim}} & \widehat{C}_s^{\mathrm{ord}} \\ {\scriptstyle \pi_{s'}^s}\downarrow & & \downarrow{\scriptstyle p^{s'-s}U(p)^{s'-s}} \\ \widehat{C}_s^{\mathrm{ord}} & = & \widehat{C}_s^{\mathrm{ord}}. \end{array}$$

and

$$\begin{array}{ccc} \widehat{D}_{s'}^{\mathrm{ord}} & \underset{\pi_{s'}^s}{\xrightarrow{\sim}} & \widehat{D}_s^{\mathrm{ord}} \\ {\scriptstyle \pi_{s,s'}^*}\uparrow & & \uparrow{\scriptstyle p^{s'-s}U(p)^{s'-s}} \\ \widehat{D}_s^{\mathrm{ord}} & = & \widehat{D}_s^{\mathrm{ord}}. \end{array}$$

Proof By $\pi_{r,s}^*$ (resp. π_s^r), we identify $\widehat{C}_s^{\mathrm{ord}}$ with $\widehat{C}_r^{\mathrm{ord}}$ (resp. $\widehat{D}_s^{\mathrm{ord}}$ with $\widehat{D}_r^{\mathrm{ord}}$) as in Lemma 6.2. Then the above two diagrams follow from (23). □

By (23), we have exact sequences

$$\begin{aligned} &0 \to C_s[p^{s-r}]^{\mathrm{ord}} \to C_s[p^\infty]^{\mathrm{ord}} \xrightarrow{\pi_s^r} C_r[p^\infty]^{\mathrm{ord}} \to 0, \\ &0 \to D_r[p^{s-r}]^{\mathrm{ord}} \to D_r[p^\infty]^{\mathrm{ord}} \xrightarrow{\pi_{r,s}^*} D_s[p^\infty]^{\mathrm{ord}} \to 0. \end{aligned} \tag{24}$$

Applying (2) to the exact sequence $\mathcal{K}_s^r(K) \hookrightarrow C_s(K) \twoheadrightarrow C_r(K)$ for $\mathcal{K}_s^r(K) = \mathrm{Ker}(\pi_s^r)(K)$ and $\mathcal{K}_{r,s}(K) \hookrightarrow C_r(K) \twoheadrightarrow D_s(K)$ for $\mathcal{K}_{r,s} = \mathrm{Ker}(\pi_{r,s}^*)$, we get the following exact sequence of fppf abelian sheaves:

$$0 \to \widehat{\mathcal{K}}_s^r \to \widehat{C}_s \xrightarrow{\pi_s^r} \widehat{C}_r \to 0,$$
$$0 \to \widehat{\mathcal{K}}_{r,s} \to \widehat{D}_r \xrightarrow{\pi_{r,s}^*} \widehat{D}_s \to 0.$$

Taking the ordinary part, we confirm exactness of

$$\begin{aligned} 0 \to C_s[p^{s-r}]^{\mathrm{ord}} &\to \widehat{C}_s^{\mathrm{ord}} \xrightarrow{\pi_s^r} \widehat{C}_r^{\mathrm{ord}} \to 0, \\ 0 \to D_r[p^{s-r}]^{\mathrm{ord}} &\to \widehat{D}_r^{\mathrm{ord}} \xrightarrow{\pi_{r,s}^*} \widehat{D}_s^{\mathrm{ord}} \to 0. \end{aligned} \tag{25}$$

Write $H^1(X) = H^1(\mathrm{Gal}(K/\kappa), X)$ for an intermediate extension $K/\kappa/k$ and $\mathrm{Gal}(K/k)$-module X and $H^1_?(X) = H^1_?(\mathrm{Spec}(\kappa), X)$ for a smooth/fppf extension for $? =$ sm or fppf. Then taking the p-adic completion, we get the following exact sequences as parts of the long exact sequences associated to (25)

$$\begin{aligned} 0 \to C_s[p^{s-r}]^{\mathrm{ord}}(\kappa) &\to \widehat{C}_s^{\mathrm{ord}}(\kappa) \underset{\pi_s^r}{\longrightarrow} \widehat{C}_r^{\mathrm{ord}}(\kappa) \to H^1_?(C_s[p^{s-r}]^{\mathrm{ord}}), \\ 0 \to D_r[p^{s-r}]^{\mathrm{ord}}(\kappa) &\to \widehat{D}_r^{\mathrm{ord}}(\kappa) \underset{\pi_{r,s}^*}{\longrightarrow} \widehat{D}_s^{\mathrm{ord}}(\kappa) \to H^1_?(D_r[p^{s-r}]^{\mathrm{ord}}) \end{aligned} \tag{26}$$

for $? =$ fppf, sm (cohomology under smooth topology) or nothing (i.e., Galois cohomology equivalent to étale cohomology in this case). Here if $? =$ fppf, κ/k is an extension of finite type, if $? =$ sm, κ/k is a smooth extension of finite type, and if ? is nothing, $K/\kappa/k$ is an intermediate field.

By Lemma 6.2, we can rewrite the first exact sequence of (24) as

$$0 \to C_r[p^{s-r}]^{\mathrm{ord}}(\kappa) \xrightarrow{\pi_{r,s}^*} \widehat{C}_s^{\mathrm{ord}}(\kappa) \xrightarrow{\pi_s^r} \widehat{C}_r^{\mathrm{ord}}(\kappa) \to H^1_?(C_r[p^{s-r}]^{\mathrm{ord}}). \tag{27}$$

This (combined with Corollary 6.3) induces the corresponding diagram for H^1, for any extension κ/k inside K,

$$\begin{array}{ccccc} H^1(C_s[p^{s'-r}]^{\mathrm{ord}}) & \xleftarrow[\pi_{r,s'}^*]{\sim} & H^1(C_r[p^{s'-r}]^{\mathrm{ord}}) & \hookleftarrow & \left(\frac{C_r(\kappa)}{p^{s'-r}C_r(\kappa)}\right)^{\mathrm{ord}} \\ \pi_{s'}^s \downarrow & & \downarrow p^{s'-s}U(p)^{s'-s} & & \downarrow p^{s'-s}U(p)^{s'-s} \\ H^1(C_s[p^{s-r}]^{\mathrm{ord}}) & \xleftarrow[\pi_{r,s}^*]{\sim} & H^1(C_r[p^{s-r}]^{\mathrm{ord}}) & \hookleftarrow & \left(\frac{C_r(\kappa)}{p^{s'-r}C_r(\kappa)}\right)^{\mathrm{ord}}. \end{array}$$

The right square is the result of Kummer theory for C_r. Passing to the projective limit with respect to s, we get a sequence

$$0 \to \varprojlim_s C_r[p^{s-r}]^{\mathrm{ord}}(\kappa) \xrightarrow{\pi^*_{r,s}} \varprojlim_s \widehat{C}^{\mathrm{ord}}_s(\kappa) \xrightarrow{\pi^r_s} \widehat{C}^{\mathrm{ord}}_r(\kappa) \to \varprojlim_s H^1(C_r[p^{s-r}]^{\mathrm{ord}}) \tag{28}$$

which is exact at left three terms up to the term $\widehat{C}^{\mathrm{ord}}_r(\kappa)$.

Proposition 6.4 *Let k be a finite extension field of $\mathbb{Q}$ or $\mathbb{Q}_l$ for a prime l. Assume (3) for $\kappa_{/k}$. Then we have the following identity*

$$\widehat{C}_\infty(\kappa)^{\mathrm{ord}} = \varprojlim_s \widehat{C}_s(\kappa)^{\mathrm{ord}} \cong \varprojlim_s C_r[p^{s-r}](\kappa)^{\mathrm{ord}} = 0$$

and exact sequences for K/k as in Sect. 2:

$$0 \to T_p C^{\mathrm{ord}}_r \to \varprojlim_s \widehat{C}_s(K)^{\mathrm{ord}} \to \widehat{C}_r(K)^{\mathrm{ord}} \to 0$$
$$0 \to T_p C^{\mathrm{ord}}_r \to \varprojlim_s C_s[p^\infty](K)^{\mathrm{ord}} \to C_r[p^\infty](K)^{\mathrm{ord}} \to 0.$$

In the last sequence, we have $\varprojlim_s C_s[p^\infty](K)^{\mathrm{ord}} \cong T_p C^{\mathrm{ord}}_r \otimes_{\mathbb{Z}} \mathbb{Q}$. By the first identity, $\widehat{C}^{\mathrm{ord}}_\infty$ as a smooth (resp. étale) sheaf vanishes if k is a number field or a local field with residual characteristic $l \neq p$ (resp. a p-adic field).

Proof By (28), we get a sequence which is exact at the first three left terms (up to the term $\widehat{C}^{\mathrm{ord}}_r(\kappa)$):

$$0 \to \varprojlim_s C_r[p^{s-r}](\kappa)^{\mathrm{ord}} \to \widehat{C}_\infty(\kappa)^{\mathrm{ord}} \xrightarrow[\pi^r_s]{} \widehat{C}_r(\kappa)^{\mathrm{ord}} \xrightarrow{\delta} \varprojlim_s H^1(C_s[p^{s-r}]^{\mathrm{ord}}).$$

Since δ is injective by Lemma 2.1 under (3), we get the first two identities. The vanishing of $\varprojlim_s C_r[p^{s-r}](\kappa)^{\mathrm{ord}}$ follows because $C_r[p^\infty]^{\mathrm{ord}}(\kappa)$ is a finite p-torsion module if κ/k is an extension of finite type.

If $\kappa = K$, we may again pass to the limit of the first exact sequence of (25) again noting $C_s[p^{s-r}](K)^{\mathrm{ord}} \cong C_r[p^{s-r}](K)^{\mathrm{ord}}$. The limit keeps exactness (as $\{C_r[p^{s-r}](K)\}_s$ is a surjective projective system), and we get the following exact sequence

$$0 \to TC_r[p^\infty](K)^{\mathrm{ord}} \to \varprojlim_s \widehat{C}_s(K)^{\mathrm{ord}} \xrightarrow[\pi^r_s]{} \widehat{C}_r(K)^{\mathrm{ord}} \to 0.$$

The divisible version can be proven taking the limit of (24). Since $C_r[p^\infty](K)^{\mathrm{ord}}$ is p-divisible and the projective system of the exact sequences $0 \to C_r[p](K)^{\mathrm{ord}} \to C_r[p^\infty](K)^{\mathrm{ord}} \xrightarrow{x \mapsto px} C_r[p^\infty](K)^{\mathrm{ord}} \to 0$ by the transition map $x \mapsto p^n U(p^n)(x)$ satisfies the Mittag–Leffler condition (as $C_r[p](K)^{\mathrm{ord}}$ is finite), $\varprojlim_s C_s[p^\infty](K)^{\mathrm{ord}}$ is a p-divisible module. Thus by the exact sequence, we have $T_p C^{\mathrm{ord}}_r \otimes_{\mathbb{Z}} \mathbb{Q} \subset \varprojlim_s C_s[p^\infty](K)^{\mathrm{ord}}$, which implies

$$T_p C_r^{\mathrm{ord}} \otimes_{\mathbb{Z}} \mathbb{Q} \cong \varprojlim_s C_s[p^\infty](K)^{\mathrm{ord}}$$

as $T_p C_r^{\mathrm{ord}} \otimes_{\mathbb{Z}} \mathbb{Q}/T_p C_r \cong C_r[p^\infty]^{\mathrm{ord}}(K)$. □

We insert here Shimura's definition of his abelian subvariety [13, Theorem 7.14] and abelian variety quotient [31] of J_s associated to a member f_P of a p-adic analytic family. Shimura mainly considered these abelian varieties associated to a primitive Hecke eigenform. Since we need those associated to old Hecke eigenforms, we give some details.

Let $P \in \mathrm{Spec}(\mathbf{h})(\overline{\mathbb{Q}}_p)$ be an arithmetic point of weight 2. Then we have a p-stabilized Hecke eigenform form f_P associated to P; i.e., $f_P|T(n) = P(T(n))f_P$ for all n (e.g., [5, Sect. 3.2]). Then $f_P^* = w_r(f_P)$ is the dual common eigenform of $T^*(n)$. If f_P is new at every prime $l|pN$, f_P^* is a constant multiple of the complex conjugate f_P^c of f_P (but otherwise, they are different). Shimura's abelian subvariety A_P (associated to f_P) is defined to be the identity connected component of $\bigcap_{\alpha \in P} J_r[\alpha]$ regarding P as a prime ideal of $h_r(\mathbb{Z})$.

The Rosati involution (induced by the canonical polarization) brings $h_r(\mathbb{Z})$ to $h_r^*(\mathbb{Z}) \subset \mathrm{End}(J_{r/\mathbb{Q}})$ isomorphically, and $\mathbf{h}$ acts on $\widehat{J}_\infty$ (resp. $\widetilde{J}_\infty$) through the identity $T(n) \mapsto T(n)$ (resp. through $T(n) \mapsto T^*(n)$). Let $f_P^*|T^*(n) = P(T(n))f_P^*$, and regard P as an algebra homomorphism $P^* : h_r^*(\mathbb{Z}) \to \overline{\mathbb{Q}}$ (so, $P^*(T^*(n)) = P(T(n))$). Identify P^* with the prime ideal $\mathrm{Ker}(P^*)$, and define A_P^* to be the identity connected component of $J_r[P^*] := \bigcap_{\alpha \in P^*} J_r[\alpha]$. Then $A_P \cong A_P^*$ by w_r over $\mathbb{Q}(\mu_{Np^r})$.

Assume that $r = r(P)$ is the minimal exponent of p in the level of f_P. For $s > r$, we write A_s (resp. A_s^*) for the abelian variety associated to f_P regarded as an old form of level p^s (resp. $w_s(f_P)$). In other words, regarding P^* as an ideal of $h_s^*(\mathbb{Z})$ via the projection $h_s^*(\mathbb{Z}) \twoheadrightarrow h_r^*(\mathbb{Z})$, we define A_s^* by the identity connected component of $J_s[P^*]$. The Albanese functoriality $\pi_* : J_s \twoheadrightarrow J_r$ induces an isogeny $A_s^* \twoheadrightarrow A_r^* = A_P^*$. Similarly the Picard functoriality $\pi^* : J_r \to J_s$ induces an isogeny $A_P = A_r \twoheadrightarrow A_s$. Since f_P^* is the complex conjugate of f_P (assuming that f_P is new), $A_P^* = A_P$ inside J_r (for $r = r(P)$). Since $w_s : A_{s/\mathbb{Q}[\zeta_{Np^s}]} \cong A^*_{s/\mathbb{Q}[\zeta_{Np^s}]}$ and A_s and A_s^* are isogenous to A_P over $\mathbb{Q}$, A_s and A_s^* are isomorphic over $\mathbb{Q}$. Consider the dual quotient $J_s \twoheadrightarrow B_s$ (resp. $J_s \twoheadrightarrow B_s^*$) of $A_s^* \hookrightarrow J_s$ (resp. $A_s \hookrightarrow J_s$). In the same manner as above, B_s and B_s^* are isomorphic over $\mathbb{Q}$. Then B_s (resp. B_s^*) is stable under $T(n)$ and $U(p)$ (resp. $T^*(n)$ and $U^*(p)$) and $\Omega_{B_s/\mathbb{C}}$ (resp. $\Omega_{B_s^*/\mathbb{C}}$) is spanned by $f_P^\sigma dz$ (resp. $g_P^\sigma dz$ for $g_P = w_s(f_P)$) for σ running over $\mathrm{Gal}(\overline{\mathbb{Q}}/\mathbb{Q})$. We mainly apply Corollary 6.3 and Proposition 6.4 taking C_s (resp. D_s) to be A_s (resp. B_s).

7 Abelian Factors of Modular Jacobians

Let k be a finite extension of $\mathbb{Q}$ inside $\overline{\mathbb{Q}}$ or a finite extension of $\mathbb{Q}_p$ over $\overline{\mathbb{Q}}_p$. We study the control theorem for $\widehat{J}_s(k)$ which is not covered in [21].

Let A_r be a group subscheme of J_r proper over k; so, A_r is an extension of an abelian scheme $A^\circ_{r/\mathbb{Q}}$ by a finite étale group. Write A_s $(s \geq r)$ for the image of A_r in J_s under the morphism $\pi^* : J_r \to J_s$ given by Picard functoriality from the projection $\pi : X_s \to X_r$. Hereafter we assume

(A) We have $\alpha \in \mathbf{h}(N)$ such that $(\gamma^{p^{r-\epsilon}} - 1) = \alpha x$ with $x \in \mathbf{h}(N)$ and that $\mathbf{h}/(\alpha)$ is free of finite rank over $\mathbb{Z}_p$. Write α_s for the image in $\mathbf{h}_s$ $(s \geq r)$ and $\mathfrak{a}_s = (\alpha_s \mathbf{h}_s \oplus (1-e)h_s(\mathbb{Z}_p)) \cap h_s(\mathbb{Z})$ and put $A_s = J_s[\mathfrak{a}_s]$ and $B_s = J_s/\mathfrak{a}_s J_s$, where $\mathfrak{a}_s J_s$ is an abelian subvariety defined over $\mathbb{Q}$ of J_s with $\mathfrak{a}_s J_s(\overline{\mathbb{Q}}) = \sum_{a \in \mathfrak{a}_s} a(J_s(\overline{\mathbb{Q}})) \subset J_s(\overline{\mathbb{Q}})$.

Here for $s > s'$, coherency of α_s means the following commutative diagram:

$$\begin{array}{ccc} \widehat{J}^{\mathrm{ord}}_{s'} & \xrightarrow{\pi^*} & \widehat{J}^{\mathrm{ord}}_{s} \\ {\scriptstyle \alpha_{s'}}\downarrow & & \downarrow{\scriptstyle \alpha_s} \\ \widehat{J}^{\mathrm{ord}}_{s'} & \xrightarrow[\pi^*]{} & \widehat{J}^{\mathrm{ord}}_{s} \end{array}$$

which is equivalent (by the self-duality of J_s) to the commutativity of

$$\begin{array}{ccc} \widehat{J}^{\mathrm{co\text{-}ord}}_{s} & \xrightarrow{\pi_*} & \widehat{J}^{\mathrm{co\text{-}ord}}_{s'} \\ {\scriptstyle \alpha^*_{s}}\downarrow & & \downarrow{\scriptstyle \alpha^*_{s'}} \\ \widehat{J}^{\mathrm{co\text{-}ord}}_{s} & \xrightarrow[\pi_*]{} & \widehat{J}^{\mathrm{co\text{-}ord}}_{s'}. \end{array}$$

The following fact is proven in [22, Lemma 5.1]:

Lemma 7.1 *Assume* (A). *Then we have* $\widehat{A}^{\mathrm{ord}}_s = \widehat{J}^{\mathrm{ord}}_s[\alpha_s]$ *and* $\widehat{A}^\circ_s = \widehat{A}_s$. *The identity connected component* A°_s $(s > r)$ *of* A_s *is the image of* A°_r *in* J_s *under the morphism* $\pi^* = \pi^*_{s,r} : J_r \to J_s$ *induced by Picard functoriality from the projection* $\pi = \pi_{s,r} : X_s \to X_r$ *and is* $\mathbb{Q}$*-isogenous to* B_s. *The morphism* $J_s \to B_s$ *factors through* $J_s \xrightarrow{\pi^r_s} J_r \to B_r$. *In addition, the sequence*

$$0 \to \widehat{A}^{\mathrm{ord}}_s \to \widehat{J}^{\mathrm{ord}}_s \xrightarrow{\alpha} \widehat{J}^{\mathrm{ord}}_s \xrightarrow{\rho_s} \widehat{B}^{\mathrm{ord}}_s \to 0 \text{ for } 0 < \epsilon \leq r \leq s < \infty$$

is an exact sequence of fppf sheaves.

This implies

Corollary 7.2 *Recall the finite set* S *of places made of prime factors of* Np *and* ∞. *Let* $R = k$ *if* k *is local, and let* R *be the* S*-integer ring of* k *(i.e., primes in* S *is inverted in* R*) if* k *is a number field. Then the sheaf* $\alpha_s(\widehat{J}^{\mathrm{ord}}_s)$ *is a* p*-divisible étale/fppf sheaf over* $\mathrm{Spec}(R)$, *and its* p*-torsion part* $\alpha_s(\widehat{J}^{\mathrm{ord}}_s)[p^\infty]$ *is a* p*-divisible Barsotti–Tate group over* R.

In particular, the Tate module $T_p\alpha(\widehat{J}_s^{\mathrm{ord}})$ is a well defined free $\mathbb{Z}_p$-module of finite rank for all $r \le s < \infty$.

Proof By the above lemma, the fppf sheaf $\alpha_s(\widehat{J}_s^{\mathrm{ord}}) = \mathrm{Ker}(\widehat{J}_s^{\mathrm{ord}} \xrightarrow{\rho_s} \widehat{B}_s^{\mathrm{ord}})$ fits into the following commutative diagram with exact rows:

$$\begin{array}{ccccc}
A_s[p^\infty]^{\mathrm{ord}} & \overset{\hookrightarrow}{\longrightarrow} & J_s[p^\infty]^{\mathrm{ord}} & \overset{\twoheadrightarrow}{\longrightarrow} & \alpha(J_s[p^\infty]^{\mathrm{ord}}) \\
\cap\downarrow & & \cap\downarrow & & \cap\downarrow \\
\widehat{A}_s^{\mathrm{ord}} & \overset{\hookrightarrow}{\longrightarrow} & \widehat{J}_s^{\mathrm{ord}} & \overset{\twoheadrightarrow}{\longrightarrow} & \alpha(\widehat{J}_s^{\mathrm{ord}}) \\
\downarrow & & \downarrow & & \downarrow \\
\widehat{A}_s^{\mathrm{ord}}/A_s[p^\infty]^{\mathrm{ord}} & \overset{\hookrightarrow}{\longrightarrow} & \widehat{J}_s^{\mathrm{ord}}/J_s[p^\infty]^{\mathrm{ord}} & \overset{\twoheadrightarrow}{\longrightarrow} & \alpha(\widehat{J}_s^{\mathrm{ord}})/\alpha(J_s[p^\infty]^{\mathrm{ord}}).
\end{array}$$

The first two terms of the bottom row are sheaves of $\mathbb{Q}_p$-vector spaces, so is the last term. Thus we conclude $\alpha(J_s[p^\infty]^{\mathrm{ord}}) = \alpha(\widehat{J}_s^{\mathrm{ord}})[p^\infty]$. Since $\widehat{A}_s = \widehat{A}_s^\circ$, $\widehat{A}_s[p^\infty]^{\mathrm{ord}}$ is a direct summand of the Barsotti–Tate group $J_s[p^\infty]^{\mathrm{ord}}$. Therefore $\alpha(J_s[p^\infty]^{\mathrm{ord}})$ is a Barsotti–Tate group as desired.

Alternatively, we can identify $\alpha_s(\widehat{J}_s^{\mathrm{ord}})[p^\infty]$ with the Barsotti–Tate p-divisible group of the abelian variety quotient J_s/A_s°. □

The condition (A) is a mild condition. Here are sufficient conditions for (α, A_s, B_s) to satisfy (A) given in [22, Proposition 5.2]:

Proposition 7.3 *Let* $\mathrm{Spec}(\mathbb{T})$ *be a connected component of* $\mathrm{Spec}(\mathbf{h})$ *and* $\mathrm{Spec}(\mathbb{I})$ *be a primitive irreducible component of* $\mathrm{Spec}(\mathbb{T})$. *Then the condition* (A) *holds for the following choices of* (α, A_s, B_s)*:*

1. *Suppose that an eigen cusp form $f = f_P$ new at each prime $l|N$ belongs to* $\mathrm{Spec}(\mathbb{T})$ *and that $\mathbb{T} = \mathbb{I}$ is regular. Writing the level of f_P as Np^r, the algebra homomorphism $\lambda : \mathbb{T} \to \overline{\mathbb{Q}}_p$ given by $f|T(l) = \lambda(T(l))f$ gives rise to a height 1 prime ideal $P = \mathrm{Ker}(\lambda)$, which is principal generated by $a \in \mathbb{T}$. This a has its image $a_s \in \mathbb{T}_s = \mathbb{T} \otimes_\Lambda \Lambda_s$ for $\Lambda_s = \Lambda/(\gamma^{p^{s-\epsilon}} - 1)$. Write $\mathbf{h}_s = \mathbf{h} \otimes_\Lambda \Lambda_s = \mathbb{T}_s \oplus 1_s\mathbf{h}_s$ as an algebra direct sum for an idempotent 1_s. Then, the element $\alpha_s = a_s \oplus 1_s \in \mathbf{h}_s$ for the identity 1_s of X_s satisfies* (A). *In this case, $\alpha = \varprojlim_s \alpha_s$.*
2. *More generally than* (1), *we pick a general connected component* $\mathrm{Spec}(\mathbb{T})$ *of* $\mathrm{Spec}(\mathbf{h})$. *Pick a (classical) Hecke eigenform $f = f_P$ (of weight 2) for $P \in$* $\mathrm{Spec}(\mathbb{T})$. *Assume that $\mathbf{h}_s$ (for every $s \ge r$) is reduced and $P = (a)$ for $a \in \mathbb{T}$, and write a_s for the image of a in $\mathbf{h}_s$. Then decomposing $\mathbf{h}_s = \mathbb{T}_s \oplus 1_s\mathbf{h}_s$, $\alpha_s = a_s \oplus 1_s$ satisfies* (A).
3. *Fix $r > 0$. Then α for a factor $\alpha|(\gamma^{p^{r-\epsilon}} - 1)$ in Λ satisfies* (A) *for $A_s = J_s[\alpha]^\circ$ (the identity connected component).*

Remark 7.4 (i) Under (1), all arithmetic points P of weight 2 in $\mathrm{Spec}(\mathbb{I})$ satisfies (A).

(ii) For a given weight 2 Hecke eigenform f, for density 1 primes $\mathfrak{p}$ of $\mathbb{Q}(f)$, f is ordinary at $\mathfrak{p}$ (i.e., $a(p, f) \not\equiv 0 \mod \mathfrak{p}$; see [19, §7]). Except for finitely many primes $\mathfrak{p}$ as above, f belongs to a connected component $\mathbb{T}$ which is regular (e.g., [14, §3.1] and [22, Theorem 5.3]); so, (1) is satisfied for such $\mathbb{T}$.

8 Mordell–Weil Groups of Modular Abelian Factors

Consider the composite morphism $\varpi_s : A_s \hookrightarrow J_s \twoheadrightarrow B_s$ of fppf abelian sheaves for triples (α_s, A_s, B_s) as in (A), and apply the results in Sect. 6 to abelian varieties $C_s = A_s$ and $D_s = B_s$. Let $C_s^{\mathrm{ord}} := (\mathrm{Ker}(\varpi_s) \otimes_{\mathbb{Z}} \mathbb{Z}_p)^{\mathrm{ord}}$ be the p-primary ordinary part of $\mathrm{Ker}(\varpi_s)$.

Recall we have written ρ_s for the morphism $J_s \to B_s$. As before, κ is an intermediate extension $K/\kappa/k$ finite over k. Define the error terms by

$$E_1^s(\kappa) := \alpha(\widehat{J}_s^{\mathrm{ord}})(\kappa)/\alpha(\widehat{J}_s^{\mathrm{ord}}(\kappa)) \text{ and } E_2^s(\kappa) := \mathrm{Coker}(\widehat{J}_s^{\mathrm{ord}}(\kappa) \xrightarrow{\rho_s} \widehat{B}_s^{\mathrm{ord}}(\kappa)) \quad (29)$$

for $\rho_s : \widehat{J}_s^{\mathrm{ord}}(\kappa) \to \widehat{B}_s^{\mathrm{ord}}(\kappa)$. Note that $E_1^s(\kappa) (\hookrightarrow H^1_?(\widehat{A}_s^{\mathrm{ord}}) = H^1_?(A_s^{\mathrm{ord}}) \otimes_{\mathbb{Z}} \mathbb{Z}_p)$ and $E_2^s(\kappa) = B_s^{\mathrm{ord}}(\kappa)/\rho_s(\widehat{J}_s^{\mathrm{ord}}(\kappa)) (\hookrightarrow H^1_?(\alpha(\widehat{J}_s^{\mathrm{ord}}))[\alpha])$ are p-torsion finite module as long as s is finite.

Lemma 8.1 *We have the following commutative diagram with exact rows and exact columns:*

$$\begin{array}{ccccc}
E_1^s(\kappa) & \hookrightarrow & H^1_?(\widehat{A}_s^{\mathrm{ord}}) & \xrightarrow{\iota_s} & H^1_?(\widehat{J}_s^{\mathrm{ord}}) \\
\text{onto}\uparrow & & \uparrow & & \uparrow \\
\dfrac{\alpha(\widehat{J}_s^{\mathrm{ord}})(\kappa)}{\alpha(\alpha(\widehat{J}_s^{\mathrm{ord}})(\kappa))} & \hookrightarrow & H^1_?(C_s^{\mathrm{ord}}) & \twoheadrightarrow & H^1_?(\alpha(\widehat{J}_s^{\mathrm{ord}}))[\alpha] \\
\overline{\alpha}_s\uparrow & & b_s\uparrow & & \uparrow\cup \\
\dfrac{\widehat{J}_s^{\mathrm{ord}}(\kappa)}{\alpha(\widehat{J}_s^{\mathrm{ord}})(\kappa)} & \underset{\hookrightarrow}{\xrightarrow{\rho_s}} & \widehat{B}_r^{\mathrm{ord}}(\kappa) & \twoheadrightarrow & E_2^s(\kappa).
\end{array} \quad (30)$$

Each term of the bottom two rows is a profinite module if either k is local or S is a finite set.

The last assertion follows as C_s is finite and $\widehat{B}_r^{\mathrm{ord}}(\kappa)$ is profinite. We will define each map in the following proof. The proof is the same in any cohomology theory: $H^1_?$ for $? =$ sm, fppf, étale and Galois cohomology. Therefore, we prove the lemma for the Galois cohomology dropping ? from the notation. This lemma is valid for the Galois cohomology for infinite S as is clear from the proof below.

Proof Exactness for the bottom row is from the definition of $E_2^s(\kappa)$, and exactness for the left column is by the definition of $E_1^s(\kappa)$. The middle column is a part of the long exact sequence attached to $0 \to C_s^{\mathrm{ord}} \to \widehat{A}_s^{\mathrm{ord}} \to \widehat{B}_r^{\mathrm{ord}} \to 0$, where $\widehat{B}_s^{\mathrm{ord}}$ is identified with $\widehat{B}_r^{\mathrm{ord}}$ by Lemma 6.2 applied to $D_s = B_s$. The right column comes from the long exact sequence attached to $0 \to \alpha(\widehat{J}_s^{\mathrm{ord}}) \to \widehat{J}_s^{\mathrm{ord}} \to \widehat{B}_r^{\mathrm{ord}} \to 0$, again $\widehat{B}_s^{\mathrm{ord}}$ is identified with $\widehat{B}_r^{\mathrm{ord}}$. The top row comes from the long exact sequence of $0 \to \widehat{A}_s^{\mathrm{ord}} \to \widehat{J}_s^{\mathrm{ord}} \xrightarrow{\alpha} \alpha(\widehat{J}_s^{\mathrm{ord}}) \to 0$.

As for the middle row, we consider the following commutative diagram (with exact rows in the category of fppf abelian sheaves):

$$\begin{array}{ccccc} \alpha(\widehat{J}_s^{\mathrm{ord}}) & \xrightarrow{\hookrightarrow} & \widehat{J}_s^{\mathrm{ord}} & \xrightarrow[\rho_s]{\twoheadrightarrow} & \widehat{B}_s^{\mathrm{ord}} \\ \cup\uparrow & & \cup\uparrow & & \uparrow\| \\ 0 \to \alpha(\widehat{J}_s^{\mathrm{ord}}) \times_{\widehat{J}_s^{\mathrm{ord}}} \widehat{A}_s^{\mathrm{ord}} & \xrightarrow[\hookrightarrow]{} & \widehat{A}_s^{\mathrm{ord}} & \xrightarrow[\twoheadrightarrow]{\varpi_s} & \widehat{B}_s^{\mathrm{ord}}. \end{array} \tag{31}$$

Under this circumstance, we have $\alpha(\widehat{J}_s^{\mathrm{ord}}) \cap \widehat{A}_s^{\mathrm{ord}} = \alpha(\widehat{J}_s^{\mathrm{ord}}) \times_{\widehat{J}_s^{\mathrm{ord}}} \widehat{A}_s^{\mathrm{ord}} = \mathrm{Ker}(\varpi_s)$ which is a finite étale p-group scheme over $\mathbb{Q}$. Since $\alpha(\widehat{J}_s^{\mathrm{ord}}) \cap \widehat{A}_s^{\mathrm{ord}}$ is equal to $\alpha(\widehat{J}_s^{\mathrm{ord}})[\alpha]$, we have $C_s^{\mathrm{ord}} = \alpha(\widehat{J}_s^{\mathrm{ord}})[\alpha]$.

Note that $\alpha^2(\widehat{J}_s^{\mathrm{ord}}) = \alpha(\widehat{J}_s^{\mathrm{ord}})$ as sheaves (as $\alpha : \alpha(\widehat{J}_s^{\mathrm{ord}}) \to \alpha(\widehat{J}_s^{\mathrm{ord}})$ is an isogeny, and hence, $\alpha(\alpha(\widehat{J}_s^{\mathrm{ord}}(K))) = \alpha(\widehat{J}_s^{\mathrm{ord}})(K))$. Thus we have a short exact sequence under ?-topology for ? = fppf, sm and ét:

$$0 \to C_s^{\mathrm{ord}}(K) \to \alpha(\widehat{J}_s^{\mathrm{ord}})(K) \xrightarrow{\alpha} \alpha(\widehat{J}_s^{\mathrm{ord}})(K) \to 0.$$

Look into the associated long exact sequence

$$\begin{aligned} 0 \to \alpha(\widehat{J}_s^{\mathrm{ord}})(k)/\alpha(\alpha(\widehat{J}_s^{\mathrm{ord}})(k)) &\to H^1(\alpha(\widehat{J}_s^{\mathrm{ord}})[\alpha]) \\ &\to H^1(\alpha(\widehat{J}_s^{\mathrm{ord}})) \xrightarrow{\alpha} H^1(\alpha^2(\widehat{J}_s^{\mathrm{ord}})) \end{aligned}$$

which shows the exactness of the middle row, taking the p-primary parts (and then the ordinary parts). □

In the diagram (30), we identify $\widehat{A}_s^{\mathrm{ord}}$ with $\widehat{A}_r^{\mathrm{ord}}$ by $\pi_{s,r}^* : J_r \to J_s$ for the projection $\pi_{s,r} : X_s \to X_r$ (Picard functoriality); so, the projective system $\{\widehat{A}_s^{\mathrm{ord}} = \widehat{A}_r^{\mathrm{ord}}, \pi_s^r\}_s$ (w-twisted Albanese functoriality) gives rise to the nontrivial maps $\pi_s^r : \widehat{A}_s^{\mathrm{ord}} = \widehat{A}_r^{\mathrm{ord}} \to \widehat{A}_r^{\mathrm{ord}}$ given by $x \mapsto U(p^{s-r})(p^{s-r}x)$. If we write $H^1(\widehat{A}_r^{\mathrm{ord}}) = (\mathbb{Q}_p/\mathbb{Z}_p)^m \oplus \Delta_r$ for a finite p-torsion group Δ_r by Lemma 2.2 (assuming that S is finite), we have

$$\varprojlim_{\pi_{s*}^r : x \mapsto p^{s-r}U(p^{s-r})(x)} H^1(\widehat{A}_r^{\mathrm{ord}}) \cong \varprojlim_{\pi_{s*}^r : x \mapsto p^{s-r}x} ((\mathbb{Q}_p/\mathbb{Z}_p)^m \oplus \Delta_r) = \mathbb{Q}_p^m. \tag{32}$$

We quote from [12, Corollary 2.7.6] the following fact (which is valid also for infinite S):

Lemma 8.2 *We have* $\varprojlim_s H^1(A_r[p^s]^{\mathrm{ord}}) = H^1(T_p A_r^{\mathrm{ord}})$.

We give a proof here for the sake of completeness.

Proof More generally, let $\{M_n\}_n$ be a projective system of finite $\mathrm{Gal}(k^S/k)$-modules with surjective transition maps. Let $B(M_n)$ (resp. $Z(M_n)$) be the module of 1-coboundaries (resp. inhomogeneous continuous 1-cocycles) $G := \mathrm{Gal}(k^S/k) \to M_n$. We have the exact sequence $0 \to B(M_n) \to Z(M_n) \to H^1(G, M_n) \to 0$. Plainly for $m > n$, the natural map $B(M_m) \to B(M_n)$ is onto. Thus the above sequences satisfies the Mittag–Leffler condition, and plainly $\varprojlim_n ?(M_n) = ?(\varprojlim_n M_n)$ for $? = B, Z$, we have

$$\varprojlim_n H^1(k^S/k, M_n) = H^1(k^S/k, \varprojlim_n M_n).$$

□

We have the following commutative diagram with exact rows:

$$\begin{array}{ccccc} C_s^{\mathrm{ord}} & \hookrightarrow & \widehat{A}_s^{\mathrm{ord}} & \twoheadrightarrow & \widehat{B}_s^{\mathrm{ord}} \\ \downarrow & & \downarrow & & \downarrow\wr \\ C_r^{\mathrm{ord}} & \hookrightarrow & \widehat{A}_r^{\mathrm{ord}} & \twoheadrightarrow & \widehat{B}_r^{\mathrm{ord}}. \end{array}$$

By the snake lemma applied to the above diagram, we get the following exact sequence:

$$0 \to A_r[p^{s-r}]^{\mathrm{ord}} \to C_s^{\mathrm{ord}} \to C_r^{\mathrm{ord}} \to 0.$$

Passing to the limit (as continuous H^1 for profinite coefficients is a projective limit of H^1 of finite coefficients; cf., [12, 2.7.6]), we have

$$T_pA = \varprojlim_s A_r[p^s]^{\mathrm{ord}} = \varprojlim_s C_s^{\mathrm{ord}}$$
$$\text{and } H^1(T_pA_r^{\mathrm{ord}}) = \varprojlim_s H^1(A_r[p^s]^{\mathrm{ord}}) = \varprojlim_s H^1(C_s^{\mathrm{ord}}). \quad (33)$$

9 Control Theorems with an Error Term

Taking the projective limit of the exact sequence $0 \to \widehat{A}_s^{\mathrm{ord}} \to \widehat{J}_s^{\mathrm{ord}} \xrightarrow{\alpha} \widehat{J}_s^{\mathrm{ord}}$, by the vanishing $\varprojlim_s \widehat{A}_s^{\mathrm{ord}}(\kappa) = 0$ in Proposition 6.4 applied to $C_s = A_s$, we get the injectivity of $\widehat{J}_\infty^{\mathrm{ord}} \xrightarrow{\alpha} \widehat{J}_\infty^{\mathrm{ord}}$.

Since all the terms of the exact sequences: $0 \to \alpha(\widehat{J}_s^{\mathrm{ord}})(\kappa) \to \widehat{J}_s^{\mathrm{ord}}(\kappa) \to \frac{\widehat{J}_s^{\mathrm{ord}}(\kappa)}{\alpha(\widehat{J}_s^{\mathrm{ord}})(\kappa)} \to 0$ are compact p-profinite groups, after taking the limit with respect to π_s^r, we still have an exact sequence

$$0 \to \varprojlim_s \alpha(\widehat{J}_s^{\mathrm{ord}})(\kappa) \to \varprojlim_s \widehat{J}_s^{\mathrm{ord}}(\kappa) \to \varprojlim_s \frac{\widehat{J}_s^{\mathrm{ord}}(\kappa)}{\alpha(\widehat{J}_s^{\mathrm{ord}})(\kappa)} \to 0$$

with $\varprojlim_s \frac{\widehat{J}_s^{\mathrm{ord}}(\kappa)}{\alpha(\widehat{J}_s^{\mathrm{ord}})(\kappa)} \hookrightarrow \widehat{B}_r^{\mathrm{ord}}(\kappa)$. Thus

$$\frac{\widehat{J}_\infty^{\mathrm{ord}}(\kappa)}{\alpha(\widehat{J}_\infty^{\mathrm{ord}})(\kappa)} := \frac{\varprojlim_s \widehat{J}_s^{\mathrm{ord}}(\kappa)}{\varprojlim_s \alpha(\widehat{J}_s^{\mathrm{ord}})(\kappa)} \cong \varprojlim_s \frac{\alpha(\widehat{J}_s^{\mathrm{ord}})(\kappa)}{\alpha(\alpha(\widehat{J}_s^{\mathrm{ord}})(\kappa))}.$$

Here the last isomorphism follows from the injectivity of α. By the same token, we have

$$\frac{\alpha(\widehat{J}_\infty^{\mathrm{ord}})(\kappa)}{\alpha(\alpha(\widehat{J}_\infty^{\mathrm{ord}})(\kappa))} := \frac{\varprojlim_s \alpha(\widehat{J}_s^{\mathrm{ord}})(\kappa)}{\varprojlim_s \alpha(\alpha(\widehat{J}_s^{\mathrm{ord}})(\kappa))} = \varprojlim_s \frac{\alpha(\widehat{J}_s^{\mathrm{ord}})(\kappa)}{\alpha(\alpha(\widehat{J}_s^{\mathrm{ord}})(\kappa))}.$$

Writing $E_j^\infty(\kappa) = \varprojlim_s E_j^s(\kappa)$ and passing to projective limit of the diagram (30), we get the following commutative diagram with exact rows:

$$\begin{array}{ccccc}
E_1^\infty(\kappa) & \xrightarrow[\text{Lemma 8.1}]{\hookrightarrow} & \varprojlim_{\substack{s:\\ x\mapsto p^{s-r}U(p^{s-r})(x)}} H^1(\widehat{A}_r^{\mathrm{ord}}) & \xrightarrow{\iota_\infty} & H^1(\varprojlim_s \widehat{J}_s^{\mathrm{ord}}(K)) \\
\text{onto}\uparrow & & \uparrow & & \uparrow \\
\frac{\alpha(\widehat{J}_\infty^{\mathrm{ord}})(\kappa)}{\alpha(\alpha(\widehat{J}_\infty^{\mathrm{ord}})(\kappa))} & \xrightarrow{\hookrightarrow} & H^1(T_pA_r(K)^{\mathrm{ord}}) & \xrightarrow{a} & \varprojlim_s H^1(\alpha(\widehat{J}_s^{\mathrm{ord}})(K))[\alpha] \\
\overline{\alpha}_\infty\uparrow & & b\uparrow\cup & & c\uparrow\cup \\
\frac{\widehat{J}_\infty^{\mathrm{ord}}(\kappa)}{\alpha(\widehat{J}_\infty^{\mathrm{ord}})(\kappa)} & \xrightarrow[\hookrightarrow]{\rho_s} & \widehat{B}_r^{\mathrm{ord}}(\kappa) & \xrightarrow{d} & E_2^\infty(\kappa).
\end{array} \qquad (34)$$

The rows are exact since projective limit is left exact. The maps a and d are onto if either S is finite or k is local (as projective limit is exact for profinite modules). By the same token, the right and left columns are also exact. Therefore $E_j^\infty(\kappa)$ $(j = 1, 2)$ is a torsion Λ-module of finite type.

To see, we look into the cohomology exact sequence of the short exact sequence: $C_s \hookrightarrow \widehat{A}_r^{\mathrm{ord}} \twoheadrightarrow \widehat{B}_r^{\mathrm{ord}}$ with transition maps $p^{s'-s}U(p)^{s'-s}$ for $\{C_s\}_s$ and $\{\widehat{A}_r^{\mathrm{ord}}\}_s$ and $U(p)^{s'-s}$ for $\{\widehat{B}_r^{\mathrm{ord}}\}_s$. Thus we have the limit sequence

$$0 \to \varprojlim_{s: x\mapsto p^{s-r}U(p^{s-r})(x)} \widehat{A}_r^{\mathrm{ord}}(\kappa)/C_s(\kappa) \to \widehat{B}_r^{\mathrm{ord}}(\kappa) \xrightarrow{b} \varprojlim_s H^1(C_s) = H^1(T_pA_r^{\mathrm{ord}}).$$

This sequence is exact as all the terms are profinite compact modules at each step. Since

$$\varprojlim_{s: x\mapsto p^{s-r}U(p^{s-r})(x)} \widehat{A}_r^{\mathrm{ord}}(\kappa)/C_s(\kappa) = 0,$$

the map b is injective.

Passing to the limit of exact sequences of profinite modules: $C_s(\kappa) \to \widehat{A}_r^{\mathrm{ord}}(\kappa) \xrightarrow{\varpi_s} \widehat{B}_r^{\mathrm{ord}}(\kappa) \twoheadrightarrow \mathrm{Coker}(\varpi_s)$, we get the limit exact sequence $0 \to \widehat{B}_r^{\mathrm{ord}}(\kappa) \cong \varprojlim_s \mathrm{Coker}(\varpi_s)$. By the left exactness of projective limit, the sequence

$$0 \to \varprojlim_s \mathrm{Coker}(\varpi_s) \to H^1(T_p A_r^{\mathrm{ord}}) \to \varprojlim_s H^1(\widehat{A}_r^{\mathrm{ord}})$$

is exact. Therefore the middle column is exact; so,

$$\overline{\alpha}_\infty \text{ is injective.} \tag{35}$$

Since $\widehat{J}_\infty(\kappa)^{\mathrm{ord}}[\alpha] = \widehat{A}_\infty^{\mathrm{ord}}(\kappa) = 0$, $\alpha : \alpha(\widehat{J}_\infty^{\mathrm{ord}})(\kappa) \to \alpha(\widehat{J}_\infty^{\mathrm{ord}})(\kappa)$ is injective.

This shows

Lemma 9.1 *Let κ be a field extension of $\mathbb{Q}$ or $\mathbb{Q}_l$ for a prime l, but we assume finiteness condition* (3) *for the extension κ/k. We allow an infinite set S of places of k when k is finite extension of $\mathbb{Q}$. Let α be as in* (A). *Then we have the following exact sequences (of p-profinite Λ-modules)*

$$0 \to \widehat{J}_\infty^{\mathrm{ord}}(\kappa) \xrightarrow{\alpha} \alpha(\widehat{J}_\infty^{\mathrm{ord}})(\kappa) \to E_1^\infty(\kappa)^{\mathrm{ord}} \to 0$$

and

$$0 \to \alpha(\widehat{J}_\infty^{\mathrm{ord}})(\kappa) \to \widehat{J}_\infty^{\mathrm{ord}}(\kappa) \xrightarrow{\rho_\infty} \widehat{B}_r^{\mathrm{ord}}(\kappa) \to E_2^\infty(\kappa) \to 0.$$

Here $E_j^\infty(\kappa)$ is a Λ-torsion module of finite type. In particular, taking $\alpha = \gamma - 1$, we conclude that the compact module $\widehat{J}_\infty(\kappa)$ is a Λ–module of finite type.

The statement of this lemma is independent of the set S (though in the proof, we used Galois cohomology groups for finite S if k is global); therefore, the lemma is valid also for an infinite set S of places of k (as long as S contains all p-adic and archimedean places and places over N).

The left column of (34) is made up of compact modules for which projective limit is an exact functor; so, left column is exact; in particular

$$\varprojlim_s \frac{\alpha(\widehat{J}_s^{\mathrm{ord}})(\kappa)}{\alpha(\alpha(\widehat{J}_s^{\mathrm{ord}})(\kappa))} \to E_1^\infty(\kappa) := \varprojlim_s E_1^s(\kappa)$$

is onto.

Take the maximal Λ-torsion module X inside $\widehat{J}_\infty^{\mathrm{ord}}(\kappa)$. Since X is unique, it is an $\mathbf{h}$-module. The module $\widehat{J}_\infty^{\mathrm{ord}}(\kappa)$ is pseudo-isomorphic to $X \oplus \Lambda^R$ for a positive integer R. Since α is injective on $\widehat{J}_\infty^{\mathrm{ord}}(\kappa)$, for the α-localization $\mathbf{h}_{(\alpha)}$, we have $X_\alpha = X \otimes_{\mathbf{h}} \mathbf{h}_{(\alpha)} = 0$. Thus $\widehat{J}_\infty^{\mathrm{ord}}(\kappa) \otimes_{\mathbf{h}} \mathbf{h}_{(\alpha)}$ is Λ_{P_α}-free, where $P_\alpha = (\alpha) \cap \Lambda$. Thus $\frac{\alpha(\widehat{J}_\infty^{\mathrm{ord}})(\kappa)}{\alpha(\alpha(\widehat{J}_\infty^{\mathrm{ord}})(\kappa))} \otimes_{\mathbb{Z}_p} \mathbb{Q}_p$ and $\frac{\widehat{J}_\infty^{\mathrm{ord}}(\kappa)}{\alpha(\widehat{J}_\infty^{\mathrm{ord}})(\kappa)} \otimes_{\mathbb{Z}_p} \mathbb{Q}_p$ have equal $\mathbb{Q}_p$-dimension. Therefore, by the injectivity of $\overline{\alpha}_\infty$ (35), $E_1^\infty(\kappa)$ is p-torsion. However by (32), this torsion module is embedded in a $\mathbb{Q}_p$-vector space by the top sequence of (34), we have $E_1^\infty(\kappa) = 0$. This shows

Theorem 9.2 *Let α be as in* (A) *and k be a finite field extension of either $\mathbb{Q}$ or $\mathbb{Q}_l$ for a prime l. Assume* (3) *for the extension κ/k. Then we have the following exact sequence (of p-profinite Λ-modules):*

$$0 \to \widehat{J}_\infty^{\mathrm{ord}}(\kappa) \xrightarrow{\alpha} \widehat{J}_\infty^{\mathrm{ord}}(\kappa) \xrightarrow{\rho_\infty} \widehat{B}_r^{\mathrm{ord}}(\kappa) \to E_2^\infty(\kappa) \to 0.$$

In particular, taking $\alpha = \gamma - 1$, we conclude that the Λ-module $\widehat{J}_\infty(\kappa)$ is a Λ–module of finite type and that $\widehat{J}_\infty(\kappa)$ does not have any pseudo-null Λ-submodule non null (i.e., $\widehat{J}_\infty(\kappa)$ has Λ-homological dimension ≤ 1).

By this theorem (applied to $\alpha = \gamma^{p^s} - 1$ for $s = 1, 2, \ldots$), the localization $\widehat{J}_\infty(\kappa)_P$ at an arithmetic prime P is Λ_P-free of finite rank, which also follows from [27, Proposition 12.7.13.4] as $\widehat{J}_\infty(\kappa)$ can be realized inside Nekovář's Selmer group by the embedding of Lemma 2.1.

10 Control Theorem for a Number Field

The following theorem is the final result of this paper for a number field k.

Theorem 10.1 *Let the notation be as in the introduction. Suppose that k is a finite extension of $\mathbb{Q}$. Let $\mathcal{A}_{\mathbb{T}}$ be the set of all principal arithmetic points of* $\mathrm{Spec}(\mathbb{T})(\overline{\mathbb{Q}}_p)$ *of weight 2 and put $\Omega_{\mathbb{T}} := \{P \in \mathcal{A}_{\mathbb{T}} | A_P$ has good reduction over $\mathbb{Z}_p[\mu_{p^\infty}]\}$. Suppose that we have a single point $P_0 \in \Omega_{\mathbb{T}}$ with finite* $\mathrm{Sel}_k(A_{P_0})^{\mathrm{ord}}$, *and write* $\mathrm{Spec}(\mathbb{I})$ *for the unique irreducible component on which P_0 lies. Let k be a finite field extension of either $\mathbb{Q}$ or $\mathbb{Q}_l$ for a prime l. Then, for almost all $P \in \Omega_{\mathbb{T}} \cap \mathrm{Spec}(\mathbb{I})$, we have the following exact sequence (of p-profinite Λ-modules):*

$$0 \to \widehat{J}_{\infty,\mathbb{T}}^{\mathrm{ord}}(k) \xrightarrow{\alpha} \widehat{J}_{\infty,\mathbb{T}}^{\mathrm{ord}}(k) \xrightarrow{\rho_\infty} \widehat{B}_P^{\mathrm{ord}}(k) \to E_2^\infty(k) \to 0$$

with finite error term $|E_2^\infty(k)| < \infty$.

Since $\mathbb{T}$ is étale at P_0 over Λ, only one irreducible component of $\mathrm{Spec}(\mathbb{T})$ contains P_0 (e.g. [6, Proposition 3.78]).

Since the root number of $L(s, A_P)$ is not equal to -1 for most points (as $\{X_r\}_r$ is the standard tower), we expect that $|\mathrm{Sel}_k(A_P)^{\mathrm{ord}}| < \infty$ for most arithmetic P; so, the assumption of the theorem is a reasonable one.

Proof The Selmer group $\mathrm{Sel}_k(A_P)^{\mathrm{ord}}$ is the one defined in [22, §8]. By [27, 12.7.13.4] or [22, Theorem A], the finiteness $|\mathrm{Sel}_k(A_{P_0})^{\mathrm{ord}}| < \infty$ for a single point $P_0 \in \Omega_{\mathbb{T}}$ implies that $\mathrm{Sel}_k(A_P)^{\mathrm{ord}}$ is finite for almost all $P \in \Omega_{\mathbb{T}} \cap \mathrm{Spec}(\mathbb{I})$. Though in [22, Theorem A], it is assumed that $\mathbb{T}$ is regular to guarantee that all arithmetic points are principal, what we need to get the result is the principality of

P_0 and P in $\mathbb{T}$; so, this holds true for $P \in \Omega_{\mathbb{T}} \cap \mathrm{Spec}(\mathbb{I})$. By the well known exact sequence

$$0 \to \widehat{B}_P^{\mathrm{ord}}(k) \otimes_{\mathbb{Z}_p} \mathbb{Q}_p/\mathbb{Z}_p \to \mathrm{Sel}_k(A_P)^{\mathrm{ord}} \to \text{Ш}_k(A_P)^{\mathrm{ord}} \to 0,$$

the finiteness of $\mathrm{Sel}_k(A_P)^{\mathrm{ord}}$ implies finiteness of $\widehat{B}_P^{\mathrm{ord}}(k)$; so, $E_2(k)$ is finite as well. □

11 Local Error Term

Now let k be an l-adic field. As before, we write $H^q(M)$ for $H^q(k, M)$. For any abelian variety $X_{/k}$, we have an exact sequence

$$\widehat{X}(k) \hookrightarrow H^1(T_pX) \twoheadrightarrow \varprojlim_n H^1(\widehat{X})[p^n]$$

by Lemma 2.1. Similarly, by Corollary 7.2, Lemma 2.1 tells us that

$$\alpha(\widehat{J}_s^{\mathrm{ord}}(k)) \hookrightarrow H^1(T_p\alpha(\widehat{J}_s^{\mathrm{ord}})) \twoheadrightarrow T_pH^1(\alpha(\widehat{J}_s^{\mathrm{ord}})) := \varprojlim_n H^1(\alpha(\widehat{J}_s^{\mathrm{ord}}))[p^n]$$

is exact. Thus we have the following commutative diagram in which the first two columns and the first three rows are exact by Lemma 8.2 and left exactness of the formation of projective limits combined (the surjectivity of the three horizontal arrows c_j ($j = 1, 2, 3$) are valid if S is finite or k is local):

$$\begin{array}{ccccc}
\alpha(\widehat{J}_s^{\mathrm{ord}})(k) & \xrightarrow{\hookrightarrow} & H^1(T_p\alpha(\widehat{J}_s^{\mathrm{ord}})) & \underset{\twoheadrightarrow}{\xrightarrow{c_1}} & T_pH^1(\alpha(\widehat{J}_s^{\mathrm{ord}})) \\
\cap\downarrow i & & a\downarrow & & \downarrow b \\
\widehat{J}_s^{\mathrm{ord}}(k) & \underset{f}{\xrightarrow{\hookrightarrow}} & H^1(T_pJ_s^{\mathrm{ord}}) & \underset{\twoheadrightarrow}{\xrightarrow{c_2}} & T_pH^1(\widehat{J}_s^{\mathrm{ord}}) \\
\rho_s\downarrow & & j\downarrow & & \downarrow h \\
\widehat{B}_r^{\mathrm{ord}}(k) & \underset{\beta}{\xrightarrow{\hookrightarrow}} & H^1(T_pB_r^{\mathrm{ord}}) & \underset{\twoheadrightarrow}{\xrightarrow{c_3}} & T_pH^1(\widehat{B}_r^{\mathrm{ord}}) \\
\text{onto}\downarrow \pi & & \varpi_s\downarrow & & \downarrow g \\
E_2^s(k) & \xrightarrow{e_s} & H^2(T_p\alpha(\widehat{J}_s^{\mathrm{ord}})) & \longrightarrow & T_pH^2(\alpha(\widehat{J}_s^{\mathrm{ord}})).
\end{array} \tag{36}$$

Assuming that S is finite, the right column is made of $\mathbb{Z}_p$-free modules, and hence, the rows are split exact sequences.

To see the existence of the map e_s, we suppose that $x = \rho_s(y) \in \mathrm{Im}(\rho_s)$. Then we have

$$\varpi_s(\beta(x)) = \varpi_s(\beta(\rho_s(y))) = \varpi_s(j(f(y))) = 0.$$

If $b \equiv b' \mod \mathrm{Im}(\rho_s)$ for $b, b' \in \widehat{B}_r^{\mathrm{ord}}(k)$, we have $\varpi_s(\beta(b)) = \varpi_s(\beta(b'))$. In other words, $\pi(b) \mapsto \varpi(\beta(b))$ is a well-defined homomorphism from $E_2^s(k) \cong \widehat{B}_r^{\mathrm{ord}}(k)/\mathrm{Im}(\rho_s)$ into $\mathrm{Im}(\varpi_s) \cong \mathrm{Coker}(j) \subset H^2(T_p\alpha(\widehat{J}_s^{\mathrm{ord}}))$, which we have written as e_s.

We have the following fact (cf. [21, Corollary 4.4]).

Lemma 11.1 *We have $H^0(T_pB_r^{\mathrm{ord}}) = H^0(T\mathcal{G}) = 0$, where $T\mathcal{G} := \mathrm{Hom}_\Lambda(\Lambda^\vee, \mathcal{G}) \cong \varprojlim_s T_pJ_s^{\mathrm{ord}}$.*

Proof We only need to prove this for a finite field extension k of $\mathbb{Q}_l$ (as this implies the result for a number field) and T_pB_r (as we can take $B_r := J_r$, which implies the result for $T\mathcal{G}$). Write $B = B_r$. By replacing k be a finite field extension, we may assume that B has either good reduction or split multiplicative reduction over the valuation ring O of k with residue field $\mathbb{F}$. If B has good reduction over O and $l \neq p$, T_pB^{ord} is unramified at l. All the eigenvalues of the action of the l-Frobenius element ϕ are a Weil l-number of positive weight; so, we conclude

$$H^0(T_pB) \subset \mathrm{Ker}(\phi - 1 : T_pB \to T_pB) = 0,$$

and the assertion follows.

Modifying B by an isogeny does not affect the outcome; so, by doing this, we may assume that $\mathrm{End}(B_{/\mathbb{Q}})$ contains the integer ring O_B of the Hecke field. Suppose that $p = l$, and take a prime factor $\mathfrak{p}|p$ in O_B such that $T_pB^{\mathrm{ord}} - T_\mathfrak{p}B := \varprojlim_n B[\mathfrak{p}^n](\overline{\mathbb{Q}})$. Then $B[\mathfrak{p}^\infty]^{\mathrm{ord}}$ extends to an ordinary Barsotti–Tate group. If B does not have complex multiplication, by [33], the connected-étale exact sequence

$$0 \to B[\mathfrak{p}]^{\circ,\mathrm{ord}} \to B[\mathfrak{p}^\infty]^{\mathrm{ord}} \to B[\mathfrak{p}^\infty]^{\text{ét}} \to 0$$

is non-split as a $\mathrm{Gal}(\overline{\mathbb{Q}}_p/k)$–module; so, $H^0(T_\mathfrak{p}B^{\mathrm{ord}}) = 0$ again. If B has complex multiplication, by the Cartier duality, we have a Galois equivariant non-degenerate pairing

$$(T_pB[\mathfrak{p}^\infty]^{\text{ét}} \otimes_{\mathbb{Z}_p} \mathbb{Q}_p) \times (T_pB[\mathfrak{p}]^{\circ,\mathrm{ord}} \otimes_{\mathbb{Z}_p} \mathbb{Q}_p) \to \mathbb{Q}_p(1).$$

On $T_pB[\mathfrak{p}^\infty]^{\text{ét}} \otimes_{\mathbb{Z}_p} \mathbb{Q}_p$, again the eigenvalues of the action of the p-Frobenius element ϕ are Weil p-numbers of positive weight. This shows $H^0(T_pB[\mathfrak{p}^\infty]^{\text{ét}}) = 0$. By duality, $H^0(T_pB[\mathfrak{p}^\infty]^{\circ,\mathrm{ord}}) = 0$. Then from the exact sequence

$$0 \to T_pB[\mathfrak{p}^\infty]^{\circ,\mathrm{ord}} \to T_pB^{\mathrm{ord}} \to T_pB[\mathfrak{p}^\infty]^{\text{ét}} \to 0,$$

we conclude $H^0(T_pB^{\mathrm{ord}}) = 0$.

If B is split multiplicative over O, this fact is a well known result of Mumford–Tate [26]. □

By the above lemma, the map a in (36) is injective.

Lemma 11.2 *Let k be either a number field or a finite extension of $\mathbb{Q}_l$ for a prime l. Then the map b in the diagram* (36) *is injective, and if k is local with $l \neq p$, we have* $\mathrm{Im}(b) = \mathrm{Ker}(h) = 0$ *in* (36) *(so the right column is exact).*

Proof Applying the snake lemma to the first two rows of (36), we find that b is injective.

Suppose that k is local. For an abelian variety X over k with $X^t := \mathrm{Pic}^0_{X/k}$, $X^t(k)$ is isomorphic to $\mathbb{Z}_l^m$ times a finite group; so, if $l \neq p$, $\widehat{X}^t(k)$ is finite p-group. By [9, I.3.4], $H^1(k, X) \cong X^t(k)^\vee$; so, $H^1(k, \widehat{X})$ is a finite group. Therefore $H^1(k, \widehat{J}_s^{\mathrm{ord}})$ and $H^1(k, \widehat{B}_r^{\mathrm{ord}})$ are finite groups, and $T_pH^1(k, \widehat{J}_s^{\mathrm{ord}}) = T_pH^1(k, \widehat{B}_r^{\mathrm{ord}}) = 0$. Since b is injective, $T_pH^1(k, \alpha(\widehat{J_s^{\mathrm{ord}}})) = 0$; so, $\mathrm{Ker}(h) = \mathrm{Im}(b) = 0$. □

We note the following fact: If k is local non-archimedean, for an abelian variety A over k,

$$H^2(k, \widehat{A}) = H^2(k, A) = 0 \text{ for any abelian variety } A \text{ over } k. \tag{37}$$

This follows from [9, Theorem I.3.2], since $H^2(k, \widehat{A}) = H^2(k, A) \otimes_{\mathbb{Z}} \mathbb{Z}_p$.

Proposition 11.3 *If k be a finite extension of $\mathbb{Q}_l$ with $l \neq p$, then $E_2^s(k) = 0$.*

Proof Since the left column of (36) by Lemma 11.2 if $l \neq p$, applying the snake lemma to the middle two exact rows of (36), we find an exact sequence

$$0 \to E_2^s(k) \xrightarrow{e_s} \mathrm{Im}(\varpi_s) \to \mathrm{Coker}(h) \to 0. \tag{38}$$

This implies $E_2^s(k) \hookrightarrow \mathrm{Im}(\varpi_s)$.

Let $X_{/k}$ be a p-divisible Barsotti–Tate group. Then we have $H^2(k, T_pX) = \varprojlim_n H^2(k, X[p^n])$ (e.g., [12, 2.7.6]). By Tate duality (e.g., [7, Theorem 4.43]), we have $H^2(k, X[p^n]) \cong X^t[p^n](k)^\vee$ for the Cartier dual $X^t := \mathrm{Hom}(T_pX, \mu_{p^\infty})$ of X. Thus we have

$$H^2(k, T_pX) = \varprojlim_n (X^t[p^n](k)^\vee) \cong (\varinjlim_n H^0(k, X^t[p^n]))^\vee,$$

since we have a canonical pairing $X[p^n] \times X^t[p^n] \to \mu_{p^n}$ (i.e., $X^t[p^n](k)^\vee \cong X[p^n](-1)(k)$).

Apply this to the complement X of $\widehat{A}_s[p^\infty]^{\mathrm{ord}}$ in $J_s[p^\infty]^{\mathrm{ord}}$; so, $X + A_s[p^\infty]^{\mathrm{ord}} = J_s[p^\infty]^{\mathrm{ord}}$ with finite $X \cap A_s[p^\infty]^{\mathrm{ord}}$. Requiring X to be stable under $\mathbf{h}_s$, for $\mathbf{h}_s(\mathbb{Q}_p) = \mathbf{h}_s \otimes_{\mathbb{Z}_p} \mathbb{Q}_p$, X is uniquely determined as $\mathbf{h}_s(\mathbb{Q}_p) = (\mathbf{h}_s(\mathbb{Q}_p)/\alpha_s\mathbf{h}_s(\mathbb{Q}_p)) \oplus 1_s\mathbf{h}_s(\mathbb{Q}_p)$ for an idempotent 1_s (so, $X = 1_sJ_s[p^\infty]^{\mathrm{ord}}$). By local Tate duality, we get $H^2(k, T_pX) \cong H^0(k, X[p^\infty]^t)^\vee$ and conclude

$$H^2(k, T_pX) \cong \varinjlim_n H^0(k, \mathrm{Hom}(X[p^n](\overline{k}), \mu_{p^n}(\overline{k})))$$
$$= \varinjlim_n X[p^n](-1)(k) = X[p^\infty](-1)(k).$$

Thus we conclude the injectivity:

$$H^2(k, T_pX) \cong X[p^\infty](-1)(k) \overset{\hookrightarrow}{\longrightarrow} J_s[p^\infty]^{\mathrm{ord}}(-1)(k) \cong H^2(k, T_pJ_s)^{\mathrm{ord}},$$

which is injective as $X \subset J_s[p^\infty]^{\mathrm{ord}}$. By definition, we have $X + A_s[p^\infty]^{\mathrm{ord}} = J_s[p^\infty]^{\mathrm{ord}}$. By the assumption (A) and the definition of X, $X = \alpha_s(J_s[p^\infty]^{\mathrm{ord}})$. Therefore we get an injection:

$$H^2(k, T_p\alpha(\widehat{J}_s^{\mathrm{ord}})) \cong H^2(k, T_p\alpha(J_s[p^\infty]^{\mathrm{ord}}))$$
$$\cong \alpha(J_s[p^\infty]^{\mathrm{ord}})(-1)(k) \underset{\hookrightarrow}{\overset{a_2}{\longrightarrow}} J_s[p^\infty]^{\mathrm{ord}}(-1)(k) \cong H^2(k, T_pJ_s)^{\mathrm{ord}}.$$

We have an exact sequence

$$H^1(k, T_p\alpha(\widehat{J}_s^{\mathrm{ord}})) \xrightarrow{\varpi_s} H^2(k, T_pA_r)^{\mathrm{ord}} \xrightarrow{a_2} H^2(k, T_pJ_s)^{\mathrm{ord}}.$$

Since a_2 is injective, we find $\mathrm{Im}(\varpi_s) = 0$; so, $E_2^s(k) = 0$ if k is l-adic with $l \neq p$. □

Here are some remarks what happens when $l = p$ for the local error terms. For simplicity, we assume that $k = \mathbb{Q}_p$; so, $W_s = \mathbb{Z}_p[\mu_{p^s}]$. For $l \neq p$, the proof of the above proposition is an argument purely of characteristic 0. In [22, §17], we studied the error term of the control of inductive limit $J_\infty^{\mathrm{ord}}(\mathbb{Q}_p) := \varinjlim_s \widehat{J}_s^{\mathrm{ord}}(\mathbb{Q}_p)$ using a result of P. Schneider [29, 30] on universal norm for abelian varieties over ramified $\mathbb{Z}_p$-extension. It works well for the inductive limit $J_\infty^{\mathrm{ord}}(\mathbb{Q}_p)$ but perhaps not for the projective limit $\widehat{J}_\infty^{\mathrm{ord}}(\mathbb{Q}_p)$ for the following reason.

This involves study of integral models of the abelian variety (in particular, its formal Lie group over W_∞). Let I_r (resp. $X_{r,0}$) be the Igusa tower of level p^r over $X_0 := X_1(N) \otimes_{\mathbb{Z}_p} \mathbb{F}_p$ containing the zero cusp (resp. the infinity cusp). Then for $P \in \Omega_{\mathbb{T}}$, if the conductor of B_P is divisible by p^r with $r > 0$, $B_P \times_{W_r} \mathbb{F}_p$ is the quotient of $\mathrm{Pic}^0_{I_r/\mathbb{F}_p} \times \mathrm{Pic}^0_{X_{r,0}/\mathbb{F}_p}$ (cf. [8, Chap. 14] or [20, §6]). On $\mathrm{Pic}^0_{I_r/\mathbb{F}_p} \times \mathrm{Pic}^0_{X_{r,0}/\mathbb{F}_p}$, $U(p)$ and $U^*(p)$ acts in a matrix form with respect to the two factors $\mathrm{Pic}^0_{I_r/\mathbb{F}_p}$ and $\mathrm{Pic}^0_{X_{r,0}/\mathbb{F}_p}$ in this order

$$U(p) = \begin{pmatrix} F & * \\ 0 & V\langle p^{(p)}\rangle \end{pmatrix} \text{ and } U^*(p) = \begin{pmatrix} V\langle p^{(p)}\rangle & 0 \\ * & F \end{pmatrix}, \tag{39}$$

where $\langle p^{(p)}\rangle$ is the diamond operator of the class of p modulo N. See [24, §3.3] or [20, (6-1)] for this formula. Since the shoulder term $*$ of the above matrix form of $U(p)$ vanishes once restricted to B_P if $r > 0$, from (39), $\widehat{B}_P^{\mathrm{ord}}(\mathbb{F})$ must be the quotient of $\mathrm{Pic}^0_{I_r/\mathbb{F}_p}$, and the ordinary part of the formal Lie group $\widehat{B}^{\circ,\mathrm{ord}}_{P/\mathbb{F}_p}$ of $\widehat{B}_P$ has to be the quotient of $\mathrm{Pic}^0_{X_{r,0}/\mathbb{F}_p}$. Similarly, $\widehat{B}^{\text{co-ord}}_P(\mathbb{F})$ must be the quotient of $\mathrm{Pic}^0_{X_{r,0}/\mathbb{F}_p}$, and the co-ordinary part of the formal Lie group $\widehat{B}^{\circ,\text{co-ord}}_{P/\mathbb{F}_p}$ of $\widehat{B}_P$ has to be the quotient of $\mathrm{Pic}^0_{I_r/\mathbb{F}_p}$.

Write B_s for the quotient of J_s corresponding to B_P. We consider the exact sequence defining $E^s_2(\mathbb{Q}_p)$:

$$0 \to \alpha(\widehat{J}^{\mathrm{ord}}_{s,\mathbb{T}})(\mathbb{Q}_p) \to \widehat{J}^{\mathrm{ord}}_{s,\mathbb{T}}(\mathbb{Q}_p) \xrightarrow{\rho_s} \widehat{B}^{\mathrm{ord}}_s(\mathbb{Q}_p) \to E^s_2(\mathbb{Q}_p) \to 0,$$

which is equivalent to, by the involution w_s over characteristic 0 field, the following exact sequence

$$0 \to \alpha^*(\widehat{J}^{\text{co-ord}}_{s,\mathbb{T}})(\mathbb{Q}_p) \to \widehat{J}^{\text{co-ord}}_{s,\mathbb{T}}(\mathbb{Q}_p) \xrightarrow{\rho^*_s} {}^t\widehat{A}^{\text{co-ord}}_s(\mathbb{Q}_p) \to \mathcal{E}^s_2(\mathbb{Q}_p) \to 0.$$

Thus we study the second exact sequence of the co-ordinary parts. Here we have used the self duality of J_s, ${}^tA_s$ is the dual abelian variety of A_s and α^* is the image of α under the Rosati involution.

Consider the complex of Néron models over W_s:

$$0 \to \alpha^*(J_s) \to J_s \to {}^tA_s \to 0$$

and its formal completion along the identity

$$0 \to \alpha^*(J^\circ_s) \to J^\circ_s \to {}^tA^\circ_s \to 0.$$

Here X° is the formal group of an abelian variety $X_{/W_s}$. These sequence might not be exact as $W_s/\mathbb{Z}_p$ is highly ramified at p (see [1, §7.5]). But just to go forward, we assume the sequence of the co-ordinary parts of the formal Lie groups are exact (and still we find some difficulties).

As explained in [22, (17.3)], taking the $\mathbb{T}^*$-component (the image of $\mathbb{T}$ under the Rosati involution), the complex

$$0 \to \alpha^*(\widehat{J}^\circ_{s,\mathbb{T}^*}) \to \widehat{J}^\circ_{s,\mathbb{T}^*} \to {}^t\widehat{A}^\circ_{s,\mathbb{T}^*} \to 0 \tag{40}$$

is, by our assumption, an exact sequence of formal Lie groups over W_s; so, the top complex of the following commutative diagram is a short exact sequence:

$$\begin{array}{ccccc}
\alpha^*(\widehat{J}^\circ_{s,\mathbb{T}^*})(W_s) & \xrightarrow{\hookrightarrow} & \widehat{J}^\circ_{s,\mathbb{T}^*}(W_s) & \xrightarrow{\twoheadrightarrow} & {}^t\widehat{A}^\circ_{s,\mathbb{T}^*}(W_s) \\
\downarrow{\scriptstyle N_{\alpha^*(J_s)}} & & \downarrow{\scriptstyle N_{J_s}} & & \downarrow{\scriptstyle N_s} \\
\alpha^*(\widehat{J}^\circ_{s,\mathbb{T}^*})(W_s)^{\mathrm{Gal}(k_s/k_r)} & \xrightarrow{\hookrightarrow} & \widehat{J}^\circ_{s,\mathbb{T}^*}(W_s)^{\mathrm{Gal}(k_s/k_r)} & \xrightarrow{\rho^*_s} & {}^t\widehat{A}^\circ_{r,\mathbb{T}^*}(W_r),
\end{array}$$

where $N_{X,s}$ is the norm map relative to k_s/k_r of an abelian variety X defined over k_r. By Schneider [30], N_s is almost onto with the index of the image bounded independent of s. However, we do not know yet ρ_s^* is surjective up to finite bounded error for the following reason:

Though ${}^t\widehat{A}_s^{\text{co-ord}} \cong {}^t\widehat{A}_r^{\text{co-ord}}$ *because* $\widehat{A}_r^{\text{ord}} \cong \widehat{A}_s^{\text{ord}}$ *as seen in* [22], *the projection map*

$${}^t\widehat{A}_s^{\circ,\text{co-ord}}(W_s) \to {}^t\widehat{A}_r^{\circ,\text{co-ord}}(W_s)$$

is not an isomorphism. After reducing modulo p, *as already remarked, the formal Lie group of* ${}^t\widehat{A}_s^{\text{co-ord}}$ *is in the identity connected component of* $\mathrm{Pic}^0_{X_{0,s}}$. *Note that* $I_s = X_{s,0}^{(p^s)}$ *(the* p^s*-power Frobenius twist); so, the projection* $I_s \to I_r$ *is given by* $F^{s-r} \circ \pi$ *for the projection* $\pi : X_{s,0} \to X_{r,0}$ ([8, Theorem 13.11.4 (1)] *or* [20, §6]) *which is purely inseparable. This shows that* $\pi_* : {}^t\widehat{A}_s^{\text{co-ord}} \to {}^t\widehat{A}_r^{\text{co-ord}}$ *is not an isomorphism.*

Thus we have two problems for proving bounded-ness of $\mathcal{E}_2^s(\mathbb{Q}_p)$ (and hence of $E_2^s(\mathbb{Q}_p)$)

1. (40) may not be exact;
2. the projection ${}^t\widehat{A}_{s/\mathbb{F}_p}^{\text{co-ord}} \to {}^t\widehat{A}_{r/\mathbb{F}_p}^{\text{co-ord}}$ is purely inseparable of degree $\geq p^{s-r}$ (as the polarization of A_s has degree of high p-power $\geq p^s$).

These problems do not appear for the pull-back map $\widehat{A}_r^{\text{ord}} \xrightarrow{\sim} \widehat{A}_s^{\text{ord}}$ even over $\mathbb{F}_p$ as the exactness of $\widehat{A}_r^{\text{ord}} \hookrightarrow \widehat{J}_s^{\text{ord}} \twoheadrightarrow \alpha(\widehat{J}_s^{\text{ord}})$ is proven by the control of the Λ-adic BT group $\mathcal{G}$ in [22, §17] and the projection $X_{s,0} \to X_{r,0}$ is étale outside the supersingular points for all s.

References

Books

1. Bosch, S., Lütkebohmert, W., Raynaud, M.: Néron Models. Springer, New York (1990)
2. Faltings, G., Wüstholtz, G., et al.: Rational Points, Aspects of Mathematics E6. Friedr. Vieweg & Sohn, Braunschweig (1992)
3. Hartshorne, R.: Algebraic Geometry, Graduate Texts in Mathematics 52. Springer, New York (1977)
4. Hida, H.: Elliptic Curves and Arithmetic Invariants. Springer Monographs in Mathematics, Springer, New York (2013)
5. Hida, H.: Geometric Modular Forms and Elliptic Curves, 2nd edn. World Scientific, Singapore (2012)
6. Hida, H.: Hilbert modular forms and Iwasawa theory. Oxford University Press (2006)
7. Hida, H.: Modular Forms and Galois Cohomology, Cambridge Studies in Advanced Mathematics 69. Cambridge University Press, Cambridge, England (2000)
8. Katz, N.M., Mazur, B.: Arithmetic moduli of elliptic curves. Ann. Math. Stud. 108, Princeton University Press, Princeton, NJ (1985)
9. Milne, J.S.: Arithmetic Duality Theorems, 2nd edn. BookSurge, LLC (2006)
10. Milne, J.S.: Étale Cohomology. Princeton University Press, Princeton, NJ (1980)
11. Miyake, T.: Modular Forms. Springer, New York-Tokyo (1989)

12. Neukirch, J., Schmidt, A., Wingberg, K.: Cohomology of Number Fields. Springer Springer Grundlehren der mathematischen Wissenschaften, 323 (2000)
13. Shimura, G.: Introduction to the Arithmetic Theory of Automorphic Functions. Princeton University Press, Princeton, NJ, and Iwanami Shoten, Tokyo (1971)

Articles

14. Fischman, A.: On the image of Λ-adic Galois representations. Ann. Inst. Fourier (Grenoble) **52**, 351–378 (2002)
15. Ghate, E., Kumar, N.: Control theorems for ordinary 2-adic families of modular forms, Automorphic representations and L-functions, 231–261, Tata Inst. Fundam. Res. Stud. Math., 22, Tata Inst. Fund. Res., Mumbai (2013)
16. Hida, H.: Iwasawa modules attached to congruences of cusp forms. Ann. Sci. Ec. Norm. Sup. 4th series 19, 231–273 (1986)
17. Hida, H.: Galois representations into $GL_2(\mathbb{Z}_p[[X]]$ attached to ordinary cusp forms. Inventiones Math. **85**, 545–613 (1986)
18. Hida, H.: Λ-adic p-divisible groups, I, II. Two lecture notes of a series of talks at the Centre de Recherches Mathématiques, Montréal (2005). http://www.math.ucla.edu/hida/CRMpaper.pdf
19. Hida, H.: Local indecomposability of Tate modules of non CM abelian varieties with real multiplication. J. Amer. Math. Soc. **26**, 853–877 (2013)
20. Hida, H.: Λ-adic Barsotti-Tate groups. Pacific J. Math. **268**, 283–312 (2014)
21. Hida, H.: Limit modular Mordell-Weil groups and their p-adic closure, Documenta Math., Extra Volume: Alexander S. Merkurjev's Sixtieth Birthday, pp. 221–264 (2015)
22. Hida, H.: Analytic variation of Tate-Shafarevich groups (preprint, 2016), 51 p. http://www.math.ucla.edu/hida/LTS.pdf
23. Mattuck, A.: Abelian varieties over p-adic ground fields. Ann. Math. **62**(2), 92–119 (1955)
24. Mazur, B., Wiles, A.: Class fields of abelian extensions of **Q**. Inventiones Math. **76**, 179–330 (1984)
25. Mazur, B., Wiles, A.: On p-adic analytic families of Galois representations. Compositio Math. **59**, 231–264 (1986)
26. Mumford, D.: An analytic construction of degenerating abelian varieties over complete rings. Compositio Math. **24**, 239–272 (1972)
27. Nekovář, J.: Selmer complexes. Astérisque **310**, viii+559 (2006)
28. Ohta, M.: Ordinary p-adic étale cohomology groups attached to towers of elliptic modular curves. Compositio Math. **115**, 241–301 (1999)
29. Schneider, P.: Iwasawa L-functions of varieties over algebraic number fields. A first approach. Invent. Math. **71**, 251–293 (1983)
30. Schneider, P.: Arithmetic of formal groups and applications. I. Universal norm subgroups. Invent. Math. **87**, 587–602 (1987)
31. Shimura, G.: On the factors of the jacobian variety of a modular function field. J. Math. Soc. Japan **25**, 523–544 (1973)
32. Tate, J.: p-divisible groups. In: Proceedings of Conference on local filds, Driebergen 1966, Springer, pp. 158–183 (1967)
33. Zhao, B.: Local indecomposability of Hilbert modular Galois representations. Annales de l'institut Fourier 64, 1521–1560 (2014)

Some Congruences for Non-CM Elliptic Curves

Mahesh Kakde

आतां कोठे धांवे मन ।
तुझे चरण देखिलिया ॥
(Sant Tukaram)

Abstract Let p be an odd prime and let G be a p-adic Lie group. The group $K_1(\Lambda(G))$, for the Iwasawa algebra $\Lambda(G)$, is well understood in terms of congruences between elements of Iwasawa algebras of abelian sub-quotients of G due to the work of Ritter-Weiss and Kato (generalised by the author). In the former one needs to work with all abelian subquotients of G whereas in Kato's approach one can work with a certain well-chosen sub-class of abelian sub-quotients of G. For instance in [11] $K_1(\Lambda(G))$ was computed for meta-abelian pro-p groups G but the congruences in this description could only be proved for p-adic L-functions of totally real fields for certain special meta-abelian pro-p groups. By changing the class of abelian subquotients a different description of $K_1(\Lambda(G))$, for a general G, was obtained in [12] and these congruences were proven for p-adic L-functions of totally real fields in all cases. In this note we propose a strategy to get an alternate description of $K_1(\Lambda(G))$ when $G = GL_2(\mathbb{Z}_p)$. For this it is sufficient to compute $K_1(\mathbb{Z}_p[GL_2(\mathbb{Z}/p^n)])$. We demonstrate how the strategy should work by explicitly computing $K_1(\mathbb{Z}_p[GL_1(\mathbb{Z}/p)])_{(p)}$, the pro-$p$ part of $K_1(\mathbb{Z}_p[GL_2(\mathbb{Z}/p)])$, which is the most interesting part.

Keywords Iwasawa theory · Whitehead groups · Group rings

To John Coates on his 70th birthday.
The author is supported by EPSRC First Grant EP/L021986/1.

M. Kakde (✉)
Department of Mathematics, King's College London, Strand Building Strand, London WC2R 2LS, UK
e-mail: mahesh.kakde@kcl.ac.uk

D. Loeffler and S.L. Zerbes (eds.), *Elliptic Curves, Modular Forms and Iwasawa Theory*, Springer Proceedings in Mathematics & Statistics 188, DOI 10.1007/978-3-319-45032-2_8

1 Introduction

We begin with a discussion of the main conjecture for non-CM elliptic curves. Let E be an elliptic curve defined over $\mathbb{Q}$. Assume that E does not admit complex multiplication. Let p be a prime and put $E[p^n]$ for the set of p^n-torsion points of E (over $\overline{\mathbb{Q}}$). Let $F_\infty := \mathbb{Q}(E[p^\infty]) := \bigcup_{n\geqslant 1} \mathbb{Q}(E[p^n])$. By a famous theorem of Serre [17] $\mathrm{Gal}(F_\infty/\mathbb{Q})$ is an open subgroup of $GL_2(\mathbb{Z}_p)$ and is in fact equal to $GL_2(\mathbb{Z}_p)$ for almost all p. From now on fix a prime $p > 3$ (so that $GL_2(\mathbb{Z}_p)$ does not have an element of order p) such that $G := \mathrm{Gal}(F_\infty/\mathbb{Q}) \cong GL_2(\mathbb{Z}_p)$ and such that E has good ordinary reduction at p. Let μ_n be the group of nth roots of 1. Put $\mathbb{Q}_\infty := \mathbb{Q}(\mu_{p^\infty}) := \bigcup_{n\geqslant 1} \mathbb{Q}(\mu_{p^n})$. By the Weil pairing $\mathbb{Q}_\infty$ is contained in F_∞. Let $\mathbb{Q}_{cyc}$ be the unique extension of $\mathbb{Q}$ (contained in $\mathbb{Q}_\infty$) such that $\Gamma := \mathrm{Gal}(\mathbb{Q}_{cyc}/\mathbb{Q})$ is isomorphic to the additive group of p-adic integers $\mathbb{Z}_p$. We put $H := \mathrm{Gal}(F_\infty/\mathbb{Q}_{cyc})$.

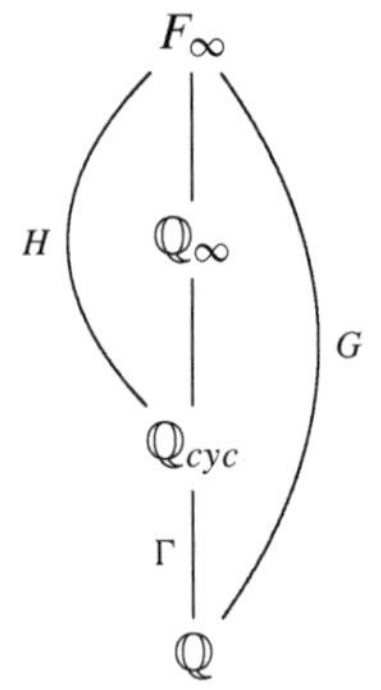

For a profinite group $P = \varprojlim_i P_i$, we put $\Lambda(P) := \varprojlim_i \mathbb{Z}_p[P_i]$ and $\Omega(P) := \varprojlim_i \mathbb{F}_p[P_i]$. Following [5] we put

$$S := \{f \in \Lambda(G) : \Lambda(G)/\Lambda(G)f \text{ is a f.g. } \Lambda(H)\text{-module}\}.$$

Put $S^* := \bigcup_{n\geqslant 0} p^n S$. By [5, Theorem 2.4] S and S^* are multiplicatively closed subsets of $\Lambda(G)$, do not contain any zero-divisors and satisfy the Ore condition. Hence we get localisations $\Lambda(G)_S$ and $\Lambda(G)_{S^*}$ of $\Lambda(G)$. Put $\mathfrak{M}_H(G)$ for the category of finitely generated S^*-torsion $\Lambda(G)$-modules i.e. all finitely generated $\Lambda(G)$-modules M such that $\Lambda(G)_{S^*} \otimes_{\Lambda(G)} M = 0$. Following [5, Sect. 5], for any algebraic extension L of $\mathbb{Q}$, we define the Selmer group

$$\mathcal{S}(E/L) := \mathrm{Ker}\left(H^1(L, E[p^\infty]) \to \prod_w H^1(L_w, E(\overline{L}_w))\right),$$

where w runs through all non-archimedian places of L and L_w denotes the union of completions at w of all finite extensions of $\mathbb{Q}$ contained in L. Also put

$$X(E/L) := \mathrm{Hom}(\mathcal{S}(E/L), \mathbb{Q}_p/\mathbb{Z}_p)$$

for the Pontrjagin dual.

Conjecture 1 (Conjecture 5.1 *[5]) The $\Lambda(G)$-module $X(E/F_\infty)$ lies in the category $\mathfrak{M}_H(G)$.*

Every continuous homomorphism $\rho : G \to GL_n(O)$, where O is the valuation ring in a finite extension L of $\mathbb{Q}_p$, induces a map (see [5, Eq. (22)])

$$K_1(\Lambda(G)_{S^*}) \to L \cup \{\infty\}, \tag{1}$$

$$x \mapsto x(\rho).$$

(Classically, this would be denoted as $\int_G \rho dx$).

Conjecture 2 (Conjecture 5.7 *[5]) There is a finite unramified extension A of $\mathbb{Z}_p$ and an element $\mathcal{L}_E \in K_1(\Lambda(G)_{S^*} \otimes_{\mathbb{Z}_p} A)$ such that, for all Artin representation ρ of G we have*

$$\mathcal{L}_E(\rho) = \frac{L_\Sigma(E, \hat{\rho}, 1)}{\Omega_+(E)^{d^+(\rho)}\Omega_-(E)^{d^-(\rho)}} \cdot e_p(\rho) \cdot \frac{P_p(\rho, u^{-1})}{P_p(\hat{\rho}, w^{-1})} \cdot u^{-f_\rho}.$$

(see [5, Sect. 5] for all unexplained notation and the paragraph before Proposition 7.5 [3] for correction to loc. cit. We thank the referee for pointing this out).

Recall the following part of the localisation sequence of K-theory

$$K_1(\Lambda(G)) \to K_1(\Lambda(G)_{S^*}) \xrightarrow{\partial} K_0(\mathfrak{M}_H(G)) \to 1.$$

The surjection of ∂ is shown in [5, Proposition 3.4]. Assuming Conjecture 1 a characteristic element of $X(E/F_\infty)$ is defined as any element of $K_1(\Lambda(G)_{S^*})$ that maps to the class of $X(E/F_\infty)$ in $K_0(\mathfrak{M}_H(G))$.

Conjecture 3 (Conjecture 5.8 *[5]) Assume Conjectures 1 and 2. Let ξ_E be a characteristic element of $X(E/F_\infty)$. Then the image of ξ_E in*

$$\frac{K_1(\Lambda(G)_{S^*} \otimes_{\mathbb{Z}_p} A)}{\textit{Image of } K_1(\Lambda(G) \otimes_{\mathbb{Z}_p} A)}$$

coincides with the image of $\mathcal{L}_E$.

For simplicity we assume that the ring A in the above Conjecture 2 is $\mathbb{Z}_p$ as we discuss the strategy for attacking the above conjecture. This strategy is due to Burns and Kato (see [1]) and is well-known by now. Put $S(G)$ for the set of all open subgroups of G. For every U in $S(G)$, there is a map

$$\theta_U : K_1(\Lambda(G)) \to K_1(\Lambda(U^{\mathrm{ab}})) \cong \Lambda(U^{\mathrm{ab}})^{\times}$$

defined as composition of the norm map $K_1(\Lambda(G)) \to K_1(\Lambda(U))$ and the map induced by natural project $\Lambda(U) \to \Lambda(U^{\mathrm{ab}})$. Hence we get a map

$$\theta := \prod_{U \in S(G)} \theta_U : K_1(\Lambda(G)) \to \prod_{U \in S(G)} \Lambda(U^{\mathrm{ab}})^{\times}$$

Similarly, there are maps

$$\theta_S : K_1(\Lambda(G)_S) \to \prod_{U \in S(G)} \Lambda(U^{\mathrm{ab}})_S^{\times},$$

$$\theta_{S^*} : K_1(\Lambda(G)_{S^*}) \to \prod_{U \in S(G)} \Lambda(U^{\mathrm{ab}})_{S^*}^{\times}.$$

As G has no element of order p by "Weierstrass preparation theorem" [3, Proposition 3.4] there is an isomorphism

$$K_1(\Lambda(G)_{S^*}) \cong K_1(\Lambda(G)_S) \oplus K_0(\Omega(G)).$$

We have the map

$$\theta_0 : K_0(\Omega(G)) \to \prod_{U \in S(G)} K_0(\Omega(U^{\mathrm{ab}}))$$

that fits into the following commutative diagram

$$\begin{array}{ccc} K_1(\Lambda(G)_{S^*}) & \xrightarrow{\cong} & K_1(\Lambda(G)_S) \oplus K_0(\Omega(G)) \\ \downarrow{\scriptstyle \theta_{S^*}} & & \downarrow{\scriptstyle (\theta_S, \theta_0)} \\ \prod_{U \in S(G)} \Lambda(U^{\mathrm{ab}})_{S^*}^{\times} & \hookleftarrow & \prod_{U \in S(G)} \Lambda(U^{\mathrm{ab}})_S^{\times} \oplus K_0(\Omega(U^{\mathrm{ab}})). \end{array}$$

Here, we abuse the notation by denoting Ore sets for U^{ab} by the same symbols S and S^*. The injection in lower row is not surjective in general. Nevertheless, an element $(x_U) \in \prod_{U \in S(G)} \Lambda(U^{\mathrm{ab}})_{S^*}^{\times}$ lies in the image of θ_{S^*} if and only if it "factorises" as $x_U = (\overline{x_U}, \mu_U)$ and $(\overline{x}_U)$ and (μ_U) lie in the images of θ_S and θ_0 respectively.

Let P be a pro-p normal subgroup of G. Put $G_0(\mathbb{F}_p[G/P])$ for the Grothendieck group of the category of finitely generated $\mathbb{F}_p[G/P]$-modules. The group is isomorphic to the group Brauer characters of G/P by [8, Proposition 17.14]. The group is also independent of the choice of the pro-p normal subgroup P (by [8, Proposition 17.16(i)]. The group $K_0(\Omega(G))$ is a Green module over the Green ring $G_0(\mathbb{F}_p[G/P])$ (this is [18, 17.1(*)(c)]. For the notion of Green rings and Green modules see [15, Chap. 11]). Moreover, there is a Cartan homomorphism [18, 15.1]

$$c : K_0(\Omega(G)) \to G_0(\mathbb{F}_p[G/P])$$

which is injective by [8, Lemma 18.22(ii)] (see also [3, 3.4.2]). Hence

$$x \in K_0(\Omega(G)) \text{ is } 0 \iff \chi(x) = 0 \; \forall \text{ Brauer characters } \chi \text{ of } G \tag{2}$$

(note that χ induces a map $\chi : K_0(\Omega(G)) \to K_0(\overline{\mathbb{F}}_p) = \mathbb{Z}$). The following lemma is a generalisation of the the 'only if' part of [7, Theorem 3.8].

Lemma 4 *The map θ_0 is injective and its image consists of all tuples (μ_U) such that for any finite collection $\{U_i\} \subset S(G)$, if there are mod p representation χ_i of U_i^{ab} and integers n_i such that $\sum_i n_i Ind_{U_i^{ab}}^G(\chi_i) = 0$, then*

$$\sum_i n_i \chi_i(\mu_{U_i}) = 0.$$

Proof There is a canonical inflation map $K_0(\Omega(U^{ab})) \to K_0(\Omega(U))$. By the Brauer induction theorem [8, Theorem 21.15] there are finitely many open subgroups U_i of G and Brauer characters χ_i of U_i^{ab} such that the trivial character 1 of G is given by

$$1 = \sum_i Ind_{U_i}^G \chi_i,$$

where we use the same symbol χ_i for the character of U_i obtained by inflating χ_i. For any tuple $(\mu_U)_U$ satisfying the condition given in the statement of the lemma define an element $\mu \in K_0(\Omega(G))$ as follows—firstly, by abuse of notation, let μ_U denote the inflation of μ_U to $K_0(\Omega(U)$. Let $i_U : K_0(\Omega(U)) \to K_0(\Omega(G))$ be the map induced by the inclusion $U \hookrightarrow G$. Define

$$\mu := \sum_i i_{U_i}(\chi_i \cdot \mu_{U_i}).$$

Here χ_i is considered as an element of $G_0(\mathbb{F}_p[U_i/P_i])$ for a suitable (i.e. χ_i is trivial on P_i) open normal pro-p subgroup P_i of U_i and $\chi_i \cdot \mu_{U_i} \in K_0(\Omega(U_i))$ is obtained by the action of $G_0(\mathbb{F}_p[U_i/P_i])$ on $K_0(\Omega(U_i))$. The element μ is independent of the choice of U_i's and χ_i's because the hypothesis on $(\mu_U)_U$ and (2). It is easy to check that this gives the inverse of θ_0 for the claimed image. □

Remark 5 The above discussion reveals two surprising consequences of S^*-torsion conjecture. Firstly, as observed in [1], the p-adic L-function in $\Lambda(U^{ab})^\times_{S^*}$ should factorise canonically into an element in $\Lambda(U^{ab})^\times_S$ and a "μ-invariant" part (even though there is no Weierstrass preparation theorem for $\Lambda(U^{ab})^\times_{S^*}$ in general). Secondly, both these parts of the p-adic L-functions should satisfy "Artin formalism" (one part satisfies Artin formalism if and only if the other does since the p-adic L-function satisfies Artin formalism).

Remark 6 Coates-Sujatha [7] proved that for $P \cong \mathbb{Z}_p \times \mathbb{Z}_p$, the S^*-torsion conjecture is equivalent to Artin formalism for μ-invariant parts of p-adic L-functions in $\Lambda(\mathbb{Z}_p \times p^n\mathbb{Z}_p)$ (in this case this is equivalent to growth of μ-invariant in a certain way). However, in general, Artin formalism for μ-invariants seems to be weaker than S^*-torsion conjecture.

In any case, to prove above conjectures at present it seems necessary to study the maps $\theta, \theta_S, \theta_{S^*}, \theta_0$ and compute their images explicitly. There are two ways to do this. One is due to Ritter-Weiss [16, 19] and other is due to Kato, generalised by the author [12–14]. The first one is very elegant to state and can be easily stated for a general p-adic Lie group (as opposed to one dimension pro-p groups treated in [16, 19]) as will be shown in [2]. However, in this approach it is necessary to use all open subgroups of G. The advantage of Kato's approach is that in special situations one can restrict to smaller classes of open subgroups of G. In this short note we propose to begin the study of $K_1(\Lambda(G))$ using a smaller class of open subgroups of G. By [9, Proposition 1.5.1] $K_1(\Lambda(G)) \cong \varprojlim_n K_1(\mathbb{Z}_p[GL_2(\mathbb{Z}/p^n)])$. Hence it is enough to compute $K_1(\mathbb{Z}_p[GL_2(\mathbb{Z}/p^n)])$. The strategy to compute $K_1(\mathbb{Z}_p[P])$ for a finite group P can roughly be stated as follows: Let $S(P)$ be a fixed class of subgroups of P. Then we have a map

$$\theta : K_1(\mathbb{Z}_p[P]) \to \prod_{U \in S(P)} \mathbb{Z}_p[U^{\mathrm{ab}}]^\times.$$

There is an additive analogue of this map

$$\psi : \mathbb{Z}_p[\mathrm{Conj}(P)] \to \prod_{U \in S(P)} \mathbb{Z}_p[U^{\mathrm{ab}}],$$

where $\mathrm{Conj}(P)$ is the set of conjugacy classes of P. For the definition of ψ see next section. Precise relation between θ and ψ via integral logarithm is often hard to describe. In short there are two ingredients to compute $K_1(\mathbb{Z}_p[P])$

(1) Explicit knowledge of $\mathrm{Conj}(P)$ so that the map ψ can be described explicitly.
(2) Explicit relation between θ and ψ via integral logarithm. Here we again need to know $\mathrm{Conj}(P)$ explicitly.

In this short note we carry out this strategy for $n = 1$. For $n > 1$ a similar computation (with a lot more blood, sweat and tears but essentially with no new ideas) should go through and will be carried out in future. In the end the congruences turn out to be rather trivial for $n = 1$ but that is expected as p-Sylow subgroups of $GL_2(\mathbb{Z}/p\mathbb{Z})$ are cyclic groups of order p. However, we hope the calculations in this simple case are illuminating and indicate how the general case will proceed.

There are several interesting aspects of this construction. For example, it is already seen from computations in meta-abelian case in [11] and in general in [12], that the shape of the congruences can be very different if one chooses different class of subgroups. It seems hard to pass from the congruences in [11] to the congruences in [12] directly in cases when they both apply. As [12] shows having an alternate description can be useful to prove the congruences. Secondly, we hope that open subgroups of G that, modulo p^m, are one of the "standard subgroups" (Borel, Cartan, Centre etc.) will suffice. These are closer to the theory of automorphic representations. Therefore one may hope that the theory of automorphic representations and automorphic forms weighs in significantly in understanding these congruences. This will only be clear after the congruences are explicitly written down in general. Lastly, in the case of elliptic curves admitting CM, the main conjecture for symmetric power representations attached to the curve has been deduced from the two variable main conjecture by Coates-Schmidt [6]. It was mentioned to the author by John Coates that a similar deduction should be possible for non-CM elliptic curves using the main conjecture stated above. This seems to be an extremely hard problem. Specially because in the main conjecture above we allow evaluation of p-adic L-function only at Artin representations. Hence one needs to understand reduction modulo powers of p of Artin representations and symmetric power representations. This study is implicit in the calculations proposed here. L-functions of symmetric power representations are extremely hard to study and the author does not claim that the calculations proposed here would provide a way to do this.

Remark 7 Throughout the paper we restrict to $\mathbb{Z}_p$ coefficients, however, the same computation should work for a more general class of coefficient rings for which integral logarithm has been constructed (for example rings considered in [4]). However, all the results we need are not stated or proven in this generality yet and a satisfactory discussion of these will take us too far off our modest goal.

2 An Additive Result

From now onwards G denotes $GL_2(\mathbb{F}_p)$. Let us denote the set of conjugacy classes of G by $\mathrm{Conj}(G)$. Fix a non-square element ϵ in $\mathbb{F}_p$. It is well-known that the conjugacy classes of G are (for example see [10])

$$i_a := \begin{pmatrix} a & 0 \\ 0 & a \end{pmatrix} \text{ for } a \in \mathbb{F}_p^\times.$$

$$c_{a,1} := \begin{pmatrix} a & 1 \\ 0 & a \end{pmatrix} \text{ for } a \in \mathbb{F}_p^\times.$$

$$t_{a,d} := \begin{pmatrix} a & 0 \\ 0 & d \end{pmatrix} \text{ for } a \neq d \in \mathbb{F}_p^\times.$$

$$k_{a,b} := \begin{pmatrix} a & \epsilon b \\ b & a \end{pmatrix} \text{ for } a \in \mathbb{F}_p \text{ and } b \in \mathbb{F}_p^\times.$$

We first describe the free $\mathbb{Z}_p$-module $\mathbb{Z}_p[Cong(G)]$ explicitly in terms of abelian subgroups of G. For this we define the following subgroups of G.

$$Z := \left\{ i_a = \begin{pmatrix} a & 0 \\ 0 & a \end{pmatrix} : a \in \mathbb{F}_p^\times \right\} = \text{ centre of } G,$$

$$C := \left\{ c_{a,b} = \begin{pmatrix} a & b \\ 0 & a \end{pmatrix} : a \in \mathbb{F}_p^\times, b \in \mathbb{F}_p \right\},$$

$$T := \left\{ t_{a,d} = \begin{pmatrix} a & 0 \\ 0 & d \end{pmatrix} : a, d \in \mathbb{F}_p^\times \right\} = \text{ split Cartan,}$$

and

$$K := \left\{ k_{a,b} = \begin{pmatrix} a & \epsilon b \\ b & a \end{pmatrix} : a, b \in \mathbb{F}_p \right\} = \text{ non-split Cartan,}$$

where ϵ is the fixed non-square element of $\mathbb{F}_p$. Note that they are all abelian. Put $S(G) := \{Z, C, T, K\}$. Define a map

$$\psi := (\psi_U)_{U \in S(G)} : \mathbb{Z}_p[\text{Conj}(G)] \to \prod_{U \in S(G)} \mathbb{Z}_p[U],$$

where (the trace map) $\psi_U : \mathbb{Z}_p[\text{Conj}(G)] \to \mathbb{Z}_p[U]$ is a $\mathbb{Z}_p$-linear map defined by

$$g \mapsto \sum_{i=1}^{n} \{h_i^{-1} g h_i : h_i^{-1} g h_i \in U\}$$

for any $g \in G$ and a fixed set $\{h_1, \ldots, h_n\}$ of left coset representatives for U in G. This map is explicitly given in the following table

	ψ_Z	ψ_C	ψ_T	ψ_K
i_a	$p(p^2-1)i_a$	$(p^2-1)i_a$	$p(p+1)i_a$	$p(p-1)i_a$
$c_{a,1}$	0	$\sum_{i=1}^{p-1} c_{a,i}$	0	0
$t_{a,d}$	0	0	$t_{a,d} + t_{d,a}$	0
$k_{a,b}$	0	0	0	$k_{a,b} + k_{a,-b}$

Definition 8 We put Ψ for the set of all tuples $(a_U) \in \prod_{U \in S(G)} \mathbb{Z}_p[U]$ satisfying the following conditions

(A1) For every $U \in S(G)$, the trace map $tr_U : \mathbb{Z}_p[U] \to \mathbb{Z}_p[Z]$ maps a_U to a_Z (as $\mathbb{Z}_p[U]$ is a free finitely generated module over $\mathbb{Z}_p[Z]$ we have the trace map tr_U).

(A2) Let $N_G U$ be the normaliser of U in G. We require that every a_U is fixed under the conjugation action of $N_G U$.
(A3) The element a_Z lies in the ideal $p\mathbb{Z}_p[Z]$ of $\mathbb{Z}_p[Z]$.

Theorem 9 (additive theorem) *The map ψ induces an isomorphism between $\mathbb{Z}_p[\mathrm{Conj}(G)]$ and Ψ.*

Proof From the above table it is clear that the image of ψ lies in Ψ. We simply define a left inverse δ of the map ψ and then show that δ is injective on Ψ. Define $\delta := \sum_{U \in S(G)} \delta_U$, with each $\delta_U : \prod_{V \in S(G)} \mathbb{Z}_p[V] \to \mathbb{Q}_p[\mathrm{Conj}(G)] = \frac{\mathbb{Q}[G]}{[\mathbb{Q}[G], \mathbb{Q}[G]]}$ defined by

$$\delta_U((a_V)) = \begin{cases} \frac{1}{[N_G U : U]} \left(a_U - \frac{1}{[U:Z]} a_Z \right) & \text{if } U \neq Z \\ \frac{1}{[G:Z]} a_Z & \text{if } U = Z. \end{cases}$$

First we show that $\delta \circ \psi = id_{\mathbb{Z}_p[\mathrm{Conj}(G)]}$. As all the maps are $\mathbb{Z}_p$-linear it is enough to check only on conjugacy classes $\mathrm{Conj}(G)$.

(1) **(For classes i_a)** It is clear from the above table that $\delta_Z(\psi_Z(i_a)) = i_a$. For every $U \in S(G)$, we have $tr_U(\psi_U(i_a)) = [U : Z]\psi_U(i_a)$ (considered as an element of $\mathbb{Z}_p[Z]$ as it already lies in that subring of $\mathbb{Z}_p[U]$). Therefore $\delta_U(\psi_U(i_a)) = 0$ for all $U \neq Z$ by (A1). Hence $\delta(\psi(i_a)) = i_a$.
(2) **(For classes $c_{a,1}$)** From the above table it is clear that $\delta_U(\psi_U(c_{a,1})) = 0$ for all $U \neq C$. Moreover, it is easy to check that $N_G C$ is the set of upper triangular matrices in G. Hence $\delta_C(\psi_C(c_{a,1})) = \frac{1}{p-1}(\sum_{h \in N_G C/C} h^{-1} c_{a,1} h) = c_{a,1}$. Hence $\delta(\psi(c_{a,1})) = c_{a,1}$.
(3) **(For classes $t_{a,d}$)** Again from the above table it is clear that $\delta_U(\psi_U(t_{a,d})) = 0$ for $U \neq T$. The normaliser

$$N_G T = \left\{ \begin{pmatrix} a & b \\ c & d \end{pmatrix} : a = d = 0 \text{ or } b = c = 0 \right\} = T \cup \begin{pmatrix} 0 & 1 \\ 1 & 0 \end{pmatrix} T.$$

Therefore $[N_G T : T] = 2$. Hence $\delta_T(\psi_T(t_{a,d})) = t_{a,d}$ (we abuse the notation and denote the conjugacy class of $t_{a,d}$ by the same symbol).
(4) **(For classes $k_{a,b}$)** Again from the above table it is clear that $\delta_U(\psi_U(k_{a,b})) = 0$ for $U \neq K$. The normaliser

$$N_G K = \left\{ \begin{pmatrix} a & \epsilon b \\ c & d \end{pmatrix} : a = d, b = c \text{ or } a = -d, b = -c \right\}$$
$$= K \cup \begin{pmatrix} 0 & -\epsilon \\ 1 & 0 \end{pmatrix} K.$$

Therefore $[N_G K : K] = 2$. Hence $\delta_K(\psi_K(k_{a,b})) = k_{a,b}$.

This show that $\delta \circ \psi = id_{\mathbb{Z}_p[\mathrm{Conj}(G)]}$.

Next we show that $\delta|_\Psi$ is injective. Let $(a_U) \in \Psi$ be such that $\delta((a_U)) = 0$. First consider $a_C = \sum_{i=0}^{p-1} \sum_{a=1}^{p-1} x_{a,i} c_{a,i}$. Then $tr_C(a_C) = [C : Z] \sum_{a=1}^{p-1} x_{a,0} c_{a,0}$ which, by (A1), is equal to a_Z. Hence $a_C - \frac{1}{[C:Z]} a_Z = \sum_{i=1}^{p-1} \sum_{a=1}^{p-1} x_{a,i} c_{a,i}$. As $c_{a,i}$, for $1 \leqslant i \leqslant p-1$, are all conjugates of $c_{a,1}$ we have, by (A2),

$$a_C - \frac{1}{[C:Z]} a_Z = \sum_{a=1}^{p-1} x_{a,1} \sum_{i=1}^{p-1} c_{a,i}.$$

Therefore $\delta_C((a_U)) = \sum_{a=1}^{p-1} x_{a,1} c_{a,1}$. Moreover these conjugacy classes $c_{a,1}$ cannot appear in the image of δ_U for any $U \neq C$. Therefore $\delta((a_U)) = 0$ implies that $x_{a,1} = 0$ for all a. Hence $a_C = \frac{1}{[C:Z]} a_Z$. Hence $\delta_C((a_U)) = 0$.

Similarly, we show that $a_T = \frac{1}{[T:Z]} a_Z$ and $a_K = \frac{1}{[K:Z]} a_Z$ and so $\delta_T((a_U)) = 0 = \delta_K((a_U))$. Hence $\delta_Z((a_U)) = 0$. Therefore $a_Z = 0$ and so $a_C = 0$, $a_T = 0$ and $a_K = 0$. This show that $\delta|_\Psi$ is injective. □

Remark 10 The above proof goes through with any coefficient ring which is a $\mathbb{Z}_{(p)}$-algebra.

3 The Main Result

We have a map

$$\theta : K_1(\mathbb{Z}_p[G]) \to \prod_{U \in S(G)} \mathbb{Z}_p[U]^\times.$$

Let χ_U be representations of U and n_U be integers such that

$$\sum_{U \in S(G)} n_U Ind_U^G \chi_U = 0.$$

This sum takes place in the group of virtual characters of G. We say that a tuple $(x_U) \in \prod_{U \in S(G)} \mathbb{Z}_p[U]^\times$ satisfies (F) if for any χ_U and n_U as above

$$\prod_U \chi_U(x_U)^{n_U} = 1.$$

It is clear that the image of θ satisfies (F).

Proposition 11 *Let $(x_U) \in \prod_U \mathbb{Z}_p[U]^\times$ satisfy (F). Then*

(M1) (x_U) satisfies the analogue of (A1), i.e. the norm map $nr : \mathbb{Z}_p[U] \to \mathbb{Z}_p[Z]$ maps x_U to x_Z for any $U \in S(G)$.

(M2) (x_U) satisfies the analogue of (A2), i.e. x_U is fixed by $N_G U$ for any $U \in S(G)$.

Proof This is an easy consequence of (F). We demonstrate (M1). Let χ be a representation of Z and $\rho := Ind_Z^U(\chi)$. Then

$$\chi(nr(x_U)) = \rho(x_U).$$

As $Ind_Z^G(\chi) = Ind_U^G(\rho)$, it is plain from (F) that $\rho(x_U) = \chi(x_Z)$. Hence $\chi(nr(x_U)) = \chi(x_Z)$. Hence $nr(x_U) = x_Z$. □

Next we observe that by Oliver [15, Proposition 12.7] $SK_1(\mathbb{Z}_p[G]) = 1$. Let P be a finite group and put J_P for the Jacobson radical of group ring $\mathbb{Z}_p[P]$. By [15, Theorem 2.10]

$$K_1(\mathbb{Z}_p[P]) \cong K_1(\mathbb{Z}_p[P]/J_P) \oplus K_1(\mathbb{Z}_p[P], J_P).$$

The group $K_1(\mathbb{Z}_p[P]/J_P)$ is a finite group of order prime to P. The group $K_1(\mathbb{Z}_p[P], J_P)$ is a $\mathbb{Z}_p$-module. Hence $K_1(\mathbb{Z}_p[P])_{(p)} = K_1(\mathbb{Z}_p[P], J_P)$. Hence the map θ induces

$$\theta|_{K_1(\mathbb{Z}_p[G]/J_G)} : K_1(\mathbb{Z}_p[G]/J_G) \to \prod_{U \in S(G)} (\mathbb{Z}_p[U]/J_U)^\times$$

and

$$\theta|_{K_1(\mathbb{Z}_p[G])_{(p)}} : K_1(\mathbb{Z}_p[G])_{(p)} \to \prod_{U \in S(G)} K_1(\mathbb{Z}_p[U])_{(p)}.$$

In this paper we will ignore the prime to p-part $K_1(\mathbb{Z}_p[G]/J_G)$ as interesting congruences come from the p-part $K_1(\mathbb{Z}_p[G])_{(p)}$.

We first recall the integral logarithm of Oliver and Taylor ([15, Chaps. 6 and 12])

$$L : K_1(\mathbb{Z}_p[G])_{(p)} \to \mathbb{Z}_p[Cong(G)]$$

defined as $L := \left(1 - \frac{\varphi}{p}\right) \circ \log$, where φ is the map induced by $g \mapsto g^p$ for every $g \in \mathrm{Conj}(G)$.

Proposition 12 *The integral logarithm on $K_1(\mathbb{Z}_p[G])$ induces an isomorphism*

$$L : K_1(\mathbb{Z}_p[G])_{(p)} \xrightarrow{\cong} \mathbb{Z}_p[\mathrm{Conj}(G)].$$

In particular, $K_1(\mathbb{Z}_p[G])_{(p)}$ is torsion-free.

Proof For a finite group P let P' denote the set representative of p-regular conjugacy classes of P. Then kernel and cokernel of integral logarithm L on $K_1(\mathbb{Z}_p[P])$ are equal to $H_1(P, \mathbb{Z}_p[P'])^{\varphi}$ and $H_1(P, \mathbb{Z}_p[P'])_{\varphi}$ respectively. Here H_1 is the Hochschild homology and P acts on the coefficients by conjugation. The operator φ is the one induced by the map $\sum a_g g \mapsto \sum a_g g^p$ on the coefficients (this is

[15, Theorem 12.9]). We apply this to the group G. Firstly note that φ is identity on p-regular elements of G. To compute the homology group $H_1(G, \mathbb{Z}_p[G'])$ notice that by the sentence after Eq. (1) on page 286 in [15]

$$H_1(G, \mathbb{Z}_p[G']) = \ker(L) = \mathrm{tor}(K_1(\mathbb{Z}_p[G])_{(p)}),$$

which by [15, Theorem 12.5(ii)] is trivial in our case (this can be computed using the explicit conjugacy classes of G given in the previous section). □

Next we find a relation between ψ and θ. Let $\eta = Ind_Z^C 1$ be a representation of C. It induces a $\mathbb{Z}_p$-linear map $\eta : \mathbb{Z}_p[C] \to \mathbb{Z}_p[C]$ given by $g \mapsto tr(\eta(g))g$. The image of this map lies in $\mathbb{Z}_p[Z] \subset \mathbb{Z}_p[C]$. The representation η also induces a map $\mathbb{Z}_p[C]^\times \to \mathbb{Z}_p[C]^\times$, that we again denote by η, given by $\sum_{g \in C} a_g g \mapsto det(\sum_g a_g \eta(g) g)$. It is easy to verify that this is just the norm map $\mathbb{Z}_p[C]^\times \to \mathbb{Z}_p[Z]^\times$.

Lemma 13 *We have the following commutative diagram*

$$\begin{array}{ccc} \mathbb{Q}_p[\mathrm{Conj}(G)] & \xrightarrow{\varphi} & \mathbb{Q}_p[\mathrm{Conj}(G)] \\ \psi \downarrow & & \downarrow \psi \\ \prod_{U \in S(G)} \mathbb{Q}_p[U] & \xrightarrow[\tilde{\varphi}]{} & \prod_{U \in S(G)} \mathbb{Q}_p[U], \end{array}$$

where the map $\tilde{\varphi} = (\tilde{\varphi}_U)$ is given by

$$\tilde{\varphi}_Z(a_Z, a_C, a_T, a_K) := \varphi(a_Z) + p(p+1)\varphi(a_C - \frac{\eta}{p}(a_C))$$

$$\tilde{\varphi}_C(a_Z, a_C, a_T, a_K) := \varphi(a_C) - p(a_C - \frac{\eta}{p}(a_C))$$

$$\tilde{\varphi}_T(a_Z, a_C, a_T, a_K) := \varphi(a_T) + \frac{p(p+1)}{p-1}\varphi(a_C - \frac{\eta}{p}(a_C))$$

$$\tilde{\varphi}_K(a_Z, a_C, a_T, a_K) := \varphi(a_K) + p\varphi(a_C - \frac{\eta}{p}(a_C)).$$

In the above we use the fact that $\varphi(a_C - \frac{\eta}{p}(a_C))$ belongs to $\mathbb{Z}_p[Z]$ and hence can be considered as an element of $\mathbb{Z}_p[U]$ for any $U \in S(G)$.

Proof This is a simple and straightforward, though somewhat tedious, calculation using the explicit description of conjugacy classes of G and the map ψ. □

Proposition 14 *The relation between the maps θ and ψ is given by*

$$\begin{array}{ccc} K_1(\mathbb{Z}_p[\mathrm{Conj}(G)])_{(p)} & \xrightarrow{L} & \mathbb{Z}_p[\mathrm{Conj}(G)] \\ \downarrow{\scriptstyle\theta} & & \downarrow{\scriptstyle\psi} \\ \prod_{U\in S(G)} \mathbb{Z}_p[U]^{\times} & \xrightarrow[\tilde{L}]{} & \prod_{U\in S(G)} \mathbb{Q}_p[U] \end{array}$$

where the map $\tilde{L} := (\tilde{L}_U)$ is given by

$$\tilde{L}_Z(x_Z, x_C, x_T, x_K) := \frac{1}{p}\log\left(\frac{x_Z^p}{\varphi(x_Z)}\cdot\frac{\varphi(\eta(x_C))^{p+1}}{\varphi(x_C)^{p(p+1)}}\right)$$

$$\tilde{L}_C(x_Z, x_C, x_T, x_K) := \frac{1}{p}\log\left(\frac{x_C^p}{\varphi(x_C)}\cdot\frac{\varphi(\eta(x_C))}{\varphi(x_C)^{p}}\right)$$

$$\tilde{L}_T(x_Z, x_C, x_T, x_K) := \frac{1}{p(p-1)}\log\left(\frac{x_T^{p(p-1)}}{\varphi(x_T)^{p-1}}\cdot\frac{\varphi(\eta(x_C))^{p+1}}{\varphi(x_C)^{p(p+1)}}\right)$$

$$\tilde{L}_K(x_Z, x_C, x_T, x_K) := \frac{1}{p}\log\left(\frac{x_K^p}{\varphi(x_K)}\cdot\frac{\varphi(\eta(x_C))}{\varphi(x_C)^{p}}\right)$$

Proof This is again a simple explicit calculation using Lemma 13,

$$L = \left(1 - \frac{\varphi}{p}\right)\circ\log$$

and the fact that $\psi\circ\log = \log\circ\theta$ (by the commutative diagram (1a) in the Proof of Theorem 6.8 in [15]). □

Remark 15 We refer the reader to the discussion on page 286 after the Proof of Theorem 12.9 in [15]. It may explain why the definition of $\tilde{L}$ is complicated.

Definition 16 Let Θ be the set of all tuples $(x_U) \in \prod_{U\in S(G)} \mathbb{Z}_p[U]^{\times}_{(p)}$ which are not torsion and such that

(1) (x_U) satisfies (F).
(2) $x_Z \equiv \varphi(x_C) (\mathrm{mod}\ p\mathbb{Z}_p[Z])$.

Lemma 17 *If $(x_Z, x_C, x_T, x_K) \in \Theta$, then $\tilde{L}_Z(x_Z, x_C, x_T, x_K)$ becomes*

$$\tilde{L}_Z(x_Z, x_C, x_T, x_K) = \log\left(\frac{x_Z\varphi(x_Z)}{\varphi(x_C)^{p+1}}\right)$$

Proof As shown in Proposition 11, condition (F) implies that $\eta(x_C) = nr(x_C) = x_Z$. □

We can now state our main theorem.

Theorem 18 *The map* θ *induces an isomorphism between* $K_1(\mathbb{Z}_p[G])_{(p)}$ *and* Θ.

Proof We prove this in two steps.

(1) First note that, as p is odd, $x_Z \equiv \varphi(x_C) (\mathrm{mod}\ p)$ is implied by

$$\left(\frac{x_Z\phi(x_Z)}{\varphi(x_C)\varphi(x_C)^p}\right) \equiv \left(\frac{x_Z\varphi(x_Z)}{\varphi(x_C)\varphi^2(x_C)}\right) \equiv \left(\frac{x_Z}{\varphi(x_C)}\right)^2 \equiv 1 (\mathrm{mod}\ p).$$

Note that -1 does not belong to $K_1(\mathbb{Z}_p[G])_{(p)}$. As log induces an isomorphism between $1 + p\mathbb{Z}_p[Z]$ and $p\mathbb{Z}_p[Z]$, Lemma 17 implies that the image of θ satisfies (C). Hence by Propositions 11 and 14 we get that the image of θ is contained in Θ.

(2) We first claim that the $\ker(\tilde{L}|_\Theta)$ is trivial. Let $(x_U)_U$ be in the kernel of $\tilde{L}|_\Theta$. As log induces an isomorphism between $1 + p\mathbb{Z}_p[C]$ and $p\mathbb{Z}_p[C]$ it follows that

$$\frac{x_C^p}{\varphi(x_C)} \cdot \frac{\varphi(\eta(x_C))}{\varphi(x_C)^p} = 1.$$

This shows that x_C^p lies in $\mathbb{Z}_p[Z]^\times$ (since $\varphi(\mathbb{Z}_p[C]^\times) \subset \mathbb{Z}_p[Z]^\times$). Hence $\eta(x_C^p) = x_C^{p^2}$. Hence $0 = p\tilde{L}_C(x_Z, x_C, x_T, x_K) = \log\left(\frac{x_C^p}{\varphi(x_C)}\right) = pL(x_C)$ (here L is the integral logarithm map on $\mathbb{Z}_p[C]^\times_{(p)}$). Whence $L(x_C) = 0$ and $x_C = 1$ by [15, Theorem 12.9] (note that x_C is not torsion by the definition of Θ). Hence $\frac{\varphi(\eta(x_C))}{\varphi(x_C)^p} = 1$ and consequently $\frac{1}{p}\log\left(\frac{x_U^p}{\varphi(x_U)}\right) = 1$ for all U. Hence $x_U = 1$ for all $U \neq C$. Therefore $\ker(\tilde{L})$ is trivial. This proves the claim. (Compare this with proof of injectivity of δ above).

Now consider the commutative diagram

$$\begin{array}{ccc} K_1(\mathbb{Z}_p[G])_{(p)} & \xrightarrow[\cong]{L} & \mathbb{Z}_p[\mathrm{Conj}(G)] \\ \Big\downarrow{\scriptstyle\theta} & & {\scriptstyle\cong}\Big\downarrow{\scriptstyle\psi} \\ \Theta & \xrightarrow[\tilde{L}]{} & \Psi. \end{array}$$

This diagram shows that $\tilde{L} : \Theta \to \Psi$ is surjective and hence an isomorphism. Therefore θ is an isomorphism. □

Acknowledgements John Coates introduced me to non-commutative Iwasawa during my PhD. He has been a constant source of encouragement and inspiration for me and I am sure will continue to be so for many years. It is my great pleasure and honour to dedicate this paper to him on the occasion of his seventieth birthday. These calculations were carried out during a visit to TIFR in fall, 2014 and I thank TIFR for its hospitality. The author would like to thank the anonymous referee for several helpful comments and careful reading of the manuscript.

References

1. Burns, D.: On main conjectures in non-commutative Iwasawa theory and related conjectures. J. Reine Angew. Math. **698**, 105–160 (2015)
2. Burns, D., Kakde, M.: Congruences in Non-commutative Iwasawa Theory (In preparation)
3. Burns, D., Venjakob, O.: On descent theory and main conjectures in non-commutative Iwasawa theory. J. Inst. Math. Jussieu **10**, 59–118 (2011)
4. Chinburg, T., Pappas, G., Taylor, M.J.: The group logarithm past and present. In: Coates, J., Schneider, P., Sujatha, R., Venjakob, O. (eds.) Noncommutative Iwasawa Main Conjectures over Totally Real Fields, vol. 29, Springer Proceedings in Mathematics and Statistics, pp. 51–78. Springer (2012)
5. Coates, J., Fukaya, T., Kato, K., Sujatha, R., Venjakob, O.: The GL_2 main conjecture for elliptic curves without complex multiplication. Publ. Math. IHES (1), 163–208 (2005)
6. Coates, J., Schmidt, C.-G.: Iwasawa theory for the symmetric square of an elliptic curve. J. Reine Angew. Math. **375**(376), 104–156 (1987)
7. Coates, J., Sujatha, R.: On the $\mathfrak{M}_H(G)$-conjecture. In: Non-abelian Fundamental Groups and Iwasawa Theory, pp. 132–161 (2012)
8. Curtis, C.W., Reiner, I.: Methods of Representation Theory with Applications to Finite Groups and Orders, vol. 1. Wiley (1981)
9. Fukaya, T., Kato, K.: A formulation of conjectures on p-adic zeta functions in non-commutative Iwasawa theory. In: Uraltseva, N.N. (ed.) Proceedings of the St. Petersburg Mathematical Society, vol. 12, pp. 1–85 (2006)
10. Fulton, W., Harris, J.: Representation theory. A first course, vol. 129, Graduate Text in Mathematics. Springer, New York (1991)
11. Kakde, M.: Proof of the main conjecture of noncommutatve Iwasawa theory for totally real number fields in certain cases. J. Algebraic Geom. **20**, 631–683 (2011)
12. Kakde, M.: The main conjecture of Iwasawa theory for totally real fields. Invent. Math. **193**(3), 539–626 (2013)
13. Kato, K.: K_1 of some non-commutative completed group rings. K-Theory **34**, 99–140 (2005)
14. Kato, K.: Iwasawa theory of totally real fields for Galois extensions of Heisenberg type. Very preliminary version (2006)
15. Oliver, R.: Whitehead Groups of Finite Groups, vol. 132, London Mathematical Society Lecture Note Series. Cambridge University Press (1988)
16. Ritter, J., Weiss, A.: On the 'main conjecture' of equivariant Iwasawa theory. J. AMS **24**, 1015–1050 (2011)
17. Serre, J.-P.: Propriétés galoisiennes des points d'ordre fini des courbes elliptiques. Invent. Math. **15**, 259–331 (1972)
18. Serre, J-P.: Linear representation of finite groups, vol. 42, Graduate Text in Mathematics. Springer (1977)
19. Venjakob, O.: On the work of Ritter and Weiss in comparison with Kakde's approach. In: Noncommutative Iwasawa Main Conjectures over Totally Real Fields, vol. 29, Springer Proceedings in Mathematics and Statistics, pp. 159–182 (2013)

Diophantine Geometry and Non-abelian Reciprocity Laws I

Minhyong Kim

> *I dive down into the depth of the ocean of forms, hoping to gain the perfect pearl of the formless...*
>
> *(Tagore)*

Abstract We use non-abelian fundamental groups to define a sequence of higher reciprocity maps on the adelic points of a variety over a number field satisfying certain conditions in Galois cohomology. The non-abelian reciprocity law then states that the global points are contained in the kernel of all the reciprocity maps.

Keywords Class field theory · Diophantine geometry · Non-abelian cohomology

Mathematics Subject Classification: 11D99 · 11R37

1 Refined Hasse Principles and Reciprocity Laws

Consider the Hasse–Minkowski theorem [11] for affine conics like

$$X: \quad ax^2 + by^2 = c$$

Dedicated to John Coates on the occasion of his 70th birthday.

M. Kim (✉)
Mathematical Institute, University of Oxford, Radcliffe Observatory Quarter, Woodstock Road, Oxford OX2 6GG, UK
e-mail: minhyong.kim@maths.ox.ac.uk

M. Kim
Korea Institute for Advanced Study, 85 Hoegiro, Seoul 02455, Dongdaemun-gu, Republic of Korea

D. Loeffler and S.L. Zerbes (eds.), *Elliptic Curves, Modular Forms and Iwasawa Theory*, Springer Proceedings in Mathematics & Statistics 188, DOI 10.1007/978-3-319-45032-2_9

stating that X has a rational point in a number field F if and only if it has a point in F_v for all places v. In spite of its great elegance, even undergraduate students are normally left with a somewhat unsatisfactory sense of the statement, having essentially to do with the fact that the theorem says nothing about the locus of

$$X(F) \subset X(\mathbb{A}_F).$$

There are various attempts to rectify the situation, the most successful of which might be the theory of the Brauer–Manin obstruction [8].

The point of view of this paper is that one should consider such problems, even for more general varieties, as that of defining a good reciprocity map. That is, let's simplify for a moment and assume $X \simeq \mathbb{G}_m$ (so that existence of a rational point is not the issue). Then we are just asking about the locus of $F^\times$ in the ideles $\mathbb{A}_F^\times$ of F. In this regard, a description of sorts is provided by Abelian class field theory [1], which gives us a map

$$\mathrm{rec}^{\mathrm{ab}} : \mathbb{A}_F^\times \longrightarrow G_F^{\mathrm{ab}},$$

with the property that

$$\mathrm{rec}^{\mathrm{ab}}(F^\times) = 0.$$

So one could well view the reciprocity map as providing a 'defining equation' for $\mathbb{G}_m(F)$ in $\mathbb{G}_m(\mathbb{A}_F)$, except for the unusual fact that the equation takes values in a group. Because F is a number field, there is also the usual complication that the kernel of rec^{ab} is not exactly equal to $\mathbb{G}_m(F)$. But the interpretation of the reciprocity law as a refined statement of Diophantine geometry is reasonable enough.

In this paper, we obtain a generalization of Artin reciprocity to an *iterative non-abelian reciprocity law* with coefficients in smooth varieties whose étale fundamental groups satisfy rather mild cohomological conditions [Coh], to be described near the end of this section. They are satisfied, for example, by any smooth curve. Given a smooth variety X equipped with a rational point $b \in X(F)$ satisfying [Coh], we define a sequence of subsets

$$X(\mathbb{A}_F) = X(\mathbb{A}_F)_1 \supset X(\mathbb{A}_F)_1^2 \supset X(\mathbb{A}_F)_2 \supset X(\mathbb{A}_F)_2^3 \supset X(\mathbb{A}_F)_3 \supset X(\mathbb{A}_F)_3^4 \supset \cdots$$

and a sequence of maps

$$\mathrm{rec}_n : X(\mathbb{A}_F)_n \longrightarrow \mathfrak{G}_n(X)$$

$$\mathrm{rec}_n^{n+1} : X(\mathbb{A}_F)_n^{n+1} \longrightarrow \mathfrak{G}_n^{n+1}(X)$$

to a sequence $\mathfrak{G}_n(X)$, $\mathfrak{G}_n^{n+1}(X)$ of profinite abelian groups in such a way that

$$X(\mathbb{A}_F)_n^{n+1} = \mathrm{rec}_n^{-1}(0)$$

and

$$X(\mathbb{A}_F)_{n+1} = (\mathrm{rec}_n^{n+1})^{-1}(0).$$

We visualize this as a diagram:

$$
\begin{array}{ccccccc}
\cdots \begin{array}{c} X(\mathbb{A}_F)_2^3 \\ = \mathrm{rec}_2^{-1}(0) \end{array} & \hookrightarrow & \begin{array}{c} X(\mathbb{A}_F)_2 \\ = (\mathrm{rec}_1^2)^{-1}(0) \end{array} & \hookrightarrow & \begin{array}{c} X(\mathbb{A}_F)_1^2 \\ = \mathrm{rec}_1^{-1}(0) \end{array} & \hookrightarrow & X(\mathbb{A}_F)_1 \\
\cdots \; \mathrm{rec}_2^3 \downarrow & & \mathrm{rec}_2 \downarrow & & \mathrm{rec}_1^2 \downarrow & & \mathrm{rec}_1 \downarrow \\
\cdots \; \mathfrak{G}_2^3(X) & & \mathfrak{G}_2(X) & & \mathfrak{G}_1^2(X) & & \mathfrak{G}_1(X)
\end{array}
$$

in which each reciprocity map is defined not on all of $X(\mathbb{A}_F)$, but only on the kernel (the inverse image of 0) of all the previous maps. We put

$$X(\mathbb{A}_F)_\infty = \bigcap_{i=1}^{\infty} X(\mathbb{A}_F)_i.$$

The non-abelian reciprocity law then states

Theorem 1.1

$$X(F) \subset X(\mathbb{A}_F)_\infty.$$

We give now a brief description of the groups $\mathfrak{G}_n$ and $\mathfrak{G}_n^{n+1}$. Let $\Delta = \pi_1(\bar{X}, b)^{(2)}$ be the profinite prime-to-2 étale fundamental group[1] [3] of $\bar{X} = X \times_{\mathrm{Spec}(F)} \mathrm{Spec}(\bar{F})$, and let $\Delta^{[n]}$ be its lower central series defined as

$$\Delta^{[1]} = \Delta$$

and

$$\Delta^{[n+1]} := \overline{[\Delta, \Delta^{[n]}]},$$

where the overline refers to the topological closure. We denote

$$\Delta_n := \Delta/\Delta^{[n+1]}$$

and

$$T_n := \Delta^{[n]}/\Delta^{[n+1]}.$$

[1] The referee has asked for an explanation for leaving out the prime 2 in the fundamental groups. To include the full profinite π_1, we would need to consider localization to Archimedean places in Poitou–Tate duality, which would then require us to include Archimedean places in the definition of restricted direct products. But then, because of non-trivial H^0 at Archimedean places, various long exact sequences in non-abelian cohomology would become problematic (cf. Lemma 4.2). We note in this regard that if the base field F had no real places, the full fundamental group could have been used in the entire paper.

Thus, we have an exact sequence

$$1 \longrightarrow T_n \longrightarrow \Delta_n \longrightarrow \Delta_{n-1} \longrightarrow 1$$

for each n, turning Δ_n into a central extension of Δ_{n-1}.

All of the objects above are equipped with canonical actions of $G_F = \mathrm{Gal}(\bar{F}/F)$. Given any topological abelian group A with continuous G_F-action, we have the continuous Galois dual

$$D(N) := \mathrm{Hom}_{ct}(A, \mu_\infty),$$

and the Pontriagin dual

$$A^\vee = \mathrm{Hom}_{ct}(A, \mathbb{Q}/\mathbb{Z}).$$

(See Appendix II for details.)

With this notation, we can define the targets of the reciprocity maps rec_n using continuous cohomology:

$$\mathfrak{G}_n(X) := H^1(G_F, D(T_n))^\vee.$$

Notice that when $X = \mathbb{G}_m$, we have $T_1 = \hat{\mathbb{Z}}^{(2)}(1)$ and $T_n = 0$ for $n > 1$. Thus, $D(T_1) = \bigoplus_{p \neq 2} \mathbb{Q}_p/\mathbb{Z}_p$ and

$$H^1(G_F, D(T_1)) = \bigoplus_{p\neq 2} \mathrm{Hom}(G_F, \mathbb{Q}_p/\mathbb{Z}_p) = \bigoplus_{p\neq 2} \mathrm{Hom}(G_F^{\mathrm{ab}}, \mathbb{Q}_p/\mathbb{Z}_p).$$

Hence, by Pontrjagin duality, there is a canonical isomorphism

$$\mathfrak{G}_1((\mathbb{G}_m)_F) \simeq G_F^{\mathrm{ab},(2)},$$

and rec_1 will agree with the prime-to-2 part of the usual reciprocity map $\mathrm{rec}^{\mathrm{ab}}$.

For the $\mathfrak{G}_n^{n+1}(X)$, we need a little more notation. Let S be a finite set of places of F and $G_F^S = \mathrm{Gal}(F_S/F)$ the Galois group of the maximal extension of F unramified outside of S. We denote by S^0 the set of non-Archimedean places in S. For a topological abelian group A with G_F^S-action, we have the kernel of localization

$$Ш^i_S(A) := \mathrm{Ker}[H^i(G_F^S, A) \xrightarrow{\mathrm{loc}_S} \prod_{v\in S^0} H^i(G_v, A)],$$

and what we might call the *strict* kernel,

$$sШ^i_S(A) := \mathrm{Ker}[H^i(G_F^S, A) \xrightarrow{\mathrm{loc}} \prod_{v\in V_F^0} H^i(G_v, A)],$$

where the localization map now goes to *all* non-Archimedean places in F. Obviously,

$$s\text{Ш}^i_S(A) \subset \text{Ш}^i_S(A).$$

For the strict kernels, whenever $S \subset T$, the restriction maps on cohomology induce maps

$$s\text{Ш}^i_S(A) \longrightarrow s\text{Ш}^i_T(A),$$

For any finite set M of odd primes, denote by Δ^M the maximal pro-M quotient of Δ, together with corresponding notation $[\Delta^M]^{[n]}$, Δ^M_n, T^M_n. Given M, we consider sets of places S of F that contain all places lying above primes of M, all Archimedean places, and all places of bad reduction for (X, b).

Then

$$\mathfrak{G}^{n+1}_n(X) := \varprojlim_M (\varinjlim_S s\text{Ш}^2_S(T^M_{n+1})).$$

We will see in Sect. 3 how to define the reciprocity maps from level of thr filtration on $X(\mathbb{A}_F)$ to all these groups.

The conditions [Coh] are the following.

- [Coh 1] For each finite set M of odd primes, T^M_n is torsion-free.
- [Coh 2] For each finite set M of odd primes and non-Archimedean place v, $H^0(G_v, T^M_n) = 0$.

They are used in suitable injectivity statements for localization in cohomology, which, in turn, feed into the inductive definition of the reciprocity maps. Also, [Coh 2] is necessary for the long exact sequence (2.28) in the restricted direct product of local cohomology.

If we pick a place v, then the projection $X(\mathbb{A}_F) \longrightarrow X(F_v)$ induces the image filtration

$$X(F_v) = X(F_v)_1 \supset X(F_v)^2_1 \supset X(F_v)_2 \supset X(F_v)^3_2 \supset X(F_v)_3 \supset X(F_v)^4_3 \supset \cdots$$

and of course,

$$X(F) \subset X(F_v)_\infty := \bigcap_n X(F_v)_n.$$

Conjecture 1.2 *When X is a proper smooth curve and v is an odd prime of good reduction, we have*

$$X(F) = X(F_v)_\infty.$$

This conjecture can be viewed as a refinement of the conjecture of Birch and Swinnerton-Dyer type made in [2]. By comparing the profinite reciprocity map here to a unipotent analogue, the computations of that paper can be viewed as evidence for (an affine analogue of) this conjecture as well. We will write more systematically about this connection and about *explicit* higher reciprocity laws in a forthcoming publication. In the meanwhile, we apologise a bit for the lack of examples in this paper, pointing out that it is possible to view the examples in [2] as illustrating the reciprocity laws here. That is, it is quite likely unfeasible to give explicit formulas for the full non-abelian reciprocity maps, in the same way such formulas are hard to come by in the abelian theory. It is rather that one should compose the reciprocity maps with natural projections or functions. Even though we do not spell it out in detail as yet, the computations in [2] arise exactly from such a process.

2 Pre-reciprocity

We will assume throughout that X is a smooth variety over F such that the conditions [Coh] are satisfied. We will denote by V_F the set of all places of F and by V_F^0 the set of non-Archimedean places.

The maps rec_n and rec_n^{n+1} will be constructed in general via non-abelian cohomology and an iterative application of Poitou–Tate duality. For this, it is important that the G_F-action on any fixed Δ^M, where M, as before, is a finite set of odd primes, factors through $G_F^S = \mathrm{Gal}(F_S/F)$ for some finite set S of places of F. Here, F_S refers to the maximal algebraic extension of F unramified outside S. If $X \hookrightarrow X'$ is a smooth compactification with a normal crossing divisor D as complement, then by [13, Theorem 2.1] it suffices to take S large enough to satisfy the conditions that

- X' has a smooth model over $\mathrm{Spec}(\mathcal{O}_F[1/S])$;
- D extends to a relative normal crossing divisor over $\mathrm{Spec}(\mathcal{O}_F[1/S])$;
- b extends to an S-integral point of the model of X, given as the complement of the closure of D in the smooth model of X';
- S contains M and all Archimedean places of F.

We will be using thereby the continuous cohomology sets and groups (Appendix I, and [5, 12])

$$H^1(G_F^S, \Delta_n^M),\ \ H^i(G_F^S, T_n^M),\ \ H^i(G_F^S, D(T_n^M)).$$

(Note that the first term is an H^1, in anticipation of the possibility that Δ_n^M is non-abelian.) Whenever this notation is employed, we assume that the finite set S has been chosen large enough so that the G_F-action factors through G_F^S. Given any topological group U with continuous action of G_F, if this action factors through G_F^S for some set S, we will call S an *admissible set* of places. For any admissible set, we denote by S^0 the non-Archimedean places in S.

For each non-Archimedean place v of F, let $G_v = \mathrm{Gal}(\bar{F}_v/F_v)$. In the following, U denotes a topologically finitely-generated profinite group that is prime to 2, in the sense that it is the inverse limit of finite groups of order prime to 2. When U has a continuous G_F-action, define

$$\prod^{S} H^i(G_v, U) := \prod_{v\in S^0} H^i(G_v, U) \times \prod_{v\in V_F^0\setminus S^0} H^i_{un}(G_v, U)$$

where

$$H^i_{un}(G_v, U) := H^i(G_v/I_v, U^{I_v})$$

and $I_v \subset G_v$ is the inertia subgroup. Here, as in the following, if U is non-abelian, we only allow $i = 1$. In any case, we have a natural map

$$H^i_{un}(G_v, U) \longrightarrow H^i(G_v, U),$$

and hence, if $T \supset S$, a natural map

$$\prod^{S} H^i(G_v, U) \longrightarrow \prod^{T} H^i(G_v, U).$$

Define the 'restricted direct product' as a direct limit

$$\prod{}' H^i(G_v, U) := \varinjlim_{S} \prod^{S} H^i(G_v, U).$$

For $i = 1$, the maps in the limit will be injective, but not in general for $i = 2$. We will also use the notation

$$\prod_{S} H^i(G_v, U) = \prod_{v\in S^0} H^i(G_v, U)$$

and

$$\prod_{T}^{S} H^i(G_v, U) = \prod_{v\in S^0} H^i(G_v, U) \times \prod_{v\in T^0\setminus S^0} H^1_{un}(G_v, U)$$

for $T \supset S$, so that

$$\prod^{S} H^i(G_v, U) = \varprojlim_{T} \prod_{T}^{S} H^i(G_v, U).$$

For each $n \geqslant 2$, we have an exact sequence

$$1 \longrightarrow T_n^M \longrightarrow \Delta_n^M \longrightarrow \Delta_{n-1}^M \longrightarrow 1.$$

of topological groups. By Appendix I, Lemma 4.4, the surjection $\Delta_n^M \longrightarrow \Delta_{n-1}^M$ is equipped with a continuous section, so that we get a long exact sequence of continuous cohomology

$$0 \longrightarrow H^1(G_F^S, T_n^M) \longrightarrow H^1(G_F^S, \Delta_n^M) \longrightarrow H^1(G_F^S, \Delta_{n-1}^M) \xrightarrow{\delta_{n-1}^g} H^2(G_F^S, T_n^M).$$

Here, the superscript in 'δ_{n-1}^g' refers to 'global'. As explained in the Appendix I, Lemmas 4.2 and 4.3, the meaning of exactness here is as follows. The group $H^1(G_F^S, T_n^M)$ acts freely on the space $H^1(G_F^S, \Delta_n^M)$. and the projection

$$p_{n-1} : H^1(G_F^S, \Delta_n^M) \longrightarrow H^1(G_F^S, \Delta_{n-1}^M)$$

identifies the orbit space with the kernel of the boundary map δ_{n-1}^g. To check that the conditions of Appendix I are satisfied, note that twisting the Galois action by a cocycle for a class $c \in H^1(G_F^S, \Delta_{n-1}^M)$ will not change the action on the graded pieces T_i^M, so that the condition [Coh 2] implies that Δ_{n-1}^M has no G_F^S-invariants. This is because the twisted action is an inner twist, which will not affect the subquotients of the lower central series.

Similarly, for each non-Archimedean local Galois group, we have exact sequences

$$0 \longrightarrow H^1(G_v, T_n^M) \longrightarrow H^1(G_v, \Delta_n^M) \longrightarrow H^1(G_v, \Delta_{n-1}^M) \xrightarrow{\delta_{n-1}} H^2(G_v, T_n^M).$$

For each n, there is a surjection

$$(T_1^M)^{\otimes n} \twoheadrightarrow T_n^M.$$

Thus, T_n^M has strictly negative weights between $-2n$ and $-n$ as a Galois representation. By [4], Theorem 3(b), we see that the localization

$$H^1(G_F^S, T_n^M) \longrightarrow \prod^S H^1(G_v, T_n^M) \subset \prod{}' H^1(G_v, T_n^M)$$

is injective.

In order to use [4], we need to make a few remarks. Firstly, there is the simple fact that

$$T_n^M = \prod_{l \in M} T_n^l,$$

so it suffices to consider l-adic representations for a fixed prime l. Next, we note that [4] proves the injectivity for the Galois representations $H^i(\bar{V}, \mathbb{Z}_l(n))/(tor)$ and

$i \neq 2n$ where V is a smooth projective variety. But an examination of the proof shows that it only uses the fact that this is torsion-free, finitely-generated, and of non-zero weight. That is to say, it is shown that

$$H^1(G_F^S, N) \longrightarrow \prod^S H^1(G_v, N) \subset \prod' H^1(G_v, N)$$

is injective for any torsion-free finitely-generated $\mathbb{Z}_l$-module of non-zero Galois weight. Note that all the T_n^l are torsion-free by condition [Coh 1].

Now, by using the exact sequences (2.8) and (2.10) and an induction over n, we get injectivity of localization

$$H^1(G_F^S, \Delta_n^M) \longrightarrow \prod^S H^1(G_v, \Delta_n^M) \subset \prod' H^1(G_v, \Delta_n^M)$$

for every n.

Of course, we can repeat the discussion with any admissible $T \supset S$. Using these natural localization maps, we will regard global cohomology simply as subsets of the $\prod^S$ or of $\prod'$.

For any U with continuous G_F^S-action such that the localization map

$$H^1(G_F^T, U) \longrightarrow \prod^T H^1(G_v, U) \subset \prod' H^1(G_v, U)$$

is injective for all admissible T, define

$$E(U) := \varinjlim_T \mathrm{loc}(H^1(G_F^T, U)) = \bigcup_T \mathrm{loc}(H^1(G_F^T, U)).$$

For admissible T, there is also the partial localization

$$H^1(G_F^T, U) \xrightarrow{\mathrm{loc}_T} \prod_T H^1(G_v, U).$$

When U is topologically finitely-generated abelian profinite group with all finite quotients prime to 2, we have the duality isomorphism (local Tate duality, [9], Chap. VII.2)

$$D : \prod_T H^1(G_v, U) \simeq \prod_T H^1(G_v, D(U))^\vee$$

that can be composed with

$$\prod_T H^1(G_v, D(U))^\vee \xrightarrow{\mathrm{loc}_T^*} H^1(G_F^T, D(U))^\vee$$

to yield a map

$$\mathrm{loc}_T^* \circ D : \prod_{v \in T} H^1(G_v, U) \longrightarrow H^1(G_F^T, D(U))^\vee$$

such that

$$\mathrm{Ker}(\mathrm{loc}_T^* \circ D) = \mathrm{loc}_T(H^1(G_F^T, U))$$

(Poitou–Tate duality, [9], Chap. VIII.6). We denote also by $\mathrm{loc}_T^* \circ D$ the map

$$\prod^T H^1(G_v, U) \longrightarrow H^1(G_F^T, D(U))^\vee$$

obtained by projecting the components in $\prod_{v \in V_F^0 \setminus T^0} H^1_{un}(G_v, U)$ to zero.

When U is abelian and $T' \supset T$, these maps fit into commutative diagrams as follows:

$$\begin{array}{ccc} \prod^T H^1(G_v, U) & \hookrightarrow & \prod^{T'} H^1(G_v, U) \\ \downarrow \mathrm{loc}_T^* \circ D & & \downarrow \mathrm{loc}_{T'}^* \circ D \\ H^1(G_F^T, D(U))^\vee & \xleftarrow{\mathrm{Inf}^*} & H^1(G_F^{T'}, D(U))^\vee \end{array}$$

where the lower arrow is the dual to inflation. The commutativity follows from the fact that $H^1_{un}(G_v, U)$ and $H^1_{un}(G_v, D(U))$ annihilate each other under duality, so that the sum of the local pairings between $\prod^T H^1(G_v, U)$ and $H^1(G_F^T, D(U))$ will be independent of the contribution from $T' \setminus T$. Hence, we get a compatible family of maps

$$\prod^T H^1(G_v, U) \longrightarrow \varprojlim_{T'} H^1(G_F^{T'}, D(U))^\vee = H^1(G_F, D(U))^\vee$$

Taking the union over T, we then get

$$\mathrm{prec}(U) : \prod{}' H^1(G_v, U) \longrightarrow \varprojlim_{T} H^1(G_F^T, D(U))^\vee = H^1(G_F, D(U))^\vee$$

(The notation prec for 'pre-reciprocity' will be placed in context below.) According to Appendix II (5.13),

Proposition 2.1 *When U is a topologically finitely-generated abelian pro-finite group with all finite quotients prime to 2, then*

$$\mathrm{Ker}(\mathrm{prec}(U)) = E(U).$$

(Recall $E(U)$ defined in (2.17).)

One distinction from the appendix is that our product runs only over non-Archimedean places. However, because we are only considering prime-to-2 coefficients, the local H^1 vanishes as all Archimedean places. The goal of this section, by and large, is to generalise this result to the coefficients Δ_n^M, which are non-abelian.

In addition to the exact sequences (2.8) and (2.10), we have exact sequences with restricted direct products

$$0 \longrightarrow \prod' H^1(G_v, T_n^M) \longrightarrow \prod' H^1(G_v, \Delta_n^M) \xrightarrow{p_{n-1}} \prod' H^1(G_v, \Delta_{n-1}^M) \xrightarrow{\delta_{n-1}} \prod' H^2(G_v, T_n^M)$$

making the second term of the first line a $\prod' H^1(G_v, T_n^M)$-torsor over the kernel of δ. To see this, let S be an admissible set of primes. Then the G_v-action for $v \notin S$ factors through G_v/I_v, so that we have an exact sequence

$$0 \longrightarrow H^1(G_v/I_v, T_n^M) \longrightarrow H^1(G_v/I_v, \Delta_n^M) \longrightarrow H^1(G_v/I_v, \Delta_{n-1}^M) \xrightarrow{\delta_{n-1}} H^2(G_v/I_v, T_n^M)$$

and hence, an exact sequence

$$\begin{aligned} 0 &\longrightarrow \prod_{v\in S^0} H^1(G_v, T_n^M) \times \prod_{v\in V_F^0\setminus S^0} H^1_{un}(G_v, T_n^M) \\ &\longrightarrow \prod_{v\in S} H^1(G_v, \Delta_n^M) \times \prod_{v\in V_F^0\setminus S^0} H^1_{un}(G_v, \Delta_n^M) \\ &\longrightarrow \prod_{v\in S} H^1(G_v, \Delta_{n-1}^M) \times \prod_{v\in V_F^0\setminus S^0} H^1_{un}(G_v, \Delta_{n-1}^M) \\ &\xrightarrow{\delta_{n-1}} \prod_{v\in S} H^2(G_v, T_n^M) \times \prod_{v\in V_F^0\setminus S^0} H^2_{un}(G_v, T_n^M). \end{aligned}$$

Taking the direct limit over S gives us the exact sequence with restricted direct products.

In the following, various local, global, and product boundary maps will occur. In the notation, we will just distinguish the level and the global boundary map, since the domain should be mostly clear from the context.

We go on to define a sequence of pre-reciprocity maps as follows. First, we let

$$\mathrm{prec}_1 := \varprojlim_M \mathrm{prec}(\Delta_1^M):$$

$$\varprojlim_M {\prod}' H^1(G_v, \Delta_1^M) \longrightarrow \varprojlim_M H^1(G_F, D(\Delta_1^M))^\vee = H^1(G_F, D(\Delta_1))^\vee$$

as above. The kernel of prec_1 is exactly $E_1 := \varprojlim_M E(\Delta_1^M)$. For $x \in E_1$, define

$$\mathrm{prec}_1^2(x) := \delta_1^g(x) \in \varprojlim_M \varinjlim_T H^2(G_F^T, T_2^M)$$

(where we identify global cohomology with its image under the injective localisation in order to apply the boundary map to elements of E_1) and

$$E_1^2 := \mathrm{Ker}(\mathrm{prec}_1^2).$$

Given $x \in E_1^2$ we will denote by x_M the projection to

$$[E_1^2]_M := \mathrm{Ker}[\delta_1^g | E(\Delta_1^M)].$$

We will be considering various inverse limits over M below, and using superscripts M in a consistent fashion.

Now define

$$W(\Delta_2^M) \subset {\prod}' H^1(G_v, \Delta_2^M)$$

to be the inverse image of $[E_1^2]_M$ under the projection map

$$p_1 : {\prod}' H^1(G_v, \Delta_2^M) \longrightarrow {\prod}' H^1(G_v, \Delta_1^M),$$

which is, therefore, a $\prod' H^1(G_v, T_2^M)$-torsor over $[E_1^2]_M$. (By (2.28) an element of $[E_1^2]_M$ is liftable to $\prod' H^1(G_v, \Delta_2^M)$.)

Consider the following diagram:

$$\begin{array}{ccc}
E(T_2^M) & \hookrightarrow & \prod' H^1(G_v, T_2^M) \\
\cap\downarrow & & \cap\downarrow \\
E(\Delta_2^M) & \hookrightarrow & W(\Delta_2^M) \\
\downarrow & & \downarrow \\
[E_1^2]_M & = & [E_1^2]_M
\end{array}$$

We see with this that $E(\Delta_2^M)$ provides a reduction of structure group for $W(\Delta_2^M)$ from $\prod' H^1(G_v, T_2^M)$ to $E(T_2^M)$. That is, $W(\Delta_2^M)$ is the torsor pushout of $E(\Delta_2^M)$ with respect to the map

$$E(T_2^M) \longrightarrow \prod{}' H^1(G_v, T_2^M).$$

Choose a set-theoretic splitting

$$s_1 : [E_1^2]_M \longrightarrow E(\Delta_2^M)$$

of the torsor in the left column. We then use this 'global' splitting to define

$$\mathrm{prec}_2^M : W(\Delta_2^M) \longrightarrow H^1(G_F, D(T_2^M))^\vee$$

by the formula

$$\mathrm{prec}_2^M(x) = \mathrm{prec}(T_2^M)(x - s_1(p_1(x)))$$

Here, we denote by $x - s_1(p_1(x))$ the unique element $z \in \prod' H^1(G_v, T_2^M)$ such that $x = s_1(p_1(x)) + z$. (We are using additive notation because the context is the action of a vector group on a torsor.)

Because $E(T_2^M)$ is killed by $\mathrm{prec}(T_2^M)$, it is easy to see that

Proposition 2.2 prec_2^M *is independent of the splitting* s_1.

Now define

$$W_2 := \varprojlim_M W_2(\Delta_2^M)$$

and

$$\mathrm{prec}_2 := \varprojlim_M \mathrm{prec}_2^M : W_2 \longrightarrow \varprojlim_M H^1(G_F, D(T_2^M))^\vee = H^1(G_F, D(T_2))^\vee.$$

In general, define

$$E_n := \varprojlim_M E(\Delta_n^M)$$

and

$$\mathrm{prec}_n^{n+1} := \delta_n^g : E_n \longrightarrow \varprojlim_M \varinjlim_T H^2(G_F^T, T_{n+1}^M).$$

Then define

$$E_n^{n+1} = \mathrm{Ker}(\delta_n^g) \subset E_n,$$

and

$$W(\Delta_{n+1}^M) = p_n^{-1}([E_n^{n+1}]_M),$$

where $[E_n^{n+1}]_M = \mathrm{Ker}(\delta^g | E(\Delta_n^M))$. As when $n = 1$, (2.28) implies that $W(\Delta_{n+1}^M)$ is a $\prod' H^1(G_v, T_{n+1}^M)$ torsor over $[E_n^{n+1}]_M$. Use a splitting s_n of

$$E(\Delta_{n+1}^M) \longrightarrow [E_n^{n+1}]_M$$

to define

$$\mathrm{prec}_{n+1}^M : W(\Delta_{n+1}^M) \longrightarrow H^1(G_F, D(T_{n+1}^M))^\vee$$

via the formula

$$\mathrm{prec}_{n+1}^M(x) = \mathrm{prec}(T_{n+1}^M)(x - s_n(p_n(x))).$$

Once again, because $E(T_{n+1}^M)$ is killed by $\mathrm{prec}(T_{n+1}^M)$, we get

Proposition 2.3 prec_n^M *is independent of the splitting* s_n.

Finally, define

$$W_{n+1} := \varprojlim_M W(\Delta_{n+1}^M)$$

and

$$\mathrm{prec}_{n+1} = \varprojlim_M \mathrm{prec}_{n+1}^M : W_{n+1} \longrightarrow \varprojlim_M H^1(G_F, D(T_{n+1}^M))^\vee = H^1(G_F, D(T_{n+1}))^\vee.$$

Then we finally have the following generalisation of Proposition 2.1.

Proposition 2.4

$$\mathrm{Ker}(\mathrm{prec}_{n+1}) = E_{n+1}.$$

Proof We have seen this already for $n = 1$. Let $x \in \mathrm{Ker}(\mathrm{prec}_{n+1})$ and x_M the projection to $\mathrm{Ker}(\mathrm{prec}^M_{n+1})$. It is clear from the definition that $E(\Delta^M_{n+1}) \subset Ker(\mathrm{prec}^M_{n+1})$. On the other hand, if $\mathrm{prec}^M_{n+1}(x_M) = 0$, then $y_M = x_M - s_n(p_n(x_M)) \in E(T^M_{n+1})$, by Proposition 2.1. Hence, $x_M = y_M + s_n(p_n(x_M)) \in E(\Delta^M_{n+1})$. Since this is true for all M, $x \in E_{n+1} = \varprojlim E(\Delta^M_{n+1})$. (The assertion is a kind of 'left exactness' of the inverse limit for pointed sets, although we are giving a direct argument.) □

3 Reciprocity

Recall the product of the local period maps

$$j_n^M : X(\mathbb{A}_F) \longrightarrow \prod{}' H^1(G_v, \Delta_n^M),$$

$$x \mapsto (\pi_1^{et}(\bar{X}; b, x_v)_n^M)_v.$$

Here,

$$\pi_1^{et}(\bar{X}; b, x_v)_n^M := \pi_1^{et}(\bar{X}; b, x_v) \times_{\pi_1^{et}(\bar{X},b)} \Delta_n^M = [\pi_1^{et}(\bar{X}; b, x_v) \times \Delta_n^M]/\pi_1^{et}(\bar{X}, b),$$

(where the $\pi_1^{et}(\bar{X}, b)$-action at the end is the diagonal one giving the pushout torsor) are torsors for Δ_n^M with compatible actions of G_v, and hence, define classes in $H^1(G_v, \Delta_n^M)$. When $v \notin S$ for S admissible and $x_v \in X(\mathcal{O}_{F_v})$, then this class belongs to $H^1_{un}(G_v, \Delta_n^M)$ ([13], Proposition 2.3). Therefore, $(\pi_1^{et}(\bar{X}; b, x_v)_n^M)_v$ defines a class in $\prod' H^1(G_v, \Delta_n^M)$. (This discussion is exactly parallel to the unipotent case [6, 7].) Clearly, we can then take the limit over M, to get the period map

$$j_n : X(\mathbb{A}_F) \longrightarrow \varprojlim_M \prod{}' H^1(G_v, \Delta_n^M).$$

The reciprocity maps will be defined by

$$\mathrm{rec}_n(x) = \mathrm{prec}_n(j_n(x)),$$

and

$$\mathrm{rec}_n^{n+1}(x) = \mathrm{prec}_n^{n+1}(j_n(x)).$$

Of course, these maps will not be defined on all of $X(\mathbb{A}_F)$. As in the introduction, define

$$X(\mathbb{A}_F)_1^2 = Ker(\text{rec}_1).$$

Then for $x \in X(\mathbb{A}_F)_1^2$, $j_1(x) \in E(\Delta_1)$, and hence, prec_1^2 is defined on $j_1(x)$. Thus, rec_1^2 is defined on $X(\mathbb{A}_F)_1^2$. Now define

$$X(\mathbb{A}_F)_2 := \text{Ker}(\text{rec}_1^2).$$

Then for $x \in X(\mathbb{A}_F)_2$, $j_1(x) \in E_1^2$, so that $j_2(x) \in W_2$. Hence, prec_2 is defined on $j_2(x)$, and rec_2 is defined on $X(\mathbb{A}_F)_2$.

In general, the following proposition is now clear.

Proposition 3.1 *Assume we have defined*

$$X(\mathbb{A}_F)_1^2 \supset X(\mathbb{A}_F)_2 \supset \cdots \supset X(\mathbb{A}_F)_{n-1}^n \supset X(\mathbb{A}_F)_n$$

as the iterative kernels of $\text{rec}_1, \text{rec}_1^2, \ldots, \text{rec}_{n-1}, \text{rec}_{n-1}^n$. *Then,* $j_n(x) \in W_n$ *for* $x \in X(\mathbb{A}_F)_n$ *so that* $\text{rec}_n = \text{prec}_n \circ j_n$ *is defined on* $X(\mathbb{A}_F)_n$ *and* rec_n^{n+1} *is defined on* $\text{Ker}(\text{rec}_n)$.

Note that prec_{n-1}^n takes values in $\varprojlim_M \varinjlim_T H^2(G_F^T, T_n^M)$. However, $j_{n-1}(x)$ lifts to $j_n(x) \in \varprojlim_M \prod' H^1(G_v, \Delta_n^M)$, and hence, is clearly in the kernel of δ_n. Therefore,

$$\text{prec}_{n-1}^n(j_{n-1}(x)) \in \varprojlim_M \varinjlim_T s\text{Ш}_T^2(T_n^M) =: \mathfrak{G}_{n-1}^n(X)$$

for all $x \in X(\mathbb{A}_F)_{n-1}$.

The global reciprocity law of Theorem 1.1,

$$X(F) \subset X(\mathbb{A}_F)_\infty,$$

now follows immediately from the commutativity of the diagram

$$\begin{array}{ccc} X(F) & \hookrightarrow & X(\mathbb{A}_F) \\ \downarrow & & \downarrow j_n \\ E(\Delta_n^M) = \varinjlim_T H^1(G_F^T, \Delta_n^M) & \hookrightarrow & \prod' H^1(G_v, \Delta_n^M) \end{array}$$

for each M.

To check compatibility with the usual reciprocity map for $X = \mathbb{G}_m$ note that the map

$$F_v^* \xrightarrow{\kappa} H^1(G_v, \hat{\mathbb{Z}}(1)^{(2)}) \xrightarrow{D} H^1(G_v, \bigoplus_{p \neq 2} \mathbb{Q}_p/\mathbb{Z}_p)^\vee = G_v^{\text{ab},(2)}$$

is the local reciprocity map ([9], Corollary (7.2.13), with the natural modification for the prime-to-2 part). Here, κ is the map given by Kummer theory, while D is local duality as before. Furthermore, the localization

$$H^1(G_F, \bigoplus_{p\neq 2} \mathbb{Q}_p/\mathbb{Z}_p) \xrightarrow{\text{loc}_v} H^1(G_v, \bigoplus_{p\neq 2} \mathbb{Q}_p/\mathbb{Z}_p)$$

is dual to the map,

$$G_v^{(2)} \longrightarrow G_F^{(2)}$$

induced by $\bar{F} \hookrightarrow \bar{F}_v$, so that the dual of localization

$$G_v^{\text{ab},(2)} = H^1(G_v, \bigoplus_{p\neq 2} \mathbb{Q}_p/\mathbb{Z}_p)^\vee \xrightarrow{\text{loc}_v^*} H^1(G_F, \bigoplus_{p\neq 2} \mathbb{Q}_p/\mathbb{Z}_p)^\vee = G_F^{\text{ab},(2)}$$

is simply the natural map we started out with. Since the global reciprocity map is the sum of local reciprocity maps followed by the inclusion of decomposition groups, we are done.

Acknowledgements It is a great pleasure and an honour to dedicate this humble paper to John Coates in gratitude for all his kind guidance and support over the years. I am also grateful to Henri Darmon, Jonathan Pridham, Romyar Sharifi, Junecue Suh, and Andrew Wiles for many helpful discussions on reciprocity laws. Finally, I am very grateful to the referee for a careful reading that gave rise to many useful corrections and clarifications.

Appendix I: A Few Lemmas on Non-abelian Cohomology

We include here some basic facts for the convenience of the reader.

Given a continuous action

$$\rho : G \times U \longrightarrow U$$

of topological group G on a topological group U, we will only need $H^0(G, U)$ and $H^1(G, U)$ in general. Of course $H^0(G, U) = U^\rho \subset U$ is the subgroup of G-invariant elements. (We will put the homomorphism ρ into the notation or not depending upon the needs of the situation.) Meanwhile, we define

$$H^1(G, U) = U \backslash Z^1(G, U).$$

Here, $Z^1(G, U)$ consists of the 1-cocycles, that is, continous maps $c : G \longrightarrow U$ such that

$$c(gh) = c(g)gc(h),$$

while the U action on it is given by

$$(uc)(g) := uc(g)g(u^{-1}).$$

We also need $H^2(G, A)$ for A abelian defined in the usual way as the 2-cocycles, that is, continuous functions $c : G \times G \longrightarrow A$ such that

$$gc(h, k) - c(gh, k) + c(g, hk) - c(g, h) = 0,$$

modulo the subgroup of elements of the form

$$df(g, h) = f(gh) - f(g) - gf(h)$$

for $f : G \longrightarrow A$ continuous. Any $H^i(G, U)$ defined in this way is pointed by the class of the constant map $G^n \longrightarrow e \in U$, even though it is a group in general only for U abelian. We denote by $[c]$ the equivalence class of a cocycle c.

Given a 1-cocycle $c \in Z^1(G, U)$, we can define the twisted action

$$\rho_c : G \longrightarrow \mathrm{Aut}(U)$$

as

$$\rho_c(g)u = c(g)\rho(g)(u)c(g)^{-1}.$$

The isomorphism class of this action depends only on the equivalence class $[c]$.

Given an exact sequence

$$0 \longrightarrow A \xrightarrow{i} B \xrightarrow{q} C \longrightarrow 0$$

of topological groups with G action such that the last map admits a continuous splitting (not necessarily a homomorphism) and A is central in B, we get the exact sequence

$$\begin{aligned} 0 \longrightarrow H^0(G, A) \longrightarrow H^0(G, B) \longrightarrow H^0(G, C) \longrightarrow \\ \longrightarrow H^1(G, A) \longrightarrow H^1(G, B) \longrightarrow H^1(G, C) \longrightarrow H^2(G, A) \end{aligned}$$

of pointed sets, in the sense that the image of one map is exactly the inverse image of the base-point for the next map ([12], Appendix to Chap. VII).

But there are several bits more of structure. Consider the fibers of the map

$$i_* : H^1(G, A) \longrightarrow H^1(G, B).$$

The group $H^0(G, C)$ will act on $H^1(G, A)$ as follows. For $\gamma \in H^0(G, C)$, choose a lift to $b \in B$. For $x \in Z^1(G, A)$, let $(bx)(g) = bx(g)g(b^{-1})$. Because γ is G-invariant, this take values in A, and defines a cocycle. Also, a different choice of b will result in an equivalent cocycle, so that the action on $H^1(G, A)$ is well-defined. From the definition, the $H^0(G, C)$-action preserves the fibers of i_*. Conversely, if $[x]$ and $[x']$ map to the same element of $H^1(G, B)$, then there is a $b \in B$ such that $x'(g) = bx(g)g(b^{-1})$ for all $g \in G$. But then by applying q, we get $1 = q(b)g(q(b)^{-1})$, that is, $q(b) \in H^0(G, C)$. We have shown:

Lemma 4.1 *The fibers of i_* are exactly the $H^0(G, C)$-orbits of $H^1(G, A)$.*

We can say more. Given $x \in Z^1(G, A)$ and $y \in Z^1(G, B)$, consider the map

$$(xy)(g) := x(g)y(g).$$

This is easily seen to be in $Z^1(G, B)$ and defines an action of $H^1(G, A)$ on $H^1(G, B)$.

Lemma 4.2 *Suppose $C^{\rho_z} = 1$ for all $[z] \in H^1(G, C)$. Then $H^1(G, A)$ acts freely on $H^1(G, B)$.*

Proof Fix an element $[y] \in H^1(G, B)$. We work out its stabilizer. We have $[x][y] = [xy] = [y]$ if and only if there is a $b \in B$ such that $x(g)y(g) = by(g)g(b^{-1})$. By composing with q, we get

$$qy(g) = q(b)qy(g)g(q(b)^{-1})$$

or

$$qy(g)g(q(b))qy(g)^{-1} = q(b).$$

This says that $q(b)$ is invariant under the G-action ρ_{qy} given by

$$c \mapsto qy(g)g(c)qy(g)^{-1}.$$

Hence, by assumption, $q(b) = 1$, and hence, $b \in A$. But then, $x(g)y(g) = bg(b^{-1})y(g)$ for all g, from which we deduce that $x(g) = bg(b^{-1})$ for all g, so that $[x] = 0$. □

On the other hand,

Lemma 4.3 *The action of $H^1(G, A)$ is transitive on the fibers of $q_* : H^1(G, B) \to H^1(G, C)$.*

Proof The action clearly preserves the fiber. Now suppose $[qy] = [qy'] \in H^1(G, C)$. Then there is a $c \in C$ such that

$$qy'(g) = cqy(g)g(c^{-1})$$

for all g. We can lift c to $b \in B$, from which we get

$$y'(g) = x(g)by(g)g(b^{-1})$$

for some $x(g) \in A$. Since y, y' and $g \mapsto by(g)g(b^{-1})$ are all cocycles and A is central, this equality implies that

$$x : G \longrightarrow A$$

is a cocycle, and $[y'] = [x][y]$. □

The existence of the continuous splitting of exact sequences that we need for applying the results above always holds in the profinite case.

Lemma 4.4 *Suppose we have an exact sequence of profinite groups*

$$0 \longrightarrow A \longrightarrow B \stackrel{p}{\longrightarrow} C \longrightarrow 0$$

where all maps are continuous. Suppose $B = \varprojlim_j B_j$*, where the* j *run over natural numbers. Then there is a continuous section to the map* $B \longrightarrow C$.

Proof If $B = \varprojlim B_j$, by replacing each B_j with the image of B if necessary, we can assume all the maps in the inverse system are surjective. Furthermore, if A_j is the image of A in B_j, and $C_j = B_j/A_j$, one gets $A = \varprojlim A_j$ (since A is closed in B) and $C = \varprojlim C_j$. That is, the exact sequence of profinite groups can be constructed as an inverse limit of exact sequences

$$0 \longrightarrow A_j \longrightarrow B_j \stackrel{p_j}{\longrightarrow} C_j \longrightarrow 0$$

indexed by the same category in such as way that all the transition maps

$$A_i \longrightarrow A_j,\ B_i \longrightarrow B_j,\ C_i \longrightarrow C_j$$

are surjective. From the commutative diagram

$$\begin{array}{ccc} B_i & \longrightarrow & C_i \\ {\scriptstyle f}\downarrow & & \downarrow{\scriptstyle g} \\ B_j & \longrightarrow & C_j \end{array}$$

we get the commutative diagram

$$\begin{array}{ccc} B_i & \xrightarrow{h} & B_j \times_{C_j} C_i \\ & \searrow & \downarrow \\ & & C_i \end{array}$$

We claim that the map h is surjective. To see this, let $c_j \in C_j$ and $b_j \in B_j$ map to c_j. We need to check that $f^{-1}(b_j)$ surjects onto $g^{-1}(c_j)$. Let $c_i \in g^{-1}(c_j)$. Choose $b'_i \in B_i$ mapping to c_i and let $b'_j = f(b'_i)$. Since b'_j and b_j both map to c_j, there is an $a_j \in A_j$ such that $b'_j = b_j + a_j$. Now choose a_i mapping to a_j and put $b_i = b'_i - a_i$. Then $b_i \in f^{-1}(b_j)$ and it still maps to c_i. This proves the claim.

For any fixed j, suppose we've chosen a section s_j of $B_j \longrightarrow C_j$. Then

$$s_j \circ g : C_i \longrightarrow B_j$$

defines a section of

$$B_j \times_{C_j} C_i \longrightarrow C_i$$

This section can then be lifted to a section s_i of $B_i \longrightarrow C_i$. Thereby, we have constructed a diagram of sections

$$\begin{array}{ccc} C_i & \xrightarrow{s_i} & B_i \\ \downarrow & & \downarrow \\ C_j & \xrightarrow{s_j} & B_j \end{array}$$

By composing s_j with the projection $C \longrightarrow C_j$, we have a compatible sequence of maps

$$C \xrightarrow{f_j = s_j \circ g_j} B_j$$

such that $p_j \circ f_j = g_j$. Thus, we get a continuous map $f : C \longrightarrow B$ such that $p \circ f = Id$. □

It has been pointed out by the referee that a more general statement can be found in [10], Proposition 2.2.2. However, we will retain the proof above for the convenience of the reader. That is, a continuous section exists in circumstances more general than countably ordered inverse limits, but we have just recalled this case since it is all we will need. This applies for example when B is the pro-M completion of a finitely-generated group: For every n, we can let $B(n) \subset B$ be the intersection of open subgroups of index $\leq n$. This is a characteristic subgroup, and still open. So the quotients defining the inverse limit can be taken as $B/B(n)$.

Appendix II: Some Complements on Duality for Galois Cohomology

When A is topological abelian group, $A^\vee$ denotes the continuous homomorphisms to the discrete group $\mathbb{Q}/\mathbb{Z}$. Thus, in the profinite case of $A = \varprojlim A/H$, where the H run over any defining system of open normal subgroups of finite index,

$$A^\vee = \varinjlim_H \mathrm{Hom}(A/H, \mathbb{Q}/\mathbb{Z})$$

with the discrete topology. If $A = \varinjlim_m A[m]$ is a torsion abelian group with the discrete topology, then

$$A^\vee = \varprojlim \mathrm{Hom}(A[m], \mathbb{Q}/\mathbb{Z})$$

with the projective limit topology. Meanwhile, if A has a continuous action of the Galois group of a local or a global field, then $D(A)$ denotes the continuous homomorphisms to the discrete group

$$\mu_\infty = \varinjlim_m \mu_m$$

with Galois action. As far as the topological group structure is concerned, $D(A)$ is of course the same as $A^\vee$.

We let F be a number field and T a finite set of places of F including the Archimedean places. We denote by G_F the Galois group $\mathrm{Gal}(\bar{F}/F)$ and by $G_F^T = \mathrm{Gal}(F_T/F)$ the Galois group of the maximal extension F_T of F unramified outside T. Let v be a place of F, and equip $G_v = \mathrm{Gal}(\bar{F}_v/F_v)$ with a choice of homomorphism $G_v \longrightarrow G_F \longrightarrow G_F^T$ given by the choice of an embedding $\bar{F} \hookrightarrow \bar{F}_v$

In the following A (with or without Galois action) will be in the abelian subcategory of all abelian groups generated by topologically finitely-generated profinite abelian groups and torsion groups A such that $A^\vee$ is topologically finitely-generated. We choose this category to give a discussion of duality.

We have local Tate duality

$$H^i(G_v, A) \simeq^D H^{2-i}(G_v, D(A))^\vee.$$

We also use the same letter D to denote the product isomorphisms

$$\prod_{v\in T'} H^i(G_v, A) \simeq^D \prod_{v\in T'} H^{2-i}(G_v, D(A))^\vee$$

for any indexing set T'.

Let $Ш^i_T(A)$ be the kernel of the localization map

$$Ш^i_T(A) := \mathrm{Ker}[H^i(G^T_F, A) \xrightarrow{\mathrm{loc}_T} \prod_{v\in T} H^i(G_v, A)]$$

and $Im^i_T(A)$, the image of the localization map

$$Im^i_T(A) := Im[H^i(G^T_F, A) \xrightarrow{\mathrm{loc}_T} \prod_{v\in T} H^i(G_v, A)].$$

Assume now that $A = \varprojlim A_n$, where T contains all the places lying above primes dividing the order of any A_n. According to Poitou–Tate duality, we have an isomorphism

$$Ш^i_T(A) \simeq Ш^{2-i}_T(D(A))^\vee,$$

and an exact sequence

$$H^i(G^T_F, A) \longrightarrow \prod_{v\in T} H^i(G_v, A) \xrightarrow{\mathrm{loc}^*_T \circ D} H^{2-i}(G^T_F, D(A))^\vee$$

Note that this is usually stated for finite coefficients.[2] But since all the groups in the exact sequence

$$H^i(G^T_F, A_n) \longrightarrow \prod_{v\in T} H^i(G_v, A_n) \xrightarrow{\mathrm{loc}^*_T \circ D} H^{2-i}(G^T_F, D(A_n))^\vee$$

are finite, we can take an inverse limit to get the exact sequence above (since the inverse limit is exact on inverse systems of finite groups).

If $T' \supset T$, since all the inertia subgroups $I_v \subset G_v$ for $v \notin T$ act trivially on A, we have

$$Im^1_{T'}(A) \cap [\prod_{v\in T} H^1(G_v, A) \times \prod_{v\in T'\setminus T} H^1(G_v/I_v, A)] = Im^1_T(A).$$

[2]It has been pointed out by the referee that this general case is well-known, for example, in work of Nekovar or Schneider.

In particular, we have an exact sequence

$$H^1(G_F^T, A) \xrightarrow{\text{loc}_{T'}} \prod_{v \in T} H^1(G_v, A) \times \prod_{v \in T' \setminus T} H^1(G_v/I_v, A) \xrightarrow{\text{loc}^*_{T'} \circ D} H^1(G_F^{T'}, D(A))^\vee$$

Taking an inverse limit over T', we get an exact sequence

$$H^1(G_F^T, A) \xrightarrow{\text{loc}} \prod_{v \in T} H^1(G_v, A) \times \prod_{v \notin T} H^1(G_v/I_v, A) \xrightarrow{\text{loc}^* \circ D} H^1(G_F, D(A))^\vee$$

References

1. Artin, E., Tate, J.: Class Field Theory. Reprinted with corrections from the 1967 original. AMS Chelsea Publishing, Providence, RI. viii+194 pp. (2009)
2. Balakrishnan, J., Dan-Cohen, I., Kim, M., Wewers, S.: A Non-abelian Conjecture of Birch and Swinnerton-Dyer Type for Hyperbolic Curves. arXiv:1209.0640
3. Grothendieck, A.: Séminaire de Géométrie Algébrique du Bois Marie—1960–61—Revêtements étales et groupe fondamental—(SGA 1) (Lecture notes in mathematics 224). Springer, Berlin, New York, xxii+447 p
4. Jannsen, U.: On the l-adic cohomology of varieties over number fields and its Galois cohomology. In: Galois Groups Over Q (Berkeley, CA, 1987), pp. 315–360, Math. Sci. Res. Inst. Publ., 16, Springer, New York (1989)
5. Kim, M.: The motivic fundamental group of $\mathbb{P}^1 \setminus \{0, 1, \infty\}$ and the theorem of Siegel. Invent. Math. **161**(3), 629–656 (2005)
6. Kim, M.: The unipotent Albanese map and Selmer varieties for curves. Publ. Res. Inst. Math. Sci. 45(1), 89–133 (2009)
7. Kim, M., Tamagawa, A.: The ℓ-component of the unipotent Albanese map. Math. Ann. **340**(1), 223–235 (2008)
8. Manin, Y. I.: Le groupe de Brauer–Grothendieck en géométrie diophantienne. Actes du Congrès International des Mathématiciens (Nice, 1970), Tome 1, pp. 401–411. Gauthier-Villars, Paris (1971)
9. Neukirch, J., Schmidt, A., Wingberg, K.: Cohomology of number fields, 2nd edn. In: Grundlehren der Mathematischen Wissenschaften, vol. 323. Springer, Berlin, xvi+825 pp. (2008)
10. Ribes, L., Zalesskii, P.: Profinite groups, 2nd edn. In: Ergebnisse der Mathematik und ihrer Grenzgebiete. 3. Folge. A Series of Modern Surveys in Mathematics, vol. 40. Springer, Berlin (2010)
11. Serre, J.-P.: A course in arithmetic. Translated from the French. In: Graduate Texts in Mathematics, No. 7. Springer, New York, Heidelberg. viii+115 pp. (1973)
12. Serre, J.-P.: Local fields. Translated from the French by Marvin Jay Greenberg. In: Graduate Texts in Mathematics, vol. 67. Springer, New York, Berlin, viii+241 pp. (1979)
13. Wojtkowiak, Z.: On the Galois Actions on Torsors of Paths. I. Descent of Galois representations. J. Math. Sci. Univ. Tokyo **14**(2), 177–259 (2007)

On p-adic Interpolation of Motivic Eisenstein Classes

Guido Kings

Abstract In this paper we prove that the motivic Eisenstein classes associated to polylogarithms of commutative group schemes can be p-adically interpolated in étale cohomology. This connects them to Iwasawa theory and generalizes and strengthens the results for elliptic curves obtained in our former work. In particular, degeneration questions can be treated easily.

Keywords Polylogarithms · étale Eisenstein cohomology · Iwasawa cohomology

1 Introduction

In this paper we prove that the motivic Eisenstein classes associated to polylogarithms of commutative group schemes can be p-adically interpolated in étale cohomology. This generalizes the results for elliptic curves obtained in our former paper [12]. Already in the one dimensional case the results obtained here are stronger and much more flexible as they allow to treat degenerating elliptic curves easily.

The interpolation of motivic Eisenstein classes connects them with Iwasawa theory and is essential for many applications. In the elliptic case for example, the interpolation was used in [11] to prove a case of the Tamagawa number conjecture for CM elliptic curves and it was one of the essential ingredients in the proof of an explicit reciprocity law for Rankin-convolutions in [13]. We hope that the general case will find similar applications.

Before we explain our results, we have to introduce the motivic Eisenstein classes (for the construction we refer to Sect. 4.2).

To John Coates, on the occasion of his 70th birthday

This research was supported by the DFG grant: SFB 1085 "Higher invariants".

G. Kings (✉)
Fakultät Für Mathematik, Universität Regensburg, 93040 Regensburg, Germany
e-mail: guido.kings@mathematik.uni-regensburg.de

D. Loeffler and S.L. Zerbes (eds.), *Elliptic Curves, Modular Forms and Iwasawa Theory*, Springer Proceedings in Mathematics & Statistics 188, DOI 10.1007/978-3-319-45032-2_10

Let $\pi : G \to S$ be a smooth commutative and connected group scheme of relative dimension d (for example a semi-abelian scheme) and denote by

$$\mathscr{H} := R^1\pi_!\mathbb{Z}_p(1)$$

the first étale homology of G/S, which is just the sheaf of relative p-adic Tate modules of G/S. We write $\mathscr{H}_{\mathbb{Q}_p}$ for the associated $\mathbb{Q}_p$-adic sheaf. Note that this is not a lisse sheaf in general. Evaluating the motivic polylogarithm at a non-zero N-torsion section $t : S \to G$ one defines motivic Eisenstein classes

$$_\alpha\mathrm{Eis}^k_{\mathrm{mot}}(t) \in H^{2d-1}_{\mathrm{mot}}(S, \mathrm{Sym}^k \mathscr{H}_{\mathbb{Q}}(d)),$$

depending on some auxiliary data α, where $\mathrm{Sym}^k \mathscr{H}_{\mathbb{Q}}(1)$ is the k-th symmetric tensor power of the motivic sheaf $\mathscr{H}_{\mathbb{Q}}$ which underlies $\mathscr{H}_{\mathbb{Q}_p}$.

In the case of an elliptic curve, the de Rham realization of $_\alpha\mathrm{Eis}^k_{\mathrm{mot}}(t)$ is the cohomology class of a holomorphic Eisenstein series, which justifies the name. These motivic Eisenstein classes in the elliptic case play a major role in Beilinson's proof of his conjectures on special values of L-functions for modular forms.

In this paper we consider the étale regulator

$$r_{\text{ét}} : H^{2d-1}_{\mathrm{mot}}(S, \mathrm{Sym}^k \mathscr{H}_{\mathbb{Q}}(d)) \to H^{2d-1}(S, \mathrm{Sym}^k \mathscr{H}_{\mathbb{Q}_p}(d))$$

which gives rise to the étale Eisenstein classes

$$_\alpha\mathrm{Eis}^k_{\mathbb{Q}_p}(t) := r_{\text{ét}}(\mathrm{Eis}^k_{\mathrm{mot}}(t)) \in H^{2d-1}(S, \mathrm{Sym}^k \mathscr{H}_{\mathbb{Q}_p}(d)).$$

In the elliptic case these classes were used by Kato in his seminal work to construct Euler systems for modular forms.

It is a natural question, whether these étale Eisenstein classes enjoy some p-adic interpolation properties, in a similar way as one can p-adically interpolate the holomorphic Eisenstein series. At first sight, this seems to be a completely unreasonable question, as for varying k the different motivic cohomology groups $H^{2d-1}_{\mathrm{mot}}(S, \mathrm{Sym}^k \mathscr{H}_{\mathbb{Q}}(1))$ are not related at all. Nevertheless, this question was answered affirmatively in the elliptic case in [12] and in this paper we will generalize this result to commutative group schemes.

To explain our answer to this question we need the *sheaf of Iwasawa-algebras* $\Lambda(\mathscr{H})$, which is defined as follows: One first defines a sheaf of "group rings" $\mathbb{Z}/p^r\mathbb{Z}[\mathscr{H}_r]$ on S, where $\mathscr{H}_r$ is the étale sheaf associated to the $[p^r]$-torsion subgroup $G[p^r]$ or alternatively the first homology with $\mathbb{Z}/p^r\mathbb{Z}$-coefficients (see Sect. 5.4 for more details). These group rings form an inverse system for varying r and hence define a pro-sheaf

$$\Lambda(\mathscr{H}) := (\mathbb{Z}/p^r\mathbb{Z}[\mathscr{H}_r])_{r\geq 0}.$$

Moreover, it is also possible to sheafify the classical moments of a measure to a morphism of pro-sheaves

$$\mathrm{mom}^k : \Lambda(\mathscr{H}) \to \Gamma_k(\mathscr{H}),$$

where $\Gamma_k(\mathscr{H})$ is the k-th graded piece of the divided power algebra $\Gamma_{\mathbb{Z}_p}(\mathscr{H})$. Thus the sheaf $\Lambda(\mathscr{H})$ p-adically interpolates the $\Gamma_k(\mathscr{H})$. For the $\mathbb{Q}_p$-sheaf $\mathscr{H}_{\mathbb{Q}_p}$ the natural map $\mathrm{Sym}^k \mathscr{H}_{\mathbb{Q}_p} \to \Gamma_k(\mathscr{H}_{\mathbb{Q}_p})$ is an isomorphism and the moment map gives rise to morphisms

$$\mathrm{mom}^k : H^{2d-1}(S, \Lambda(\mathscr{H})(d)) \to H^{2d-1}(S, \mathrm{Sym}^k \mathscr{H}_{\mathbb{Q}_p}(d)).$$

To understand this better, it is instructive to consider the case of an abelian scheme $\pi : A \to S$ over a scheme S which is of finite type over $\mathrm{Spec}\, \mathbb{Z}$ (see also Sect. 6.5). Then

$$H^{2d-1}(S, \Lambda(\mathscr{H})(d)) = \varprojlim_r H^{2d-1}(A[p^r], \mathbb{Z}/p^r\mathbb{Z}(d))$$

where the inverse limit is taken with respect to the trace maps along $A[p^r] \to A[p^{r-1}]$. The right hand side is obviously an Iwasawa theoretic construction. In the one dimensional case $d = 1$, the right hand side has an interpretation as an inverse limit of units via Kummer theory.

Our main result can now be formulated as follows:

Main Theorem (see Theorem 7.3.3) *There exists a cohomology class*

$${}_{\alpha}\mathcal{EI}(t)_N \in H^{2d-1}(S, \Lambda(\mathscr{H})(d))$$

called the Eisenstein–Iwasawa class, *such that*

$$\mathrm{mom}^k({}_{\alpha}\mathcal{EI}(t)_N) = N^k {}_{\alpha}\mathrm{Eis}^k_{\mathbb{Q}_p}(t).$$

This interpolation result in the elliptic case is one of the key ingredients in the proof of an explicit reciprocity law for Rankin-convolutions of modular forms in [13].

The use of this theorem also considerably simplifies the computations of the degeneration of polylogarithm in [12]. We hope to treat this at another occasion.

We would also like to point out an important open problem: In the one-dimensional torus or the elliptic curve case, the Eisenstein–Iwasawa class has a direct description in terms of cyclotomic units or Kato's norm compatible elliptic units respectively. Unfortunately, we do not have a similar description of the Eisenstein–Iwasawa class in the higher dimensional case.

2 Notations and Set up

2.1 The Category of $\mathbb{Z}_p$-sheaves

All schemes will be separated of finite type over a noetherian regular scheme of dimension 0 or 1. Let X be such a scheme and let p be a prime invertible on X. We work in the category of constructable $\mathbb{Z}_p$-sheaves $\mathscr{S}(X)$ on X in the sense of [15, Exposé V].

Recall that a constructible $\mathbb{Z}_p$-sheaf is an inverse system $\mathscr{F} = (\mathscr{F}_n)_{n\geqslant 1}$ where $\mathscr{F}_n$ is a constructible $\mathbb{Z}/p^n\mathbb{Z}$-sheaf and the transition maps $\mathscr{F}_n \to \mathscr{F}_{n-1}$ factor into isomorphisms

$$\mathscr{F}_n \otimes_{\mathbb{Z}/p^n\mathbb{Z}} \mathbb{Z}/p^{n-1}\mathbb{Z} \cong \mathscr{F}_{n-1}.$$

The $\mathbb{Z}_p$-sheaf is *lisse*, if each $\mathscr{F}_n$ is locally constant. If X is connected and $x \in X$ is a geometric point, then the category of lisse sheaves is equivalent to the category of finitely generated $\mathbb{Z}_p$-modules with a continous $\pi_1(X, x)$-action. For a general $\mathbb{Z}_p$-sheaf there exists a finite partition of X into locally closed subschemes X_i, such that $\mathscr{F}|_{X_i}$ is lisse (see [4, Rapport, Proposition 2.4., 2.5.]).

For a $\mathbb{Z}_p$-sheaf $\mathscr{F}$ we denote by $\mathscr{F} \otimes \mathbb{Q}_p$ its image in the category of $\mathbb{Q}_p$-sheaves, i.e., the quotient category modulo $\mathbb{Z}_p$-torsion sheaves.

We also consider the "derived" category $\mathbf{D}(X)$ of $\mathscr{S}(X)$ in the sense of Ekedahl [5]. This is a triangulated category with a t-structure whose heart is the category of constructible $\mathbb{Z}_p$-sheaves. By loc. cit. Theorem 6.3 there is a full 6 functor formalism on these categories.

Recall that an inverse system $A := (A_r)_{r\geqslant 0}$ (in any abelian category $\mathcal{A}$) satisfies the Mittag-Leffler condition (resp. is Mittag-Leffler zero), if for each r the image of $A_{r+s} \to A_r$ is constant for all sufficiently big s (is zero for some $s \geqslant 1$). If A satisfies the Mittag-Leffler condition and $\mathcal{A}$ satisfies $AB4^*$ (i.e. products exists and products of epimorphisms are epimorphisms) then $\varprojlim_r^1 A_r = 0$ (see [14, Proposition 1]). If A is Mittag-Leffler zero, then for each left exact functor $h : \mathcal{A} \to \mathcal{B}$ one has $R^i\varprojlim_r h(A_r) = 0$ for all $i \geqslant 0$ ([9, Lemma 1.11.]).

For a pro-system of étale sheaves $\mathscr{F} = (\mathscr{F}_r)_{r\geqslant 0}$ on X we work with Jannsen's continuous étale cohomology $H^i(X, \mathscr{F})$ which is the i-th derived functor of $\mathscr{F} \mapsto \varprojlim_r H^0(X, \mathscr{F}_n)$. By [9, 3.1] one has an exact sequence

$$0 \to \varprojlim_r{}^1 H^{i-1}(X, \mathscr{F}_r) \to H^i(X, \mathscr{F}) \to \varprojlim_r H^i(X, \mathscr{F}_r) \to 0. \tag{1}$$

Note in particular, that if $H^{i-1}(X, \mathscr{F}_r)$ is finite for all r, one has

$$H^i(X, \mathscr{F}) = \varprojlim_r H^i(X, \mathscr{F}_r). \tag{2}$$

For $\mathscr{F} = (\mathscr{F}_r)$ Mittag-Leffler zero, one has for all $i \geqslant 0$

$$H^i(X, \mathscr{F}) = 0. \tag{3}$$

2.2 The Divided Power Algebra

Let A be a commutative ring and M be an A-module. Besides the usual symmetric power algebra $\mathrm{Sym}_A(M)$ we need also the divided power algebra $\Gamma_A(M)$ (see [1, Appendix A] for more details).

The A-algebra $\Gamma_A(M)$ is a graded augmented algebra with $\Gamma_0(M) = A$, $\Gamma_1(M) = M$ and augmentation ideal $\Gamma^+(M) := \bigoplus_{k\geqslant 1} \Gamma_k(M)$. For each element $m \in M$ one has the divided power $m^{[k]} \in \Gamma_k(M)$ with the property that $m^k = k!m^{[k]}$ where m^k denotes the k-th power of m in $\Gamma_A(M)$. Moreover, one has the formula

$$(m+n)^{[k]} = \sum_{i+j=k} m^{[i]} n^{[j]}.$$

In the case where M is a free A-module with basis $m_1, \ldots, m_r$ the A-module $\Gamma_k(M)$ is free with basis $\{m_1^{[i_1]} \cdots m_r^{[i_r]} \mid \sum i_j = k\}$. Further, for M free, there is an A-algebra isomorphism

$$\Gamma_A(M) \cong \mathrm{TSym}_A(M)$$

with the algebra of symmetric tensors ($\mathrm{TSym}_A^k(M) \subset \mathrm{Sym}_A^k(M)$ are the invariants of the symmetric group), which maps $m^{[k]}$ to $m^{\otimes k}$. Also, by the universal property of $\mathrm{Sym}_A(M)$, one has an A-algebra homomorphism

$$\mathrm{Sym}_A(M) \to \Gamma_A(M) \tag{4}$$

which maps m^k to $k!m^{[k]}$. In particular, if A is a $\mathbb{Q}$-algebra, this map is an isomorphism.

If M is free and $M^\vee := \mathrm{Hom}_A(M, A)$ denotes the A-dual one has in particular

$$\mathrm{Sym}^k(M^\vee) \cong \Gamma_k(M)^\vee \cong \mathrm{TSym}_A^k(M)^\vee.$$

The algebra $\Gamma_A(M)$ has the advantage over $\mathrm{TSym}_A(M)$ of being compatible with arbitrary base change

$$\Gamma_A(M) \otimes_A B \cong \Gamma_B(M \otimes_A B)$$

and thus sheafifies well. Recall from [8, I 4.2.2.6.] that if $\mathscr{F}$ is an étale sheaf of $\mathbb{Z}_p$-modules, then $\Gamma_{\mathbb{Z}_p}(\mathscr{F})$ is defined to be the sheaf associated to the presheaf

$$U \mapsto \Gamma_{\mathbb{Z}_p(U)}(\mathscr{F}(U)). \tag{5}$$

Definition 2.2.1 We denote by

$$\widehat{\Gamma}_A(M) := \varprojlim_r \Gamma_A(M)/\,\Gamma^+(M)^{[r]}$$

the completion of $\Gamma_A(M)$ with respect to the divided powers of the augmentation ideal.

Note that $\Gamma^+(M)^{[r]} = \bigoplus_{k\geqslant r}\Gamma_k(M)$ so that as A-module one has $\widehat{\Gamma}_A(M) \cong \prod_{k\geqslant 0}\Gamma_k(M)$.

In the same way we define the completion of $\mathrm{Sym}_A(M)$ with respect to the augmentation ideal $\mathrm{Sym}_A^+(M)$ to be

$$\widehat{\mathrm{Sym}_A}(M) := \varprojlim_k \mathrm{Sym}_A(M)/(\mathrm{Sym}_A^+(M))^k \tag{6}$$

2.3 Unipotent Sheaves

Let $\Lambda = \mathbb{Z}/p^r\mathbb{Z}, \mathbb{Z}_p$ or $\mathbb{Q}_p$ and let $\pi : X \to S$ be a separated scheme of finite type, with X, S as in Sect. 2.1. A Λ-sheaf $\mathscr{F}$ on X is *unipotent of length n*, if it has a filtration $0 = \mathscr{F}^{n+1} \subset \mathscr{F}^n \subset \ldots \subset \mathscr{F}^0 = \mathscr{F}$ such that $\mathscr{F}^i/\mathscr{F}^{i+1} \cong \pi^*\mathscr{G}^i$ for a Λ-sheaf $\mathscr{G}^i$ on S.

The next lemma is taken from [7], where it is stated in the setting of $\mathbb{Q}_p$-sheaves.

Lemma 2.3.1 *Let $\Lambda = \mathbb{Z}/p^r\mathbb{Z}, \mathbb{Z}_p$ or $\mathbb{Q}_p$ and let $\pi_1 : X_1 \to S$ and $\pi_2 : X_2 \to S$ be smooth of constant fibre dimension d_1 and d_2. Let $f : X_1 \to X_2$ be an S-morphism. Let $\mathscr{F}$ be a unipotent Λ-sheaf. Then*

$$f^!\mathscr{F} = f^*\mathscr{F}(d_1 - d_2)[2d_1 - 2d_2].$$

Proof Put $c = d_1 - d_2$ the relative dimension of f. We start with the case $\mathscr{F} = \pi_2^*\mathscr{G}$. In this case

$$\begin{aligned} f^!\mathscr{F} &= f^!\pi_2^*\mathscr{G} = f^!\pi_2^!\mathscr{G}(-d_2)[-2d_2] = \pi_1^!\mathscr{G}(-d_2)[-2d_2] \\ &= \pi_1^*\mathscr{G}(c)[2c] = f^*\pi_2^*\mathscr{G}(c)[2c] = f^*\mathscr{F}\otimes\Lambda(c)[2c]. \end{aligned}$$

In particular, $f^!\Lambda = \Lambda(c)[2c]$ and we may rewrite the formula as

$$f^*\mathscr{F}\otimes f^!\Lambda = f^!(\mathscr{F}\otimes\Lambda).$$

There is always a map from the left to right via adjunction from the projection formula

$$Rf_!(f^*\mathscr{F}\otimes f^!\Lambda) = \mathscr{F}\otimes Rf_!f^!\Lambda \to \mathscr{F}\otimes\Lambda.$$

Hence we can argue on the unipotent length of $\mathscr{F}$ and it suffices to consider the case $\mathscr{F} = \pi^*\mathscr{G}$. This case was settled above. □

The next lemma is also taken from [7]. Let $X \to S$ be a smooth scheme with connected fibres and $e : S \to X$ a section. Homomorphisms of unipotent sheaves are completely determined by their restriction to S via e^*:

Lemma 2.3.2 *Let $\pi : X \to S$ be smooth with connected fibres and $e : S \to X$ a section of π. Let $\Lambda = \mathbb{Z}/p^r\mathbb{Z}, \mathbb{Z}_p$ or $\mathbb{Q}_p$ and $\mathscr{F}$ a unipotent Λ-sheaf on X. Then*

$$e^* : \mathrm{Hom}_X(\Lambda, \mathscr{F}) \to \mathrm{Hom}_S(\Lambda, e^*\mathscr{F})$$

is injective.

Proof Let $0 \to \mathscr{F}_1 \to \mathscr{F}_2 \to \mathscr{F}_3 \to 0$ be a short exact sequence of unipotent Λ-sheaves on G. As e^* is exact and Hom left exact, we get a commutative diagram of exact sequences

$$\begin{array}{ccccccc}
0 & \longrightarrow & \mathrm{Hom}_X(\Lambda, \mathscr{F}_1) & \longrightarrow & \mathrm{Hom}_X(\Lambda, \mathscr{F}_2) & \longrightarrow & \mathrm{Hom}_X(\Lambda, \mathscr{F}_3) \\
 & & \downarrow & & \downarrow & & \downarrow \\
0 & \longrightarrow & \mathrm{Hom}_S(\Lambda, e^*\mathscr{F}_1) & \longrightarrow & \mathrm{Hom}_S(\Lambda, e^*\mathscr{F}_2) & \longrightarrow & \mathrm{Hom}_S(\Lambda, e^*\mathscr{F}_3).
\end{array}$$

Suppose that the left and right vertical arrows are injective, then the middle one is injective as well and it is enough to show the lemma in the case where $\mathscr{F} = \pi^*\mathscr{G}$. But the isomorphism

$$\mathrm{Hom}_X(\Lambda, \pi^*\mathscr{G}) \cong \mathrm{Hom}_X(\pi^!\Lambda, \pi^!\mathscr{G}) \cong \mathrm{Hom}_S(R\pi_!\pi^!\Lambda, \mathscr{G})$$

factors through

$$\mathrm{Hom}_X(\pi^!\Lambda, \pi^!\mathscr{G}) \xrightarrow{e^!} \mathrm{Hom}_S(\Lambda, \mathscr{G}) \to \mathrm{Hom}_S(R\pi_!\pi^!\Lambda, \mathscr{G})$$

where the last map is induced by the trace map $R\pi_!\pi^!\Lambda \to \Lambda$. This proves the claim. □

2.4 The Geometric Situation

We recall the geometric set up from [7] using as much as possible the notations from loc. cit. Let

$$\pi : G \to S$$

be a smooth separated commutative group scheme with connected fibres of relative dimension d. We denote by $e : S \to G$ the unit section and by $\mu : G \times_S G \to G$ the multiplication. Let $j : U \to G$ be the open complement of $e(S)$.

Let $\iota_D : D \to G$ be a closed subscheme with structural map $\pi_D : D \to S$. Typically π_D will be étale and contained in the c-torsion of G for some $c \geqslant 1$. We note in passing, that for c invertible on S the c-torsion points of G, i.e. the kernel of the c-multiplication $G[c]$, is quasi-finite and étale over S. Denote by $j_D : U_D = G \setminus D \to G$ the open complement of D. We summarize the situation in the basic diagram

$$\begin{array}{ccccc} U_D := G \setminus D & \xrightarrow{j_D} & G & \xleftarrow{\iota_D} & D \\ & \searrow & \downarrow{\scriptstyle \pi} & \swarrow{\scriptstyle \pi_D} & \\ & & S & & \end{array}$$

We will also consider morphisms $\phi : G_1 \to G_2$ of S-group schemes as above. In this case we decorate all notation with an index 1 or 2, e.g., d_1 for the relative dimension of G_1/S.

3 The Logarithm Sheaf

3.1 Homology of G

The basic sheaf in our constructions is the relative first $\mathbb{Z}_p$-homology $\mathscr{H}_G$ of G/S, which we define as follows:

Definition 3.1.1 For the group scheme $\pi : G \to S$ we let

$$\mathscr{H} := \mathscr{H}_G := R^{2d-1}\pi_!\mathbb{Z}_p(d) = R^{-1}\pi_!\pi^!\mathbb{Z}_p.$$

We write $\mathscr{H}_r := \mathscr{H} \otimes \mathbb{Z}/p^r\mathbb{Z}$ and $\mathscr{H}_{\mathbb{Q}_p} := \mathscr{H} \otimes \mathbb{Q}_p$ for the associated $\mathbb{Q}_p$-sheaf.

Note that $\mathscr{H}$ is not a lisse $\mathbb{Z}_p$-sheaf in general, but the stalks are free $\mathbb{Z}_p$-modules of finite rank, which follows for example from Lemma 3.1.2 below.

The sheaf $\mathscr{H}$ and more generally $R^i\pi_!\mathbb{Z}_p$ is covariant functorial for any map of S-schemes $f : G \to X$ using the adjunction $f_!f^!\mathbb{Z}_p \to \mathbb{Z}_p$. In particular, the group multiplication $\mu : G \times_S G \to G$ induces a product

$$R^i\pi_!\mathbb{Z}_p(d) \otimes R^j\pi_!\mathbb{Z}_p(d) \to R^{i+j-2d}\pi_!\mathbb{Z}_p(d)$$

and the diagonal $\Delta : G \to G \times_S G$ induces a coproduct

$$R^i\pi_!\mathbb{Z}_p(d) \to \bigoplus_j R^j\pi_!\mathbb{Z}_p(d) \otimes R^{2d+i-j}\pi_!\mathbb{Z}_p(d)$$

on $R^{\cdot}\pi_!\mathbb{Z}_p$, which gives it the structure of a Hopf algebra and one has

$$R^i\pi_!\mathbb{Z}_p(d) \cong \bigwedge^{2d-i} \mathscr{H} \tag{7}$$

(this follows by base change to geometric points and duality from [2, Lemma 4.1.]). The same result holds for $\mathbb{Z}/p^r\mathbb{Z}$-coefficients.

Lemma 3.1.2 *Let $G[p^r]$ be the kernel of the p^r-multiplication $[p^r] : G \to G$. Then there is a canonical isomorphism of étale sheaves*

$$G[p^r] \cong R^{-1}\pi_!\pi^!\mathbb{Z}/p^r\mathbb{Z} = \mathscr{H}_r.$$

In particular, $\mathscr{H}_G$ is the p-adic Tate-module of G.

Proof This is standard and we only sketch the proof: Consider $G[p^r]$ as an étale sheaf on S. The Kummer sequence is a $G[p^r]$-torsor on G, hence gives a class in

$$\begin{aligned} H^1(G, \pi^*G[p^r]) &\cong \operatorname{Ext}^1_G(\pi^*\mathbb{Z}/p^r\mathbb{Z}, \pi^*G[p^r]) \cong \operatorname{Ext}^1_G(\pi^!\mathbb{Z}/p^r\mathbb{Z}, \pi^!G[p^r]) \cong \\ &\cong \operatorname{Ext}^1_S(R\pi_!\pi^!\mathbb{Z}/p^r\mathbb{Z}, G[p^r]) \cong \operatorname{Hom}_S(R^{-1}\pi_!\pi^!\mathbb{Z}/p^r\mathbb{Z}, G[p^r]). \end{aligned}$$

Thus the Kummer torsor induces a map $R^{-1}\pi_!\pi^!\mathbb{Z}/p^r\mathbb{Z} \to G[p^r]$ and one can perform a base change to geometric points $\bar{s} \in S$ to show that this is an isomorphism. But this follows then from Poincaré-duality and the isomorphism $\operatorname{Hom}_{\bar{s}}(G[p^r], \mu_{p^r}) \cong H^1(G, \mu_{p^r})$ shown in [2, Lemma 4.2.]. □

3.2 The First Logarithm Sheaf

Consider the complex $R\pi_!\pi^!\mathbb{Z}_p$ calculating the homology of $\pi : G \to S$ and its canonical filtration whose associated graded pieces are the $R^i\pi_!\pi^!\mathbb{Z}_p$. We apply this to

$$R\operatorname{Hom}_G(\pi^!\mathbb{Z}_p, \pi^!\mathscr{H}) \cong R\operatorname{Hom}_S(R\pi_!\pi^!\mathbb{Z}_p, \mathscr{H}).$$

Then the resulting hypercohomology spectral sequence gives rise to the five term sequence

$$\begin{aligned} 0 \to \operatorname{Ext}^1_S(\mathbb{Z}_p, \mathscr{H}) \xrightarrow{\pi^!} \operatorname{Ext}^1_G(\pi^!\mathbb{Z}_p, \pi^!\mathscr{H}) \to \operatorname{Hom}_S(\mathscr{H}, \mathscr{H}) \to \\ \to \operatorname{Ext}^2_S(\mathbb{Z}_p, \mathscr{H}) \xrightarrow{\pi^!} \operatorname{Ext}^2_G(\pi^!\mathbb{Z}_p, \pi^!\mathscr{H}) \end{aligned}$$

and the maps $\pi^!$ are injective because they admit the splitting $e^!$ induced by the unit section e. This gives

$$0 \to \mathrm{Ext}^1_S(\mathbb{Z}_p, \mathscr{H}) \xrightarrow{\pi^!} \mathrm{Ext}^1_G(\pi^!\mathbb{Z}_p, \pi^!\mathscr{H}) \to \mathrm{Hom}_S(\mathscr{H}, \mathscr{H}) \to 0. \tag{8}$$

Note that $\mathrm{Ext}^1_G(\pi^!\mathbb{Z}_p, \pi^!\mathscr{H}) \cong \mathrm{Ext}^1_G(\mathbb{Z}_p, \pi^*\mathscr{H})$. The same construction is also possible with the base ring $\Lambda_r := \mathbb{Z}/p^r\mathbb{Z}$ and $\mathscr{H}_r$ and gives the exact sequence

$$0 \to \mathrm{Ext}^1_S(\Lambda_r, \mathscr{H}_r) \xrightarrow{\pi^!} \mathrm{Ext}^1_G(\pi^!\Lambda_r, \pi^!\mathscr{H}_r) \to \mathrm{Hom}_S(\mathscr{H}_r, \mathscr{H}_r) \to 0. \tag{9}$$

Definition 3.2.1 The *first logarithm sheaf* $(\mathcal{L}og^{(1)}, \mathbf{1}^{(1)})$ on G consists of an extension class

$$0 \to \pi^*\mathscr{H} \to \mathcal{L}og^{(1)} \to \mathbb{Z}_p \to 0$$

such that its image in $\mathrm{Hom}_S(\mathscr{H}, \mathscr{H})$ is the identity together with a fixed splitting $\mathbf{1}^{(1)} : e^*\mathbb{Z}_p \to e^*\mathcal{L}og^{(1)}$. In exactly the same way one defines $\mathcal{L}og^{(1)}_{\Lambda_r}$. We denote by $\mathcal{L}og^{(1)}_{\mathbb{Q}_p}$ the associated $\mathbb{Q}_p$-sheaf.

The existence and uniqueness of $(\mathcal{L}og^{(1)}, \mathbf{1}^{(1)})$ follow directly from (8). The automorphisms of $\mathcal{L}og^{(1)}$ form a torsor under $\mathrm{Hom}_G(\mathbb{Z}_p, \pi^*\mathscr{H})$. In particular, the pair $(\mathcal{L}og^{(1)}, \mathbf{1}^{(1)})$ admits no automorphisms except the identity.

It is obvious from the definition that one has

$$\mathcal{L}og^{(1)} \otimes_{\mathbb{Z}_p} \Lambda_r \cong \mathcal{L}og^{(1)}_{\Lambda_r} \tag{10}$$

so that $\mathcal{L}og^{(1)} = (\mathcal{L}og^{(1)}_{\Lambda_r})_{r \geqslant 0}$. Moreover, $\mathcal{L}og^{(1)}$ is compatible with arbitrary base change. If

$$\begin{array}{ccc} G_T & \xrightarrow{f_T} & G \\ {\scriptstyle \pi_T}\downarrow & & \downarrow{\scriptstyle \pi} \\ T & \xrightarrow{f} & S \end{array} \tag{11}$$

is a cartesian diagram one has $f_T^*\mathcal{L}og^{(1)}_G \cong \mathcal{L}og^{(1)}_{G_T}$ and $f_T^*(\mathbf{1}^{(1)})$ defines a splitting.

Let

$$\varphi : G_1 \to G_2$$

be a homomorphism of group schemes of relative dimension d_1, d_2, respectively and write $c := d_1 - d_2$.

Theorem 3.2.2 *For $\varphi : G_1 \to G_2$ as above, there is a unique morphism of sheaves*

$$\varphi_\# : \mathcal{L}og^{(1)}_{G_1} \to \varphi^*\mathcal{L}og^{(1)}_{G_2} \cong \varphi^!\mathcal{L}og^{(1)}_{G_2}(-c)[-2c]$$

such that $\varphi_\#(\mathbf{1}^{(1)}_{G_1}) = \mathbf{1}^{(1)}_{G_2}$. Moreover, if φ is an isogeny of degree prime to p, then $\varphi_\#$ is an isomorphism.

Proof Pull-back of $\mathcal{L}og^{(1)}_{G_2}$ gives an exact sequence

$$0 \to \pi_1^* \mathcal{H}_{G_2} \to \varphi^* \mathcal{L}og^{(1)}_{G_2} \to \mathbb{Z}_p \to 0$$

and push-out of $\mathcal{L}og^{(1)}_{G_1}$ by $\pi_1^* \mathcal{H}_{G_1} \to \pi_1^* \mathcal{H}_{G_2}$ induces a map

$$\begin{array}{ccccccccc} 0 & \longrightarrow & \pi_1^* \mathcal{H}_{G_1} & \longrightarrow & \mathcal{L}og^{(1)}_{G_1} & \longrightarrow & \mathbb{Z}_p & \longrightarrow & 0 \\ & & \downarrow & & \downarrow h & & \| & & \\ 0 & \longrightarrow & \pi_1^* \mathcal{H}_{G_2} & \longrightarrow & \varphi^* \mathcal{L}og^{(1)}_{G_2} & \longrightarrow & \mathbb{Z}_p & \longrightarrow & 0. \end{array} \tag{12}$$

If φ is an isogeny and $\deg \varphi$ is prime to p, then $\pi_1^* \mathcal{H}_{G_1} \to \pi_1^* \mathcal{H}_{G_2}$ is an isomorphism, hence also h. By uniqueness there is a unique isomorphism of the pair $(\mathcal{L}og^{(1)}_{G_2}, e_1^*(h) \circ \mathbf{1}^{(1)}_{G_1})$ with $(\mathcal{L}og^{(1)}_{G_2}, \mathbf{1}^{(1)}_{G_2})$. The composition of this isomorphism with h is the desired map. If $h' : \mathcal{L}og^{(1)}_{G_1} \to \varphi^! \mathcal{L}og^{(1)}_{G_2}$ is another map with this property, the difference $h - h' : \mathbb{Z}_p \to \pi_1^* \mathcal{H}_{G_2}$ is uniquely determined by its pull-back $e^*(h - h') : \mathbb{Z}_p \to e_2^* \mathcal{L}og^{(1)}_{G_2}$ according to Lemma 2.3.2. If both, h and h' are compatible with the splittings, then $e^*(h - h') = 0$ and hence $h = h'$. □

Corollary 3.2.3 (Splitting principle) *Let $\varphi : G_1 \to G_2$ be an isogeny of degree prime to p. Then if $t : S \to G_1$ is in the kernel of φ, then*

$$t^* \mathcal{L}og^{(1)}_{G_1} \cong t^* \varphi^* \mathcal{L}og^{(1)}_{G_2} \cong e_1^* \varphi^* \mathcal{L}og^{(1)}_{G_2} \cong e_1^* \mathcal{L}og^{(1)}_{G_1}.$$

Proof Apply t^* to $\varphi_\#$. □

3.3 The $\mathbb{Q}_p$-logarithm Sheaf

We are going to define the $\mathbb{Q}_p$-logarithm sheaf, which has been studied extensively in [7].

Definition 3.3.1 We define

$$\mathcal{L}og^{(k)}_{\mathbb{Q}_p} := \operatorname{Sym}^k(\mathcal{L}og^{(1)}_{\mathbb{Q}_p})$$

and denote by

$$\mathbf{1}^{(k)} := \frac{1}{k!} \operatorname{Sym}^k(\mathbf{1}^{(1)}) : \mathbb{Q}_p \to \mathcal{L}og^{(k)}_{\mathbb{Q}_p}$$

the splitting induced by $\mathbf{1}^{(1)}$.

We note that $\mathcal{L}og^{(k)}_{\mathbb{Q}_p}$ is unipotent of length k and that the splitting $\mathbf{1}^{(k)}$ induces an isomorphism

$$e^*\mathcal{L}og^{(k)}_{\mathbb{Q}_p} \cong \prod_{i=0}^{k} \mathrm{Sym}^i \mathscr{H}_{\mathbb{Q}_p}. \tag{13}$$

To define transition maps

$$\mathcal{L}og^{(k)}_{\mathbb{Q}_p} \to \mathcal{L}og^{(k-1)}_{\mathbb{Q}_p} \tag{14}$$

consider the morphism $\mathcal{L}og^{(1)}_{\mathbb{Q}_p} \to \mathbb{Q}_p \oplus \mathcal{L}og^{(1)}_{\mathbb{Q}_p}$ given by the canonical projection and the identity. Then we have

$$\begin{aligned}
\mathcal{L}og^{(k)}_{\mathbb{Q}_p} &= \mathrm{Sym}^k(\mathcal{L}og^{(1)}_{\mathbb{Q}_p}) \\
&\to \mathrm{Sym}^k(\mathbb{Q}_p \oplus \mathcal{L}og^{(1)}_{\mathbb{Q}_p}) \cong \bigoplus_{i+j=k} \mathrm{Sym}^i(\mathbb{Q}_p) \otimes \mathrm{Sym}^j(\mathcal{L}og^{(1)}_{\mathbb{Q}_p}) \\
&\to \mathrm{Sym}^1(\mathbb{Q}_p) \otimes \mathrm{Sym}^{k-1}(\mathcal{L}og^{(1)}_{\mathbb{Q}_p}) \cong \mathcal{L}og^{(k-1)}_{\mathbb{Q}_p}.
\end{aligned}$$

A straightforward computation shows that $\mathbf{1}^{(k)} \mapsto \mathbf{1}^{(k-1)}$ under this transition map.

3.4 *Main Properties of the $\mathbb{Q}_p$-logarithm Sheaf*

The logarithm sheaf has three main properties: functoriality, vanishing of cohomology and a universal mapping property for unipotent sheaves. Functoriality follows trivially from Theorem 3.2.2. We review the others briefly, referring for more details to [7].

Let $\varphi : G_1 \to G_2$ be a homomorphism of group schemes of relative dimension d_1, d_2, respectively and let $c := d_1 - d_2$ be the relative dimension of the homomorphism.

Theorem 3.4.1 (Functoriality) *For $\varphi : G_1 \to G_2$ as above there is a unique homomorphism of sheaves*

$$\varphi_\# : \mathcal{L}og_{\mathbb{Q}_p,G_1} \to \varphi^*\mathcal{L}og_{\mathbb{Q}_p,G_2} \cong \varphi^!\mathcal{L}og_{\mathbb{Q}_p,G_2}(-c)[-2c]$$

such that $\mathbf{1}_{G_1}$ maps to $\mathbf{1}_{G_2}$. Moreover, if φ is an isogeny, the $\varphi_\#$ is an isomorphism.

Proof This follows directly from Theorem 3.2.2 and the fact that $\deg\varphi$ is invertible in $\mathbb{Q}_p$. □

Corollary 3.4.2 (Splitting principle) *Let $\varphi : G_1 \to G_2$ be an isogeny. Then if $t : S \to G_1$ is in the kernel of φ, one has*

$$\varrho_t : t^*\mathcal{L}og_{\mathbb{Q}_p,G_1} \cong t^*\varphi^*\mathcal{L}og_{\mathbb{Q}_p,G_2} \cong e_1^*\varphi^*\mathcal{L}og_{\mathbb{Q}_p,G_2} \cong e_1^*\mathcal{L}og_{\mathbb{Q}_p,G_1} \cong \prod_{k\geqslant 0} \mathrm{Sym}^k \mathscr{H}_{\mathbb{Q}_p,G_1}.$$

More generally, if $\iota : \ker\varphi \to G_1$ is the closed immersion, one has

$$\iota^*\mathcal{L}og_{G_1} \cong \pi\ |_{\ker\varphi}^* \prod_{k\geqslant 0} \mathrm{Sym}^k \mathscr{H}_{\mathbb{Q}_p,G_1},$$

where $\pi\ |_{\ker\varphi}\colon \ker\varphi \to S$ is the structure map.

Proof Apply t^* to both sides of the isomorphism $\varphi_\#$ and use (13). For the second statement make the base change to $\ker\varphi$ and apply the first statement to the tautological section of $\ker\varphi$. □

Theorem 3.4.3 (Vanishing of cohomology) *One has*

$$R^i\pi_!\mathcal{L}og_{\mathbb{Q}_p} \cong \begin{cases} \mathbb{Q}_p(-d) & \text{if } i = 2d \\ 0 & \text{if } i \neq 2d. \end{cases}$$

More precisely, the transition maps $R^i\pi_!\mathcal{L}og^{(k)}_{\mathbb{Q}_p} \to R^i\pi_!\mathcal{L}og^{(k-1)}_{\mathbb{Q}_p}$ are zero for $i < 2d$ and one has an isomorphism $R^{2d}\pi_!\mathcal{L}og^{(k)}_{\mathbb{Q}_p} \simeq \mathbb{Q}_p(-d)$ compatible with the transition maps.

Proof This is Theorem 3.3.1. in [7]. □

Let $\mathscr{F}$ be a unipotent sheaf of finite length n on G. Consider the homomorphism

$$\pi_*\underline{\mathrm{Hom}}_G(\mathcal{L}og_{\mathbb{Q}_p}, \mathscr{F}) \to e^*\mathscr{F} \tag{15}$$

defined as the composition of

$$\pi_*\underline{\mathrm{Hom}}_G(\mathcal{L}og_{\mathbb{Q}_p}, \mathscr{F}) \to \pi_*e_*e^*\underline{\mathrm{Hom}}_G(\mathcal{L}og_{\mathbb{Q}_p}, \mathscr{F}) \to \underline{\mathrm{Hom}}_S(e^*\mathcal{L}og_{\mathbb{Q}_p}, e^*\mathscr{F})$$

with

$$\underline{\mathrm{Hom}}_S(e^*\mathcal{L}og_{\mathbb{Q}_p}, e^*\mathscr{F}) \xrightarrow{(\mathbf{1})^*} \underline{\mathrm{Hom}}_S(\mathbb{Q}_p, e^*\mathscr{F}) \cong e^*\mathscr{F}.$$

Theorem 3.4.4 (Universal property) *Let $\mathscr{F}$ be a unipotent sheaf of finite length. Then the map* (15) *induces an isomorphism*

$$\pi_*\underline{\mathrm{Hom}}(\mathcal{L}og_{\mathbb{Q}_p}, \mathscr{F}) \cong e^*\mathscr{F}.$$

Proof This is Theorem 3.3.2. in [7]. □

4 The $\mathbb{Q}_p$-polylogarithm and Eisenstein Classes

4.1 Construction of the $\mathbb{Q}_p$-polylogarithm

Fix an auxiliary integer $c > 1$ invertible on S and consider the c-torsion subgroup $D := G[c] \subset G$. We write $U_D := G \setminus D$ and consider

$$U_D \stackrel{j_D}{\to} G \stackrel{\iota_D}{\leftarrow} D.$$

We also write $\pi_D : D \to S$ for the structure map.

For any sheaf $\mathscr{F}$ the localization triangle defines a connecting homomorphism

$$R\pi_! Rj_{D*}j_D^*\mathscr{F}[-1] \to R\pi_!\iota_{D!}\iota_D^!\mathscr{F}. \tag{16}$$

As $\mathcal{L}og_{\mathbb{Q}_p}^{(k)}(d)[2d]$ is unipotent we may use Lemma 2.3.1 to replace $\iota_D^!$ by ι_D^*. Using Corollary 3.4.2 one gets

$$\pi_{D!}\iota_D^!\mathcal{L}og_{\mathbb{Q}_p}^{(k)}(d)[2d] \cong \prod_{i=0}^{k} \pi_{D!}\pi_D^* \operatorname{Sym}^i \mathscr{H}_{\mathbb{Q}_p}.$$

Putting everything together and taking the limit over the transition maps $\mathcal{L}og_{\mathbb{Q}_p}^{(k)} \to \mathcal{L}og_{\mathbb{Q}_p}^{(k-1)}$ gives the *residue map*

$$\mathrm{res} : H^{2d-1}(S, R\pi_! Rj_{D*}j_D^*\mathcal{L}og_{\mathbb{Q}_p}(d)) \to H^0(S, \prod_{k\geqslant 0} \pi_{D!}\pi_D^* \operatorname{Sym}^k \mathscr{H}_{\mathbb{Q}_p}). \tag{17}$$

Proposition 4.1.1 *The localization triangle induces a short exact sequence*

$$0 \to H^{2d-1}(S, R\pi_! Rj_{D*}j_D^*\mathcal{L}og_{\mathbb{Q}_p}(d)) \stackrel{\mathrm{res}}{\to} H^0(S, \prod_{k\geqslant 0} \pi_{D!}\pi_D^* \operatorname{Sym}^k \mathscr{H}_{\mathbb{Q}_p}) \to H^0(S, \mathbb{Q}_p) \to 0.$$

Proof This is an immediate consequence from the localization triangle and the computation of $R\pi_! \mathcal{L}og$ in Theorem 3.4.3. □

Definition 4.1.2 Let

$$\mathbb{Q}_p[D]^0 := \ker(H^0(S, \pi_{D!}\mathbb{Q}_p) \to H^0(S, \mathbb{Q}_p))$$

where the map is induced by the trace $\pi_{D!}\mathbb{Q}_p \to \mathbb{Q}_p$.

Note that

$$\mathbb{Q}_p[D]^0 \subset \ker\left(H^0(S, \prod_{k\geqslant 0} \pi_{D!}\pi_D^* \operatorname{Sym}^k \mathscr{H}_{\mathbb{Q}_p}) \to H^0(S, \mathbb{Q}_p)\right).$$

Definition 4.1.3 Let $\alpha \in \mathbb{Q}_p[D]^0$. Then the unique class

$${}_\alpha\mathrm{pol}_{\mathbb{Q}_p} \in H^{2d-1}(S, R\pi_! Rj_{D*}j_D^* \mathcal{L}og_{\mathbb{Q}_p}(d))$$

with $\mathrm{res}({}_\alpha\mathrm{pol}_{\mathbb{Q}_p}) = \alpha$ is called the *polylogarithm class associated to* α. We write ${}_\alpha\mathrm{pol}^k_{\mathbb{Q}_p}$ for the image of ${}_\alpha\mathrm{pol}_{\mathbb{Q}_p}$ in $H^{2d-1}(S, R\pi_! Rj_{D*}j_D^* \mathcal{L}og^{(k)}_{\mathbb{Q}_p}(d))$.

4.2 Eisenstein Classes

Recall that $D = G[c]$ and fix an integer $N > 1$ invertible on S, such that $(N, c) = 1$ and let $t : S \to U_D = G \setminus D$ be an N-torsion section. Consider the composition

$$R\pi_! Rj_{D*}j_D^* \mathcal{L}og(d) \to R\pi_! Rj_{D*}j_D^* Rt_* t^* \mathcal{L}og(d) \cong R\pi_! Rt_* t^* \mathcal{L}og(d) \cong t^* \mathcal{L}og(d) \tag{18}$$

induced by the adjunction $\mathrm{id} \to Rt_* t^*$, the fact that $Rt_* = Rt_!$ and because $\pi \circ t = \mathrm{id}$. Together with the splitting principle from Corollary 3.4.2 and the projection to the k-th component one gets an evaluation map

$$H^{2d-1}(S, R\pi_! Rj_{D*}j_D^* \mathcal{L}og^{(k)}_{\mathbb{Q}_p}(d)) \xrightarrow{\varrho_t \circ t^*} H^{2d-1}(S, \prod_{i=0}^{k} \operatorname{Sym}^i \mathscr{H}_{\mathbb{Q}_p}) \xrightarrow{\mathrm{pr}_k} H^{2d-1}(S, \operatorname{Sym}^k \mathscr{H}_{\mathbb{Q}_p}). \tag{19}$$

Definition 4.2.1 Let $\alpha \in \mathbb{Q}_p[D]$. The image of ${}_\alpha\mathrm{pol}_{\mathbb{Q}_p}$ under the evaluation map (19)

$${}_\alpha\mathrm{Eis}^k_{\mathbb{Q}_p}(t) \in H^{2d-1}(S, \operatorname{Sym}^k \mathscr{H}_{\mathbb{Q}_p})$$

is called the *k-th étale* $\mathbb{Q}_p$*-Eisenstein class for* G.

Remark 4.2.2 The normalization in [12, Definition 12.4.6] is different. There we had an additional factor of $-N^{k-1}$ in front of ${}_{\alpha}\mathrm{Eis}^k_{\mathbb{Q}_p}(t)$. This has the advantage to make the residues of ${}_{\alpha}\mathrm{Eis}^k_{\mathbb{Q}_p}(t)$ at the cusps integral, but is very unnatural from the point of view of the polylogarithm.

Recall from [7, Theorem 5.2.1] that the polylogarithm ${}_{\alpha}\mathrm{pol}^k_{\mathbb{Q}_p}$ is motivic, i.e., there exists a class in motivic cohomology

$${}_{\alpha}\mathrm{pol}^k_{\mathrm{mot}} \in H^{2d-1}_{\mathrm{mot}}(S, R\pi_! Rj_{D*}j_D^*\mathcal{L}og^{(k)}_{\mathrm{mot}}(d)),$$

the *motivic polylogarithm*, which maps to ${}_{\alpha}\mathrm{pol}^k{}_{\mathbb{Q}_p}$ under the étale regulator

$$r_{\text{ét}} : H^{2d-1}_{\mathrm{mot}}(S, R\pi_! Rj_{D*}j_D^*\mathcal{L}og^{(k)}_{\mathrm{mot}}(d)) \to H^{2d-1}(S, R\pi_! Rj_{D*}j_D^*\mathcal{L}og^{(k)}_{\mathbb{Q}_p}(d)).$$

With the motivic analogue of the evaluation map (19) one can define exactly in the same way as in the étale case motivic Eisenstein classes for $\alpha \in \mathbb{Q}[D]^0$

$${}_{\alpha}\mathrm{Eis}^k_{\mathrm{mot}}(t) \in H^{2d-1}_{\mathrm{mot}}(S, \mathrm{Sym}^k \mathscr{H}_{\mathbb{Q}}). \tag{20}$$

The next proposition is obvious from the fact that the evaluation map is compatible with the étale regulator.

Proposition 4.2.3 *For $\alpha \in \mathbb{Q}[D]^0$ the image of the motivic Eisenstein class ${}_{\alpha}\mathrm{Eis}^k_{\mathrm{mot}}(t)$ under the étale regulator*

$$r_{\text{ét}} : H^{2d-1}_{\mathrm{mot}}(S, \mathrm{Sym}^k \mathscr{H}_{\mathbb{Q}}) \to H^{2d-1}(S, \mathrm{Sym}^k \mathscr{H}_{\mathbb{Q}_p})$$

is the étale $\mathbb{Q}_p$-Eisenstein class ${}_{\alpha}\mathrm{Eis}^k_{\mathbb{Q}_p}(t)$.

5 Sheaves of Iwasawa Algebras

5.1 Iwasawa Algebras

Let $X = \varprojlim_r X_r$ be a profinite space with transition maps $\lambda_r : X_{r+1} \to X_r$ and

$$\Lambda_r[X_r] := \mathrm{Map}(X_r, \mathbb{Z}/p^r\mathbb{Z})$$

the $\mathbb{Z}/p^r\mathbb{Z}$-module of maps from X_r to $\mathbb{Z}/p^r\mathbb{Z}$. For each x_r we write $\delta_{x_r} \in \Lambda_r[X_r]$ for the map which is 1 at x_r and 0 else. It is convenient to interpret $\Lambda_r[X_r]$ as the space of $\mathbb{Z}/p^r\mathbb{Z}$-valued measures on X_r and δ_{x_r} as the delta measure at x_r. Then the push-forward along $\lambda_r : X_{r+1} \to X_r$ composed with reduction modulo p^r induces $\mathbb{Z}_p$-module maps

$$\lambda_{r*} : \Lambda_{r+1}[X_{r+1}] \to \Lambda_r[X_r] \tag{21}$$

which are characterized by $\lambda_{r*}(\delta_{x_{r+1}}) = \delta_{\lambda_r(x_r)}$.

Definition 5.1.1 The *module of* $\mathbb{Z}_p$*-valued measures on* X is the inverse limit

$$\Lambda(X) := \varprojlim_r \Lambda_r[X_r]$$

of $\Lambda_r[X_r]$ with respect to the transition maps from (21).

Let $x = (x_r)_{r\geqslant 0} \in X$. We define $\delta_x := (\delta_{x_r})_{r\geqslant 0} \in \Lambda(X)$, which provides a map

$$\delta : X \to \Lambda(X).$$

For each continuous map $\varphi : X \to Y$ of profinite spaces we get a homomorphism

$$\varphi_* : \Lambda(X) \to \Lambda(Y) \tag{22}$$

"push-forward of measures" with the property $\varphi_*(\delta_x) = \delta_{\varphi(x)}$. Obviously, one has $\Lambda_r[X_r \times Y_r] \cong \Lambda_r[X_r] \otimes \Lambda_r[Y_r]$ so that

$$\Lambda(X \times Y) \cong \Lambda(X)\widehat{\otimes}\Lambda(Y) := \varprojlim_r \Lambda_r[X_r] \otimes \Lambda_r[Y_r].$$

In particular, if $X = G = \varprojlim_r G_r$ is a profinite group, the group structure $\mu : G \times G \to G$ induces a $\mathbb{Z}_p$-algebra structure on $\Lambda(G)$, which coincides with the $\mathbb{Z}_p$-algebra structure induced by the inverse limit of group algebras $\varprojlim_r \Lambda_r[G_r]$.

Definition 5.1.2 If $G = \varprojlim_r G_r$ is a profinite group, we call

$$\Lambda(G) := \varprojlim_r \Lambda_r[G_r]$$

the *Iwasawa algebra of* G.

More generally, if G acts continuously on the profinite space X, one gets a map

$$\Lambda(G)\widehat{\otimes}\Lambda(X) \to \Lambda(X)$$

which makes $\Lambda(X)$ a $\Lambda(G)$-module. If X is a principal homogeneous space under G, then $\Lambda(X)$ is a free $\Lambda(G)$-module of rank 1.

5.2 *Properties of the Iwasawa Algebra*

In this section we assume that H is a finitely generated free $\mathbb{Z}_p$-module. We let

$$H_r := H \otimes_{\mathbb{Z}_p} \mathbb{Z}_p/p^r\mathbb{Z}_p$$

so that $H = \varprojlim_r H_r$ with the natural transition maps $H_{r+1} \to H_r$.

In the case $H = \mathbb{Z}_p$, the so called *Amice transform* of a measure $\mu \in \Lambda(\mathbb{Z}_p)$

$$\mathcal{A}_\mu(T) := \sum_{n=0}^{\infty} T^n \int_{\mathbb{Z}_p} \binom{x}{n} \mu(x)$$

induces a ring isomorphism $\mathcal{A} : \Lambda(\mathbb{Z}_p) \cong \mathbb{Z}_p[[T]]$ (see [3, Sect. 1.1.]). A straightforward generalization shows that $\Lambda(H)$ is isomorphic to a power series ring in $\operatorname{rk} H$ variables. On the other hand one has the so called *Laplace transform* of μ (see loc. cit.)

$$\mathcal{L}_\mu(t) := \sum_{n=0}^{\infty} \frac{t^n}{n!} \int_{\mathbb{Z}_p} x^n \mu(x).$$

This map is called the *moment map* in [10] and we will follow his terminology. In the next section, we will explain this map from an abstract algebraic point of view. For this we interpret $\frac{t^n}{n!}$ as $t^{[n]}$ in the divided power algebra $\Gamma_{\mathbb{Z}_p}(\mathbb{Z}_p)$.

5.3 *The Moment Map*

We return to the case of a free $\mathbb{Z}_p$-module H of finite rank.

Proposition 5.3.1 *Let H be a free $\mathbb{Z}_p$-module of finite rank and $H_r := H \otimes_{\mathbb{Z}_p} \mathbb{Z}/p^r\mathbb{Z}$. Then*

$$\widehat{\Gamma}_{\mathbb{Z}_p}(H) \cong \varprojlim_r \widehat{\Gamma}_{\mathbb{Z}/p^r\mathbb{Z}}(H_r).$$

Proof As each $\Gamma_{\mathbb{Z}_p}(H)/\Gamma^+(H)^{[k]}$ is a finitely generated free $\mathbb{Z}_p$-module, this follows by the compatibility with base change of $\Gamma_{\mathbb{Z}_p}(H)$ and the fact that one can interchange the inverse limits. □

By the universal property of the finite group ring $\Lambda_r[H_r]$, the group homomorphism

$$\begin{aligned} H_r &\to \widehat{\Gamma}_{\mathbb{Z}/p^r\mathbb{Z}}(H_r)^\times \\ h_r &\mapsto \sum_{k\geqslant 0} h_r^{[k]} \end{aligned}$$

induces a homomorphism of $\mathbb{Z}/p^r\mathbb{Z}$-algebras

$$\mathrm{mom}_r : \Lambda_r[H_r] \to \widehat{\Gamma}_{\mathbb{Z}/p^r\mathbb{Z}}(H_r).$$

Corollary 5.3.2 *The maps* mom_r *induce in the inverse limit a* $\mathbb{Z}_p$*-algebra homomorphism*

$$\mathrm{mom} : \Lambda(H) \to \widehat{\Gamma}_{\mathbb{Z}_p}(H).$$

which is functorial in H.

Definition 5.3.3 We call $\mathrm{mom} : \Lambda(H) \to \widehat{\Gamma}_{\mathbb{Z}_p}(H)$ the *moment map* and the composition with the projection to $\Gamma_k(H)$

$$\mathrm{mom}^k : \Lambda(H) \to \Gamma_k(H)$$

the k*-th moment map*.

5.4 Sheafification of the Iwasawa Algebras

Let X be a separated noetherian scheme of finite type as in Sect. 2.1 and $\mathscr{X} := (p_r : \mathscr{X}_r \to X)_r$ be an inverse system of quasi-finite étale schemes over X with étale transition maps $\lambda_r : \mathscr{X}_r \to \mathscr{X}_{r-1}$. We often write

$$\Lambda_r := \mathbb{Z}/p^r\mathbb{Z}. \tag{23}$$

The adjunction $\lambda_{r!}\lambda_r^! \to \mathrm{id}$ defines a homomorphism

$$p_{r+1!}\Lambda_{r+1} = p_{r!}\lambda_{r!}\lambda_r^!\Lambda_{r+1} \to p_{r!}\Lambda_{r+1},$$

because λ_r is étale. If one composes this with reduction modulo p^r, one gets a trace map

$$\mathrm{Tr}_{r+1} : p_{r+1,!}\Lambda_{r+1} \to p_{r,!}\Lambda_r. \tag{24}$$

Definition 5.4.1 We define an étale sheaf on X by

$$\Lambda_r[\mathscr{X}_r] := p_{r!}\Lambda_r.$$

With the trace maps $\mathrm{Tr}_{r+1} : \Lambda_{r+1}[\mathscr{X}_{r+1}] \to \Lambda_r[\mathscr{X}_r]$ as transition morphisms we define the pro-sheaf

$$\Lambda(\mathscr{X}) := (\Lambda_r[\mathscr{X}_r])_{r\geqslant 0}.$$

This definition is functorial in $\mathscr{X}$. If $(\varphi_r)_r : (\mathscr{X}_r)_r \to (\mathscr{Y}_r)_r$ is a morphism of inverse system of quasi-finite étale schemes over X, then the adjunction $\varphi_{r!}\varphi_r^! \to \mathrm{id}$ defines a morphism

$$\varphi_{r!} : \Lambda_r[\mathscr{X}_r] \to \Lambda_r[\mathscr{Y}_r]$$

compatible with the transition maps, and hence a morphism of pro-sheaves

$$\Lambda(\mathscr{X}) \to \Lambda(\mathscr{Y}).$$

Moreover, the formation of $\Lambda(\mathscr{X})$ is compatible with base change: if $\mathscr{X}_{r,T} := \mathscr{X}_r \times_S T$ for an S-scheme $f : T \to S$, then by proper base change one has

$$f^*\Lambda_r[\mathscr{X}_r] \cong \Lambda[\mathscr{X}_{r,T}].$$

By the Künneth formula, one has

$$\Lambda_r[\mathscr{X}_r \times_X \mathscr{Y}_r] \cong \Lambda_r[\mathscr{X}_r] \otimes \Lambda_r[\mathscr{Y}_r]$$

and hence $\Lambda(\mathscr{X} \times_X \mathscr{Y}) \cong \Lambda(\mathscr{X})\widehat{\otimes}\Lambda(\mathscr{Y})$ by taking the inverse limit. In particular, in the case where $\mathscr{X} = \mathscr{G}$ is an inverse system of quasi-finite étale group schemes $\mathscr{G}_r$, the group structure $\mu_r : \mathscr{G}_r \times_X \mathscr{G}_r \to \mathscr{G}_r$ induces a ring structure

$$\Lambda(\mathscr{G})\widehat{\otimes}\Lambda(\mathscr{G}) \to \Lambda(\mathscr{G})$$

on $\Lambda(\mathscr{G})$. Similarly, if

$$\mathscr{G} \times_X \mathscr{X} \to \mathscr{X}$$

is a group action of inverse systems, i.e., a compatible family of actions $\mathscr{G}_r \times_X \mathscr{X}_r \to \mathscr{X}_r$, then $\Lambda(\mathscr{X})$ becomes a $\Lambda(\mathscr{G})$-module.

The next lemma shows that the above construction indeed sheafifies the Iwasawa algebras considered before.

Lemma 5.4.2 *Let $\overline{x} \in X$ be a geometric point and write $p_{r,\overline{x}} : \mathscr{X}_{r,\overline{x}} \to \overline{x}$ for the base change of $\mathscr{X}_r$ to $\overline{x}$ considered as a finite set. Then*

$$\Lambda_r[\mathscr{X}_r]_{\overline{x}} \cong \Lambda_r[X_r].$$

Proof This follows directly from the base change property of $\Lambda_r[\mathscr{X}_r]$ and the fact that $p_{r,\overline{x},!}\Lambda_r \cong \Lambda_r[X_r]$ over an algebraically closed field. □

We return to our basic set up, where $\pi : G \to S$ is a separated smooth commutative group scheme with connected fibres. Recall from Lemma 3.1.2 that $\mathscr{H}_r$ is the sheaf associated to $G[p^r]$, which is quasi-finite and étale over S.

Definition 5.4.3 Define the *sheaf of Iwasawa algebras* $\Lambda(\mathscr{H})$ on S to be the pro-sheaf

$$\Lambda(\mathscr{H}) := (\Lambda_r[\mathscr{H}_r])_{r\geqslant 0}.$$

5.5 Sheafification of the Moment Map

We keep the notation of the previous section. In particular, we consider the étale sheaf $\mathscr{H}_r$ and the sheaf $\Lambda_r[\mathscr{H}_r]$.

Over $G[p^r]$ the sheaf $[p^r]^*\mathscr{H}_r$ has the tautological section $\tau_r \in \Gamma(G[p^r], [p^r]^*\mathscr{H}_r)$ corresponding to the identity map $G[p^r] \to \mathscr{H}_r$. This gives rise to the section

$$\tau_r^{[k]} \in \Gamma(G[p^r], [p^r]^*\Gamma_k(\mathscr{H}_r)) \tag{25}$$

of the k-th divided power of $\mathscr{H}_r$. Using the chain of isomorphisms (note that $[p^r]^* = [p^r]^!$ as $[p^r]$ is étale)

$$\begin{aligned}\Gamma(G[p^r], [p^r]^*\Gamma_k(\mathscr{H}_r)) &\cong \mathrm{Hom}_{G[p^r]}(\mathbb{Z}/p^r\mathbb{Z}, [p^r]^*\Gamma_k(\mathscr{H}_r))\\ &\cong \mathrm{Hom}_S([p^r]_!\mathbb{Z}/p^r\mathbb{Z}, \Gamma_k(\mathscr{H}_r)),\end{aligned}$$

the section $\tau_r^{[k]}$ gives rise to a morphism of sheaves

$$\mathrm{mom}_r^k : \Lambda_r[\mathscr{H}_r] \to \Gamma_k(\mathscr{H}_r). \tag{26}$$

Lemma 5.5.1 *There is a commutative diagram*

$$\begin{array}{ccc} \Lambda_r[\mathscr{H}_r] & \xrightarrow{\mathrm{mom}_r^k} & \Gamma_k(\mathscr{H}_r) \\ {\scriptstyle \mathrm{Tr}_r}\downarrow & & \downarrow \\ \Lambda_{r-1}[\mathscr{H}_{r-1}] & \xrightarrow{\mathrm{mom}_{r-1}^k} & \Gamma_k(\mathscr{H}_{r-1}) \end{array}$$

where the right vertical map is given by the reduction map

$$\Gamma_k(\mathscr{H}_r) \to \Gamma_k(\mathscr{H}_r) \otimes_{\mathbb{Z}/p^r\mathbb{Z}} \mathbb{Z}/p^{r-1}\mathbb{Z} \cong \Gamma_k(\mathscr{H}_{r-1}).$$

Proof Denote by $\lambda_r : \mathscr{H}_r \to \mathscr{H}_{r-1}$ the transition map. Reduction modulo p^{r-1} gives a commutative diagram

$$\begin{array}{ccc} [p^r]_!\mathbb{Z}/p^r\mathbb{Z} & \xrightarrow{\mathrm{mom}_r^k} & \Gamma_k(\mathscr{H}_r) \\ \downarrow & & \downarrow \\ [p^r]_!\lambda_r^*\mathbb{Z}/p^{r-1}\mathbb{Z} & \xrightarrow{\mathrm{mom}_r^k \otimes \mathbb{Z}/p^{r-1}\mathbb{Z}} & \Gamma_k(\mathscr{H}_{r-1}). \end{array}$$

As the image of the tautological class $\tau_r^{[k]} \in \Gamma(G[p^r], [p^r]^*\Gamma_k(\mathscr{H}_r))$ under the reduction map gives the the pull-back of the tautological class

$$\begin{aligned} \lambda_r^*\tau_{r-1}^{[k]} \in \Gamma(G[p^r], [p^r]^*\Gamma_k(\mathscr{H}_{r-1})) &\cong \mathrm{Hom}_{G[p^r]}(\lambda_r^*\mathbb{Z}/p^{r-1}\mathbb{Z}, [p^r]^*\Gamma_k(\mathscr{H}_{r-1})) \\ &\cong \mathrm{Hom}_S([p^r]_!\lambda_r^*\mathbb{Z}/p^{r-1}\mathbb{Z}, \Gamma_k(\mathscr{H}_{r-1})) \end{aligned}$$

one concludes that $\mathrm{mom}_r^k \otimes \mathbb{Z}/p^{r-1}\mathbb{Z}$ coincides with the map given by $\lambda_r^*\tau_{r-1}^{[k]}$. This means that $\mathrm{mom}_r^k \otimes \mathbb{Z}/p^{r-1}\mathbb{Z}$ has to factor through Tr_r, i.e., the diagram

$$\begin{array}{ccc} [p^{r-1}]_!\lambda_{r!}\lambda_r^*\mathbb{Z}/p^{r-1}\mathbb{Z} & \xrightarrow{\mathrm{mom}_r^k \otimes \mathbb{Z}/p^{r-1}\mathbb{Z}} & \Gamma_k(\mathscr{H}_{r-1}) \\ & {\scriptstyle \mathrm{Tr}_r} \searrow \quad \nearrow {\scriptstyle \mathrm{mom}_{r-1}^k} & \\ & [p^{r-1}]_!\mathbb{Z}/p^{r-1}\mathbb{Z} & \end{array}$$

commutes, which gives the desired result. □

With this result we can now define the moment map for the sheaf of Iwasawa algebras $\Lambda(\mathscr{H})$.

Definition 5.5.2 We define the *k-th moment map* to be the map of pro-sheaves

$$\mathrm{mom}^k : \Lambda(\mathscr{H}) \to \Gamma_k(\mathscr{H})$$

defined by $(\mathrm{mom}_r^k)_{r \geqslant 0}$ and

$$\mathrm{mom} : \Lambda(\mathscr{H}) \to \widehat{\Gamma}_{\mathbb{Z}_p}(\mathscr{H})$$

by taking mom^k in the k-th component.

Remark 5.5.3 In each stalk the the map mom^k coincides with the map mom^k defined in Definition 5.3.3 (see [12, Lemma 12.2.14]).

6 The Integral Logarithm Sheaf

6.1 Definition of the Integral Logarithm Sheaf

We now define a pro-sheaf $\mathcal{L}$ on G of modules over $\pi^*\Lambda(\mathscr{H})$, which will give a $\mathbb{Z}_p$-structure of the logarithm sheaf $\mathcal{L}og_{\mathbb{Q}_p}$. For this write $G_r := G$ considered as a quasi-finite étale G-scheme via the p^r-multiplication

$$[p^r] : G_r = G \to G. \tag{27}$$

Note that this is a $G[p^r]$-torsor

$$0 \to G[p^r] \to G_r \xrightarrow{[p^r]} G \to 0$$

over G. Let $\lambda_r : G_r \to G_{r-1}$ be the transition map, which is just the $[p]$-multiplication in this case. Then, as in (24), we have trace maps

$$\mathrm{Tr}_r : \Lambda_r[G_r] \to \Lambda_{r-1}[G_{r-1}].$$

We will also need the following variant. Let $\Lambda_s := \mathbb{Z}/p^s\mathbb{Z}$ and write

$$\Lambda_s[G_r] := [p^r]_!\Lambda_s. \tag{28}$$

Then the adjunction $\lambda_{r!}\lambda_r^! \to \mathrm{id}$ defines transition morphisms

$$\lambda_{r!} : \Lambda_s[G_r] \to \Lambda_s[G_{r-1}]. \tag{29}$$

Definition 6.1.1 With the above transition maps we can define the pro-sheaves

$$\mathcal{L} := (\Lambda_r[G_r])_{r\geqslant 0} \qquad \text{and} \qquad \mathcal{L}_{\Lambda_s} := (\Lambda_s[G_r])_{r\geqslant 0}.$$

We call $\mathcal{L}$ the *integral logarithm sheaf.*

Note that the reduction modulo p^{s-1} gives transition maps $\mathcal{L}_{\Lambda_s} \to \mathcal{L}_{\Lambda_{s-1}}$ and that we have an isomorphism of pro-sheaves

$$\mathcal{L} \cong (\mathcal{L}_{\Lambda_s})_{s\geqslant 0}. \tag{30}$$

By the general theory outlined above, $\mathcal{L}$ is a module over $\pi^*\Lambda(\mathscr{H})$ which is free of rank 1.

Let $t : S \to G$ be a section and denote by $G[p^r]\langle t\rangle$ the $G[p^r]$-torsor defined by the cartesian diagram

$$\begin{array}{ccc} G[p^r]\langle t\rangle & \longrightarrow & G_r \\ \downarrow & & \downarrow{\scriptstyle [p^r]} \\ S & \xrightarrow{t} & G. \end{array} \tag{31}$$

We denote by $\mathscr{H}_r\langle t\rangle$ the étale sheaf defined by $G[p^r]\langle t\rangle$ and by $\mathscr{H}\langle t\rangle := (\mathscr{H}_r\langle t\rangle)$ the pro-system defined by the trace maps. We write

$$\Lambda(\mathscr{H}\langle t\rangle) := (\Lambda_r[\mathscr{H}_r\langle t\rangle])_{r\geqslant 0}$$

for the sheaf of Iwasawa modules defined by $\mathscr{H}\langle t\rangle$.

Lemma 6.1.2 *There is an canonical isomorphism*

$$t^*\mathcal{L} \cong \Lambda(\mathscr{H}\langle t\rangle).$$

In particular, for the unit section $e : S \to G$ one has

$$e^*\mathcal{L} \cong \Lambda(\mathscr{H})$$

and hence a section $\mathbf{1} : \mathbb{Z}_p \to e^*\mathcal{L}$ given by mapping 1 to 1.

Proof This follows directly from the fact that $\mathcal{L}$ is compatible with base change and the definitions. □

6.2 Basic Properties of the Integral Logarithm Sheaf

The integral logarithm sheaf enjoys the same properties as its $\mathbb{Q}_p$-counterpart, namely functoriality, vanishing of cohomology and a universal property for unipotent sheaves.

Let $\varphi : G_1 \to G_2$ be a homomorphism of group schemes of relative dimension d_1 and d_2 over S. Denote by $\mathcal{L}_1$ and $\mathcal{L}_2$ the integral logarithm sheaves on G_1 and G_2 respectively.

Theorem 6.2.1 (Functoriality) *Let $c := d_1 - d_2$. Then there is a canonical map*

$$\varphi_\# : \mathcal{L}_1 \to \varphi^*\mathcal{L}_2 \cong \varphi^!\mathcal{L}_2(-c)[-2c].$$

Moreover, if φ is an isogeny of degree prime to p, then $\varphi_\# : \mathcal{L}_1 \cong \varphi^*\mathcal{L}_2$ is an isomorphism.

Proof The homomorphism φ induces a homomorphism of group schemes over G_1

$$\varphi : G_{1,r} \to G_{2,r} \times_{G_2} G_1 \tag{32}$$

which induces by adjunction $\varphi_!\varphi^! \to \mathrm{id}$ and the base change property of $\Lambda_r[G_{2,r}]$ a morphism of sheaves

$$\varphi_\# : \Lambda_r[G_{1,r}] \to \varphi^*\Lambda_r[G_{2,r}] = \varphi^!\Lambda_r[G_{2,r}](-c)[-2c].$$

Passing to the limit gives the required map. If φ is an isogeny of degree prime to p, then the map in (32) is an isomorphism. Hence this is also true for $\varphi_\#$. □

Corollary 6.2.2 (Splitting principle) *Let c be an integer prime to p and let $t : S \to G$ be a c-torsion section. Then there is an isomorphism*

$$[c]_\# : t^*\mathcal{L} \cong \Lambda(\mathscr{H}).$$

More generally, if $D := G[c]$ *with* $(c, p) = 1$ *then*

$$\iota_D^*\mathcal{L} \cong \pi_D^*\Lambda(\mathcal{H}),$$

where $\iota_D : D \to G$ *and* $\pi_D : D \to S$ *is the structure map.*

Proof Apply t^* respectively, ι_D^* to the isomorphism $[c]_\# : \mathcal{L} \to [c]^*\mathcal{L}$. □

Theorem 6.2.3 (Vanishing of cohomology) *Recall that* $2d$ *is the relative dimension of* $\pi : G \to S$. *Then the pro-sheaves*

$$R^i\pi_!\mathcal{L} \text{ for } i < 2d$$

are Mittag-Leffler zero (see Sect. 2.1) and

$$R^{2d}\pi_!\mathcal{L}(d) \cong \mathbb{Z}_p.$$

We start the proof of this theorem with a lemma:

Lemma 6.2.4 *The endomorphism* $[p^r]_! : R^i\pi_!\mathbb{Z}/p^s\mathbb{Z} \to R^i\pi_!\mathbb{Z}/p^s\mathbb{Z}$ *is given by multiplication with* $p^{r(2d-i)}$.

Proof By Lemma 3.1.2 we see that $[p^r]_!$ is given by p^r-multiplication on $\mathcal{H}_s$. The result follows from this and the $\mathbb{Z}/p^s\mathbb{Z}$-version of the isomorphism (7) □

Proof of Theorem 6.2.3. Consider the transition map $\Lambda_s[G_{r+j}] \to \Lambda_s[G_r]$. If we apply $R^i\pi_!$ we get the homomorphism

$$[p^j]_! : R^i\pi_{r+j,!}\Lambda_s \to R^i\pi_{r,!}\Lambda_s,$$

where $\pi_r = \pi : G_r \to S$ is the structure map of $G_r = G$. By Lemma 6.2.4, the map $[p^j]_!$ acts by multiplication with $p^{j(2d-i)}$ on $R^i\pi_{r+j,!}\Lambda_s$. In particular, this is zero for $i \neq 2d$ and $j \geqslant s$ and the identity for $i = 2d$. This proves the theorem, because $R^{2d}\pi_!\Lambda_s(d) \cong \Lambda_s$. □

The sheaf $\mathcal{L}$ satisfies also a property analogous to Theorem 3.4.4. To formulate this properly, we first need a property of unipotent $\mathbb{Z}/p^s\mathbb{Z}$-sheaves.

Lemma 6.2.5 *Let* $\mathcal{F}$ *be a unipotent* $\Lambda_s = \mathbb{Z}/p^s\mathbb{Z}$*-sheaf of length* n *on* G. *Then* $[p^{ns}]^*\mathcal{F}$ *is trivial on* G_{ns} *in the sense that there exists a* Λ_s*-sheaf* $\mathcal{G}$ *on* S *such that*

$$[p^{ns}]^*\mathcal{F} \cong \pi_{ns}^*\mathcal{G},$$

where $\pi_{ns} : G_{ns} \to S$ *is the structure map.*

Proof We show this by induction. For $n = 0$ there is nothing to show. So let

$$0 \to \mathcal{F}' \to \mathcal{F} \to \pi^*\mathcal{G}'' \to 0$$

be an exact sequence with $\mathscr{F}'$ unipotent of length $n-1$, so that by induction hypotheses $[p^{(n-1)s}]^*\mathscr{F}' \cong \pi^*\mathscr{G}'$ on $G_{(n-1)s}$. Thus it suffices to show that for an extension $\mathscr{F} \in \operatorname{Ext}^1_G(\pi^*\mathscr{G}'', \pi^*\mathscr{G}')$, the sheaf $[p^s]^*\mathscr{F}$ is trivial on G_s. One has

$$\operatorname{Ext}^1_G(\pi^*\mathscr{G}'', \pi^*\mathscr{G}') \cong \operatorname{Ext}^1_G(\pi^!\mathscr{G}'', \pi^!\mathscr{G}') \cong \operatorname{Ext}^1_S(R\pi_!\pi^!\mathscr{G}'', \mathscr{G}')$$

and the pull-back by $[p^s]^*$ on the first group is induced by the trace map $[p^s]_! : R\pi_![p^s]_![p^s]^!\pi^!\mathscr{G}'' \to R\pi_!\pi^!\mathscr{G}''$ on the last group. By the projection formula we have $R\pi_!\pi^!\mathscr{G}'' \cong R\pi_!\Lambda_s(d)[2d] \otimes \mathscr{G}''$ and the triangle

$$\tau_{<2d}R\pi_!\Lambda_s(d)[2d] \to R\pi_!\Lambda_s(d)[2d] \to R^{2d}\pi_!\Lambda_s(d) \cong \Lambda_s$$

gives rise to a long exact sequence of Ext-groups

$$\begin{aligned} \ldots \to \operatorname{Ext}^1_S(\mathscr{G}'', \mathscr{G}') &\to \operatorname{Ext}^1_S(R\pi_!\Lambda_s(d)[2d] \otimes \mathscr{G}'', \mathscr{G}') \\ &\to \operatorname{Ext}^1_S(\tau_{<2d}R\pi_!\Lambda_s(d)[2d] \otimes \mathscr{G}'', \mathscr{G}') \to \ldots \end{aligned}$$

If we pull-back by $[p^s]^*$ and use Lemma 6.2.4 the resulting map on the module $\operatorname{Ext}^1_S(\tau_{<2d}R\pi_!\Lambda_s(d)[2d] \otimes \mathscr{G}'', \mathscr{G}')$ is zero, which shows that $[p^s]^*\mathscr{F}$ is in the image of

$$\operatorname{Ext}^1_S(\mathscr{G}'', \mathscr{G}') \xrightarrow{[p^s]^*\pi^*} \operatorname{Ext}^1_G([p^s]^*\pi^*\mathscr{G}'', [p^s]^*\pi^*\mathscr{G}').$$

This is the desired result. □

Exactly as in (15) one can define for each Λ_s-sheaf $\mathscr{F}$ and each r a homomorphism

$$\pi_*\underline{\operatorname{Hom}}_G(\Lambda_s[G_r], \mathscr{F}) \to e^*\mathscr{F} \tag{33}$$

as the composition

$$\begin{aligned} \pi_*\underline{\operatorname{Hom}}_G(\mathcal{L}_{\Lambda_s,r}, \mathscr{F}) &\to \pi_*e_*e^*\underline{\operatorname{Hom}}_G(\mathcal{L}_{\Lambda_s,r}, \mathscr{F}) \\ &\to \underline{\operatorname{Hom}}_S(e^*\mathcal{L}_{\Lambda_s,r}, e^*\mathscr{F}) \xrightarrow{\mathbf{1}^*} \underline{\operatorname{Hom}}_S(\Lambda_s, e^*\mathscr{F}) \end{aligned}$$

The next theorem corrects and generalizes [12, Proposition 4.5.3], which was erroneously stated for all $\mathbb{Z}/p^s\mathbb{Z}$-sheaves and not just for unipotent ones.

Theorem 6.2.6 (Universal property) *Let $\mathscr{F}$ be a unipotent Λ_s-sheaf of length n. Then the homomorphism* (33)

$$\pi_*\underline{\operatorname{Hom}}_G(\Lambda_s[G_{ns}], \mathscr{F}) \cong e^*\mathscr{F}$$

is an isomorphism.

Proof Let $\mathscr{F}$ be unipotent of length n. Then we know from Lemma 6.2.5 that there is a Λ_s-sheaf $\mathscr{G}$ on S such that $[p^{ns}]^*\mathscr{F} \cong \pi_{ns}^*\mathscr{G}$, where $\pi_{ns} : G_{ns} \to S$ is the structure map. Similarly, we write e_{ns} for the unit section of G_{ns}. Then one has

$$e^*\mathscr{F} \cong e_{ns}^*[p^{ns}]^*\mathscr{F} \cong e_{ns}^*\pi_{ns}^*\mathscr{G} \cong \mathscr{G}.$$

Further, one has the following chain of isomorphisms

$$\begin{aligned}\pi_*\underline{\mathrm{Hom}}_G(\Lambda_s[G_{ns}], \mathscr{F}) = \pi_*\underline{\mathrm{Hom}}_G([p^{ns}]_!\Lambda_s, \mathscr{F}) &\cong \pi_{ns*}\underline{\mathrm{Hom}}_{G_{ns}}(\Lambda_s, [p^{ns}]^*\mathscr{F})\\ &\cong \pi_{ns*}\underline{\mathrm{Hom}}_{G_{ns}}(\Lambda_s, \pi_{ns}^*\mathscr{G})\\ &\cong \underline{\mathrm{Hom}}_S(R\pi_{ns!}\Lambda_s(d)[2d], \mathscr{G})\\ &\cong \underline{\mathrm{Hom}}_S(R^{2d}\pi_{ns!}\Lambda_s(d), \mathscr{G})\\ &\cong \mathscr{G} \cong e^*\mathscr{F},\end{aligned}$$

which prove the theorem. □

6.3 The Integral étale Poylogarithm

In this section we define in complete analogy with the $\mathbb{Q}_p$-case the integral étale polylogarithm.

We recall the set-up from Sect. 4.1. Denote by $c > 1$ an integer invertible on S and prime to p and let $D := G[c]$ be the c-torsion subgroup. Then the localization triangle for $j_D : U_D \subset G$ and $\iota_D : D \to G$ reads

$$R\pi_!\mathcal{L}(d)[2d-1] \to R\pi_!Rj_{D*}j_D^*\mathcal{L}(d)[2d-1] \to \pi_{D!}\iota_D^!\mathcal{L}(d).$$

By relative purity and the splitting principle $\iota_D^!\mathcal{L}(d)[2d] \cong \iota_D^*\mathcal{L} \cong \pi_D^*\Lambda(\mathscr{H})$. We apply the functor $H^j(S, -)$ to this triangle. As the $R^i\pi_!\mathcal{L}$ are Mittag-Leffler zero for $i \neq 2d$ by Theorem 6.2.3 one gets with (3):

Proposition 6.3.1 *In the above situation there is a short exact sequence*

$$0 \to H^{2d-1}(S, R\pi_!Rj_{D*}j_D^*\mathcal{L}(d)) \xrightarrow{\mathrm{res}} H^0(S, \pi_{D!}\pi_D^*\Lambda(\mathscr{H})) \to H^0(S, \mathbb{Z}_p) \to 0.$$

As in the $\mathbb{Q}_p$-case we define

$$\mathbb{Z}_p[D]^0 := \ker\left(H^0(S, \pi_{D!}\pi_D^*\mathbb{Z}_p) \to H^0(S, \mathbb{Z}_p)\right)$$

so that one has

$$\mathbb{Z}_p[D]^0 \subset \ker\left(H^0(S, \pi_{D!}\pi_D^*\Lambda(\mathscr{H})) \to H^0(S, \mathbb{Z}_p)\right).$$

With these preliminaries we can define the integral polylogarithm.

Definition 6.3.2 The *integral étale polylogarithm* associated to $\alpha \in \mathbb{Z}_p[D]^0$ is the unique class

$$_\alpha\mathrm{pol} \in H^{2d-1}(S, R\pi_! Rj_{D*} j_D^* \mathcal{L}(d))$$

such that $\mathrm{res}(_\alpha\mathrm{pol}) = \alpha$.

6.4 The Eisenstein–Iwasawa Class

Recall that $D = G[c]$ and let $t : S \to U_D = G \setminus D$ be an N-torsion section with $(N, c) = 1$ but N not necessarily prime to p. The same chain of maps as in (18) gives a map

$$H^{2d-1}(S, R\pi_! Rj_{D*} j_D^* \mathcal{L}(d)) \to H^{2d-1}(S, t^* \mathcal{L}(d)) \cong H^{2d-1}(S, \Lambda(\mathscr{H}\langle t\rangle)(d)). \quad (34)$$

By functoriality the N-multiplication induces a homomorphism

$$[N]_\# : \Lambda(\mathscr{H}\langle t\rangle) \to \Lambda(\mathscr{H}).$$

Definition 6.4.1 Let $\alpha \in \mathbb{Z}_p[D]^0$ and $t : S \to U_D$ be an N-torsion section. Then the image

$$_\alpha\mathcal{EI}(t) \in H^{2d-1}(S, \Lambda(\mathscr{H}\langle t\rangle)(d))$$

of $_\alpha\mathrm{pol}$ under the map (34) is called the *Eisenstein–Iwasawa class*. We write

$$_\alpha\mathcal{EI}(t)_N := [N]_\#(_\alpha\mathcal{EI}(t)) \in H^{2d-1}(S, \Lambda(\mathscr{H})(d)).$$

Remark 6.4.2 Note that $_\alpha\mathcal{EI}(t)_N$ depends on N and not on t alone. The class $_\alpha\mathcal{EI}(t)_{NM}$ differs from $_\alpha\mathcal{EI}(t)_N$.

The k-th moment map induces a homomorphism of cohomology groups

$$\mathrm{mom}^k : H^{2d-1}(S, \Lambda(\mathscr{H})(d)) \to H^{2d-1}(S, \Gamma_k(\mathscr{H})(d)). \quad (35)$$

Definition 6.4.3 The class

$$_\alpha\mathrm{Eis}_N^k(t) := \mathrm{mom}^k(_\alpha\mathcal{EI}_N) \in H^{2d-1}(S, \Gamma_k(\mathscr{H})(d))$$

is called the *integral étale Eisenstein class*.

These Eisenstein classes are interpolated by the Eisenstein–Iwasawa class by definition. We will see later how they are related to the $\mathbb{Q}_p$-Eisenstein class, which are motivic, i.e., in the image of the étale regulator from motivic cohomology.

6.5 The Eisenstein–Iwasawa Class for Abelian Schemes

It is worthwhile to consider the case of abelian schemes in more detail. In this section we let $G = A$ be an abelian scheme over S, so that in particular $\pi : A \to S$ is proper and we can write $R\pi_*$ instead of $R\pi_!$.

The first thing to observe is the isomorphism

$$H^{2d-1}(S, R\pi_! Rj_{D*} j_D^* \mathcal{L}og(d)) \cong H^{2d-1}(U_D, \mathcal{L}og(d)),$$

so that the $\mathbb{Q}_p$-polylogarithm is a class

$$_\alpha\mathrm{pol}_{\mathbb{Q}_p} \in H^{2d-1}(U_D, \mathcal{L}og(d)).$$

Evaluation at the N-torsion section $t : S \to U_D$ is just the pull-back with t^*

$$t_\alpha^* \mathrm{pol}_{\mathbb{Q}_p} \in H^{2d-1}(S, t^* \mathcal{L}og(d)) \cong H^{2d-1}(S, \prod_{k \geqslant 0} \mathrm{Sym}^k \mathscr{H}_{\mathbb{Q}_p}(d))$$

and the k-th component of $t_\alpha^* \mathrm{pol}_{\mathbb{Q}_p}$ is $_\alpha\mathrm{Eis}_{\mathbb{Q}_p}^k(t)$.

There is one specific choice of α which is particularly important, which we define next. Consider the finite étale morphism $\pi_D : G[c] \to S$ and the unit section $e : S \to G[c]$. These induce

$$e_* : H^0(S, \mathbb{Q}_p) \to H^0(S, \pi_{D*}\mathbb{Q}_p)$$

(coming from $\pi_{D*} e_! e^! \mathbb{Q}_p \to \pi_{D*}\mathbb{Q}_p$) and

$$\pi_D^* : H^0(S, \mathbb{Q}_p) \to H^0(S, \pi_{D*}\mathbb{Q}_p).$$

One checks easily that $e_*(1) - \pi_D^*(1)$ is in the kernel of $H^0(S, \pi_{D*}\mathbb{Q}_p) \to H^0(S, \mathbb{Q}_p)$.

Definition 6.5.1 Let $\alpha_c \in \mathbb{Q}_p[D]^0$ be the class

$$\alpha_c := e_*(1) - \pi_D^*(1).$$

We write $_c\mathrm{pol}_{\mathbb{Q}_p}$ and $_c\mathrm{Eis}_{\mathbb{Q}_p}^k(t)$ for the polylogarithm and the Eisenstein class defined with α_c.

We now assume that S is of finite type over $\operatorname{Spec}\mathbb{Z}$. Then $H^{2d-1}(A_r \setminus A_r[cp^r], \mathbb{Z}/p^r\mathbb{Z}(d))$ is finite, so that one has by (2)

$$H^{2d-1}(S, R\pi_! Rj_{D*} j_D^* \mathcal{L}(d)) \cong H^{2d-1}(A \setminus A[c], \mathcal{L}(d))$$
$$\cong \varprojlim_r H^{2d-1}(A_r \setminus A_r[cp^r], \mathbb{Z}/p^r\mathbb{Z}(d))$$

where, as before, $[p^r] : A_r = A \to A$ is the p^r-multiplication and the transition maps are given by the trace maps. The integral étale polylogarithm is then a class

$$ {}_\alpha\mathrm{pol} \in \varprojlim_r H^{2d-1}(A_r \setminus A_r[cp^r], \mathbb{Z}/p^r\mathbb{Z}(d)). $$

In the special case where $A = E$ is an elliptic curve over S it is shown in [12, Theorem 12.4.21] that

$$ {}_c\mathrm{pol} \in \varprojlim_r H^1(E_r \setminus E_r[cp^r], \mathbb{Z}/p^r\mathbb{Z}(d)) $$

is given by the inverse limit of Kato's norm compatible elliptic units ${}_c\vartheta_E$. Unfortunately, we do not have such a description even in the case of abelian varieties of dimension $\geqslant 2$. If we write $A[p^r]\langle t\rangle$ for the $A[p^r]$-torsor defined by diagram (31), then

$$ {}_\alpha\mathcal{EI}(t) \in H^{2d-1}(S, t^*\mathcal{L}(d)) = \varprojlim_r H^{2d-1}(A[p^r]\langle t\rangle, \mathbb{Z}/p^r\mathbb{Z}(d)) $$

where the inverse limit is again over the trace maps.

7 Interpolation of the $\mathbb{Q}_p$-Eisenstein Classes

7.1 An Integral Structure on $\mathcal{L}og^{(k)}_{\mathbb{Q}_p}$

For the comparison between the integral $\mathcal{L}$ and the $\mathbb{Q}_p$-polylogarithm $\mathcal{L}og_{\mathbb{Q}_p}$ we need an intermediate object, which we define in this section. This is purely technical. The reason for this is as follows: In general a unipotent $\mathbb{Q}_p$-sheaf does not necessarily have a $\mathbb{Z}_p$-lattice which is again a unipotent sheaf. In the case of $\mathcal{L}og^{(k)}_{\mathbb{Q}_p}$ however, it is even possible to construct a $\mathbb{Z}_p$-structure $\mathcal{L}og^{(k)}$ such that

$$ \mathcal{L}og^{(k)}_{\Lambda_r} := \mathcal{L}og^{(k)} \otimes_{\mathbb{Z}_p} \Lambda_r $$

is a unipotent $\Lambda_r = \mathbb{Z}/p^r\mathbb{Z}$-sheaf.

Let $\mathcal{L}og^{(1)}$ be the $\mathbb{Z}_p$-sheaf defined in Definition 3.2.1

$$ 0 \to \mathscr{H} \to \mathcal{L}og^{(1)} \to \mathbb{Z}_p \to 0 \tag{36} $$

and denote by $\mathbf{1}^{(1)} : \mathbb{Z}_p \to e^*\mathcal{L}og^{(1)}$ a fixed splitting.

Definition 7.1.1 We define

$$ \mathcal{L}og^{(k)} := \Gamma_k(\mathcal{L}og^{(1)}) $$

as the k-th graded piece of the divided power algebra $\Gamma_{\mathbb{Z}_p}(\mathcal{L}og^{(1)})$. We further denote by

$$\mathbf{1}^{(k)} := \Gamma_k(\mathbf{1}^{(1)}) : \mathbb{Z}_p \to \mathcal{L}og^{(k)}$$

the splitting induced by $\mathbf{1}^{(1)}$.

As $\mathbb{Z}_p$ and $\mathscr{H}$ are flat $\mathbb{Z}_p$-sheaves (all stalks are $\mathbb{Z}_p$-free), the k-th graded piece of the divided power algebra $\Gamma_k(\mathcal{L}og^{(1)})$ has a filtration with graded pieces $\pi^*\Gamma_i(\mathscr{H}) \otimes \Gamma_{k-i}(\mathbb{Z}_p)$ (see [8, V 4.1.7]). In particular, the $\Gamma_k(\mathcal{L}og^{(1)})$ are unipotent $\mathbb{Z}_p$-sheaves of length k. By base change the same is true for the Λ_r-sheaf

$$\mathcal{L}og^{(k)}_{\Lambda_r} := \mathcal{L}og^{(k)} \otimes_{\mathbb{Z}_p} \Lambda_r. \tag{37}$$

To define transition maps

$$\mathcal{L}og^{(k)} \to \mathcal{L}og^{(k-1)} \tag{38}$$

we proceed as in Sect. 3.3. Consider $\mathcal{L}og^{(1)} \to \mathbb{Z}_p \oplus \mathcal{L}og^{(1)}$ given by the canonical projection and the identity. Then we define

$$\mathcal{L}og^{(k)} = \Gamma_k(\mathcal{L}og^{(1)}) \to \Gamma_k(\mathbb{Z}_p \oplus \mathcal{L}og^{(1)}) \cong \bigoplus_{i+j=k} \Gamma_i(\mathbb{Z}_p) \otimes \Gamma_j(\mathcal{L}og^{(1)}) \to$$
$$\to \Gamma_1(\mathbb{Z}_p) \otimes \Gamma_{k-1}(\mathcal{L}og^{(1)}) \cong \mathcal{L}og^{(k-1)}$$

where we identify $\Gamma_1(\mathbb{Z}_p) \cong \mathbb{Z}_p$. A straightforward computation shows that $\mathbf{1}^{(k)} \mapsto \mathbf{1}^{(k-1)}$ under the transition map.

Definition 7.1.2 We denote by $\mathcal{L}og$ the pro-sheaf $(\mathcal{L}og^{(k)})_{k\geqslant 0}$ with the above transition maps and let $\mathbf{1} : \mathbb{Z}_p \to e^*\mathcal{L}og$ be the splitting defined by $(\mathbf{1}^{(k)})_{k\geqslant 0}$.

Remark 7.1.3 We would like to point out that, contrary to the $\mathbb{Q}_p$-situation, the pro-sheaf $(\mathcal{L}og^{(k)})_{k\geqslant 0}$ is *not* the correct definition of the $\mathbb{Z}_p$-logarithm sheaf. In fact, the correct integral logarithm sheaf is $\mathcal{L}$.

Proposition 7.1.4 *Denote by $\mathcal{L}og^{(k)} \otimes \mathbb{Q}_p$ the $\mathbb{Q}_p$-sheaf associated to $\mathcal{L}og^{(k)}$. Then there is a canonical isomorphism*

$$\mathcal{L}og^{(k)}_{\mathbb{Q}_p} \cong \mathcal{L}og^{(k)} \otimes \mathbb{Q}_p$$

which maps $\mathbf{1}^{(k)}_{\mathbb{Q}_p}$ to $\mathbf{1}^{(k)}$.

Proof First note that the canonical map $\operatorname{Sym}^k \mathcal{L}og^{(1)}_{\mathbb{Q}_p} \to \Gamma_k(\mathcal{L}og^{(1)}_{\mathbb{Q}_p})$ is an isomorphism. This can be checked at stalks, where it follows from (4) as $\mathcal{L}og^{(1)}_{\mathbb{Q}_p}$ is a sheaf of $\mathbb{Q}_p$-modules. The claim in the proposition then follows from the isomorphisms

$$\mathcal{L}og^{(k)}_{\mathbb{Q}_p} = \operatorname{Sym}^k \mathcal{L}og^{(1)}_{\mathbb{Q}_p} \cong \Gamma_k(\mathcal{L}og^{(1)}_{\mathbb{Q}_p}) \cong \Gamma_k(\mathcal{L}og^{(1)}) \otimes \mathbb{Q}_p = \mathcal{L}og^{(k)} \otimes \mathbb{Q}_p$$

and the claim about the splitting follows from the explicit formula for the map $\mathrm{Sym}^k \mathcal{L}og^{(1)}_{\mathbb{Q}_p} \to \Gamma_k(\mathcal{L}og^{(1)}_{\mathbb{Q}_p})$ given after (4). □

Corollary 7.1.5 *For all i there are isomorphisms*

$$H^i(S, R\pi_! Rj_{D*}j_D^* \mathcal{L}og^{(k)}(d)) \otimes_{\mathbb{Z}_p} \mathbb{Q}_p \cong H^i(S, R\pi_! Rj_{D*}j_D^* \mathcal{L}og^{(k)}_{\mathbb{Q}_p}(d))$$

$$H^i(S, \pi_{D!}\pi_D^* \prod_{i=0}^{k} \Gamma_i(\mathscr{H})) \otimes_{\mathbb{Z}_p} \mathbb{Q}_p \cong H^i(S, \pi_{D!}\pi_D^* \prod_{i=0}^{k} \mathrm{Sym}^i \mathscr{H}_{\mathbb{Q}_p})$$

$$H^i(S, R\pi_! \mathcal{L}og^{(k)}(d)[2d]) \otimes_{\mathbb{Z}_p} \mathbb{Q}_p \cong H^i(S, R\pi_! \mathcal{L}og^{(k)}_{\mathbb{Q}_p}(d)[2d])$$

Proof The first and the third follow directly from the proposition and the definition of the cohomology of a $\mathbb{Q}_p$-sheaf. For the second one observes that the canonical map

$$\mathrm{Sym}^k \mathscr{H}_{\mathbb{Q}_p} \cong \mathrm{Sym}^k \mathscr{H} \otimes \mathbb{Q}_p \to \Gamma_k(\mathscr{H}) \otimes \mathbb{Q}_p \cong \Gamma_k(\mathscr{H}_{\mathbb{Q}_p})$$

is an isomorphism. This can be checked on stalks, where it follows again from (4). □

7.2 Comparison of Integral and $\mathbb{Q}_p$-polylogarithm

In this section we want to compare $\mathcal{L}$ and $\mathcal{L}og_{\mathbb{Q}_p}$. We first compare $\mathcal{L}$ with the sheaves $\mathcal{L}og^{(k)}$ defined in Definition 7.1.1.

Define a comparison map

$$\mathrm{comp}^k : \mathcal{L} \to \mathcal{L}og^{(k)}$$

as follows. By Theorem 6.2.6 one has for the sheaves $\mathcal{L}og^{(k)}_{\Lambda_r}$ from (37) the isomorphism

$$\mathrm{Hom}_G(\Lambda_r[G_{rk}], \mathcal{L}og^{(k)}_{\Lambda_r}) \cong H^0(S, e^* \mathcal{L}og^{(k)}_{\Lambda_r}),$$

so that the splitting $\mathbf{1}^{(k)} \otimes \Lambda_r : \Lambda_r \to e^* \mathcal{L}og^{(k)}_{\Lambda_r}$ defines a morphism of sheaves on G

$$\mathrm{comp}^k_r : \Lambda_r[G_{rk}] \to \mathcal{L}og^{(k)}_{\Lambda_r}, \tag{39}$$

which is obviously compatible with the transition maps and functorial in G. Passing to the pro-systems over $r \geqslant 0$, this defines a homomorphism

$$\mathrm{comp}^k : \mathcal{L} \to \mathcal{L}og^{(k)}. \tag{40}$$

Taking also the pro-system in the k-direction leads to a comparison map

$$\mathrm{comp} : \mathcal{L} \to \mathcal{L}og. \tag{41}$$

For each k applying comp^k to the localization triangle for $D \hookrightarrow G \hookleftarrow U_D$ gives

$$\begin{array}{ccccc} R\pi_! Rj_{D*}j_D^*\mathcal{L}(d)[2d-1] & \longrightarrow & \pi_{D!}\pi_D^*\Lambda(\mathscr{H}) & \longrightarrow & R\pi_!\mathcal{L}(d)[2d] \\ \big\downarrow{\scriptstyle \mathrm{comp}^k} & & \big\downarrow{\scriptstyle \mathrm{comp}^k} & & \big\downarrow{\scriptstyle \mathrm{comp}^k} \\ R\pi_! Rj_{D*}j_D^*\mathcal{L}og^{(k)}(d)[2d-1] & \longrightarrow & \pi_{D!}\pi_D^*\mathcal{L}og^{(k)} & \longrightarrow & R\pi_!\mathcal{L}og^{(k)}(d)[2d] \end{array} \tag{42}$$

compatible with the transition maps $\mathcal{L}og^{(k)} \to \mathcal{L}og^{(k-1)}$.

Proposition 7.2.1 *There is a commutative diagram with short exact columns*

$$\begin{array}{ccc} 0 & & 0 \\ \big\downarrow & & \big\downarrow \\ H^{2d-1}(S, R\pi_! Rj_{D*}j_D^*\mathcal{L}(d)) & \xrightarrow{\mathrm{comp}} & H^{2d-1}(S, R\pi_! Rj_{D*}j_D^*\mathcal{L}og_{\mathbb{Q}_p}(d)) \\ \big\downarrow{\scriptstyle \mathrm{res}} & & \big\downarrow{\scriptstyle \mathrm{res}} \\ H^0(S, \pi_{D!}\pi_D^*\Lambda(\mathscr{H})) & \xrightarrow{e^*\,\mathrm{comp}} & H^0(S, \pi_{D!}\pi_D^* \prod_{k\geqslant 0} \mathrm{Sym}^k \mathscr{H}_{\mathbb{Q}_p}(d)) \\ \big\downarrow & & \big\downarrow \\ H^0(S, \mathbb{Z}_p) & \longrightarrow & H^0(S, \mathbb{Q}_p) \\ \big\downarrow & & \big\downarrow \\ 0 & & 0 \end{array}$$

Proof Take the long exact cohomology sequence of the commutative diagram in (42), tensor the lower horizontal line with $\mathbb{Q}_p$ and then pass to the inverse limit over k. Using the isomorphisms in Corollary 7.1.5 gives the commutative diagram as stated. □

Corollary 7.2.2 *Let $\alpha \in \mathbb{Z}_p[D]^0$, with $D = G[c]$ as before. Then one has*

$$\mathrm{comp}({}_\alpha\mathrm{pol}) = {}_\alpha\mathrm{pol}_{\mathbb{Q}_p}$$

in $H^{2d-1}(S, R\pi_! Rj_{D*}j_D^*\mathcal{L}og_{\mathbb{Q}_p}(d))$. In particular, for every N-torsion section $t : S \to U_D$ one has

$$\mathrm{comp}({}_\alpha\mathcal{EI}(t)) = t^*({}_\alpha\mathrm{pol}_{\mathbb{Q}_p}).$$

Proof Immediate from the definition of ${}_\alpha\mathrm{pol}$ and ${}_\alpha\mathrm{pol}_{\mathbb{Q}_p}$ and the commutative diagram in the proposition. The second statement follows from the first as comp is compatible with the evaluation map at t. □

7.3 Interpolation of the $\mathbb{Q}_p$-Eisenstein Classes

For our main result, we first have to relate the comparison map comp^k with the moment map mom^k.

Proposition 7.3.1 *The composition*

$$\Lambda(\mathscr{H}) \xrightarrow{e^*(\mathrm{comp}^k)} e^*\mathcal{L}og^{(k)} \xrightarrow{\mathrm{pr}_k} \Gamma_k(\mathscr{H})$$

coincides with the moment map mom^k.

Proof By the definitions of mom^k and comp^k it suffices to prove this statement for Λ_r-coefficients. Consider

$$\mathrm{comp}_r^k : \Lambda_r[G_{rk}] \to \mathcal{L}og_{\Lambda_r}^{(k)}$$

from (39). This comes by adjunction from a map

$$\beta_r : \Lambda_r \to [p^{rk}]^*\mathcal{L}og_{\Lambda_r}^{(k)},$$

on G_{rk} which has by definition the property that its pull-back $e_{rk}^*(\beta_r)$ coincides with $\mathbf{1}^{(k)} : \Lambda_r \to e^*\mathcal{L}og_{\Lambda_r}^{(k)}$. By Lemma 2.3.2 the map β_r is uniquely determined by this property. As $\mathcal{L}og_{\Lambda_r}^{(k)}$ is unipotent of length k, the pull-back $[p^{rk}]^*\mathcal{L}og_{\Lambda_r}^{(k)}$ is trivial by Lemma 6.2.5 and is hence of the form

$$[p^{rk}]^*\mathcal{L}og_{\Lambda_r}^{(k)} \cong \pi_{rk}^* e^*\mathcal{L}og_{\Lambda_r}^{(k)} \cong \pi_{rk}^* \prod_{i=0}^{k} \Gamma_i(\mathscr{H}_r),$$

where the last isomorphism is obtained by the splitting $\mathbf{1}^{(k)}$. Thus the map

$$\Lambda_r \to [p^{rk}]^*\mathcal{L}og_{\Lambda_r}^{(k)} \cong \pi_{rk}^* \prod_{i=0}^{k} \Gamma_i(\mathscr{H}_r) \qquad 1 \mapsto \sum_{i=0}^{k} \tau_r^{[i]},$$

where $\tau_r^{[i]}$ is the i-th divided power of the tautological section from (25), has the property that its pull-back by e_{rk}^* coincides with $\mathbf{1}^{(k)}$. It follows that this map equals β_r and by definition of the moment map in (26) the projection to the k-th component coincides also with the moment map. □

Let $t : S \to U_D$ be an N-torsion section. We need a compatibility between the composition

$$\mathrm{mom}_N^k := \mathrm{mom}^k \circ [N]_\# : \Lambda(\mathscr{H}\langle t\rangle) \to \Lambda(\mathscr{H}\langle t\rangle) \to \Gamma_k(\mathscr{H})$$

and the map ϱ_t in the splitting principle Corollary 3.4.2 composed with the projection onto the k-th component

$$\mathrm{pr}_k \circ \varrho_t : t^*\mathcal{L}og^{(k)}_{\mathbb{Q}_p} \cong \prod_{i=0}^{k} \mathrm{Sym}^k \mathcal{H}_{\mathbb{Q}_p} \to \mathrm{Sym}^k \mathcal{H}_{\mathbb{Q}_p}.$$

Proposition 7.3.2 *There is a commutative diagram*

$$\begin{array}{ccc} H^{2d-1}(S, \Lambda(\mathcal{H}\langle t\rangle)(d)) & \xrightarrow{\mathrm{mom}^k_N} & H^{2d-1}(S, \Gamma_k(\mathcal{H})(d)) \\ {\scriptstyle t^*\mathrm{comp}^k}\downarrow & & \downarrow \\ H^{2d-1}(S, t^*\mathcal{L}og^{(k)}_{\mathbb{Q}_p}(d)) & \xrightarrow{N^k \mathrm{pr}_k \circ \varrho_t} & H^{2d-1}(S, \mathrm{Sym}^k \mathcal{H}_{\mathbb{Q}_p}(d), \end{array}$$

where $\mathrm{mom}^k_N = \mathrm{mom}^k \circ [N]_\#$ *and* $\varrho_t = [N]_\#^{-1} \circ [N]_\#$.

Proof The commutative diagram

$$\begin{array}{ccc} H^{2d-1}(S, \Lambda(\mathcal{H}\langle t\rangle)(d)) & \xrightarrow{[N]_\#} & H^{2d-1}(S, \Lambda(\mathcal{H})(d)) \\ {\scriptstyle t^*\mathrm{comp}^k}\downarrow & & \downarrow{\scriptstyle e^*\mathrm{comp}^k} \\ H^{2d-1}(S, t^*\mathcal{L}og^{(k)}_{\mathbb{Q}_p}(d)) & \underset{\cong}{\xrightarrow{[N]_\#}} & H^{2d-1}(S, e^*\mathcal{L}og^{(k)}_{\mathbb{Q}_p}(d)) \end{array}$$

coming from functoriality of comp^k and the isomorphisms

$$H^{2d-1}(S, t^*\mathcal{L}og^{(k)}(d)) \otimes_{\mathbb{Z}_p} \mathbb{Q}_p \cong H^{2d-1}(S, t^*\mathcal{L}og^{(k)}_{\mathbb{Q}_p}(d))$$
$$H^{2d-1}(S, e^*\mathcal{L}og^{(k)}(d)) \otimes_{\mathbb{Z}_p} \mathbb{Q}_p \cong H^{2d-1}(S, e^*\mathcal{L}og^{(k)}_{\mathbb{Q}_p}(d))$$

reduces the proof of the proposition to show the commutativity of the diagram

$$\begin{array}{ccc} H^{2d-1}(S, \Lambda(\mathcal{H})(d)) & \xrightarrow{\mathrm{mom}^k} & H^{2d-1}(S, \Gamma_k(\mathcal{H})(d)) \\ {\scriptstyle e^*\mathrm{comp}^k}\downarrow & & \downarrow \\ H^{2d-1}(S, e^*\mathcal{L}og^{(k)}_{\mathbb{Q}_p}(d)) & \xrightarrow{N^k \mathrm{pr}_k \circ [N]_\#^{-1}} & H^{2d-1}(S, \mathrm{Sym}^k \mathcal{H}_{\mathbb{Q}_p}(d). \end{array}$$

The isogeny $[N]$ acts by N-multiplication on $\mathcal{H}$, hence by multiplication with N^k on $\mathrm{Sym}^k \mathcal{H}_{\mathbb{Q}_p}$, which means that

$$\mathrm{pr}_k \circ [N]_\#^{-1} = [N]_\#^{-1} \circ \mathrm{pr}_k = N^{-k} \mathrm{pr}_k .$$

Thus it remains to show that the diagram

$$\begin{array}{ccc} H^{2d-1}(S, \Lambda(\mathscr{H})(d)) & \xrightarrow{\mathrm{mom}^k} & H^{2d-1}(S, \Gamma_k(\mathscr{H})(d)) \\ {\scriptstyle e^*\,\mathrm{comp}^k}\downarrow & & \downarrow \\ H^{2d-1}(S, e^*\mathcal{L}og^{(k)}_{\mathbb{Q}_p}(d)) & \xrightarrow{\mathrm{pr}_k} & H^{2d-1}(S, \mathrm{Sym}^k \mathscr{H}_{\mathbb{Q}_p}(d)) \end{array}$$

commutes, which follows from Proposition 7.3.1 and the isomorphism

$$H^{2d-1}(S, \Gamma_k(\mathscr{H})(d)) \otimes_{\mathbb{Z}_p} \mathbb{Q}_p \cong H^{2d-1}(S, \mathrm{Sym}^k \mathscr{H}_{\mathbb{Q}_p}(d))$$

which was obtained in Corollary 7.1.5. □

Recall from Definition 6.4.1 the Eisenstein–Iwasawa class

$${}_\alpha\mathcal{EI}(t)_N = [N]_\#({}_\alpha\mathcal{EI}(t)) \in H^{2d-1}(S, \Lambda(\mathscr{H})(d))$$

and from Definition 4.2.1 the $\mathbb{Q}_p$-Eisenstein class

$${}_\alpha\mathrm{Eis}^k_{\mathbb{Q}_p}(t) \in H^{2d-1}(S, \mathrm{Sym}^k \mathscr{H}_{\mathbb{Q}_p}).$$

We consider its image under the k-th moment map

$$\mathrm{mom}^k : H^{2d-1}(S, \Lambda(\mathscr{H})(d)) \to H^{2d-1}(S, \Gamma_k(\mathscr{H})(d)).$$

The main result of this paper can now be formulated as follows:

Theorem 7.3.3 (Interpolation of $\mathbb{Q}_p$-Eisenstein classes) *The image of* ${}_\alpha\mathcal{EI}(t)_N$ *under the k-th moment map is given by*

$$\mathrm{mom}^k({}_\alpha\mathcal{EI}(t)_N) = N^k{}_\alpha\mathrm{Eis}^k_{\mathbb{Q}_p}(t).$$

Proof This follows by combining Corollaries 7.2.2, 7.3.2 and the definition of the $\mathbb{Q}_p$-Eisenstein class Definition 4.2.1. □

Remark 7.3.4 For comparison with [12, Theorem 12.4.21] we point out again that the normalization of ${}_\alpha\mathrm{Eis}^k_{\mathbb{Q}_p}(t)$ in loc. cit. is *different*. We had there a factor of $-N^{k-1}$ in front of the Eisenstein series.

Acknowledgements This paper is a sequel to [12], which was written as a response to John Coates' wish to have an exposition of the results in [6] at a conference in Pune, India. This triggered a renewed interest of mine in the p-adic interpolation of motivic Eisenstein classes which goes back to [11]. In this sense the paper would not exist without the persistence of John Coates. The paper [7] with Annette Huber created the right framework to treat these questions. It is a pleasure to thank them both and to dedicate this paper to John Coates on the occasion of his seventieth birthday.

References

1. Berthelot, P., Ogus, A.: Notes on crystalline cohomology. University of Tokyo Press, Tokyo, Princeton University Press, Princeton (1978)
2. Brion, M., Szamuely, T.: Prime-to-p étale covers of algebraic groups and homogeneous spaces. Bull. Lond. Math. Soc. **45**(3), 602–612 (2013). doi:10.1112/blms/bds110
3. Colmez, P.: Fonctions L p-adiques. Astérisque **266**, Exp. No. 851, 3, pp. 21–58 (2000). Séminaire Bourbaki, vol. 1998/99. http://www.numdam.org/item?id=SB_1998-1999__41__21_0
4. Deligne, P.: Cohomologie étale. In: Lecture Notes in Mathematics, vol. 569. Springer, Berlin-New York (1977)
5. Ekedahl, T.: On the adic formalism. In: The Grothendieck Festschrift, vol. II, Progr. Math., vol. 87, pp. 197–218. Birkhäuser Boston, Boston, MA (1990)
6. Huber, A., Kings, G.: Degeneration of l-adic Eisenstein classes and of the elliptic polylog. Invent. Math. **135**(3), 545–594 (1999). doi:10.1007/s002220050295
7. Huber, A., Kings, G.: Polylogarithm for Families of Commutative Group Schemes (2015) (Preprint)
8. Illusie, L.: Complexe Cotangent et Déformations I. In: Lecture Notes in Mathematics, 239, vol. 239. Springer (1971)
9. Jannsen, U.: Continuous étale cohomology. Math. Ann. **280**(2), 207–245 (1988). doi:10.1007/BF01456052
10. Katz, N.M.: P-adic interpolation of real analytic Eisenstein series. Ann. Math. (2)**104**(3), 459–571 (1976). doi:10.2307/1970966
11. Kings, G.: The Tamagawa number conjecture for CM elliptic curves. Invent. Math. **143**(3), 571–627 (2001). doi:10.1007/s002220000115
12. Kings, G.: Eisenstein classes, elliptic Soulé elements and the ℓ-adic elliptic polylogarithm. In: The Bloch–Kato conjecture for the Riemann zeta function, *London Math. Soc. Lecture Note Ser.*, vol. 418, pp. 237–296. Cambridge University Press (2015)
13. Kings, G., Loeffler, D., Zerbes, S.: Rankin-Eisenstein classes and explicit reciprocity laws (2015) (Preprint)
14. Roos, J.E.: Sur les foncteurs dérivés de $\varprojlim$ applications. C. R. Acad. Sci. Paris **252**, 3702–3704 (1961)
15. Cohomologie l-adique et fonctions L. Lecture Notes in Mathematics, vol. 589. Springer, Berlin-New York. Séminaire de Géometrie Algébrique du Bois-Marie 1965–1966 (SGA 5), Edité par Luc Illusie (1977)

Vanishing of Some Galois Cohomology Groups for Elliptic Curves

Tyler Lawson and Christian Wuthrich

Abstract Let $E/\mathbb{Q}$ be an elliptic curve and p be a prime number, and let G be the Galois group of the extension of $\mathbb{Q}$ obtained by adjoining the coordinates of the p-torsion points on E. We determine all cases when the Galois cohomology group $H^1\big(G, E[p]\big)$ does not vanish, and investigate the analogous question for $E[p^i]$ when $i > 1$. We include an application to the verification of certain cases of the Birch and Swinnerton-Dyer conjecture, and another application to the Grunwald–Wang problem for elliptic curves.

Keywords Elliptic curves · Galois cohomology · Grunwald-Wang problem · Birch and Swinnerton-Dyer conjecture

1 Introduction

Let E be an elliptic curve over $\mathbb{Q}$ and p a prime number. Denote by K the Galois extension of $\mathbb{Q}$ obtained by adjoining the coordinates of the p torsion points on E and let G be the Galois group of $K/\mathbb{Q}$. The Galois action on the p-torsion points $E[p]$ identifies G with a subgroup of $\mathrm{GL}\big(E[p]\big) \cong \mathrm{GL}_2(\mathbb{F}_p)$ via the representation $\rho\colon \mathrm{Gal}\big(\bar{\mathbb{Q}}/\mathbb{Q}\big) \to \mathrm{GL}\big(E[p]\big)$. A celebrated theorem of Serre [21] shows that G is equal to the full group $\mathrm{GL}_2(\mathbb{F}_p)$ for all but finitely many primes p when the curve is fixed.

We are interested in the vanishing of the Galois cohomology group $H^1\big(G, E[p]\big)$; see [22] or [19] for the basic definitions of Galois cohomology. This specific

Tyler Lawson's work is partially supported by NSF DMS-1206008.

T. Lawson
Vincent Hall 323, Department of Mathematics, University of Minnesota,
Vincent Hall, 206 Church St SE, Minneapolis, MN 55455, USA
e-mail: tlawson@math.umn.edu

C. Wuthrich (✉)
School of Mathematical Sciences, University of Nottingham,
University Park, Nottingham NG7 2RD, UK
e-mail: christian.wuthrich@nottingham.ac.uk

D. Loeffler and S.L. Zerbes (eds.), *Elliptic Curves, Modular Forms and Iwasawa Theory*, Springer Proceedings in Mathematics & Statistics 188, DOI 10.1007/978-3-319-45032-2_11

cohomology group appears as an obstruction in various contexts. For instance, Kolyvagin's work uses the vanishing of this group in the case G is equal to $\mathrm{GL}_2(\mathbb{F}_p)$ (see Proposition 9.1 in [14]). The following first theorem characterizes completely when this cohomology group does not vanish, answering a question at [15].

Theorem 1 *Fix a prime p. Let $E/\mathbb{Q}$ be an elliptic curve, $K = \mathbb{Q}(E[p])$, and G the Galois group of $K/\mathbb{Q}$. Then $H^1\big(G, E[p]\big)$ is trivial except in the following cases:*

- *$p = 3$, there is a rational point of order 3 on E, and there are no other isogenies of degree 3 from E that are defined over $\mathbb{Q}$.*
- *$p = 5$ and the quadratic twist of E by $D = 5$ has a rational point of order 5, but no other isogenies of degree 5 defined over $\mathbb{Q}$.*
- *$p = 11$ and E is the curve labeled as 121c2 in Cremona's tables [6], given by the global minimal equation $y^2 + xy = x^3 + x^2 - 3632\,x + 82757$.*

In each of these cases, $H^1\big(G, E[p]\big)$ has p elements.

Partial results on this question have appeared in various sources. For instance, Lemma 10 in [5] by Coates shows that $H^1\big(G, E[p]\big)$ vanishes when $E[p]$ is irreducible as a Galois module. Section 3 in [4] also treats related questions.

The above result extends to elliptic curves E over more general number fields F if we assume that $F \cap \mathbb{Q}(\mu_p) = \mathbb{Q}$, where $\mathbb{Q}(\mu_p)$ is the field generated by p-th roots of unity. Rather than a single elliptic curve for $p > 5$, one finds possibly infinitely many exceptions for $p = 11$ and $p = 17$, but only finitely many further exceptions for each $p > 17$ and none for all p such that $p \equiv 1 \pmod 3$. See Theorem 11 for a precise statement.

Next, we address the analogous question for $E[p^i]$ for $i > 1$, but assuming that $p > 3$.

Theorem 2 *Fix a prime $p > 3$. Let $E/\mathbb{Q}$ be an elliptic curve, $K_i = \mathbb{Q}(E[p^i])$ the extension of $\mathbb{Q}$ obtained by adjoining the coordinates of all p^i-torsion points, and G_i the Galois group of $K_i/\mathbb{Q}$. Then $H^1\big(G_2, E[p^2]\big)$ is trivial if and only if $H^1\big(G_i, E[p^i]\big)$ is trivial for all $i \geqslant 2$. This vanishing holds if and only if (E, p) is not among the following cases:*

- *$p = 5$ or $p = 7$ and E contains a rational p-torsion point.*
- *$p = 5$ and there is an isogeny $\varphi\colon E \to E'$ of degree 5 defined over $\mathbb{Q}$ and the quadratic twist by $D = 5$ of E contains a rational 5-torsion point.*
- *$p = 5$ and there is an isogeny $\varphi\colon E \to E'$ of degree 5 defined over $\mathbb{Q}$ but none of degree 25 and the quadratic twist by $D = 5$ of E' contains a rational 5-torsion point.*
- *$p = 5$ and E admits an isogeny $E \to E' \to E''$ of degree 25 defined over $\mathbb{Q}$ and E' contains a rational 5-torsion point.*
- *$p = 11$ and E is 121c1 or 121c2.*

Again, we will also obtain some results that are valid over more general base fields F with $F \cap \mathbb{Q}(\mu_p) = \mathbb{Q}$, and some that are valid for $p = 3$. See Sect. 6.

This more general question has also been investigated before, and Cha has obtained results in this direction in [3]. He proved the vanishing of $H^1\big(G_i, E[p^i]\big)$ when $p > 3$, the curve has semi-stable reduction at an unramified place above p, and E does not have a rational p-torsion point. He also describes when this cohomology group vanishes for $p = 3$ under his assumptions. The method of proof is similar.

The results in Theorem 2 can be applied to the Grunwald–Wang problem for elliptic curves as formulated by Dvornicich and Zannier in [9]. In Proposition 25, we give an example of an elliptic curve $E/\mathbb{Q}$ with a point $P \in E(\mathbb{Q})$ divisible by $m = 9$ in $E(\mathbb{Q}_\ell)$ for almost all primes ℓ but not divisible by 9 in $E(\mathbb{Q})$. Previously, the only known examples [10] were with $m = 4$. In Theorem 24, we also give a simplified proof of the result in [20] that it is impossible to find such a point P when $m = p^2$ and $p > 3$.

The paper is structured as follows. We begin with some background in Sect. 2, both establishing notation and reducing to cases where the Galois group G does not contain a nontrivial homothety. In Sect. 3 we prove a general form of Theorem 1. Section 4 establishes a vanishing result for H^2. In Sect. 5 we give an application to verifying cases of the Birch and Swinnerton-Dyer conjecture, correcting an oversight in [13]. Our main results classifying the vanishing of $H^1(G_i, E[p^i])$ are then discussed in Sect. 6, and some supplementary numerical computations for $H^1(G_2, E[p^2])$ are included in Sect. 7. Finally, in Sect. 8 we give the application to the Grunwald–Wang problem for elliptic curves.

2 Preliminaries and Notation

Throughout this paper E will be an elliptic curve defined over a number field F and p will be a prime number. We will denote by $K = F\big(E[p]\big)$ the number field obtained by adjoining the coordinates of the p-torsion points to F. Let G be the Galois group of K/F. More generally, for $i \geqslant 1$ we let $K_i = F\big(E[p^i]\big)$ and $G_i = \mathrm{Gal}(K_i/F)$. The faithful actions of G_i on $E[p^i]$ give embeddings $G_i \hookrightarrow \mathrm{Aut}(E[p^i]) \cong \mathrm{GL}_2(\mathbb{Z}/p^i)$, and so we may regard them as subgroups.

We will also use the groups $H_i = \mathrm{Gal}(K_{i+1}/K_i)$ and $M_i = \mathrm{Gal}(K_i/K)$. We note that, as H_i is the kernel of the map $G_{i+1} \to G_i$, it is identified with a subgroup of

$$\ker\Big(\mathrm{GL}_2(\mathbb{Z}/p^{i+1}) \to \mathrm{GL}_2(\mathbb{Z}/p^i)\Big) \cong \mathrm{Mat}_2(\mathbb{F}_p),$$

where the conjugation action of $G_{i+1} \subset \mathrm{GL}_2(\mathbb{Z}/p^{i+1})$ is by the adjoint representation. Therefore, all elements in H_i have order p and commute with the elements of M_{i+1} inside G_{i+1}.

In summary, we have the following situation:

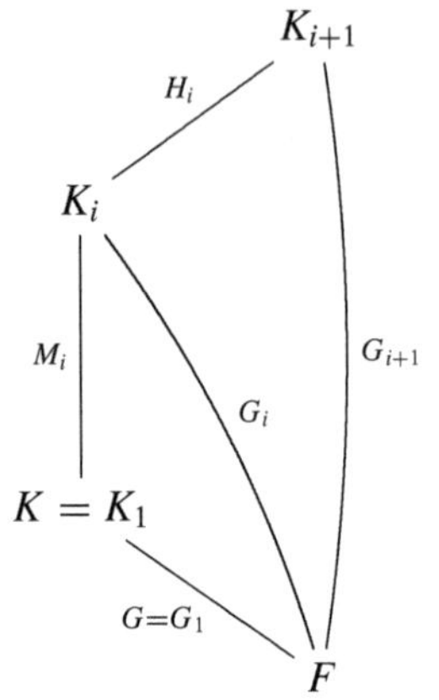

We will later use the inflation-restriction sequence

$$0 \longrightarrow H^1\big(G_i, E[p^j]\big) \xrightarrow{\text{inf}} H^1\big(G_{i+1}, E[p^j]\big) \xrightarrow{\text{res}} H^1\big(H_i, E[p^j]\big)^{G_i} \longrightarrow H^2\big(G_i, E[p^j]\big) \tag{1}$$

which is valid for all $1 \leqslant j \leqslant i$. In inductive arguments, we will also use that the short exact sequence

$$0 \longrightarrow E[p] \longrightarrow E[p^j] \longrightarrow E[p^{j-1}] \longrightarrow 0$$

gives a long exact sequence

$$E(F)[p^{j-1}] \longrightarrow H^1\big(G_i, E[p]\big) \longrightarrow H^1\big(G_i, E[p^j]\big) \longrightarrow H^1\big(G_i, E[p^{j-1}]\big). \tag{2}$$

As mentioned in the introduction, these cohomology groups only start to be interesting when $E[p]$ is reducible. The following argument for this is given in [3] as Theorem 7.

Lemma 3 *If G contains a non-trivial homothety, then $H^1\big(G_i, E[p^i]\big) = 0$.*

Proof Let g be a non-trivial homothety. Since g is central, $\langle g \rangle$ is a normal subgroup in G. Consider the inflation-restriction sequence

$$0 \longrightarrow H^1\big(G/\langle g \rangle, E[p]^{g=1}\big) \longrightarrow H^1\big(G, E[p]\big) \longrightarrow H^1\big(\langle g \rangle, E[p]\big)$$

The homothety g cannot have fixed points in $E[p]$; in particular $E(F)[p] = 0$. The left-hand side cohomology group in the above sequence is therefore trivial. The right-hand side is also trivial because $\langle g \rangle$ is of order coprime to p.

We assume by induction that $H^1\big(G_i, E[p^i]\big)$ and $H^1\big(G_i, E[p]\big)$ are both trivial. By assumption, the restriction maps

$$H^1\big(G_{i+1}, E[p^i]\big) \longrightarrow H^1\big(H_i, E[p^i]\big)^{G_i} \cong \mathrm{Hom}\big(H_i, E[p^i]\big)^{G_i}$$

$$H^1\big(G_{i+1}, E[p]\big) \longrightarrow H^1\big(H_i, E[p]\big)^{G_i} \cong \mathrm{Hom}\big(H_i, E[p]\big)^{G_i}$$

from (1) are both injective. Note that the target groups are actually equal because all elements in H_i have order p. Since M_{i+1} and H_i commute, the action of G_i on H_i factors through G, so the target in both cases is $\mathrm{Hom}\big(H_i, E[p]\big)^G$.

The homothety g acts trivially on H_i and non-trivially on any non-zero point in $E[p]$. Therefore, there are no homomorphisms from H_i to $E[p]$ which are fixed by g. It follows that $H^1\big(G_{i+1}, E[p^i]\big)$ and $H^1\big(G_{i+1}, E[p]\big)$ are both trivial. The exact sequence (2) now implies that $H^1\big(G_{i+1}, E[p^{i+1}]\big)$ is also trivial. □

Lemma 4 *Suppose $p > 2$. Assume that G does not contain a non-trivial homothety and $F \cap \mathbb{Q}(\mu_p) = \mathbb{Q}$. Then G is contained in a Borel subgroup.*

Proof By the Weil pairing, the determinant of ρ is the Teichmüller character ω describing the action of Galois on the p-th roots of unity μ_p. The assumption $F \cap \mathbb{Q}(\mu_p) = \mathbb{Q}$ implies that $\det\colon G \to \mathbb{F}_p^\times$ must be surjective.

Assume first that $p > 3$. We fix a basis of $E[p]$ and view G as a subgroup of $\mathrm{GL}_2(\mathbb{F}_p)$. By the classification of maximal subgroups of $\mathrm{GL}_2(\mathbb{F}_p)$, we have to show that the following cases can not occur: G is a subgroup of the normalizer of a split Cartan group, G is a subgroup of the normalizer of a non-split Cartan group, or G maps to an exceptional group A_4, A_5 or S_4 in $\mathrm{PGL}_2(\mathbb{F}_p)$.

Suppose G is a subgroup of the group of diagonal and anti-diagonal matrices, which is the normalizer of a split Cartan subgroup. Suppose moreover that G is not a subgroup of the diagonal matrices. The square of $\left(\begin{smallmatrix} 0 & b \\ c & 0 \end{smallmatrix}\right) \in G$ is the homothety by bc. Therefore, all anti-diagonal elements in G must be of the form $\left(\begin{smallmatrix} 0 & c^{-1} \\ c & 0 \end{smallmatrix}\right)$. Multiplying this with a diagonal element $\left(\begin{smallmatrix} u & 0 \\ 0 & v \end{smallmatrix}\right)$ in G then shows that all diagonal elements must have determinant 1. Hence the determinant would not be surjective for $p > 3$.

Next, suppose that G is a subgroup of the normalizer of a non-split Cartan group. Since G contains no non-trivial homothety, the image of G in $\mathrm{PGL}_2(\mathbb{F}_p)$ is isomorphic to G. In other words, G must be a subgroup of a dihedral group of order $2(p+1)$. No such group could have a surjective map onto $\mathbb{F}_p^\times$ if $p > 3$.

Finally, assume that G is exceptional. As before, our hypothesis implies that G is isomorphic to a subgroup of A_4, A_5 or S_4. However the only case in which we could have a surjective map onto $\mathbb{F}_p^\times$ with $p > 3$ is when $p = 5$ and G is a cyclic group of order 4 in S_4. However, as $\mathbb{F}_5$ contains the fourth roots of unity μ_4, all such subgroups are diagonalizable in $\mathrm{GL}_2(\mathbb{F}_5)$.

We now return to the case $p = 3$. By assumption, G is isomorphic to its image in $\mathrm{PGL}_2(\mathbb{F}_p)$, which is the full symmetric group on the four elements $\mathbb{P}^1(\mathbb{F}_3)$. Since the determinant is surjective, the image of G cannot be contained in the alternating group. Therefore it is not transitive on $\mathbb{P}^1(\mathbb{F}_3)$, and G is contained in a Borel subgroup. □

From now on we will suppose that $\varphi\colon E \to E'$ is an isogeny of degree p defined over F, and write $E[\varphi]$ for its kernel. The dual isogeny is denoted by $\hat{\varphi}\colon E' \to E$. We will now also fix a basis of $E[p]$ with the property that the first point belongs to $E[\varphi]$. In this basis, the Galois representation $\rho\colon \operatorname{Gal}(\bar{F}/F) \to \operatorname{GL}(E[p]) \cong \operatorname{GL}_2(\mathbb{F}_p)$ now takes values in the Borel subgroup of upper triangular matrices. We will write $\chi\colon \operatorname{Gal}(\bar{F}/F) \to \mathbb{F}_p^\times$ for the character of the Galois group on $E'[\hat{\varphi}]$. Then the character on $E[\varphi]$ is $\omega\chi^{-1}$, where ω is the Teichmüller character introduced above. The representation now is of the form $\rho = \left(\begin{smallmatrix} \omega\chi^{-1} & * \\ 0 & \chi \end{smallmatrix}\right)$.

Corollary 5 *Suppose $p > 2$. If $F \cap \mathbb{Q}(\mu_p) = \mathbb{Q}$ and the group $H^1\big(G, E[p]\big)$ is non-trivial, then E admits exactly one isogeny $\varphi\colon E \to E'$ of degree p that is defined over F.*

Proof By Lemma 3, we know that there is no non-trivial homothety in G. Then Lemma 4 implies that G is contained in a Borel subgroup. Hence there is a subgroup of order p in $E[p]$ fixed by the Galois group. If there were a second subgroup of order p fixed by the Galois group, then in a suitable basis of $E[p]$ the group G would consist of diagonal matrices. It would follow that G has order coprime to p and therefore that the cohomology group is trivial. Therefore, there is a unique isogeny defined over F of degree p. □

3 Proof of Theorem 1

We begin by assuming that E is defined over a number field F such that $F \cap \mathbb{Q}(\mu_p) = \mathbb{Q}$.

Lemma 6 *The cohomology group $H^1(G; E[2])$ always vanishes.*

Proof The group $\operatorname{GL}_2(\mathbb{F}_2)$ is isomorphic to the symmetric group on 3 letters. For any cyclic subgroup of $\operatorname{GL}_2(\mathbb{F}_2)$ of order 2 generated by h, we may compute $H^1\big(\langle h\rangle, E[2]\big)$ as the quotient of the kernel of the norm $N_G = 1 + h$ on $E[2]$ modulo the image of $h - 1$. Because $p = 2$, this group is trivial.

For a general subgroup $G \leqslant \operatorname{GL}_2(\mathbb{F}_2)$, let H be the intersection of G with the normal subgroup of order 3. We have $H^1\big(H, E[2]\big) = 0$ because the order of H is coprime to 2. We also have $H^1\big(G/H, E[2]^H\big) = 0$ because H is either of order 3 and only fixes 0 in $E[2]$, or H is trivial and this group is $H^1\big(\langle h\rangle, E[2]\big) = 0$. By the inflation-restriction sequence, we conclude that $H^1\big(G, E[2]\big) = 0$. □

Lemma 7 *Let $H < \operatorname{GL}_2(\mathbb{F}_p)$ be the subgroup generated by $h = \left(\begin{smallmatrix} 1 & 1 \\ 0 & 1 \end{smallmatrix}\right)$. We have an isomorphism*

$$H^1\big(H, E[p]\big) \cong \mathbb{F}_p,$$

and the action of an element $g = \left(\begin{smallmatrix} u & w \\ 0 & v \end{smallmatrix}\right)$ *in the normalizer* $N(H)$ *of* H *on this cohomology group is multiplication by* $u^{-1}v^2$.

Proof The cohomology of the cyclic group H is computed to be

$$H^1(H, E[p]) \cong \frac{\ker\left(\sum_{a=0}^{p-1} h^a\right)}{\operatorname{im}(h-1)} = \frac{\ker(0)}{\operatorname{im}\left(\begin{smallmatrix} 0 & 1 \\ 0 & 0 \end{smallmatrix}\right)} \cong \mathbb{F}_p.$$

The explicit isomorphism $i\colon H^1(H, E[p]) \to \mathbb{F}_p$ sends a cocycle $\xi : H \to E[p]$ to the second coordinate of $\xi(h)$. Let now $g = \left(\begin{smallmatrix} u & w \\ 0 & v \end{smallmatrix}\right)$ be an element of $N(H)$ with $u, v \in \mathbb{F}_p^\times$. Then the action of g on $\xi \in H^1(H, E[p])$ is as follows.

$$\begin{aligned}
(g \star \xi)(h) &= g\,\xi\left(g^{-1}hg\right) \\
&= g\,\xi\left(h^{u^{-1}v}\right) \\
&= g\left(h^{u^{-1}v-1} + \cdots + h + 1\right)\xi(h) \\
&= \left(\begin{smallmatrix} u & 0 \\ 0 & v \end{smallmatrix}\right)\left(\begin{smallmatrix} u^{-1}v & * \\ 0 & u^{-1}v \end{smallmatrix}\right)\left(\begin{smallmatrix} * \\ i(\xi) \end{smallmatrix}\right)
\end{aligned}$$

Here the terms denoted by $*$ are unknown entries which do not alter the result that

$$i(g \star \xi) = u^{-1}\,v^2\,i(\xi). \qquad \square$$

For the remainder of this section we will assume that $p > 2$ and E satisfies $H^1(G, E[p]) \neq 0$; we wish to show that we fall into one of the cases listed in the Theorem 1.

Lemma 8 *Suppose* $p > 2$. *Then* G *satisfies* $H^1(G, E[p]) \neq 0$ *if and only if* $p \not\equiv 1 \pmod 3$ *and there exists a basis of* $E[p]$ *such that* G *consists of all matrices of the form* $\left(\begin{smallmatrix} v^2 & w \\ 0 & v \end{smallmatrix}\right)$ *with* $v \in \mathbb{F}_p^\times$ *and* $w \in \mathbb{F}_p$. *In this case, the cohomology group is isomorphic to* $\mathbb{F}_p$ *and the representation* ρ *is of the form* $\left(\begin{smallmatrix} \chi^2 & * \\ 0 & \chi \end{smallmatrix}\right)$, *where* χ^3 *is the Teichmüller character* ω.

Proof By Corollary 5, we may view G as a group of upper triangular matrices containing the subgroup H generated by element $h = \left(\begin{smallmatrix} 1 & 1 \\ 0 & 1 \end{smallmatrix}\right)$ of order p.

Since H is a normal subgroup of G, we can use the inflation-restriction sequence to show that

$$H^1(G, E[p]) \longrightarrow H^1(H, E[p])^{G/H}$$

is an isomorphism because G/H is of order coprime to p. Because we assumed that $H^1(G, E[p])$ is non-trivial, by Lemma 7 we must have that G/H acts trivially on $H^1(H, E[p])$, that $H^1(G, E[p])$ has precisely p elements, and that all elements in G must be of the form $\left(\begin{smallmatrix} v^2 & w \\ 0 & v \end{smallmatrix}\right)$ with $w \in \mathbb{F}_p$ and $v \in \mathbb{F}_p^\times$.

Recall that the character χ is such that $\rho = \left(\begin{smallmatrix} \omega\chi^{-1} & * \\ 0 & \chi \end{smallmatrix}\right)$. We now deduce that $\chi^2 = \omega\chi^{-1}$ and hence $\chi^3 = \omega$. Since we assumed $F \cap \mathbb{Q}(\mu_p) = \mathbb{Q}$, the determinant ω

from G to $\mathbb{F}_p^\times$ must be surjective. As the determinant of the typical element in G is v^3 with $v \in \mathbb{F}_p^\times$, we must conclude that either $p = 3$ or $p \equiv 2 \pmod 3$, and that G is equal to the group of all matrices of the form $\begin{pmatrix} v^2 & w \\ 0 & v \end{pmatrix}$. □

Corollary 9 *If $p = 3$, we have $H^1(G, E[3]) \neq 0$ if and only if E has a 3-torsion point and no other isogenies defined over F.*

Proof This can only occur if the group G is the group of matrices of the form $\begin{pmatrix} 1 & w \\ 0 & v \end{pmatrix}$ of order 6. This is precisely the case when $E(F)[3]$ is of order 3 and no other isogenies are defined over F. □

Lemma 10 *If $p = 5$, we have $H^1(G, E[5]) \neq 0$ if and only if the quadratic twist of E by $D = 5$ has a 5-torsion point and no other isogenies defined over F.*

Proof This happens precisely when we have

$$\rho = \begin{pmatrix} \omega^2 & * \\ 0 & \omega^{-1} \end{pmatrix}$$

Here ω^2 is the quadratic character corresponding to the non-trivial extension $F(\sqrt{5})/F$ contained in $F(\mu_5)$. Let $E^\dagger$ be the quadratic twist of E by $D = 5$. Then we have the desired form of representation ρ if and only if the representation $\rho^\dagger$ on $E^\dagger[5]$ is now of the form $\begin{pmatrix} 1 & * \\ 0 & \omega \end{pmatrix}$. We conclude that this occurs if and only if $E^\dagger(F)[5]$ has five points and $E^\dagger$ has no other isogenies of degree 5 defined over F. □

Theorem 11 *Let E be an elliptic curve defined over a number field F with $F \cap \mathbb{Q}(\mu_p) = \mathbb{Q}$. Let $K = F(E[p])$ and $G = \mathrm{Gal}(F/K)$. Then $H^1(G, E[p]) = 0$ except in the following cases:*

- *$p = 3$, there is a rational 3-torsion point in $E(F)$, and there are no other 3-isogenies from E defined over F.*
- *$p = 5$ and the quadratic twist of E by $D = 5$ has a rational point of order 5, but no other isogenies of degree 5 defined over F.*
- *$p \geqslant 11$, $p \equiv 2 \pmod 3$, there is a unique isogeny $\varphi\colon E \to E'$ of degree p defined over F, its kernel $E[\varphi]$ acquires a rational point over $F \cdot \mathbb{Q}(\mu_p)^+$, and $E[\varphi] \cong \mu_p^{\otimes (p+1)/3}$.*

There are only finitely many cases for each prime p with $p > 17$.

Proof The only remaining cases to prove are those where $p > 5$. As we may assume $p \equiv 2 \pmod 3$, one sees that

$$\rho = \begin{pmatrix} \omega^{\frac{p+1}{3}} & * \\ 0 & \omega^{\frac{2-p}{3}} \end{pmatrix}.$$

This explains the condition in the cases $p \geqslant 11$ in the above list.

The curve E and its unique isogeny φ of degree p defined over F represent a point on the modular curve $Y_0(p)$ defined over F. For $p = 11$ and $p = 17$, the curve $Y_0(p)$ is of genus 1; for all larger primes $p \equiv 2 \pmod 3$ it is of genus at least two. Therefore there are only finitely many $\bar{\mathbb{Q}}$-isomorphism classes of curves E/F with an isogeny of degree p defined over F. Only a single twist in each class can have ρ of the above shape. Hence there are only finitely many exceptions for $p > 17$. □

We specialize now to the field $F = \mathbb{Q}$ where the points on $Y_0(p)$ are well-known.

Lemma 12 *If $F = \mathbb{Q}$ and $p > 5$, we have $H^1\big(G, E[p]\big) \neq 0$ if and only if E is the curve labeled as 121c2 in Cremona's tables.*

Proof For all those p, there are only a finite number of $\bar{\mathbb{Q}}$-isomorphism classes of elliptic curves E with a p-isogeny defined over $\mathbb{Q}$. Mazur's theorem [16] shows that there are no rational points on $Y_0(p)$ except for three points on $Y_0(11)$ and two points on $Y_0(17)$. All of these five examples have no other automorphisms than ± 1. Hence, all elliptic curves $E/\mathbb{Q}$ representing one of them are quadratic twists of each other.

Let us first look at $p = 11$. The j-invariants of the three families are -121, -32768, and -24729001, and the representation ρ must now be of the form $\left(\begin{smallmatrix} \omega^4 & * \\ 0 & \omega^7 \end{smallmatrix}\right)$. We start with the last. The curve 121c2 is an example of an elliptic curve with j-invariant -24729001. Using SageMath [23], we find a point P of order 11 in $E\big(\mathbb{Q}(\mu_{11})\big)$. Its x-coordinate in the global minimal model given above is $11\zeta^9 + 11\zeta^8 + 22\zeta^7 + 22\zeta^6 + 22\zeta^5 + 22\zeta^4 + 11\zeta^3 + 11\zeta^2 + 39$, where ζ is a primitive 11-th root of unity. One finds that $\sigma(P) = 5P$ for the Galois element with $\sigma(\zeta) = \zeta^2$. Therefore the action of Galois on the group generated by P is given by ω^4. The isogeny with P in its kernel is defined over $\mathbb{Q}$ and it is the only isogeny on E defined over $\mathbb{Q}$. Therefore the group G is precisely of the form required. Hence $H^1\big(G, E[p]\big)$ has p elements. No quadratic twist of E could have the same property.

With similar computation one finds that the group G for the curve 121b1 with j-invariant -32768 is of the form $\left(\begin{smallmatrix} \omega^8 & * \\ 0 & \omega^3 \end{smallmatrix}\right)$ and for the curve 121c1 with j-invariant -121 it is $\left(\begin{smallmatrix} \omega^7 & * \\ 0 & \omega^4 \end{smallmatrix}\right)$. No quadratic twist of these curves could have the required form for G.

For $p = 17$, the representation ρ must now be of the form $\left(\begin{smallmatrix} \omega^6 & * \\ 0 & \omega^{11} \end{smallmatrix}\right)$. In particular, for any prime $\ell \neq 11$ of good reduction for E, the Frobenius element is sent to a matrix of the form $\left(\begin{smallmatrix} \ell^6 & * \\ 0 & \ell^{11} \end{smallmatrix}\right)$. We conclude that we must have $\ell^6 + \ell^{11} \equiv a_\ell \pmod p$, where a_ℓ is the trace of Frobenius. This gives an easy criterion to rule out specific curves.

There are two j-invariants of elliptic curves that admit a 17-isogeny over $\mathbb{Q}$: $-297756989/2$ and $-882216989/131072$. In fact, these values were computed by Vélu and published on page 80 of [1]. We pick a curve E for each of these j-invariants. The curves 14450p1 and 14450n1 are examples. Now for both curves, it is easy to show that $\pm 3^6 \pm 3^{11} \not\equiv a_3 \pmod{17}$ for any choice of the signs as

$a_3 = \pm 2$. Therefore no quadratic twist of E will satisfy the congruence that we need. Thus $H^1\big(G, E[p]\big) = 0$ for all curves with a degree-17 isogeny. Similar computations were done by Greenberg in Remark 2.1.2 in [11]. $\square$

This concludes the proof of Theorem 1.

4 Vanishing of the Second Cohomology

We continue to assume that E is defined over a number field F such that $F \cap \mathbb{Q}(\mu_p) = \mathbb{Q}$.

Lemma 13 *Let p be a prime. Then $H^2\big(G, E[p]\big) = 0$ except if $p > 2$, E admits a p-isogeny $\varphi\colon E \to E'$ and no other p-isogenies over F, and $E'[\hat{\varphi}]$ contains an F-rational p-torsion point. If this cohomology group is non-zero then it contains p elements.*

We could also write the condition in the lemma as either that $E[\varphi] \cong \mu_p$ or that χ is trivial.

Proof As before, only the cases when p divides the order of G are of interest.

We again discuss the case $p = 2$ separately. A Sylow subgroup of G is a cyclic group of order 2 generated by h, and the restriction $H^2\big(G, E[p]\big) \to H^2\big(\langle h\rangle, E[p]\big)$ is an inclusion. However, $H^2\big(\langle h\rangle, E[p]\big)$ can be computed as the Tate cohomology group $\hat{H}^0\big(\langle h\rangle, E[p]\big)$, which is zero.

For $p > 2$, we have to deal with the cases when G contains $\mathrm{SL}\big(E[p]\big)$ and when G is contained in a Borel subgroup.

In the first case, G is actually the full group $\mathrm{GL}\big(E[p]\big)$ as the Weil pairing forces the determinant to be surjective. If Z is the center of G, then $H^i\big(Z, E[p]\big) = 0$ for all $i \geqslant 0$. The Hochschild-Serre spectral sequence implies that $H^i\big(G, E[p]\big) = 0$ for all $i \geqslant 0$.

Now we may assume that G is contained in the Borel subgroup of upper-triangular matrices. If there is more that one isomorphism class of p-isogeny leaving E which is defined over F, then G is of order coprime to p and hence $H^2\big(G, E[p]\big) = 0$. Therefore, we may assume that G contains the unique p-Sylow H generated by $h = \left(\begin{smallmatrix} 1 & 1 \\ 0 & 1 \end{smallmatrix}\right)$. Since H is normal and G/H is of order coprime to p, the restriction

$$H^2\big(G, E[p]\big) \cong H^2\big(H, E[p]\big)^{G/H}$$

is an isomorphism.

Fix an injective homomorphism $\psi : H \to \mathbb{Q}/\mathbb{Z}$. Let $\delta\colon H^1\big(H, \mathbb{Q}/\mathbb{Z}\big) \to H^2\big(H, \mathbb{Z}\big)$ be the connecting homomorphism. Then we have an isomorphism $\hat{H}^0\big(H, E[p]\big) \to H^2\big(H, E[p]\big)$ given by sending a point $P \in E[p]$ to the cup product $\delta\psi \cup P$. For

$p > 2$, the Tate cohomology group $\hat{H}^0(H, E[p])$ is equal to the usual cohomology group $H^0(H, E[p]) = E[\varphi]$, which has p elements.

Let $g = \left(\begin{smallmatrix} u & w \\ 0 & v \end{smallmatrix}\right) \in G$. On the one hand, it acts on P by multiplication by u. On the other hand, it acts on ψ by multiplication by $u^{-1}v$ because

$$(g \star \psi)(h) = g\,\psi\left(g^{-1}hg\right) = \psi\left(h^{u^{-1}v}\right) = u^{-1}v\,\psi(h).$$

It follows that g acts on the generator of $H^2(H, E[p])$ by multiplication by $uu^{-1}v = v$. Unless all such $g \in G$ have $v = 1$, we conclude that the second cohomology group vanishes. Otherwise it has p elements, and this occurs if and only if $E'[\hat{\varphi}]$ contains a rational p-torsion point. □

5 Application to the Conjecture of Birch and Swinnerton-Dyer and p-descent

The vanishing of the Galois cohomology group we consider is used when trying to extend Kolyvagin's results to find a sharper bound on the Birch and Swinnerton-Dyer conjecture for elliptic curves of analytic rank at most 1. This was the original motivation in Cha's work [3]. In [13], the authors attempt to extend Cha's results, but there is a mistake in the proof of their Lemma 5.4 and consequently their Theorem 3.5 is not correct. The latter is also copied as Theorem 5.3 in [17]. Using our results above, we can now state and prove a corrected version of Theorem 3.5 in [13]. We refer to the original paper for the notations.

Theorem 14 *Let $E/\mathbb{Q}$ be an elliptic curve of analytic rank at most* 1. *Let p be an odd prime. Let F a quadratic imaginary field satisfying the Heegner hypothesis and suppose p does not ramify in $F/\mathbb{Q}$. Suppose that (E, p) does not appear in the list of Theorem 1 and that E is not isogenous to an elliptic curve over $\mathbb{Q}$ such that the dual isogeny contains a rational p-torsion point. Then the p-adic valuation of the order of the Tate-Shafarevich group is bounded by twice the index of the Heegner point.*

Proof In their proof, only the vanishing of $H^1(G, E[p])$ and $H^2(G, E[p])$ are needed for the argument. Under our assumptions they both vanish by Theorem 1 and Lemma 13. One has also to note that, as pointed out in [17], the assumption in their theorem that E does not admit complex multiplication is not used in the proof. Finally, the paper [13] needs that F is not included in $K = \mathbb{Q}(E[p])$ to conclude that $H^i(\mathrm{Gal}(F(E[p])/F), E[p])$ also vanishes for $i = 1$ and 2. This is guaranteed by the Heegner hypothesis and the assumption that p does not ramify in F, as F and K then have disjoint sets of ramified primes. □

The following is a short-cut in the usual p-descent for $E =$121c2 and $p = 11$. It is not a new result as it appears already in [18] as Example 7.4. However it illustrates that the non-trivial class in $H^1(G, E[p])$ can be of use.

Proposition 15 *The Tate-Shafarevich group of the curve 121c2 does not contain any non-trivial elements of order* 11. *The full Birch and Swinnerton-Dyer conjecture holds for this curve.*

Proof Set $p = 11$. Let $\varphi\colon E \to E'$ be the p-isogeny defined over $\mathbb{Q}$. We saw before that $E[\varphi] \cong \mathbb{F}_p(4)$ and $E'[\hat{\varphi}] \cong \mathbb{F}_p(7)$ where $\mathbb{F}_p(k)$ is the 1-dimensional $\mathbb{F}_p$-vector space with the Galois group acting by the character ω^k.

Let F be the maximal extension of $\mathbb{Q}$ which is unramified at all finite places $\ell \neq p$. Write $\mathcal{G} = \mathrm{Gal}\big(F/\mathbb{Q}\big)$ and $\mathcal{H} = \mathrm{Gal}\big(F/\mathbb{Q}(\zeta)\big)$ where ζ is a primitive p-th root of 1. Let $\Gamma = \mathcal{G}/\mathcal{H}$. Since $|\Gamma|$ is coprime to p, we have an isomorphism $H^1\big(\mathcal{G}, E[\varphi]\big) \cong H^1(\mathcal{H}, E[\varphi])^{\Gamma}$. Now Dirichlet's unit theorem can be used to compute

$$H^1\big(\mathcal{H}, \mathbb{F}_p(1)\big) = H^1\big(\mathcal{H}, \mu_p\big) \cong \mathbb{F}_p(1) \oplus \bigoplus_{i=0}^{4} \mathbb{F}_p(2\,i)$$

as a $\mathbb{F}_p[\Gamma]$-module; see for instance Corollary 8.6.12 (or 8.7.3 in the second edition) in [19]. Since $H^1\big(\mathcal{H}, \mathbb{F}_p(k)\big) \cong H^1\big(\mathcal{H}, \mathbb{F}_p(1)\big)(k-1)$, the group $H^1\big(\mathcal{G}, \mathbb{F}_p(k)\big)$ is a sum of copies of $\mathbb{F}_p$ corresponding to the copies of $\mathbb{F}_p(1-k)$ in $H^1\big(\mathcal{H}, \mathbb{F}_p(1)\big)$. We deduce that $H^1\big(\mathcal{G}, E[\varphi]\big)$ is trivial and that $H^1\big(\mathcal{G}, E'[\hat{\varphi}]\big)$ is 1-dimensional.

Since $K/\mathbb{Q}$ is only ramified at p, we have an inflation map $H^1\big(G, E'[\hat{\varphi}]\big) \to H^1\big(\mathcal{G}, E'[\hat{\varphi}]\big)$. By Theorem 1 and the above, this is now an isomorphism and our explicit cocycle ξ can be viewed as a generator for $H^1\big(\mathcal{G}, E'[\hat{\varphi}]\big)$.

The $\hat{\varphi}$-Selmer group $\mathrm{Sel}^{\hat{\varphi}}$ is defined to be the kernel of the map

$$H^1\big(\mathcal{G}, E'[\hat{\varphi}]\big) \to H^1\big(\mathbb{Q}_p, E'\big)[\hat{\varphi}].$$

An explicit local computation shows that $\hat{\varphi}\colon E'(\mathbb{Q}_p) \to E(\mathbb{Q}_p)$ is surjective. Therefore $H^1\big(\mathbb{Q}_p, E'\big)[\hat{\varphi}] \cong H^1\big(\mathbb{Q}_p, E'[\hat{\varphi}]\big)$. Since $K/\mathbb{Q}$ is totally ramified at p, the decomposition group of $K/\mathbb{Q}$ at the unique place above p in K is equal to G. Therefore ξ also inflates to a non-trivial element in $H^1\big(\mathbb{Q}_p, E'[\hat{\varphi}]\big)$. It follows that the generator of $H^1\big(\mathcal{G}, E'[\hat{\varphi}]\big)$ does not lie in the Selmer group. Therefore $\mathrm{Sel}^{\hat{\varphi}}$ is trivial.

Since $H^1\big(\mathcal{G}, E[\varphi]\big) = 0$, the φ-Selmer group Sel^{φ} is trivial. The usual exact sequence

$$\mathrm{Sel}^{\varphi} \longrightarrow \mathrm{Sel}^{p}(E/\mathbb{Q}) \longrightarrow \mathrm{Sel}^{\hat{\varphi}}$$

shows now that the p-Selmer group $\mathrm{Sel}^p(E/\mathbb{Q})$ is trivial. Therefore the rank of E is zero and the p-primary part of the Tate-Shafarevich group $\text{Ш}(E/\mathbb{Q})$ is trivial.

As explained in Theorem 8.5 in [17], the only prime at which one has to check the Birch and Swinnerton-Dyer conjecture after the Heegner point computations done there is $p = 11$. Therefore, this completes the proof of the conjecture for this specific elliptic curve. □

The main result of [17] (based on [8, 18]) by Miller and his collaborators states that the Birch and Swinnerton-Dyer conjecture holds for all elliptic curves of conductor at most 5000 and analytic rank at most 1. As a consequence of the error in [13], the verification for some curves in this list is not complete. The following is a description how we performed the necessary computations to fill in the gaps for all these curves. See also [24] for the correction of the corresponding bug in SageMath.

From the change in Theorem 14, it follows that only curves E contained in the list of Theorem 1 could have been affected when verifying the p-part of the conjecture. The case 121c2 was verified in Proposition 15. The exceptional cases with $p = 3$ are already dealt with in Theorem 9.1 in [18], as they were already considered exceptional cases there. That only leaves the curves with non-vanishing $H^1\big(G, E[5]\big)$. For the following list of curves, we had to perform a 5-descent to verify the conjecture: 50a3, 50a4, 75a2, 150b3, 150b4, 175c2, 275b1, 325d2, 550b1, 550f3, 775c1, 950a1, 1050d2, 1425b1, 1450a1, 1650b1, 1650b2, 1650c2, 1650d2, 1950b2, 1975d1, 2175f2, 2350e2, 2550f2, 2850a1, 2850a2, 2850g2, 2950a1, 3075d1, 3075g2, 3325c1, 3550d1, 3850k2, 3950a1, 4350a1, 4350a2, 4425c1, 4450a1, 4450f2, 4650e1, 4650k2, 4650m2. The methods in [18] are sufficient in all these cases. If the rank is 1, then even the weaker bound in their Corollary 7.3 is enough. Otherwise, if the rank is 0, the Selmer groups for φ and $\hat{\varphi}$ are trivial as one finds quickly by looking at a few local conditions.

6 Results for $i > 1$

We now turn to the question of finding all cases of elliptic curves E/F and primes p such that the group $H^1\big(G_i, E[p^i]\big)$ does not vanish for some $i > 1$. We continue to assume that $F \cap \mathbb{Q}(\mu_p) = \mathbb{Q}$ and we will assume now that $p > 2$.

By Lemmas 3 and 4, we know that all these groups vanish unless there is an isogeny $\varphi\colon E \to E'$ defined over F. Therefore, we may continue to assume the existence of φ and that the group G is contained in the Borel subgroup of upper triangular matrices. This fixes (up to scalar) the first basis element of $E[p]$ and we still have some flexibility about the second; if there is a second subgroup of $E[p]$ fixed by the Galois group, we will choose the second basis element in there. Unlike in the case $i = 1$, we may not yet assume that p divides the order of G.

In what follows we will write expressions like $G = \left(\begin{smallmatrix} 1 & * \\ 0 & * \end{smallmatrix}\right)$. By this we mean that G is equal to the group of all matrices of this form in $\mathrm{GL}_2(\mathbb{F}_p)$, so $*$ on the diagonal can take any non-zero value and $*$ in the top right corner can be any value in $\mathbb{F}_p$.

Let M be the additive group of 2×2-matrices with coefficients in $\mathbb{F}_p$. Then $G \leqslant \mathrm{GL}_2(\mathbb{F}_p)$ acts on M by conjugation. We would like to determine $\mathrm{Hom}_G\big(M, E[p]\big)$. We do so by computing first $\mathrm{Hom}_G\big(M, E[\varphi]\big)$ and $\mathrm{Hom}_G\big(M, E'[\hat{\varphi}]\big)$.

Lemma 16 *Suppose first* $p > 3$. *The group* $\mathrm{Hom}_G\big(M, E[\varphi]\big)$ *is trivial except in the following cases.*

- *If* $G = \left(\begin{smallmatrix} 1 & 0 \\ 0 & * \end{smallmatrix}\right)$ *in a suitable basis of* $E[p]$, *then* $\mathrm{Hom}_G\big(M, E[\varphi]\big)$ *has dimension* 2 *over* $\mathbb{F}_p$.
- *If* $G = \left(\begin{smallmatrix} * & 0 \\ 0 & 1 \end{smallmatrix}\right)$ *in a suitable basis of* $E[p]$, *this group has dimension* 1.
- *If* $G = \left(\begin{smallmatrix} 1 & * \\ 0 & * \end{smallmatrix}\right)$, *this group has dimension* 1.

- *If* $G \leqslant \{\left(\begin{smallmatrix} u & w \\ 0 & u^2 \end{smallmatrix}\right) \mid u \in \mathbb{F}_p^\times, w \in \mathbb{F}_p\}$, *this group has dimension* 1.

If $p = 3$, *the list is the same with one modification to the second and to the last case above.*

- *If* $G = \left(\begin{smallmatrix} * & 0 \\ 0 & 1 \end{smallmatrix}\right)$ *in a suitable basis of* $E[3]$, *this group has dimension* 2.

Proof If $f : M \to E[\varphi]$ is fixed by $g \in G$, then $f(m) = g \cdot f(g^{-1}mg)$ for all $m \in M$. Let α, β, γ, and δ be the images in $E[\varphi]$ under f of $\left(\begin{smallmatrix} 1 & 0 \\ 0 & 0 \end{smallmatrix}\right)$, $\left(\begin{smallmatrix} 0 & 1 \\ 0 & 0 \end{smallmatrix}\right)$, $\left(\begin{smallmatrix} 0 & 0 \\ 1 & 0 \end{smallmatrix}\right)$, and $\left(\begin{smallmatrix} 0 & 0 \\ 0 & 1 \end{smallmatrix}\right)$ respectively. Then the above equation for m being one of the these four matrices yields four equations that have to hold for all $g = \left(\begin{smallmatrix} u & w \\ 0 & v \end{smallmatrix}\right) \in G$:

$$\begin{aligned} \alpha &= u \cdot (\alpha + u^{-1} w\, \beta) \\ \beta &= u \cdot (u^{-1} v\, \beta) \\ \gamma &= u \cdot (-v^{-1} w\, \alpha - u^{-1} v^{-1} w^2\, \beta + u v^{-1}\, \gamma + v^{-1} w\, \delta) \\ \delta &= u \cdot (\delta - u^{-1} w\, \beta) \end{aligned} \tag{3}$$

From these equations, we deduce the following:

$$\begin{aligned} f \text{ is fixed by } \left(\begin{smallmatrix} 1 & 1 \\ 0 & 1 \end{smallmatrix}\right) &\iff \beta = 0 \text{ and } \alpha = \delta \\ f \text{ is fixed by } \left(\begin{smallmatrix} 1 & 0 \\ 0 & v \end{smallmatrix}\right) \text{ for some } v \neq 1 &\iff \beta = \gamma = 0 \\ f \text{ is fixed by } \left(\begin{smallmatrix} u & 0 \\ 0 & 1 \end{smallmatrix}\right) \text{ for some } u \neq \pm 1 &\iff \alpha = \gamma = \delta = 0 \\ f \text{ is fixed by } \left(\begin{smallmatrix} -1 & 0 \\ 0 & 1 \end{smallmatrix}\right) &\iff \alpha = \delta = 0 \end{aligned}$$

Assume first that G is contained in $\left(\begin{smallmatrix} 1 & * \\ 0 & * \end{smallmatrix}\right)$. Since the determinant must be surjective, G is either $\left(\begin{smallmatrix} 1 & * \\ 0 & * \end{smallmatrix}\right)$ or $\left(\begin{smallmatrix} 1 & 0 \\ 0 & * \end{smallmatrix}\right)$, after choosing a suitable second basis element for $E[p]$. In both cases, the above allows us to verify the statements in the lemma. The case when G is contained in $\left(\begin{smallmatrix} * & * \\ 0 & 1 \end{smallmatrix}\right)$ is very similar, except that when $p = 3$, in which case we are in the group of matrices with $v = u^2$ and we can only apply the fourth equation instead of the third.

Assume now that G contains an element $\left(\begin{smallmatrix} u & w \\ 0 & v \end{smallmatrix}\right)$ with $v \neq 1$ and one with $u \neq 1$. Then $\beta = 0$ by the second equation in (3). From the last two equations, we deduce that $\alpha = \delta = 0$. Now the equations (3) simplify to one equation $(1 - u^2 v^{-1})\gamma = 0$. Therefore, if G is contained in the group of matrices with $v = u^2$, then the dimension of $\mathrm{Hom}_G\big(M, E[\varphi]\big)$ is 1 and $p > 3$ as otherwise all $g \in G$ have $v = 1$, otherwise the space is trivial. □

Recall that $E'[\hat{\varphi}]$ is the kernel of the dual isogeny.

Lemma 17 *Suppose* $p > 2$. *The group* $\mathrm{Hom}_G\big(M, E'[\hat{\varphi}]\big)$ *is trivial except in the following cases. If* G *is contained in* $\left(\begin{smallmatrix} 1 & * \\ 0 & * \end{smallmatrix}\right)$ *or if* $G = \left(\begin{smallmatrix} * & * \\ 0 & 1 \end{smallmatrix}\right)$ *or if* $G = \{\left(\begin{smallmatrix} v^2 & 0 \\ 0 & v \end{smallmatrix}\right) \mid v \in \mathbb{F}_p^\times\}$ *in a suitable basis for* $E[p]$, *then* $\mathrm{Hom}_G\big(M, E'[\hat{\varphi}]\big)$ *has dimension* 1. *If* $G = \left(\begin{smallmatrix} * & 0 \\ 0 & 1 \end{smallmatrix}\right)$ *in a suitable basis of* $E[p]$, *then it has dimension* 2.

Proof This is analogous to the proof of the previous lemma. The equations (3) become equations where the u at the start of the right hand side of each equation is replaced by a v. This new set of equations can be rewritten as follows.

$$\begin{aligned} (1-v)\alpha &= u^{-1}vw\,\beta \\ (1-u^{-1}v^2)\beta &= 0 \\ (1-u)\gamma &= w(\delta-\alpha) - u^{-1}w^2\beta \\ (1-v)\delta &= -u^{-1}vw\,\beta \end{aligned} \tag{4}$$

From here, the computations are again straightforward for the cases when G is contained in $\left(\begin{smallmatrix} * & * \\ 0 & 1 \end{smallmatrix}\right)$ or $\left(\begin{smallmatrix} 1 & * \\ 0 & * \end{smallmatrix}\right)$. If G is contained in the group $\left\{\left(\begin{smallmatrix} v^2 & w \\ 0 & v \end{smallmatrix}\right) \mid v \in \mathbb{F}_p^\times, w \in \mathbb{F}_p\right\}$, it is either equal to this group, in which case the cohomology group in question is trivial, or it is equal to a subgroup of order $p-1$. In the latter case, we may change the choice of basis of $E[p]$ to get G to be equal to $\left\{\left(\begin{smallmatrix} v^2 & 0 \\ 0 & v \end{smallmatrix}\right) \mid v \in \mathbb{F}_p^\times\right\}$, in which case $\alpha = \gamma = \delta = 0$, but β is free. In all other cases it is trivial. □

The exact sequence

$$0 \longrightarrow \mathrm{Hom}_G\big(M, E[\varphi]\big) \longrightarrow \mathrm{Hom}_G\big(M, E[p]\big) \xrightarrow{\varphi} \mathrm{Hom}_G\big(M, E'[\hat{\varphi}]\big) \tag{5}$$

connects the results from the previous two lemmas.

Proposition 18 *If $p > 3$, the group $\mathrm{Hom}_G\big(M, E[p]\big)$ vanishes except when, for some choice of basis of $E[p]$, it is one of the following subgroups.*

G	$= \left(\begin{smallmatrix} 1 & 0 \\ 0 & * \end{smallmatrix}\right)$	$= \left(\begin{smallmatrix} * & 0 \\ 0 & 1 \end{smallmatrix}\right)$	$= \left(\begin{smallmatrix} 1 & * \\ 0 & * \end{smallmatrix}\right)$	$\leqslant \left(\begin{smallmatrix} u & * \\ 0 & u^2 \end{smallmatrix}\right)$	$= \left(\begin{smallmatrix} v^2 & 0 \\ 0 & v \end{smallmatrix}\right)$
$\dim_{\mathbb{F}_p} \mathrm{Hom}_G\big(M, E[p]\big)$	3	3	2	1	1

If $p = 3$, the group $\mathrm{Hom}_G\big(M, E[p]\big)$ vanishes except when, for some choice of basis of $E[p]$, it is one of the following subgroups.

G	$= \left(\begin{smallmatrix} 1 & 0 \\ 0 & * \end{smallmatrix}\right)$	$= \left(\begin{smallmatrix} * & 0 \\ 0 & 1 \end{smallmatrix}\right)$	$= \left(\begin{smallmatrix} 1 & * \\ 0 & * \end{smallmatrix}\right)$	$= \left(\begin{smallmatrix} * & * \\ 0 & 1 \end{smallmatrix}\right)$
$\dim_{\mathbb{F}_3} \mathrm{Hom}_G\big(M, E[3]\big)$	3	4	2	1

Here we have chosen a suitable second basis element in $E[p]$ as in the previous lemmas. Of course, the first two cases are in fact the same when the basis elements are swapped.

Proof If $\mathrm{Hom}_G\big(M, E'[\hat{\varphi}]\big) = 0$, then the exact sequence (5) reduces this to Lemma 16. Otherwise, we have to check if the homomorphisms $f\colon M \to E'[\hat{\varphi}]$ lift to homomorphisms $e\colon M \to E[p]$ that are G-equivariant. In the following four cases, they all lift indeed. We will just give the explicit map which form a basis of $\mathrm{Hom}_G(M, E[p])$ modulo the image from $\mathrm{Hom}_G(M, E[\varphi])$. One can verify without difficult that they are G-equivariant.

G	$=\begin{pmatrix}1&0\\0&*\end{pmatrix}$	$=\begin{pmatrix}*&0\\0&1\end{pmatrix}$	$=\begin{pmatrix}1&*\\0&*\end{pmatrix}$	$=\begin{pmatrix}v^2&0\\0&v\end{pmatrix}$
$e\begin{pmatrix}a&b\\c&d\end{pmatrix}$	$\begin{pmatrix}a\\c\end{pmatrix}$	$\begin{pmatrix}0\\a\end{pmatrix}$ and $\begin{pmatrix}0\\d\end{pmatrix}$	$\begin{pmatrix}a\\c\end{pmatrix}$	$\begin{pmatrix}0\\b\end{pmatrix}$

There is only the case $G = \begin{pmatrix}*&*\\0&1\end{pmatrix}$ left to treat. The generator of $\mathrm{Hom}_G(M, E'[\hat{\varphi}])$ is given by $f\begin{pmatrix}a&b\\c&d\end{pmatrix} = a + d$. We will show that f does not lift to a map $e\colon M \to E[p]$. Denote by $\begin{pmatrix}\alpha\\1\end{pmatrix}$ the image of $\begin{pmatrix}1&0\\0&0\end{pmatrix}$ under such an e and by $\begin{pmatrix}\beta\\0\end{pmatrix}$ the image of $\begin{pmatrix}0&1\\0&0\end{pmatrix}$. Then we must have for all $u \neq 1$ and w in $\mathbb{F}_p$ that

$$\begin{pmatrix}\beta\\0\end{pmatrix} = \begin{pmatrix}u&w\\0&1\end{pmatrix} e\begin{pmatrix}0&u^{-1}w\\0&0\end{pmatrix} = \begin{pmatrix}u&w\\0&1\end{pmatrix}\begin{pmatrix}u^{-1}w\beta\\0\end{pmatrix} = \begin{pmatrix}w\beta\\0\end{pmatrix}.$$

Hence $\beta = 0$. Again for all u and w, we should have that

$$\begin{pmatrix}\alpha\\1\end{pmatrix} = \begin{pmatrix}u&w\\0&1\end{pmatrix} e\begin{pmatrix}1&u^{-1}w\\0&0\end{pmatrix} = \begin{pmatrix}u&w\\0&1\end{pmatrix}\begin{pmatrix}\alpha\\1\end{pmatrix} = \begin{pmatrix}u\alpha+w\\1\end{pmatrix}.$$

However, this cannot hold for all choices no matter what α is. □

Definition Let E/F be an elliptic curve. We will say that G_i is *greatest possible* if it consists of all the matrices in $\mathrm{GL}_2(\mathbb{Z}/p^i\mathbb{Z})$ that reduce to a matrix in G modulo p. Equivalently, M_i is the kernel of the map $\mathrm{GL}_2(\mathbb{Z}/p^i) \to \mathrm{GL}_2(\mathbb{F}_p)$.

We will show that if $p > 2$ and $i > 1$, then G_i is greatest possible if and only if G_2 is greatest possible: Since $G_i \to G_{i-1}$ is surjective, by induction it suffices to prove that the kernel H_{i-1} contains all matrices of the form $1 + p^{i-1}A \in \mathrm{GL}_2(\mathbb{Z}/p^i)$. If G_2 is greatest possible, then for any $A \in M_2(\mathbb{F}_p)$ there exists an element $g \in G_i$ whose image in $\mathrm{GL}_2(\mathbb{Z}/p^2)$ is $1 + pA$. Then $g^{p^{i-2}}$ has image $(1 + p^{i-1}A)$ in $\mathrm{GL}_2(\mathbb{Z}/p^i)$ by taking binomial expansions, and so G_i contains all of H_{i-1}.

Proposition 19 *Let $p > 2$ be a prime and let E/F be an elliptic curve. Suppose G lies in the Borel subgroup of upper triangular matrices and that G_2 is greatest possible. If G is not among the exceptional cases in Theorem 1 or in Proposition 18, then $H^1(G_i, E[p^j]) = 0$ for all $i \geqslant j \geqslant 1$.*

Proof The short exact sequence (2) implies that if the groups $H^1(G_{i+1}, E[p^{j-1}])$ and $H^1(G_{i+1}, E[p])$ are zero, then so is $H^1(G_{i+1}, E[p^j])$. By induction on j, it suffices to prove the proposition in the case $j = 1$.

For $i = 1$, the statement follows from Theorem 1. We assume now that it holds for $i \geqslant 1$. By assumption M_{i+1} is isomorphic to the group $(1 + p\,\mathrm{Mat}_2(\mathbb{Z}/p^i)) \subset \mathrm{GL}_2(\mathbb{Z}/p^{i+1})$ and H_i is isomorphic to the group M of all matrices with coefficients in $\mathbb{F}_p$. Using Proposition 18, we find

$$H^1(M_{i+1}, E[p])^G = \mathrm{Hom}_G(M_{i+1}, E[p]) \cong \mathrm{Hom}_G(M, E[p]) = 0$$

because all elements in the kernel of the map $M_{i+1} \to M$ are p'th powers. (Note that this requires $p > 2$.) Now considering the inflation-restriction sequence

$$0 \longrightarrow H^1\big(G, E[p]\big) \longrightarrow H^1\big(G_{i+1}, E[p]\big) \longrightarrow H^1\big(M_{i+1}, E[p]\big)^G \qquad (6)$$

yields that $H^1\big(G_{i+1}, E[p]\big) = 0$. □

Lemma 20 *Let $p > 2$ be a prime and E/F an elliptic curve such that G_2 is greatest possible. If G is among the exceptional cases in Theorem 1 or in Proposition 18 then $H^1\big(G_i, E[p^i]\big) \neq 0$ for all $i \geqslant 2$.*

Proof We claim that the sequence (6) is part of a short exact sequence. The next term in the sequence is $H^2\big(G, E[p]\big)$, and so it suffices to show that the map $H^1(M_{i+1}, E[p])^G \to H^2\big(G, E[p]\big)$ is zero. By Lemma 13, the target group is trivial unless $G = \left(\begin{smallmatrix} * & * \\ 0 & 1 \end{smallmatrix}\right)$. If $p > 3$ and $G = \left(\begin{smallmatrix} * & * \\ 0 & 1 \end{smallmatrix}\right)$, then the source group $H^1(M_{i+1}, E[p])^G$ vanishes. If $p = 3$ and $G = \left(\begin{smallmatrix} * & * \\ 0 & 1 \end{smallmatrix}\right)$, the source is cyclic and generated by a cocycle $\xi : M_{i+1} \to E[3]$ such that $\left(\begin{smallmatrix} a & b \\ c & d \end{smallmatrix}\right) \mapsto \left(\begin{smallmatrix} c/3 \\ 0 \end{smallmatrix}\right)$. The image of ξ in $H^2\big(G, E[p]\big)$ is zero because the formula $\left(\begin{smallmatrix} a & b \\ c & d \end{smallmatrix}\right) \mapsto \left(\begin{smallmatrix} ac/3 \\ 0 \end{smallmatrix}\right)$ lifts it to a cocycle $G_{i+1} \to E[3]$.

Therefore, the dimension of $H^1(G_{i+1}, E[p])$ is the sum of the dimensions of the two groups surrounding it in the sequence (6). In all cases, this dimension is strictly larger than the dimension of the group of p-torsion points of E defined over F.

Now we turn to sequence (2) with $i \geqslant 2$. In all cases, the dimension of $H^1\big(G_i, E[p]\big)$ is strictly larger than the dimension of $E(F)[p^{i-1}]/pE(F)[p^i]$ (which is at most 1 because $E(F)[p^{i-1}]$ must be cyclic by the assumption on F). We conclude that $H^1\big(G_i, E[p^i]\big)$ is non-trivial. □

So far we have been able to treat all cases in which G_i is greatest possible and $F \cap \mathbb{Q}(\mu_p) = \mathbb{Q}$. We will now restrict our attention to $F = \mathbb{Q}$ and $p > 3$. Luckily, for the large majority of elliptic curves over $\mathbb{Q}$ the groups G_i are indeed greatest possible. The following is a summary of the results in [11, 12].

Theorem 21 *Let $p > 3$ and $i > 1$. Let $E/\mathbb{Q}$ be an elliptic curve with an isogeny of degree p defined over $\mathbb{Q}$. Then G_i is greatest possible except in two cases:*

- *when $p = 7$ and the curve is the quadratic twist of a curve of conductor 49, or*
- *when $p = 5$ and there is an isogeny $\psi : E \to E''$ of degree 25 defined over $\mathbb{Q}$.*

We will now treat the two exceptional cases, starting with $p = 5$.

Lemma 22 *Let $E/\mathbb{Q}$ be an elliptic curve and suppose there is an cyclic isogeny $\psi : E \to E' \to E''$ of degree $p^2 = 25$ defined over $\mathbb{Q}$. Then $H^1\big(G_2, E[p^2]\big) = 0$ if and only if $H^1\big(G_i, E[p^i]\big) = 0$ for all $i > 1$. This vanishing holds except if $G = \left(\begin{smallmatrix} 1 & * \\ 0 & * \end{smallmatrix}\right)$, if $G = \left(\begin{smallmatrix} * & * \\ 0 & 1 \end{smallmatrix}\right)$, or if E appears in Theorem 1 as an exception.*

For instance, it is non-vanishing if E admits a rational 5-torsion point or if E' admits a rational 5-torsion point. The curves 11a3 and 11a2 are examples of these two situations where $H^1\big(G, E[p]\big) = 0$, yet $H^1\big(G_i, E[p^i]\big) \neq 0$ for all $i > 1$ because there are two 5-isogenies 11a3 $\to$ 11a1 $\to$ 11a2 with only 11a3 and 11a1 having a rational 5-torsion point. The cohomology group $H^1\big(G_2, E[25]\big)$ is also non-trivial for 11a1 by Proposition 19.

Proof Note that there are no elliptic curves with rational points of order 25 and there are no cyclic isogenies over $\mathbb{Q}$ of degree $p^3 = 125$. Greenberg shows in Theorem 2 in [11] that the index of G_2 in $\mathrm{GL}_2\big(\mathbb{Z}/25\big)$ is divisible by 5 but not 25. Hence the group G_2 can be identified with a subgroup of the upper triangular matrices modulo p^2, but the top left entry is not constant 1 modulo p^2 and the top right corner is not constant zero modulo p. Since the index is only divisible by 5 once, the group G_2 consists of all the upper triangular matrices that reduce to an element of G.

We wish to use the same strategy as in the proof of Proposition 19, but we have to show that G-fixed part of $H^1\big(M_2, E[p]\big)$ is still zero despite $M_2 \neq M$. This time M_2 can be identified with upper triangular matrices modulo p and the computations are slightly easier. One finds that $\mathrm{Hom}_G\big(M_2, E[\varphi]\big)$ has dimension 2 if $G = \left(\begin{smallmatrix}1 & * \\ 0 & *\end{smallmatrix}\right)$ and 0 in all other cases. Similarly, the dimension of $\mathrm{Hom}_G\big(M_2, E'[\hat{\varphi}]\big)$ is equal to 2 if $G = \left(\begin{smallmatrix}* & * \\ 0 & 1\end{smallmatrix}\right)$ and zero otherwise. (Alternatively, it is not too hard to show by direct calculation that the dimension of $\mathrm{Hom}_G\big(M_2, E[p]\big)$ is 2 if $G = \left(\begin{smallmatrix}1 & * \\ 0 & *\end{smallmatrix}\right)$, 1 if $G = \left(\begin{smallmatrix}* & * \\ 0 & 1\end{smallmatrix}\right)$, and 0 otherwise.)

Hence if we assume that neither $G = \left(\begin{smallmatrix}* & * \\ 0 & 1\end{smallmatrix}\right)$ nor $G = \left(\begin{smallmatrix}1 & * \\ 0 & *\end{smallmatrix}\right)$ nor $G = \left\{\left(\begin{smallmatrix}v^2 & w \\ 0 & v\end{smallmatrix}\right) \,\middle|\, v \in \mathbb{F}_p^{\times}, w \in \mathbb{F}_p\right\}$, then $H^1\big(G_i, E[p^i]\big) = 0$ for all $i \geqslant 1$ with the same proof as in Proposition 19.

If $G = \left(\begin{smallmatrix}1 & * \\ 0 & *\end{smallmatrix}\right)$, then one can show as in Lemma 20 that $H^1\big(G_i, E[p^i]\big)$ is non-zero for all $i > 1$. Similarly for $G = \left\{\left(\begin{smallmatrix}v^2 & w \\ 0 & v\end{smallmatrix}\right) \,\middle|\, v \in \mathbb{F}_p^{\times}, w \in \mathbb{F}_p\right\}$.

Finally if $G = \left(\begin{smallmatrix}* & * \\ 0 & 1\end{smallmatrix}\right)$, then one may compute $H^1\big(G_2, E[p^2]\big)$ directly: the group G_2 consists of all upper triangular matrices modulo p^2 whose lower right entry is congruent to 1 modulo p. Let H be the subgroup generated by $\left(\begin{smallmatrix}1 & 1 \\ 0 & 1\end{smallmatrix}\right)$. Then the method used in the proof of Theorem 1 shows that the subgroup of $H^1\big(H, E[p^2]\big)$ fixed by the action of G_2/H is trivial. However $H^1\big(G_2/H, E[p^2]^H\big) \cong \mathbb{Z}/p\mathbb{Z}$ implies then that $H^1\big(G_2, E[p^2]\big) \cong \mathbb{Z}/p\mathbb{Z}$ where an explicit isomorphism sends a cocycle ξ to the first coordinate of $\xi\big(\left(\begin{smallmatrix}1 & 0 \\ 0 & 1+p\end{smallmatrix}\right)\big)$ in $p\mathbb{Z}/p^2\mathbb{Z}$. From the exact sequence (2), one deduces that $H^1\big(G_2, E[p]\big)$ is non-trivial and again this implies that all $H^1\big(G_i, E[p^i]\big)$ are non-zero for $i > 1$. □

Lemma 23 *Let $E/\mathbb{Q}$ be a quadratic twist of a curve of conductor* 49 *and let $p = 7$. Then $H^1\big(G_i, E[p^i]\big) = 0$ for all $i \geqslant 1$.*

Proof Assume first that E is one of the curves of conductor 49. By assumption E has complex multiplication by O, where O is either $\mathbb{Z}[\sqrt{-7}]$ or the ring of integers in $\mathbb{Q}(\sqrt{-7})$. Since $\mathbb{Q}(\sqrt{-7}) \subset K$, the subgroup $\mathrm{Gal}(K/\mathbb{Q}(\sqrt{-7})) < G$ acts by O-linear endomorphisms on $E[7^i]$. By scaling with the period, we may choose points p and $\sqrt{-7} \cdot p$ as a basis for $E[7^i]$. Any lift of these forms a $\mathbb{Z}_7$-basis of the Tate module T_7E. The endomorphism $a + b\sqrt{-7}$ with $a, b \in \mathbb{Q} \cap \mathbb{Z}_7$ acts via $\left(\begin{smallmatrix}a & b \\ -7b & a\end{smallmatrix}\right)$ on $T_7(E)$.

The Frobenius element $\mathrm{Fr}_\ell \in \mathrm{GL}(T_7E)$ for $\ell = 347$ has trace $a_\ell = 4$ for all four curves of conductor 49. Since ℓ splits in $\mathbb{Q}(\sqrt{-7})$, the Frobenius Fr_ℓ in $\mathrm{GL}_2(\mathbb{Z}_7)$ is a matrix of the above shape with trace 4 and determinant 347. We find that it is congruent to $\left(\begin{smallmatrix}2&0\\0&2\end{smallmatrix}\right)$ modulo 7. Since G contains now the homotheties by 2 and 4, the result follows from Lemma 3.

Let now E be a quadratic twist of a curve of conductor 49. Then it is the quadratic twist of one of them by an integer D coprime to 7. The above homotheties are multiplied by a non-zero scalar and hence Lemma 3 also implies the result for E. □

Proof of Theorem 2 We combine the results from Proposition 19, Lemmas 20, 22 and 23. From these we conclude immediately that $H^1\big(G_i, E[p^i]\big) = 0$ if and only if $H^1\big(G_2, E[p^2]\big) = 0$.

We are now left with making the list in Theorem 2 match with the non-vanishing cases. We start by verifying that the cohomology groups are non-vanishing in each of the five special cases in the theorem.

- First, if E contains a rational point of order p, then $G = \left(\begin{smallmatrix}1&0\\0&*\end{smallmatrix}\right)$, $G = \left(\begin{smallmatrix}*&0\\0&1\end{smallmatrix}\right)$, or $G = \left(\begin{smallmatrix}1&*\\0&*\end{smallmatrix}\right)$. In all these cases, the cohomology groups in question do not vanish by Lemmas 20 and 22.
- In the second point in the list of Theorem 2, $p = 5$ and the quadratic twist by $D = 5$ of E has a rational 5-torsion point. Then G is contained in $\left\{\left(\begin{smallmatrix}v^2&*\\0&v\end{smallmatrix}\right) \mid v \in \mathbb{F}_p^\times\right\}$. If G is equal to that group, then Theorem 1 and Lemmas 20 or 22 imply the non-vanishing of $H^1\big(G_2, E[p^2]\big)$. Otherwise we may choose the basis of $E[p]$ so that G is contained in the diagonal matrices of this form, in which case Lemma 20 proves the assertion.
- In the third point, $p = 5$ and the quadratic twist by $D = 5$ of E' has a rational 5-torsion point. Then G is contained in $\left\{\left(\begin{smallmatrix}u&*\\0&u^2\end{smallmatrix}\right) \mid u \in \mathbb{F}_p^\times\right\}$. There is no isogeny of degree 25 defined over $\mathbb{Q}$ leaving from E, hence G_2 is greatest possible; therefore Lemma 20 proves the desired non-vanishing.
- If we are in the situation of the fourth point in Theorem 2, we are in the situation of Lemma 22 and $G = \left(\begin{smallmatrix}*&*\\0&1\end{smallmatrix}\right)$. Therefore $H^1\big(G_2, E[p^2]\big) \neq 0$.
- In the final point, if $p = 11$ and E is 121c2, then Theorem 1 and Lemma 20 shows the desired non-vanishing. If the curve is 121c1 instead, then $G = \left\{\left(\begin{smallmatrix}u&*\\0&u^2\end{smallmatrix}\right) \mid u \in \mathbb{F}_p^\times\right\}$ and Lemma 20 treats this case too.

Next, we have to check that every case when the group $H^1\big(G_2, E[p^2]\big)$ is non-trivial is among the exceptional cases of Theorem 2 above.

Let us assume first that G_2 is greatest possible and consider the cases in Lemma 20. If $G = \left(\begin{smallmatrix}1&0\\0&*\end{smallmatrix}\right)$ or $G = \left(\begin{smallmatrix}1&*\\0&*\end{smallmatrix}\right)$, then E has a rational p-torsion point. By Mazur's Theorem on the torsion point on elliptic curves over $\mathbb{Q}$, we know that this can only occur if $p = 5$ or $p = 7$ and we fall under the first point in the list of Theorem 2. If (E, p) appears as an exception in Theorem 1, then either $p = 5$ and we are in the situation of the second point, or $p = 11$ and we are in the last point on the list. If G is the group of all matrices of the form $\left(\begin{smallmatrix}v^2&0\\0&v\end{smallmatrix}\right)$, then the quadratic twist by $D = 5$ has a rational 5-torsion point and we are in the situation of the second point. Finally, assume G is contained in the group $\left\{\left(\begin{smallmatrix} u & * \\ 0 & u^2 \end{smallmatrix}\right) \mid u \in \mathbb{F}_p^\times\right\}$. Then $p \equiv 2 \pmod 3$. If $p = 5$, then the quadratic twist by $D = 5$ of E' has a rational 5-torsion point and we are in the third case. If $p \geqslant 11$, then the proof that there is only one curve, namely 121c1, is very analogous to Lemma 12.

Finally, we consider the cases in Lemma 22. If $G = \left(\begin{smallmatrix} 1 & * \\ 0 & * \end{smallmatrix}\right)$, then E has a rational 5-torsion point and we are in the first point in the list. If $G = \left(\begin{smallmatrix} * & * \\ 0 & 1 \end{smallmatrix}\right)$, then E' admits a rational 5-torsion point, which is the fourth point on the list. If (E, p) appear as exceptions in Theorem 1, then we fall into the second point on the list. □

7 Numerical Computations

We used Magma [2] to perform, for small primes p, the numerical computation of our cohomology group $H^1(G_2, V_2)$ for various subgroup $G_2 \leqslant \mathrm{GL}_2(\mathbb{Z}/p^2)$, where V_2 is the natural rank 2 module over $\mathbb{Z}/p^2$ on which G_2 acts. We restricted our attention to groups with surjective determinants and we only considered groups up to conjugation in $\mathrm{GL}_2(\mathbb{Z}/p^2)$.

We will continue to write M_2 for the kernel of reduction $G_2 \to \mathrm{GL}_2(\mathbb{F}_p)$ and G for its image.

7.1 $p = 2$

For the prime $p = 2$, the groups $H^1(G_2, V_2)$ are non-zero for 36 conjugacy classes of subgroup $G_2 \leqslant \mathrm{GL}_2(\mathbb{Z}/4)$ with surjective determinant. The possible cohomology groups are $(\mathbb{Z}/2)^k$ for $0 \leqslant k \leqslant 6$ and $\mathbb{Z}/4$. Non-trivial cohomology groups appear for all dimensions $1 \leqslant d \leqslant 4$ of M_2.

7.2 $p = 3$

There are 41 groups G_2 with non-vanishing $H^1(G_2, V_2)$. For thirteen of them the cohomology group is $\mathbb{Z}/3 \oplus \mathbb{Z}/3$, for one it is $\mathbb{Z}/3 \oplus \mathbb{Z}/3 \oplus \mathbb{Z}/3$ and for all others it is just $\mathbb{Z}/3$. In all non-vanishing cases the image $G \leqslant \mathrm{GL}_2(\mathbb{F}_3)$ of reduction has either non-trivial $H^0(G, V)$ or non-trivial $H^2(G, V)$, where V is the 2-dimensional vector space over $\mathbb{F}_3$ with its natural action by $G \leqslant \mathrm{GL}_2(\mathbb{F}_3)$. In other words, these numerical computations show that if the group $H^1(G_2, E[9])$ is non-trivial for an elliptic curve $E/\mathbb{Q}$, there is an isogeny $\varphi\colon E \to E'$ defined over $\mathbb{Q}$ of degree 3 such that either φ or its dual $\hat{\varphi}$ has a rational 3-torsion point in its kernel.

The maximal order of the cohomology group appears for the group G_2 consisting of all matrices in $\mathrm{GL}_2(\mathbb{Z}/9)$ with reduction $\left(\begin{smallmatrix} 1 & 0 \\ 0 & 1 \end{smallmatrix}\right)$ or $\left(\begin{smallmatrix} 1 & 0 \\ 0 & -1 \end{smallmatrix}\right)$ modulo 3.

7.3 $p = 5$

There are 39 groups G_2 with non-vanishing $H^1(G_2, V_2)$. For two of them, the group is $\mathbb{Z}/5 \oplus \mathbb{Z}/5$, for one it is $\mathbb{Z}/25$ and for all others it is $\mathbb{Z}/5$. If we restrict to those groups for which M_2 has dimension 4, then there are five cases as found in Sect. 6:

G	$\begin{pmatrix} v^2 & 0 \\ 0 & v \end{pmatrix}$	$\begin{pmatrix} 1 & 0 \\ 0 & * \end{pmatrix}$	$\begin{pmatrix} u & * \\ 0 & u^2 \end{pmatrix}$	$\begin{pmatrix} v^2 & * \\ 0 & v \end{pmatrix}$	$\begin{pmatrix} 1 & * \\ 0 & * \end{pmatrix}$
$\lvert G \rvert$	4	4	20	20	20
$H^1(G_2, V_2)$	$\mathbb{Z}/5$	$\mathbb{Z}/5 \oplus \mathbb{Z}/5$	$\mathbb{Z}/5$	$\mathbb{Z}/5$	$\mathbb{Z}/5$

This determines what the non-vanishing cohomology groups can be for this specific prime.

7.4 $p = 7$

Here we restricted our attention to the subgroups G_2 for which M_2 has dimension 4. Then, as previously found, there are only two cases. The group G can be of the form $\begin{pmatrix} 1 & 0 \\ 0 & * \end{pmatrix}$ or $\begin{pmatrix} 1 & * \\ 0 & * \end{pmatrix}$. In the first case the cohomology group $H^1(G_2, V_2)$ is $\mathbb{Z}/7 \oplus \mathbb{Z}/7$; in the latter it is $\mathbb{Z}/7$.

8 Applications to Local and Global Divisibility of Rational Points

The cohomology groups that we have discussed in this paper also appear in the analogue of the Grunwald–Wang problem for elliptic curves. This question was raised by Dvornicich and Zannier in [9].

Grunwald–Wang problem for elliptic curves. Let $E/\mathbb{Q}$ be an elliptic curve, $P \in E(\mathbb{Q})$, and $m > 1$. If P is divisible by m in $E(\mathbb{Q}_\ell)$ for almost all ℓ, is it true that P is divisible by m in $E(\mathbb{Q})$?

By the Chinese remainder theorem, it is sufficient to restrict to the case when $m = p^i$ is a prime power. The answer is positive if m is prime. The explicit example in [10] shows that the answer is negative for $m = 4$. In [20], it is shown that the answer is positive for all $m = p^2$ with p a prime larger than 3. To our knowledge, the case $m = 9$ has not been determined.

This question connects to our cohomology groups through the following reinterpretation. Suppose $m = p^i$ for our fixed prime p. Let Σ be a finite set of places in $\mathbb{Q}$. Let

$$D(E/\mathbb{Q}) = \ker\Big(E(\mathbb{Q})/p^i E(\mathbb{Q}) \to \prod_{v \notin \Sigma} E(\mathbb{Q}_v)/p^i E(\mathbb{Q}_v)\Big)$$

be the group that measures if there are points P that are locally divisible by p^i, but not globally. Let

$$L(E/\mathbb{Q}) = L(G_i) = \ker\Big(H^1(G_i, E[p^i]) \to \prod_{\substack{C \leqslant G_i \\ C \text{ cyclic}}} H^1(C, E[p^i])\Big) \qquad (7)$$

be the kernel of reduction to all the cyclic subgroups of G_i. We now assume that Σ contains all places above p and all bad places. By Chebotarev's theorem, $L(E/\mathbb{Q})$ is also the kernel of localization from $H^1(G_i, E[p^i])$ to all $H^1(D_{w|v}, E[p^i])$ where $D_{w|v}$ is the decomposition group in $K_i/\mathbb{Q}$ of a place w above v. Hence a natural notation for $L(E/\mathbb{Q})$ could be $\text{Ш}^1(U, E[p^i])$ with U the complement of Σ in $\operatorname{Spec}(\mathbb{Z})$. The sequence

$$0 \longrightarrow D(E/\mathbb{Q}) \longrightarrow L(E/\mathbb{Q}) \longrightarrow H^1(G_i, E(K_i))$$

is exact. Hence the answer is positive for $m = p^i$ if $H^1(G_i, E[p^i])$ vanishes. Note that the description of $L(E/\mathbb{Q})$ in (7) is now entirely group-theoretic, and can be computed numerically with the methods described in the previous section.

Theorem 24 *Let p a prime and $i \geqslant 1$. Then the Grunwald–Wang problem for local-global divisibility by $m = p^i$ admits a positive answer for all elliptic curves $E/\mathbb{Q}$ if and only if $p > 3$ or $m = 2$ or $m = 3$.*

Proof If we find a point P of infinite order that is a counter-example for $m = p^i$, then $p^j P$ is a counter-example for $m = p^{i+j}$ for any $j > 0$. As mentioned before, the negative answer for $m = 4$ is explained in [10]. This settles also all higher powers of 2 as their examples are points of infinite order. Counter-examples when m is a power of 3 were first found by Creutz in [7]. We will give below in Proposition 25 a new counter-example of infinite order for $m = 9$. For $p \geqslant 5$ the theorem follows from [20]. However, we wish to give a slightly simplified proof with our methods.

Assume therefore $p \geqslant 5$. We will now show that the kernel of localization $L(G_i)$ is zero. Note that by Greenberg's result in Theorem 21 and the work done in the exceptional cases in Lemmas 22 and 23, we may assume that G_i is greatest possible or that $i = 2$ and M_2 consists of all matrices m such that $m - 1$ is upper triangular. In both cases the elements in $E[p^i]$ fixed by M_i are just $E[p]$. We get an exact sequence

$$0 \longrightarrow H^1(G, E[p]) \xrightarrow{\text{inf}} H^1(G_i, E[p^i]) \longrightarrow H^1(M_i, E[p^i]).$$

First assume that $L(G_i)$ contains a non-trivial element which belongs to the image of the inflation map from $H^1(G, E[p])$. Since the latter must now be non-trivial, G must contain the element $\bar{h} = \left(\begin{smallmatrix} 1 & 1 \\ 0 & 1 \end{smallmatrix}\right)$. By the description of M_i, we find that G_i contains the element $h = \left(\begin{smallmatrix} 1 & 1 \\ 0 & 1 \end{smallmatrix}\right)$. Let C be the cyclic group generated by h and let $\bar{C}$ be its image in G. Our computations for proving Theorem 1 showed that $H^1(G, E[p]) \to H^1(\bar{C}, E[p])$ is a bijection. Next, both maps in the composition

$$H^1(\bar{C}, E[p]) \longrightarrow H^1(\bar{C}, E[p^i]^{C\cap M_i}) \longrightarrow H^1(C, E[p^i])$$

are injective: for the latter it is because any inflation map is injective, and for the first it can be read off the long exact sequence associated to the inclusion $E[p] \to E[p^i]^{C\cap M_i}$. We conclude that $H^1(G, E[p]) \to H^1(C, E[p^i])$ is injective. This now contradicts the assumption that $L(G_i)$ contained a non-trivial element from $H^1(G, E[p])$.

Therefore, $L(G_i)$ injects into to

$$L(M_i) = \ker\Big(H^1(M_i, E[p^i]) \to \prod_{\substack{C\leqslant M_i\\ \text{cyclic}}} H^1(C, E[p^i])\Big),$$

where the product now runs over all cyclic subgroups of M_i. We will now prove by induction on i that $L(M_i)$ is trivial. It is known for $i = 1$.

Recall that the group H_i acts trivially on $E[p^i]$. We consider the following diagram with exact rows:

$$\begin{array}{ccccccc} 0 & \longrightarrow & H^1(M_i, E[p^i]) & \longrightarrow & H^1(M_{i+1}, E[p^i]) & \longrightarrow & H^1(H_i, E[p^i]) \\ & & \downarrow & & \downarrow & & \downarrow \\ 0 & \to & \prod_C H^1(C/C\cap H_i, E[p^i]^{C\cap H_i}) & \to & \prod_C H^1(C, E[p^i]) & \to & \prod_C H^1(C\cap H_i, E[p^i]) \end{array} \tag{8}$$

where the products run over all cyclic subgroups C of M_{i+1}. Now the vertical map on the right hand side has the same kernel as

$$H^1(H_i, E[p^i]) = \mathrm{Hom}(H_i, E[p^i]) \longrightarrow \prod_{\substack{D\leqslant H_i\\ \text{cyclic}}} \mathrm{Hom}(D, E[p^i])$$

and this map is clearly injective. Since $C \cap H_i$ fixes $E[p^i]$ the vertical map on the right in the above diagram (8) is injective by induction hypothesis because $C/C\cap H_i \cong CH_i/H_i$ will run through all cyclic subgroups of M_i at least once. Therefore the middle vertical map in (8) is injective, too.

Next, consider the following diagram with exact rows:

$$\begin{array}{ccccccccc} & & 0 & \longrightarrow & E[p] & \xrightarrow{\delta} & \mathrm{Hom}(M_{i+1}, E[p]) & \xrightarrow{\iota} & H^1(M_{i+1}, E[p^{i+1}]) \xrightarrow{[p]} H^1(M_{i+1}, E[p^i]) \\ & & & & & & \downarrow & & \downarrow \qquad\qquad \downarrow \\ E[p^{i+1}]^C & \xrightarrow[{[p]}]{} & E[p^i]^C & \xrightarrow[\delta_C]{} & & & \mathrm{Hom}(C, E[p]) & \longrightarrow & H^1(C, E[p^{i+1}]) \longrightarrow H^1(C, E[p^i]) \end{array}$$

Here C is any cyclic subgroup of M_{i+1}. The zero at the top left corner is a consequence from the fact that the M_{i+1}-fixed points in $E[p^j]$ are exactly $E[p]$ for all $1 \leqslant j \leqslant i+1$.

If $\xi \in L(M_{i+1})$, then its image under $[p]$ in $H^1(M_{i+1}, E[p^i])$ must be trivial by what we have shown for the middle vertical map in (8). Therefore ξ is the image under ι of an element f in $\mathrm{Hom}(M_{i+1}, E[p])$. Since $E[p]$ is p-torsion, we can identify $\mathrm{Hom}(M_{i+1}, E[p])$ with $\mathrm{Hom}(M_2, E[p])$. To say that ξ restricts to zero for a cyclic group $C \leqslant M_{i+1}$ forces $f\colon M_2 \to E[p]$ to be in the image of the map $\delta_C\colon E[p] \to \mathrm{Hom}(C, E[p])$ for all cyclic subgroups C of M_2.

Now we identify M_2 with the additive subgroup $\tilde{M}_2 \leqslant \mathrm{Mat}_2(\mathbb{F}_p)$ as before. Under this identification the map δ sends a p-torsion point $T \in E[p] = \mathbb{F}_p^2$ to the map f sending a matrix $m \in \tilde{M}_2$ to $m(T)$. Thus, the restriction of f to $\mathrm{Hom}(\langle m\rangle, \mathbb{F}_p^2)$ is in the image of $\delta_{\langle m\rangle}$ for a particular $m \in \tilde{M}_2$ if and only if $f(m) \in \mathbb{F}_p^2$ belongs to the image of m. Therefore, we have shown that

$$L(M_{i+1}) = \frac{\left\{f \in \mathrm{Hom}(\tilde{M}_2, \mathbb{F}_p^2) \;\middle|\; f(m) \in \mathrm{im}(m)\ \forall m \in \tilde{M}_2\right\}}{\left\{f(m) = m(T) \text{ for some } T \in \mathbb{F}_p^2\right\}}. \tag{9}$$

We wish to show that $L(M_{i+1})$ is trivial if $\tilde{M}_2$ is the full matrix group or the upper triangular matrices. Assume first that $\tilde{M}_2$ is the full matrix group. Then f is determined by its image on the matrices with only one non-zero entry. However, the local condition of being in the image of δ_C for these matrices and the matrices $\left(\begin{smallmatrix}1&0\\1&0\end{smallmatrix}\right)$ and $\left(\begin{smallmatrix}0&1\\0&1\end{smallmatrix}\right)$ forces $f(m)$ to be just $m(T)$ for $T = f\left(\left(\begin{smallmatrix}1&0\\0&0\end{smallmatrix}\right)\right) + f\left(\left(\begin{smallmatrix}0&0\\0&1\end{smallmatrix}\right)\right)$. Therefore $L(M_{i+1})$ is trivial. The case when $\tilde{M}_2$ is the group of upper triangular matrices is very similar. □

The result about the vanishing of $L(G_2)$ for $p > 3$ in the above proof is reminiscent of Proposition 3.2.ii in [9]. We have reproved part of this result with a more conceptual approach. The main reason for doing so is that the general statement there is slightly incorrect. The case $\dim(M_2) = 3$ assumes that $\left(\begin{smallmatrix}1&1\\0&1\end{smallmatrix}\right)$ belongs to G_2. However, for $p = 3$, the group G_2 generated by $\left(\begin{smallmatrix}7&8\\3&1\end{smallmatrix}\right)$ and the group of all matrices m with $m - 1$ upper-triangular is a counterexample. This group does not contain any elements of order 9 and one can compute that $L(G_2)$ is isomorphic to $\mathbb{Z}/3$.

We include here a new counter-example for $m = 9$; the method is quite different from [7] where a first such example was found.

Proposition 25 *Let E be the elliptic curve labeled 243a2, given by the global minimal equation $y^2 + y = x^3 + 20$, and let $P = (-2, 3)$. Then $3P$ is divisible by 9 in $E(\mathbb{Q}_\ell)$ for all primes $\ell \neq 3$, but it is not divisible by 9 in $E(\mathbb{Q})$.*

Proof Since P is a generator of the free part of this curve of rank 1, it is clear that $3P$ is not divisible by 9 in $E(\mathbb{Q})$.

Let k be the unique subfield of $\mathbb{Q}(\mu_9)$ of degree 3 over $\mathbb{Q}$ and let ζ be a primitive 9-th root of unity. Then $P' = (3\zeta^5 + 3\zeta^4 + 3, 9\zeta^4 - 9\zeta^2 + 9\zeta + 4) \in E(k)$ satisfies

$3P' = P$. Thus, if $\ell \equiv \pm 1 \pmod 9$, then ℓ splits in k and hence P is divisible by 3 in $E(\mathbb{Q}_\ell)$. As a consequence, $3P$ is divisible by 9 over $\mathbb{Q}_\ell$.

For this curve and $p = 3$, the group $G = \left(\begin{smallmatrix} 1 & * \\ 0 & * \end{smallmatrix}\right)$ is of order 6 and $K = \mathbb{Q}(\theta)$ with $\theta^6 + 3 = 0$. Factoring the 9-division polynomial, one finds that K_2/K is an extension of degree 3. The field of definition of the points of order 9 is K_2 except for those points T with $3T \in E(\mathbb{Q})[3]$. Those are defined instead over a non-Galois extension of degree 3 over k.

Let $\ell \not\equiv \pm 1 \pmod 9$ and $\ell \neq 3$. Then the Frobenius element Fr_ℓ in G_2 cannot belong to $\mathrm{Gal}(K_2/k)$. Therefore Fr_ℓ does not fix any point of order 9. It follows that $\tilde{E}(\mathbb{F}_\ell)[9] = \tilde{E}(\mathbb{F}_\ell)[3]$. Consider the following commutative diagram, whose lower row is exact.

$$\begin{array}{ccccccc} & & & & P \in E(\mathbb{Q})/3E(\mathbb{Q}) & \xrightarrow{[3]} & E(\mathbb{Q})/9E(\mathbb{Q}) \\ & & & & \downarrow & & \downarrow \\ E(\mathbb{Q}_\ell)[9] & \xrightarrow{[3]} & E(\mathbb{Q}_\ell)[3] & \xrightarrow{\delta} & E(\mathbb{Q}_\ell)/3E(\mathbb{Q}_\ell) & \xrightarrow{[3]} & E(\mathbb{Q}_\ell)/9E(\mathbb{Q}_\ell) \end{array}$$

Since $\ell \neq 3$, the reduction of E at ℓ is good and hence [3] is an isomorphism on the kernel of reduction $E(\mathbb{Q}_\ell) \to \tilde{E}(\mathbb{F}_\ell)$. It follows that $E(\mathbb{Q}_\ell)[3] \cong \tilde{E}(\mathbb{F}_\ell)[3]$ and $E(\mathbb{Q}_\ell)/3E(\mathbb{Q}_\ell) \cong \tilde{E}(\mathbb{F}_\ell)/3\tilde{E}(\mathbb{F}_\ell)$ have the same size. By the above argument δ is an injective map between two groups of the same size. Thus δ is a bijection. This implies that $3P$ is divisible by 9 in $E(\mathbb{Q}_\ell)$. □

This is the counter-example of smallest conductor for $m = 9$; here is how we found that this curve is a likely candidate.

Consider curves E with a 3 isogeny where either the kernel has a rational 3-torsion point or where the kernel of the dual isogeny has a rational 3-torsion point. On the one hand, we computed (for a few thousand primes $\ell \neq 3$ of good reduction) the pairs $(a_\ell(E), \ell)$ modulo 9. On the other hand, we may determine all subgroups $G_2 \leqslant \mathrm{GL}_2(\mathbb{Z}/9)$ with surjective determinant to find the examples for which the kernel (7) is non-trivial. There are 13 such groups. The dimension of M_2 in these cases is 1, 2 or 3. For each of them we may list pairs $(\mathrm{tr}(g), \det(g))$ when g runs through all matrices $g \in G_2$.

Now, if the list of possible pairs $(a_\ell(E), \ell)$ modulo 9 agrees with one of the lists above, then G_2 could be among the groups for which the localization kernel is non-trivial. Furthermore, it is easy to check local divisibility for primes $\ell < 1000$ for all possible candidates in $3E(\mathbb{Q})/9E(\mathbb{Q})$. The above curve 243a2 was the first to pass all these tests.

Here are a few more candidates. Note that we have not formally proved that local divisibility holds by 9 holds for all primes ℓ of good reduction.

The point $P = (6, 17)$ on the curve 9747f1 gives a point $3P$ which is likely to be locally divisible by 9 for *all* primes, but not divisible by 9 globally. In this example G_2 has 54 elements.

On the curve 972d2 the point $3P$ with $P = (13, 35)$ is likely to be locally divisible by 9 for all places $\ell \neq 3$, yet not globally so. This is a curve without a rational 3-torsion point and G_2 having 54 elements again.

All the above examples have complex multiplication by the maximal order in $\mathbb{Q}(\sqrt{-3})$. The curve 722a1, with a point P having x-coordinate $\frac{27444}{169}$, is an example without complex multiplication and $|G_2| = 162$. Again, it is likely that $3P$ is locally divisible by 9 at all places $\ell \neq 19$, but $3P$ is not globally divisible by 9. The group G_2 here is probably conjugate to the one mentioned earlier as a counter-example to Proposition 3.2.ii in [9].

We have also done numerical calculation of the kernel in (7) for other primes. For $p = 5$, there are only three subgroups G_2 in $\mathrm{GL}_2(\mathbb{Z}/25)$ with non-trivial localization kernel. They all have $\dim(M_2) = 2$ and $|G| = 4$.

For $p = 2$, there are twelve cases. The dimensions of M_2 can be 1, 2, or 3. In only one of these cases is the localization kernel is $\mathbb{Z}/2 \oplus \mathbb{Z}/2$; otherwise it is $\mathbb{Z}/2$.

Acknowledgements It is our pleasure to thank Jean Gillibert and John Coates for interesting comments and suggestions. We are also grateful to Brendan Creutz for pointing us to [7].

References

1. Birch, B.J., Kuyk, W. (eds.): Modular functions of one variable. IV. In: Lecture Notes in Mathematics, vol. 476. Springer, Berlin, New York (1975)
2. Bosma, W., Cannon, J., Playoust, C.: The Magma algebra system. I. The user language. J. Symbolic Comput. **24**(3-4), 235–265 (1997). http://dx.doi.org/10.1006/jsco.1996.0125. Computational algebra and number theory (London, 1993)
3. Cha, B.: Vanishing of some cohomology groups and bounds for the Shafarevich-Tate groups of elliptic curves. J. Number Theory **111**(1), 154–178 (2005)
4. Çiperiani, M., Stix, J.: Weil-Châtelet divisible elements in Tate-Shafarevich groups II: On a question of Cassels. To appear in Journal für die Reine und Angewandte Mathematik
5. Coates, J.: An application of the division theory of elliptic functions to diophantine approximation. Invent. Math. **11**, 167–182 (1970)
6. Cremona, J.E.: Algorithms for Modular Elliptic Curves, 2nd edn. Cambridge University Press, Cambridge (1997)
7. Creutz, B.: On the local-glocal principle for divisibility in the cohomology of elliptic curves (2013). http://arxiv.org/abs/1305.5881
8. Creutz, B., Miller, R.L.: Second isogeny descents and the Birch and Swinnerton-Dyer conjectural formula. J. Algebra **372**, 673–701 (2012). http://dx.doi.org/10.1016/j.jalgebra.2012.09.029
9. Dvornicich, R., Zannier, U.: Local-global divisibility of rational points in some commutative algebraic groups. Bull. Soc. Math. France **129**(3), 317–338 (2001)
10. Dvornicich, R., Zannier, U.: An analogue for elliptic curves of the Grunwald–Wang example. C. R. Math. Acad. Sci. Paris **338**(1), 47–50 (2004). http://dx.doi.org/10.1016/j.crma.2003.10.034
11. Greenberg, R.: The image of Galois representations attached to elliptic curves with an isogeny. Amer. J. Math. **134**(5), 1167–1196 (2012). http://dx.doi.org/10.1353/ajm.2012.0040
12. Greenberg, R., Rubin, K., Silverberg, A., Stoll, M.: On elliptic curves with an isogeny of degree 7. Amer. J. Math. **136**(1), 77–109 (2014). http://dx.doi.org/10.1353/ajm.2014.0005

13. Grigorov, G., Jorza, A., Patrikis, S., Stein, W., Tarniţă, C.: Computational verification of the Birch and Swinnerton-Dyer conjecture for individual elliptic curves. Math. Comp. **78**(268), 2397–2425 (2009). http://dx.doi.org/10.1090/S0025-5718-09-02253-4
14. Gross, B.H.: Kolyvagin's work on modular elliptic curves. In: L-functions and arithmetic (Durham, 1989), *London Math. Soc. Lecture Note Ser.*, vol. 153, pp. 235–256. Cambridge University Press, Cambridge (1991). http://dx.doi.org/10.1017/CBO9780511526053.009
15. Matar, A.: For an elliptic curve $E/\mathbb{Q}$ can the cohomology group $H^1\left(\mathrm{Gal}(\mathbb{Q}(E[p])/\mathbb{Q}), E[p]\right)$ be nontrivial? (2014). http://mathoverflow.net/questions/186807
16. Mazur, B.: Rational isogenies of prime degree (with an appendix by D. Goldfeld). Invent. Math. **44**(2), 129–162 (1978)
17. Miller, R.L.: Proving the Birch and Swinnerton-Dyer conjecture for specific elliptic curves of analytic rank zero and one. LMS J. Comput. Math. **14**, 327–350 (2011). http://dx.doi.org/10.1112/S1461157011000180
18. Miller, R.L., Stoll, M.: Explicit isogeny descent on elliptic curves. Math. Comp. **82**(281), 513–529 (2013). http://dx.doi.org/10.1090/S0025-5718-2012-02619-6
19. Neukirch, J., Schmidt, A., Wingberg, K.: Cohomology of number fields (Grundlehren der Mathematischen Wissenschaften), vol. 323. Springer (2000)
20. Paladino, L., Ranieri, G., Viada, E.: On the minimal set for counterexamples to the local-global principle. J. Algebra **415**, 290–304 (2014). http://dx.doi.org/10.1016/j.jalgebra.2014.06.004
21. Serre, J.P.: Propriétés galoisiennes des points d'ordre fini des courbes elliptiques. Invent. Math. **15**(4), 259–331 (1972)
22. Serre, J.P.: Cohomologie Galoisienne. In: Lecture Notes in Mathematics, vol. 5. Springer, Berlin-New York (1973)
23. Stein, W.A., et al.: Sage Mathematics Software (Version 6.4). The Sage Development Team (2014). http://www.sagemath.org
24. Wuthrich, C.: prove_BSD for elliptic curve uses an incorrect lemma (2015). Bug report and fixing patch for SageMath. http://trac.sagemath.org/ticket/17869

Coates–Wiles Homomorphisms and Iwasawa Cohomology for Lubin–Tate Extensions

Peter Schneider and Otmar Venjakob

Abstract For the p-cyclotomic tower of $\mathbb{Q}_p$ Fontaine established a description of local Iwasawa cohomology with coefficients in a local Galois representation V in terms of the ψ-operator acting on the attached etale (φ, Γ)-module $D(V)$. In this chapter we generalize Fontaine's result to the case of arbitrary Lubin–Tate towers L_∞ over finite extensions L of $\mathbb{Q}_p$ by using the Kisin–Ren/Fontaine equivalence of categories between Galois representations and (φ_L, Γ_L)-modules and extending parts of [20, 33]. Moreover, we prove a kind of explicit reciprocity law which calculates the Kummer map over L_∞ for the multiplicative group twisted with the dual of the Tate module T of the Lubin–Tate formal group in terms of Coleman power series and the attached (φ_L, Γ_L)-module. The proof is based on a generalized Schmid–Witt residue formula. Finally, we extend the explicit reciprocity law of Bloch and Kato [3] Theorem 2.1 to our situation expressing the Bloch–Kato exponential map for $L(\chi_{LT}^r)$ in terms of generalized Coates–Wiles homomorphisms, where the Lubin–Tate character χ_{LT} describes the Galois action on T.

Keywords p-adic Hodge theory · Explicit reciprocity law · Coates–Wiles homomorphisms · Lubin–Tate formal groups · Ramified Witt vectors · Artin–Schreier pairing · Schmidt–Witt residue formula · Coleman power series

MSCs: 11Sxx

To John Coates on the occasion of his 70th birthday.

P. Schneider
Mathematisches Institut, Universität Münster, Einsteinstr. 62, 48291 Münster, Germany
e-mail: pschnei@uni-muenster.de
URL: http://www.uni-muenster.de/u/schneider/

O. Venjakob (✉)
Mathematisches Institut, Universität Heidelberg, Im Neuenheimer Feld 288, 69120 Heidelberg, Germany
e-mail: venjakob@mathi.uni-heidelberg.de
URL: http://www.mathi.uni-heidelberg.de/~venjakob/

D. Loeffler and S.L. Zerbes (eds.), *Elliptic Curves, Modular Forms and Iwasawa Theory*, Springer Proceedings in Mathematics & Statistics 188, DOI 10.1007/978-3-319-45032-2_12

1 Introduction

The invention of Coates–Wiles homomorphisms and Coleman power series [7, 8] were the starting point of a range of new and important developments in arithmetic. They have not lost their impact and fascination up to now. In order to recall it we shall first introduce some notation.

Consider a finite extension L of $\mathbb{Q}_p$ and fix a Lubin–Tate formal group LT over the integers o_L (with uniformizer π_L). By $\eta = (\eta_n)$ we denote a generator of the Tate module T of LT as o_L-module. The π_L^n-division points generate a tower of Galois extensions $L_n = L(LT[\pi_L^n])$ of L the union of which we denote by L_∞ with Galois group Γ_L. Coleman assigned to any norm compatible system of units $u = (u_n) \in \varprojlim_n L_n^\times$ a Laurent series $g_{u,\eta} \in o_L((Z))^\times$ such that $g_{u,\eta}(\eta_n) = u_n$ for all n. If ∂_{inv} denotes the invariant derivation with respect to LT, then, for $r \geq 1$, the rth Coates–Wiles homomorphism is given by

$$\psi_{CW}^r : \varprojlim_n L_n^\times \to L(\chi_{LT}^r),$$

$$u \mapsto \frac{1}{r!}\partial_{\mathrm{inv}}^r \log g_{u,\eta}(Z)|_{Z=0} := \frac{1}{r!}\partial_{\mathrm{inv}}^{r-1} \frac{\partial_{\mathrm{inv}} g_{u,\eta}(Z)}{g_{u,\eta}(Z)}|_{Z=0} ,$$

it is Galois invariant and satisfies at least heuristically—setting $t_{LT} = \log_{LT}(Z)$—the equation

$$\log g_{u,\eta}(Z) = \sum_r \psi_{CW}^r(u) t_{LT}^r$$

the meaning of which in p-adic Hodge theory has been crucially exploited, e.g. [5, 13, 16].

Explicitly or implicitly this mysterious map plays—classically for the multiplicative group over $\mathbb{Q}_p$—a crucial role in the Bloch–Kato (Tamagawa number) conjecture [3], in the study of special L-values [6], in explicit reciprocity laws [23, 37] and even in the context of the cyclotomic trace map from K-theory into topological cyclic homology for $\mathbb{Z}_p$ [2].

In this context one motivation for the present work is to understand Kato's (and hence Wiles's) explicit reciprocity law in terms of (φ_L, Γ_L)-modules. Since in the classical situation a successful study of explicit reciprocity laws has been achieved by Colmez, Cherbonnier/Colmez, Benois and Berger using Fontaine's work on (φ, Γ)-modules and Herr's calculation of Galois cohomology by means of them, the plan for this chapter is to firstly use Kisin–Ren/Fontaine's equivalence of categories (recalled in Sect. 4) to find a description of Iwasawa cohomology $H^i_{\mathrm{Iw}}(L_\infty/L, V)$ for the tower L_∞ and a (finitely generated o_L-module) representation V of G_L in terms of a ψ-operator acting on the etale (φ_L, Γ_L)-module $D_{LT}(V)$. To this aim we have to generalize parts of [20] in Sect. 3, in particular the residue pairing, which we relate to Pontrjagin duality. But instead of using Herr complexes (which one also could define easily in this context) we use local Tate duality $H^i_{\mathrm{Iw}}(L_\infty/L, V) \cong H^{2-i}(L_\infty, V^\vee(1))$

for V being an o_L-module of finite length and an explicit calculation of the latter groups in terms of (φ_L, Γ_L)-modules inspired by [15, 16, 33]. Using the key observation that $D_{LT}(V)^\vee \cong D_{LT}(V^\vee(\chi_{LT}))$ (due to the residue pairing involving differentials $\Omega^1 \cong D_{LT}(o_L)(\chi_{LT})$ and the compatibility of inner Homs under the category equivalence) we finally establish in Theorem 5.13 the following exact sequence

$$0 \longrightarrow H^1_{\mathrm{Iw}}(L_\infty/L, V) \longrightarrow D_{LT}(V(\tau^{-1})) \xrightarrow{\psi_L - 1} D_{LT}(V(\tau^{-1})) \longrightarrow H^2_{\mathrm{Iw}}(L_\infty/L, V) \longrightarrow 0$$

as one main result of this article, where the twist by $\tau = \chi_{LT}^{-1}\chi_{cyc}$ is a new phenomenon (disappearing obviously in the cyclotomic case) arising from the joint use of Pontrjagin and local Tate duality. The second main result is the explicit calculation of the twisted (by the o_L-dual T^* of T) Kummer map

$$\varprojlim_n L_n^\times \otimes_{\mathbb{Z}} T^* \xrightarrow{\kappa \otimes_{\mathbb{Z}_p} T^*} H^1_{\mathrm{Iw}}(L_\infty/L, \mathbb{Z}_p(1)) \otimes_{\mathbb{Z}_p} T^* \cong H^1_{\mathrm{Iw}}(L_\infty/L, o_L(\tau))$$

in terms of Coleman series (recalled in Sect. 2) and (φ_L, Γ_L)-modules, see Theorem 6.2, which generalizes the explicit reciprocity laws of Benois and Colmez. Inspired by [15, 16] we reduce its proof to an explicit reciprocity law, Proposition 6.3, in characteristic p, which in turn is proved by the Schmid–Witt residue formula which we generalize to our situation in Sect. 7, see Theorem 7.16. In Sect. 8 we generalize the approach sketched in [16] to prove in Theorem 8.6 a generalization of the explicit reciprocity law of Bloch and Kato [3] Theorem 2.1: again in this context the Bloch–Kato exponential map is essentially given by the Coates–Wiles homomorphism. As a direct consequence we obtain as Corollary 8.7 a new proof for a special case of Kato's explicit reciprocity law for Lubin–Tate formal groups.

It is a great honour and pleasure to dedicate this work to John Coates who has been a source of constant inspiration for both of us.

We thank L. Berger for pointing Lemma 8.2.i out to us and for a discussion about Proposition 9.1 and G. Kings for making available to us a copy of [25]. We also acknowledge support by the DFG Sonderforschungsbereich 878 at Münster and by the DFG research unit 1920 "Symmetrie, Geometrie und Arithmetik" at Heidelberg/Darmstadt.

Notation

Let $\mathbb{Q}_p \subseteq L \subset \mathbb{C}_p$ be a field of finite degree d over $\mathbb{Q}_p$, o_L the ring of integers of L, $\pi_L \in o_L$ a fixed prime element, $k_L = o_L/\pi_L o_L$ the residue field, and $q := |k_L|$. We always use the absolute value $|\ |$ on $\mathbb{C}_p$ which is normalized by $|\pi_L| = q^{-1}$.

We fix a Lubin–Tate formal o_L-module $LT = LT_{\pi_L}$ over o_L corresponding to the prime element π_L. We always identify LT with the open unit disk around zero, which gives us a global coordinate Z on LT. The o_L-action then is given by formal power series $[a](Z) \in o_L[[Z]]$. For simplicity the formal group law will be denoted by $+_{LT}$.

The power series $\frac{\partial(X+_{LT}Y)}{\partial Y}_{|(X,Y)=(0,Z)}$ is a unit in $o_L[[Z]]$ and we let $g_{LT}(Z)$ denote its inverse. Then $g_{LT}(Z)dZ$ is, up to scalars, the unique invariant differential form on LT ([19] §5.8). We also let $\log_{LT}(Z) = Z + \dots$ denote the unique formal power series in $L[[Z]]$ whose formal derivative is g_{LT}. This $\log_{LT}$ is the logarithm of LT ([26] 8.6). In particular, $g_{LT}dZ = d\log_{LT}$. The invariant derivation ∂_{inv} corresponding to the form $d\log_{LT}$ is determined by

$$f'dZ = df = \partial_{\text{inv}}(f)d\log_{LT} = \partial_{\text{inv}}(f)g_{LT}dZ$$

and hence is given by

$$\partial_{\text{inv}}(f) = g_{LT}^{-1}f' . \tag{1}$$

For any $a \in o_L$ we have

$$\log_{LT}([a](Z)) = a \cdot \log_{LT} \qquad \text{and hence} \qquad ag_{LT}(Z) = g_{LT}([a](Z)) \cdot [a]'(Z) \tag{2}$$

([26] 8.6 Lemma 2).

Let T be the Tate module of LT. Then T is a free o_L-module of rank one, and the action of $G_L := \text{Gal}(\overline{L}/L)$ on T is given by a continuous character $\chi_{LT} : G_L \longrightarrow o_L^\times$. Let T' denote the Tate module of the p-divisible group Cartier dual to LT, which again is a free o_L-module of rank one. The Galois action on T' is given by the continuous character $\tau := \chi_{cyc} \cdot \chi_{LT}^{-1}$, where χ_{cyc} is the cyclotomic character.

For $n \geq 0$ we let L_n/L denote the extension (in $\mathbb{C}_p$) generated by the π_L^n-torsion points of LT, and we put $L_\infty := \bigcup_n L_n$. The extension L_∞/L is Galois. We let $\Gamma_L := \text{Gal}(L_\infty/L)$ and $H_L := \text{Gal}(\overline{L}/L_\infty)$. The Lubin–Tate character χ_{LT} induces an isomorphism $\Gamma_L \xrightarrow{\cong} o_L^\times$.

2 Coleman Power Series

We recall the injective ring endomorphism

$$\begin{aligned} \varphi_L : o_L[[Z]] &\longrightarrow o_L[[Z]] \\ f(Z) &\longmapsto f([\pi_L](Z)) . \end{aligned}$$

In order to characterize its image we let LT_1 denote the group of π_L-torsion points of LT. According to [8] Lemma 3 we have

$$\text{im}(\varphi_L) = \{f \in o_L[[Z]] : f(Z) = f(a +_{LT} Z) \text{ for any } a \in LT_1\}.$$

This leads to the existence of a unique o_L-linear endomorphism ψ_{Col} of $o_L[[Z]]$ such that

$$\varphi_L \circ \psi_{Col}(f)(Z) = \sum_{a \in LT_1} f(a +_{LT} Z) \qquad \text{for any } f \in o_L[[Z]]$$

([8] Theorem 4 and Corollary 5) as well as of a unique multiplicative map $\mathscr{N}: o_L[[Z]] \longrightarrow o_L[[Z]]$ such that

$$\varphi_L \circ \mathscr{N}(f)(Z) = \prod_{a \in LT_1} f(a +_{LT} Z) \qquad \text{for any } f \in o_L[[Z]]$$

([8] Theorem 11).

The group Γ_L acts continuously on $o_L[[Z]]$ via

$$\begin{aligned} \Gamma_L \times o_L[[Z]] &\longrightarrow o_L[[Z]] \\ (\gamma, f) &\longmapsto f([\chi_{LT}(\gamma)](Z)) \end{aligned} \tag{3}$$

([8] Theorem 1).

Remark 2.1 i. $\psi_{Col} \circ \varphi_L = q$.
ii. $\psi_{Col}([\pi_L] \cdot f) = Z\psi_{Col}(f)$ for any $f \in o_L[[Z]]$.
iii. $\mathcal{N}([\pi_L]) = Z^q$.

Proof Because of the injectivity of φ_L it suffices in all three cases to verify the asserted identity after applying φ_L. i. We compute

$$\begin{aligned} \varphi_L \circ \psi_{Col} \circ \varphi_L(f) &= \sum_{a \in LT_1} (\varphi_L f)(a +_{LT} Z) = \sum_{a \in LT_1} f([\pi_L](a +_{LT} Z)) \\ &= \sum_{a \in LT_1} f([\pi_L](Z)) = \varphi_L(qf). \end{aligned}$$

ii. We compute

$$\begin{aligned} (\varphi_L \circ \psi_{Col})([\pi_L]f) &= \sum_{a \in LT_1} [\pi_L](a +_{LT} Z) f(a +_{LT} Z) \\ &= [\pi_L](Z) \sum_{a \in LT_1} f(a +_{LT} Z) = \varphi_L(Z)(\varphi_L \circ \psi_{Col})(f) \\ &= \varphi_L(Z\psi_{Col}(f)) . \end{aligned}$$

iii. We omit the entirely analogous computation. □

We observe that for any $f \in o_L((Z)) = o_L[[Z]][Z^{-1}]$ there is an $n(f) \geq 1$ such that $[\pi_L]^{n(f)} \cdot f \in o_L[[Z]]$. The above remark therefore allows to extend ψ_{Col} to an o_L linear endomorphism

$$\begin{aligned} \psi_{Col} : o_L((Z)) &\longrightarrow o_L((Z)) \\ f &\longmapsto Z^{-n(f)} \psi_{Col}([\pi_L]^{n(f)} f) \end{aligned}$$

and to extend $\mathcal{N}$ to a multiplicative map

$$\begin{aligned} \mathcal{N} : o_L((Z)) &\longrightarrow o_L((Z)) \\ f &\longmapsto Z^{-qn(f)}\mathcal{N}([\pi_L]^{n(f)}f) \ . \end{aligned}$$

We choose an o_L-generator η of T. This is a sequence of elements $\eta_n \in \pi_{L_n} o_{L_n}$ such that $[\pi_L](\eta_{n+1}) = \eta_n$ for $n \geq 1$, $[\pi_L](\eta_1) = 0$, and $\eta_1 \neq 0$.

Theorem 2.2 (Coleman) *For any norm-coherent sequence $u = (u_n)_n \in \varprojlim_n L_n^\times$ there is a unique Laurent series $g_{u,\eta} \in (o_L((Z))^\times)^{\mathcal{N}=1}$ such that $g_{u,\eta}(\eta_n) = u_n$ for any $n \geq 1$. This defines a multiplicative isomorphism*

$$\begin{aligned} \varprojlim_n L_n^\times &\xrightarrow{\cong} (o_L((Z))^\times)^{\mathcal{N}=1} \\ u &\longmapsto g_{u,\eta} \ . \end{aligned}$$

Proof See [8] Theorem A and Corollary 17. □

Remark 2.3 i. The map $(o_L((Z))^\times)^{\mathcal{N}=1} \xrightarrow{\cong} k_L((Z))^\times$ given by reduction modulo π_L is an isomorphism; hence

$$\begin{aligned} \varprojlim_n L_n^\times &\xrightarrow{\cong} k_L((Z))^\times \\ u &\longmapsto g_{u,\eta} \bmod \pi_L \end{aligned}$$

is an isomorphism of groups.

ii. If $\vartheta = c\eta$ is a second o_L-generator of T then $g_{u,\vartheta}([c](Z)) = g_{u,\eta}(Z)$ for any $u \in \varprojlim_n L_n^\times$.

Proof i. [8] Corollary 18. ii. This is immediate from the characterizing property of $g_{u,\eta}$ in the theorem. □

We now introduce the "logarithmic" homomorphism

$$\begin{aligned} \Delta_{LT} : o_L[[Z]]^\times &\longrightarrow o_L[[Z]] \\ f &\longmapsto \frac{\partial_{\mathrm{inv}}(f)}{f} = g_{LT}^{-1}\frac{f'}{f} \ , \end{aligned}$$

whose kernel is $o_L^\times$.

Lemma 2.4 *i.* $\Delta_{LT} \circ \varphi_L = \pi_L \varphi_L \circ \Delta_{LT}$.

ii. $\psi_{Col} \circ \Delta_{LT} = \pi_L \Delta_{LT} \circ \mathcal{N}$.

Proof We begin with a few preliminary observations. From (2) we deduce

$$g_{LT} = \frac{[\pi_L]'}{\pi_L}\varphi_L(g_{LT}) \,. \tag{4}$$

Secondly we have

$$\tfrac{d}{dZ}\varphi_L(f(Z)) = \tfrac{d}{dZ}f([\pi_L](Z)) = f'([\pi_L](Z))[\pi_L]'(Z) = \varphi_L(f')[\pi_L]' \tag{5}$$

for any $f \in o_L[[Z]]$. Finally, the fact that $g_{LT}(Z)dZ$ is an invariant differential form implies that

$$g_{LT} = \tfrac{d}{dZ}\log_{LT}(a +_{LT} Z) \qquad \text{for any } a \in LT_1. \tag{6}$$

For i. we now compute

$$\Delta_{LT} \circ \varphi_L(f) = \frac{1}{g_{LT}}\frac{\frac{d}{dZ}\varphi_L(f(Z))}{\varphi_L(f)} = \frac{[\pi_L]'}{g_{LT}}\frac{\varphi_L(f')}{\varphi_L(f)} = \frac{\pi_L}{\varphi_L(g_{LT})}\frac{\varphi_L(f')}{\varphi_L(f)} = \pi_L\varphi_L \circ \Delta_{LT}(f)\,,$$

where the second, resp. the third, identity uses (5), resp. (4). For ii. we compute

$$\begin{aligned}
\varphi_L \circ \psi_{Col} \circ \Delta_{LT}(f) &= \sum_{a\in LT_1} \frac{1}{g_{LT}(a+_{LT}Z)}\frac{f'}{f}(a +_{LT} Z) \\
&= \sum_{a\in LT_1} \frac{1}{g_{LT}(a+_{LT}Z)}\frac{\frac{d}{dZ}f(a+_{LT}Z)}{f(a+_{LT}Z)}\frac{1}{\frac{d}{dZ}(a+_{LT}Z)} \\
&= \sum_{a\in LT_1} \frac{1}{\frac{d}{dZ}\log_{LT}(a+_{LT}Z)}\frac{\frac{d}{dZ}f(a+_{LT}Z)}{f(a+_{LT}Z)} \\
&= \sum_{a\in LT_1} \Delta_{LT}(f(a +_{LT} Z)) = \Delta_{LT}(\prod_{a\in LT_1} f(a +_{LT} Z)) \\
&= \Delta_{LT} \circ \varphi_L \circ \mathcal{N}(f) = \pi_L\varphi_L \circ \Delta_{LT} \circ \mathcal{N}(f) \\
&= \varphi_L(\pi_L\Delta_{LT} \circ \mathcal{N}(f))\,,
\end{aligned}$$

where the fourth, resp. the seventh, identity uses (6), resp. part i. of the assertion. □

It follows that Δ_{LT} restricts to a homomorphism

$$\Delta_{LT} : (o_L[[Z]]^\times)^{\mathcal{N}=1} \longrightarrow o_L[[Z]]^{\psi_{Col}=\pi_L}\,.$$

Its kernel is the subgroup $\mu_{q-1}(L)$ of $(q-1)$th roots of unity in $o_L^\times$.

On the other hand Δ_{LT} obviously extends to the homomorphism

$$\begin{aligned} \Delta_{LT} : o_L((Z))^\times &\longrightarrow o_L((Z)) \\ f &\longmapsto g_{LT}^{-1}\frac{f'}{f} , \end{aligned}$$

with the same kernel $o_L^\times$.

Lemma 2.5 *The identity $\psi_{Col} \circ \Delta_{LT} = \pi_L \Delta_{LT} \circ \mathcal{N}$ holds true on $o_L((Z))^\times$.*

Proof Let $f \in o_L((Z)^\times$ be any element. It can be written $f = Z^{-n} f_0$ with $f_0 \in o_L[[Z]]^\times$. Then

$$\psi_{Col} \circ \Delta_{LT}(f) = -n\psi_{Col}(\frac{1}{Zg_{LT}}) + \psi_{Col} \circ \Delta_{LT}(f_0)$$

and

$$\pi_L \Delta_{LT} \circ \mathcal{N}(f) = -n\pi_L \Delta_{LT}(\mathcal{N}(Z)) + \pi_L \Delta_{LT} \circ \mathcal{N}(f_0) .$$

The second summands being equal by Lemma 2.4.ii we see that we have to establish that

$$\psi_{Col}(\frac{1}{Zg_{LT}}) = \pi_L \Delta_{LT}(\mathcal{N}(Z)) .$$

By definition the left hand side is $Z^{-1}\psi_{Col}(\frac{[\pi_L]}{Zg_{LT}})$ and the right hand side is $\frac{\pi_L}{g_{LT}} \frac{\frac{d}{dZ}\mathcal{N}(Z)}{\mathcal{N}(Z)}$. Hence we are reduced to proving the identity

$$g_{LT}\mathcal{N}(Z)\psi_{Col}(\frac{[\pi_L]}{Zg_{LT}}) = \pi_L Z \frac{d}{dZ}\mathcal{N}(Z) ,$$

which is an identity in $o_L[[Z]]$ and therefore can be checked after applying φ_L. On the left hand side we obtain

$$\begin{aligned} &\varphi_L(g_{LT}) \prod_{a \in LT_1} (a +_{LT} Z) \sum_{b \in LT_1} \frac{[\pi_L](b +_{LT} Z)}{(b +_{LT} Z) g_{LT}(b +_{LT} Z)} \\ &= \varphi_L(g_{LT})\varphi_L(Z) \prod_{a \in LT_1} (a +_{LT} Z) \sum_{b \in LT_1} \frac{1}{(b +_{LT} Z) g_{LT}(b +_{LT} Z)} \\ &= g_{LT}\frac{\pi_L}{[\pi_L]'}\varphi_L(Z) \prod_{a \in LT_1} (a +_{LT} Z) \sum_{b \in LT_1} \frac{1}{(b +_{LT} Z) g_{LT}(b +_{LT} Z)} , \end{aligned}$$

where the second equality uses (4). On the right hand side, using (5), we have

$$\begin{aligned}\pi_L\varphi_L(Z)\varphi_L(\tfrac{d}{dZ}\mathcal{N}(Z)) &= \frac{\pi_L}{[\pi_L]'}\varphi_L(Z)\tfrac{d}{dZ}\varphi_L(\mathcal{N}(Z)) \\ &= \frac{\pi_L}{[\pi_L]'}\varphi_L(Z)\tfrac{d}{dZ}\prod_{a\in LT_1}(a+_{LT}Z)\,.\end{aligned}$$

This further reduces us to proving that

$$g_{LT}\sum_{b\in LT_1}\frac{1}{(b+_{LT}Z)g_{LT}(b+_{LT}Z)} = \frac{\frac{d}{dZ}\prod_{a\in LT_1}(a+_{LT}Z)}{\prod_{a\in LT_1}(a+_{LT}Z)}\,.$$

The invariance of $g_{LT}(Z)dZ$ implies

$$\tfrac{d}{dZ}(a+_{LT}Z) = \frac{g_{LT}(Z)}{g_{LT}(a+_{LT}Z)}\,. \tag{7}$$

We see that the above right hand side, indeed, is equal to

$$\begin{aligned}\frac{\frac{d}{dZ}\prod_{a\in LT_1}(a+_{LT}Z)}{\prod_{a\in LT_1}(a+_{LT}Z)} &= \sum_{a\in LT_1}\frac{\frac{d}{dZ}(a+_{LT}Z)}{a+_{LT}Z} \\ &= g_{LT}\sum_{a\in LT_1}\frac{1}{(a+_{LT}Z)g_{LT}(a+_{LT}Z)}.\end{aligned}$$

□

Hence we even have the homomorphism

$$\Delta_{LT} : (o_L((Z))^\times)^{\mathcal{N}=1} \longrightarrow o_L((Z))^{\psi_{Col}=\pi_L}$$

with kernel $\mu_{q-1}(L)$.

3 Etale (φ_L, Γ_L)-modules

We define the ring $\mathscr{A}_L$ to be the π_L-adic completion of $o_L[[Z]][Z^{-1}]$ and we let $\mathscr{B}_L := \mathscr{A}_L[\pi_L^{-1}]$ denote the field of fractions of $\mathscr{A}_L$. The ring endomorphism φ_L of $o_L[[Z]]$ maps Z to $[\pi_L](Z)$. Since $[\pi_L](Z) \equiv Z^q \bmod \pi_L$ the power series $[\pi_L](Z)$ is a unit in $\mathscr{A}_L$. Hence φ_L extends to a homomorphism $o_L[[Z]][Z^{-1}] \longrightarrow \mathscr{A}_L$ and then by continuity to a ring endomorphism φ_L of $\mathscr{A}_L$ and finally to an embedding of fields $\varphi_L : \mathscr{B}_L \longrightarrow \mathscr{B}_L$. Similarly the invariant derivation ∂_{inv} first extends algebraically to $o_L[[Z]][Z^{-1}]$, then by continuity to $\mathscr{A}_L$, and finally by linearity to $\mathscr{B}_L$. Evidently we still have (1) for any $f \in \mathscr{B}_L$.

Remark 3.1 $1, Z, \ldots, Z^{q-1}$ is a basis of $\mathscr{B}_L$ as a $\varphi_L(\mathscr{B}_L)$-vector space.

Proof See [17] Remark before Lemma 2.1 or [31] Proposition 1.7.3. □

This remark allows us to introduce the unique additive endomorphism ψ_L of $\mathscr{B}_L$ which satisfies

$$\varphi_L \circ \psi_L = \pi_L^{-1} \cdot trace_{\mathscr{B}_L/\varphi_L(\mathscr{B}_L)} .$$

By the injectivity of φ_L and the linearity of the field trace we have the projection formula

$$\psi_L(\varphi_L(f_1)f_2) = f_1\psi_L(f_2) \qquad \text{for any } f_i \in \mathscr{B}_L$$

as well as the formula

$$\psi_L \circ \varphi_L = \frac{q}{\pi_L} \cdot \mathrm{id} .$$

Correspondingly, we consider the unique multiplicative map $\mathcal{N}_L : \mathscr{B}_L \longrightarrow \mathscr{B}_L$ which satisfies

$$\varphi_L \circ \mathcal{N}_L = \mathrm{Norm}_{\mathscr{B}_L/\varphi_L(\mathscr{B}_L)} . \tag{8}$$

Remark 3.2 i. $\psi_L(\mathscr{A}_L) \subseteq \mathscr{A}_L$ and $\mathcal{N}_L(\mathscr{A}_L) \subseteq \mathscr{A}_L$.
ii. On $o_L((Z))$ we have $\psi_L = \pi_L^{-1} \cdot \psi_{Col}$ and $\mathcal{N}_L = \mathcal{N}$.
iii. $\varphi_L \circ \psi_L \circ \partial_{\mathrm{inv}} = \partial_{\mathrm{inv}} \circ \varphi_L \circ \psi_L$ on $\mathscr{B}_L$.
iv. $\mathcal{N}_L(f)([c](Z)) = \mathcal{N}_L(f([c](Z)))$ for any $c \in o_L^\times$ and $f \in \mathscr{B}_L$.
v. $\mathcal{N}_L(f) \equiv f \bmod \pi_L\mathscr{A}_L$ for any $f \in \mathscr{A}_L$.
vi. If $f \in \mathscr{A}_L$ satisfies $f \equiv 1 \bmod \pi_L^m\mathscr{A}_L$ for some $m \geq 1$ then $\mathcal{N}_L(f) \equiv 1 \bmod \pi_L^{m+1}\mathscr{A}_L$.
vii. $(o_L((Z))^\times)^{\mathcal{N}=1} = (\mathscr{A}_L^\times)^{\mathcal{N}_L=1}$.

Proof i. The homomorphism φ_L induces on $\mathscr{A}_L/\pi_L\mathscr{A}_L = k_L((Z))$ the injective q-Frobenius map. It follows that $\{f \in \mathscr{B}_L : \varphi_L(f) \in \mathscr{A}_L\} = \mathscr{A}_L$. Hence the assertion reduces to the claim that

$$trace_{\mathscr{B}_L/\varphi_L(\mathscr{B}_L)}(\mathscr{A}_L) \subseteq \pi_L\mathscr{A}_L . \tag{9}$$

But the trace map $trace_{\mathscr{B}_L/\varphi_L(\mathscr{B}_L)}$ induces the trace map for the purely inseparable extension $k_L((Z))/k_L((Z^q))$, which is the zero map.

ii. For any $a \in LT_1$ we have the ring homomorphism

$$\begin{aligned} \sigma_a : o_L[[Z]] &\longrightarrow o_{L_1}[[Z]] \subseteq \mathscr{A}_{L_1} \\ f(Z) &\mapsto f(a +_{LT} Z) . \end{aligned}$$

Since $\sigma_a(Z) = a +_{LT} Z \equiv a + Z \bmod \deg 2$ we have $\sigma_a(Z) \in \mathscr{A}_{L_1}^\times$, so that σ_a extends to $o_L[[Z]][Z^{-1}]$. By continuity σ_a further extends to $\mathscr{A}_L$ and then by linearity to an embedding of fields

$$\sigma_a : \mathscr{B}_L \longrightarrow \mathscr{B}_{L_1} = \mathscr{B}_L L_1 .$$

Clearly these σ_a are pairwise different. Moreover, for any $f \in o_L[[Z]]$, we have

$$\sigma_a \circ \varphi_L(f)(Z) = f([\pi_L](a +_{LT} Z)) = f(Z) .$$

We conclude, by continuity, that $\sigma_a | \varphi_L(\mathscr{B}_L) = \mathrm{id}$. It follows that $\prod_{a \in LT_1}(X - \sigma_a(f))$, for any $f \in \mathscr{B}_L$, is the characteristic polynomial of f over $\varphi_L(\mathscr{B}_L)$. Hence

$$trace_{\mathscr{B}_L/\varphi_L(\mathscr{B}_L)}(f) = \sum_{a \in LT_1} \sigma_a(f) , \tag{10}$$

which proves the assertion for $f \in o_L[[Z]]$. For general $f \in o_L((Z))$, using the notation and definition before Theorem 2.2 we compute

$$\begin{aligned}
\varphi_L \circ \psi_{Col}(f) &= \varphi_L \left(Z^{-n(f)} \psi_{Col}([\pi_L]^{n(f)} f)\right) \\
&= \varphi_L(Z)^{-n(f)} \sum_{a \in LT_1} \sigma_a([\pi_L]^{n(f)} f) \\
&= \sum_{a \in LT_1} \sigma_a(f) = trace_{\mathscr{B}_L/\varphi_L(\mathscr{B}_L)}(f) \\
&= \varphi_L \circ \pi_L \psi_L(f) .
\end{aligned}$$

The proof for $\mathcal{N}_L$ is completely analogous.

iii. By the invariance of ∂_{inv} we have $\partial_{\mathrm{inv}} \circ \sigma_a = \sigma_a \circ \partial_{\mathrm{inv}}$ on $o_L[[Z]][Z^{-1}]$, whence on $\mathscr{B}_L$ by continuity and linearity. Therefore, the claim follows from (10).

iv. We compute

$$\begin{aligned}
\varphi_L(\mathcal{N}_L(f)([c](Z))) &= \mathcal{N}_L(f)([c]([\pi_L](Z))) = \mathcal{N}_L(f)([\pi_L]([c](Z))) \\
&= \varphi_L(\mathcal{N}_L(f))([c](Z)) = \prod_{a \in LT_1} \sigma_a(f)([c](Z)) \\
&= \prod_{a \in LT_1} \sigma_{[c^{-1}](a)}(f([c](Z))) = \prod_{a \in LT_1} \sigma_a(f([c](Z))) \\
&= \varphi_L(\mathcal{N}_L(f([c](Z)))) .
\end{aligned}$$

v. We have

$$\begin{aligned}
\varphi_L \circ \mathcal{N}_L(f) \bmod \pi_L \mathscr{A}_L &= \mathrm{Norm}_{k_L((Z))/k_L((Z^q))}(f \bmod \pi_L \mathscr{A}_L) \equiv f^q \bmod \pi_L \mathscr{A}_L \\
&= \varphi_L(f) \bmod \pi_L \mathscr{A}_L .
\end{aligned}$$

vi. Let $f = 1 + \pi_L^m g$ with $g \in \mathscr{A}_L$. We compute

$$\begin{aligned}
\varphi_L(\mathcal{N}_L(1 + \pi_L^m g)) &= \prod_{a \in LT_1} 1 + \pi_L^m \sigma_a(g) \equiv 1 + \pi_L^m (\sum_{a \in LT_1} \sigma_a(g)) \bmod \pi_L^{m+1} \mathscr{A}_L \\
&\equiv 1 \bmod \pi_L^{m+1} \mathscr{A}_L \equiv \varphi_L(1) \bmod \pi_L^{m+1} \mathscr{A}_L
\end{aligned}$$

where the third identity uses (9). The assertion follows since φ_L remains injective modulo π_L^j for any $j \geq 1$.

vii. We have the commutative diagram

$$\begin{array}{ccc} (o_L((Z))^\times)^{\mathcal{N}=1} & \xrightarrow{\subseteq} & (\mathscr{A}_L^\times)^{\mathcal{N}_L=1} \\ & {\searrow}^{\cong} \quad \swarrow & \\ & k_L((Z))^\times & \end{array}$$

where the oblique arrows are given by reduction modulo π_L. The left one is an isomorphism by Remark 2.3.i. The right one is injective as a consequence of the assertion vi. Hence all three maps must be bijective. □

Due to Remark 3.2.vii we may view the Coleman isomorphism in Theorem 2.2 as an isomorphism

$$\varprojlim_n L_n^\times \xrightarrow{\cong} (\mathscr{A}_L^\times)^{\mathcal{N}_L=1} . \tag{11}$$

We always equip $\mathscr{A}_L$ with the weak topology, for which the o_L-submodules $\pi_L^m \mathscr{A}_L + Z^m o_L[[Z]]$, for $m \geq 1$, form a fundamental system of open neighbourhoods of zero. The weak topology on any finitely generated $\mathscr{A}_L$-module M is defined to be the quotient topology, with respect to any surjective homomorphism $\mathscr{A}_L^n \twoheadrightarrow M$, of the product topology on $\mathscr{A}_L^n$; this is independent of the choice of this homomorphism. We have the following properties (cf. [32] Lemmas 8.2 and 8.22 for a detailed discussion of weak topologies):

- $\mathscr{A}_L$ is a complete Hausdorff topological o_L-algebra (with jointly continuous multiplication).
- $\mathscr{A}_L$ induces on $o_L[[Z]]$ its compact topology.
- M is a complete Hausdorff topological module (with jointly continuous scalar multiplication).
- $M/\pi_L^m M$, for any $m \geq 1$, is locally compact.

Remark 3.3 The endomorphisms φ_L and ψ_L of $\mathscr{A}_L$ are continuous for the weak topology.

Proof For φ_L see [31] Proposition 1.7.8.i. For ψ_L see [17] Proposition 2.4(b) (note that their ψ is our $\frac{\pi_L}{q}\psi_L$). □

Let $\Omega^1 = \Omega^1_{\mathscr{A}_L} = \mathscr{A}_L dZ$ denote the free rank one $\mathscr{A}_L$-module of differential forms. Obviously the residue map

$$\begin{aligned} \mathrm{Res} : \qquad \Omega^1 &\longrightarrow o_L \\ (\sum_i a_i Z^i) dZ &\longmapsto a_{-1} \end{aligned}$$

is continuous. Later on in Sect. 7 it will be a very important fact that this map does not depend on the choice of the variable Z. For the convenience of the reader we explain the argument (cf. [15] A2.2.3). First of all we have to extend the maps d and Res by linearity to maps

$$\mathscr{B}_L \xrightarrow{d} \Omega^1_{\mathscr{B}_L} := L \otimes_{o_L} \Omega^1_{\mathscr{A}_L} \xrightarrow{\text{Res}} L\,.$$

Only for the purposes of the subsequent remark we topologize $\mathscr{B}_L$ by taking as a fundamental system of open neighbourhoods of zero the $o_L[[Z]]$-submodules

$$\pi_L^m \mathscr{A}_L + L \otimes_{o_L} Z^m o_L[[Z]]) \qquad \text{for } m \geq 1.$$

Using the isomorphism $\Omega^1_{\mathscr{B}_L} = \mathscr{B}_L dZ \cong \mathscr{B}_L$ we also make $\Omega^1_{\mathscr{B}_L}$ into a topological o_L-module. It is easy to see that the maps d and Res are continuous.

Remark 3.4 i. $d(\mathscr{B}_L)$ is dense in $\ker(\text{Res})$.
ii. $\text{Res} = \text{Res}_Z$ does not depend on the choice of the variable Z, i.e., if Z' is any element in $\mathscr{A}_L$ whose reduction modulo π_L is a uniformizing element in $k((Z))$, then $\text{Res}_Z(\omega) = \text{Res}_{Z'}(\omega)$ for all $\omega \in \Omega^1_{\mathscr{B}_L}$.

Proof i. On the one hand $L[Z, Z^{-1}] \cap \ker(\text{Res})$ is dense in $\ker(\text{Res})$. On the other hand we have $L[Z, Z^{-1}] \cap \ker(\text{Res}) \subseteq d(\mathscr{B}_L)$. ii. As a consequence of i. both maps Res_Z and $\text{Res}_{Z'}$ have the same kernel. It therefore suffices to show that $\text{Res}_Z(\frac{dZ'}{Z'}) = 1$. We have $Z' = cZv(1 + \pi_L\alpha)$ with $c \in o_L^\times$, $v \in 1 + Zo_L[[Z]]$, and $\alpha \in \mathscr{A}_L$. Hence

$$\frac{dZ'}{Z'} = \frac{dZ}{Z} + \frac{dv}{v} + \frac{d(1+\pi_L\alpha)}{1+\pi_L\alpha}\,.$$

Clearly $\text{Res}_Z(\frac{dv}{v}) = 0$. Furthermore, if $m \geq 1$ is sufficiently big, then $\log(1 + \pi_L^m\beta)$, for any $\beta \in \mathscr{A}_L$, converges in $\mathscr{A}_L$. Since $(1 + \pi_L^j\mathscr{A}_L)/(1 + \pi_L^{j+1}\mathscr{A}_L) \cong \mathscr{A}_L/\pi_L\mathscr{A}_L$, for any $j \geq 1$, we have $(1 + \pi_L\alpha)^{p^m} = 1 + \pi_L^m\beta$ for some $\beta \in \mathscr{A}_L$. It follows that $p^m \frac{d(1+\pi_L\alpha)}{1+\pi_L\alpha} = \frac{d(1+\pi_L^m\beta)}{1+\pi_L^m\beta} = d(\log(1 + \pi_L^m\beta))$ and therefore that $\text{Res}_Z(\frac{d(1+\pi_L\alpha)}{1+\pi_L\alpha}) = 0$. □

Since Ω^1 is a topological $\mathscr{A}_L$-module it follows that the residue pairing

$$\begin{aligned} \mathscr{A}_L \times \Omega^1 &\longrightarrow o_L \\ (f, \omega) &\longmapsto \text{Res}(f\omega) \end{aligned} \tag{12}$$

is jointly continuous. It induces, for any $m \geq 1$, the continuous pairing

$$\begin{aligned} \mathscr{A}_L/\pi_L^m\mathscr{A}_L \times \Omega^1/\pi_L^m\Omega^1 &\longrightarrow L/o_L \\ (f, \omega) &\longmapsto \pi_L^{-m}\text{Res}(f\omega) \bmod o_L \end{aligned}$$

and hence (cf. [4] X.28 Theorem 3) the continuous o_L-linear map

$$\begin{aligned} \Omega^1/\pi_L^m\Omega^1 &\longrightarrow \mathrm{Hom}^c_{o_L}(\mathscr{A}_L/\pi_L^m\mathscr{A}_L, L/o_L) \\ \omega &\longmapsto [f \mapsto \pi_L^{-m}\mathrm{Res}(f\omega) \bmod o_L] , \end{aligned} \tag{13}$$

where $\mathrm{Hom}^c_{o_L}$ denotes the module of continuous o_L-linear maps equipped with the compact-open topology. For the convenience of the reader we recall the following well known fact.

Lemma 3.5 *The map* (13) *is an isomorphism of topological o_L-modules.*

Proof Let $R := o_L/\pi_L^m o_L$. It is convenient to view the map in question as the map

$$\begin{aligned} R((Z))dZ &\longrightarrow \mathrm{Hom}^c_R(R((Z)), R) \\ \omega &\longmapsto \ell_\omega(f) := \mathrm{Res}(f\omega) . \end{aligned}$$

One easily checks that $\omega = \sum_i \ell_\omega(Z^{-i-1})Z^i dZ$. Hence injectivity is clear. If ℓ is an arbitrary element in the right hand side we put $\omega := \sum_i \ell(Z^{-i-1})Z^i dZ$. The continuity of ℓ guarantees that $\ell(Z^i) = 0$ for any sufficiently big i. Hence ω is a well defined preimage of ℓ in the left hand side. Finally, the map is open since

$$\{f \in R((Z)) : \mathrm{Res}(fZ^n R[[Z]]dZ) = 0\} = Z^{-n}R[[Z]]$$

is compact for any $n \geq 1$. □

For an arbitrary $\mathscr{A}_L$-module N we have the adjunction isomorphism

$$\begin{aligned} \mathrm{Hom}_{\mathscr{A}_L}(N, \mathrm{Hom}_{o_L}(\mathscr{A}_L, L/o_L)) &\xrightarrow{\cong} \mathrm{Hom}_{o_L}(N, L/o_L) \\ F &\longmapsto F(.)(1) . \end{aligned} \tag{14}$$

Lemma 3.6 *For any finitely generated $\mathscr{A}_L/\pi_L^m\mathscr{A}_L$-module M the adjunction* (14) *together with* (13) *induces the topological isomorphism*

$$\begin{aligned} \mathrm{Hom}_{\mathscr{A}_L}(M, \Omega^1/\pi_L^m\Omega^1) &\xrightarrow{\cong} \mathrm{Hom}^c_{o_L}(M, L/o_L) \\ F &\longmapsto \pi_L^{-m}\mathrm{Res}(F(.)) \bmod o_L . \end{aligned}$$

Proof It is clear that (14) restricts to an injective homomorphism

$$\mathrm{Hom}_{\mathscr{A}_L}(M, \mathrm{Hom}^c_{o_L}(\mathscr{A}_L/\pi_L^m\mathscr{A}_L, L/o_L)) \longrightarrow \mathrm{Hom}^c_{o_L}(M, L/o_L) . \tag{15}$$

The inverse of (14) sends $\ell \in \mathrm{Hom}_{o_L}(M, L/o_L)$ to $F(m)(f) := \ell(fm)$ and visibly restricts to an inverse of (15). By inserting (13) we obtain the asserted algebraic isomorphism. To check that it also is a homeomorphism we first clarify that on the left hand side we consider the weak topology of $\mathrm{Hom}_{\mathscr{A}_L}(M, \Omega^1/\pi_L^m\Omega^1)$ as a finitely generated $\mathscr{A}_L$-module. The elementary divisor theorem for the discrete valuation ring $\mathscr{A}_L$ implies that M is isomorphic to a finite direct product of modules of the form $\mathscr{A}_L/\pi_L^n\mathscr{A}_L$ with $1 \le n \le m$. It therefore suffices to consider the case $M = \mathscr{A}_L/\pi_L^n\mathscr{A}_L$. We then have the commutative diagram of isomorphisms

$$\begin{array}{ccc}
\mathrm{Hom}_{\mathscr{A}_L}(\mathscr{A}_L/\pi_L^n\mathscr{A}_L, \Omega^1/\pi_L^m\Omega^1) & & \\
{\scriptstyle =}\big\downarrow & \searrow & \\
\pi_L^{m-n}\Omega^1/\pi_L^m\Omega^1 & & \mathrm{Hom}^c_{o_L}(\mathscr{A}_L/\pi_L^n\mathscr{A}_L, L/o_L). \\
{\scriptstyle \pi_L^{m-n}\cdot}\big\uparrow & \nearrow_{(13)} & \\
\Omega^1/\pi_L^n\Omega^1 & &
\end{array}$$

By Lemma 3.5 all maps in this diagram except possibly the upper oblique arrow, which is the map in the assertion, are homeomorphisms. Hence the oblique arrow must be a homeomorphism as well. □

The Γ_L-action (3) on $o_L[[Z]]$ extends, by the same formula, to a Γ_L-action on $\mathscr{A}_L$ which, moreover, is continuous for the weak topology (see [31] Proposition 1.7.8.ii).

Definition 3.7 A (φ_L, Γ_L)-module M (over $\mathscr{A}_L$) is a finitely generated $\mathscr{A}_L$-module M together with

- a Γ_L-action on M by semilinear automorphisms which is continuous for the weak topology, and
- a φ_L-linear endomorphism φ_M of M which commutes with the Γ_L-action.

It is called etale if the linearized map

$$\begin{aligned}
\varphi_M^{lin} : \mathscr{A}_L \otimes_{\mathscr{A}_L,\varphi_L} M &\xrightarrow{\cong} M \\
f \otimes m &\longmapsto f\varphi_M(m)
\end{aligned}$$

is bijective. We let $\mathfrak{M}^{et}(\mathscr{A}_L)$ denote the category of etale (φ_L, Γ_L)-modules M over $\mathscr{A}_L$.

Remark 3.8 Let $\alpha : \mathscr{A}_L \longrightarrow \mathscr{A}_L$ be a continuous ring homomorphism, and let $\beta : M \longrightarrow M$ be any α-linear endomorphism of a finitely generated $\mathscr{A}_L$-module M; then β is continuous for the weak topology on M.

Proof The map

$$\begin{aligned}
\beta^{lin} : \mathscr{A}_L \otimes_{\mathscr{A}_L,\alpha} M &\longrightarrow M \\
f \otimes m &\longmapsto f\beta(m)
\end{aligned}$$

is $\mathscr{A}_L$-linear. We pick a free presentation $\lambda : \mathscr{A}_L^n \twoheadrightarrow M$. Then we find an $\mathscr{A}_L$-linear map $\widetilde{\beta}$ such that the diagram

$$\begin{array}{ccccc} \mathscr{A}_L^n & \xrightarrow{\alpha^n} & \mathscr{A}_L^n = \mathscr{A}_L \otimes_{\mathscr{A}_L,\alpha} \mathscr{A}_L^n & \xrightarrow{\widetilde{\beta}} & \mathscr{A}_L^n \\ \downarrow{\scriptstyle \lambda} & & \downarrow{\scriptstyle \mathrm{id}\otimes\lambda} & & \downarrow{\scriptstyle \lambda} \\ M & \xrightarrow{m \mapsto 1\otimes m} & \mathscr{A}_L \otimes_{\mathscr{A}_L,\alpha} M & \xrightarrow{\beta^{lin}} & M \end{array}$$

$$M \xrightarrow{\ \beta\ } M$$

is commutative. All maps except possibly the lower left horizontal arrow are continuous. The universal property of the quotient topology then implies that β must be continuous as well. □

Remarks 3.3 and 3.8 imply that the endomorphism φ_M of a (φ_L, Γ_L)-module M is continuous.

On any etale (φ_L, Γ_L)-module M we have the o_L-linear endomorphism

$$\begin{aligned} \psi_M : M \xrightarrow{(\varphi_M^{lin})^{-1}} \mathscr{A}_L \otimes_{\mathscr{A}_L,\varphi_L} M &\longrightarrow M \\ f \otimes m &\longmapsto \psi_L(f)m \ , \end{aligned}$$

which, by construction, satisfies the projection formulas

$$\psi_M(\varphi_L(f)m) = f\psi_M(m) \qquad \text{and} \qquad \psi_M(f\varphi_M(m)) = \psi_L(f)m \ ,$$

for any $f \in \mathscr{A}_L$ and $m \in M$, as well as the formula

$$\psi_M \circ \varphi_M = \frac{q}{\pi} \cdot \mathrm{id}_M \ .$$

Remark 3.3 is easily seen to imply that ψ_M is continuous for the weak topology.

For technical purposes later on we need to adapt part of Colmez's theory of treillis to our situation. We will do this in the following setting. Let M be a finitely generated $\mathscr{A}_L$-module (always equipped with its weak topology) such that $\pi_L^n M = 0$ for some $n \geq 1$; we also assume that M is equipped with a φ_L-linear endomorphism φ_M which is etale, i.e., such that φ_M^{lin} is bijective.

Definition 3.9 A treillis N in M is an $o_L[[Z]]$-submodule $N \subseteq M$ which is compact and such that its image in $M/\pi_L M$ generates this $k_L((Z))$-vector space.

Remark 3.10 i. If $e_1, \ldots, e_d$ are $\mathscr{A}_L$-generators of M then $o_L[[Z]]e_1 + \cdots + o_L[[Z]]e_d$ is a treillis in M.

ii. A compact $o_L[[Z]]$-submodule N of M is a treillis if and only if it is open.

iii. For any two treillis $N_0 \subseteq N_1$ in M the quotient N_1/N_0 is finite; in particular, any intermediate $o_L[[Z]]$-submodule $N_0 \subseteq N \subseteq N_1$ is a treillis as well.

Proof Part i. is obvious from the compactness of $o_L[[Z]]$. For ii. and iii. see [11] Proposition I.1.2(i). □

Following Colmez we define

$$M^{++} := \{m \in M : \varphi_M^i(m) \xrightarrow{i\to\infty} 0\}.$$

Since $o_L[[Z]]$ is compact it is easily seen that M^{++} is an $o_L[[Z]]$-submodule of M. Obviously M^{++} is φ_M-invariant.

Lemma 3.11 *i. M^{++} is a treillis.*
ii. $\varphi_M - 1$ is an automorphism of M^{++}.

Proof i. Using Remark 3.10.i/iii this follows from [11] Lemma II.2.3. This lemma is stated and proved there in the cyclotomic situation. But the only property of φ_L, besides being etale, which is used is that $\varphi_L(Z) \in Z^2 o_L[[Z]] + \pi_L Z o_L[[Z]]$.

ii. Obviously $m = 0$ is the only element in M^{++} which satisfies $\varphi_M(m) = m$. Now let $m \in M^{++}$ be an arbitrary element. Since M is complete the series $m' := \sum_i \varphi_M^i(m)$ converges and satisfies $(\varphi_M - 1)(-m') = m$. But M^{++} is open and hence closed in M so that $-m' \in M^{++}$. □

The following lemma is a slight generalization of a result of Fontaine (cf. [20] Proposition 2.4.1).

Lemma 3.12 *On any etale (φ_L, Γ_L)-module M such that $\pi_L^n M = 0$ for some $n \geq 1$ the map $\varphi_M - 1$ is open and, in particular, is topologically strict.*

Proof As M^{++}, by Lemma 3.11.i and Remark 3.10.ii, is compact and open in M we first see, using Lemma 3.11.ii, that $\varphi_M - 1$ is a homeomorphism on M^{++} and then that $\varphi_M - 1$ is an open map. □

The category $\mathfrak{M}^{et}(\mathscr{A}_L)$ has an internal Hom-functor. For any two modules M and N in $\mathfrak{M}^{et}(\mathscr{A}_L)$ the $\mathscr{A}_L$-module $\mathrm{Hom}_{\mathscr{A}_L}(M, N)$ is finitely generated. It is a (φ_L, Γ_L)-module with respect to

$$\gamma(\alpha) := \gamma \circ \alpha \circ \gamma^{-1} \quad \text{and} \quad \varphi_{\mathrm{Hom}_{\mathscr{A}_L}(M,N)}(\alpha) := \varphi_N^{lin} \circ (\mathrm{id}_{\mathscr{A}_L} \otimes \alpha) \circ (\varphi_M^{lin})^{-1}$$

for any $\gamma \in \Gamma_L$ and any $\alpha \in \mathrm{Hom}_{\mathscr{A}_L}(M, N)$. We need to verify that the Γ_L-action, indeed, is continuous. This is a consequence of the following general facts.

Remark 3.13 For any two finitely generated $\mathscr{A}_L$-modules M and N we have:

i. The weak topology on $\mathrm{Hom}_{\mathscr{A}_L}(M, N)$ coincides with the topology of pointwise convergence.
ii. The bilinear map

$$\begin{aligned} \mathrm{Hom}_{\mathscr{A}_L}(M, N) \times M &\longrightarrow N \\ (\alpha, m) &\longmapsto \alpha(m) \end{aligned}$$

is continuous for the weak topology on all three terms.

Proof Since any finitely generated module over the discrete valuation ring $\mathscr{A}_L$ is a direct sum of modules of the form $\mathscr{A}_L$ or $\mathscr{A}_L/\pi_L^j\mathscr{A}_L$ for some $j \geq 1$, it suffices to consider the case that M and N both are such cyclic modules. In fact, we may even assume that $M = \mathscr{A}_L$ and $N = \mathscr{A}_L =: \mathscr{A}_L/\pi_L^\infty\mathscr{A}_L$ or $N = \mathscr{A}_L/\pi_L^j\mathscr{A}_L$. We then have the isomorphism of $\mathscr{A}_L$-modules

$$\begin{aligned} \mathrm{ev}_1 : \mathrm{Hom}_{\mathscr{A}_L}(\mathscr{A}_L, \mathscr{A}_L/\pi_L^j\mathscr{A}_L) &\xrightarrow{\cong} \mathscr{A}_L/\pi_L^j\mathscr{A}_L \\ \alpha &\longmapsto \alpha(1) . \end{aligned}$$

For i. we have to show that this map is a homeomorphism for the topology of pointwise convergence and the weak topology on the left and right term, respectively. The topology of pointwise convergence is generated by the subsets $C(f, U) := \{\alpha \in \mathrm{Hom}_{\mathscr{A}_L}(\mathscr{A}_L, \mathscr{A}_L/\pi_L^j\mathscr{A}_L) : \alpha(f) \in U\}$, where $f \in \mathscr{A}_L$ and where U runs over open subsets $U \subseteq \mathscr{A}_L/\pi_L^j\mathscr{A}_L$ (for the weak topology). For the open subset $U_f := \{n \in \mathscr{A}_L/\pi_L^j\mathscr{A}_L : fn \in U\}$ we have $C(f, U) = C(1, U_f)$. We see that $\mathrm{ev}_1(C(f, U)) = U_f$.

Using the topological isomorphism ev_1 the bilinear map in ii. becomes the multiplication map $\mathscr{A}_L/\pi_L^j\mathscr{A}_L \times \mathscr{A}_L \longrightarrow \mathscr{A}_L/\pi_L^j\mathscr{A}_L$. It is continuous since, as noted earlier, $\mathscr{A}_L$ is a topological algebra for the weak topology. □

Let $(\gamma_i)_{i\in\mathbb{N}}$ in Γ_L, resp. $(\alpha_i)_{i\in\mathbb{N}}$ in $\mathrm{Hom}_{\mathscr{A}_L}(M, N)$, be a sequence which converges to $\gamma \in \Gamma_L$, resp. to $\alpha \in \mathrm{Hom}_{\mathscr{A}_L}(M, N)$ for the weak topology. We have to show that the sequence $(\gamma_i(\alpha_i))_i$ converges to $\gamma(\alpha)$ for the weak topology. By Remark 3.13.i it, in fact, suffices to check pointwise convergence. Let therefore $m \in M$ be an arbitrary element. As Γ_L acts continuously on M, we have $\lim_{i\to\infty} \gamma_i^{-1}(m) = \gamma^{-1}(m)$. The Remark 3.13.ii then implies that $\lim_{i\to\infty} \alpha_i(\gamma_i^{-1}(m)) = \alpha(\gamma^{-1}(m)$. By the continuity of the Γ_L-action on N we finally obtain that $\lim_{i\to\infty} \gamma_i(\alpha_i(\gamma_i^{-1}(m))) = \gamma(\alpha(\gamma^{-1}(m)))$.

In order to check etaleness we use the linear isomorphisms φ_M^{lin} and φ_N^{lin} to identify $\mathrm{Hom}_{\mathscr{A}_L}(M, N)$ and $\mathrm{Hom}_{\mathscr{A}_L}(\mathscr{A}_L \otimes_{\mathscr{A}_L,\varphi_L} M, \mathscr{A}_L \otimes_{\mathscr{A}_L,\varphi_L} N) = \mathrm{Hom}_{\mathscr{A}_L}(M, \mathscr{A}_L \otimes_{\mathscr{A}_L,\varphi_L} N)$. Then the linearized map $\varphi_{\mathrm{Hom}_{\mathscr{A}_L}(M,N)}^{lin}$ becomes the map

$$\begin{aligned} \mathscr{A}_L \otimes_{\mathscr{A}_L,\varphi_L} \mathrm{Hom}_{\mathscr{A}_L}(M, N) &\longrightarrow \mathrm{Hom}_{\mathscr{A}_L}(M, \mathscr{A}_L \otimes_{\mathscr{A}_L,\varphi_L} N) \\ f \otimes \alpha &\longmapsto [m \mapsto f \otimes \alpha(m)] . \end{aligned}$$

To see that the latter map is bijective we may use, because of the flatness of φ_L as an injective ring homomorphism between discrete valuation rings, a finite presentation of the module M in order to reduce to the case $M = \mathscr{A}_L$, in which the bijectivity is obvious. Hence $\mathrm{Hom}_{\mathscr{A}_L}(M, N)$ is an etale (φ_L, Γ_L)-module (cf. [15] A.1.1.7). One easily checks the validity of the formula

$$\varphi_{\mathrm{Hom}_{\mathscr{A}_L}(M,N)}(\alpha)(\varphi_M(m)) = \varphi_N(\alpha(m)) . \tag{16}$$

As a basic example we point out that Ω^1 naturally is an etale (φ_L, Γ_L)-module via

$$\gamma(dZ) := [\chi_{LT}(\gamma)]'(Z)dZ = d[\chi_{LT}(\gamma)](Z)$$
$$\varphi_{\Omega^1}(dZ) := \pi_L^{-1}[\pi_L]'(Z)dZ = \pi_L^{-1}d[\pi_L](Z) .$$

Note that the congruence $[\pi_L](Z) \equiv \pi_L Z + Z^q \bmod \pi_L$ indeed implies that the derivative $[\pi_L]'(Z)$ is divisible by π_L. The simplest way to see that Ω^1 is etale is to identify it with another obviously etale (φ_L, Γ_L)-module.

If $\chi : \Gamma_L \longrightarrow o_L^\times$ is any continuous character with representation module $W_\chi = o_L t_\chi$ then, for any M in $\mathfrak{M}^{et}(\mathscr{A}_L)$, we have the twisted module $M(\chi)$ in $\mathfrak{M}^{et}(\mathscr{A}_L)$ where $M(\chi) := M \otimes_{o_L} W_\chi$ as $\mathscr{A}_L$-module, $\varphi_{M(\chi)}(m \otimes w) := \varphi_M(m) \otimes w$, and $\gamma|M(\chi)(m \otimes w) := \gamma|M(m) \otimes \gamma|W_\chi(w) = \chi(\gamma) \cdot \gamma|M(m) \otimes w$ for $\gamma \in \Gamma_L$. It follows that $\psi_{M(\chi)}(m \otimes w) = \psi_M(m) \otimes w$. For the character χ_{LT} we take $W_{\chi_{LT}} = T = o_L\eta$ and $W_{\chi_{LT}^{-1}} = T^* = o_L\eta^*$ as representation module, where T^* denotes the o_L-dual with dual basis η^* of η.

Lemma 3.14 *The map*

$$\mathscr{A}_L(\chi_{LT}) \xrightarrow{\cong} \Omega^1$$
$$f \otimes \eta \longmapsto fd\log_{LT} = f g_{LT} dZ$$

is an isomorphism of (φ_L, Γ_L)*-modules.*

Proof Since g_{LT} is a unit in $o_L[[Z]]$ it is immediately clear that the map under consideration is well defined and bijective. The equivariance follows from (2). □

Remark 3.15 For later applications we want to point out that for $\hat{u} \in (o_L((Z))^\times)^{\mathcal{N}=1}$ the differential form $\frac{d\hat{u}}{\hat{u}}$ is ψ_{Ω^1}-invariant: In fact using Lemma 3.14, Remark 3.2.ii, and Lemma 2.5 for the second, fourth, and fifth identity, respectively, we compute

$$\begin{aligned}\psi_{\Omega^1}(\frac{d\hat{u}}{\hat{u}}) &= \psi_{\Omega^1}(\Delta_{LT}(\hat{u})d\log_{LT}) = \psi_{\Omega^1}(\Delta_{LT}(\hat{u})\varphi_{\Omega^1}(d\log_{LT}))\\ &= \psi_L(\Delta_{LT}(\hat{u}))d\log_{LT} = \pi_L^{-1}\psi_{Col}(\Delta_{LT}(\hat{u}))d\log_{LT}\\ &= \Delta_{LT}(\hat{u})d\log_{LT} = \frac{d\hat{u}}{\hat{u}} .\end{aligned}$$

Lemma 3.16 *The map* $d : \mathscr{A}_L \longrightarrow \Omega^1$ *satisfies:*

i. $\pi_L \cdot \varphi_{\Omega^1} \circ d = d \circ \varphi_L$;
ii. $\gamma \circ d = d \circ \gamma$ *for any* $\gamma \in \Gamma_L$;
iii. $\pi_L^{-1} \cdot \psi_{\Omega^1} \circ d = d \circ \psi_L$.

Proof i. For $f \in \mathscr{A}_L$ we compute

$$\begin{aligned}\varphi_{\Omega^1}(df) &= \varphi_{\Omega^1}(f'dZ)\\ &= \pi_L^{-1}f'([\pi_L](Z))[\pi_L]'(Z)dZ\\ &= \pi_L^{-1}d(f([\pi_L](Z)))\\ &= \pi_L^{-1}d(\varphi_L(f)) .\end{aligned}$$

ii. The computation is completely analogous to the one for i.

iii. Since φ_{Ω^1} is injective, the asserted identity is equivalent to $\varphi_{\Omega^1} \circ \psi_{\Omega^1} \circ d = d \circ \varphi_L \circ \psi_L$ by i. Lemma 3.14 implies that $(\varphi_L \circ \psi_L(f))g_{LT}dZ = \varphi_{\Omega^1} \circ \psi_{\Omega^1}(fg_{LT}dZ)$. Using this, (1), and Remark 3.2.iii in the second, first and fourth, and third identity, respectively, we compute

$$\begin{aligned}\varphi_{\Omega^1} \circ \psi_{\Omega^1}(df) &= \varphi_{\Omega^1} \circ \psi_{\Omega^1}(\partial_{\mathrm{inv}}(f)g_{LT}dZ)\\ &= (\varphi_L \circ \psi_L(\partial_{\mathrm{inv}}(f))g_{LT}dZ = \partial_{\mathrm{inv}}(\varphi_L \circ \psi_L(f))g_{LT}dZ\\ &= d(\varphi_L \circ \psi_L(f)) .\end{aligned}$$

□

Proposition 3.17 *The residue map* $\mathrm{Res} : \Omega^1 \longrightarrow L$ *satisfies:*

i. $\mathrm{Res} \circ \varphi_{\Omega^1} = \pi_L^{-1}q \cdot \mathrm{Res}$*;*
ii. $\mathrm{Res} \circ \gamma = \mathrm{Res}$ *for any* $\gamma \in \Gamma_L$*;*
iii. $\mathrm{Res} \circ \psi_{\Omega^1} = \mathrm{Res}$.

Proof Of course, exact differential forms have zero residue. Let now α denote any of the endomorphisms φ_{Ω^1}, γ, or ψ_{Ω^1} of Ω^1. Using Lemma 3.16 we have $(m+1)\mathrm{Res}(\alpha(Z^m dZ)) = \mathrm{Res}(\alpha(d(Z^{m+1}))) = \pi_L^{\epsilon}\mathrm{Res}(d\alpha(Z^{m+1})) = 0$ with $\epsilon \in \{-1, 0, 1\}$ and hence $\mathrm{Res}(\alpha(Z^m dZ)) = 0$ for any $m \neq -1$. Since Res is continuous it follows that α preserves the kernel of Res. This reduces us to showing the asserted identities on the differential form $Z^{-1}dZ$. In other words we have to check that $\mathrm{Res}(\alpha(Z^{-1}dZ)) = \pi_L^{-1}q$, 1, 1, respectively, in the three cases.

i. We have $\varphi_{\Omega^1}(Z^{-1}dZ) = \pi_L^{-1}\frac{[\pi_L]'(Z)}{[\pi_L](Z)}dZ$. But $[\pi_L](Z) = Z^q(1 + \pi_L v(Z))$ with $v \in \mathscr{A}_L$. Hence $\varphi_{\Omega^1}(Z^{-1}dZ) = \pi_L^{-1}qZ^{-1}dZ + \pi_L^{-1}\frac{d(1+\pi_L v)}{1+\pi_L v}$. In the proof of Remark 3.4.ii we have seen that $\frac{d(1+\pi_L v)}{1+\pi_L v}$ has zero residue.

ii. Here we have $\gamma(Z^{-1}dZ) = \frac{[\chi_{LT}(\gamma)]'(Z)}{[\chi_{LT}(\gamma)](Z)}dZ$. But $[\chi_{LT}(\gamma)](Z) = Zu(Z)$ with a unit $u \in o_L[[Z]]^\times$. It follows that $\gamma(Z^{-1}dZ) = Z^{-1}dZ + \frac{u'}{u}dZ$. The second summand has zero residue, of course.

iii. The identity in i. implies that $\mathrm{Res} \circ \varphi_{\Omega^1} = \pi_L^{-1}q \cdot \mathrm{Res} = \mathrm{Res} \circ \psi_{\Omega^1} \circ \varphi_{\Omega^1}$. Hence the identity in iii. holds on the image of φ_{Ω^1} (as well as on the kernel of Res). But in the course of the proof of i. we have seen that $Z^{-1}dZ \in \mathrm{im}(\varphi_{\Omega^1}) + \ker(\mathrm{Res})$. □

Corollary 3.18 *The residue pairing satisfies*

$$\mathrm{Res}(f\psi_{\Omega^1}(\omega)) = \mathrm{Res}(\varphi_L(f)\omega) \quad \textit{for any } f \in \mathscr{A}_L \textit{ and } \omega \in \Omega^1.$$

Proof By the projection formula the left hand side of the asserted equality is equal to $\mathrm{Res}(\psi_{\Omega^1}(\varphi_L(f)\omega))$. Hence the assertion follows from Proposition 3.17.iii. □

For a finitely generated $\mathscr{A}_L/\pi_L^n\mathscr{A}_L$-module M the isomorphism in Lemma 3.6 induces the following pairing

$$\begin{aligned}[\ ,\ ] = [\ ,\ ]_M : M \times \mathrm{Hom}_{\mathscr{A}_L}(M, \Omega^1/\pi_L^n\Omega^1) &\longrightarrow L/o_L \\ (m, F) &\longmapsto \pi_L^{-n}\mathrm{Res}(F(m)) \bmod o_L .\end{aligned}$$

Since M is locally compact it is (jointly) continuous by [4] X.28 Theorem 3. Note that this pairing (and hence also the isomorphism in Lemma 3.6) is Γ_L-invariant by Proposition 3.17.ii.

The map $\mathrm{Hom}_{\mathscr{A}_L}(\mathscr{A}_L/\pi_L^n\mathscr{A}_L, \Omega^1/\pi_L^n\Omega^1) \xrightarrow{\cong} \Omega^1/\pi_L^n\Omega^1$ which sends F to $F(1)$ is an isomorphism of (etale) (φ_L, Γ_L)-modules. Corollary 3.18 then implies that

$$[\varphi_{\mathscr{A}_L/\pi_L^n\mathscr{A}_L}(f), F]_{\mathscr{A}_L/\pi_L^n\mathscr{A}_L} = [f, \psi_{\mathrm{Hom}_{\mathscr{A}_L}(\mathscr{A}_L/\pi_L^n\mathscr{A}_L, \Omega^1/\pi_L^n\Omega^1)}(F)]_{\mathscr{A}_L/\pi_L^n\mathscr{A}_L}$$

for all $f \in \mathscr{A}_L/\pi_L^n\mathscr{A}_L$ and $F \in \mathrm{Hom}_{\mathscr{A}_L}(\mathscr{A}_L/\pi_L^n\mathscr{A}_L, \Omega^1/\pi_L^n\Omega^1)$. More generally, we show:

Proposition 3.19 *Let M be an etale (φ_L, Γ_L)-module such that $\pi_L^n M = 0$ for some $n \geq 1$; we have:*

i. The operator ψ_M is left adjoint to $\varphi_{\mathrm{Hom}_{\mathscr{A}_L}(M, \Omega^1/\pi_L^n\Omega^1)}$ under the pairing $[\ ,\ ]$, i.e.,

$$[\psi_M(m), F] = [m, \varphi_{\mathrm{Hom}_{\mathscr{A}_L}(M, \Omega^1/\pi_L^n\Omega^1)}(F)]$$

for all $m \in M$ and all $F \in \mathrm{Hom}_{\mathscr{A}_L}(M, \Omega^1/\pi_L^n\Omega^1)$;

ii. the operator φ_M is left adjoint to $\psi_{\mathrm{Hom}_{\mathscr{A}_L}(M, \Omega^1/\pi_L^n\Omega^1)}$ under the pairing $[\ ,\ ]$, i.e.,

$$[\varphi_M(m), F] = [m, \psi_{\mathrm{Hom}_{\mathscr{A}_L}(M, \Omega^1/\pi_L^n\Omega^1)}(F)]$$

for all $m \in M$ and all $F \in \mathrm{Hom}_{\mathscr{A}_L}(M, \Omega^1/\pi_L^n\Omega^1)$.

Proof For notational simplicity we abbreviate the subscript $\mathrm{Hom}_{\mathscr{A}_L}(M, \Omega^1/\pi_L^n\Omega^1)$ to Hom.

i. Since M is etale it suffices to check the asserted identity on elements of the form $f\varphi_M(m)$ with $f \in \mathscr{A}_L$ and $m \in M$. By the projection formula for ψ_M the left hand side then becomes $[\psi_L(f)m, F]$. We compute the right hand side:

$$\begin{aligned}
[f\varphi_M(m), \varphi_{\mathrm{Hom}}(F)] &\equiv \pi_L^{-n}\mathrm{Res}(\varphi_{\mathrm{Hom}}(F)(f\varphi_M(m))) \\
&\equiv \pi_L^{-n}\mathrm{Res}(f\varphi_{\mathrm{Hom}}(F)(\varphi_M(m))) \\
&\equiv \pi_L^{-n}\mathrm{Res}(f\varphi_{\Omega^1/\pi_L^n\Omega^1}(F(m))) \\
&\equiv \pi_L^{-n}\mathrm{Res}(\psi_{\Omega^1/\pi_L^n\Omega^1}(f\varphi_{\Omega^1/\pi_L^n\Omega^1}(F(m)))) \\
&\equiv \pi_L^{-n}\mathrm{Res}(\psi_L(f)F(m)) \\
&\equiv \pi_L^{-n}\mathrm{Res}(F(\psi_L(f)m)) \\
&\equiv [\psi_L(f)m, F] \mod o_L ;
\end{aligned}$$

here the first and last identities are just the definition of [,], the second and sixth use $\mathscr{A}_L$-linearity, the third the formula (16), the fourth Proposition 3.17 iii., and the fifth the projection formula for $\psi_{\Omega^1/\pi_L^n\Omega^1}$.

ii. Correspondingly we compute

$$\begin{aligned}
[\varphi_M(m), f\varphi_{\mathrm{Hom}}(F)] &\equiv \pi_L^{-n}\mathrm{Res}(f\varphi_{\mathrm{Hom}}(F)(\varphi_M(m))) \\
&\equiv \pi_L^{-n}\mathrm{Res}(f\varphi_{\Omega^1/\pi_L^n\Omega^1}(F(m))) \\
&\equiv \pi_L^{-n}\mathrm{Res}(\psi_{\Omega^1/\pi_L^n\Omega^1}(f\varphi_{\Omega^1/\pi_L^n\Omega^1}(F(m)))) \\
&\equiv \pi_L^{-n}\mathrm{Res}(\psi_L(f)F(m)) \\
&\equiv \pi_L^{-n}\mathrm{Res}((\psi_{\mathrm{Hom}}(f\varphi_{\mathrm{Hom}}(F)))(m)) \\
&\equiv [m, \psi_{\mathrm{Hom}}(f\varphi_{\mathrm{Hom}}(F))] \mod o_L .
\end{aligned}$$

□

Remark 3.20 Similarly, for any etale (φ_L, Γ_L)-module M such that $\pi_L^n M = 0$ one can consider the Γ_L-invariant (jointly) continuous pairing

$$\begin{aligned}
[\,,\,\rangle = [\,,\,\rangle_M : M \times \mathrm{Hom}_{\mathscr{A}_L}(M, \mathscr{A}_L(\chi_{LT})/\pi_L^n\mathscr{A}_L(\chi_{LT})) &\longrightarrow L/o_L \\
(m, F\otimes\eta) &\longmapsto \pi_L^{-n}\mathrm{Res}(F(m)d\log_{LT}) \bmod o_L
\end{aligned}$$

which arises from $[\,,\,]_M$ by plugging in the isomorphism from Lemma 3.14. Clearly, it has adjointness properties analogous to the ones in Proposition 3.19.

4 The Kisin–Ren Equivalence

Let $\widetilde{\mathbf{E}}^+ := \varprojlim o_{\mathbb{C}_p}/po_{\mathbb{C}_p}$ with the transition maps being given by the Frobenius $\phi(a) = a^p$. We may also identify $\widetilde{\mathbf{E}}^+$ with $\varprojlim o_{\mathbb{C}_p}/\pi_L o_{\mathbb{C}_p}$ with the transition maps being given by the q-Frobenius $\phi_q(a) = a^q$. Recall that $\widetilde{\mathbf{E}}^+$ is a complete valuation ring with residue field $\overline{\mathbb{F}_p}$ and its field of fractions $\widetilde{\mathbf{E}} = \varprojlim \mathbb{C}_p$ being algebraically closed of characteristic p. Let $\mathfrak{m}_{\widetilde{\mathbf{E}}}$ denote the maximal ideal in $\widetilde{\mathbf{E}}^+$.

The q-Frobenius ϕ_q first extends by functoriality to the rings of the Witt vectors $W(\widetilde{\mathbf{E}}^+) \subseteq W(\widetilde{\mathbf{E}})$ and then o_L-linearly to $W(\widetilde{\mathbf{E}}^+)_L := W(\widetilde{\mathbf{E}}^+) \otimes_{o_{L_0}} o_L \subseteq W(\widetilde{\mathbf{E}})_L := W(\widetilde{\mathbf{E}}) \otimes_{o_{L_0}} o_L$, where L_0 is the maximal unramified subextension of L. The Galois group G_L obviously acts on $\widetilde{\mathbf{E}}$ and $W(\widetilde{\mathbf{E}})_L$ by automorphisms commuting with ϕ_q. This G_L-action is continuous for the weak topology on $W(\widetilde{\mathbf{E}})_L$ (cf. [31] Lemma 1.5.3). Let $\mathbb{M}_L$ denote the ideal in $W(\widetilde{\mathbf{E}}^+)_L$ which is the preimage of $\mathfrak{m}_{\widetilde{\mathbf{E}}}$ under the residue class map.

Evaluation of the global coordinate Z of LT at π_L-power torsion points induces a map (not a homomorphism of abelian groups) $\iota : T \longrightarrow \widetilde{\mathbf{E}}^+$. Namely, if $t = (z_n)_{n\geq 1} \in T$ with $[\pi_L](z_{n+1}) = z_n$ and $[\pi_L](z_1) = 0$, then $z_{n+1}^q \equiv z_n \bmod \pi_L$ and hence $\iota(t) := (z_n \bmod \pi_L)_n \in \widetilde{\mathbf{E}}^+$.

Lemma 4.1 *The image of the map ι is contained in $\mathfrak{m}_{\widetilde{\mathbf{E}}}$. The map*

$$\iota_{LT} : T \longrightarrow \mathbb{M}_L$$
$$t \longmapsto \lim_{n\to\infty} ([\pi_L] \circ \phi_q^{-1})^n([\iota(t)])\ ,$$

where $[\iota(t)]$ denotes the Teichmüller representative of $\iota(t)$, is well defined and satisfies:

1. $[a](\iota_{LT}(t)) = \iota_{LT}(at)$ *for any* $a \in o_L$;
2. $\phi_q(\iota_{LT}(t)) = \iota_{LT}(\pi_L t) = [\pi_L](\iota_{LT}(t))$;
3. $\sigma(\iota_{LT}(t)) = [\chi_{LT}(\sigma)](\iota_{LT}(t))$ *for any* $\sigma \in G_L$.

Proof This is [24] Lemma 1.2, which refers to [9] Lemma 9.3. For full details see [31] §2.1. □

As before we fix an o_L-generator η of T and put $\omega_{LT} := \iota_{LT}(\eta)$. By sending Z to ω_{LT} we obtain an embedding of rings

$$o_L[[Z]] \longrightarrow W(\widetilde{\mathbf{E}}^+)_L\ .$$

As explained in [24] (1.3) it extends to embeddings of rings

$$\mathscr{A}_L \longrightarrow W(\widetilde{\mathbf{E}})_L \quad \text{and} \quad \mathscr{B}_L \longrightarrow L \otimes_{o_L} W(\widetilde{\mathbf{E}})_L\ .$$

The left map, in fact, is a topological embedding for the weak topologies on both sides ([31] Proposition 2.1.16.i). The Galois group G_L acts through its quotient Γ_L on $\mathscr{B}_L$ by $(\sigma, f) \longmapsto f([\chi_{LT}(\sigma)](Z))$. Then, by Lemma 4.1.c, the above embeddings are G_L-equivariant. Moreover, the q-Frobenius ϕ_q on $L \otimes_{o_L} W(\widetilde{\mathbf{E}})_L$, by Lemma 4.1.b, restricts to the endomorphism $f \longmapsto f \circ [\pi_L]$ of $\mathscr{B}_L$ which we earlier denoted by φ_L.

We define $\mathbf{A}_L$ to be the image of $\mathscr{A}_L$ in $W(\widetilde{\mathbf{E}})_L$. It is a complete discrete valuation ring with prime element π_L and residue field the image $\mathbf{E}_L$ of $k_L((Z)) \hookrightarrow \widetilde{\mathbf{E}}$. As a consequence of Lemma 4.1.a this subring $\mathbf{A}_L$ is independent of the choice of η. As explained above each choice of η gives rise to an isomorphism

$$(\mathscr{A}_L, \varphi_L, \Gamma_L, \text{weak topology}) \xrightarrow{\cong} (\mathbf{A}_L, \phi_q, \Gamma_L, \text{weak topology}) \tag{17}$$

between the o_L-algebras $\mathscr{A}_L$ and $\mathbf{A}_L$ together with their additional structures. By literally repeating the Definition 3.7 we have the notion of (etale) (ϕ_q, Γ_L)-modules over $\mathbf{A}_L$ as well as the category $\mathfrak{M}^{et}(\mathbf{A}_L)$. In the same way as for Remark 3.2.i we may define the operator ψ_L on $\mathbf{A}_L$ and then on any etale (ϕ_q, Γ_L)-module over $\mathbf{A}_L$. The above algebra isomorphism gives rise to an equivalence of categories

$$\mathfrak{M}^{et}(\mathscr{A}_L) \xrightarrow{\simeq} \mathfrak{M}^{et}(\mathbf{A}_L) , \tag{18}$$

which also respects the ψ_L-operators. Using the norm map for the extension $\mathbf{A}_L/\phi_q(\mathbf{A}_L)$ we define, completely analogously as in (8), a multiplicative norm operator $\mathcal{N} : \mathbf{A}_L \longrightarrow \mathbf{A}_L$. Then, using also Lemma 4.1.b, $\mathcal{N}_L$ and $\mathcal{N}$ correspond to each other under the isomorphism (17). In particular, (17) (for any choice of η) induces an isomorphism

$$(\mathscr{A}_L^\times)^{\mathcal{N}_L=1} \xrightarrow{\cong} (\mathbf{A}_L^\times)^{\mathcal{N}=1} . \tag{19}$$

We form the maximal integral unramified extension (= strict Henselization) of $\mathbf{A}_L$ inside $W(\widetilde{\mathbf{E}})_L$. Its p-adic completion $\mathbf{A}$ still is contained in $W(\widetilde{\mathbf{E}})_L$. Note that $\mathbf{A}$ is a complete discrete valuation ring with prime element π_L and residue field the separable algebraic closure $\mathbf{E}_L^{sep}$ of $\mathbf{E}_L$ in $\widetilde{\mathbf{E}}$. By the functoriality properties of strict Henselizations the q-Frobenius ϕ_q preserves $\mathbf{A}$. According to [24] Lemma 1.4 the G_L-action on $W(\widetilde{\mathbf{E}})_L$ respects $\mathbf{A}$ and induces an isomorphism $H_L = \ker(\chi_{LT}) \xrightarrow{\cong} \mathrm{Aut}^{cont}(\mathbf{A}/\mathbf{A}_L)$.

Let $\mathrm{Rep}_{o_L}(G_L)$ denote the abelian category of finitely generated o_L-modules equipped with a continuous linear G_L-action. The following result is established in [24] Theorem 1.6.

Theorem 4.2 *The functors*

$$V \longmapsto D_{LT}(V) := (\mathbf{A} \otimes_{o_L} V)^{\ker(\chi_{LT})} \quad \text{and} \quad M \longmapsto (\mathbf{A} \otimes_{\mathbf{A}_L} M)^{\phi_q \otimes \varphi_M = 1}$$

are exact quasi-inverse equivalences of categories between $\mathrm{Rep}_{o_L}(G_L)$ *and* $\mathfrak{M}^{et}(\mathbf{A}_L)$.

For the convenience of the reader we discuss a few properties, which will be used later on, of the functors in the above theorem.

First of all we recall that the tensor product $M \otimes_{o_L} N$ of two linear-topological o_L-modules M and N is equipped with the linear topology for which the o_L-submodules

$$\mathrm{im}(U_M \otimes_{o_L} N \to M \otimes_{o_L} N) + \mathrm{im}(M \otimes_{o_L} U_N \to M \otimes_{o_L} N) \subseteq M \otimes_{o_L} N ,$$

where U_M and U_N run over the open submodules of M and N, respectively, form a fundamental system of open neighbourhoods of zero. One checks that, if a profinite group H acts linearly and continuously on M and N, then its diagonal action on $M \otimes_{o_L} N$ is continuous as well.

In our situation we consider $M = \mathbf{A}$ with its weak topology induced by the weak topology of $W(\widetilde{\mathbf{E}})_L$ and $N = V$ in $\mathrm{Rep}_{o_L}(G_L)$ equipped with its π_L-adic topology. The diagonal action of G_L on $\mathbf{A} \otimes_{o_L} V$ then, indeed, is continuous. In addition, since the π_L-adic topology on $\mathbf{A}$ is finer than the weak topology, any open o_L-submodule of $\mathbf{A}$ contains $\pi_L^j \mathbf{A}$ for a sufficiently big j. Hence $\{\mathrm{im}(U \otimes_{o_L} V \to \mathbf{A} \otimes_{o_L} V) : U \subseteq \mathbf{A}$ any open o_L -submodule$\}$ is a fundamental system of open neighbourhoods of zero in $\mathbf{A} \otimes_{o_L} V$. This implies that the tensor product topology on $\mathbf{A} \otimes_{o_L} V$ is nothing else than its weak topology as a finitely generated $\mathbf{A}$-module.

Remark 4.3 For any V in $\mathrm{Rep}_{o_L}(G_L)$ the tensor product topology on $\mathbf{A} \otimes_{o_L} V$ induces the weak topology on $D_{LT}(V)$. In particular, the residual Γ_L-action on $D_{LT}(V)$ is continuous.

Proof The finitely generated $\mathbf{A}_L$-module $D_{LT}(V)$ is of the form $D_{LT}(V) \cong \oplus_{i=1}^r \mathbf{A}_L / \pi_L^{n_i} \mathbf{A}_L$ with $1 \leq n_i \leq \infty$. Using the isomorphism in the subsequent Proposition 4.4.ii we obtain that $\mathbf{A} \otimes_{o_L} V \cong \oplus_{i=1}^r \mathbf{A}/\pi_L^{n_i}\mathbf{A}$. We see that the inclusion $D_{LT}(V) \subseteq \mathbf{A} \otimes_{o_L} V$ is isomorphic to the direct product of the inclusions $\mathbf{A}_L/\pi_L^{n_i}\mathbf{A}_L \subseteq \mathbf{A}/\pi_L^{n_i}\mathbf{A}$, which clearly are compatible with the weak topologies. □

Proposition 4.4 *i. The functor D_{LT} is exact.*

ii. For any V in $\mathrm{Rep}_{o_L}(G_L)$ the natural map $\mathbf{A} \otimes_{\mathbf{A}_L} D_{LT}(V) \xrightarrow{\cong} \mathbf{A} \otimes_{o_L} V$ is an isomorphism (compatible with the G_L-action and the Frobenius on both sides).

Proof We begin with three preliminary observations.

(1) As $\mathbf{A}$ is o_L-torsion free, the functor $\mathbf{A} \otimes_{o_L} -$ is exact.

(2) The functor D_{LT} restricted to the full subcategory of finite length objects V in $\mathrm{Rep}_{o_L}(G_L)$ is exact. This follows immediately from 1) and the vanishing of $H^1(H_L, \mathbf{A} \otimes_{o_L} V)$ in Lemma 5.2 below.

(3) For any V in $\mathrm{Rep}_{o_L}(G_L)$ we have $D_{LT}(V) = \varprojlim_n D_{LT}(V/\pi_L^n V)$. To see this we compute

$$\begin{aligned}
\varprojlim_n D_{LT}(V/\pi_L^n V) &= \varprojlim_n (\mathbf{A} \otimes_{o_L} V/\pi_L^n V)^{H_L} = (\varprojlim_n (\mathbf{A} \otimes_{o_L} V/\pi_L^n V))^{H_L} \\
&= (\mathbf{A} \otimes_{o_L} \varprojlim_n V/\pi_L^n V)^{H_L} = (\mathbf{A} \otimes_{o_L} V)^{H_L} \\
&= D_{LT}(V) .
\end{aligned}$$

Here the third identity becomes obvious if one notes that V as an o_L-module is a finite direct sum of modules of the form $o_L/\pi_L^j o_L$ with $1 \leq j \leq \infty$. In this case $\varprojlim_n (\mathbf{A} \otimes_{o_L} o_L/\pi_L^{j+n} o_L) = \varprojlim_n \mathbf{A}/\pi_L^{j+n}\mathbf{A} = \mathbf{A}/\pi_L^j\mathbf{A} = \mathbf{A} \otimes_{o_L} o_L/\pi_L^j o_L$.

i. Let

$$0 \longrightarrow V_1 \longrightarrow V_2 \longrightarrow V_3 \longrightarrow 0$$

be an exact sequence in $\mathrm{Rep}_{o_L}(G_L)$. By (2) we obtain the exact sequences of projective systems of finitely generated $\mathbf{A}_L$-modules

$$D_{LT}(\ker(\pi_L^n|V_3)) \to D_{LT}(V_1/\pi_L^n V_1) \to D_{LT}(V_2/\pi_L^n V_2) \to D_{LT}(V_3/\pi_L^n V_3) \to 0\,.$$

Since $\mathbf{A}_L$ is a noetherian pseudocompact ring taking projective limits is exact. By (3) the resulting exact sequence is

$$\varprojlim_n D_{LT}(\ker(\pi_L^n|V_3)) \longrightarrow D_{LT}(V_1) \longrightarrow D_{LT}(V_2) \longrightarrow D_{LT}(V_3) \longrightarrow 0\,.$$

But since the torsion subgroup of V_3 is finite and the transition maps in the projective system $(\ker(\pi_L^n|V_3))_n$ are multiplication by π_L, any composite of sufficiently many transition maps in this projective system and hence also in the projective system $(D_{LT}(\ker(\pi_L^n|V_3)))_n$ is zero. It follows that $\varprojlim_n D_{LT}(\ker(\pi_L^n|V_3)) = 0$.

ii. The compatibility properties are obvious from the definition of the map. To show its bijectivity we may assume, by devissage and (3), that $\pi_L V = 0$. In this case our assertion reduces to the bijectivity of the natural map $\mathbf{E}_L^{sep} \otimes_{\mathbf{E}_L} (\mathbf{E}_L^{sep} \otimes_k V)^{\mathrm{Gal}(\mathbf{E}_L^{sep}/\mathbf{E}_L)} \longrightarrow \mathbf{E}_L^{sep} \otimes_k V$. But this is a well known consequence of the vanishing of the Galois cohomology group $H^1(\mathrm{Gal}(\mathbf{E}_L^{sep}/\mathbf{E}_L), \mathrm{GL}_d(\mathbf{E}_L^{sep}))$ where $d := \dim_k V$. □

Lemma 4.5 *The above equivalence of categories D_{LT} is compatible with the formation of inner Hom-objects, i.e., there are canonical isomorphisms*

$$\mathrm{Hom}_{\mathbf{A}_L}(D_{LT}(V_1), D_{LT}(V_2)) = D_{LT}(\mathrm{Hom}_{o_L}(V_1, V_2))$$

for every V_1, V_2 in $\mathrm{Rep}_{o_L}(G_L)$. We also have

$$\psi_{\mathrm{Hom}_{\mathbf{A}_L}(D_{LT}(V_1), D_{LT}(V_2))} = \psi_{D_{LT}(\mathrm{Hom}_{o_L}(V_1,V_2))}\,.$$

Proof We have

$$\begin{aligned}
D_{LT}(\mathrm{Hom}_{o_L}(V_1, V_2)) &= (\mathbf{A} \otimes_{o_L} \mathrm{Hom}_{o_L}(V_1, V_2))^{H_L} \\
&= \mathrm{Hom}_{\mathbf{A}}(\mathbf{A} \otimes_{o_L} V_1, \mathbf{A} \otimes_{o_L} V_2)^{H_L} \\
&= \mathrm{Hom}_{\mathbf{A}}(\mathbf{A} \otimes_{\mathbf{A}_L} D_{LT}(V_1), \mathbf{A} \otimes_{o_L} V_2)^{H_L} \\
&= \mathrm{Hom}_{\mathbf{A}_L}(D_{LT}(V_1), \mathbf{A} \otimes_{o_L} V_2)^{H_L} \\
&= \mathrm{Hom}_{\mathbf{A}_L}(D_{LT}(V_1), (\mathbf{A} \otimes_{\mathbf{A}_L} V_2)^{H_L}) \\
&= \mathrm{Hom}_{\mathbf{A}_L}(D_{LT}(V_1), D_{LT}(V_2))\,.
\end{aligned}$$

Here the second identity is clear for V_1 being free, the general case follows by choosing a finite presentation of V_1 (as o_L-module neglecting the group action). The third identity uses Proposition 4.4.ii, while the fourth one comes from the adjointness of base extension and restriction. The fifth one uses the fact that H_L acts trivially on $D_{LT}(V_1)$.

One easily checks that the above sequence of identities is compatible with the Γ_L-actions (which are induced by the diagonal G_L-action on $\mathbf{A} \otimes_{o_L} -$).

The compatibility with Frobenius can be seen as follows. First of all we abbreviate $\varphi_{D_{LT}(\mathrm{Hom})} := \varphi_{D_{LT}(\mathrm{Hom}_{o_L}(V_1,V_2))}$ and $\varphi_{\mathrm{Hom}} := \varphi_{\mathrm{Hom}_{\mathbf{A}_L}(D_{LT}(V_1),D_{LT}(V_2))}$. An element $\beta = \sum_i a_i \otimes \alpha_i \in (\mathbf{A} \otimes_{o_L} \mathrm{Hom}_{o_L}(V_1, V_2))^{H_L}$ becomes, under the above identifications, the map $\iota_\beta : D_{LT}(V_1) \to (\mathbf{A} \otimes_{o_L} V_2)^{H_L}$ which sends $\sum_j c_j \otimes v_j$ to $\sum_{i,j} a_i c_j \otimes \alpha_i(v_j)$. Assuming that $c_j = \phi_q(c_j')$ we compute

$$\begin{aligned}
\iota_{\varphi_{D_{LT}(\mathrm{Hom})}(\beta)}(\varphi_{D_{LT}(V_1)}(\sum_j c_j' \otimes v_j) &= \iota_{\sum_i \phi_q(a_i)\otimes\alpha_i}(\sum_j \phi_q(c_j') \otimes v_j) \\
&= \sum_{i,j} \phi_q(a_i)\phi_q(c_j') \otimes \alpha_i(v_j) \\
&= \varphi_{D_{LT}(V_2)}(\sum_{i,j} a_i c_j' \otimes \alpha_i(v_j)) \\
&= \varphi_{D_{LT}(V_2)}(\iota_\beta(\sum_j c_j' \otimes v_j)) \\
&= \varphi_{\mathrm{Hom}}(\iota_\beta)(\varphi_{D_{LT}(V_1)}(\sum_j c_j' \otimes v_j)) ,
\end{aligned}$$

where the last identity comes from (16). Using the etaleness of $D_{LT}(V_1)$ we deduce that $\iota_{\varphi_{D_{LT}(\mathrm{Hom})}(\beta)} = \varphi_{\mathrm{Hom}}(\iota_\beta)$ for any $\beta \in D_{LT}(\mathrm{Hom}_{o_L}(V_1, V_2))$. The additional formula for the ψ-operators is a formal consequence of the compatibility of the φ-operators. □

Remark 4.6 For any V in $\mathrm{Rep}_{o_L}(G_L)$ and any continuous character $\chi : \Gamma_L \longrightarrow o_L^\times$ with representation module $W_\chi = o_L t_\chi$ the twisted G_L-representation $V(\chi)$ is defined to be $V(\chi) = V \otimes W_\chi$ as o_L-module and $\sigma_{|V(\chi)}(v \otimes w) = \sigma_{|V}(v) \otimes \sigma_{|W_\chi}(w) = \chi(\sigma) \cdot \sigma_{|V}(v) \otimes w$. One easily checks that $D_{LT}(V(\chi)) = D_{LT}(V)(\chi)$. If $V = o_L/\pi_L^n o_L$, $1 \le n \le \infty$ is the trivial representation, we usually identify $V(\chi)$ and W_χ. Recall that for the character χ_{LT} we take $W_{\chi_{LT}} = T = o_L\eta$ and $W_{\chi_{LT}^{-1}} = T^* = o_L\eta^*$ as representation module, where T^* denotes the o_L-dual with dual basis η^* of η.

Defining $\Omega^1 := \Omega^1_{\mathbf{A}_L} = \mathbf{A}_L d\omega_{LT}$ any choice of η defines an isomorphism $\Omega^1_{\mathscr{A}_L} \cong \Omega^1_{\mathbf{A}_L}$ by sending $f(Z)dZ$ to $f(\omega_{LT})d\omega_{LT}$; moreover $\Omega^1_{\mathscr{A}_L}$ and $\Omega^1_{\mathbf{A}_L}$ correspond to each other under the equivalence of categories (18). Due to the isomorphism (17) we obtain a residue pairing

$$\begin{aligned}
\mathbf{A}_L \times \Omega^1_{\mathbf{A}_L} &\longrightarrow o_L \\
(f(\omega_{LT}), g(\omega_{LT})d\omega_{LT}) &\longmapsto \mathrm{Res}(f(Z)g(Z)dZ)
\end{aligned} \tag{20}$$

which satisfies

$$\mathrm{Res}(f\psi_{\Omega^1}(\omega)) = \mathrm{Res}(\phi_q(f)\omega) \qquad \text{for any } f \in \mathbf{A}_L \text{ and } \omega \in \Omega^1_{\mathbf{A}_L} \tag{21}$$

(by Corollary 3.18) and which is independent of the choice of η, i.e., ω_{LT}, by Remark 3.4.ii and Lemma 4.1.a. In particular, we have a well defined map $\mathrm{Res} : \Omega^1_{\mathbf{A}_L} \to o_L$. In this context Remark 3.15 together with Remark 3.2.vii tell us that

$$\psi_{\Omega^1}(\frac{d\hat{u}}{\hat{u}}) = \frac{d\hat{u}}{\hat{u}} \tag{22}$$

holds true for every $\hat{u} \in (\mathbf{A}_L^\times)^{\mathcal{N}=1}$.

Moreover, Lemma 3.6 translates into the (existence of the) topological isomorphism

$$\begin{aligned} \mathrm{Hom}_{\mathbf{A}_L}(M, \Omega^1_{\mathbf{A}_L}/\pi_L^n\Omega^1_{\mathbf{A}_L}) &\xrightarrow{\cong} \mathrm{Hom}^c_{o_L}(M, L/o_L) \\ F &\longmapsto \pi_L^{-n}\mathrm{Res}(F(.)) \bmod o_L , \end{aligned} \tag{23}$$

for any M annihilated by π_L^n. Lemma 3.14 implies the isomorphism

$$\begin{aligned} \mathbf{A}_L(\chi_{LT}) = \mathbf{A}_L \otimes_{o_L} T &\xrightarrow{\cong} \Omega^1_{\mathbf{A}_L} \\ f(\iota_{LT}(\eta)) \otimes \eta &\longmapsto f(\iota_{LT}(\eta))g_{LT}(\iota_{LT}(\eta))d\iota_{LT}(\eta) . \end{aligned}$$

Using Lemma 4.1.a as well as the second identity in (2) one verifies that this isomorphism (unlike its origin in Lemma 3.14) does not depend on the choice of η. We use it in order to transform (23) into the topological isomorphism

$$\mathrm{Hom}_{\mathbf{A}_L}(M, \mathbf{A}_L/\pi_L^n\mathbf{A}_L(\chi_{LT})) \xrightarrow{\cong} \mathrm{Hom}^c_{o_L}(M, L/o_L). \tag{24}$$

Finally we obtain the following analogues of Proposition 3.19 and Remark 3.20.

Remark 4.7 For any etale (φ_L, Γ_L)-module M such that $\pi_L^n M = 0$ one has the Γ_L-invariant (jointly) continuous pairing

$$\begin{aligned} [\ ,\ \rangle = [\ ,\ \rangle_M : M \times \mathrm{Hom}_{\mathbf{A}_L}(M, \mathbf{A}_L/\pi_L^n\mathbf{A}_L(\chi_{LT})) &\longrightarrow L/o_L \\ (m, F) &\longmapsto \pi_L^{-n}\mathrm{Res}(F(m)d\log_{LT}(\omega_{LT})) \bmod o_L \end{aligned}$$

with adjointness properties analogous to the ones in Proposition 3.19. Again, it is independent of the choice of η.

5 Iwasawa Cohomology

For any V in $\mathrm{Rep}_{o_L}(G_L)$ we also write $H^i(K, V) = H^i(G_K, V)$, for any algebraic extension K of L, and we often abbreviate $\varphi := \varphi_{D_{LT}(V)}$.

Remark 5.1 For any V in $\mathrm{Rep}_{o_L}(G_L)$ the sequence

$$0 \longrightarrow V \xrightarrow{\subseteq} \mathbf{A} \otimes_{o_L} V \xrightarrow{\phi_q \otimes \mathrm{id} - 1} \mathbf{A} \otimes_{o_L} V \longrightarrow 0 \tag{25}$$

is exact.

Proof We clearly have the exact sequence $0 \to k_L \to \mathbf{E}_L^{sep} \xrightarrow{x \mapsto x^q - x} \mathbf{E}_L^{sep} \to 0$. By devissage we deduce the exact sequence $0 \to o_L/\pi_L^n o_L \to \mathbf{A}/\pi_L^n \mathbf{A} \xrightarrow{\phi_q - 1} \mathbf{A}/\pi_L^n \mathbf{A} \to 0$ for any $n \geq 1$. Since the projective system $\{o_L/\pi_L^n o_L\}_n$ has surjective transition maps, passing to the projective limit is exact and gives the exact sequence

$$0 \to o_L \to \mathbf{A} \xrightarrow{\phi_q - 1} \mathbf{A} \to 0. \tag{26}$$

Finally, $\mathbf{A}$ is o_L-torsion free and hence flat over o_L. It follows that tensoring by V is exact. □

Since $\mathbf{A}$ is the π_L-adic completion of an unramified extension of $\mathbf{A}_L$ with Galois group H_L the H_L-action on $\mathbf{A}/\pi_L^n\mathbf{A}$, for any $n \geq 1$, and hence on $\mathbf{A} \otimes_{o_L} V$, whenever $\pi_L^n V = 0$ for some $n \geq 1$, is continuous for the discrete topology. We therefore may, in the latter case, pass from (25) to the associated long exact Galois cohomology sequence with respect to H_L.

Lemma 5.2 *Suppose that $\pi_L^n V = 0$ for some $n \geq 1$; we then have:*

i. $H^i(H_L, \mathbf{A} \otimes_{o_L} V) = 0$ for any $i \geq 1$;
ii. the long exact cohomology sequence for (25) *gives rise to an exact sequence*

$$0 \longrightarrow H^0(L_\infty, V) \longrightarrow D_{LT}(V) \xrightarrow{\varphi - 1} D_{LT}(V) \xrightarrow{\partial_\varphi} H^1(L_\infty, V) \longrightarrow 0\,. \tag{27}$$

Proof i. Since $\mathbf{A} \otimes_{o_L} V \cong \mathbf{A} \otimes_{\mathbf{A}_L} D_{LT}(V)$ by Proposition 4.4.ii, it suffices to show the vanishing of $H^i(H_L, \mathbf{A}/\pi_L^n\mathbf{A}) = 0$ for $i \geq 1$. This reduces, by devissage, to the case $n = 1$, i.e., to $H^i(H_L, \mathbf{E}_L^{sep}) = 0$, which is a standard fact of Galois cohomology. ii. follows immediately from i. □

In order to derive from this a computation of Iwasawa cohomology in terms of (φ_L, Γ_L)-modules we first have to recall Pontrjagin duality and local Tate duality in our setting. The trace pairing

$$\begin{aligned} L \times L &\longrightarrow \mathbb{Q}_p \\ (x, y) &\longmapsto \mathrm{Tr}_{L/\mathbb{Q}_p}(xy) \end{aligned}$$

gives rise to the inverse different

$$\begin{aligned} \mathfrak{D}_{L/\mathbb{Q}_p}^{-1} = \{x \in L : \mathrm{Tr}_{L/\mathbb{Q}_p}(x o_L) \subseteq \mathbb{Z}_p\} &\xrightarrow{\cong} \mathrm{Hom}_{\mathbb{Z}_p}(o_L, \mathbb{Z}_p) \\ y &\longmapsto [x \mapsto \mathrm{Tr}_{L/\mathbb{Q}_p}(yx)]\,. \end{aligned}$$

Let $\mathfrak{D}_{L/\mathbb{Q}_p} = \pi_L^s o_L$. We fix once and for all the o_L-linear isomorphism

$$\begin{aligned} o_L &\xrightarrow{\cong} \mathrm{Hom}_{\mathbb{Z}_p}(o_L, \mathbb{Z}_p) \\ y &\longmapsto [x \mapsto \mathrm{Tr}_{L/\mathbb{Q}_p}(\pi_L^{-s} xy)] \,. \end{aligned} \tag{28}$$

By tensoring with $\mathbb{Q}_p/\mathbb{Z}_p$ it induces the isomorphism of torsion o_L-modules

$$\Xi : L/o_L \cong \mathrm{Hom}_{\mathbb{Z}_p}(o_L, \mathbb{Z}_p) \otimes_{\mathbb{Z}_p} \mathbb{Q}_p/\mathbb{Z}_p \cong \mathrm{Hom}_{\mathbb{Z}_p}(o_L, \mathbb{Q}_p/\mathbb{Z}_p) \,.$$

Now let M be any topological o_L-module. Since $\mathrm{Hom}_{\mathbb{Z}_p}(o_L, -)$ is right adjoint to scalar restriction from o_L to $\mathbb{Z}_p$ and by using Ξ^{-1} in the second step, we have a natural isomorphism

$$\mathrm{Hom}_{\mathbb{Z}_p}(M, \mathbb{Q}_p/\mathbb{Z}_p) \cong \mathrm{Hom}_{o_L}(M, \mathrm{Hom}_{\mathbb{Z}_p}(o_L, \mathbb{Q}_p/\mathbb{Z}_p)) \cong \mathrm{Hom}_{o_L}(M, L/o_L) \,. \tag{29}$$

Lemma 5.3 *The isomorphism* (29) *restricts to an isomorphism*

$$\mathrm{Hom}^c_{\mathbb{Z}_p}(M, \mathbb{Q}_p/\mathbb{Z}_p) \cong \mathrm{Hom}^c_{o_L}(M, L/o_L)$$

of topological groups between the subgroups of continuous homomorphisms endowed with the compact-open topology.

Proof Coming from an isomorphism between the targets the second isomorphism in (29) obviously restricts to a topological isomorphism

$$\mathrm{Hom}^c_{o_L}(M, \mathrm{Hom}_{\mathbb{Z}_p}(o_L, \mathbb{Q}_p/\mathbb{Z}_p)) \cong \mathrm{Hom}^c_{o_L}(M, L/o_L) \,.$$

The first isomorphism is induced by the homomorphism $\lambda \mapsto \lambda(1)$ between the targets and therefore, at least restricts to a continuous injective map

$$\mathrm{Hom}^c_{o_L}(M, \mathrm{Hom}_{\mathbb{Z}_p}(o_L, \mathbb{Q}_p/\mathbb{Z}_p)) \longrightarrow \mathrm{Hom}^c_{\mathbb{Z}_p}(M, \mathbb{Q}_p/\mathbb{Z}_p) \,.$$

Let $\ell : M \to \mathbb{Q}_p/\mathbb{Z}_p$ be a continuous homomorphism. Then the composite map

$$M \times o_L \xrightarrow{(m,a) \mapsto am} M \xrightarrow{\ell} \mathbb{Q}_p/\mathbb{Z}_p$$

is continuous. Therefore the preimage $F_\ell \in \mathrm{Hom}_{o_L}(M, \mathrm{Hom}_{\mathbb{Z}_p}(o_L, \mathbb{Q}_p/\mathbb{Z}_p))$ of ℓ, which is given by $F_\ell(m)(a) := \ell(am)$, is continuous by [4] X.28 Theorem 3. Finally let $A \subseteq M$ be any compact subset and $V \subseteq \mathbb{Q}_p/\mathbb{Z}_p$ be any subset. Define $B := \{a \in o_L : aA \subseteq A\}$. Then $1 \in B$, and, since A is closed, also B is closed and hence compact. Put $\widetilde{V} := \{\lambda \in \mathrm{Hom}_{\mathbb{Z}_p}(o_L, \mathbb{Q}_p/\mathbb{Z}_p) : \lambda(B) \subseteq V\}$. One easily checks that $\ell(A) \subseteq V$ if and only if $F_\ell(A) \subseteq \widetilde{V}$. This means that the inverse bijection $\ell \mapsto F_\ell$ is continuous as well. □

In the following we shall use the notation

$$M^\vee := \mathrm{Hom}^c_{o_L}(M, L/o_L) ,$$

always equipped with the compact-open topology. The following version of Pontrjagin duality should be well known. Since we could not find a reference we will sketch a proof for the convenience of the reader.

Proposition 5.4 (Pontrjagin duality) *The functor $-^\vee$ defines an involutory contravariant autoequivalence of the category of (Hausdorff) locally compact linear-topological o_L-modules. In particular, for such a module M, the canonical map*

$$M \xrightarrow{\cong} (M^\vee)^\vee$$

is an isomorphism of topological o_L-modules.

Proof We recall that a topological o_L-module M is called linear-topological if it has a fundamental system of open zero neighbourhoods consisting of o_L-submodules. If M is linear-topological and locally compact one easily checks that it has a fundamental system of open zero neighbourhoods consisting of compact open o_L-submodules.

Classical Pontrjagin duality $M \longmapsto \mathrm{Hom}^c(M, \mathbb{R}/\mathbb{Z})$, the right hand side being the group of all continuous group homomorphisms equipped with the compact-open topology, is an autoequivalence of the category of locally compact abelian groups M. We first compare this, for any locally compact linear-topological $\mathbb{Z}_p$-module M, with the group $\mathrm{Hom}^c_{\mathbb{Z}_p}(M, \mathbb{Q}_p/\mathbb{Z}_p)$, as always equipped with the compact-open topology. There is an obvious injective and continuous map

$$\mathrm{Hom}^c_{\mathbb{Z}_p}(M, \mathbb{Q}_p/\mathbb{Z}_p) \longrightarrow \mathrm{Hom}^c(M, \mathbb{R}/\mathbb{Z}) . \tag{30}$$

Step 1: The map (30) *is bijective.* Let $\ell : M \to \mathbb{R}/\mathbb{Z}$ be any continuous group homomorphism. We have to show that $\mathrm{im}(\ell) \subseteq \mathbb{Q}_p/\mathbb{Z}_p$ and that ℓ is continuous for the discrete topology on $\mathbb{Q}_p/\mathbb{Z}_p$. We fix a compact-open $\mathbb{Z}_p$-submodule $U \subseteq M$. Then $\ell(U)$ is a compact subgroup of $\mathbb{R}/\mathbb{Z}$, and hence is either equal to $\mathbb{R}/\mathbb{Z}$ or is finite. Since U is profinite the former cannot occur. We conclude that $\ell(U)$ is a finite subgroup of $\mathbb{Q}_p/\mathbb{Z}_p$. In particular, there is an $r \in \mathbb{N}$ such that $p^r \cdot \ell|U = 0$. The quotient module M/U is discrete and $\mathbb{Z}_p$-torsion. It follows that $p^r \cdot \ell(M) \subseteq \mathbb{Q}_p/\mathbb{Z}_p$ and hence that $\ell(M) \subseteq \mathbb{Q}_p/\mathbb{Z}_p$. Since $\mathbb{R}/\mathbb{Z}$ induces the discrete topology on the finite subgroup $\ell(U)$ the restricted homomorphism $\ell : U \to \mathbb{Q}_p/\mathbb{Z}_p$ is continuous. Since U is open in M this suffices for the continuity of $\ell : M \to \mathbb{Q}_p/\mathbb{Z}_p$. That ℓ then is a homomorphism of $\mathbb{Z}_p$-modules is automatic.

Step 2: The map (30) *is open.* As usual, $C(A, V)$, for a compact subset $A \subseteq M$ and an arbitrary subset $V \subseteq \mathbb{Q}_p/\mathbb{Z}_p$, denotes the open subset of all $\ell \in \mathrm{Hom}^c_{\mathbb{Z}_p}(M, \mathbb{Q}_p/\mathbb{Z}_p)$ such that $\ell(A) \subseteq V$. First one checks that the $C(m + U, V)$, for $m \in M$, $U \subseteq M$ a compact-open $\mathbb{Z}_p$-submodule, and V arbitrary, form a subbase of the compact-open topology. For any $\ell_0 \in C(m + U, V)$ we have $\ell_0 \in C(m + U, \ell_0(m) + \ell_0(U)) \subseteq$

$C(m+U, V)$, where $\ell_0(U)$ is a finite subgroup of $\mathbb{Q}_p/\mathbb{Z}_p$. These observations reduce us to showing that the sets $C(m+U, v+\frac{1}{p^n}\mathbb{Z}/\mathbb{Z})$ are open in $\mathrm{Hom}^c(M, \mathbb{R}/\mathbb{Z})$. We fix a point $\ell_0 \in C(m+U, v+\frac{1}{p^n}\mathbb{Z}/\mathbb{Z})$, and we let p^t be the order of m modulo U. Note that we have $\ell_0(m) \in v+\frac{1}{p^n}\mathbb{Z}/\mathbb{Z}$ and hence

$$\ell_0(U) \subseteq \tfrac{1}{p^n}\mathbb{Z}/\mathbb{Z} \qquad \text{and} \qquad \ell_0(m) \in \tfrac{1}{p^{n+t}}\mathbb{Z}/\mathbb{Z}\,. \tag{31}$$

We use the open subsets

$$V_1(\ell_0) := \big((-\tfrac{1}{2p^{n+t+1}}, \tfrac{1}{2p^{n+t+1}}) + \tfrac{1}{p^n}\mathbb{Z}\big)/\mathbb{Z} \subseteq V_2(\ell_0) := V_1(\ell_0) - V_1(\ell_0) \subseteq \mathbb{R}/\mathbb{Z}\,.$$

They satisfy:

(a) $V_2(\ell_0) \cap \frac{1}{p^{n+t+1}}\mathbb{Z}/\mathbb{Z} = V_1(\ell_0) \cap \frac{1}{p^{n+t+1}}\mathbb{Z}/\mathbb{Z} = \frac{1}{p^n}\mathbb{Z}/\mathbb{Z}$.
(b) $V_1(\ell_0) + \frac{1}{p^n}\mathbb{Z}/\mathbb{Z} \subseteq V_1(\ell_0)$.

We claim that $\ell_0 \in C(m+U, v+V_1(\ell_0)) \subseteq C(m+U, v+\frac{1}{p^n}\mathbb{Z}/\mathbb{Z})$. This means that in $\mathrm{Hom}^c(M, \mathbb{R}/\mathbb{Z})$ we have found an open neighbourhood of ℓ_0 which is contained in $C(m+U, v+\frac{1}{p^n}\mathbb{Z}/\mathbb{Z})$. Hence $C(m+U, v+\frac{1}{p^n}\mathbb{Z}/\mathbb{Z})$ is open in $\mathrm{Hom}^c(M, \mathbb{R}/\mathbb{Z})$. Since $\frac{1}{p^n}\mathbb{Z}/\mathbb{Z} \subseteq V_1(\ell_0)$ we certainly have $\ell_0 \in C(m+U, v+V_1(\ell_0))$. Let now $\ell \in C(m+U, v+V_1(\ell_0))$ be an arbitrary element. We have $\ell(U) = \frac{1}{p^j}\mathbb{Z}/\mathbb{Z}$ for some $j \geq 0$ and consequently $\ell(m) \in \frac{1}{p^{j+t}}\mathbb{Z}/\mathbb{Z}$.

Case 1: $j \leq n$. Using (31) we then have $\ell_0(m) - v \in \frac{1}{p^n}\mathbb{Z}/\mathbb{Z}$ and $\ell(m) - \ell_0(m) \in \frac{1}{p^{n+t}}\mathbb{Z}/\mathbb{Z}$ and hence

$$\ell(m) - v = \ell(m) - \ell_0(m) + \ell_0(m) - v \in \tfrac{1}{p^{n+t}}\mathbb{Z}/\mathbb{Z}\,.$$

Since also $\ell(m) - v \in V_1(\ell_0)$ we deduce from (a) that $\ell(m) - v \in \frac{1}{p^n}\mathbb{Z}/\mathbb{Z}$ and therefore

$$\ell(m+U) = \ell(m) + \ell(U) \subseteq v + \tfrac{1}{p^n}\mathbb{Z}/\mathbb{Z} + \tfrac{1}{p^j}\mathbb{Z}/\mathbb{Z} = v + \tfrac{1}{p^n}\mathbb{Z}/\mathbb{Z}\,.$$

Case 2: $j > n$. We obtain

$$\begin{aligned}
\tfrac{1}{p^{n+1}}\mathbb{Z}/\mathbb{Z} \subseteq \tfrac{1}{p^j}\mathbb{Z}/\mathbb{Z} = \ell(U) &\subseteq -\ell(m) + v + V_1(\ell_0) \\
&= -(\ell(m) - \ell_0(m)) - \ell_0(m) + v + V_1(\ell_0) \\
&\subseteq -(\ell(m) - \ell_0(m)) + \tfrac{1}{p^n}\mathbb{Z}/\mathbb{Z} + V_1(\ell_0) \\
&\subseteq -(\ell(m) - \ell_0(m)) + V_1(\ell_0) \\
&\subseteq -(V_1(\ell_0) + \tfrac{1}{p^n}\mathbb{Z}/\mathbb{Z}) + V_1(\ell_0) \\
&\subseteq V_1(\ell_0) - V_1(\ell_0) \\
&\subseteq V_2(\ell_0)\,,
\end{aligned}$$

where the fourth and the sixth inclusion use (b). This is in contradiction to (b). We deduce that this case, in fact, cannot occur.

At this point we have shown that (30) is a topological isomorphism of locally compact abelian groups. From now on we assume that M is a locally compact linear-topological o_L-module. By combining this latter isomorphism with the isomorphism in Lemma 5.3 we obtain a topological isomorphism of locally compact abelian groups

$$\mathrm{Hom}^c_{o_L}(M, L/o_L) \cong \mathrm{Hom}^c(M, \mathbb{R}/\mathbb{Z}) , \tag{32}$$

which is natural in M. Of course, $\mathrm{Hom}^c_{o_L}(M, L/o_L)$ naturally is an o_L-module again.

Step 3: The o_L-module $\mathrm{Hom}^c_{o_L}(M, L/o_L)$ is linear-topological. It is straightforward to check that the $C(U, \{0\})$ with U running over all compact open o_L-submodules of M form a fundamental system of open zero neighbourhoods in $M^\vee$. Each such $C(U, \{0\})$ evidently is an o_L-submodule.

Hence the topological isomorphism (32) also applies to $M^\vee$ instead of M. One checks that the natural map $M \to \mathrm{Hom}^c(\mathrm{Hom}^c(M, \mathbb{R}/\mathbb{Z}), \mathbb{R}/\mathbb{Z})$ corresponds under this isomorphism to the natural map $M \to (M^\vee)^\vee$. We finally see that the classical Pontrjagin duality implies that $M \xrightarrow{\cong} (M^\vee)^\vee$ is a topological isomorphism; it, of course, is o_L-linear. □

Remark 5.5 Let $M_0 \xrightarrow{\alpha} M \xrightarrow{\beta} M_1$ be a sequence of locally compact linear-topological o_L-modules such that $\mathrm{im}(\alpha) = \ker(\beta)$ and β is topologically strict with closed image; then the dual sequence $M_1^\vee \xrightarrow{\beta^\vee} M^\vee \xrightarrow{\alpha^\vee} M_0^\vee$ is exact as well, i.e., we have $\mathrm{im}(\beta^\vee) = \ker(\alpha^\vee)$.

Proof We have $\ker(\alpha^\vee) = (M/\mathrm{im}(\alpha))^\vee \cong \mathrm{im}(\beta)^\vee$, where the second isomorphism uses the assumption that β is topologically strict. The assertion therefore reduces to the claim that the closed immersion $\mathrm{im}(\beta) \subseteq M_1$ induces a surjection between the corresponding Pontrjagin duals. For this see, for example, [21] Theorem 24.11. □

We recall that the weak topology on a finitely generated $\mathbf{A}_L$-module M is o_L-linear; moreover, it is locally compact if M is annihilated by some power of π_L. Suppose that M is a finitely generated $\mathbf{A}_L/\pi_L^n\mathbf{A}_L$-module. From (23) and (24) we have topological isomorphisms $M^\vee \cong \mathrm{Hom}_{\mathbf{A}_L}(M, \Omega^1/\pi_L^n\Omega^1) \cong \mathrm{Hom}_{\mathbf{A}_L}(M, \mathbf{A}_L/\pi_L^n\mathbf{A}_L(\chi_{LT}))$. By Proposition 5.4 they dualize into topological isomorphisms $M \cong \mathrm{Hom}_{\mathbf{A}_L}(M, \Omega^1/\pi_L^n\Omega^1)^\vee \cong \mathrm{Hom}_{\mathbf{A}_L}(M, \mathbf{A}_L/\pi_L^n\mathbf{A}_L(\chi_{LT}))^\vee$. If M actually is an etale (φ_q, Γ_L)-module then we see that in the adjoint pairs of maps $(\psi_M, \varphi_{\mathrm{Hom}_{\mathbf{A}_L}(M,\Omega^1/\pi_L^n\Omega^1)})$ and $(\varphi_M, \psi_{\mathrm{Hom}_{\mathbf{A}_L}(M,\Omega^1/\pi_L^n\Omega^1)})$ from Remark 4.7 each map determines the other uniquely.

Remark 5.6 Let V be an object in $\mathrm{Rep}_{o_L}(G_L)$ of finite length. Then, the pairing $[\ ,\ \rangle_{D_{LT}(V)}$ from Remark 4.7, the Remark 4.6, and the compatibility of the functor $D_{LT}(-)$ with internal Hom's by Lemma 4.5 induce, for n sufficiently large, a natural isomorphism of topological groups

$$\begin{aligned} D_{LT}(V)^\vee &\cong \mathrm{Hom}_{\mathbf{A}_L}(D_{LT}(V), \mathbf{A}_L/\pi_L^n\mathbf{A}_L(\chi_{LT})) \\ &\cong \mathrm{Hom}_{\mathbf{A}_L}(D_{LT}(V), D_{LT}((o_L/\pi_L^n o_L)(\chi_{LT}))) \\ &= D_{LT}(\mathrm{Hom}_{o_L}(V, (o_L/\pi_L^n o_L)(\chi_{LT}))) = D_{LT}(V^\vee(\chi_{LT}))\ , \end{aligned}$$

which is independent of the choice of n and under which $\psi_{D_{LT}(V^\vee(\chi_{LT}))}$ identifies with $\varphi_{D_{LT}(V)}^\vee$ by Remark 4.7.

Proposition 5.7 (Local Tate duality) *Let V be an object in $\mathrm{Rep}_{o_L}(G_L)$ of finite length, and K any finite extension of L. Then the cup product and the local invariant map induce perfect pairings of finite o_L-modules*

$$H^i(K, V) \times H^{2-i}(K, \mathrm{Hom}_{\mathbb{Z}_p}(V, \mathbb{Q}_p/\mathbb{Z}_p(1))) \longrightarrow H^2(K, \mathbb{Q}_p/\mathbb{Z}_p(1)) = \mathbb{Q}_p/\mathbb{Z}_p$$

and

$$H^i(K, V) \times H^{2-i}(K, \mathrm{Hom}_{o_L}(V, L/o_L(1))) \longrightarrow H^2(K, L/o_L(1)) = L/o_L$$

where—(1) *denotes the Galois twist by the cyclotomic character. In other words, there are canonical isomorphisms*

$$H^i(K, V) \cong H^{2-i}(K, V^\vee(1))^\vee\ .$$

Proof Note that the isomorphism $H^2(K, L/o_L(1)) = L/o_L$ arises from the isomorphism $H^2(K, \mathbb{Q}_p/\mathbb{Z}_p(1)) = \mathbb{Q}_p/\mathbb{Z}_p$ by tensoring with o_L over $\mathbb{Z}_p$. The first pairing is the usual version of local Tate duality (cf. [35] II.5.2 Theorem 2). It induces the first isomorphism in

$$\begin{aligned} H^i(K, V) &\cong \mathrm{Hom}_{\mathbb{Z}_p}(H^{2-i}(K, \mathrm{Hom}_{\mathbb{Z}_p}(V, \mathbb{Q}_p/\mathbb{Z}_p(1))), \mathbb{Q}_p/\mathbb{Z}_p) \\ &\cong \mathrm{Hom}_{o_L}(H^{2-i}(K, \mathrm{Hom}_{\mathbb{Z}_p}(V, \mathbb{Q}_p/\mathbb{Z}_p(1))), L/o_L) \\ &\cong \mathrm{Hom}_{o_L}(H^{2-i}(K, \mathrm{Hom}_{o_L}(V, L/o_L(1))), L/o_L)\ , \end{aligned}$$

while the second and third are induced by Lemma 5.3. To obtain the second pairing it remains to check that the above composite isomorphism is given by the cup product again. By the functoriality properties of the cup product this reduces to the following formal fact. Let $\xi : L/o_L \longrightarrow \mathbb{Q}_p/\mathbb{Z}_p$ be any group homomorphism. Then the diagram

$$\begin{array}{ccc} H^2(K, L/o_L(1)) & \xrightarrow{=} & L/o_L \\ \downarrow{\scriptstyle H^2(K,\xi(1))} & & \downarrow{\scriptstyle \xi} \\ H^2(K, \mathbb{Q}_p/\mathbb{Z}_p(1)) & \xrightarrow{=} & \mathbb{Q}_p/\mathbb{Z}_p \end{array}$$

commutes, where the horizontal maps are the local invariant maps. This in turn is an easy consequence of the $\mathbb{Z}_p$-linearity of the local invariant map if one uses the following description of ξ viewed as a map $L/o_L = \mathbb{Q}_p/\mathbb{Z}_p \otimes_{\mathbb{Z}_p} o_L \xrightarrow{\xi} \mathbb{Q}_p/\mathbb{Z}_p$. Let $\zeta : o_L \longrightarrow \mathrm{Hom}_{\mathbb{Z}_p}(\mathbb{Q}_p/\mathbb{Z}_p, \mathbb{Q}_p/\mathbb{Z}_p) = \mathbb{Z}_p$ be the homomorphism which sends a to $c \mapsto \xi(c \otimes a)$. Then $\xi(c \otimes a) = \zeta(a)c$. $\square$

For any V in $\mathrm{Rep}_{o_L}(G_L)$ we define the generalized Iwasawa cohomology of V by

$$H^*_{\mathrm{Iw}}(L_\infty/L, V) := \varprojlim_K H^*(K, V)$$

where K runs through the finite Galois extensions of L contained in L_∞ and the transition maps in the projective system are the cohomological corestriction maps.[1]

Shapiro's lemma for cohomology gives natural isomorphisms

$$H^*(K, V) = H^*(G_L, o_L[G_L/G_K] \otimes_{o_L} V)$$

where, on the right hand side, G_L acts diagonally on the coefficients. In this picture the corestriction map, for $K \subseteq K'$, becomes the map induced on cohomology by the map $\mathrm{pr} \otimes \mathrm{id}_V : o_L[G_L/G_{K'}] \otimes_{o_L} V \longrightarrow o_L[G_L/G_K] \otimes_{o_L} V$.

Lemma 5.8 $H^*_{\mathrm{Iw}}(L_\infty/L, V) = H^*(G_L, o_L[[\Gamma_L]] \otimes_{o_L} V)$ *(where the right hand side refers to cohomology with continuous cochains).*

Proof On the level of continuous cochain complexes we compute

$$\begin{aligned}\varprojlim_K C^\bullet(G_L, o_L[G_L/G_K] \otimes_{o_L} V) &= C^\bullet(G_L, \varprojlim_K(o_L[G_L/G_K] \otimes_{o_L} V)) \\ &= C^\bullet(G_L, o_L[[\Gamma_L]] \otimes_{o_L} V)\,.\end{aligned}$$

The second identity comes from the isomorphism $\varprojlim_K(o_L[G_L/G_K] \otimes_{o_L} V) \cong o_L[[\Gamma_L]] \otimes_{o_L} V$ which is easily seen by using a presentation of the form $0 \to o_L^s \to o_L^r \to V \to 0$. Since the transition maps in this projective system of complexes are surjective the first hypercohomology spectral sequence for the composite functor $\varprojlim \circ H^0(G_L, .)$ degenerates so that the second hypercohomology spectral sequence becomes

$$R^i \varprojlim_K H^j(G_L, o_L[G_L/G_K] \otimes_{o_L} V) \Longrightarrow H^{i+j}(G_L, o_L[[\Gamma_L]] \otimes_{o_L} V)\,.$$

[1] Note that, for any finite extension K/L contained in L_∞ the definition $H^*_{\mathrm{Iw}}(L_\infty/K, V) := \varprojlim_{K \subseteq K' \subseteq L_\infty} H^*(K', V)$ produces the same o_L-modules. Our notation indicates that we always consider these groups as Γ_L-modules.

Due to the countability of our projective system we have $R^i \varprojlim_K = 0$ for $i \geq 2$. Hence this spectral sequence degenerates into short exact sequences

$$0 \longrightarrow R^1 \varprojlim_K H^{i-1}(G_L, o_L[G_L/G_K] \otimes_{o_L} V) \longrightarrow$$
$$H^i(G_L, o_L[[\Gamma_L]] \otimes_{o_L} V) \longrightarrow \varprojlim_K H^i(G_L, o_L[G_L/G_K] \otimes_{o_L} V) \longrightarrow 0\,.$$

It is well known that the Galois cohomology groups $H^*(G_L, o_L[G_L/G_K] \otimes_{o_L} V)$ are finitely generated o_L-modules (cf. [35] II.5.2 Proposition 14). It therefore follows from [22] Theorem 8.1 that the above $R^1 \varprojlim_K$-terms vanish. □

Lemma 5.9 *$H^*_{\mathrm{Iw}}(L_\infty/L, V)$ is a δ-functor on* $\mathrm{Rep}_{o_L}(G_L)$.

Proof Let $0 \to V_1 \to V_2 \to V_3 \to 0$ be a short exact sequence in $\mathrm{Rep}_{o_L}(G_L)$. Then the sequence of topological G_L-modules $0 \to o_L[[\Gamma_L]] \otimes_{o_L} V_1 \to o_L[[\Gamma_L]] \otimes_{o_L} V_2 \to o_L[[\Gamma_L]] \otimes_{o_L} V_3 \to 0$ is short exact as well. In view of Lemma 5.8 our assertion therefore follows from [30] Lemma 2.7.2 once we show that

1. the topology of $o_L[[\Gamma_L]] \otimes_{o_L} V_1$ is induced by the topology of $o_L[[\Gamma_L]] \otimes_{o_L} V_2$ and
2. the (surjective continuous) map $o_L[[\Gamma_L]] \otimes_{o_L} V_2 \to o_L[[\Gamma_L]] \otimes_{o_L} V_3$ has a continuous section as a map of topological spaces.

Each $o_L[[\Gamma_L]] \otimes_{o_L} V_i$ is a profinite (hence compact) abelian group with a countable base of the topology, which therefore is metrizable by [4] IX.21 Proposition 16. One easily deduces 1. and that the map in 2. is open. The property 2. then follows from [29] Corollary 1.4. □

Remark 5.10 For any V_0 in $\mathrm{Rep}_{o_L}(G_L)$ which is o_L-free and on which G_L acts through its factor Γ_L there is a natural isomorphism $H^*_{\mathrm{Iw}}(L_\infty/L, V \otimes_{o_L} V_0) \cong H^*_{\mathrm{Iw}}(L_\infty/L, V) \otimes_{o_L} V_0$.

Proof In view of Lemma 5.8 the asserted isomorphism is induced by the G_L-equivariant isomorphism on coefficients

$$o_L[[\Gamma_L]] \otimes_{o_L} V \otimes_{o_L} V_0 \xrightarrow{\cong} o_L[[\Gamma_L]] \otimes_{o_L} V \otimes_{o_L} V_0$$
$$\gamma \otimes v \otimes v_0 \longmapsto \gamma \otimes v \otimes \gamma^{-1} v_0\,;$$

on the left G_L acts diagonally on all three factors, whereas on the right it acts trivially on the third factor. □

Remark 5.11 Let V be in $\mathrm{Rep}_{o_L}(G_L)$ of finite length; in particular V is discrete. Then local Tate duality (Proposition 5.7) induces an isomorphism

$$H^i_{\mathrm{Iw}}(L_\infty/L, V) \cong H^{2-i}(L_\infty, V^\vee(1))^\vee .$$

Proof Use Proposition 5.7 over the layers L_n and take limits. □

We point out that $H^*_{\mathrm{Iw}}(L_\infty/L, V) = H^*(G_L, o_L[[\Gamma_L]] \otimes_{o_L} V)$ is a left $o_L[[\Gamma_L]]$-module through the action of $\gamma \in \Gamma_L$ by right multiplication with γ^{-1} on the factor $o_L[[\Gamma_L]]$.

Lemma 5.12 *i.* $H^*_{\mathrm{Iw}}(L_\infty/L, V) = 0$ *for* $* \neq 1, 2$.
ii. $H^2_{\mathrm{Iw}}(L_\infty/L, V)$ *is finitely generated as* o_L*-module.*
iii. $H^1_{\mathrm{Iw}}(L_\infty/L, V)$ *is finitely generated as* $o_L[[\Gamma_L]]$*-module.*

Proof i. In case $* > 2$ the assertion follows from the fact that the groups G_K have cohomological p-dimension 2 ([35] II.4.3 Proposition 12). The vanishing of $H^0_{\mathrm{Iw}}(L_\infty/L, V) = \varprojlim_K V^{G_K}$ is clear if V is finite. Hence we may assume that V is finitely generated free over o_L. Note that the identity $H^0_{\mathrm{Iw}}(L_\infty/L, V) = \varprojlim_K V^{G_K}$ shows that $H^0_{\mathrm{Iw}}(L_\infty/L, V)$ is a profinite o_L-module. On the other hand we then have the exact sequence

$$0 \longrightarrow H^0_{\mathrm{Iw}}(L_\infty/L, V) \xrightarrow{\pi_L \cdot} H^0_{\mathrm{Iw}}(L_\infty/L, V) \longrightarrow H^0_{\mathrm{Iw}}(L_\infty/L, V/\pi_L V).$$

Since we observed already that the last term vanishes it follows that $H^0_{\mathrm{Iw}}(L_\infty/L, V)$ is an L-vector space. Both properties together enforce the vanishing of $H^0_{\mathrm{Iw}}(L_\infty/L, V)$.

ii. We have

$$H^2_{\mathrm{Iw}}(L_\infty/L, V) = \varprojlim_K H^2(K, V) = \varprojlim_K H^0(K, V^\vee(1))^\vee = (\bigcup_K V^\vee(1)^{G_K})^\vee ,$$

which visibly is a finitely generated o_L-module.

iii. *Case 1: V is finite.* By Remark 5.11 $H^1_{\mathrm{Iw}}(L_\infty/L, V) = H^1(L_\infty, V^\vee(1))^\vee$ is the Pontrjagin dual of a discrete torsion module and hence is a compact $o_L[[\Gamma_L]]$-module. The compact Nakayama lemma (cf. [30] Lemma 5.2.18) therefore reduces us to showing that the Pontrjagin dual $(H^1_{\mathrm{Iw}}(L_\infty/L, V)_\Gamma)^\vee = H^1(L_\infty, V^\vee(1))^\Gamma$ of the Γ-coinvariants of $H^1_{\mathrm{Iw}}(L_\infty/L, V)$ is cofinitely generated over o_L; here Γ is a conveniently chosen open subgroup of Γ_L. The Hochschild–Serre spectral sequence for the extension L_∞/K, where $K := L_\infty^\Gamma$, gives us an exact sequence

$$H^1(K, V^\vee(1)) \longrightarrow H^1(L_\infty, V^\vee(1))^\Gamma \longrightarrow H^2(\Gamma, V^\vee(1)^{H_L}) .$$

The first group is finite by local Galois cohomology. At this point we choose Γ to be isomorphic to $\mathbb{Z}_p^d$. Then $H^2(\Gamma, \mathbb{F}_p)$ is finite. Since $\mathbb{F}_p$ is the only simple $\mathbb{Z}_p[[\Gamma]]$-module it follows by devissage that $H^2(\Gamma, V^\vee(1)^{H_L})$ is finite.

Case 2: V is o_L-free. As pointed out above $o_L[[\Gamma_L]] \otimes_{o_L} V$ is a $o_L[[\Gamma_L]]$-module, which is finitely generated free and on which G_L acts continuously and $o_L[[\Gamma_L]]$-linearly. In view of Lemma 5.8 we therefore may apply [18] Proposition 1.6.5 and obtain that $H^*_{\mathrm{Iw}}(L_\infty/L, V)$, as a $o_L[[\Gamma_L]]$-module, is isomorphic to the cohomology of a bounded complex of finitely generated projective $o_L[[\Gamma_L]]$-modules. The ring $o_L[[\Gamma_L]]$ is noetherian. Hence the cohomology of such a complex is finitely generated.

The general case follows by using Lemma 5.9 and applying the above two special cases to the outer terms of the short exact sequence $0 \to V_{tor} \to V \to V/V_{tor} \to 0$, where V_{tor} denotes the torsion submodule of V. □

Theorem 5.13 *Let V in $\mathrm{Rep}_{o_L}(G_L)$. Then, with $\psi = \psi_{D_{LT}(V(\tau^{-1}))}$, we have a short exact sequence*

$$0 \longrightarrow H^1_{\mathrm{Iw}}(L_\infty/L, V) \longrightarrow D_{LT}(V(\chi_{LT}\chi_{cyc}^{-1})) \xrightarrow{\psi-1} D_{LT}(V(\chi_{LT}\chi_{cyc}^{-1})) \\ \longrightarrow H^2_{\mathrm{Iw}}(L_\infty/L, V) \longrightarrow 0\,, \quad (33)$$

which is functorial in V.

Proof In the sense of Remark 4.6 we take $T^* \otimes_{\mathbb{Z}_p} \mathbb{Z}_p(1)$ and $T \otimes_{\mathbb{Z}_p} \mathrm{Hom}_{\mathbb{Z}_p}(\mathbb{Z}_p(1), \mathbb{Z}_p)$ as representation module for τ and τ^{-1}, respectively.

We first assume that V has finite length. Then the exact sequence (27) is a sequence of locally compact linear-topological o_L-modules. In fact, the first term is finite and the last term is cofinitely generated over o_L. In particular, the first and the last term carry the discrete topology and the first map is a closed immersion. Moreover, the map $\varphi - 1$ in the middle, by Lemma 3.12, is topologically strict with open image. The latter implies that ∂_φ induces an isomorphism of discretely topologized o_L-modules $D_{LT}(V)/(\varphi-1)D_{LT}(V) \xrightarrow{\cong} H^1(L_\infty, V)$. In particular, the map ∂_φ is topologically strict as well.

Therefore, using Remark 5.5, we obtain that the dual sequence

$$0 \longrightarrow H^1(L_\infty, V)^\vee \longrightarrow D_{LT}(V)^\vee \xrightarrow{\varphi^\vee-1} D_{LT}(V)^\vee \longrightarrow H^0(L_\infty, V)^\vee \longrightarrow 0$$

is exact. If we identify the terms in this latter sequence according to Remarks 5.6 and 5.11 then the result is the exact sequence in the assertion.

Now let V be arbitrary and put $V_n := V/\pi_L^n V$. We have the exact sequence of projective systems

$$0 \to H^1_{\mathrm{Iw}}(L_\infty/L, V_n) \to D_{LT}(V_n(\tau^{-1})) \xrightarrow{\psi-1} D_{LT}(V_n(\tau^{-1})) \\ \to H^2_{\mathrm{Iw}}(L_\infty/L, V_n) \to 0\,.$$

Since the functor D_{LT} is exact (Proposition 4.4.i) we have

$$\varprojlim_n D_{LT}(V_n(\tau^{-1})) = \varprojlim_n D_{LT}(V(\tau^{-1}))/\pi_L^n D_{LT}(V(\tau^{-1})) = D_{LT}(V(\tau^{-1}))\,.$$

Moreover,

$$\varprojlim_n H^i_{\mathrm{Iw}}(L_\infty/L, V_n) = \varprojlim_n \varprojlim_K H^i(K, V_n) = \varprojlim_K \varprojlim_n H^i(K, V_n) = \varprojlim_K H^i(K, V)$$
$$= H^i_{\mathrm{Iw}}(L_\infty/L, V) .$$

Therefore it remains to show that passing to the projective limit in the above exact sequence of projective systems is exact. For this it suffices to show that $R^1 \varprojlim$ of the two projective systems $\{H^1_{\mathrm{Iw}}(L_\infty/L, V_n)\}_n$ and $\{(\psi - 1)D_{LT}(V_n(\tau^{-1}))\}_n$ vanishes. Because of $D_{LT}(V_n(\tau^{-1})) = D_{LT}(V(\tau^{-1}))/\pi_L^n D_{LT}(V(\tau^{-1}))$ the transition maps in the second projective system are surjective, which guarantees the required vanishing. For the first projective system we choose an open pro-p subgroup Γ in Γ_L, so that $o_L[[\Gamma]]$ is a complete local noetherian commutative ring. From Lemma 5.12.iii we know that $\{H^1_{\mathrm{Iw}}(L_\infty/L, V_n)\}_n$ is a projective system of finitely generated $o_L[[\Gamma]]$-modules. Hence [22] Theorem 8.1 applies and gives the required vanishing. □

Remark 5.14 Each map in the exact sequence (33) is continuous and $o_L[[\Gamma_L]]$-equivariant.

Proof Continuity and Γ_L-equivariance follow from the construction. Since the weak topology on $D_{LT}(V)$ is o_L-linear and complete we may apply [28] Theorem II.2.2.6 (which is valid for any profinite group) and obtain that the continuous Γ_L-action extends, by continuity, uniquely to an $o_L[[\Gamma_L]]$-action on $D_{LT}(V)$. □

6 The Kummer Map

We consider the Kummer isomorphism

$$\kappa : A(L_\infty) := \varprojlim_{n,m} L_n^\times / L_n^{\times p^m} \xrightarrow{\cong} H^1_{\mathrm{Iw}}(L_\infty/L, \mathbb{Z}_p(1)) .$$

Recall that we have fixed a generator η of the Tate module $T = o_L\eta$. Correspondingly we have the dual generator η^* of the o_L-dual $T^* = o_L\eta^*$ of T. This leads to the twisted Kummer isomorphism

$$A(L_\infty) \otimes_{\mathbb{Z}_p} T^* \xrightarrow[\cong]{\kappa \otimes_{\mathbb{Z}_p} T^*} H^1_{\mathrm{Iw}}(L_\infty/L, \mathbb{Z}_p(1)) \otimes_{\mathbb{Z}_p} T^* \cong H^1_{\mathrm{Iw}}(L_\infty/L, o_L(\tau))$$

where the second isomorphism comes from Remark 5.10. On the other hand, by Theorem 5.13, we have a natural isomorphism

$$Exp^* : H^1_{\mathrm{Iw}}(L_\infty/L, o_L(\tau)) \xrightarrow{\cong} D_{LT}(o_L)^{\psi=1} = \mathbf{A}_L^{\psi=1} .$$

Finally we have the homomorphism

$$\nabla : (\varprojlim_n L_n^\times) \otimes_{\mathbb{Z}} T^* \longrightarrow \mathbf{A}_L^{\psi=1}$$

$$u \otimes a\eta^* \longmapsto a \frac{\partial_{\mathrm{inv}}(g_{u,\eta})}{g_{u,\eta}}(\iota_{LT}(\eta)) \ .$$

It is well defined by Theorem 2.2, the last sentence in Sect. 2, and Remark 3.2.ii.

Remark 6.1 The map ∇ is independent of the choice of η.

Proof Let $u \otimes a\eta^* \in (\varprojlim_n L_n^\times) \otimes_{\mathbb{Z}} T^*$. We temporarily write ∇_η instead of ∇, and we let ϑ be a second generator of T, so that $\vartheta = c\eta$ for some $c \in o_L^\times$. Then $u \otimes a\eta^* = u \otimes ac\vartheta^*$, and by inserting the definitions in the first line we compute

$$\begin{aligned}
\nabla_\vartheta(u \otimes a\eta^*) = \nabla_\vartheta(u \otimes ac\vartheta^*) &= \frac{ac}{g_{LT}(\iota_{LT}(\vartheta))} \frac{g'_{u,\vartheta}(\iota_{LT}(\vartheta))}{g_{u,\vartheta}(\iota_{LT}(\vartheta))} \\
&= \frac{ac}{g_{LT}(\iota_{LT}(c\eta))} \frac{g'_{u,c\eta}(\iota_{LT}(c\eta))}{g_{u,c\eta}(\iota_{LT}(c\eta))} \\
&= \frac{ac}{g_{LT}([c](\iota_{LT}(\eta)))} \frac{g'_{u,c\eta}([c](\iota_{LT}(\eta)))}{g_{u,c\eta}([c](\iota_{LT}(\eta)))} \\
&= \frac{a}{g_{LT}(\iota_{LT}(\eta))} \frac{(g_{u,c\eta} \circ [c])'(\iota_{LT}(\eta))}{(g_{u,c\eta} \circ [c])(\iota_{LT}(\eta))} \\
&= \frac{a}{g_{LT}(\iota_{LT}(\eta))} \frac{g'_{u,\eta}(\iota_{LT}(\eta))}{g_{u,\eta}(\iota_{LT}(\eta))} \\
&= \nabla_\eta(u \otimes a\eta^*) \ ,
\end{aligned}$$

where we use Lemma 4.1.a for the fourth identity, (2) for the fifth one, and Remark 2.3.ii for the sixth one. □

Generalizing ([5] Proposition V.3.2.iii) (see also [10] Theorem 7.4.1) we will establish the following kind of reciprocity law.

Theorem 6.2 *The diagram*

$$\begin{array}{ccc}
(\varprojlim_n L_n^\times) \otimes_{\mathbb{Z}} T^* & \xrightarrow{\kappa \otimes T^*} & H^1_{\mathrm{Iw}}(L_\infty/L, o_L(\tau)) \\
 & {\scriptstyle \nabla} \searrow \qquad \swarrow {\scriptstyle Exp^*}\ \cong & \\
 & \mathbf{A}_L^{\psi=1} &
\end{array} \tag{34}$$

is commutative.

In a first step we consider, for any $n \geq 1$, the diagram

$$\begin{array}{ccccc}
\big((\mathbf{A}_L/\pi_L^n\mathbf{A}_L)(\chi_{LT})\big)/\operatorname{im}(\varphi-1) & \times & (\mathbf{A}_L/\pi_L^n\mathbf{A}_L)^{\psi=1} & \xrightarrow{[\;,\;\rangle} & L/o_L \\
\partial_\varphi \big\downarrow \cong & & \cong \big\uparrow Exp^* & & \big\| \\
H^1(L_\infty, o_L/\pi_L^n o_L(\chi_{LT})) & \times & H^1_{\mathrm{Iw}}(L_\infty/L, o_L/\pi_L^n o_L(\chi_{cyc}\chi_{LT}^{-1})) & \longrightarrow & L/o_L \\
\big\| & & \big\uparrow \kappa\otimes_{\mathbb{Z}_p}\mathrm{id} & & \big\| \\
\big\| & & (\varprojlim_n L_n^\times)\otimes_{\mathbf{Z}} o_L/\pi_L^n o_L(\chi_{LT}^{-1}) & & \big\| \\
\big\| & & \big\downarrow rec\otimes_{\mathbb{Z}_p}\mathrm{id} & & \big\| \\
\operatorname{Hom}(H_L, o_L/\pi_L^n o_L)(\chi_{LT}) & \times & H_L^{ab}(p)\otimes_{\mathbb{Z}_p} o_L/\pi_L^n o_L(\chi_{LT}^{-1}) & \longrightarrow & L/o_L,
\end{array}$$

where the second pairing is induced by local Tate duality and the third pairing is the obvious one. By $rec : (\varprojlim_n L_n^\times) \longrightarrow H_L^{ab}(p)$ we denote the map into the maximal abelian pro-p quotient $H_L^{ab}(p)$ of H_L induced by the reciprocity homomorphisms of local class field theory for the intermediate extensions L_m. Note that $\mathrm{Gal}(L_\infty^{ab}/L_\infty) = \varprojlim_m \mathrm{Gal}(L_m^{ab}/L_\infty) = \varprojlim_m \mathrm{Gal}(L_m^{ab}/L_m)$, where $L_?^{ab}$ denotes the maximal abelian extension of $L_?$. The upper half of the diagram is commutative by the construction of the map Exp^*. The commutativity of the lower half follows from [30] Corollary 7.2.13. All three pairings are perfect in the sense of Pontrjagin duality.

In order to prove Theorem 6.2 we have to show that, for any $u \in \varprojlim_n L_n^\times$ and any $a \in o_L$, we have

$$[z \otimes \eta, Exp^*(\kappa(u) \otimes a\eta^*)\rangle \equiv \operatorname{Res}(za\frac{\partial_{\mathrm{inv}}(g_{u,\eta})}{g_{u,\eta}} d\log_{LT}) \mod \pi_L^n$$

for any $z \in \mathbf{A}_L$ and any $n \geq 1$. Due to the commutativity of the above diagram the left hand side is equal to $a\partial_\varphi(z \otimes \eta)(rec(u) \otimes \eta^*) = a\partial_\varphi(z)(rec(u))$. On the other hand the right hand side, by (1), is equal to $\operatorname{Res}(za(\frac{g'_{u,\eta}}{g_{u,\eta}}dZ)_{|Z=\iota_{LT}(\eta)}) = \operatorname{Res}(za\frac{d(g_{u,\eta}(\iota_{LT}(\eta)))}{g_{u,\eta}(\iota_{LT}(\eta))})$. By the o_L-bilinearity of all pairings involved we may assume that $a = 1$. Hence we are reduced to proving that

$$\operatorname{Res}(z\frac{d(g_{u,\eta}(\iota_{LT}(\eta)))}{g_{u,\eta}(\iota_{LT}(\eta))}) = \partial_\varphi(z)(rec(u))$$

holds true for any $z \in \mathbf{A}_L$ and $u \in \varprojlim_n L_n^\times$. According to the theory of fields of norms we have the natural identification $\varprojlim_n L_n^\times = \mathbf{E}_L^\times$ (cf. [24] Lemma 1.4). Under this identification, by [27] Theorem 3.2.2, $rec(u)$ coincides with the image $rec_{\mathbf{E}_L}(u)$ of u under the reciprocity homomorphism $rec_{\mathbf{E}_L} : \mathbf{E}_L^\times \longrightarrow H_L^{ab}(p)$ in characteristic p. Furthermore, $g_{u,\eta}(\iota_{LT}(\eta)) \in (\mathbf{A}_L^\times)^{\mathcal{N}=1}$, is, by Remark 2.3.i, Remark 3.2.vii, and

(19), a lift of $u \in \mathbf{E}_L^\times$. This reduces the proof of Theorem 6.2 further to the following proposition which generalizes the explicit reciprocity law in [15] Proposition 2.4.3.

Proposition 6.3 *For any $z \in \mathbf{A}_L$ and any $u \in \mathbf{E}_L^\times$ with (unique) lift $\hat{u} \in (\mathbf{A}_L^\times)^{\mathcal{N}=1}$ we have*

$$\mathrm{Res}(z\frac{d\hat{u}}{\hat{u}}) = \partial_\varphi(z)(rec_{\mathbf{E}_L}(u)) ,$$

where ∂_φ is the connecting homomorphism in (27).

Obviously the connecting homomorphism ∂_φ for $V = o_L$ induces, by reduction modulo $\pi^n o_L$, the corresponding connecting homomorphism for $V = o_L/\pi_L^n o_L$. Hence we may prove the identity in Proposition 6.3 as a congruence modulo $\pi^n o_L$ for any $n \geq 1$. Recall that for $\hat{u} \in (\mathbf{A}_L^\times)^{\mathcal{N}=1}$ the differential form $\frac{d\hat{u}}{\hat{u}}$ is ψ_{Ω^1}-invariant by (22). Hence, by the adjointness of ψ_{Ω^1} and φ_L cf. (21), we obtain the equality

$$\mathrm{Res}(\varphi_L^m(z)\frac{d\hat{u}}{\hat{u}}) = \mathrm{Res}(z\frac{d\hat{u}}{\hat{u}})$$

for any $m \geq 1$. This reduces Proposition 6.3 and consequently Theorem 6.2 to proving the congruence

$$\mathrm{Res}(\varphi_L^{n-1}(z)\frac{d\hat{u}}{\hat{u}}) \equiv \partial_\varphi(z)(rec_{\mathbf{E}_L}(u)) \mod \pi_L^n o_L \tag{35}$$

for all $n \geq 1$. This will be the content of the next section (cf. Lemma 7.18).

7 The Generalized Schmid–Witt Formula

The aim of this section is to generalize parts of Witt's seminal paper [38] (see also the detailed accounts [36] and [25] of Witt's original article) to the case of *ramified* Witt vectors.

First of all we need to recall a few facts about ramified Witt vectors $W(B)_L$ for o_L-algebras B. Details of this construction can be found in [19]. But we will use [31] where a much more straightforward approach is fully worked out. We denote by $\Phi_B = (\Phi_0, \Phi_1, \ldots) : W(B)_L \longrightarrow B^{\mathbb{N}_0}$ the homomorphism of o_L-algebras, called the ghost map, given by the polynomials $\Phi_n(X_0, \ldots, X_n) = X_0^{q^n} + \pi_L X_1^{q^{n-1}} + \ldots \pi_L^n X_n$. On the other hand, the multiplicative Teichmüller map $B \longrightarrow W(B)_L$ is given by $b \mapsto [b] := (b, 0, \ldots)$ (cf. [31] Lemma 1.1.15). If B is a k_L-algebra then the Frobenius endomorphism $F = \phi_q$ of $W(B)_L$ has the form $\phi_q(b_0, \ldots, b_n, \ldots) = (b_0^q, \ldots, b_n^q, \ldots)$ (cf. [31] Proposition 1.1.18.i).

For a perfect k_L-algebra B we have $W_n(B)_L = W(B)_L/\pi_L^n W(B)_L$ for any $n \geq 1$ and, in particular, $\ker(\Phi_0) = \pi_L W(B)_L$; moreover, any $b = (b_0, b_1, \ldots) \in W(B)_L$ has the unique convergent expansion $b = \sum_{m=1}^\infty \pi_L^m [b_m^{q^{-m}}]$ (cf. [31] Proposition 1.1.19).

Proposition 7.1 *Suppose that π_L is not a zero divisor in B and that B has an endomorphism of o_L-algebras σ such that $\sigma(b) \equiv b^q \bmod \pi_L B$ for any $b \in B$. Then there is a unique homomorphism of o_L-algebras*

$$s_B : B \longrightarrow W(B)_L \quad \text{such that } \Phi_i \circ s_B = \sigma^i \text{ for any } i \geq 0.$$

Moreover, we have:

i. s_B is injective;
ii. for any $n \geq 1$ there is a unique homomorphism of o_L-algebras $s_{B,n} : B/\pi_L^n B \longrightarrow W_n(B/\pi_L B)_L$ such that the diagram

$$\begin{array}{ccccc} B & \xrightarrow{s_B} & W(B)_L & \xrightarrow{W(\mathrm{pr})_L} & W(B/\pi_L B)_L \\ \downarrow{\scriptstyle \mathrm{pr}} & & & & \downarrow{\scriptstyle \mathrm{pr}} \\ B/\pi_L^n B & & \xrightarrow{\quad s_{B,n} \quad} & & W_n(B/\pi_L B)_L \end{array}$$

is commutative;
iii. if $B/\pi_L B$ is perfect then $s_{B,n}$, for any $n \geq 1$, is an isomorphism.

Proof See [31] Proposition 1.1.23. □

Lemma 7.2 *For any perfect k_L-algebra B we have:*

i. The diagram

$$\begin{array}{ccc} W_n(W(B)_L)_L & \xrightarrow{\Phi_{n-1}} & W(B)_L \\ \downarrow{\scriptstyle W_n(\mathrm{pr})_L} & & \downarrow{\scriptstyle \mathrm{pr}} \\ W_n(B)_L & \xrightarrow{\phi_q^{n-1}} & W_n(B)_L \end{array}$$

is commutative for any $n \geq 1$.
ii. The composite map

$$W(B)_L \xrightarrow{s_{W(B)_L}} W(W(B)_L)_L \xrightarrow{W(\mathrm{pr})_L} W(B)_L$$

is the identity.

Proof i. Let $(\mathbf{b}_0, \ldots, \mathbf{b}_{n-1}) \in W_n(W(B)_L)_L$ with $\mathbf{b}_j = (b_{j,0}, b_{j,1}, \ldots)$. As $\ker(\Phi_0) = \pi_L W(B)_L$ we have $\mathbf{b}_j \equiv [b_{j,0}] \bmod \pi_L W(B)_L$. Hence [31] Lemma 1.1.1 implies that $\mathbf{b}_j^{q^m} \equiv [b_{j,0}^{q^m}] \bmod \pi_L^{m+1} W(B)_L$ for any $m \geq 0$. Using this as well as [31] Lemma 1.1.13.i we now compute

$$
\begin{aligned}
\Phi_{n-1}(\mathbf{b}_0, \ldots, \mathbf{b}_{n-1}) &= \sum_{m=0}^{n-1} \pi_L^m \mathbf{b}_m^{q^{n-1-m}} \\
&\equiv \sum_{m=0}^{n-1} \pi_L^m [b_{m,0}^{q^{n-1-m}}] \mod \pi_L^n W(B)_L \\
&= (b_{0,0}^{q^{n-1}}, \ldots, b_{n-1,0}^{q^{n-1}}, 0, \ldots) \\
&= \phi_q^{n-1}(b_{0,0}, \ldots, b_{n-1,0}, 0, \ldots) \\
&= \phi_q^{n-1} \circ W_n(\mathrm{pr})_L(\mathbf{b}_0, \ldots, \mathbf{b}_{n-1}) .
\end{aligned}
$$

ii. First of all we note that the Frobenius on $W(B)_L$ is the qth power map modulo π_L. Hence the homomorphism $s_{W(B)_L}$ exists.

Let $\mathbf{b} = (\mathbf{b}_0, \ldots, \mathbf{b}_j, \ldots) \in W(W(B)_L)_L$ with $\mathbf{b}_j = (b_{j,0}, b_{j,1}, \ldots) \in W(B)_L$ be the image under $s_{W(B)_L}$ of some $b = (b_0, b_1, \ldots) \in W(B)_L$. We have to show that $b_i = b_{i,0}$ for any $i \geq 0$. By the characterizing property of $s_{W(B)_L}$ we have $\Phi_i(\mathbf{b}) = (b_0^{q^i}, \ldots, b_j^{q^i}, \ldots)$. On the other hand the computation in the proof of i. shows that $\Phi_i(\mathbf{b}) = (b_{0,0}^{q^i}, \ldots, b_{i,0}^{q^i}, \ldots)$. Hence $b_i^{q^i} = b_{i,0}^{q^i}$ and therefore $b_i = b_{i,0}$. □

By construction (and [31] Proposition 1.1.26) we have $\mathbf{A}_L \subseteq W(\widetilde{\mathbf{E}})_L$. But there is the following observation.

Remark 7.3 Let $A \subseteq W(\widetilde{\mathbf{E}})_L$ be a ϕ_q-invariant o_L-subalgebra such that $A/\pi_L A \subseteq \widetilde{\mathbf{E}}$; we then have $A \subseteq W(A/\pi_L A)_L$.

Proof We consider the diagram

$$
\begin{array}{ccccc}
A & \xrightarrow{s_A} & W(A)_L & \xrightarrow{W(\mathrm{pr})_L} & W(A/\pi_L A)_L \\
\downarrow \subseteq & & \downarrow \subseteq & & \downarrow \subseteq \\
W(\widetilde{\mathbf{E}})_L & \xrightarrow{s_{W(\widetilde{\mathbf{E}})_L}} & W(W(\widetilde{\mathbf{E}})_L)_L & \xrightarrow{W(\mathrm{pr})_L} & W(\widetilde{\mathbf{E}})_L .
\end{array}
$$

For the commutative left square we apply Proposition 7.1 (with $\sigma := \phi_q$). The right hand square is commutative by naturality. By Lemma 7.2.ii the composite map in the bottom row is the identity. Hence the composite map in the top row must be an inclusion. □

This applies to $\mathbf{A}_L$ and shows that

$$\mathbf{A}_L \subseteq W(\mathbf{E}_L)_L \subseteq W(\widetilde{\mathbf{E}})_L$$

holds true. In particular, we have the commutative diagram (cf. Proposition 7.1.ii)

$$\begin{array}{ccc} \mathbf{A}_L & \xrightarrow{\subseteq} & W(\mathbf{E}_L)_L \\ {\scriptstyle \mathrm{pr}}\downarrow & \overset{\alpha_n}{\searrow} & \downarrow{\scriptstyle \mathrm{pr}} \\ \mathbf{A}_L/\pi_L^n\mathbf{A}_L & \xrightarrow{\overline{\alpha}_n := s_{\mathbf{A}_L,n}} & W_n(\mathbf{E}_L)_L \end{array} \tag{36}$$

for any $n \geq 1$, where α_n by definition is the composite of the outer maps. For later use before Lemma 7.18 we note that Remark 7.3 also applies to $\mathbf{A}$ showing that $\mathbf{A} \subseteq W(\mathbf{E}_L^{sep})_L$.

Lemma 7.4 *For any $n \geq 1$ the diagram*

$$\begin{array}{ccc} W_n(\mathbf{A}_L)_L & \xrightarrow{\Phi_{n-1}} & \mathbf{A}_L \\ {\scriptstyle W_n(\mathrm{pr})_L}\downarrow & & \downarrow{\scriptstyle \alpha_n} \\ W_n(\mathbf{E}_L)_L & \xrightarrow{\phi_q^{n-1}} & W_n(\mathbf{E}_L)_L \end{array}$$

is commutative.

Proof We consider the diagram

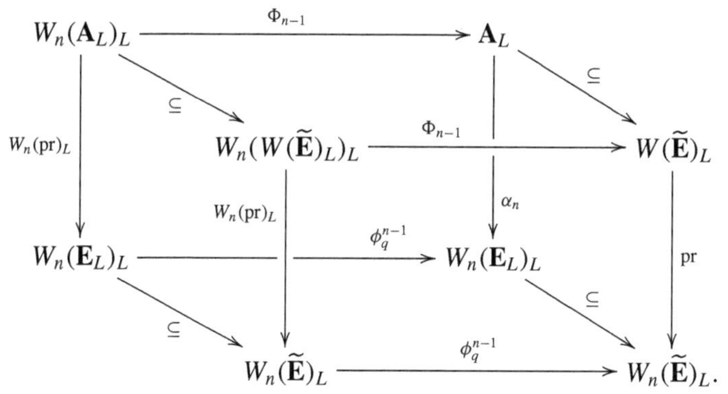

The front square is commutative by Lemma 7.2.i. The top and bottom squares are commutative by the naturality of the involved maps and the side squares for trivial reasons. Hence the back square is commutative as claimed. □

Lemma 7.5 *For any $n \geq 1$ the map $\overline{\alpha}_n : \mathbf{A}_L/\pi_L^n\mathbf{A}_L \longrightarrow W_n(\mathbf{E}_L)_L$ is injective.*

Proof We have to show that $V_n(\mathbf{E}_L)_L \cap \mathbf{A}_L = \pi_L^n\mathbf{A}_L$. We know already that $\pi_L^n\mathbf{A}_L \subseteq V_n(\mathbf{E}_L)_L \cap \mathbf{A}_L = V_n(\widetilde{\mathbf{E}})_L \cap \mathbf{A}_L = \pi_L^n W(\widetilde{\mathbf{E}})_L \cap \mathbf{A}_L$. But $\mathbf{A}_L \subseteq W(\widetilde{\mathbf{E}})_L$ both are discrete valuation rings with the prime element π_L. Therefore we must have equality. □

The above two lemmas together with the surjectivity of $W_n(\alpha_1)_L$ imply that, for any $n \geq 1$, there is a unique homomorphism of o_L-algebras $w_{n-1} : W_n(\mathbf{E}_L)_L \longrightarrow \mathbf{A}_L/\pi_L^n\mathbf{A}_L$ such that the diagram

$$\begin{array}{ccc} W_n(\mathbf{A}_L)_L & \xrightarrow{\Phi_{n-1}} & \mathbf{A}_L \\ {\scriptstyle W_n(\mathrm{pr})_L}\downarrow & & \downarrow{\scriptstyle \mathrm{pr}} \\ W_n(\mathbf{E}_L)_L & \xrightarrow{w_{n-1}} & \mathbf{A}_L/\pi_L^n\mathbf{A}_L \end{array} \tag{37}$$

is commutative. Furthermore, we have

$$\overline{\alpha}_n \circ w_{n-1} = \phi_q^{n-1}{}_{|W_n(\mathbf{E}_L)_L} \quad \text{and} \quad w_{n-1} \circ \overline{\alpha}_n = \phi_q^{n-1}{}_{|\mathbf{A}_L/\pi_L^n\mathbf{A}_L} . \tag{38}$$

We also may apply Proposition 7.1 to o_L itself (with $\sigma := \mathrm{id}$) and obtain analogous commutative diagrams as well as the corresponding maps

$$\begin{array}{ccccc} o_L & & & & \\ {\scriptstyle \mathrm{pr}}\downarrow & \searrow^{\alpha_n} & & & \\ o_L/\pi_L^n o_L & \xrightarrow{\overline{\alpha}_n} & W_n(k_L)_L & \xrightarrow{w_{n-1}} & o_L/\pi_L^n o_L . \end{array}$$

But here $\overline{\alpha}_n$ and w_{n-1} are isomorphisms which are inverse to each other (cf. Proposition 7.1.iii). Of course these maps for o_L and $\mathbf{A}_L$ are compatible with respect to the inclusions $o_L \subseteq \mathbf{A}_L$ and $k_L \subseteq \mathbf{E}_L$.

For the rest of this section let K denote any local field isomorphic to $k((Z))$ with $k = k_L$ (such an isomorphism depending on the choice of an uniformizing element Z of K), K^{sep} any separable closure of it and $H = \mathrm{Gal}(K^{sep}/K)$ its Galois group. Furthermore we write $\overline{K}$ for an algebraic closure of K^{sep}, ϕ_q for the qth power Frobenius, and $\wp := \phi_q - 1$ for the corresponding Artin–Schreier operator. By induction with respect to n one easily proves the following fact.

Lemma 7.6 *We have the short exact sequences*

$$0 \longrightarrow W_n(k)_L \longrightarrow W_n(K^{sep})_L \xrightarrow{\wp} W_n(K^{sep})_L \longrightarrow 0$$

and

$$0 \longrightarrow W_n(k)_L \longrightarrow W_n(\overline{K})_L \xrightarrow{\wp} W_n(\overline{K})_L \longrightarrow 0 .$$

From the H-group cohomology long exact sequence associated with the first sequence above we obtain a homomorphism

$$W_n(K)_L = (W_n(K^{sep})_L)^H \xrightarrow{\partial} H^1(H, W_n(k)_L) \xrightarrow{rec_K^*} \mathrm{Hom}^{cont}(K^\times, W_n(k)_L) ,$$

which induces the generalized, bilinear Artin–Schreier–Witt pairing

$$\begin{aligned} [\,,\,) := [\,,\,)_K : W_n(K)_L \times K^\times &\longrightarrow W_n(k)_L \\ (x, a) &\longmapsto [x, a) := \partial(x)(rec_K(a)) , \end{aligned}$$

i.e., $[x, a) = rec_K(a)(\alpha) - \alpha$ for any $\alpha \in W_n(K^{sep})_L$ with $\wp(\alpha) = x$. It is bilinear in the sense that it is o_L-linear in the first and additive in the second variable.

Remark 7.7 Let K^{rad} be the perfect closure of K in $\overline{K}$. Then one can use the second exact sequence to extend the above pairing to $W_n(K^{rad})_L \times K^\times$.

For a separable extension F of K we obtain similarly by taking $\mathrm{Gal}(K^{sep}/F)$-invariants (instead of H-invariants) an Artin–Schreier–Witt pairing for F

$$[\ ,\ )_F : W_n(F)_L \times F^\times \longrightarrow W_n(k)_L$$
$$(x, a) \longmapsto [x, a) := \partial(x)(rec_F(a)) ,$$

(with respect to the same q!) satisfying

$$[x, a)_F = [x, \mathrm{Norm}_{F/K}(a))_K \qquad \text{for } x \in W_n(K)_L \text{ and } a \in F^\times$$

—and similarly for any pair of separable extensions F and F'—by the functoriality of class field theory.

Although $W_n(k)_L$ is not a cyclic group in general, many aspects of Kummer/Artin–Schreier theory still work. In particular, for any $\alpha = (\alpha_0, \ldots, \alpha_{n-1}) \in W_n(K^{sep})_L$ with $\wp(\alpha) = x \in W_n(K)_L$ the extension $K(\alpha) := K(\alpha_0, \ldots, \alpha_{n-1}) = (K^{sep})^{H_x} = K(\wp^{-1}(x))$ of K is Galois with abelian Galois group $\mathrm{Gal}(K(\alpha)/K)$ contained in $W_n(k)_L$ via sending σ to $\chi_x(\sigma) := \sigma(\alpha) - \alpha$; here $H_x \subseteq H$ denotes the stabilizer of α, which also is the stabilizer of $\wp^{-1}(x)$.

We also need the injective additive map

$$\tau : \quad W_n(B)_L \longrightarrow W_{n+1}(B)_L$$
$$(x_0, \ldots, x_{n-1}) \longmapsto (0, x_0, \ldots, x_{n-1})$$

induced in an obvious way by the additive Verschiebung V (cf. [31] Proposition 1.1.10). If B is a k_L-algebra then

$$\tau \circ \phi_q = \phi_q \circ \tau \quad \text{and, in particular,} \quad \wp \circ \tau = \tau \circ \wp \tag{39}$$

(cf. [31] Proposition 1.1.18.i).

Lemma 7.8 *Let $K \subseteq F \subseteq K^{sep}$ be a finite extension. Then, for any $a \in F^\times$, $x \in W_n(F)_L$, and $\alpha \in W_n(K^{sep})_L$ with $\wp(\alpha) = x$ we have:*

i. $[\tau x, a)_F = \tau[x, a)_F$ *(where we use the same notation for the pairing at level $n + 1$ and n, respectively!);*
ii. *if a belongs to $(F^\times)^{p^n}$, then* $[x, a)_F = 0$;
iii. $[x, a)_F = 0$ *if and only if* $a \in \mathrm{Norm}_{F(\alpha)/F}(F(\alpha)^\times)$.

Proof i. By (39) we have $\wp(\tau\alpha) = \tau x$. Therefore $[\tau x, a) = rec_K(a)(\tau\alpha) - \tau\alpha = \tau(rec_K(a)(\alpha) - \alpha) = \tau[x, a)$. ii. Since $p^n W_n(k)_L = 0$ (this is not sharp with regard to p^n!) this is immediate from the bilinearity of the pairing. iii. Because of $[x, a)_F = \chi_x(rec_F(a))$ this is clear from local class field theory. □

For any subset S of an o_L-algebra R we define the *subset*

$$W_n(S)_L := \{(s_0, \ldots, s_{n-1}) \in W_n(R)_L : s_i \in S \text{ for all } i\}$$

of $W_n(R)_L$ as well as $VW_n(S)_L := V(W_n(S)_L)$. If $I \subseteq R$ is an ideal then $W_n(I)_L$ is an ideal in $W_n(R)_L$, and we have the exact sequence

$$0 \longrightarrow W_n(I)_L \longrightarrow W_n(R)_L \longrightarrow W_n(R/I)_L \longrightarrow 0 . \tag{40}$$

If $R' \subseteq R$ is an o_L-subalgebra (not necessarily with a unit), then $W_n(R')_L \subseteq W_n(R)_L$ forms a subgroup and there is an exact sequence of abelian groups

$$0 \longrightarrow VW_n(R')_L \longrightarrow W_n(R')_L \xrightarrow{\Phi_0} R' \longrightarrow 0 . \tag{41}$$

We apply this to $R' = ak[a]$ for $a \in K^\times$.

Proposition 7.9 *For any $x \in W_n(ak[a])_L$ we have $[x, a) = 0$.*

Proof We prove by induction on n that for any finite separable extension F of K and for any $a \in F^\times$ the corresponding statement holds true with $R' = ak[a]$.

For both, $n = 1$ (trivially) and $n > 1$ (by induction hypothesis and Lemma 7.8.i), we know the implication

$$x \in VW_n(ak[a])_L \Longrightarrow [x, a) = 0 .$$

Therefore, for arbitrary $x \in W_n(ak[a])_L$ we have $[x, a) = [[x_0], a)$ by the bilinearity of the pairing and Lemma 1.1.13.i in [31]—we constantly will make use of the fact that the first component of Witt vectors behaves additively. Moreover, again by the additivity of the pairing and using (41) it suffices to prove (for all $n > 0$) that

$$[[ra^l], a) = 0 \text{ for all } r \in k^\times, l \geq 1 .$$

Writing $l = l'p^m$ with l' and p coprime and denoting by r' the p^mth root of r we see that

$$\begin{aligned} l'[[ra^l], a) &= [[(r'a^{l'})^{p^m}], a^{l'}) = [[(r'a^{l'})^{p^m}], a^{l'}) + [[(r'a^{l'})^{p^m}], r') \\ &= [[(r'a^{l'})^{p^m}], r'a^{l'}) \end{aligned}$$

by Lemma 7.8.ii because $r' \in (k^\times)^{p^n}$. Noting that l' is a unit in $W_n(k)_L$ we are reduced to the case $x = [a^{p^m}]$ for $m \geq 0$.

To this end let $\alpha_0 \in F^{sep}$ be in $\wp^{-1}(a)$ and define $\tilde{\alpha_0} := \prod_{\xi \in k/\operatorname{im}(\chi_a)}(\alpha_0 + \xi)$. Then

$$\begin{aligned} \operatorname{Norm}_{F(\alpha_0)/F}(\tilde{\alpha_0}) &= \prod_{\sigma \in \operatorname{Gal}(F(\alpha_0)/F)} \sigma\tilde{\alpha_0} \\ &= \prod_{\xi' \in \operatorname{im}(\chi_a)} \prod_{\xi \in k/im(\chi_a)} (\alpha_0 + \xi' + \xi) \\ &= \prod_{\xi \in k}(\alpha_0 + \xi) = a\,, \end{aligned} \tag{42}$$

since $\alpha_0 + \xi$, $\xi \in k$, are precisely the zeros of $X^q - X - a$.

Now let β be in $\wp^{-1}([a^{p^m}])$ with $\beta_0 = \alpha_0^{p^m}$. Then we have $F(\alpha_0)(\beta) = F(\alpha_0, \beta_0, \ldots, \beta_{n-1}) = F(\alpha_0)(\beta - [\beta_0])$ and

$$\wp(\beta - [\beta_0]) = [a^{p^m}] - \wp([\alpha_0^{p^m}])$$

belongs to $V W_n(\alpha_0 k[\alpha_0])_L$ because $a^{p^m} = (\alpha_0^q - \alpha_0)^{p^m}$ belongs to $\alpha_0 k[\alpha_0]$ as well as $\wp([\alpha_0^{p^m}]) = [(\alpha_0^{p^m})^q] - [\alpha_0^{p^m}] \in W_n(\alpha_0 k[\alpha_0])_L$ and $[a^{p^m}]_0 = a^{p^m} = \wp([\alpha_0^{p^m}])_0$.

Note that for $n = 1$ we have $F(\alpha_0)(\beta) = F(\alpha_0)$ (i.e., the last consideration is not needed) and since a is a norm with respect to the extension $F(\alpha_0)/F$ the claim follows from Lemma 7.8.iii.

Now let $n > 1$. Then, the induction hypothesis for $F' := F(\alpha_0)$ and $a' := \alpha_0$ implies that $[[a^{p^m}] - \wp([\alpha_0^{p^m}]), \alpha_0) = 0$, i.e., that $\alpha_0 \in \operatorname{Norm}_{F(\alpha_0)(\beta)/F(\alpha_0)}(F(\alpha_0)(\beta)^\times)$ by Lemma 7.8.iii.

Replacing α_0 by $\alpha_0 + \xi$, for $\xi \in k$, we see that we also have $\alpha_0 + \xi \in \operatorname{Norm}_{F(\alpha_0)(\beta)/F(\alpha_0)}(F(\alpha_0)(\beta)^\times)$ (note that $F(\alpha_0) = F(\alpha_0 + \xi)$ and the composite $F(\alpha_0)(\beta) = F(\alpha_0)F(\beta)$ does not depend on the choices involved above). By the multiplicativity of the norm we obtain that $\tilde{\alpha_0}$ lies in $\operatorname{Norm}_{F(\alpha_0)(\beta)/F(\alpha_0)}(F(\alpha_0)(\beta)^\times)$, whence by transitivity of the norm and (42) a belongs to $\operatorname{Norm}_{F(\alpha_0)(\beta)/F}(F(\alpha_0)(\beta)^\times)$ and thus also to $\operatorname{Norm}_{F(\beta)/F}(F(\beta)^\times)$ because $F(\beta) \subseteq F(\alpha_0)(\beta)$. Thus $[[a^{p^m}], a) = 0$, again by Lemma 7.8.iii, as had to be shown. □

Now we will define a second bilinear pairing

$$(\ ,\ ) : W_n(K)_L \times K^\times \longrightarrow W_n(k)_L$$

by using the residue pairing (cf. (12))

$$\operatorname{Res} : \mathscr{A}_L \times \Omega^1_{\mathscr{A}_L} \longrightarrow o_L\,.$$

To this end we choose an isomorphism $\mathscr{A}_L/\pi_L\mathscr{A}_L = k((Z)) \cong K$ and remark that our construction will not depend on this choice of a prime element of K by Remark 3.4. Consider the map $d\log : o_L((Z))^\times \longrightarrow \Omega^1_{\mathscr{A}_L}$, sending f to $\frac{df}{f}$. We define the upper pairing in

$$\begin{array}{ccccc} W_n(\mathscr{A}_L)_L & \times & o_L((Z))^\times & \xrightarrow{\{\,,\,\}} & o_L \\ \downarrow{\scriptstyle \Phi_{n-1}} & & \downarrow{\scriptstyle d\log} & & \| \\ \mathscr{A}_L & \times & \Omega^1_{\mathscr{A}_L} & \xrightarrow{\mathrm{Res}} & o_L \end{array}$$

via the commutativity of the diagram.

Lemma 7.10 *There is a unique well defined bilinear pairing* (,) *such that the diagram*

$$\begin{array}{ccccc} W_n(\mathscr{A}_L)_L & \times & o_L((Z))^\times & \xrightarrow{\{\,,\,\}} & o_L \\ \downarrow{\scriptstyle W_n(\alpha_1)_L} & & \downarrow{\scriptstyle \mathrm{mod}\ \pi_L} & & \downarrow{\scriptstyle \alpha_n} \\ W_n(K)_L & \times & K^\times & \xrightarrow{(\,,\,)} & W_n(k)_L, \end{array}$$

is commutative.

Proof (Note that the reduction map $o_L((Z))^\times \to K^\times$ indeed is surjective.) We need to show that

$$\{\ker(W_n(\mathrm{pr})_L), o_L((Z))^\times\} \subseteq \pi_L^n o_L$$
$$\text{and}\quad \{W_n(\mathscr{A}_L)_L, \ker(o_L((Z))^\times \to K^\times)\} \subseteq \pi_L^n o_L\ .$$

For $a = (a_0, \ldots, a_{n-1}) \in W_n(\mathrm{pr})_L$ such that $a_i \in \pi_L \mathscr{A}_L$ we obviously have $\Phi_{n-1}(a) \in \pi_L^n \mathscr{A}_L$. Hence $\{a, o_L((Z))^\times\} \subseteq \pi_L^n o_L$.

For the second inclusion we first observe that $\ker(o_L((Z))^\times \to K^\times) = 1 + \pi_L o_L[[Z]]$. Hence we have to prove that

$$\mathrm{Res}(\Phi_{n-1}(f) d\log(1 + \pi_L h)) \in \pi_L^n o_L \tag{43}$$

holds true for all $f = (f_0, \ldots, f_{n-1}) \in \mathscr{A}_L^n$ and $h \in o_L[[Z]]$. We observe that sending Z to $Z' := Z(1 + \pi_L h)$ defines a ring automorphism first of $o_L[[Z]]$, then by localization of $o_L((Z))$, and finally by π_L-adic completion of $\mathscr{A}_L$. We write $f_i(Z) = g_i(Z')$ and $g := (g_0, \ldots, g_{n-1})$, and we compute

$$\begin{aligned} &\mathrm{Res}_Z(\Phi_{n-1}(f) d\log(1 + \pi_L h)) \\ &\qquad = \mathrm{Res}_Z(\Phi_{n-1}(g(Z')) d\log(Z')) - \mathrm{Res}_Z(\Phi_{n-1}(f(Z)) d\log(Z)) \\ &\qquad = \mathrm{Res}_{Z'}(\Phi_{n-1}(g(Z')) d\log(Z')) - \mathrm{Res}_Z(\Phi_{n-1}(f(Z)) d\log(Z)) \\ &\qquad = \mathrm{Res}_Z([\Phi_{n-1}(g(Z)) - \Phi_{n-1}(f(Z))] d\log(Z))\ . \end{aligned}$$

Here the second equality uses the fact that the residue does not depend on the choice of the variable (cf. Remark 3.4) while in the third equality we just rename the variable Z' into Z both in the argument and the index of Res, which of course does not change the value. Now note that, since $Z' \equiv Z \bmod \pi_L o_L[[Z]]$, we have the congruences

$$f_i(Z) = g_i(Z') \equiv g_i(Z) \mod \pi_L \mathscr{A}_L \qquad \text{for any } 0 \leq i \leq n-1.$$

This implies that

$$\Phi_{n-1}(g(Z)) - \Phi_{n-1}(f(Z)) \equiv 0 \mod \pi_L^n o_L$$

(cf. [31] Lemma 1.1.2.i) whence the claim (43) by the o_L-linearity of the residue. □

Remark 7.11 Alternatively, one can define similarly a pairing by using the full ghost map $\Phi = (\Phi_0, \ldots, \Phi_{n-1})$ via commutativity of the diagram

$$\begin{array}{ccccc} W_n(\mathscr{A}_L)_L & \times & o_L((Z))^\times & \longrightarrow & W_n(o_L)_L \\ \downarrow{\scriptstyle \Phi} & & \downarrow{\scriptstyle d\log} & & \downarrow{\scriptstyle \Phi} \\ \mathscr{A}_L^n & \times & \Omega^1_{\mathscr{A}_L} & \xrightarrow{\text{Res}} & o_L^n \end{array}$$

and by showing that for all $f \in W_n(\mathscr{A}_L)_L$ and $h \in o_L((Z))^\times$ the residue vector $(\text{Res}(\Phi_i(f)\frac{dh}{h}))_i$ belongs to the image of (the right hand) Φ. We leave it to the interested reader to check that this induces the same pairing as $(\ ,\ )$ above by applying $W_n(\text{pr})_L$ to the target. For unramified Witt vectors this is done in [36] Proposition 3.5.[2]

Our aim is to show that the two pairings $[\ ,\ )$ and $(\ ,\ )$, in fact, coincide. This generalizes a result of Witt ([38] Satz 18), which we learned from [15, 16]. The strategy is to reduce this to the comparison of the restrictions of the two pairings to $W_n(k)_L \times K^\times$.

For an element $x = \sum_j x_j Z^j \in K$ with $x_j \in k$ and $x_j = 0$ for $j < v_Z(x)$ (the valuation of K) we set $x^+ := \sum_{j \geq 1} x_j Z^j$ and $x^- := \sum_{j<0} x_j Z^j$. Then, for $x = (x_0, \ldots, x_{n-1}) \in W_n(K)_L$, with arbitrary $n \geq 1$, we define iteratively elements (the 'constant term' and the plus and negative parts of x with respect to the variable Z)

$$\Omega_Z^n(x) \in W_n(k)_L, \ x^+ \in W_n(Zk[[Z]])_L, \ \text{and } x^- \in W_n(Z^{-1}k[Z^{-1}])_L$$

[2]Another alternative formulation for the definition of $(\ ,\ )$ goes as follows: The residue pairing

$$\text{Res} : \mathscr{A}_L/\pi_L^n \mathscr{A} \times \Omega^1_{\mathscr{A}_L/\pi_L^n \mathscr{A}} \longrightarrow o_L/\pi_L^n o_L$$

induces the pairing

$$\begin{array}{ccccc} \text{im}(\Phi_{n-1}) + \pi_L^n \mathscr{A}_L/\pi_L^n \mathscr{A} & \times & \Omega^1_{\mathscr{A}_L/\pi_L^n \mathscr{A}}/d\log(1+\pi_L o_L[[Z]]) & \longrightarrow & o_L/\pi_L^n o_L \\ \uparrow{\scriptstyle w_{n-1}} & & \uparrow & & \uparrow{\scriptstyle w_{n-1}} \\ W_n(K)_L & \times & K^\times & \xrightarrow{(\ ,\ )} & W_n(k)_L, \end{array}$$

where the middle vertical map is induced by $d\log$ and the inverse of the isomorphism $o_L((Z))^\times/(1 + \pi_L o_L[[Z]]) \cong K^\times$.

such that

$$x = \Omega_Z^n(x) + x^+ + x^- .$$

For $n = 1$ put

$$\Omega_Z^1(x) := [x_0 - x_0^+ - x_0^-],\ x^+ := [x_0^+],\ \text{and}\ x^- := [x_0^-]$$

and for $n > 1$ define

$$\Omega_Z^n(x) := [x_0 - x_0^+ - x_0^-] + \tau\Omega_Z^{n-1}(y),\ x^+ := [x_0^+] + \tau y^+,\ \text{and}\ x^- := [x_0^-] + \tau y^- ,$$

where $y \in W_{n-1}(K)_L$ satisfies $\tau y = x - [x_0 - x_0^+ - x_0^-] - [x_0^+] - [x_0^-]$.

Remark 7.12 $x = \Omega_Z^n(x) + x^+ + x^-$ is the unique decomposition of $x \in W_n(K)_L$ such that the three summands lie in $W_n(k)_L$, $W_n(Zk[[Z]])_L$, and $W_n(Z^{-1}k[Z^{-1}])_L$, respectively.

Proof Let $x = a + a_+ + a_-$ be any decomposition such that $a \in W_n(k)_L$, $a_+ \in W_n(Zk[[Z]])_L$, and $a_- \in W_n(Z^{-1}k[Z^{-1}])_L$. Since the projection onto the zeroth component is additive we immediately obtain that $a = [x_0 - x_0^+ - x_0^-] + \tau b$, $a_+ = [x_0^+] + \tau b_+$, and $x_- = [x_0^-] + \tau b_-$ for (uniquely determined) elements $b \in W_{n-1}(k)_L$, $b_+ \in W_{n-1}(Zk[[Z]])_L$, and $b_- \in W_{n-1}(Z^{-1}k[Z^{-1}])_L$. We put $y := b + b_+ + b_-$ and obtain $x = [x_0 - x_0^+ - x_0^-] + [x_0^+] + [x_0^-] + \tau y$. Hence y is the element in the above inductive construction for x. By induction with respect to n we have $b = \Omega_Z^{n-1}(y)$, $b_+ = y^+$ and $b_- = y^-$. It follows that $a = \Omega_Z^n(x)$, $a_+ = x^+$, and $a_- = x^-$. □

Lemma 7.13 *For any prime element Z in K and any $x \in W_n(K)_L$ we have*

$$(x, Z) = \Omega_Z^n(x) .$$

Proof For $x \in W_n(Zk[[Z]])_L \cup W_n(Z^{-1}k[Z^{-1}])_L$ we may choose the lift f of x to lie in $W_n(Zo_L[[Z]])_L$ and $W_n(Z^{-1}o_L[Z^{-1}])_L$, respectively. It is straightforward to see that then $\{f, Z\} = \mathrm{Res}(\Phi_{n-1}(f)d\log Z) = \mathrm{Res}(\Phi_{n-1}(f)\frac{dZ}{Z}) = 0$. By Remark 7.12 we have $\Omega_Z^n(x) = 0$ as well.

By the additivity of (,) in the first component it therefore remains to treat the case that $x \in W_n(k)_L$. Let $\tilde{x} \in W_n(W(k)_L)_L \subseteq W_n(\mathscr{A}_L)_L$ be any lift of x. Then we have that $\{\tilde{x}, Z\} = \mathrm{Res}(\Phi_{n-1}(\tilde{x})\frac{dZ}{Z}) = \Phi_{n-1}(\tilde{x})$. But $\alpha_n(\Phi_{n-1}(\tilde{x})) = \overline{\alpha}_n \circ w_{n-1}(x) = \phi_q^{n-1}(x) = x$ by (37) and (38) for o_L. Hence $(x, Z) = \alpha_n(\{\tilde{x}, Z\}) = \alpha_n(\Phi_{n-1}(\tilde{x})) = x = \Omega_Z^n(x)$, the last identity again by Remark 7.12. □

Lemma 7.14 *For any prime element Z in K and any $x \in W_n(k)_L$ we have $[x, Z) = (x, Z)$.*

Proof We choose $\alpha \in W_n(k^{sep})_L$ such that $\wp(\alpha) = x$. Then $K(\alpha) \subseteq k(\alpha)((Z))$ is an unramified extension of $K = k((Z))$. From local class field theory we therefore obtain that $rec_K(Z) = \phi_q$. It follows that $[x, Z) = rec_K(Z)(\alpha) - \alpha = \wp(\alpha) = x$. On the other hand Lemma 7.13 implies that $(x, Z) = \Omega_Z^n(x) = x$ as well. □

Proposition 7.15 *For any prime element Z in K, any $a \in K^\times$, and any $x \in W_n(K)_L$ we have*

$$[x, a) = [(x, a), Z) \quad \textit{and} \quad (x, a) = ((x, a), Z)\ .$$

Proof As $\Omega_Z^n((x, a)) = (x, a)$ the second identity is a consequence of Lemma 7.13.

For the fist identity we first consider the *special case* $a = Z$. We will compare the decompositions

$$[x, Z) = [\Omega_Z^n(x), Z) + [x^+, Z) + [-x^-, Z^{-1}) \ \text{ and}$$
$$[(x, Z), Z) = [(\Omega_Z^n(x), Z), Z) + [(x^+, Z), Z) + [(-x^-, Z^{-1}), Z)$$

term by term. By Lemma 7.13 the two first terms coincide and the remaining terms in the second decomposition vanish. The last term in the first decomposition vanishes by Proposition 7.9. Hence it remains to show that $[x^+, Z) = 0$. For this it suffices to check that $W_n(Zk[[Z]])_L \subseteq \wp(W_n(K)_L)$. Indeed, we claim that for $y \in W_n(Zk[[Z]])_L$ the series $\sum_{i=0}^{\infty} \phi_q^i(y)$ converges in $W_n(k[[Z]])_L$ (componentwise in the Z-adic topology). We observe that, for $x \in W_n(k[[Z]])_L$ and $z \in W_n(Z^l k[[Z]])_L$ with $l \geq 0$, one has, by (40), the congruence

$$(x + z)_i \equiv x_i \mod Z^l k[[Z]]$$

for the components of the respective Witt vectors. It follows that each component of the sequence of partial sums $\sum_{i=0}^{m} \phi_q^i(y)$ forms a Cauchy sequence. Since ϕ_q, being the componentwise qth power map, obviously is continuous for the topology under consideration we obtain $\wp(-\sum_{i=0}^{\infty} \phi_q^i(y)) = y$.

For a *general* a we find a $\nu \in \mathbb{Z}$ and another prime element $Z' \in K$ such that $a = Z^\nu Z'$. Using bilinearity and the special case (for Z as well as Z') we compute

$$\begin{aligned}[x, a) &= \nu[x, Z) + [x, Z') = \nu[(x, Z), Z) + [(x, Z'), Z') \\ &= [(x, Z), Z^\nu) + [(x, Z'), Z) = [(x, Z^\nu Z'), Z) \\ &= [(x, a), Z)\ ;\end{aligned}$$

the third equality uses Lemma 7.14. □

Theorem 7.16 (Schmid–Witt formula) *The pairings* [,) *and* (,) *coincide.*

Proof This now is an immediate consequence of Lemma 7.14 and Proposition 7.15. □

Corollary 7.17 *For all $z \in W_n(K)_L$ and $\hat{u} \in o_L((Z))^\times$ any lift of $u \in K^\times$ we have*

$$Res(w_{n-1}(z)\frac{d\hat{u}}{\hat{u}}) = w_{n-1}(\partial(z)(rec_K(u))).$$

Proof Let $f \in W_n(\mathscr{A}_L)_L$ be a lift of z. Theorem 7.16 implies that

$$\partial(z)(rec_K(u)) = \alpha_n(\mathrm{Res}(\Phi_{n-1}(f)\frac{d\hat{u}}{\hat{u}}))$$

holds true. Applying w_{n-1} (for o_L) we obtain

$$w_{n-1}(\partial(z)(rec_K(u))) = \mathrm{Res}(\Phi_{n-1}(f)\frac{d\hat{u}}{\hat{u}}) \mod \pi_L^n o_L\ .$$

On the other hand, by (37), the element $\Phi_{n-1}(f)$ module π_L^n is equal to $w_{n-1}(z)$ (with w_{n-1} for $\mathscr{A}_L \cong \mathbf{A}_L$). □

We finally are able to establish the congruence (35). First note that since $\mathbf{A} \subseteq W(\mathbf{E}_L^{sep})_L$ we obtain the commutative diagram

$$\begin{array}{ccccccccc} 0 & \longrightarrow & o_L/\pi_L^n o_L & \longrightarrow & \mathbf{A}/\pi_L^n\mathbf{A} & \xrightarrow{\phi_q - 1} & \mathbf{A}/\pi_L^n\mathbf{A} & \longrightarrow & 0 \\ & & \cong \downarrow w_{n-1}^{-1} = \overline{\alpha_n} & & \downarrow & & \downarrow & & \\ 0 & \longrightarrow & W_n(k)_L & \longrightarrow & W_n(\mathbf{E}_L^{sep})_L & \xrightarrow{\phi_q - 1} & W_n(\mathbf{E}_L^{sep})_L & \longrightarrow & 0. \end{array}$$

We recall that ∂_φ and ∂ denote the connecting homomorphisms arising from the upper and lower exact sequence, respectively. We obviously have the identity $(\overline{\alpha}_n)_* \circ \partial_\varphi = \partial \circ \overline{\alpha}_n$ for the map $\overline{\alpha}_n$ which was defined in (36).

Lemma 7.18 *For any $z \in \mathbf{A}_L$ and $\hat{u} \in o_L((\omega_{LT}))^\times \subseteq \mathbf{A}_L^\times$ any lift of $u \in \mathbf{E}_L^\times$ we have*

$$Res(\varphi_L^{n-1}(z)\frac{d\hat{u}}{\hat{u}}) \equiv \partial_\varphi(z)(rec_{\mathbf{E}_L}(u)) \mod \pi_L^n o_L\ . \tag{44}$$

Proof We use the identity in Corollary 7.17 for the element $z' := \alpha_n(z)$ with respect to $K = \mathbf{E}_L$. Since $w_{n-1} \circ \overline{\alpha}_n = \varphi_L^{n-1}$ by (38) its left hand side becomes the left hand side of the assertion. For the right hand sides we compute

$$w_{n-1}(\partial(\alpha_n(z))(rec_{\mathbf{E}_L}(u))) = w_{n-1} \circ \alpha_n(\partial_\varphi(z)(rec_{\mathbf{E}_L}(u))) = \partial_\varphi(z)(rec_{\mathbf{E}_L}(u))\ .$$

□

As explained at the end of Sect. 6 this last lemma implies Proposition 6.3. The proof of Theorem 6.2 therefore now is complete.

8 Bloch and Kato's as Well as Kato's Explicit Reciprocity Law Revisited

In this section we give a proof of a generalization of the explicit reciprocity law of Bloch and Kato ([3] Theorem 2.1) as well as of (a special case of) Kato's explicit reciprocity law ([23] Theorem II.2.1.7) replacing his method of syntomic cohomology by generalizing the method of Fontaine in [16] from the cyclotomic to the general Lubin–Tate case.

First we recall some definitions and facts from [9]. (This reference assumes that the power series $[\pi_L](Z)$ is a polynomial. But, by some additional convergence considerations, the results can be seen to hold in general (cf. [31] §2.1 for more details).) The ideal $\mathbb{I}_L \subseteq W(\widetilde{\mathbf{E}}^+)_L$ is defined to be the preimage of $\pi_L o_{\mathbb{C}_p}$ under the surjective homomorphism of o_L-algebras $\theta : W(\widetilde{\mathbf{E}}^+)_L \longrightarrow o_{\mathbb{C}_p}$ (cf. [31] Lemma 1.4.17). [9] §8.5 introduces the π_L-adic completion $A_{max,L}$ of the subalgebra $W(\widetilde{\mathbf{E}}^+)_L[\frac{1}{\pi_L}\mathbb{I}_L] \subseteq W(\widetilde{\mathbf{E}}^+)_L[\frac{1}{\pi_L}]$ as well as $B^+_{max,L} := A_{max,L}[\frac{1}{\pi_L}] \subseteq B^+_{dR}$. The important point is that the Frobenius ϕ_q naturally extends to $A_{max,L} \subseteq B^+_{max,L}$ (but not to B^+_{dR}). Let $\overline{\mathbf{E}}_L$ denote the algebraic closure of $\mathbf{E}_L$ in $\widetilde{\mathbf{E}}$ and $\overline{\mathbf{E}}^+_L$ its ring of integers. Setting $\omega_L := \omega_{LT}$, $\omega_1 := \phi_q^{-1}(\omega_{LT}) \in W(\overline{\mathbf{E}}^+)_L$, $\xi_L := \omega_L\omega_1^{-1} \in W(\overline{\mathbf{E}}^+_L)_L$ (compare [31] Lemmas 2.1.13 and 2.1.18), $t_L := \log_{LT}(\omega_L) \in A_{max,L}$, and $B_{max,L} := B^+_{max,L}[\frac{1}{t_L}] \subseteq B_{dR}$ we have the following properties by Propositions 9.10 and 9.6 in loc. cit.:

$$(B^+_{max,L})^{G_L} = L, \tag{45}$$

$$\phi_q(t_L) - \pi_L t_L \quad \text{and} \quad \frac{t_L}{\omega_L} = 1 + \sum_{k\geq 2} e_k\omega_L^{k-1} \in A^\times_{max,L} \quad \text{with } e_k \in \pi_L^{-l_q(k)} o_L, \tag{46}$$

where $l_q(k)$ denotes the maximal integer l such that $q^l \leq k$,

$$W(\widetilde{\mathbf{E}}^+)_L\xi_L = \ker\left(W(\widetilde{\mathbf{E}}^+)_L \xrightarrow{\theta} o_{\mathbb{C}_p}\right), \tag{47}$$

$$W(\widetilde{\mathbf{E}}^+)_L\omega_L = \{x \in W(\widetilde{\mathbf{E}}^+)_L : \theta(\phi_q^i(x)) = 0 \text{ for all } i \geq 0\},$$

$$B^+_{max,L}t_L = \{x \in B^+_{max,L} : \theta(\phi_q^i(x)) = 0 \text{ for all } i \geq 0\}. \tag{48}$$

By (46) we see that

$$\left(\frac{t_L}{\omega_L}\right)^{-r} =: \sum_{m\geq 0} \lambda_{m,r}\omega_L^m$$

belongs to $L[[\omega_L]] \subseteq B^+_{dR}$ for $r \geq 0$.

Lemma 8.1 *i.* $\sum_{m\geq 0}\lambda_{m,r}\omega_L^m$ *converges in* $A_{max,L}$.
ii. $\sum_{m\geq 1}\lambda_{m+r,r}\omega_L^{m-1}$ *converges in* $B^+_{max,L}$.

Proof First of all we note that the π_L-adic completion R of the polynomial ring $o_L[Z]$ is the subring of all power series in $o_L[[Z]]$ whose coefficients tend to zero. Using the geometric series we see that $1+\pi_L R\subseteq R^\times$.

According to (46) and the proof of [9] Proposition 9.10 there exists a $g(Z)\in R$ such that

$$\left(\frac{t_L}{\omega_L}\right)^r = 1-\pi_L g(u)\ ,$$

where $u=\frac{\omega_L}{\pi_L}\in\pi_L^{-1}\mathbb{I}_L$ for $q\neq 2$ and $u=\frac{\omega_L}{\pi_L^2}\in\pi_L^{-2}\mathbb{I}_L^2$ for $q=2$, respectively. By the initial observation we have

$$(1-\pi_L g(Z))^{-1}=\sum_{m\geq 0} b_m Z^m\in R\ .$$

Thus

$$\left(\frac{t_L}{\omega_L}\right)^{-r}=\sum_{m\geq 0} b_m u^m=\sum_{m\geq 0}\lambda_{m,r}\omega_L^m$$

converges in $A_{max,L}$. For the second part of the assertion it remains to note that

$$\sum_{m\geq 1}\lambda_{m+r,r}\omega_L^{m-1}=\pi_L^{-(r+1)}\sum_{m\geq 0} b_{m+r+1}u^m\ .$$

□

Setting $\tau_r' := \sum_{m=0}^{r}\lambda_{m,r}\omega_L^m$ and $\tau_r := \omega_L^{-r}\tau_r'\in L[\frac{1}{\omega_L}]\subseteq W(\overline{\mathbf{E}}_L)_L[\frac{1}{\pi_L}]$ we have

$$\tau_r - t_L^{-r}\in L[[\omega_L]]\subseteq B^+_{dR}\ . \tag{49}$$

By [9] Proposition 9.25 (SEF 3E) we have the exact sequence

$$0\longrightarrow L\longrightarrow (B_{max,L})^{\phi_q=1}\longrightarrow B_{dR}/B^+_{dR}\longrightarrow 0\ . \tag{50}$$

We define

$$Fil^r B^+_{max,L} := B^+_{max,L}\cap t_L^r B^+_{dR}\qquad \text{for } r\geq 0.$$

Lemma 8.2 *i. For* $r\geq 1$ *the sequence*

$$0\longrightarrow Lt_L^r\longrightarrow Fil^r B^+_{max,L}\xrightarrow{\pi_L^{-r}\phi_q-1} B^+_{max,L}\longrightarrow 0$$

is exact.

ii. For $r = 0$ the sequence

$$0 \longrightarrow L \longrightarrow B^+_{max,L} \xrightarrow{\phi_q - 1} (\phi_q - 1)B^+_{max,L} \longrightarrow 0$$

is exact, and $((\phi_q - 1)B^+_{max,L})^{G_L} = L$.

iii. $\phi_q - 1$ is bijective on $\omega_L B^+_{max,L}$.

Proof i. and ii. By [9] Proposition 9.22 we have, for any $r \geq 0$, the exact sequence

$$0 \longrightarrow Lt_L^r \longrightarrow (B^+_{max,L})^{\phi_q = \pi_L^r} \longrightarrow B^+_{dR}/t_L^r B^+_{dR} \longrightarrow 0\,.$$

First we deduce that $Lt_L^r = (B^+_{max,L})^{\phi_q=\pi_L^r} \cap t_L^r B^+_{dR} = (Fil^r B^+_{max,L})^{\phi_q=\pi_L^r}$. Secondly it implies that $B^+_{dR} = (B^+_{max,L})^{\phi_q=\pi_L^r} + t_L^r B^+_{dR}$ and hence $B^+_{max,L} = (B^+_{max,L})^{\phi_q=\pi_L^r} + Fil^r B^+_{max,L}$. In the proof of [9] Proposition 9.25 it is shown that, for $r \geq 1$, the map $B^+_{max,L} \xrightarrow{\pi_L^{-r}\phi_q - 1} B^+_{max,L}$ is surjective. It follows that $B^+_{max,L} = (\pi_L^{-r}\phi_q - 1)Fil^r B^+_{max,L}$ for $r \geq 1$.

It remains to verify the second part of ii. By (45) we have $((\phi_q - 1)B^+_{max,L})^{G_L} \subseteq L$. For the reverse inclusion it suffices to consider any $a \in o_L \subseteq W(\overline{\mathbf{E}}_L^+)_L$. Since $\overline{\mathbf{E}}_L^+$ is integrally closed the map $\phi_q - 1$ on $W(\overline{\mathbf{E}}_L^+)_L$ is surjective. Hence we find a $y \in W(\overline{\mathbf{E}}_L^+)_L \subseteq B^+_{max,L}$ such that $(\phi_q - 1)y = a$.

iii. First of all we note that $\phi_q(\omega_L B^+_{max,L}) \subseteq \phi_q(\omega_1 \xi) B^+_{max,L} \subseteq \omega_L B^+_{max,L}$, so that, indeed, $\phi_q - 1$ restricts to an endomorphism of $\omega_L B^+_{max,L}$. By ii. we have $(\omega_L B^+_{max,L})^{\phi_q=1} \subseteq (B^+_{max,L})^{\phi_q=1} = L$. But $\omega_L B^+_{max,L} \cap L = 0$ by (47). This proves the injectivity. It suffices to establish surjectivity on $\omega_L A_{max,L}$. Let $\omega_L a \in \omega_L A_{max,L}$. We let $\omega_L = t_L u$ with $u \in A^\times_{max,L}$ and compute

$$\begin{aligned}\phi_q^n(\omega_L a) = \phi_q^n(t_L a u) = \phi_q^n(t_L)\phi_q^n(au) &= \pi_L^n t_L \phi_q^n(au) \\ &\in \pi_L^n t_L A_{max,L} = \omega_L \pi_L^n A_{max,L}\,.\end{aligned}$$

It follows that the series $-\sum_{n\geq 0} \phi_q^n(\omega_L a)$ converges (π_L-adically) to some element $\omega_L c \in \omega_L A_{max,L}$ such that $(\phi_q - 1)(\omega_L c) = \omega_L a$. □

The sequences in Lemma 8.2.i/ii induce, for any $r \geq 0$, the connecting homomorphism in continuous Galois cohomology

$$L = (B^+_{max,L})^{G_L} \xrightarrow{\partial^r} H^1(L, Lt_L^r)\,,$$

Note[3] that as a G_L-representation Lt_L^r is isomorphic to $V := L \otimes_{o_L} T$ (cf. Lemma 4.1.c and (2)). We introduce the composite homomorphism

[3] Setting $L^r_{adm} := L \cap (\pi_L^{-r}\phi_q - 1)(Fil^r B^+_{max,L})$ we still may define

$$L^r_{adm} \xrightarrow{\partial^r} H^1(L, Lt_L^r)$$

$$\delta^r : L \xrightarrow{\partial^r} H^1(L, Lt_L^r) \xrightarrow{\text{res}} \text{Hom}_{\Gamma_L}(H_L, Lt_L^r) .$$

By Lemma 7.6 we also have the connecting homomorphism

$$\partial_\varphi : W(\overline{\mathbf{E}}_L^{H_L})_L[\tfrac{1}{\pi_L}] \longrightarrow H^1(H_L, L) .$$

Proposition 8.3 *$\delta^r(a) = \partial_\varphi(\tau_r a)t_L^r$ for any $a \in L$.*

Proof Let $n \geq 0$ such that $\pi_L^n \tau_r \in o_L[\frac{1}{\omega_L}]$, and assume without loss of generality that a belongs to $\pi_L^n o_L$, i.e., that $\tau_r' a \in o_L[\omega_L] \subseteq W(\overline{\mathbf{E}}_L^+)_L$. In order to compute $\delta^r(a)$ we choose any $\alpha \in Fil^r B_{max,L}^+$ such that $(\pi_L^{-r}\phi_q - 1)(\alpha) = a$. Then

$$\delta^r(a)(g) = (g-1)\alpha \quad \text{for all } g \in H_L.$$

In fact, choosing α is equivalent to choosing $\beta := t_L^{-r}\alpha \in t_L^{-r}B_{max,L}^+ \cap B_{dR}^+$ such that

$$(\phi_q - 1)(\beta) = b := t_L^{-r} a .$$

On the other hand, to compute $\partial_\varphi(\tau_r a)$ we note that $\tau_r' a$ and ξ_L^r belong to $W(\overline{\mathbf{E}}_L^+)_L$ and that, since $\overline{\mathbf{E}}_L^+$ is integrally closed, the map $\phi_q - \xi_L^r : W(\overline{\mathbf{E}}_L^+)_L \to W(\overline{\mathbf{E}}_L^+)_L$ is surjective (argue inductively with respect to the length of Witt vectors). Hence we find a $y \in W(\overline{\mathbf{E}}_L^+)_L$ such that

$$\varphi_L(y) - \xi_L^r y = \tau_r' a .$$

By using (46) we see that $\omega_1^{-r} = \xi_L^r \omega_L^{-r} \in t_L^{-r} B_{max,L}^+$. It follows that the element $\beta_0 := \omega_1^{-r} y$ belongs to $t_L^{-r}B_{max,L}^+$ as well as to $W(\overline{\mathbf{E}}_L)_L$ and satisfies

$$\begin{aligned}(\phi_q - 1)(\beta_0) &= \phi_q(\omega_1)^{-r}\phi_q(y) - \omega_1^{-r} y \\ &= \omega_L^{-r}(\phi_q(y) - \xi_L^r y) \\ &= \omega_L^{-r}\tau_r' a = \tau_r a .\end{aligned}$$

Hence

$$\partial_\varphi(\tau_r a)(g) = (g-1)\beta_0 \quad \text{for all } g \in H_L.$$

We observe that $y \in W(\overline{\mathbf{E}}_L^+)_L \subseteq B_{dR}^+$, that ω_1 is a unit in B_{dR}^+ (since $\theta(\omega_1) \neq 0$ by the first sentence in the proof of [9] Proposition 9.6), and hence that $\beta_0 \in t_L^{-r}B_{max,L}^+ \cap B_{dR}^+$. At this point we are reduced to finding an element $\gamma \in (\omega_L B_{max,L}^+)^{H_L} \subseteq B_{max,L}^+ \subseteq t_L^{-r}B_{max,L}^+ \cap B_{dR}^+$ such that $(\phi_q - 1)(\gamma) = (t_L^{-r} - \tau_r)a$. We then put $\beta := \beta_0 + \gamma$ and obtain

---

(Footnote 3 continued)
without knowing the right hand surjectivity in Lemma 8.2.i and define ∂^r with source L_{adm}^r instead. In the course of the next Proposition one can then shown that $L_{adm}^r = L$.

$$\delta^r(a)(g) = (g-1)\beta \otimes t_L^r = (g-1)(\beta_0+\gamma) \otimes t_L^r = (g-1)\beta_0 \otimes t_L^r = \partial_\varphi(\tau_r a)(g) \otimes t_L^r$$

for any $g \in H_L$. In order to find γ it suffices, because of Lemma 8.2.iii, to observe that

$$t_L^{-r} - \tau_r = \omega_L\Big(\sum_{m\geq 1} \lambda_{m+r,r}\omega_L^{m-1}\Big) \in (\omega_L B^+_{max,L})^{H_L}$$

by Lemma 8.1.ii. □

Now we define the Coates–Wiles homomorphisms in this context for $r \geq 1$ and $m \geq 0$ by[4]

$$\psi^r_{CW,m} : \varprojlim_n o_{L_n}^\times \longrightarrow L_m$$
$$u \longmapsto \frac{1}{r!\pi_L^{rm}}\left(\partial_{\mathrm{inv}}^{r-1}\Delta_{LT}g_{u,\eta}\right)_{|Z=\eta_m} .$$

Then the map

$$\Psi^r_{CW,m} : \varprojlim_n o_{L_n}^\times \longrightarrow L_m t_L^r$$
$$u \longmapsto \psi^r_{CW,m}(u)t_L^r$$

is G_L-equivariant (it depends on the choice of η). In the following we abbreviate $\psi^r_{CW} := \psi^r_{CW,0}$ and $\Psi^r_{CW} := \Psi^r_{CW,0}$. One might think about these maps in terms of the formal identity

$$\log g_{u,\eta}(\omega_{LT}) = \sum_r \psi^r_{CW}(u)t_L^r = \sum_r \Psi^r_{CW}(u) \qquad \text{in } L[[t_L]] \subseteq B_{dR}.$$

But instead of justifying in which sense we may insert $g_{u,\eta}(\omega_{LT}(t_L))$[5] into the logarithm series, we shall only explain (and below use) the following identity

$$d\log g_{u,\eta}(\omega_{LT}) = \frac{dg_{u,\eta}(\omega_{LT})}{g_{u,\eta}(\omega_{LT})} = \sum_{r\geq 1} r\psi^r_{CW}(u)t_L^{r-1}dt_L .$$

Indeed, $t_L = \log_{LT}(\omega_{LT})$ implies $\frac{d}{dt_L}\omega_{LT} = g_{LT}(\omega_{LT})^{-1}$ and hence

$$\frac{d}{dt_L}f(\omega_{LT}) = g_{LT}(\omega_{LT})^{-1}\frac{d}{d\omega_{LT}}f(\omega_{LT}) = \partial_{\mathrm{inv}}(f)(Z)_{|Z=\omega_{LT}} .$$

---

[4] For $m > 0$ one can extend the definition to $\varprojlim_n L_n^\times$ while for $m = 0$ one cannot evaluate at $\eta_0 = 0$!

[5] This power series has a constant term: see [16] for a technical solution.

We calculate

$$\frac{1}{(r-1)!}\left((\tfrac{d}{dt_L})^{r-1}\frac{1}{g_{u,\eta}(\omega_{LT})}\frac{dg_{u,\eta}(\omega_{LT})}{dt_L}\right)_{|``t_L=0''}$$
$$= \frac{1}{(r-1)!}((\partial_{\mathrm{inv}}^{r-1}\Delta_{LT}g_{u,\eta}(Z))_{|Z=\omega_{LT}})_{|``t_L=0''}$$
$$= \frac{1}{(r-1)!}(\partial_{\mathrm{inv}}^{r-1}\Delta_{LT}g_{u,\eta}(Z))_{|Z=0}$$
$$= r\psi_{CW}^{r}(u)\ .$$

Proposition 8.4 *For all $a \in L$, $r \geq 1$, and $u \in \varprojlim_n o_{L_n}^{\times}$ we have*

$$ar\psi_{CW}^{r}(u) = \partial_{\varphi}(\tau_r a)(rec(u))\ .$$

Proof Using $L[[t_L]]=L[[\omega_L]]\subseteq B_{dR}^{+}$ we obtain from (49) that $\tau_r - t_L^{-r} \in L[[t_L]]$. By the discussion before Proposition 6.3 we therefore obtain

$$\begin{aligned}\partial_{\varphi}(\tau_r a)(rec(u)) &= \mathrm{Res}_{\omega_L}(\tau_r a d\log g_{u,\eta}(\omega_{LT}))\\ &= \mathrm{Res}_{t_L}(\tau_r a d\log g_{u,\eta}(\omega_{LT}))\\ &= \mathrm{Res}_{t_L}(a t_L^{-r} d\log g_{u,\eta}(\omega_{LT}))\\ &= \mathrm{Res}_{t_L}(a t_L^{-r}\sum_{n\geq 1} n\psi_{CW}^{n}(u)t_L^{n-1}dt_L)\\ &= ar\psi_{CW}^{r}(u)\ .\end{aligned}$$

□

With (50) also the sequence

$$0 \longrightarrow L \xrightarrow{diag} B_{max,L}^{\phi_q=1} \oplus B_{dR}^{+} \xrightarrow{(x,y)\mapsto x-y} B_{dR} \longrightarrow 0 \tag{51}$$

is exact. Tensoring with $V = L \otimes_{o_L} T$ over L gives the upper exact sequence in the commutative diagram

$$\begin{array}{ccccccccc} 0 \longrightarrow & V^{\otimes r} & \longrightarrow & (B_{max,L}^{\phi_q=1} \oplus B_{dR}^{+}) \otimes_L V^{\otimes r} & \longrightarrow & B_{dR} \otimes_L V^{\otimes r} & \longrightarrow 0 \\ & \uparrow{\scriptstyle j(at_L^r):=a\eta^{\otimes r}}\ \cong & & \cong\ \uparrow{\scriptstyle (x,y)\mapsto (xt_L^{-r},yt_L^{-r})\otimes\eta^{\otimes r}} & & \| & \\ 0 \longrightarrow & Lt_L^r & \xrightarrow{diag} & B_{max,L}^{\phi_q=\pi_L^r} \oplus t_L^r B_{dR}^{+} & \xrightarrow{(x,y)\mapsto (x-y)t_L^{-r}\otimes\eta^{\otimes r}} & B_{dR} \otimes_L V^{\otimes r} & \longrightarrow 0. \end{array} \tag{52}$$

Passing to continuous G_L-cohomology gives rise to the connecting isomorphism

$$\begin{array}{ccc} & \xrightarrow{\exp:=\exp_{L,V^{\otimes r},\mathrm{id}}} & H^1(L, V^{\otimes r}) \\ (B_{dR} \otimes_L V^{\otimes r})^{G_L} = tan_L(V^{\otimes r}) & & \uparrow \cong \\ & \longrightarrow & H^1(L, Lt_L^r), \end{array}$$

which is the identity component (see §9) of the Bloch–Kato exponential map over L for $V^{\otimes r}$. We introduce the composite map

$$\exp_r : L \xrightarrow[\cong]{a \mapsto at_L^{-r} \otimes \eta^{\otimes r}} tan_L(V^{\otimes r}) \xrightarrow{\exp_{L,V^{\otimes r},\mathrm{id}}} H^1(L, V^{\otimes r}) \xrightarrow{\mathrm{res}} \mathrm{Hom}_{\Gamma_L}(H_L, V^{\otimes r}) .$$

Proposition 8.5 *For all $a \in L$ we have*

$$j^{-1} \circ \exp_r(a) = -\delta^r((\pi_L^{-r} - 1)a) .$$

Proof By (52) we find $(x, y) \in B_{max,L}^{\phi_q=\pi_L^r} \oplus t_L^r B_{dR}^+$ such that $x - y = a$. Then

$$(j^{-1} \circ \exp_r(a))(g) = (g - 1)x = (g - 1)y \qquad \text{for all } g \in H_L.$$

We *claim* that y belongs to $Fil^r B_{max,L}^+$. For this it suffices to prove that y lies in $B_{max,L}^+$ (because it is contained in $t_L^r B_{dR}^+$ by assumption). We know that $y = x - a \in B_{max,L} = \bigcup_{s \geq 0} t_L^{-s} B_{max,L}^+$. Let s be minimal with respect to the property that $y \in t_L^{-s} B_{max,L}^+$, i.e., that $t_L^s y \in B_{max,L}^+$. We want to show that $s = 0$. Assume to the contrary that $s > 0$. Then $B_{max,L}^+ \ni \phi_q^i(t_L^s y) = \pi_L^{is} t_L^s \phi_q^i(y)$, for any $i \geq 0$, belongs to $Fil^s B_{max,L}^+ \subseteq \ker(\theta)$ because

$$\phi_q^i(y) = \phi_q^i(x - a) = \pi_L^{ri} x - a = \pi_L^{ri} y + \pi_L^{ri} a - a \in B_{dR}^+ .$$

By (48) we obtain $t_L^s y = t_L y'$ for some $y' \in B_{max,L}^+$. Hence $t_L^{s-1} y$ already belongs to $B_{max,L}^+$, which is a contradiction. The above claim follows.

In particular, by the definition of δ^r and using that $y = x - a$ we see that

$$(j^{-1} \circ \exp_r(a))(g) = (g - 1)y = \delta^r((\pi_L^{-r}\phi_q - 1)(y))(g) = -\delta^r((\pi_L^{-r} - 1)a)(g)$$

for $g \in H_L$, because $(\pi_L^{-r}\phi_q - 1)(x) = 0$ as x belongs to $B_{max,L}^{\phi_q=\pi_L^r}$. □

Putting the previous results together we obtain the following generalization of the explicit reciprocity law of Bloch and Kato ([3] Theorem 2.1) from the cyclotomic to the general Lubin–Tate case. In particular, this confirms partly the speculations in

[14] §11: de Shalit had suggested to find a replacement for $B_{max,\mathbb{Q}_p}$ (or rather B_{cris} which was used at that time) in the context of general Lubin–Tate formal groups and it is precisely Colmez' $B_{max,L}$ which has this function (although the path in (loc. cit.) is slightly different from the one chosen here).

Theorem 8.6 *For all $u \in \varprojlim_n o_{L_n}^\times$, $a \in L$, and $r \geq 1$ we have the identities*

$$\delta^r(a)(rec(u)) = ar\Psi^r_{CW}(u)$$

and

$$\begin{aligned}(j^{-1} \circ \exp_r(a))(rec(u)) &= -(\pi_L^{-r} - 1)ar\Psi^r_{CW}(u) \\ &= \tfrac{1}{(r-1)!}(1 - \pi_L^{-r})a\partial^r_{\mathrm{inv}} \log g_{u,\eta}(Z)_{|Z=0}t_L^r .\end{aligned}$$

Finally we consider the following commutative diagram

$$\begin{array}{ccccc}
H_L^{ab}(p) \otimes_{\mathbb{Z}_p} V^{\otimes -r} & \times & \mathrm{Hom}^c(H_L, V^{\otimes r}) & \longrightarrow & L \\
\uparrow {\scriptstyle rec\otimes\mathrm{id}} & & & & \| \\
\varprojlim_n L_n^\times \otimes_{\mathbb{Z}_p} V^{\otimes -r} & & \uparrow {\scriptstyle \mathrm{res}} & & \| \\
\downarrow {\scriptstyle \mathrm{cores}(-\kappa\otimes\mathrm{id})} & & & & \| \\
H^1(L, V^{\otimes -r}(1)) & \times & H^1(L, V^{\otimes r}) & \xrightarrow{\cup} & H^2(L, L(1)) = L \\
\downarrow {\scriptstyle \exp^*} & & \uparrow {\scriptstyle \exp} & & {\scriptstyle \cong}\downarrow {\scriptstyle c\mapsto ct_{\mathbb{Q}_p}^{-1}\otimes\eta^{cyc}} \\
D^0_{dR,L}(V^{\otimes -r}(1)) & \times & tan_L(V^{\otimes r}) & \longrightarrow & D_{dR,L}(L(1)) \\
{\scriptstyle a\mapsto at_L^r t_{\mathbb{Q}_p}^{-1}\otimes(\eta^{\otimes -r}\otimes\eta^{cyc})}\uparrow {\scriptstyle \cong} & & {\scriptstyle \cong}\uparrow {\scriptstyle b\mapsto bt_L^{-r}\otimes\eta^{\otimes r}} & & {\scriptstyle \cong}\uparrow {\scriptstyle c\mapsto ct_{\mathbb{Q}_p}^{-1}\otimes\eta^{cyc}} \\
L & \times & L & \xrightarrow{(a,b)\mapsto ab} & L.
\end{array}$$

Here η^{cyc} is a generator of the cyclotomic Tate module $\mathbb{Z}_p(1)$, and $t_{\mathbb{Q}_p} := \log_{\mathbb{G}_m}([\iota(\eta^{cyc}) + 1] - 1)$. The commutativity of the upper part can be shown by taking inverse limits (on both sides) of a similar diagram with appropriate torsion coefficients and afterwards tensoring with L over o_L. Its middle part is the definition of the dual exponential map $\exp^*$. The commutativity of the lower part is easily checked. Note also that the composite of the middle maps going up is nothing else than $\exp_r$ by definition. Thus setting $\mathbf{d}_r := t_L^r t_{\mathbb{Q}_p}^{-1} \otimes (\eta^{\otimes -r} \otimes \eta^{cyc})$ we obtain the following consequence.

Corollary 8.7 (A special case of Kato's explicit reciprocity law) *For $r \geq 1$ the diagram*

$$\begin{array}{ccc}
\varprojlim_n o_{L_n}^{\times} \otimes_{\mathbb{Z}} T^{\otimes -r} & & \\
\Big\downarrow {\scriptstyle -\kappa \otimes \mathrm{id}} & \searrow {\scriptstyle ''(1-\pi_L^{-r}) r \psi_{CW}^r(_)\mathbf{d}_r''} & \\
H^1_{\mathrm{Iw}}(L_\infty/L, T^{\otimes -r}(1)) & & \\
\Big\downarrow {\scriptstyle \mathrm{cores}} & & \\
H^1(L, T^{\otimes -r}(1)) & \xrightarrow{\exp^*} & D^0_{dR,L}(V^{\otimes -r}(1)) = L\mathbf{d}_r,
\end{array}$$

commutes, i.e., the diagonal map sends $u \otimes a\eta^{\otimes -r}$ to

$$a(1 - \pi_L^{-r}) r \psi_{CW}^r(u)\mathbf{d}_r = a\frac{1 - \pi_L^{-r}}{(r-1)!} \partial_{\mathrm{inv}}^r \log g_{u,\eta}(Z)_{|Z=0}\mathbf{d}_r \, .$$

9 Appendix: p-adic Hodge Theory

For a continuous representation of G_K on a finite dimensional $\mathbb{Q}_p$-vector space V we write as usual

$$D_{dR,K}(V) := (B_{dR} \otimes_{\mathbb{Q}_p} V)^{G_K} \supseteq D^0_{dR,K}(V) := (B^+_{dR} \otimes_{\mathbb{Q}_p} V)^{G_K} \quad \text{and}$$
$$D_{cris,K}(V) := (B_{max,\mathbb{Q}_p} \otimes_{\mathbb{Q}_p} V)^{G_K} \, .$$

The quotient $tan_K(V) := D_{dR,K}(V)/D^0_{dR,K}(V)$ is called the tangent space of V.

Henceforth we assume that V is de Rham. Then the usual Bloch–Kato exponential map $\exp_{K,V} : tan_K(V) \to H^1(K, V)$ can be defined as follows. Apply the tensor functor $- \otimes_{\mathbb{Q}_p} V$ to the exact sequence

$$0 \to \mathbb{Q}_p \to B_{max,\mathbb{Q}_p}^{\phi_p = 1} \to B_{dR}/B^+_{dR} \to 0 \tag{53}$$

and take the (first) connecting homomorphism in the associated G_K-cohomology sequence.[6] Note that by [3] Lemma 3.8.1 we have $tan_K(V) = (B_{dR}/B^+_{dR} \otimes_{\mathbb{Q}_p} V)^{G_K}$. Furthermore, the dual exponential map $exp^*_{K,V}$ is defined by the commutativity of the following diagram

[6] It follows from [12, Proposition III.3.1] that this sequence splits in the category of topological $\mathbf{Q}_p$-vector spaces. Since the p-adic topology on $\mathbf{Q}_p$ coincides with the induced topology from $B_{max,\mathbf{Q}_p}$ the existence of the transition map is granted by [30, Lem. 2.7.2].

$$\begin{array}{ccc} H^1(K,V) & \xrightarrow{\exp^*_{K,V}} & D^0_{dR,K}(V) \\ \cong\downarrow & & \downarrow\cong \\ H^1(K,V^*(1))^* & \xrightarrow{(\exp_{K,V^*(1)})^*} & (D_{dR,K}(V^*(1))/D^0_{dR,K}(V^*(1)))^*, \end{array} \tag{54}$$

where the left, resp. right, perpendicular isomorphism comes from local Tate duality, resp. from the perfect pairing

$$D_{dR,K}(V) \times D_{dR,K}(\mathrm{Hom}_{\mathbb{Q}_p}(V, \mathbb{Q}_p(1))) \longrightarrow D_{dR,K}(\mathbb{Q}_p(1)) \cong K, \tag{55}$$

in which the $D^0_{dR,K}$-subspaces are orthogonal to each other. Note that the isomorphism $K \cong D_{dR,K}(\mathbb{Q}_p(1))$ sends a to $at^{-1}_{\mathbb{Q}_p} \otimes \eta^{cyc}$. Also, $(-)^*$ here means the $\mathbb{Q}_p$-dual.

Now assume that V is in $Rep_L(G_K)$ and consider $K = L$ in the following. Tensoring (28) with $\mathbb{Q}_p$ gives the isomorphism of L-vector spaces

$$\tilde{\Xi} : L \cong \mathrm{Hom}_{\mathbb{Z}_p}(o_L, \mathbb{Z}_p) \otimes_{\mathbb{Z}_p} \mathbb{Q}_p \cong \mathrm{Hom}_{\mathbb{Q}_p}(L, \mathbb{Q}_p) .$$

Since $\mathrm{Hom}_{\mathbb{Q}_p}(L, -)$ is right adjoint to scalar restriction from L to $\mathbb{Q}_p$, and by using $\tilde{\Xi}^{-1}$ in the second step, we have a natural isomorphism

$$\mathrm{Hom}_{\mathbb{Q}_p}(V, \mathbb{Q}_p) \cong \mathrm{Hom}_L(V, \mathrm{Hom}_{\mathbb{Q}_p}(L, \mathbb{Q}_p)) \cong \mathrm{Hom}_L(V, L) . \tag{56}$$

Combined with (55) we obtain the perfect pairing

$$D_{dR,L}(V) \times D_{dR,L}(\mathrm{Hom}_L(V, L(1))) \longrightarrow L \tag{57}$$

with an analogous orthogonality property. Furthermore, similarly as in Proposition 5.7 local Tate duality can be seen as a perfect pairing of finite dimensional L-vector spaces

$$H^i(K, V) \times H^{2-i}(K, \mathrm{Hom}_L(V, L(1))) \longrightarrow H^2(K, L(1)) = L . \tag{58}$$

Altogether we see that, for such a V, the dual Bloch–Kato exponential map can also be defined by an analogous diagram as (54) involving the pairings (57) and (58) and in which $(-)^*$ means taking the L-dual.

Since B_{dR} contains the algebraic closure $\overline{L}$ of L we have the isomorphism

$$B_{dR} \otimes_{\mathbb{Q}_p} V = (B_{dR} \otimes_{\mathbb{Q}_p} L) \otimes_L V \xrightarrow{\cong} \prod_{\sigma \in G_{\mathbb{Q}_p}/G_L} B_{dR} \otimes_{\sigma,L} V$$

which sends $b \otimes v$ to $(b \otimes v)_\sigma$. The tensor product in the factor $B_{dR} \otimes_{\sigma,L} V$ is formed with respect to L acting on B_{dR} through σ. With respect to the G_L-action the right hand side decomposes according to the double cosets in $G_L \backslash G_{\mathbb{Q}_p} / G_L$. It follows, in particular, that $D^{\mathrm{id}}_{dR,L}(V) := (B_{dR} \otimes_L V)^{G_L}$ is a direct summand of $D_{dR,L}(V)$. Similarly, $tan_{L,\mathrm{id}}(V) := (B_{dR}/B^+_{dR} \otimes_L V)^{G_L}$ is a direct summand of $tan_L(V)$. We then have the composite map

$$\widetilde{\exp}_{L,V,\mathrm{id}} : tan_{L,\mathrm{id}}(V) \overset{\subseteq}{\to} tan_L(V) \xrightarrow{\exp_{L,V}} H^1(L, V),$$

the identity component of $\exp_{L,V}$. On the other hand, applying the tensor functor $- \otimes_L V$ to the exact sequence (50)

$$0 \to L \to B^{\phi_q=1}_{max,L} \to B_{dR}/B^+_{dR} \to 0$$

and taking the (first) connecting homomorphism[7] in the associated G_L-cohomology sequence gives rise to a map

$$\exp_{L,V,\mathrm{id}} : tan_{L,\mathrm{id}}(V) \to H^1(L, V) .$$

Suppose that V is even L-analytic, i.e., that the Hodge–Tate weights of V at all embeddings $\mathrm{id} \neq \sigma : L \to \overline{L}$ are zero. We then have $tan_L(V) = tan_{L,\mathrm{id}}(V)$ and the following fact.

Proposition 9.1 *If V is L-analytic, the maps $\exp_{L,V}$, $\widetilde{\exp}_{L,V,\mathrm{id}}$ and $\exp_{L,V,\mathrm{id}}$ coincide.*

Because of this fact we call also $\exp_{L,V,\mathrm{id}}$ the identity component of $\exp_{L,V}$ in the situation of the Proposition. We remark that by [34] III.A4 Proposition 4 and Lemma 2(b) the character $V = L(\chi_{LT})$ is L-analytic.

Proof of Proposition 9.1 Let $L_0 \subseteq L$ be the maximal unramified subextension and let $f := [L_0 : \mathbb{Q}_p]$. As explained at the beginning of §9.7 in [9] we have $B_{max,L} \subseteq B_{max,\mathbb{Q}_p} \otimes_{L_0} L$ and hence $B^{\phi_q=1}_{max,L} \subseteq B^{\phi_p^f=1}_{max,\mathbb{Q}_p} \otimes_{L_0} L$. We claim that

$$B^{\phi_q=1}_{max,\mathbb{Q}_p} = B^{\phi_p^f=1}_{max,\mathbb{Q}_p} = B^{\phi_p=1}_{max,\mathbb{Q}_p} \otimes_{\mathbb{Q}_p} L_0 .$$

The left hand side contains L_0. Let $\Delta := \mathrm{Gal}(L_0/\mathbb{Q}_p)$ with Frobenius generator δ. For any $x \in B^{\phi_p^f=1}_{max,\mathbb{Q}_p}$ we have the finite dimensional L_0-vector space $V_x := \sum_{i=0}^{f-1} L_0\phi_p^i(x)$ on which the Galois group Δ acts semilinearly by sending δ to ϕ_p. Hilbert 90 therefore implies that $V_x = L_0 \otimes_{\mathbb{Q}_p} V_x^\Delta$. This proves the claim.

[7] Analogous arguments as in Footnote 6 grant the existence of this connecting homomorphism.

It follows that we have the commutative diagram

$$\begin{array}{ccccccccc}
0 & \longrightarrow & L & \longrightarrow & B^{\phi_p=1}_{max,\mathbb{Q}_p} \otimes_{\mathbb{Q}_p} L & \longrightarrow & B_{dR}/B^+_{dR} \otimes_{\mathbb{Q}_p} L & \longrightarrow & 0 \\
 & & \downarrow & & \downarrow{=} & & \downarrow{mult} & & \\
0 & \longrightarrow & C & \longrightarrow & B^{\phi_q=1}_{max,\mathbb{Q}_p} \otimes_{L_0} L & \longrightarrow & B_{dR}/B^+_{dR} & \longrightarrow & 0 \\
 & & \uparrow & & \uparrow{\subseteq} & & \uparrow{=} & & \\
0 & \longrightarrow & L & \longrightarrow & B^{\phi_q=1}_{max,L} & \longrightarrow & B_{dR}/B^+_{dR} & \longrightarrow & 0,
\end{array}$$

in which the upper exact sequence is induced by tensoring (53) by L over $\mathbf{Q}_p$ while the lower one is (50). C is defined to be the kernel in the middle horizontal sequence which is therefore also exact. Note that the two vertical maps $L \to C$ both coincide as their composites into the middle term each sends $l \in L$ to $1 \otimes l$. By tensoring this diagram with V over L and forming the cohomology sequences we conclude that the composites of $\exp_{L,V}$ and $\exp_{L,V,\mathrm{id}}$ with $H^1(G_L, V) \to H^1(G_L, C \otimes_L V)$ coincide whence the claim shall follow from the injectivity of the latter map. The snake lemma applied to the upper part of the diagram (tensored with V) leads to the exact sequence[8]

$$0 \to V \to C \otimes_L V \to \prod_{\sigma \neq \mathrm{id}} B_{dR}/B^+_{dR} \otimes_{\sigma,L} V \to 0 \,,$$

which in turn, by forming continuous G_L-cohomology, induces the exact sequence

$$0 = \ker(tan_L(V) \to tan_{L,\mathrm{id}}(V)) \to H^1(G_L, V) \to H^1(G_L, C \otimes_L V) \,,$$

i.e., we obtain the desired injectivity. □

References

1. Benois, D.: On Iwasawa theory of crystalline representations. Duke Math. J. **104**, 211–267 (2000)
2. Bentzen, S., Madsen, I.: Trace maps in algebraic K-theory and the Coates-Wiles homomorphism. J. Reine Angew. Math. **411**, 171–195 (1990)
3. Bloch, S., Kato, K.: L-functions and Tamagawa numbers of motives. In: The Grothendieck Festschrift, vol. I, 333–400, Progress Math., 86, Birkhäuser Boston (1990)
4. Bourbaki, N.: Topologie Générale. Chaps. 1–10. Springer (2007)

[8]Using the facts from Footnote 6 one checks that this sequence again satisfies the conditions of [30, Lem. 2.7.2] whence the existence of the long exact cohomology sequence below is granted.

5. Cherbonnier, F., Colmez, P.: Théorie d'Iwasawa des représentations p-adiques d'un corps local. J. AMS **12**, 241–268 (1999)
6. Coates J., Sujatha, R.: Cyclotomic fields and zeta values. Springer (2006)
7. Coates, J., Wiles, A.: On p-adic L-functions and elliptic units. J. Austral. Math. Soc. Ser. A **26**(1), 1–25 (1978)
8. Coleman, R.: Division values in local fields. Invent. math. **53**, 91–116 (1979)
9. Colmez, P.: Espaces de Banach de dimension finie. J. Inst. Math. Jussieu **1**, 331–439 (2002)
10. Colmez, P.: Fontaine's rings and p-adic L-functions. Lecture Notes at Tsinghua Univ. (2004)
11. Colmez, P.: (φ, Γ)-modules et représentations du mirabolique de $GL_2(\mathbb{Q}_p)$. In: Berger, L., Breuil, C., Colmez, P. (eds.) Représentations p-adiques de groupes p-adiques, vol. II. Astérisque 330, 61–153 (2010)
12. Colmez, P.: Théorie d'Iwasawa des représentations de de Rham d'un corps local. Ann. Math. **148**(2), 485–571 (1998)
13. Colmez, P.: A generalization of Coleman's isomorphism. In: Algebraic Number Theory and Related Topics (Kyoto, 1997). Srikaisekikenkysho kkyroku **1026**, 110–112 (1998)
14. de Shalit, E.: The explicit reciprocity law of Bloch–Kato. Columbia University Number Theory Seminar (New York, 1992). Astérisque **228**(4), 197–221 (1995)
15. Fontaine, J.-M.: Répresentations p-adiques des corps locaux. In: The Grothendieck Festschrift, vol. II, 249–309, Birkhäuser (1990)
16. Fontaine, J.-M.: Appendice: Sur un théorème de Bloch et Kato (lettre à B. Perrin-Riou). Invent. Math. **115**, 151–161 (1994)
17. Fourquaux, L., Xie, B.: Triangulable O_F-analytic (φ_q, Γ)-modules of rank 2. Algebra Number Theory **7**(10), 2545–2592 (2013)
18. Fukaya, T., Kato, K.: A formulation of conjectures on p-adic zeta functions in non-commutative Iwasawa theory. In: Proceedings of St. Petersburg Math. Soc., vol. XII, AMS Transl. Ser. 2, vol. 219, 1–86 (2006)
19. Hazewinkel, M.: Formal Groups and Applications. Academic Press (1978)
20. Herr, L.: Sur la cohomologie galoisienne des corps p-adiques. Bull. Soc. Math. France **126**, 563–600 (1998)
21. Hewitt, E., Ross, K.: Abstract Harmonic Analysis, vol. I. Springer (1994)
22. Jensen, C.U.: Les Foncteurs Dérivés de $\varprojlim$ et leurs Applications en Théorie des Modules. Springer Lect. Notes Math., vol. 254 (1972)
23. Kato, K.: Lectures on the approach to Iwasawa theory for Hasse-Weil L-functions via B_{dR}. I. Arithmetic algebraic geometry (Trento, 1991), Springer. Lect. Notes Math. **1553**, 50–163 (1993)
24. Kisin, M., Ren, W.: Galois representations and Lubin-Tate groups. Documenta Math. **14**, 441–461 (2009)
25. Kölcze P.: Ein Analogon zum Hilbertsymbol für algebraische Funktionen und Witt-Vektoren solcher Funktionen. Diplomarbeit (Betreuer: J. Neukirch) Universität Regensburg (1990)
26. Lang, S.: Cyclotomic Fields. Springer (1978)
27. Laubie, F.: Extensions de Lie et groupes d'automorphismes de corps locaux. Compositio Math. **67**, 165–189 (1988)
28. Lazard, M.: Groupes analytiques p-adiques. Publ. Math. IHES **26**, 389–603 (1965)
29. Michael, E.: Continuous selections II. Ann. Math. **64**, 562–580 (1956)
30. Neukirch, J., Schmidt, A., Wingberg, K.: Cohomology of Number Fields. 2nd edn. Springer (2008)
31. Schneider, P.: Galois representations and (φ, Γ)-modules. Lecture Notes, Münster (2015). http://wwwmath.uni-muenster.de/u/schneider/publ/lectnotes/index.html
32. Schneider, P., Vigneras, M.-F.: A functor from smooth o-torsion representations to (φ, Γ)-modules. In: Arthur, Cogdell, ... (eds.) On Certain L-Functions. Clay Math. Proc., vol. 13, 525–601, AMS-CMI (2011)
33. Scholl, A. J.: Higher fields of norms and (ϕ, Γ)-modules. Documenta Math. 2006, Extra Vol., pp. 685–709 (2006)
34. Serre, J.-P.: Abelian l-Adic Representations and Elliptic Curves. Benjamin, W.A (1968)

35. Serre, J.-P.: Cohomologie Galoisienne. Springer Lect. Notes Math., vol. 5 (1973)
36. Thomas, L.: Ramification groups in Artin-Schreier-Witt extensions. J. Théorie des Nombres de Bordeaux **17**, 689–720 (2005)
37. Wiles, A.: Higher explicit reciprocity laws. Ann. Math. **107**(2), 235–254 (1978)
38. Witt, E.: Zyklische Körper und Algebren der Charakteristik p vom Grad p^n. Struktur diskret bewerteter perfekter Körper mit vollkommenem Restklassenkörper der Charakteristik p. J. Reine Angew. Math. **176**, 126–140 (1936)

Bigness in Compatible Systems

Andrew Snowden and Andrew Wiles

Abstract Clozel, Harris and Taylor have recently proved a modularity lifting theorem of the following general form: if ρ is an ℓ-adic representation of the absolute Galois group of a number field for which the residual representation $\overline{\rho}$ comes from a modular form then so does ρ. This theorem has numerous hypotheses; a crucial one is that the image of $\overline{\rho}$ must be "big," a technical condition on subgroups of GL_n. In this paper we investigate this condition in compatible systems. Our main result is that in a sufficiently irreducible compatible system the residual images are big at a density one set of primes. This result should make some of the work of Clozel, Harris and Taylor easier to apply in the setting of compatible systems.

Keywords Galois representations · Bigness · Compatible systems

Mathematics Subject Classification: 11F80

Authors' note: This paper was written in 2008 but has only been available on the arxiv. The notion of bigness was later superseded by the notion of adequacy, but the focus on compatible systems which are irreducible on open subgroups has increased in importance (see the remark added to this version at the end of the introduction) and we believe the techniques used here will continue to be useful. It is a pleasure to dedicate this paper to John Coates on the occasion of his 70th birthday, both for the beauty of his work and his inspiration as a teacher.

A. Snowden was partially supported by NSF fellowship DMS-0902661. A. Wiles was supported by an NSF grant.

A. Snowden
Department of Mathematics, University of Michigan, 2074 East Hall,
530 Church Street, Ann Arbor, MI 48109-1043, USA
e-mail: asnowden@math.umich.edu

A. Wiles (✉)
Mathematical Institute, University of Oxford, Andrew Wiles Building,
Radcliffe Observatory Quarter, Woodstock Road, Oxford OX2 6GG, UK
e-mail: Andrew.Wiles@maths.ox.ac.uk

D. Loeffler and S.L. Zerbes (eds.), *Elliptic Curves, Modular Forms and Iwasawa Theory*, Springer Proceedings in Mathematics & Statistics 188, DOI 10.1007/978-3-319-45032-2_13

1 Introduction

Let k be a finite field of characteristic ℓ, let V be a finite dimensional vector space over k and let G be a subgroup of $\mathrm{GL}(V)$. For an endomorphism g of V and an element α of k we let $V_{g,\alpha}$ denote the generalized eigenspace of g with eigenvalue α. It is naturally a sub and a quotient of V. Following Clozel, Harris and Taylor (see [6, Def. 2.5.1]), we say that G is *big* if the following four conditions hold:

(B1) The group G has no non-trivial quotient of ℓ-power order.
(B2) The space V is absolutely irreducible as a G-module.
(B3) We have $H^1(G, \mathrm{ad}^\circ V) = 0$.
(B4) For every irreducible G-submodule W of ad V we can find $g \in G, \alpha \in k$ and $f \in W$ such that $V_{g,\alpha}$ is one dimensional and the composite

$$V_{g,\alpha} \hookrightarrow V \xrightarrow{f} V \twoheadrightarrow V_{g,\alpha}$$

is non-zero.

The bigness condition is important in the work of Clozel, Harris and Taylor [6]. They prove modularity lifting theorems of the following general form: if ρ is an ℓ-adic representation of the absolute Galois group G_F of a number field F such that $\overline{\rho}$ comes from a modular form then so does ρ. There are several hypotheses in these theorems, but one crucial one is that the image of $\overline{\rho}$ must be big. In this paper, we investigate the bigness condition in compatible systems and show that it automatically holds at a density one set of primes, assuming the system is sufficiently irreducible. Thus the theorems of [6] should become easier to apply in the setting of compatible systems. Precisely, our main theorem is the following:

Theorem 1.1 *Let F be a number field. Let L be a set of prime numbers of Dirichlet density one and for each $\ell \in L$ let $\rho_\ell : G_F \to \mathrm{GL}_n(\mathbf{Q}_\ell)$ be a continuous representation of G_F. Assume that the ρ_ℓ form a compatible system and that each ρ_ℓ is absolutely irreducible when restricted to any open subgroup of G_F. Then there is a subset $L' \subset L$ of density one such that $\overline{\rho}_\ell(G_F)$ is a big subgroup of $\mathrm{GL}_n(\mathbf{F}_\ell)$ for all $\ell \in L'$.*

Here $\overline{\rho}_\ell$ denotes the semi-simplified mod ℓ reduction of ρ_ℓ. For our definition of "compatible system" see §7. We in fact prove a more general result, allowing for compatible systems with coefficients in a number field and for F to be a function field; see §8 for details.

1.1 Outline of Proof

Broadly speaking, the proof of Theorem 1.1 has three steps:

1. We first show that if $G/\mathbf{F}_\ell$ is a reductive group and $\rho : G \to \mathrm{GL}_n$ is an absolutely irreducible algebraic representation of G such that ℓ is large compared to n and the weights appearing in ρ then the group $\rho(G(\mathbf{F}_\ell))$ is big.
2. Using this, we show that if $\rho : \Gamma \to \mathrm{GL}_n(\mathbf{Q}_\ell)$ is an ℓ-adic representation of a profinite group such that (a) ℓ is large compared to n; (b) ρ is absolutely irreducible when restricted to any open subgroup of Γ; and (c) $\rho(\Gamma)$ is close to being a hyperspecial subgroup of its Zariski closure, then $\overline{\rho}(\Gamma)$ is a big subgroup of $\mathrm{GL}_n(\mathbf{F}_\ell)$.
3. Finally, we combine the above with results of Serre and Larsen on compatible systems to deduce Theorem 1.1.

1.2 Examples

We should point out that one can construct compatible systems which satisfy the hypotheses of the theorem. Let F be a totally real number field (resp. imaginary CM field) and let π be a cuspidal automorphic representation of $\mathrm{GL}_n(\mathbf{A}_F)$ satisfying the following conditions:

(C1) π is *regular algebraic*. This means that π_∞ has the same infinitesimal character as some irreducible algebraic representation of the restriction of scalars from F to $\mathbf{Q}$ of GL_n.

(C2) π is *essentially self-dual* (resp. *conjugate self-dual*). When F is totally real this means that $\pi^\vee = \chi \otimes \pi$ for some character χ of the idele group of F for which $\chi_v(-1)$ is independent of v, as v varies over the infinite places of F. When F is imaginary CM, this means that $\pi^\vee = \pi^c$, where c denotes complex conjugation.

(C3) There is some finite place v_0 of F such that π_{v_0} is a twist of the Steinberg representation.

Under these conditions, we can associate to π a compatible system of semi-simple representations $\{\rho_w\}$ of G_F with coefficients in some number field E. The system is indexed by the places w of E. For more precise statements, see [6, Propositions 3.2.1 and 3.3.1].

Let w be a place of E with residue characteristic different from that of v_0. Assume that F is imaginary CM. By the main result of [19] the (Frobenius semi-simplification of the) representation $\rho_w|_{G_{F,v_0}}$ corresponds to π_{v_0} under the local Langlands correspondence (we write G_{F,v_0} for the decomposition group at v_0). As π_{v_0} is a twist of the Steinberg representation, we find that $\rho_w|_{G_{F,v_0}}$ is absolutely indecomposable, and remains so after restricting to any open subgroup of G_{F,v_0}. It follows that ρ_w is absolutely indecomposable when restricted to any open subgroup of G_F. Since ρ_w

is semi-simple, we conclude that it is in fact absolutely irreducible when restricted to any open subgroup of G_F. When F is totally real we can still conclude that ρ_w has this property by making an appropriate abelian base change to an imaginary CM field and appealing to the above argument.

We thus see that all but finitely many members of the compatible system $\{\rho_w\}$ are absolutely irreducible on any open subgroup of G_F. By the more general version of the main theorem given in §8, we conclude that there is a set of primes P of $\mathbf{Q}$ of Dirichlet density $1/[E : \mathbf{Q}]$, all of which split in E, such that $\overline{\rho}_w(G_F)$ is big subgroup of $\mathrm{GL}_n(\mathbf{F}_\ell)$ for all w which lie above a prime $\ell \in P$.

1.3 Remark

In view of Patrikis' thesis [16, Proposition 1.1.15], every absolutely irreducible ρ_ℓ is of the form $\rho_\ell = \mathrm{Ind}_K^F(\rho'_\ell \otimes \Psi)$ where K is a finite extension of F, Ψ is an Artin representation of G_K, and ρ'_ℓ is an ℓ-adic representation of G_K that is absolutely irreducible on every open subgroup. We would conjecture that if ρ_ℓ is part of a semi-simple compatible system then so is ρ'_ℓ, and that K and Ψ may be chosen independent of ℓ. A similar statement should hold for compatible systems with coefficients in a number field.

1.4 Notation and Conventions

Reductive groups over fields are connected. A semi-simple group G over a field k is called simply connected if the root datum of $G_{\overline{k}}$ is simply connected (i.e., coroots span the coweight lattice). If G is a semi-simple group over k then there is a simply connected group G^{sc} and an isogeny $G^{\mathrm{sc}} \to G$ whose kernel is central. The group G^{sc} and the map $G^{\mathrm{sc}} \to G$ are unique up to isomorphism. We call G^{sc} the universal cover of G. For an arbitrary algebraic group G over k we let G° denote the connected component of the identity, G^{ad} the adjoint group of the quotient of G° by its radical, which is a semi-simple group, and G^{sc} the universal cover of G^{ad}. We also write G^{der} for the derived subgroup of G°, which is semi-simple if G° is reductive. For a vector space V we denote by ad V the space of endomorphisms of V and by ad° V the subspace of traceless endomorphisms. More definitions are given in the body of the paper.

2 Elementary Properties of Bigness

In this section we establish some elementary properties of bigness. Throughout this section, k denotes a finite field of characteristic ℓ, V a finite dimensional vector space over k and G a subgroup of $\mathrm{GL}(V)$.

Proposition 2.1 *Let H be a normal subgroup of G. If H satisfies (B2), (B3) and (B4) then G does as well. In particular, if H is big and the index $[G : H]$ is prime to ℓ then G is big.*

Proof Assume H satisfies (B2), (B3) and (B4). Since V is absolutely irreducible for H, it is for G as well, and so G satisfies (B2). We have an exact sequence

$$H^1(G/H, (\mathrm{ad}^\circ V)^H) \to H^1(G, \mathrm{ad}^\circ V) \to H^1(H, \mathrm{ad}^\circ V)^{G/H}.$$

Since H satisfies (B3), $H^1(H, \mathrm{ad}^\circ V) = 0$ and so the rightmost term vanishes. Since V is absolutely irreducible for H we have $(\mathrm{ad}^\circ V)^H = 0$ and so the leftmost term vanishes. Thus $H^1(G, \mathrm{ad}^\circ V) = 0$ and G satisfies (B3). Now let W be a G-irreducible submodule of $\mathrm{ad}\, V$. Let W' be an H-irreducible submodule of W. Since H satisfies (B4), we can find $g \in H$, $\alpha \in k$ and $f \in W'$ such that $V_{g,\alpha}$ is one dimensional and $f(V_{g,\alpha})$ has non-zero projection to $V_{g,\alpha}$. Of course, g also belongs to G and f also belongs to W. Thus G satisfies (B4) as well.

Now say that H is big and $[G : H]$ is prime to ℓ. The above arguments show that G satisfies (B2), (B3) and (B4), so to show that G is big we need only verify (B1). Let K be an ℓ-power order quotient of G. Since H has no ℓ-power order quotient, its image in K is trivial. Thus K is a quotient of G/H. But this group has prime-to-ℓ order, and so $K = 1$. This shows that the only ℓ-power order quotient of G is the trivial group, and so G satisfies (B1). □

Proposition 2.2 *The group G is big if and only if $k^\times G$ is, where $k^\times$ denotes the group of scalar matrices in* $\mathrm{GL}(V)$.

Proof Since G is a normal subgroup of $k^\times G$ of prime-to-ℓ index, the bigness of the former implies that of the latter by Proposition 2.1. Now assume that $k^\times G$ is big. Let K be an ℓ-power order quotient of G. Since $k^\times \cap G$ is prime to ℓ, its image in K is trivial. Thus K is a quotient of the group $G/(G \cap k^\times) = k^\times G/k^\times$. By assumption, $k^\times G$ has no non-trivial quotient of ℓ-power order. Thus $K = 1$ and G satisfies (B1).

Since V is absolutely irreducible for $k^\times G$ it is for G as well. Thus G satisfies (B2).

We have an exact sequence

$$1 \to G \to k^\times G \to H \to 1$$

where H is a quotient of $k^\times$. We thus have an exact sequence

$$H^1(k^\times G, \mathrm{ad}^\circ V) \to H^1(G, \mathrm{ad}^\circ V)^H \to H^2(H, (\mathrm{ad}^\circ V)^G).$$

The group on the left vanishes by hypothesis. The group on the right vanishes since $(\mathrm{ad}^\circ\ V)^G = 0$. Thus the group in the middle vanishes. Now, the action of H on $H^1(G, \mathrm{ad}^\circ V)$ is trivial. (Proof: Let $f : G \to \mathrm{ad}^\circ\ V$ be a 1-cocycle representing a cohomology class $[f]$ and let h be an element of H. Then $h \cdot [f]$ is represented by the 1-cocycle $g \mapsto \widetilde{h} f(\widetilde{h}^{-1} g \widetilde{h})$ for any lift $\widetilde{h}$ of h. We can pick a lift $\widetilde{h}$ of h which belongs to $k^\times$. Thus $\widetilde{h}$ acts trivially on G by conjugation and acts trivially on $\mathrm{ad}^\circ\ V$. Therefore $h \cdot [f] = [f]$.) It thus follows that $H^1(G, \mathrm{ad}^\circ\ V)$ vanishes and so G satisfies (B3).

As for the last condition, let W be an irreducible G-submodule of ad V. Then it is also an irreducible $k^\times G$-module. Thus we can find $g \in k^\times G$, $\alpha \in k$ and $f \in W$ such that $V_{g,\alpha}$ is one dimensional and $f(V_{g,\alpha})$ has non-zero projection to $V_{g,\alpha}$. We can write $g = zg'$ where z belongs to $k^\times$ and g' belongs to G. Put $\alpha' = \alpha z^{-1}$. Then $V_{g',\alpha'} = V_{g,\alpha}$. Thus this space is one dimensional and $f(V_{g',\alpha'})$ has non-zero projection to $V_{g',\alpha'}$. We have therefore shown that G satisfies (B4). Thus G is big. □

The following result will not be used, but is good to know.

Proposition 2.3 *Let k'/k be a finite extension and put $V' = V \otimes k'$. If G is a big subgroup of* $\mathrm{GL}(V)$ *then it is a big subgroup of* $\mathrm{GL}(V')$.

Proof Conditions (B1), (B2) and (B3) for G as a subgroup of $\mathrm{GL}(V')$ are immediate. We prove (B4). Let S be the set of pairs $(g, \alpha) \in G \times k^\times$ such that $V_{g,\alpha}$ is one dimensional. Consider the natural map

$$\Phi : \mathrm{ad}\ V \to \bigoplus_{(g,\alpha)\in S} \mathrm{End}(V_{g,\alpha}).$$

The map Φ is G-equivariant, using the natural action of G on the target. Let W be an irreducible submodule of ad V. Since $G \subset \mathrm{GL}(V)$ satisfies (B4), we can find $f \in W$ such that the image of f in $\mathrm{End}(V_{g,\alpha})$ is non-zero, for some $(g, \alpha) \in S$. Clearly then, $\Phi(f) \neq 0$. It follows that $\Phi(W) \neq 0$, and therefore (since W is irreducible), $\Phi|_W$ is injective. As this holds for every irreducible submodule of ad V, it follows that Φ is injective. Tensoring Φ with k', we find that the natural map

$$\Phi' : \mathrm{ad}\ V' \to \bigoplus_{(g,\alpha)\in S} \mathrm{End}(V'_{g,\alpha})$$

is injective. (Note that $V_{g,\alpha} \otimes_k k' = V'_{g,\alpha}$.) Let W be an irreducible submodule of ad V'. Given any non-zero $f \in W$ the image of f under Φ' is non-zero. It follows that there exists some $(g, \alpha) \in S$ such that the image of f in $\mathrm{End}(V'_{g,\alpha})$ is non-zero. This shows that $G \subset \mathrm{GL}(V')$ satisfies (B4). □

3 Background on Representations of Algebraic Groups

In this section we review some representation theory of algebraic groups. To a representation V of an algebraic group G we attach a non-negative integer $\|V\|$ which measures the size of the weights appearing in V. The key principle we need is: over a field of positive characteristic, the representations of G with $\|V\|$ small behave in many ways like representations in characteristic 0. We will give several precise statements of this type.

3.1 Borel–Weil Type Representations

Let S be a scheme. A group scheme G/S is *reductive* (resp. *semi-simple*) if it is smooth, affine and its geometric fibers are reductive (resp. semi-simple). This implies that its geometric fibers are connected, by our conventions. Such a group is a *torus* if it is fppf locally isomorphic to $\mathbf{G}_m^n$. A torus is *split* if it is (globally) isomorphic to $\mathbf{G}_m^n$ (it may be best to allow n to be a locally constant function on the base S; this will not be an issue for us). By a *maximal torus* in G we mean a subtorus which is maximal in each geometric fiber. Similarly, by a *Borel subgroup* we mean a closed subgroup which is smooth over S and a Borel subgroup in each geometric fiber. A reductive group G/S is *split* if it has a split maximal torus T such that the weight spaces of T on $\mathrm{Lie}(G)$ are free coherent sheaves on S. See [7, Exp. XIX] for the general theory.

Let G/S be a split reductive group over a connected locally noetherian base S. Let B be a Borel subgroup of G and let $T \subset B$ be a split maximal torus. Let λ be a dominant weight of T and let $\mathcal{L}_S(\lambda)$ be the natural G-equivariant line bundle on G/B associated to λ. Put $V_{S,\lambda} = f_*\mathcal{L}_S(\lambda)$, where $f : G/B \to S$ is the structure map. Thus $V_{S,\lambda}$ is a coherent sheaf on S with a natural action of G. We call these sheaves "Borel–Weil representations." We omit the S from the notation if it is clear from context. Consider a cartesian diagram

$$\begin{array}{ccc} (G/B)_{S'} & \xrightarrow{g'} & G/B \\ {\scriptstyle f'}\downarrow & & \downarrow{\scriptstyle f} \\ S' & \xrightarrow{g} & S. \end{array}$$

Note that formation of $\mathcal{L}_S(\lambda)$ commutes with pull-back, that is, $(g')^*\mathcal{L}_S(\lambda) = \mathcal{L}_{S'}(\lambda)$. Kempf's vanishing theorem [9, Proposition II.4.5] states that if S' is a geometric point of S then $R^i f'_*\mathcal{L}_{S'}(\lambda) = 0$ for $i > 0$. (When S' has characteristic 0 this is part of the classical Borel–Weil–Bott theorem.) Thus, using a combination of the formal functions theorem and the proper base change theorem (see also the chapter "Cohomology and base change" in [14]), we see that $V_{S,\lambda}$ is a locally free sheaf on

S and its formation commutes with base change, that is, for any diagram as above we have $V_{S',\lambda} = g^* V_{S,\lambda}$.

Assume now that $S = \mathrm{Spec}(k)$ with k an algebraically closed field. If k has characteristic zero then V_λ is an irreducible representation of G. Furthermore, every irreducible representation of G is isomorphic to a unique V_λ. This is the classical Borel–Weil theorem. If k does not have characteristic zero then V_λ may not be irreducible. However, it has a unique irreducible submodule $\mathrm{soc}(V_\lambda)$ and any irreducible representation of G is isomorphic to a unique $\mathrm{soc}(V_\lambda)$ (see [9, Corollary II.2.7]). The representation $\mathrm{soc}(V_\lambda)$ is the unique irreducible with λ as its highest weight.

3.2 *The Norm of a Representation*

Let G/k be a reductive group over a field k. Assume for the moment that G is split and pick a split maximal torus T of G. For a weight λ of T we let $\|\lambda\|$ be the maximum value of $|\langle\lambda, \alpha^\vee\rangle|$ as α varies over the roots of G with respect to T. Let V be a representation of G. We let $\|V\|$ be the maximum value of the $\|\lambda\|$ among the weights λ appearing in V. The value of $\|V\|$ is independent of the choice of the torus T; furthermore, if k'/k is a field extension then $\|V_{k'}\| = \|V\|$. Now drop the assumption that G is split. For a representation V of G we define $\|V\|$ to be $\|V_{k'}\|$ where k'/k is an extension over which G splits. It is clear that if V is an extension of V_1 by V_2 then $\|V\| = \max(\|V_1\|, \|V_2\|)$. We also have the following:

Proposition 3.1 *Let $f : G' \to G$ be a map of reductive groups over a field k and let V be a representation of G. Assume one of the following holds:*

1. *f is a central isogeny.*
2. *f is the projection onto a direct factor.*
3. *f is the inclusion of the derived subgroup G' of G.*
4. *f is a surjection and $\ker f/(Z' \cap \ker f)$ is smooth, where Z' is the center of G'.*

Then $\|f^*V\| = \|V\|$. (Note (4) subsumes (1) and (2).)

Proof We may prove the proposition after passing to the closure of k. We thus assume k that is closed, and therefore, that G and G' are split. Although (4) subsumes (1) and (2) we will use (1) and (2) in the proof of (4), and so prove them separately.

(1) Let T be a split maximal torus of G. Then $T' = f^{-1}(T)$ is a split maximal torus of G'. Every weight of f^*V is of the form $f^*\lambda$ where λ is a weight of V. For a coroot $\alpha^\vee$ of G' we have the identity $\langle f^*\lambda, \alpha^\vee\rangle = \langle\lambda, f_*\alpha^\vee\rangle$. The push-forward $f_*\alpha^\vee$ is a coroot of G, and every coroot arises in this manner. We thus find $\|f^*V\| = \|V\|$.

(2) Write $G' = G \times G''$ so that f is the projection onto G. Let T be a split maximal torus of G and T'' a split maximal torus of G'' so that $T' = T \times T''$ is a split maximal torus of G'. The weights of f^*V coincide with the weights of V in the obvious manner. The coroots of G' are the union of the coroots of G and G''. As any coroot of G'' pairs to zero with a weight of T, we find $\|f^*V\| = \|V\|$.

(3) Let $G' = G^{\mathrm{der}}$ and let $f : G' \to G$ be the natural inclusion. Let T be a split maximal torus of G. Then the reduced subscheme of the connected component of the identity of $f^{-1}(T)$ is a group (by the lemma following this proof) and is a split maximal torus T' of G'. If $\alpha^\vee$ is a coroot of G' then $f_*\alpha^\vee$ is a coroot of G and all coroots arise in this manner. We also have an adjointness between f^* on weights and f_* on coweights. The proof now proceeds as in part (1).

(4) We have a diagram

$$\begin{array}{ccccc} \widetilde{G}' & \xrightarrow{p'} & (G')^{\mathrm{der}} & \xrightarrow{i'} & G' \\ \downarrow{\scriptstyle \widetilde{f}} & & \downarrow{\scriptstyle f'} & & \downarrow{\scriptstyle f} \\ \widetilde{G} & \xrightarrow{p} & G^{\mathrm{der}} & \xrightarrow{i} & G \end{array}$$

Here $\widetilde{G}$ is the universal cover of G^{der} and similarly for $\widetilde{G}'$. The map $\widetilde{f}$ is a lift of f'. Let $H = \ker \widetilde{f}$ and let H_{red} be the reduced subscheme of H; it is a normal subgroup of $\widetilde{G}'$ by the lemma following this proof. Of course, H_{red} is smooth since it is reduced. Let $K = H \cap \widetilde{Z}'$. Then $H/K = \ker f/(\ker f \cap Z')$ is smooth by hypothesis. The map $H_{\mathrm{red}} \to H/K$ is between smooth groups of the same dimension and is surjective on connected components; it is therefore surjective. We thus have $H = K H_{\mathrm{red}}$. The map $\widetilde{f}$ can now be factored as

$$\widetilde{G}' \to \widetilde{G}'/H^\circ_{\mathrm{red}} \to \widetilde{G}'/H_{\mathrm{red}} \to \widetilde{G}'/H = \widetilde{G}$$

The kernel of the first map is H°_{red}, which is a direct factor of $\widetilde{G}'$ since it is smooth, connected and normal. The kernel of the second map is $\pi_0(H_{\mathrm{red}})$, which is étale and therefore central. The kernel of the third map is H/H_{red}, which is the image of K in $\widetilde{G}'/H_{\mathrm{red}}$, and therefore central. We thus see that $\widetilde{f}$ is a composition of a projection onto a direct factor with two central isogenies. It follows from (1) and (2) that $\|\widetilde{f}^*W\| = \|W\|$ for any representation W of $\widetilde{G}$. We now have:

$$\|f^*V\| = \|(i')^*f^*V\| = \|(p')^*(i')^*f^*V\| = \|\widetilde{f}^*p^*i^*V\| = \|p^*i^*V\| = \|i^*V\| = \|V\|.$$

The first equality uses (3), the second (1), the third the commutativity of the diagram, the fourth the fact that $\widetilde{f}^*$ preserves norm, the fifth (1), the sixth (3). □

Lemma 3.2 *Let G be an affine group over a field k and let G_{red} be the reduced subscheme of G.*

1. *If k is perfect then G_{red} is a subgroup of G.*
2. *If G is a closed normal subgroup of a smooth affine group H then G_{red} is stable under conjugation by H.*

Thus if k is perfect and G is a closed normal subgroup of a smooth affine group then G_{red} is a closed normal subgroup of G.

Proof (1) Since k is perfect, the product $G_{\text{red}} \times G_{\text{red}}$ (fiber product over k) is reduced. Therefore the composite

$$G_{\text{red}} \times G_{\text{red}} \to G \times G \to G$$

factors through the inclusion $G_{\text{red}} \to G$. This shows that G_{red} is a subgroup of G.

(2) Given $h \in H(k)$, the map $G \to G$ given by conjugation by h induces a map $G_{\text{red}} \to G_{\text{red}}$. If k is infinite then $H(k)$ is dense in H, and so G_{red} is stable under conjugation by H. If k is finite then it is perfect, and one may therefore verify that G_{red} is stable by conjugation after passing to the closure; since the closure is infinite, the previous argument applies. □

We thank Brian Conrad for informing us of counterexamples to the above statements when the hypotheses are not in place.

3.3 *Representations of Small Norm*

We let $\text{Rep}(G)$ be the category of representations of G. For an integer n we let $\text{Rep}^{(n)}(G)$ be the full subcategory of $\text{Rep}(G)$ on those representations V which satisfy $\|V\| < n$. Both $\text{Rep}(G)$ and $\text{Rep}^{(n)}(G)$ are abelian categories. Furthermore, if

$$0 \to V' \to V \to V'' \to 0$$

is an exact sequence in $\text{Rep}(G)$ then V belongs to $\text{Rep}^{(n)}(G)$ if and only if both V' and V'' do. In other words, $\text{Rep}^{(n)}(G)$ is a Serre subcategory of $\text{Rep}(G)$.

Proposition 3.3 *Let G/k be a reductive group over an algebraically closed field k. Assume* char k *is zero or large compared to n and* $\dim G$. *Then:*

1. *The category* $\text{Rep}^{(n)}(G)$ *is semi-simple.*
2. *The simple objects of* $\text{Rep}^{(n)}(G)$ *are Borel–Weil representations.*

In other words, any representation V of G with $\|V\|$ small compared to char k *is a direct sum of V_λ's.*

Proof The statements are well-known in characteristic zero, so we assume k has positive characteristic. We prove (2) first. The simple objects of $\text{Rep}(G)$ are exactly the $\text{soc}(V_\lambda)$. Now, $\|\text{soc}(V_\lambda)\| \geqslant \|\lambda\|$ as λ occurs as a weight in $\text{soc}(V_\lambda)$. Thus if $\text{soc}(V_\lambda)$ belongs to $\text{Rep}^{(n)}(G)$ then $\|\lambda\| < n$. On the other hand, it is known that for char k large compared to $\dim G$ and $\|\lambda\|$ the representation V_λ is irreducible (see [17]). This proves (2).

We now prove (1). First note that the simple objects of $\text{Rep}^{(n)}(G)$ have dimension bounded in terms of n and $\dim G$. Indeed, the group G is the pull-back to k of a unique split reductive group over $\mathbf{Z}$, which we still call G. The simple $V_{k,\lambda}$ is just $V_{\mathbf{Z},\lambda} \otimes k$. Thus the dimension of $V_{k,\lambda}$ is the same as the dimension of $V_{\mathbf{C},\lambda}$. This dimension can then be bounded in terms of $\dim G$ and $\|\lambda\|$ using the Weyl dimension formula [8, Corollary 24.6] and the fact that there are only finitely many root data of a given rank.

Now, it is known (see [10, 12]) that any representation of G with small dimension compared to char k is semi-simple. Since char k is large, we thus find that if A and B are two simples of $\mathrm{Rep}^{(n)}(G)$ then any extension of A by B is semi-simple, and therefore $\mathrm{Ext}^1(A, B) = 0$. This shows that $\mathrm{Rep}^{(n)}(G)$ is semi-simple. This completes the proof of the proposition. □

Corollary 3.4 *Let G/k be a reductive group over an algebraically closed field k of characteristic ℓ and let V be a representation of G such that* $\dim V$ *and* $\|V\|$ *are small compared to ℓ. Then V is semi-simple and a direct sum of simple Borel–Weil representations.*

Proof Let H be the kernel of $G \to \mathrm{GL}(V)$, let H' be the reduced subscheme of H° and let $G' = G/H'$. Then the map $G' \to \mathrm{GL}(V)$ has finite kernel. Thus $\dim G'$ is bounded by $\dim V$ and is therefore small compared to ℓ. By Proposition 3.1, $\|V\|$ is the same for G and G'. By Proposition 3.3, V is semi-simple for G' and a direct sum of simple Borel–Weil representations. It follows that the same holds for G. (The restriction of a Borel–Weil representation along a surjection is still Borel–Weil.) □

Proposition 3.5 *Let $K/\mathbf{Q}_\ell$ be an extension with ring of integers $\mathcal{O}_K$ and residue field k. Let $G/\mathcal{O}_K$ be a reductive group and let V be a finite free $\mathcal{O}_K$-module with a representation of G. Then $\|V_k\| = \|V_K\|$ and this is bounded in terms of* $\mathrm{rk}\, V$. *If the representation of G_K on V_K is absolutely irreducible and ℓ is large compared to* $\mathrm{rk}\, V$ *then the representation of G_k on V_k is absolutely irreducible.*

Proof By enlarging K if necessary we can assume that G is split. Let T be a split maximal torus in G. As maps of tori are rigid, the weights of T in V_K and V_k are the same. This shows that their norms agree. The fact that $\|V_K\|$ is bounded in terms of $\mathrm{rk}\, V$ is a fact about representations of complex Lie groups and can be proved using the Weyl dimension formula [8, Corollary 24.6].

Now, assume that $V_{\overline{K}}$ is irreducible for the action of $G_{\overline{K}}$ and that ℓ is large compared to $\mathrm{rk}\, V$. By the first paragraph, ℓ is large compared to $\|V_k\|$. It thus follows from Corollary 3.4 that we can write $V_{\overline{k}} = \bigoplus V_{\overline{k},\lambda_i}$ with each $V_{\overline{k},\lambda_i}$ irreducible. The representations $V_{\overline{k},\lambda_i}$ lift to $\mathcal{O}_{\overline{K}}$. By the first paragraph, we see that $V_{\overline{K}}$ and $\bigoplus V_{\overline{K},\lambda_i}$ have the same weights, and are thus isomorphic. Since $V_{\overline{K}}$ is irreducible, there must therefore be only one term in the sum, and so $V_{\overline{k}}$ must be irreducible as well. □

3.4 Representations of $G(k)$

Let k be a finite field and let G/k be a reductive group. We denote by $\mathrm{Rep}(G(k))$ the category of representations of the finite group $G(k)$ on $\overline{k}$-vector spaces.

Proposition 3.6 *Let k be a finite field and G/k a semi-simple simply connected group. Assume* char k *is sufficiently large compared to* $\dim G$ *and n. Then the functor* $\mathrm{Rep}^{(n)}(G_{\overline{k}}) \to \mathrm{Rep}(G(k))$ *is fully faithful and the essential image is a Serre subcategory of* $\mathrm{Rep}(G(k))$.

Proof The functor $\mathrm{Rep}^{(n)}(G_{\overline{k}}) \to \mathrm{Rep}(G(k))$ is clearly faithful and exact. The desired properties now follow from the fact that $\mathrm{Rep}^{(n)}(G_{\overline{k}})$ is semi-simple and the fact that if V is an irreducible representation of $G_{\overline{k}}$ with norm small compared to char k then it stays irreducible when restricted to $G(k)$ (see [11, §1.13]). □

3.5 *Representations of* Lie(G)

We will need the following result:

Proposition 3.7 *Let k be the algebraic closure of a finite field, G/k a semi-simple group and V an irreducible representation of G. Pick a system of positive roots P in* $\mathfrak{g} = \mathrm{Lie}(G)$. *Assume* char k *is large compared to* $\| V \|$ *and* $\dim G$. *Then the subspace of V annihilated by P is one dimensional. (This subspace is the highest weight space of V with respect to P.)*

Proof Denote by G still the unique split group over $\mathbf{Z}$ giving rise to G/k. By our hypotheses we have $V = V_{k,\lambda}$ for some dominant weight λ. We know that $V_{k,\lambda} = V_{\mathbf{Z},\lambda} \otimes k$ and similarly $V_{\mathbf{C},\lambda} = V_{\mathbf{Z},\lambda} \otimes \mathbf{C}$. Now, since G is split over $\mathbf{Z}$ for each $r \in P$ we can find $X_r \in \mathfrak{g}_{\mathbf{Z}}$ which generates the r root space of $\mathfrak{g}_{\mathbf{Z}}$. Consider the map

$$V_{\mathbf{Z},\lambda} \to \bigoplus_P V_{\mathbf{Z},\lambda}, \qquad v \mapsto (X_r \cdot v)_{r \in P}.$$

This is a linear map of finite free $\mathbf{Z}$-modules. After tensoring with $\mathbf{C}$ the kernel of this map is the subspace of $V_{\mathbf{C},\lambda}$ annihilated by P. This is one dimensional by the usual highest weight theory over $\mathbf{C}$. It thus follows that for ℓ sufficiently large, the reduction of the map modulo ℓ will have one dimensional kernel. This proves the proposition. □

4 Highly Regular Elements of Semi-simple Groups

Fix a finite field k of cardinality q. The purpose of this section is to demonstrate the following result:

Proposition 4.1 *Let G/k be a semi-simple group and let $T \subset G$ be a maximal torus defined over k. Let n be an integer and assume q is large compared to* $\dim G$ *and n. Then there exists an element $g \in T(k)$ for which the map*

$$\{\lambda \in X(T_{\overline{k}}) \mid \|\lambda\| < n\} \to \overline{k}^{\times}, \qquad \lambda \mapsto \lambda(g)$$

is injective.

Before proving the proposition we give a few lemmas.

Lemma 4.2 *Let T/k be a torus of rank r. Then $(q-1)^r \leqslant \#T(k) \leqslant (q+1)^r$.*

Proof We have $\#T(k) = \det((q-F)|X(T_{\overline{k}}))$, where F is the Frobenius in $\mathrm{Gal}(\overline{k}/k)$. (For a proof of this, see [15, §1.5].) Since T splits over a finite extension, the action of F on $X(T_{\overline{k}})$ has finite order and so its eigenvalues are roots of unity. We thus have $\#T(k) = \prod_{i=1}^{r}(q-\zeta_i)$ where each ζ_i is a root of unity, from which the lemma easily follows. □

Lemma 4.3 *Let G/k be a semi-simple group and let $T \subset G$ be a maximal torus defined over k. Then $\#\{\lambda \in X(T_{\overline{k}}) \mid \|\lambda\| < n\}$ is bounded in terms of* $\dim G$ *and n.*

Proof The quantity $\#\{\lambda \in X(T_{\overline{k}}) \mid \|\lambda\| < n\}$ depends only on n and the root datum associated to $(G_{\overline{k}}, T_{\overline{k}})$. Since there are only finitely many semi-simple root data of a given dimension, the result follows. □

Lemma 4.4 *Let G/k be a semi-simple group of rank r, $T \subset G$ a maximal torus defined over k and $\lambda \in X(T_{\overline{k}})$ a non-zero character satisfying $\|\lambda\| < n$. Then the kernel of the map $\lambda : T(k) \to \overline{k}^{\times}$ has order at most $C(q+1)^{r-1}$ for some constant C depending only on n and* $\dim G$.

Proof For a subset S of $\{\lambda \in X(T_{\overline{k}}) \mid \|\lambda\| < n\}$ let $C(S)$ denote the cardinality of the torsion of the quotient of $X(T_{\overline{k}})$ by the subgroup generated by S. Let C be the least common multiple of the $C(S)$ over all S. Since C only depends upon n and the root datum associated to $(G_{\overline{k}}, T_{\overline{k}})$, it can be bounded in terms n and $\dim G$.

Now, say the character λ of $T_{\overline{k}}$ is defined over the extension k'/k. Then λ defines a map $T \to \mathrm{Res}_{k'/k}(\mathbf{G}_m)$, where Res denotes restriction of scalars. The kernel of λ is a diagonalizable group scheme whose character group is the cokernel of the map $f : \mathbf{Z}[\mathrm{Gal}(k'/k)] \to X(T_{\overline{k}})$ given by $\sigma \mapsto \sigma \cdot \lambda$ (note that $\mathbf{Z}[\mathrm{Gal}(k'/k)]$ is the character group of $\mathrm{Res}_{k'/k}(\mathbf{G}_m)$). The image of f is spanned by the $\mathrm{Gal}(\overline{k}/k)$ orbit of λ. Since $\|\cdot\|$ is Galois invariant, it follows that the torsion in the cokernel of f has order at most C. Furthermore, since λ is non-zero, we see that the rank of the cokernel of f is at most $r-1$.

We have thus shown that the kernel of $T \to \mathrm{Res}_{k'/k}(\mathbf{G}_m)$ is an extension of a finite group scheme of order at most C by a torus of rank at most $r-1$. It follows from Lemma 4.2 that the set of k-points of the kernel — which is identified with the kernel of $\lambda : T(k) \to \overline{k}^{\times}$ — has cardinality at most $C(q+1)^{r-1}$, as was to be shown. □

We now prove the proposition.

Proof of Proposition 4.1 Let S be the set of all non-zero $\lambda \in X(T_{\overline{k}})$ such that $\|\lambda\| < 2n$ and let N be the cardinality of S. We first claim that

$$T(k) \not\subset \bigcup_{\lambda \in S} \ker \lambda.$$

Of course, this is equivalent to $T(k) \neq \bigcup_{\lambda \in S} (\ker \lambda \cap T(k))$. To see this, we look at the cardinality of each side. The right side is a union of N sets, each of which has cardinality at most $C(q+1)^{r-1}$, while the left side has cardinality at least $(q-1)^r$ by Lemma 4.2. Since N and C are small compared to q (by Lemmas 4.3 and 4.4), the claim follows.

Now, pick an element $g \in T(k)$ such that $g \notin \bigcup_{\lambda \in S} \ker \lambda$. Let λ and λ' be distinct elements of $X(T_{\overline{k}})$ each of which has $\| \cdot \| < n$. Then $\lambda - \lambda'$ belongs to S. Thus $(\lambda/\lambda')(g) \neq 1$ and so $\lambda(g) \neq \lambda'(g)$. Therefore, $\lambda \mapsto \lambda(g)$ is injective on those λ with $\| \cdot \| < n$. □

5 Bigness for Algebraic Representations

In this section we prove that "small" algebraic representations have big image. The main result is the following:

Proposition 5.1 *Let k be a finite field, let G/k be a reductive group and let ρ be an absolutely irreducible representation of G on a k-vector space V. Assume $\ell =$ char k is large compared to* $\dim V$ *and* $\|V\|$. *Then $\rho(G(k))$ is a big subgroup of* $\mathrm{GL}(V)$.

Proof Let $G_1 = G^{\mathrm{der}}$, a semi-simple group. The group $G_1(k)$ is a normal subgroup of $G(k)$. The quotient is a subgroup of $(G/G_1)(k)$, which is prime to ℓ since G/G_1 is a torus. Thus it suffices by Proposition 2.1 to show that $\rho(G_1(k))$ is big.

Let H be the kernel of $\rho|_{G_1}$ and let H' be the reduced subscheme of the identity component of H. Then H' is a closed normal subgroup of G_1 by Lemma 3.2 and is smooth, since it is reduced. Put $G_2 = G_1/H'$. The map ρ factors through G_2. By Lang's theorem, the natural map $G_1(k) \to G_2(k)$ is surjective. Thus $\rho(G_1(k)) = \rho(G_2(k))$ and so it is enough to show that $\rho(G_2(k))$ is big. Note that the kernel of $\rho|_{G_2}$ is finite, and so the dimension of G_2 can be bounded in terms of the dimension of V.

Let G_3 be the universal cover of G_2. The image of $G_3(k)$ in $G_2(k)$ is normal and the quotient is a subgroup of $H^1(k, Z)$, where $Z = \ker(G_3 \to G_2)$. Now, the order of Z divides the order of the center of $(G_3)_{\overline{k}}$, which can be computed in terms of the root datum of $(G_3)_{\overline{k}}$. Since the dimension of G_3 is bounded in terms of that of V and there are only finitely many root data of a given dimension, it follows that the order of Z can be bounded in terms of the dimension of V. Thus since ℓ is large compared to the dimension of V, we find that the order of Z is prime to ℓ. It follows that the index of $G_3(k)$ in $G_2(k)$ is prime to ℓ. Thus it is enough to show that $\rho(G_3(k))$ is big.

We have thus shown that if $\rho(G_3(k))$ is big then so is $\rho(G(k))$. Now, since ρ is an absolutely irreducible representation of G the center of G acts by a character under ρ. Thus the restriction of ρ to G_1 is still absolutely irreducible. Therefore

ρ defines an absolutely irreducible representation of G_3. Furthermore, the norm of V as a representation of G is equal to the norm of V as a representation of G_3 by Proposition 3.1. We have thus shown that it suffices to prove the proposition when G is a simply connected, semi-simple group and $\ker \rho$ is a finite subgroup of G. We now begin the proof proper.

As G is semi-simple and simply connected, it is a product of simple simply connected groups G_i. Each G_i, being simple and simply connected, is of the form $\mathrm{Res}_{k_i/k}(G_i')$ where k_i is a finite extension of k and G_i' is an absolutely simple simply connected group over k_i (this is explained in §6.21(ii) of [3]). We have $G_i(k) = G_i'(k_i)$. Let Z_i' be the center of G_i'. By [5, Theorem 11.1.2] and [5, Theorem 14.4.1] the group $G_i'(k_i)/Z_i'(k_i)$ is simple and non-abelian. As we have previously explained, the order of Z_i' can be bounded by $\dim G_i'$, which is in turn bounded by $\dim V$. Thus, by our assumptions, the order of Z_i' is small compared to ℓ. We therefore find that the Jordan–Hölder constituents of $G(k)$ are all simple groups of Lie type and abelian groups of the form $\mathbf{Z}/p\mathbf{Z}$ with p a prime that is small compared to ℓ. In particular, $\mathbf{Z}/\ell\mathbf{Z}$ is not a Jordan–Hölder constituent of $G(k)$ and therefore not a constituent of the quotient $\rho(G(k))$. Thus $\rho(G(k))$ does not have a quotient of ℓ-power order, as such a quotient would be solvable and have a quotient isomorphic to $\mathbf{Z}/\ell\mathbf{Z}$. This shows that $\rho(G(k))$ satisfies (B1).

Proposition 3.6 shows that V is absolutely irreducible as a representation of $G(k)$ and so $\rho(G(k))$ satisfies (B2).

We now examine $H^1(\rho(G(k)), \mathrm{ad}^\circ V)$. By Propositions 3.3 and 3.6, we have $H^1(G(k), \mathrm{ad}\, V) = 0$ since this group classifies self-extensions of V and any such extension is semi-simple. Since ℓ is large compared to $\dim V$ this implies $H^1(G(k), \mathrm{ad}^\circ V) = 0$, as $\mathrm{ad}^\circ (V)$ is a summand of $\mathrm{ad}\, V$. Let H be the kernel of ρ. We have an exact sequence

$$1 \to H(k) \to G(k) \to \rho(G(k)) \to 1$$

and thus we get an injection

$$H^1(\rho(G(k)), (\mathrm{ad}^\circ V)^{H(k)}) \to H^1(G(k), \mathrm{ad}^\circ V).$$

The group on the right vanishes and so the group on the left does too. Since $H(k)$ acts trivially on V, it acts trivially on $\mathrm{ad}^\circ V$. Thus $H^1(\rho(G(k)), \mathrm{ad}^\circ V) = 0$ and so $\rho(G(k))$ satisfies (B3).

We now turn to condition (B4). As every reductive group over a finite field is quasi-split (see [2, Proposition 16.6]), we can pick a Borel subgroup B of G defined over k. Let T be a maximal torus of B, which is automatically a maximal torus of G, and let U be the unipotent radical of B. By Proposition 4.1 we can pick an element g of $T(k)$ such that $\lambda(g) \neq \lambda'(g)$ for any two distinct characters λ and λ' of $T_{\overline{k}}$ which are weights of $V_{\overline{k}}$. Now, the space $V_0 = V^U$ is one dimensional and stable under the action of T. Let $\lambda_0 : T \to \mathbf{G}_m$ give the action of T on V_0. Then λ_0 is the highest weight of $V_{\overline{k}}$, and thus occurs as a weight in this representation with multiplicity one. Put $\alpha = \lambda_0(g)$, an element of $k^\times$. We then have $V_{g,\alpha} = V_0$, and so the α-generalized eigenspace of g is one dimensional. To show that the image of

$G(k)$ is big, it thus suffices to show that any irreducible $G(k)$-submodule of ad V has non-zero projection to ad V_0.

Thus let W be an irreducible $G(k)$-submodule of ad V. To show that the image of W in ad V_0 is non-zero it suffices to show that the image of $\overline{W}$ in ad $\overline{V}_0$ is, where the bar denotes $- \otimes \overline{k}$. Let $\overline{U}$ be an irreducible $G(k)$-submodule of $\overline{W}$. It is enough, of course, to show that the image of $\overline{U}$ in ad $\overline{V}_0$ is non-zero. Now, $\overline{U}$ is an irreducible $G(k)$-submodule of ad $\overline{V}$, and so, by Proposition 3.6, it is an irreducible G-submodule of ad $\overline{V}$. Thus to show that the image of $G(k)$ in $\mathrm{GL}(V)$ satisfies (B4) it is enough to prove the following: *every irreducible G-submodule of* ad $\overline{V}$ *has non-zero image in* ad $\overline{V}_0$. This is established in the following lemma. □

Lemma 5.2 *Let k be the algebraic closure of a finite field, let G/k be a semi-simple group and let (ρ, V) be an irreducible representation of G with* dim V *and* $\|V\|$ *small compared to* char k. *Let V_0 be the highest weight space of V. Then every irreducible submodule of* ad V *has non-zero projection to* ad V_0.

Proof By the same reductions used in the proof of the proposition we can assume that ker ρ is finite, and so dim G is bounded in terms of dim V. Pick a maximal torus of G and a system of positive roots. For a weight λ let V_λ denote the λ-weight space of V. Let λ_0 be the highest weight for V and let V_0 be the λ_0-weight space. For a root α we pick an element X_α of $\mathrm{Lie}(G)$ which spans the α root space. Any positive element λ of the root lattice has a unique expression $\lambda = \sum n_i\alpha_i$ where the n_i are non-negative integers and the α_i are the simple roots. We let len λ be the sum of the n_i.

By a *simple tuple* we mean an ordered tuple $\underline{\alpha} = (\alpha_1, \dots, \alpha_n)$ consisting of simple roots. For such a tuple $\underline{\alpha}$ we put $|\underline{\alpha}| = \sum \alpha_i$. Note that len $|\underline{\alpha}| = n$. We let $X_{\underline{\alpha}}$ (resp. $Y_{\underline{\alpha}}$) denote the product $X_{\alpha_1} \cdots X_{\alpha_n}$ (resp. $X_{-\alpha_1} \cdots X_{-\alpha_n}$), regarded as an element of the universal enveloping algebra.

Given a weight λ for which V_λ is non-zero the difference $\lambda_0 - \lambda$ is positive and lies in the root lattice. For a simple tuple $\underline{\alpha}$ with $|\underline{\alpha}| = \lambda_0 - \lambda$ the operator $X_{\underline{\alpha}}$ maps V_λ into V_0. The resulting map

$$V_\lambda \to \bigoplus_{|\underline{\alpha}|=\lambda_0-\lambda} V_0$$

is injective. (Proof: By Proposition 3.7, the only vector annihilated by all of the X_α is the highest weight vector. Thus if $\lambda \neq \lambda_0$ and v belongs to V_λ then we can find some α such that $X_\alpha v$ is non-zero. We can thus move v closer to the λ_0-weight space. By induction on len$(\lambda_0 - \lambda)$ we can therefore find $\alpha_1, \dots, \alpha_n$ such that $X_{\alpha_n} \cdots X_{\alpha_1} v$ is non-zero and belongs to V_0.) We can thus pick $m = \dim V_\lambda$ simple tuples $\underline{\alpha}_1, \dots, \underline{\alpha}_m$ for which the resulting map is injective. We can then pick a basis $\{v_i\}$ of V_λ such that v_i belongs to the kernel of $X_{\underline{\alpha}_j}$ whenever $i \neq j$ but does not belong to the kernel of $X_{\underline{\alpha}_i}$. We call such a basis *admissible*. Note that in V^* the space V_0^* is a lowest weight space. The same process as above, but with $X_{\underline{\alpha}}$ replaced by $Y_{\underline{\alpha}}$, yields the notion of an admissible basis for V_α^*.

Let W be an irreducible submodule of ad V. Let $p : W \to V_0 \otimes V^*$ be the natural projection. We first show that $p(W)$ is non-zero. Among those weights λ for which the projection $W \to V_\lambda \otimes V^*$ is non-zero, pick one for which $\operatorname{len}(\lambda_0 - \lambda)$ is minimal. Let w be an element of W which has non-zero projection to $V_\lambda \otimes V^*$ and write

$$w = \left(\sum v_i \otimes v_i^*\right) + v'$$

where $\{v_i\}$ is an admissible basis of V_λ, the v_i^* belong to V^* and v' belongs to the complement of $V_\lambda \otimes V^*$. Let $\underline{\alpha}_i$ be the simple tuples yielding the basis v_i. Let 1 denote an index such that v_1^* is non-zero. We then have

$$p(X_{\underline{\alpha}_1} w) = (X_{\underline{\alpha}_1} v_1) \otimes v_1^* \tag{1}$$

(explained below). Since the right side is non-zero, it follows that $p(W)$ is non-zero.

We now explain why (1) holds. Recall that if X is an element of $\operatorname{Lie}(G)$ then the formula for how X acts on a pure tensor is

$$X(v \otimes w) = (Xv) \otimes w + v \otimes (Xw).$$

Thus when we apply $X_{\underline{\alpha}_1}$ to a pure tensor $v \otimes w$ we get a sum of terms and in each term some $X_{\alpha_{1,i}}$ land on v and some land on w. We now examine $X_{\underline{\alpha}_1} v$. First consider the v' part. Write $v' = \sum v_i' \otimes u_i'$ where v_i' has weight μ_i. If $\underline{\alpha}'$ is any sub-sequence of $\underline{\alpha}_1$ then $X_{\underline{\alpha}'} v_i'$ lands in the $\mu_i + |\underline{\alpha}'|$ weight space. If $\underline{\alpha}'$ is not all of $\underline{\alpha}_1$ then this cannot equal λ_0 for length reasons. Even if $\underline{\alpha}'$ is all of $\underline{\alpha}_1$ this is not equal to λ_0 since $\lambda_0 = |\underline{\alpha}_1| + \lambda$ and no μ_i is equal to λ. Thus $p(X_{\underline{\alpha}_1} v') = 0$. We now consider the first term in v. The same length argument shows that the only way to land in $V_0 \otimes V^*$ is to have all of $X_{\underline{\alpha}_1}$ land on the first factor. However, $X_{\underline{\alpha}_1}$ kills v_i for $i \neq 1$. We have thus proved (1).

We now show that the image of the projection $q : W \to V_0 \otimes V_0^*$ is non-zero. Among those weights λ for which the projection $W \to V_0 \otimes V_\lambda^*$ is non-zero, pick one for which $\operatorname{len}(\lambda_0 - \lambda)$ is minimal. (Such a weight exists by the previous paragraphs.) Let w be an element of W which has non-zero projection to $V_0 \otimes V_\lambda^*$. We may as well assume that w has weight $\lambda_0 - \lambda$. We can thus write

$$w = \left(\sum v_i \otimes v_i^*\right) + v'$$

where the v_i belong to V_0, $\{v_i^*\}$ is an admissible basis of V_λ^* and v' belongs to the complement of $V_0 \otimes V^*$. Let $\underline{\alpha}_i$ be the simple tuples yielding the basis v_i^*. Let 1 denote an index such that v_1 is non-zero. We then have

$$q(Y_{\underline{\alpha}_1} w) = v_1 \otimes (Y_{\underline{\alpha}_1} v_1^*) \tag{2}$$

(explained below). Since the right side is non-zero, it follows that $q(W)$ is non-zero, proving the proposition.

We now explain (2). The point is that, since $Y_{\underline{\alpha}_1}$ is a lowering operator, the only way for a term of $Y_{\underline{\alpha}_1} w$ to have its first factor in V_0 is if $Y_{\underline{\alpha}_1}$ lands entirely on the second factor. Of course, none of the terms in v' have their first factor in V_0 to begin with, so they will not after applying $Y_{\underline{\alpha}_1}$. As for the first term, $Y_{\underline{\alpha}_1}$ kills v_i^* for $i \neq 1$. This proves (2). □

6 Bigness for Nearly Hyperspecial Groups

Throughout this section K denotes a finite extension of $\mathbf{Q}_\ell$, $\mathscr{O}_K$ its ring of integers and k its residue field.

We begin by recalling some definitions. Let G/K be a reductive group. The group G is *quasi-split* if it has a Borel subgroup. It is *unramified* if it is quasi-split and it splits over an unramified extension of K. A subgroup $\Gamma \subset G(K)$ is *hyperspecial* if there exists a reductive group $\widetilde{G}/\mathscr{O}_K$ with generic fiber G such that $\Gamma = \widetilde{G}(\mathscr{O}_K)$. Hyperspecial subgroups of $G(K)$ are maximal compact subgroups. The group $G(K)$ has a hyperspecial subgroup if and only if G is unramified. Let G^{ad} be the adjoint group of G and let G^{sc} be the simply connected cover of G^{ad}. We have maps

$$G \xrightarrow{\sigma} G^{\mathrm{ad}} \xleftarrow{\tau} G^{\mathrm{sc}}.$$

We say that a subgroup $\Gamma \subset G(K)$ is *nearly hyperspecial* if $\tau^{-1}(\sigma(\Gamma))$ is a hyperspecial subgroup of $G^{\mathrm{sc}}(K)$. (This is not a standard term.)

The purpose of this section is to prove the following proposition:

Proposition 6.1 *Let $\rho : \Gamma \to \mathrm{GL}_n(K)$ be a continuous representation of the profinite group Γ. Assume:*

- *The characteristic ℓ of k is large compared to n.*
- *The restriction of ρ to any open subgroup of Γ is absolutely irreducible.*
- *The index of G° in G is small compared to ℓ, where G is the Zariski closure of $\rho(\Gamma)$.*
- *The subgroup $\rho(\Gamma) \cap G^\circ(K)$ of $G^\circ(K)$ is nearly hyperspecial.*

Then $\overline{\rho}(\Gamma)$ is a big subgroup of $\mathrm{GL}_n(k)$.

We remark that the second condition in the proposition, that the restriction of ρ to any open subgroup remain absolutely irreducible, is equivalent to the condition that the representation of G° on V be absolutely irreducible. We need some auxiliary lemmas to prove the proposition. We begin with the following one.

Lemma 6.2 *Let $\widetilde{G}/\mathscr{O}_K$ be a simply connected semi-simple group and let σ be an automorphism of the generic fiber $G = \widetilde{G}_K$ such that σ maps $\widetilde{G}(\mathscr{O}_K)$ into itself. Then for any tamely ramified finite extension L/K the automorphism σ maps $\widetilde{G}(\mathscr{O}_L)$ into itself.*

Proof The group $\widetilde{G}(\mathcal{O}_K)$ fixes a point x on the building $B(G, K)$ by [18, §2.3.1] or [4, §4.6.31] which is known to be unique. Similarly, the group $\widetilde{G}(\mathcal{O}_L)$ fixes a unique point x' on the building $B(G, L)$. Furthermore $\widetilde{G}(\mathcal{O}_K)$ (resp. $\widetilde{G}(\mathcal{O}_L)$) is the full stabilizer of x (resp. x') since $\widetilde{G}(\mathcal{O}_K)$ (resp. $\widetilde{G}(\mathcal{O}_L)$) is maximal compact ([18, §3.2]). We now claim that under the natural inclusion $B(G, K) \to B(G, L)$ the point x is identified with x'. To see this, first note that if τ is an element of $\mathrm{Gal}(L/K)$ then $\widetilde{G}(\mathcal{O}_L)$ fixes $\tau x'$ and so $\tau x' = x'$ by the uniqueness of x'. Thus x' is fixed by $\mathrm{Gal}(L/K)$ and therefore belongs to $B(G, K)$ by [18, §2.6.1] (this uses the hypothesis that L/K is tamely ramified). Since x' is fixed by $\widetilde{G}(\mathcal{O}_L)$ it is certainly also fixed by the subgroup $\widetilde{G}(\mathcal{O}_K)$. By the uniqueness of x we conclude $x = x'$.

Now, the automorphism σ of G acts on $B(G, K)$ and $B(G, L)$ and respects the inclusion map. As σ carries $\widetilde{G}(\mathcal{O}_K)$ into itself it must fix x. It therefore also fixes x' and so must carry its stabilizer, $\widetilde{G}(\mathcal{O}_L)$, into itself. This proves the lemma. □

We can now prove the following:

Lemma 6.3 *Let Γ be a profinite group and let ρ be an absolutely irreducible representation of Γ on a K-vector space V. Assume that the Zariski closure G of $\rho(\Gamma)$ is connected and that $\rho(\Gamma)$ is a nearly hyperspecial subgroup of $G(K)$. Then we can find:*

- *a Γ-stable lattice Λ in V;*
- *a semi-simple group $\widetilde{G}/\mathcal{O}_K$ with generic fiber equal to G^{sc}; and*
- *a representation $r : \widetilde{G} \to \mathrm{GL}(\Lambda)$ which induces the natural map $G^{\mathrm{sc}} \to G$ on the generic fiber,*

such that $\mathcal{O}_K^{\times} \cdot r(\widetilde{G}(\mathcal{O}_K))$ is an open normal subgroup of $\mathcal{O}_K^{\times} \cdot \rho(\Gamma)$, the index of which can be bounded in terms of dim V. *Necessarily, the generic fiber of r is an absolutely irreducible representation of $\widetilde{G}_K$ on V.*

Proof The group G is a reductive (and in particular connected) group, by hypothesis. Since ρ is absolutely irreducible, the center Z of G is contained in the center of $\mathrm{GL}(V)$. We have maps

$$\begin{array}{ccccc} G & \xrightarrow{\sigma} & G^{\mathrm{ad}} & \xleftarrow{\tau} & G^{\mathrm{sc}} \\ & \nwarrow & \uparrow & \swarrow & \\ & & G^{\mathrm{der}}. & & \end{array}$$

By hypothesis, $\tau^{-1}(\sigma(\rho(\Gamma)))$ is a hyperspecial subgroup of G^{sc}. Thus we can find a semi simple group $\widetilde{G}/\mathcal{O}_K$ with generic fiber G^{sc} such that $\widetilde{G}(\mathcal{O}_K) = \tau^{-1}(\sigma(\rho(\Gamma)))$.

Let $r : G^{\mathrm{sc}} \to G$ be the natural map; this factors through G^{der} in the above diagram. Let U be the image of $G^{\mathrm{sc}}(K)$ under τ. It is an open normal subgroup of $G^{\mathrm{ad}}(K)$, the index of which can be bounded in terms of $\dim G$ and thus $\dim V$ (by arguments similar to those used in the third paragraph of the proof of Proposition 5.1). Now, we have

$$\sigma(r(\widetilde{G}(\mathcal{O}_K))) = \tau(\widetilde{G}(\mathcal{O}_K)) = \sigma(\rho(\Gamma)) \cap U.$$

Applying σ^{-1}, we find

$$K^\times \cdot r(\widetilde{G}(\mathscr{O}_K)) = K^\times \cdot (\rho(\Gamma) \cap \sigma^{-1}(U)).$$

Since $r(\widetilde{G}(\mathscr{O}_K))$ and $\rho(\Gamma) \cap \sigma^{-1}(U)$ are both compact, it follows that

$$\mathscr{O}_K^\times \cdot r(\widetilde{G}(\mathscr{O}_K)) = \mathscr{O}_K^\times \cdot (\rho(\Gamma) \cap \sigma^{-1}(U)).$$

Thus $\mathscr{O}_K^\times \cdot r(\widetilde{G}(\mathscr{O}_K))$ is an open normal subgroup of $\mathscr{O}_K^\times \cdot \rho(\Gamma)$, the index of which can be bounded in terms of $\dim V$.

We now claim that for any finite unramified extension L/K the group $\rho(\Gamma)$ normalizes $\mathscr{O}_L^\times \cdot r(\widetilde{G}(\mathscr{O}_L))$. To see this, let γ be an element of $\rho(\Gamma)$. Write $\overline{\gamma}$ for the image of γ in $G^{\mathrm{ad}}(K)$ under σ. Thus $\overline{\gamma}$ gives an automorphism of G^{sc}, which we denote by $x \mapsto \overline{\gamma}x\overline{\gamma}^{-1}$. Now, let x be an element of $G^{\mathrm{sc}}(L)$. We then have

$$\gamma r(x)\gamma^{-1} = z \cdot r(\overline{\gamma}x\overline{\gamma}^{-1}),$$

for some $z \in \mathscr{O}_L^\times$, as is easily seen by applying σ. It thus suffices to show that conjugation by $\overline{\gamma}$ carries $\widetilde{G}(\mathscr{O}_L)$ into itself. By Lemma 6.2 it suffices to show that $\overline{\gamma}$ carries $\widetilde{G}(\mathscr{O}_K)$ into itself. Thus let x be an element of $\widetilde{G}(\mathscr{O}_K)$. Using the above formula and the fact that γ normalizes $\mathscr{O}_K^\times \cdot r(\widetilde{G}(\mathscr{O}_K))$, we can find an element y of $\widetilde{G}(\mathscr{O}_K)$ and an element z of $\mathscr{O}_K^\times$ such that $r(\overline{\gamma}x\overline{\gamma}^{-1}) = zr(y)$. It thus follows that $\overline{\gamma}x\overline{\gamma}^{-1} = z'y$ for some element z' of the K-points of center of G^{sc}. However, z' must be contained in $\widetilde{G}(\mathscr{O}_K)$ since it belongs to a compact central group and $\widetilde{G}(\mathscr{O}_K)$ is maximal compact. Thus $\overline{\gamma}x\overline{\gamma}^{-1}$ belongs to $\widetilde{G}(\mathscr{O}_K)$.

Now, the group $\widetilde{G}(\mathscr{O}_K^{\mathrm{un}})$ is bounded in the sense of [18, §2.2.1]. Thus, arguing as in [11, §1.12], we can find a lattice $\Lambda' \subset V$ such that $\Lambda' \otimes \mathscr{O}_K^{\mathrm{un}}$ is stable under the action of $\widetilde{G}(\mathscr{O}_K^{\mathrm{un}})$. Now, we have shown that $\mathscr{O}_K^\times \cdot r(\widetilde{G}(\mathscr{O}_K))$ has finite index in $\mathscr{O}_K^\times \cdot \rho(\Gamma)$. Let $\gamma_1, \dots, \gamma_n$ be coset representatives and put

$$\Lambda = \sum_{i=1}^{n} \gamma_i \cdot \Lambda'.$$

Thus Λ is a lattice in V. It is easy to see that Γ maps Λ into itself and $\widetilde{G}(\mathscr{O}_K^{\mathrm{un}})$ maps $\Lambda \otimes \mathscr{O}_K^{\mathrm{un}}$ into itself. Following the argument in [11, §1.12] once again, we see that $r : \widetilde{G}_K \to \mathrm{GL}(V)$ lifts to a map $\widetilde{G} \to \mathrm{GL}(\Lambda)$, which we still call r. This completes the proof of the proposition. □

We can now prove the proposition.

Proof of Proposition 6.1 Let Γ, ρ and G be as in the statement of the proposition, and let $V = K^n$ be the representation space of ρ. We must show that $\overline{\rho}(\Gamma)$ is big. Let $\Gamma^\circ = \Gamma \cap G^\circ(K)$. Then Γ° is a normal subgroup of Γ of prime to ℓ index (since the number of components of G is assumed small compared to ℓ). It is therefore enough, by Proposition 2.1, to show that $\overline{\rho}(\Gamma^\circ)$ is big. Replacing Γ by Γ°, it thus suffices to prove the proposition under the assumption that the Zariski closure G of $\rho(\Gamma)$ is connected.

Let $\widetilde{G}/\mathscr{O}_K$, Λ and r be as in Lemma 6.3. Let $\overline{\rho}$ be the representation of Γ on $U = \Lambda \otimes_{\mathscr{O}_K} k$. By Proposition 3.5, the representation of $\widetilde{G}_k$ on U is absolutely irreducible and its norm is bounded in terms of $\dim V$. It thus follows from Proposition 5.1 that $r(\widetilde{G}(k))$ is a big subgroup of $\mathrm{GL}(U)$. Now, $\mathscr{O}_K^\times \rho(\Gamma)$ contains $\mathscr{O}_K^\times r(\widetilde{G}(\mathscr{O}_K))$ as a normal subgroup of index prime to ℓ. Taking the image of each group in $\mathrm{GL}(U)$, we find that $k^\times \overline{\rho}(\Gamma)$ contains $k^\times r(\widetilde{G}(k))$ as a normal subgroup of index prime to ℓ. (Note that the image of $r(\widetilde{G}(\mathscr{O}_K))$ in $\mathrm{GL}(U)$ is equal to $r(\widetilde{G}(k))$ since $\widetilde{G}$ is smooth over $\mathscr{O}_K$.) Since $r(\widetilde{G}(k))$ is big, we conclude the same for $\overline{\rho}(\Gamma)$ by Propositions 2.1 and 2.2. □

7 Groups with Frobenii and Compatible Systems

A *group with Frobenii* is a pair $(\Gamma, \mathscr{F})$ consisting of a profinite group Γ and a dense set of elements $\mathscr{F}$ of Γ indexed by a set P. The elements of $\mathscr{F}$ are called "Frobenius elements." The motivating example of a group with Frobenii is the Galois group of a global field. Let F be a global field (that is, a finite extension of $\mathbf{F}_p(t)$ or of $\mathbf{Q}$) and let Γ be its absolute Galois group. For each place v of F choose a Frobenius element Frob_v and let $\mathscr{F}$ be the set of all the Frob_v. Then $(\Gamma, \mathscr{F})$ is a group with Frobenii.

Let Γ be a group with Frobenii, let E be a number field, let L be a set of places of E and for each $w \in L$ let $\rho_w : \Gamma \to \mathrm{GL}_n(E_w)$ be a continuous representation. We say that the ρ_w form a *compatible system* (with coefficients in E) if for each Frobenius element $F \in \mathscr{F}$ there exists a finite set of places $L_F \subset L$ (the "bad places" for F) such that the following conditions hold:

- The characteristic polynomial of F has coefficients in E and is independent of w for good w. Precisely, given $F \in \mathscr{F}$ there is a polynomial p with coefficients in E such that for all places $w \in L \setminus L_F$ the characteristic polynomial of $\rho_w(F)$ is equal to p.
- For any finite subset L' of L the Frobenii for which all primes in L' are good form a dense set in Γ. That is, for any such L' the set $\{F \in \mathscr{F} \mid L' \cap L_F = \emptyset\}$ is dense.

By a "compatible system of semi-simple representations" we simply mean a compatible system in which each ρ_w is semi-simple. We call a set L of places of E *full* if there exists a set of rational primes P of Dirichlet density one such that for all $\ell \in P$ all places of E over ℓ belong to L.

Proposition 7.1 *Let Γ be a group with Frobenii and let $\{\rho_w\}_{w \in L}$ be a compatible system of n dimensional semi-simple representations of Γ with coefficients in E, with L a full set of places. We assume that E is Galois over $\mathbf{Q}$. Let G_w be the Zariski closure of $\rho_w(\Gamma)$ in $\mathrm{GL}_n(E_w)$ and let G_w° be its identity component. Then there is a finite index subgroup Γ° of Γ and a set of primes P of $\mathbf{Q}$ of Dirichlet density*

$1/[E : \mathbf{Q}]$, *all of which split completely in* E, *such that if* $w \in L$ *lies over a prime in* P *then:*

1. *The Zariski closure of* $\rho_w(\Gamma^\circ)$ *is* G_w°.
2. *The group* $\rho_w(\Gamma^\circ)$ *is a nearly hyperspecial subgroup of* $G_w^\circ(E_w)$.

Proof[1] When $E = \mathbf{Q}$, the first statement is due to Serre (see [13, Proposition 6.14]) and the second to Larsen (see [11]). We will deduce the statement for arbitrary E from the $E = \mathbf{Q}$ case. Let P_0 be the set of rational primes ℓ such that all places of E above ℓ belong to L. Then P_0 has Dirichlet density one since L is full. For $\ell \in P$ define $\sigma_\ell = \bigoplus_{w|\ell} \rho_w$, where here ρ_w is regarded as a $\mathbf{Q}_\ell$ representation of dimension $n[E_w : \mathbf{Q}_\ell]$. Then σ_ℓ is a $\mathbf{Q}_\ell$ representation of Γ of dimension nm, where $m = [E : \mathbf{Q}]$. One easily sees that $\{\sigma_\ell\}_{\ell \in P}$ forms a compatible system.

Let H_ℓ be the Zariski closure of the image of σ_ℓ. Applying the $E = \mathbf{Q}$ case of the proposition, we can find a set of primes $P_1 \subset P_0$ of Dirichlet density one and a finite index subgroup Γ° of Γ such that for all $\ell \in P$ we have: (1) the Zariski closure of $\sigma_\ell(\Gamma^\circ)$ is H_ℓ°; and (2) $\sigma_\ell(\Gamma^\circ)$ is a nearly hyperspecial subgroup of $H_\ell^\circ(\mathbf{Q}_\ell)$. Let P be the set of primes in P_1 which split completely in E. Let $\ell \in P$ and pick $w \mid \ell$. Since $E_w = \mathbf{Q}_\ell$, the representation ρ_w is an n-dimensional $\mathbf{Q}_\ell$ representation, and as such a summand of σ_ℓ. Thus $\sigma_\ell(\Gamma^\circ)$ surjects onto $\rho_w(\Gamma^\circ)$, and so H_ℓ° surjects onto the Zariski closure of $\rho_w(\Gamma^\circ)$. It follows that the Zariski closure of $\rho_w(\Gamma^\circ)$ is connected. Since $\rho_w(\Gamma^\circ)$ has finite index in $\rho_w(\Gamma)$, the Zariski closure of the former must be the connected component of the Zariski closure of the latter, namely G_w°. The following lemma shows that $\rho_w(\Gamma^\circ)$ is nearly hyperspecial in $G_w^\circ(\mathbf{Q}_\ell)$. □

Lemma 7.2 *Let* $K/\mathbf{Q}_\ell$ *be a finite extension, let* $f : G \to H$ *be a surjection of reductive groups over* K *and let* Γ *be a nearly hyperspecial subgroup of* $G(K)$. *Then* $f(\Gamma)$ *is a nearly hyperspecial subgroup of* $H(K)$.

Proof Consider the diagram

$$\begin{array}{ccccc} G^{\mathrm{sc}} & \xrightarrow{\tau} & G^{\mathrm{ad}} & \xleftarrow{\sigma} & G \\ \downarrow{\scriptstyle f''} & & \downarrow{\scriptstyle f'} & & \downarrow{\scriptstyle f} \\ H^{\mathrm{sc}} & \xrightarrow{\tau'} & H^{\mathrm{ad}} & \xleftarrow{\sigma'} & H \end{array}$$

where f'' is the lift of f'. Let $\Gamma' = f(\Gamma)$, $\Delta = \tau^{-1}(\sigma(\Gamma))$ and $\Delta' = (\tau')^{-1}(\sigma'(\Gamma'))$. We are given that Δ is hyperspecial and we want to show that Δ' is hyperspecial. One easily sees that $f''(\Delta) \subset \Delta'$ and that Δ' is compact. Now, since G^{sc} and H^{sc} are simply connected semi-simple groups the map f'' is a projection onto a direct factor. It follows that $G^{\mathrm{sc}} = H^{\mathrm{sc}} \times H'$ for some group H'. The following lemma shows that $\Delta = \Delta_1 \times \Delta_2$ where Δ_1 is a hyperspecial subgroup of $H^{\mathrm{sc}}(K)$ and Δ_2 is a hyperspecial subgroup of $H'(K)$. We thus find $f''(\Delta) = \Delta_1 \subset \Delta'$. Since Δ' is compact and Δ_1 is maximal compact, we have $\Delta' = \Delta_1$ and so Δ' is hyperspecial. □

[1] An argument similar to the one given here appeared in [1].

Lemma 7.3 *Let $K/\mathbf{Q}_\ell$ be a finite extension, let H_1 and H_2 be reductive groups over K and let Δ be a hyperspecial subgroup of $H_1(K) \times H_2(K)$. Then $\Delta = \Delta_1 \times \Delta_2$ where Δ_i is a hyperspecial subgroup of $H_i(K)$.*

Proof We thank Brian Conrad for this argument. Let $\Delta = \widetilde{G}(\mathcal{O}_K)$ where $\widetilde{G}/\mathcal{O}_K$ is a reductive group with generic fiber G. We wish to find $\widetilde{G}_i$ such that $\Delta_i = \widetilde{G}_i(\mathcal{O}_K)$. If $\widetilde{G}_i$ exists then it is necessarily the Zariski closure of G_i in $\widetilde{G}$ and thus unique. To establish the existence of $\widetilde{G}_i$ we may therefore (by descent theory) work étale locally on $\mathcal{O}_K$. We may therefore replace $\mathcal{O}_K$ by a cover and assume that $\widetilde{G}$ is split. Let $\widetilde{T}$ be a split maximal torus of $\widetilde{G}$. Then the root datum for $(\widetilde{G}, \widetilde{T})$ is canonically identified with that for (G, T), where T is the generic fiber of $\widetilde{T}$. As the latter is a product, so is the former. Thus $\widetilde{G} = \widetilde{G}_1 \times \widetilde{G}_2$ where the generic fiber of $\widetilde{G}_i$ is G_i. This establishes the lemma. □

8 Bigness for Compatible Systems

We can now prove our main theorem:

Theorem 8.1 *Let Γ be a group with Frobenii, let E be a Galois extension of $\mathbf{Q}$, let L be a full set of places of E and for each $w \in L$ let $\rho_w : \Gamma \to \mathrm{GL}_n(E_w)$ be a continuous representation. Assume that $\{\rho_w\}_{w\in L}$ forms a compatible system and that each ρ_w is absolutely irreducible when restricted to any open subgroup of Γ. Then there is a set of primes P of $\mathbf{Q}$ of Dirichlet density $1/[E : \mathbf{Q}]$, all of which split completely in E, such that $\overline{\rho}_w(\Gamma)$ is a big subgroup of $\mathrm{GL}_n(\mathbf{F}_\ell)$ for any $w \in L$ lying over a prime $\ell \in P$.*

Proof Let G_w be the Zariski closure of $\rho_w(\Gamma)$ in $\mathrm{GL}_n(E_w)$. Let P_0 be the set of primes provided by Proposition 7.1. Then as w varies amongst places of L lying over elements of P_0 the index of G_w° in G_w is bounded. Thus by Proposition 6.1, $\overline{\rho}_w(\Gamma)$ is a big subgroup of $\mathrm{GL}_n(\mathbf{F}_\ell)$ if $w \in L$ lies over $\ell \in P_0$ and ℓ is sufficiently large. It follows that we can take P to be the set of all sufficiently large elements of P_0. □

We expect that one should be able to take the set P of primes in the above theorem to have density one, but we have not proved this. Applying the theorem in the case where Γ is the absolute Galois group of a number field and $E = \mathbf{Q}$ gives Theorem 1.1 from the introduction.

Acknowledgements We would like to thank Thomas Barnet-Lamb, Bhargav Bhatt, Brian Conrad, Alireza Salehi Golsefidy and Jiu-Kang Yu for useful discussions. We would also like to thank an anonymous referee for some helpful comments.

References

1. Barnet-Lamb, T., Gee, T., Geraghty, D., Taylor, R.: Potential automorphy and change of weight. Ann. Math **179**, 501–609 (2014)
2. Borel, A.: Linear algebraic groups, 2nd edn. In: Graduate Texts in Mathematics, vol. 126. Springer, New York (1991)
3. Borel, A., Tits, J.: Groupes réductifs. Publ. Math. de l'IHÉS **27**, 55–151 (1965)
4. Bruhat, F., Tits, J.: Groupes réductifs sur un corps local: II. Schémas en groupes. Existence d'une donnée radicielle valuée. Publ. Math. de l'IHÉS **60**, 5–184 (1984)
5. Carter, R.W.: Simple Groups of Lie Type. Wiley, London (1972)
6. Clozel, L., Harris, M., Taylor, R.: Automorphy for some ℓ-adic lifts of automorphic mod ℓ Galois representations. Publ. Math. de l'IHÉS **108**, 1–181 (2008)
7. Demazure, M., Grothendieck, A.: Séminaire de Géométrie Algébrique du Bois Marie — Schémas en groups, vols. 1, 2 and 3, Lecture notes in Math. vols. 151, 152 and 153. Springer, New York (1970)
8. Fulton, W., Harris, J.: Representation Theory: A First Course. Graduate Texts in Mathematics, vol. 129. Springer, New York (1991)
9. Jantzen, J.C.: Representations of Algebraic Groups. Academic Press Inc, Orlando (1987)
10. Jantzen, J.C.: Low-dimensional representations of reductive groups are semisimple. In: Richardson, R.W. (ed.) Algebraic Groups and Lie Groups: A Volume of Papers in Honour of the Late, Aust. Math. Soc. Lect. Ser., vol. 9. Cambridge University Press, Cambridge (1997)
11. Larsen, M.: Maximality of Galois actions for compatible systems. Duke Math. J. **80**, 601–630 (1995)
12. Larsen, M.: On the semisimplicity of low-dimensional representations of semisimple groups in characteristic p. J. Algebra **173**(2), 219–236 (1995)
13. Larsen, M., Pink, R.: On ℓ-independence of algebraic monodromy groups in compatible systems of representations. Invent. Math. **107**, 603–636 (1992)
14. Mumford, D.: Abelian varieties. Tata Institute of Fundamental Research Studies in Math. no. 5. Oxford Univ. Press (1970)
15. Oesterlé, J.: Nombres de Tamagawa et groupes unipotents en caractéristique p. Invent. Math. **78**(1), 13–88 (1984)
16. Patrikis, S.: Variations on a Theorem of Tate (preprint). arXiv:1207.6724
17. Springer, T.A.: Weyl's character formula for algebraic groups. Invent. Math. **5**, 85–105 (1968)
18. Tits, J.: Reductive groups over local fields. In: Automorphic forms, representations and L-functions. In: Proceedings of Symposium Pure Math., vol. 33. Amer. Math. Soc., Providence, pp. 29–69 (1979)
19. Taylor, R., Yoshida, T.: Compatibility of local and global Langlands correspondences. J. Amer. Math. Soc. **20–2**, 467–493 (2007)

MIX
Papier aus verantwortungsvollen Quellen
Paper from responsible sources
FSC® C105338

If you have any concerns about our products,
you can contact us on
ProductSafety@springernature.com

In case Publisher is established outside the EU,
the EU authorized representative is:
Springer Nature Customer Service Center GmbH
Europaplatz 3, 69115 Heidelberg, Germany

Printed by Libri Plureos GmbH
in Hamburg, Germany